ADVANCED SIGNAL PROCESSING

Theory and Implementation for Sonar, Radar, and Non-Invasive Medical Diagnostic Systems

SECOND EDITION

THE ELECTRICAL ENGINEERING
AND APPLIED SIGNAL PROCESSING SERIES
Edited by Alexander D. Poularikas

The Transform and Data Compression Handbook
K.R. Rao and P.C. Yip

Handbook of Antennas in Wireless Communications
Lal Chand Godara

Handbook of Neural Network Signal Processing
Yu Hen Hu and Jenq-Neng Hwang

Optical and Wireless Communications: Next Generation Networks
Matthew N.O. Sadiku

Noise Reduction in Speech Applications
Gillian M. Davis

Signal Processing Noise
Vyacheslav P. Tuzlukov

Digital Signal Processing with Examples in MATLAB®
Samuel Stearns

Applications in Time-Frequency Signal Processing
Antonia Papandreou-Suppappola

The Digital Color Imaging Handbook
Gaurav Sharma

Pattern Recognition in Speech and Language Processing
Wu Chou and Biing-Hwang Juang

Propagation Handbook for Wireless Communication System Design
Robert K. Crane

Nonlinear Signal and Image Processing: Theory, Methods, and Applications
Kenneth E. Barner and Gonzalo R. Arce

Smart Antennas
Lal Chand Godara

Mobile Internet: Enabling Technologies and Services
Apostolis K. Salkintzis and Alexander D. Poularikas

Soft Computing with MATLAB®
Ali Zilouchian

Signal and Image Processing in Navigational Systems
Vyacheslav P. Tuzlukov

Medical Image Analysis Methods
Lena Costaridou

MIMO System Technology for Wireless Communications
George Tsoulos

Signals and Systems Primer with MATLAB®
Alexander D. Poularikas

Adaptation in Wireless Communications - 2 volume set
Mohamed Ibnkahla

Handbook of Multisensor Data Fusion: Theory and Practice, Second Edition
Martin E. Liggins, David L. Hall, and James Llinas

Discrete Random Signal Processing and Filtering Primer with MATLAB®
Alexander D. Poularikas

Advanced Signal Processing: Theory and Implementation for Sonar, Radar, and Non-Invasive Medical Diagnostic Systems, Second Edition
Stergios Stergiopoulos

THE ELECTRICAL ENGINEERING
AND APPLIED SIGNAL PROCESSING SERIES

ADVANCED SIGNAL PROCESSING

Theory and Implementation for Sonar, Radar, and Non-Invasive Medical Diagnostic Systems

SECOND EDITION

Edited by

STERGIOS STERGIOPOULOS

CRC Press
Taylor & Francis Group
Boca Raton   London   New York

CRC Press is an imprint of the
Taylor & Francis Group, an **Informa** business

CRC Press
Taylor & Francis Group
6000 Broken Sound Parkway NW, Suite 300
Boca Raton, FL 33487-2742

First issued in paperback 2017

ISBN 13: 978-1-138-11356-5 (pbk)
ISBN 13: 978-1-4200-6238-0 (hbk)

This book contains information obtained from authentic and highly regarded sources. Reasonable efforts have been made to publish reliable data and information, but the author and publisher cannot assume responsibility for the valid-ity of all materials or the consequences of their use. The authors and publishers have attempted to trace the copyright holders of all material reproduced in this publication and apologize to copyright holders if permission to publish in this form has not been obtained. If any copyright material has not been acknowledged please write and let us know so we may rectify in any future reprint.

Library of Congress Cataloging-in-Publication Data

Advanced signal processing : theory and implementation for sonar, radar, and non-invasive medical
    diagnostic systems / editor, Stergios Stergiopoulos. -- 2nd ed.
      p. cm. -- (The electrical engineering and applied signal
  processing series)
  "A CRC title."
  Includes bibliographical references and index.
  ISBN 978-1-4200-6238-0 (hardcover : alk. paper)
    1. Signal processing--Digital techniques. 2. Diagnostic imaging--Digital techniques. 3. Image
processing--Digital techniques. 4. Sonar. 5. Radar. I. Stergiopoulos, Stergios. II. Title. III. Series.

TK5102.9.A383 2009
621.382'2--dc22
                                                    2009003579

Visit the Taylor & Francis Web site at
http://www.taylorandfrancis.com

and the CRC Press Web site at
http://www.crcpress.com

To my son Sotirios and my daughter Erene

Contents

Section III   Medical Diagnostic System Applications

Preface to Second Edition

This handbook emerged from the Editor's most recent investigations at Defence R&D Canada (DRDC) in implementing defence oriented research from sonar and radar system applications into non-invasive medical diagnostic R&D to address difficult diagnostic problems relevant with the protection of members of the Canadian Forces serving in hostile fields of operations. The development of these medical technologies has proven to be equally applicable for civilian use and as a result a private company has been established to commercialize them. Thus, the writing of this handbook was prompted by a desire to bring together some of the most recent theoretical developments on advanced signal processing; and to provide a glimpse on how modern technology can be applied to the development of current and next generation real-time sonar and medical diagnostic systems.

The first edition focused on advances in digital signal processing algorithms and on their implementation in PC-based computing architectures that can provide the ability to produce real-time systems that have capabilities far exceeding those of a few years ago. It included also a generic concept for implementing successfully adaptive schemes with near-instantaneous convergence in 2-dimensional (2D) and 3-dimensional (3D) array of sensors, such as planar, circular, cylindrical and spherical arrays of sensors.

The present edition preserves the form of the more mathematical part of the previous edition, but contains a number of major changes and much new material, especially in the first and third parts of the handbook that focus on the emerging medical technologies in the areas of non-invasive tomography imaging, biometrics, and monitoring vital signs. The earlier chapters give essential background theory on the basic elements of signal processing for practical system applications that are the scope of the handbook. More specifically, the material in Chapters 2 through 4 on 'Adaptive Systems in Signal Processing', 'Advanced Beamformers' and 'Volume Visualization Methods in Medicine' remain the same since it was found to be helpful to the general graduate students of engineering and applied physics and to system engineers specializing on sonar and medical technology developments. However, Chapters 5 and 6 are new and their material addresses topics on recent advances on image segmentation, registration and fusion techniques for 3D/4D ultrasound and other tomography imaging modalities. The material in Chapters 7 and 8 is also new and expands to include aspects of the handbook in the areas of diffraction computed tomography for non-destructive 3D tomography imaging and biometrics, respectively, for security screening applications.

In the second part of the handbook on sonar and radar system applications, apart from only minor changes in Chapters 10 and 11 on sonar systems, Chapter 9 on phased array radar signal processing has been completely re-written to address issues of space—time adaptive processing and detection of targets in interference intense backgrounds comprised of clutter and jamming.

Finally, the changes in the third part of the handbook on non-invasive medical diagnostic system applications have been considerable, resulting in a substantially different set of chapters from the first edition. In particular, Chapter 12 introduces the concept of a fully digital 3D/(4D: 3D+time) ultrasound system technology, computing architecture requirements and the relevant implementation issues for a set of 3D adaptive ultrasound beamformers that have been discussed in detail in Chapter 3. The material of Chapters 13 and 14 on magnetic resonance imaging (MRI) and on organ motion correction issues for single slice CT scanners, respectively, remains the same, while the new material of Chapter 15 extends the topic of cardiac motion correction in the emerging field of multi-slice X-ray CT imaging. Further new topics on vital signs technologies and dispersive ultrasound for monitoring non-visible traumatic brain injuries have been included in Chapter 16. The next two chapters (Chapters 17 and 18) introduce new topics on MRI diagnostic applications. More specifically, Chapter 17 focuses on contrast agent kinetic analysis that has applications in medical diagnostics by helping to characterize the functional state of a tissue and in drug discovery by offering insight into the behavior of the contrast agent itself. Contrast agent kinetic analysis is used often in conjunction with an imaging device, such as MRI, that can measure non-invasively the concentration of the contrast agent, at one or more locations, as a function of time. Chapter 18 discusses arterial spin labeling (ASL) methods that can be interpreted using the same classical tracer kinetic theory to generate quantitative cerebral blood flow MRI diagnostic applications; and the final Chapter 19 presents the importance of automatic diagnosis of microcalcifications in early detection and diagnostic procedures for breast cancer using computer aided diagnosis (CAD) in mammography.

I wish to thank my colleagues and contributing authors for their valuable assistance in forming the material of this handbook.

Stergios Stergiopoulos
Toronto, Ontario, Canada

Preface to First Edition

Recent advances in digital signal processing algorithms and computer technology have combined to provide the ability to produce real-time systems that have capabilities far exceeding those of a few years ago. The writing of this handbook was prompted by a desire to bring together some of the recent theoretical developments on advanced signal processing; and to provide a glimpse of how modern technology can be applied to the development of current and next generation active and passive real-time systems.

The handbook is intended to serve as an introduction to the principles and applications of advanced signal processing. It will focus on the development of a generic processing structure that exploits the great degree of processing concept similarities existing among the radar, sonar, and medical imaging systems. A high-level view of the above real-time systems consists of a high-speed *Signal Processor* to provide mainstream signal processing for detection and initial parameter estimation, a *Data Manager* which supports the data and information processing functionality of the system and a *Display Sub-System* through which the system operator can interact with the data structures in the data manager to make the most effective use of the resources at his command.

The *Signal Processor* normally incorporates a few fundamental operations. For example, the sonar and radar signal processors include beamforming, 'matched' filtering, data normalization and image processing. The first two processes are used to improve both the signal-to-noise ratio (SNR) and parameter estimation capability through spatial and temporal processing techniques. Data normalization is required to map the resulting data into the dynamic range of the display devices in a manner which provides a CFAR (constant false alarm rate) capability across the analysis cells.

The processing algorithms for spatial and temporal spectral analysis in real-time systems are based on conventional FFT and vector dot product operations because they are computationally cheaper and very robust than the modern non-linear high resolution adaptive methods. However, these non-linear algorithms trade robustness for improved array gain performance. Thus, the challenge is to develop a concept which allows an appropriate mixture of these algorithms to be implemented in practical real-time systems.

The non-linear processing schemes are adaptive and synthetic aperture beamformers that have been shown experimentally to provide improvements in array gain for signals embedded in partially correlated noise fields. Using system image outputs, target tracking and localization results as performance criteria, the impact and merits of these techniques are contrasted with those obtained using the conventional processing schemes. The reported real data results show that the advaned processing schemes provide improvements in array gain for signals embedded in anisotropic noise fields. However, the same set of results demonstrates that these processing schemes are not adequate enough to be considered as a replacement for conventional processing. This restriction adds an additional element in our generic signal processing structure, in that the conventional and the advanced signal processing schemes should run in parallel in a real-time system in order to achieve optimum use of the advanced signal processing schemes of this study.

The handbook will include also a generic concept for implementing successfully adaptive schemes with near-instantaneous convergence in 2-dimensional (2D) and 3-dimensional (3D) arrays of sensors, such as planar, circular, cylindrical, and spherical arrays of sensors. It will be shown that the basic step is to minimize the number of degrees of freedom associated with the adaptation process. This step will minimize the adaptive scheme's convergence period and achieve near-instantaneous convergence for integrated active and passive sonar applications. The reported results are part of a major research project, which includes the definition of a generic signal processing structure that allows the implementation of adaptive and synthetic aperture signal processing schemes in real-time radar, sonar, and medical tomography (CT, MRI, ultrasound) systems that have 2D and 3D arrays of sensors.

The material in the handbook will bridge a number of related fields: detection and estimation theory; filter theory (finite impulse response filters); 1D, 2D, and 3D sensor array processing that includes conventional, adaptive, synthetic aperture beamforming and imaging; spatial and temporal spectral analysis; and data normalization. Emphasis will be be placed on topics that have been found to be particularly useful in practice. These are several inter-related topics of interest such as the influence of medium on array gain system performance, detection and estimation theory, filter theory, spaceñtime processing, conventional, adaptive processing and model-based signal processing concepts. Moreover, the system concept similarities between sonar and ultrasound problems are identified in order to exploit the use of advanced sonar and model-based signal processing concepts in ultrasound systems.

Furthermore, issues of information post-processing functionality supported by the data manager and the display units of real-time systems of interest are addressed in the relevant chapters that discuss normalizers, target tracking, target motion analysis, image post-processing and volume visualization methods.

The presentation of the subject matter has been influenced by the authors' practical experience, and it is hoped that the volume will be useful to scientists and system engineers as a textbook for a graduate course on sonar, radar, and medical imaging digital signal processing. In particular, a number of chapters summarize the state-of-the-art application of advanced processing concepts in sonar, radar, and medical imaging X-ray CT scanners, magnetic resonance imaging, 2D and 3D ultrasound systems. The focus of these chapters is to point out their applicability, benefits, and potential in the sonar, radar, and medical environments. Although an all-encompassing general approach to a subject is mathematically elegant, practical insight and understanding may be sacrificed. To avoid this problem and to keep the handbook to a reasonable size, only a modest introduction is provided. In consequence, the reader is expected to be familiar with the basics of linear and sampled systems and the principles of probability theory. Furthermore, since modern real-time systems entail sampled signals that are digitized at the sensor level, our signals are assumed to be discrete in time and the subsystems that perform the processing are assumed to be digital.

It has been a pleasure for me to edit this book and to have the relevant technical exchanges with so many experts on advanced signal processing. I take this opportunity to thank all authors for their response to my invitation to contribute. I am also greatful to CRC Press LLC and in particular to Bob Stern for his truly professional cooperation; and to Dr. George Metakides, Director in the IST-Program of the European Commission, for his efforts to initiate the European Canadian exchange that resulted in a number of major collaborative R&D projects. The preparation of this handbook is the result of this initiative. Finally, the support by the European Commission is acknowledged for awarding Professor Uzunonglu and myself with the Fourier Euroworkshop grant (HPCF-1999-00034) to organize two workshops that enabled the contributing authors to refine and coherently integrate the material of their chapters as a handbook on advanced signal processing for sonar, radar, and medical imaging system applications.

Stergios Stergiopoulos

Editor

Stergios Stergiopoulos received a BSc (Hon.) from the University of Athens in 1976 and the MSc and PhD in Physics in 1977 and 1982, respectively, from York University, Toronto, Canada. Presently, he is a defence scientist at the Defence R&D Canada (DRDC) Toronto, an adjunct professor at the Edward S. Rogers Sr. Department of Electrical and Computer Engineering of the University of Toronto, and the main innovator of the Defence R&D Canada (DRDC) medical diagnostic technologies and patents that have been licensed to a Canadian company for commercialization. These innovation include a number of non-invasive 3D imaging (i.e. cardiac 3D CT, portable 3D/4D ultrasound) and vital signs monitoring (i.e. motion and noise tolerant automated blood pressure and intracranial dispersive-ultrasound) technologies. To complete their development and their commercialization process, Dr. Stergiopoulos raised approximately $11 million from private investors and Government grants. He has an extensive background in science and research. Since 1991 he is with DRDC, a Research Agency for the Canadian Department of National Defence. From 1988 to 1991, he was with the NATO SACLANT Centre in La Spezia, Italy, where he performed both theoretical and experimental research in sonar signal processing. At SACLANTCEN, he developed jointly with Dr. Edmund J. Sullivan from NUWC an acoustic synthetic aperture technique that has been patented by the US Navy. From 1984 to 1988, he developed an underwater fixed array surveillance system for the Hellenic Navy in Greece and there, he was appointed also senior advisor to the Greek Minister of Defence. From 1982 to 1984, he worked as a research associate at York University and in collaboration with the US Army Ballistic Research Lab (BRL), Aberdeen, MD, on projects related to the stability of liquid filled spin stabilized projectiles. In 1984, he was awarded a US NRC Research Fellowship for BRL. He was associate editor for the *IEEE Journal of Oceanic Engineering* and for this journal he has prepared two special issues on Acoustic Synthetic Aperture and Sonar System Technology. He has published numerous scientific articles and a handbook (i.e. CRC Press) in the areas of advanced signal processing for sonar and medical non-invasive system applications. His present interests are associated with the implementation of advanced processing schemes in multi-dimensional arrays of sensors for sonar and medical tomography (CT, MRI, and ultrasound) systems. His research activities are supported by Canadian-DND Grants, by Research and Strategic Grants (NSERC-CANADA), the Ontario Challenge Fund and NATO Collaborative Research Grant. He has been awarded with European Commission-IST grants as technical manager of several projects that included as project partners major European corporations and institutes (i.e. Siemens, Nucletron, Philips, Sema Group, Esaote, Atmel, Fraunhofer). These project were entitled 'New Roentgen', 'MITTUG', 'ADUMS', 'MRI-MARCB', 'DUST' and Euroworkshop 'Fourier', with an average budget level of the order of €1.5 million per project. Dr. Stergiopoulos is a fellow of the Acoustical Society of America and a senior member of the IEEE. He has been a consultant to a number of companies, including Atlas Elektronik in Germany, Hellenic Arms Industry, and Hellenic Aerospace Industry.

Contributors

Bhashyam Balaji
Defence R&D Canada Ottawa
Ottawa, Ontario, Canada

G. Clifford Carter
Naval Undersea Warfare Center
Newport, Rhode Island

Ian Cunningham
The John Robarts Research Institute
University of Western Ontario
London, Ontario, Canada

Anthony Damini
Defence R&D Canada Ottawa,
Ottawa, Ontario, Canada

Amar Dhanantwari
Defence R&D Canada Toronto
Toronto, Ontario, Canada

Geoffrey Edelson
Advanced Systems & Technology
Nashua, New Hampshire

Marius Erdt
Department of Cognitive Computing
& Medical Imaging
Fraunhofer Institute for Computer
Graphics
Darmstadt, Germany

Andreas Freibert
Defence R&D Canada Toronto,
Toronto, Ontario, Canada

Julius Grodski
Defence R&D Canada Toronto
Toronto, Ontario, Canada

Dimitrios Hatzinakos
The Edward S. Rogers Sr. Department
of ECE
University of Toronto
Toronto, Ontario, Canada

David Havelock
National Research Council
Ottawa, Ontario, Canada

Simon Haykin
Communications Research Laboratory
McMaster University
Hamilton, Ontario, Canada

Michael Jerosch-Herold
Department of Radiology
Brigham & Women's Hospital
Boston, Massachussets

Sung Ho Jin
Multimedia Group
Information Communication University
Yuseong-gu
Daejon, Korea

Grigorios Karangelis
Department of Cognitive Computing and
Medical Imaging
Fraunhofer Institute for Computer Graphics
Darmstadt, Germany

Benoit Lewden
Lawson Health Research Institute
London, Ontario, Canada

Haiping Lu
The Edward S. Rogers Sr. Department
of ECE
University of Toronto
Toronto, Ontario, Canada

George K. Matsopoulos
Department of Electrical and Computer
Engineering
National Technical University of
Athens
Athens, Greece

Michael K. McDonald
Defence R&D Canada Ottawa
Ottawa, Ontario, Canada

Bernard E. McTaggart
Naval Undersea Warfare Center
Newport, Rhode Island

Sanjay K. Mehta
Naval Undersea Warfare Center
Newport, Rhode Island

Arnulf Oppelt
Siemens AG
Erlangen, Germany

Daron G. Owen
Imaging Division
Lawson Health Research Institute
London, Ontario, Canada

and

Department of Medical Biophysics
University of Western Ontario
London, Ontario, Canada

Konstantinos N. Plataniotis
The Edward S. Rogers Sr. Department
    of ECE
University of Toronto
Toronto, Ontario, Canada

Andreas Pommert
Institute of Mathematics and Computer
    Science in Medicine (IMDM)
University Hospital Eppendorf
Hamburg, Germany

Frank S. Prato
Imaging Division
Lawson Health Research Institute
London, Ontario, Canada

Department of Medical Biophysics
University of Western Ontario
London, Ontario, Canada

Department of Diagnostic Radiology and
    Nuclear Medicine
University of Western Ontario
London, Ontario, Canada

and

Department of Diagnostic Imaging
St Joseph's Health Care
London, Ontario, Canada

Yong Man Ro
Multimedia Group
Information Communication University
Yuseong-gu
Daejon, Korea

Eric Sabondjian
Lawson Health Research Institute
University of Western Ontario
London, Ontario, Canada

Georgios Sakas
Department of Cognitive Computing
    and Medical Imaging
Fraunhofer Institute for Computer
    Graphics
Darmstadt, Germany

    and

Technical University Darmstadt
Darmstadt, Germany

Keith St. Lawrence
Imaging Division
Lawson Health Research Institute
London, Ontario, Canada

Department of Medical Biophysics
University of Western Ontario
London, Ontario, Canada

and

Department of Diagnostic Radiology
    and Nuclear Medicine
University of Western Ontario
London, Ontario, Canada

Stergios Stergiopoulos
Defence R&D Canada Toronto,
Toronto, Ontario, Canada

The Edward S. Rogers Sr. Department
    of ECE
University of Toronto
Toronto, Ontario, Canada

and

Department of Medical Biophysics
Schulich School of Medicine
University of Western Ontario
London, Ontario, Canada

Robert Z. Stodilka
University of Western Ontario
Lawson Health Research Institute
London, Ontario, Canada

Jie Wang
Epson Edge
Toronto, Ontario, Canada

Waheed A. Younis
Defence R&D Canada Toronto
Toronto, Ontario, Canada

Jason Zhang
Defence R&D Canada Toronto
Toronto, Ontario, Canada

1

Signal Processing Concept Similarities among Sonar, Radar, and Noninvasive Medical Diagnostic Systems

Stergios Stergiopoulos

Defence R&D Canada Toronto

University of Toronto

University of Western Ontario

CONTENTS

1.1 Introduction

Several review articles on sonar [1,3,4,5], radar [2,3] and medical imaging [3,6–14] system technologies have provided a detailed description of the mainstream signal processing functions along with their associated implementation considerations. The attempt of this handbook is to extend the scope of these articles by introducing an implementation effort of nonmainstream processing schemes in real-time systems. To a large degree, work in the area of *sonar and radar system technology* has traditionally been funded either directly or indirectly by governments and military agencies in an attempt to improve the capability of antisubmarine warfare (ASW) sonar and radar systems. A secondary aim of this handbook is to promote, where possible, wider dissemination of this military-inspired research.

1.2 Overview of a Real-Time System

In order to provide a context for the material contained in this handbook, it would seem appropriate to briefly review the basic requirements of a high-performance real-time system. Figure 1.1 shows one possible high-level view of a generic system [15]. It consists of an array of sensors and/or sources; a high-speed *signal processor* to provide mainstream signal processing for detection and initial parameter estimation; a *data manager*, which supports the data and information processing functionality of the system; and a *display subsystem* through which the system operator can interact with the data structures in the data manager to make the most effective use of the resources at his command.

In this handbook, we will be limiting our attention to the *signal processor*, the *data manager*, and *display subsystem*, which consist of the algorithms and the processing architectures required for their implementation. *Arrays of sources and sensors* include devices of varying degrees of complexity that illuminate the medium of interest and sense the existence of signals of interest. These devices are arrays of transducer with cylindrical, spherical, plane or line geometric configurations, depending on the application of interest. Quantitative estimates of the various benefits that result from the deployment of arrays of transducers are obtained by the *array gain* term, which will be the subject of our discussion in Chapters 3, 9, and 10. Sensor array design concepts, however, are beyond the scope of this handbook and readers interested in transducers can refer to other publications on the topic [16–19].

The *signal processor* is probably the single most important component of a real-time system of interest for this handbook. In order to satisfy the basic requirements, the processor normally incorporates the following fundamental operations:

- Multidimensional beamforming
- Matched filtering

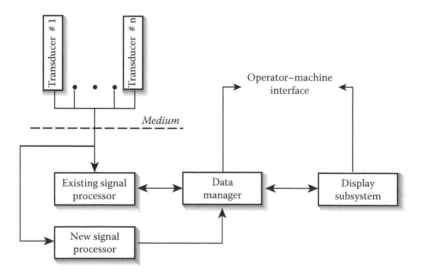

FIGURE 1.1
Overview of a generic real-time system. It consists of an array of transducers, a *signal processor* to provide mainstream signal processing for detection and initial parameter estimation, a *data manager*, which supports the data, information processing functionality, and data fusion; and a *display subsystem* through which the system operator can interact with the manager to make the most effective use of the information available at his command.

- Temporal and spatial spectral analysis
- Tomography image reconstruction processing
- Multidimensional image processing

The first three processes are used to improve both the signal-to-noise ratio (SNR) and parameter estimation capability through spatial and temporal processing techniques. The next two operations are image reconstruction and processing schemes associated mainly with image processing applications. As indicated in Figure 1.1, the replacement of the *existing signal processor* with a *new signal processor* that would include advanced processing schemes, could lead into improved performance functionality of a real-time system of interest, while the associated development cost could be significantly lower than using other hardware (H/W) alternatives. In a sense, this statement highlights the future trends of state of the art investigations on advanced real-time signal processing functionalities that are the subject of the handbook.

Furthermore, postprocessing of the information provided by the previous operations includes mainly:

- Signal tracking and target motion analysis (TMA)
- Image postprocessing and data fusion
- Data normalization
- OR-ing

These operations form the functionality of the *data manager* of sonar and radar systems. However, identification of the processing concept similarities between sonar, radar and medical imaging systems may be valuable in identifying the implementation of the above operations in other medical imaging system applications. In particular, the operation of data normalization in sonar and radar systems is required to map the resulting data into the dynamic range of the display devices in a manner which provides a constant false alarm rate (CFAR) capability across the analysis cells. The same operation, however, is required in the display functionality of medical ultrasound imaging systems as well.

In what follows, each subsystem, shown in Figure 1.1, is examined briefly by associating the evolution of its functionality and characteristics with the corresponding signal processing technological developments.

1.3 Signal Processor

The implementation of signal processing concepts in real-time systems is heavily dependent on the computing architecture characteristics, and, therefore, it is limited by the progress made in this field. While the mathematical foundations of the signal processing algorithms have been known for many years, it was the introduction of the microprocessor and high-speed multiplier-accumulator devices in the early 1970s which heralded the turning point in the development of digital systems. The first systems were primarily fixed-point machines with limited dynamic range and hence were constrained to use conventional beamforming and filtering techniques [1,4,15]. As floating-point central processing units (CPUs) and supporting memory devices were introduced in the mid to late 1970s, multiprocessor digital systems and modern signal processing algorithms could be considered for implementation in real-time systems. This major breakthrough expanded in the 1980s into massively parallel architectures supporting multisensor requirements.

The limitations associated with these massively parallel architectures became evident by the fact that they allow only fast-Fourier-transform (FFT) vector-based processing schemes because of their very efficient implementation and of their very cost-effective through-put characteristics. Thus, nonconventional schemes (i.e., adaptive, synthetic-aperture and high-resolution processing) could not be implemented in these types of real-time systems of interest, even though their theoretical and experimental developments suggest that they have advantages over existing conventional processing approaches [2,3,15,20–25]. It is widely believed that these advantages can address the requirements associated with the difficult operational problems that next generation real-time sonar, radar and noninvasive medical diagnostic systems will have to solve.

New scalable computing architectures, however, which support both scalar and vector operations satisfying high input/output bandwidth requirements of large multisensor systems, are becoming available [15]. Recent frequent announcements include successful developments of super-scalar and massively parallel signal processing computers that have throughput capabilities of hundred of billions of floating point operation per second (GFLOPS) [31]. This resulted in a resurgence of interest in algorithm development of new covariance-based high-resolution, adaptive [15,20–22,25] and synthetic-aperture beam-forming algorithms [15,23], and time-frequency analysis techniques [24].

A fundamental question, however, that must be addressed at this point is whether it is worthwhile to attempt to develop a dedicated architecture that can compete with a multiprocessor using stock microprocessors, as defined for ultrasound imaging applications in Chapter 12. However, the experience gained from sonar computing architecture developments [15] suggests that a cost effective approach in that direction is to develop a highly parallelized PC-based computing architecture that will be based on the rapidly evolving microprocessor technology of PCs.

Then, and assuming that the signal processing flow of advanced processing schemes that include both scalar and vector operations is well established, such as in Figures 3.13, 3.19 through 3.21 of Chapter 3, the distribution of the signal processing flow in a number of parallel CPU's can be straightforward, as discussed in Chapter 12.

The Chapters 2, 3, 9, 11 and 12 of this handbook discuss in some detail the recent developments in adaptive, high-resolution and synthetic aperture array signal processing and their advantages for real-time system applications. In particular, Chapter 2 reviews the basic issues involved in the study of adaptive systems for signal processing. The virtues of this approach to statistical signal processing may be summarized as follows:

- The use of an adaptive filtering algorithm, which enables the system to adjust its free parameters (in a supervised or unsupervised manner) in accordance with the underlying statistics of the environment in which the system operates. Hence, the need for determining the statistical characteristics of the environment is avoided.

- Tracking capability, which permits the system to follow statistical variations (i.e., nonstationarity) of the environment.

- The availability of many different adaptive filtering algorithms, both linear and nonlinear, which can be used to deal with a wide variety of signal-processing applications in radar, sonar and biomedical imaging.

- Digital implementation of the adaptive filtering algorithms, which can be carried out in hardware or software form.

In other cases these advanced algorithms, introduced in Chapters 2 and 3, trade robustness for improved performance [15,25,26]. Furthermore, the improvements achieved are generally not uniform across all signal and noise environments of operational scenarios. The challenge is to develop a concept which allows an appropriate mixture of these algorithms to be implemented in practical real-time systems. The advent of new adaptive processing techniques is only the first step in the utilization of *a priori* information as well as more detailed information for the mediums of the propagating signals of interest. Of particular interest is the rapidly growing field of matched field processing (MFP) [26]. The use of linear models will also be challenged by techniques that utilize higher-order statistics [24], neural networks [27], fuzzy systems [28], chaos, and other nonlinear approaches. Although these concerns have been discussed [27] in a special issue of the *IEEE Journal of Oceanic Engineering* devoted to sonar system technology, it should be noted that a detailed examination of MFP can be found also in the July 1993 issue of the same journal, which has been devoted to detection and estimation of MFP [29].

Moreover, the system concept similarities between sonar and ultrasound systems are analyzed in Chapters 3 and 12 in order to exploit the use of model-based sonar signal processing concepts in ultrasound problems.

Thus, Chapters 2, 3, and 10 through 12 address a major issue: the implementation of advanced processing schemes in real-time systems of interest. The starting point will be to identify the signal processing concept similarities among radar, sonar and medical diagnostic systems by defining a generic signal processing structure integrating the processing functionalities of the real-time systems of interest. The definition of a generic signal processing structure for a variety of systems will address the above continuing interest that is supported by the fact that synthetic-aperture and adaptive processing techniques provide new gain [2,15,20,21,23]. This kind of improvement in array gain is equivalent with improvements in system performance.

In general, improvements in system performance or array gain improvements are required when the noise environment of an operational system is nonisotropic, such as the noise environment of (1) atmospheric noise or clutter (radar applications); (2) cluttered coastal waters and areas with high shipping density in which sonar systems operate (sonar applications); and (3) the complexity of the human body (medical imaging applications). An alternative approach to improve the array gain of a real-time system requires the deployment of very large aperture arrays, which leads to technical and operational implications. Thus, the implementation of nonconventional signal processing schemes in operational systems will minimize very costly hardware requirements associated with array gain improvements.

Figure 1.2 shows the configuration of a generic signal processing scheme integrating the functionality of radar, sonar, ultrasound and medical tomography CT/X-ray and MRI imaging systems. There are five major and distinct processing blocks in the generic structure. Moreover, reconfiguration of the different processing blocks of Figure 1.2 allows the application of the proposed concepts to a variety of active or passive digital signal processing (DSP) systems [42].

The first point of the generic processing flow configuration is that its implementation is in the frequency domain. The second point is that with proper selection of filtering weights and careful data partitioning, the frequency domain outputs of conventional or advanced processing schemes can be made equivalent to the FFT of the broadband outputs. This equivalence corresponds to implementing finite impulse response (FIR) filters via circular convolution with the FFT, and it allows spatial-temporal processing of narrowband and broadband type of signals [2,15,30], as defined in Chapter 3. Thus, each processing block in the generic DSP structure provides continuous time series; and this is the central

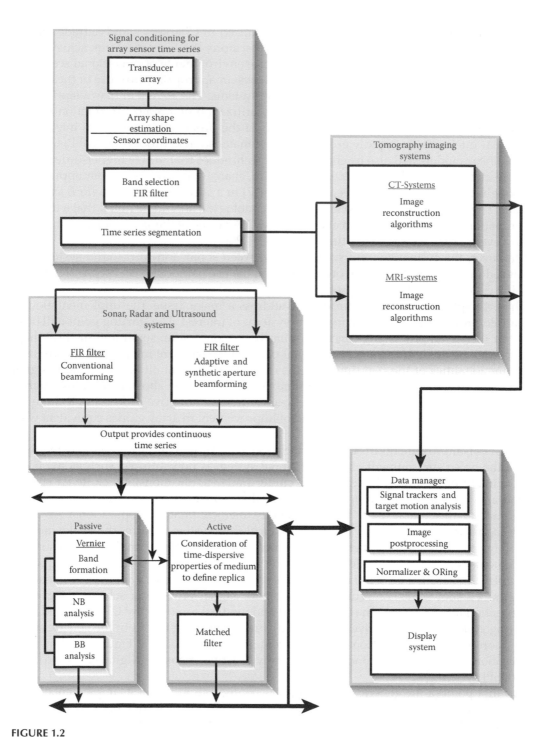

FIGURE 1.2
A generic signal processing structure integrating the signal processing functionalities of sonar, radar, ultrasound, CT/X-ray, and MRI medical imaging systems.

point of the implementation concept that allows the integration of quite diverse processing schemes, such as those shown in Figure 1.2.

More specifically, the details of the generic processing flow of Figure 1.2, are discussed briefly in the following sections.

1.3.1 Signal Conditioning of Array Sensor Time Series

The block titled "Signal conditioning for array sensor time series" in Figure 1.2 includes the partitioning of the time series from the receiving sensor array, their initial spectral FFT, the selection of the signal's frequency band of interest via band-pass FIR filters, and downsampling. The output of this block provides continuous time series at reduced sampling rate for improved temporal spectral resolution. In many system applications including moving arrays of sensors, array shape estimation or the sensor coordinates would be required to be integrated with the signal processing functionality of the system, as shown in this block.

Typical system requirements of this kind are towed array sonars [15], which are discussed in Chapters 3, 9, and 10; CT tomography imaging systems [6–8], which are analyzed in Chapters 7, 14 and 15; and ultrasound systems deploying long line or planar arrays [8–10] which are discussed in Chapters 3, 12 and 16.

The processing details of this block will be illustrated in schematic diagrams in Chapter 3. The FIR band selection processing of this block is typical in all the real-time systems of interest. As a result, its output can be provided as input to the blocks named: "Sonar, radar and ultrasound systems", or "Tomography imaging CT/X-ray and MRI systems".

1.3.2 Tomography Imaging X-ray, CT and MRI Systems

The block at the right-hand side of Figure 1.2, which is titled "Tomography imaging CT/X-ray and MRI systems", includes image reconstruction algorithms for acoustic diffraction and X-ray tomography imaging CT and MRI systems. The processing details of these algorithms will be discussed in Chapters 7, 13 through 15. In general, image reconstruction algorithms [6,7,11–13] are distinct processing schemes and their implementation is practically efficient in CT and MRI applications. However, tomography imaging and the associated image reconstruction algorithms can be applied in other system applications such as, diffraction tomography using ultrasound sources [8] and acoustic tomography of the ground or building interiors using various acoustic frequency regimes or GPR low frequencies. Diffraction tomography is not practical for medical imaging applications because of the very poor image resolution and the very high absorption rate of the acoustic energy by the bone structure of the human body. In geophysical applications, however, seismic waves can be used in tomographic imaging procedures to detect and classify very large buried objects. On the other hand, in working with higher acoustic frequencies, a better image resolution would allow detection and classification of small shallow buried objects such as anti-personnel land mines [41], which is a major humanitarian issue that has attracted the interest of the UN and the highly industrialized countries in North America and Europe. The rule of thumb in acoustic tomography imaging applications is that higher frequency regimes in radiated acoustic energy would provide better image resolution at the expense of higher absorption rates for the radiated energy penetrating the medium of interest. All these issues on diffraction computed tomography (CT) imaging are discussed in Chapter 7.

Most recently, implementation of tomography imaging using the low frequency regime of ground penetrating radars (GPR), has grown to become an emerging issue for security applications, such as through the wall cameras and surveillance of shallow buried objects to detect improvised explosive devices (IED). However, the performance characteristics of a few experimental systems, in terms of false alarm rates and detection capabilities of deeply buried IEDs, are not yet well established. Tomography imaging for surveillance security applications is a challenging problem and a potential system solution may have to integrate the tomography imaging capability of GPR and passive infra-red (IR) technologies to effectively increase detection of IEDs while minimizing through image fusion false alarm rates. Furthermore, implementation of the 3D beamforming imaging concept using planar arrays of sensors, outlined in Chapter 12 for 3D ultrasound imaging applications, has the potential to be implemented also in the low GPR frequency regime to allow for nondestructive 3D visualization and surveillance of interiors of building structures and subsurface buried objects. Moreover, the 3D visualization, segmentation and fusion capabilities of medical imaging technologies, discussed in Chapters 4 through 6 can find a direct implementation in the above security surveillance applications for IED classification of detected buried objects. Most definitely, an illumination process of buried objects by GPR combined with IR imaging through fusion can enhance the identification process of detected objects and minimizes false alarm rates, and this indicates a potential cross-fertilization of tomography imaging technologies, discussed in this handbook, in the emerging field of security surveillance.

1.3.3 Sonar, Radar, and Ultrasound Systems

The underlying signal processing functionality in sonar, radar, and modern ultrasound imaging systems deploying linear, planar, cylindrical or spherical arrays, is beamforming. Thus, the block in Figure 1.2 titled 'Sonar, radar and ultrasound systems, includes such subblocks as FIR filter/conventional beamforming and FIR filter/adaptive and synthetic-aperture beamforming for multidimensional arrays with linear, planar, circular, cylindrical and spherical geometric configurations. The output of this block provides continuous directional beam time series by using the FIR implementation scheme of the spatial filtering via circular convolution. The segmentation and overlap of the time series at the input of the beamformers take care of the wraparound errors that arise in fast-convolution signal processing operations. The overlap size is equal to the effective FIR filter's length [15,30]. Chapter 3 will discuss in detail the conventional, adaptive and synthetic aperture beamformers that can be implemented in this block of the generic processing structure of Figure 1.2. Moreover, Chapters 3 and 10 provide some real data output results from sonar systems deploying linear or cylindrical arrays.

1.3.4 Active and Passive Systems

The blocks named "Passive" and "Active" in the generic structure of Figure 1.2 are the last major processes that are included in most of the DSP systems. Inputs to these blocks are continuous beam time series, which are the outputs of the conventional and advanced beamformers of the previous block. However, continuous sensor time series from the first block titled "Signal conditioning of array sensor time series" can be provided as the input of the active and passive blocks for temporal spectral analysis. The block titled "Active" includes a matched filter subblock for the processing of active signals. The option here is to include the medium's propagation characteristics in the replica of the active signal considered in the matched filter in order to improve detection and gain [15,26]. The subblocks "Vernier/band formation" and "NB (narrowband), BB (broadband) analysis" include the final processing steps of a temporal spectral analysis for the beam time series. The inclusion of the Vernier subblock is to allow the option for improved frequency resolution. Chapters 10 and 11 discuss the signal processing functionality and system-oriented applications associated with active and passive sonar systems. Chapter 9 examines the radar detection problem as it applied to the wide area surveillance of targets, and analyzes a series of strategies to address difficult target detection scenarios using space–time-adaptive-processing (STAP). Implementation of adaptive processing (STAP) in radars has resulted in improvements in system robustness and in detection capability by suppressing both clutter and jammer interference, thus increasing the probability of target detection. Similarly, in the case of fixed surface radar platforms, two-dimensional (2D) array processing introduces improved gains in the area of target detection in the presence of spatially localized interference. Furthermore, Chapter 12 extends the discussion to address the signal processing issues relevant with three-dimensional (3D) ultrasound medical imaging systems deploying planar array probes.

In summary, the strength of the generic processing structure of Figure 1.2 is that it identifies and exploits the processing concept similarities among radar, sonar and medical imaging systems. Moreover, it enables the implementation of nonlinear signal processing methods, adaptive and synthetic aperture, as well as the equivalent conventional approaches. This kind of parallel functionality for conventional and advanced processing schemes allows for a very cost-effective evaluation of any type of improvement during the concept demonstration phase.

As stated above, the derivation of the effective filter length of an FIR adaptive and synthetic aperture filtering operation is essential for any type of application that will allow simultaneous NB and BB signal processing. This is a nontrivial problem because of the dynamic characteristics of the adaptive algorithms; and it has not as yet been addressed.

In the past, attempts to implement matrix-based signal processing methods such as adaptive processing, were based on the development of systolic array hardware because systolic arrays allow large amounts of parallel computation to be performed efficiently since communications occur locally. Unfortunately systolic arrays have been much less successful in practice than in theory. Systolic arrays big enough for real problems cannot fit on one board, much less on one chip, and interconnects have problems. A 2D systolic array implementation will be even more difficult. Recent announcements, however, include successful developments of super-scalar and massively parallel signal processing computers that have throughput capabilities of hundred of billions of floating point operation per second (GFLOPS) [40,42], and as discussed in Section 1.3, this raises a fundamental question whether it is worthwhile to attempt to develop a dedicated architecture that can compete with a PC-based cluster of microprocessors, as defined for ultrasound imaging applications in Chapter 12. It is anticipated that these recent computing architecture developments will address the computationally intensive scalar and matrix-based operations of advanced signal processing schemes for next generation real-time systems.

Finally, the block "Data manager", in Figure 1.2, includes the display system, normalizers, TMA, image postprocessing and ORing operations to map the output results into the dynamic range of the display devices; and this will be discussed in the next section.

1.4 Data Manager and Display Subsystem

Processed data at the output of the mainstream signal processing system must be stored in a temporary database before they are presented to the system operator for analysis. Until very recently, owing to the physical size and cost associated with constructing large databases, the *data manager* played a relatively small role in the overall capability of the aforementioned systems. However, with the dramatic drop in the cost of solid-state memories and the introduction of powerful microprocessors in the 1980s, the role of the *data manager* has now been expanded to incorporate postprocessing of the signal processor's output data. Thus, postprocessing operations, in addition to the traditional display data management functions, may include:

- For sonar and radar systems
  - Normalization and ORing
  - Signal tracking
  - Localization
  - Data fusion
  - Classification functionality
- For medical imaging systems
  - Image postprocessing
  - Normalizing operations
  - Registration, segmentation and image fusion

It is apparent from the above discussion that for a next-generation DSP system, emphasis should be placed on the degree of interaction between the operator and the system, through an operator–machine interface (OMI) as shown schematically in Figure 1.1. Through this interface, the operator may selectively proceed with localization, tracking, diagnosis and classification tasks.

A high-level view of the generic requirements and the associated technologies of the data manager of a next-generation DSP system reflecting the above concerns could be as shown in Figures 1.3 and 1.4. The central points of these two figures are the two kinds of displays (the processed-information and tactical displays) through a continuous interrogation procedure. In response to an operator's request, the units in the *data manager* and *display subsystem* have a continuous interaction including data-flow and requests for processing that include localization, tracking, classification for sonar-radar systems (Figure 1.3), and diagnostic images for medical imaging systems (Figure 1.4). Even though the processing steps of radar and airborne systems associated with localization, tracking and classification have conceptual similarities with those of a sonar system, the processing techniques that have been successfully applied in airborne systems have not been successful with sonar systems. This is a typical situation that indicates how hostile, in terms of signal propagation characteristics, the underwater environment is with respect to the atmospheric environment. However, technologies associated with *data fusion, neural networks, knowledge-based systems and automated parameter estimation* will provide solutions to the very difficult operational sonar problem regarding localization, tracking and classification. These issues are discussed in detail in Chapters 9 and 10.

1.4.1 Postprocessing for Sonar and Radar Systems

To provide a better understanding of these differences, let us examine the levels of information required by the data management of sonar and radar systems. Normally, for sonar and radar systems, the processing and integration of information from sensor level to a

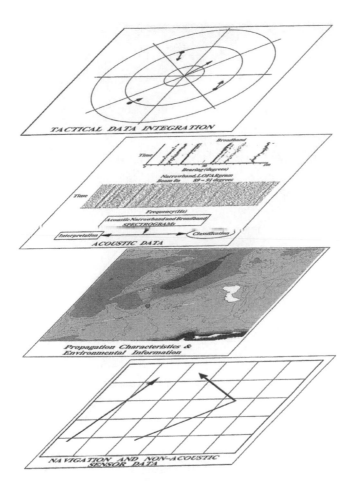

FIGURE 1.3
A simplified overview of integration of different levels of information from sensor level to a command and control level for a sonar or radar system. These levels consist mainly of (1) navigation; (2) environmental information to access the medium's influence on sonar or radar system performance; (3) signal processing of received array sensor signals that provides parameter estimation in terms of bearing, range and temporal spectral estimates for detected signals; and (4) signal following (tracking) and localization of detected targets. (Reprinted from IEEE ©1998. With permission.)

command and control level includes a few distinct processing steps. Figure 1.3 shows a simplified overview of the integration of four different levels of information for a sonar or radar system. These levels consist mainly of:

- Navigation and nonsensor array data
- Environmental information and estimation of propagation characteristics in order to assess the medium's influence on sonar or radar system performance
- Signal processing of received sensor signals that provide parameter estimation in terms of bearing, range and temporal spectral estimates for detected signals
- Signal following (tracking) and localization that monitors the time evolution of a detected signal's estimated parameters

This last tracking and localization capability [32,33] allows the sonar or radar operator to rapidly assess the data from a multisensor system and carry out the processing required to

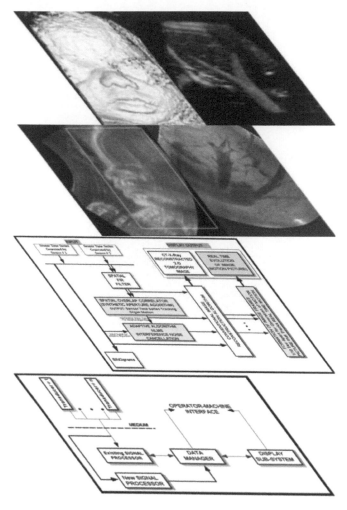

FIGURE 1.4
A simplified overview of integration of different levels of information from the sensor level to a command and control level for a medical imaging system. These levels consist mainly of (1) sensor array configuration, (2) computing architecture, (3) signal processing structure, and (4) reconstructed image to assist medical diagnosis.

develop an array sensor-based tactical picture for integration into the platform level command and control system, as shown later by Figure 1.8.

In order to allow the databases to be searched effectively, a high-performance OMI is required. These interfaces are beginning to draw heavily on modern workstation technology through the use of windows, on-screen menus etc. Large, flat panel displays driven by graphic engines which are equally adept at pixel manipulation as they are with 3D object manipulation will be critical components in future systems. It should be evident by now that the term data manager describes a level of functionality which is well beyond simple data management. The data manager facility applies technologies ranging from relational databases, neural networks [26], fuzzy systems [27], to expert systems [15,26]. The problems it addresses can be variously characterized as signal, data or information processing.

1.4.2 Postprocessing for Medical Imaging Systems

Let us examine the different levels of information to be integrated by the data manager of a medical diagnostic system. Figure 1.4 provides a simplified overview of the levels of information to be integrated by a current medical imaging system. These levels include:

- The system structure in terms of array-sensor configuration and computing architecture
- Sensor time series signal processing structure
- Image processing structure
- Postprocessing for reconstructed image to assist medical diagnosis

In general, current medical imaging systems include very limited postprocessing functionality to enhance the images that may result from main stream image reconstruction processing. It is anticipated, however, that next generation medical imaging systems will enhance their capabilities in postprocessing functionality by including image postprocessing algorithms that are discussed in Chapters 4 through 6 and 19.

More specifically, although modern medical imaging modalities such as CT, MRA, MRI, nuclear medicine, 3D-ultrasound and laser con-focal microscopy provide "slices of the body", significant differences exist between the image content of each modality. Postprocessing, in this case, is essential with special emphasis on data structures, segmentation, image registration, fusion and surface- and volume-based rendering for visualizing volumetric data. To address these issues, the first part of Chapter 4 focuses less in explaining algorithms and rendering techniques, but rather to point out their applicability, benefits, and potential in the medical environment. Moreover, in the second part of Chapter 5, applications are illustrated from the areas of craniofacial surgery, traumatology, neurosurgery, radiotherapy, and medical education. Furthermore, some new applications of volumetric methods are presented: 3D ultrasound, laser con-focal data-sets, and 3D-reconstruction of cardiological data-sets, i.e., vessels as well as ventricles. These new volumetric methods are currently under development but due to their enormous application potential they are expected to be clinically accepted within the coming years.

As an example, Figures 1.5 and 1.6 present the results of image enhancement by means of postprocessing on images that have been acquired by current CT/X-ray and ultrasound systems. The left-hand side image of Figure 1.5 shows a typical X-ray image of a human skull provided by a current type of CT/X-ray imaging system. The right-hand side image of Figure 1.5 is the result of postprocessing the original X-ray image. It is apparent from these results that the right-hand side image includes imaging details that can be valuable to medical staff in minimizing diagnostic errors and interpretation of image-results. Moreover, this kind of postprocessing image functionality may assist in cognitive operations associated with medical diagnostic applications.

Ultrasound medical imaging systems are characterized by poor image resolution capabilities. The three images in Figure 1.6 (top left and right images, bottom left-hand-side image) provide pictures of the skull of a fetus as provided by a conventional ultrasound imaging system. The bottom right-hand-side image of Figure 1.6, presents the resulting 3D postprocessed image by applying the processing algorithms discussed in Chapter 4. The 3D features and characteristics of the skull of the fetus are very pronounced in this case, although the clarity is not as good as in the case of the CT/X-ray image in Figure 1.6. Nevertheless, the image resolution characteristics and 3D features that have been reconstructed in both cases, shown in Figures 1.5 and 1.6, provide an example of the potential

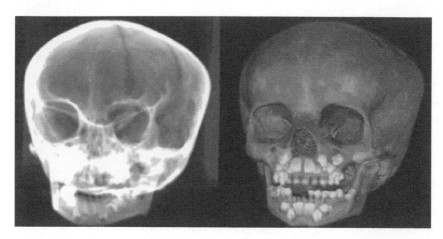

FIGURE 1.5
Left-hand-side is an X-ray image of a human skull. The right-hand-side image is the result of image enhancement by means of postprocessing the original X-ray image. (Courtesy of Prof. G. Sakas, Fraunhofer IDG, Durmstadt, Germany. With permission.)

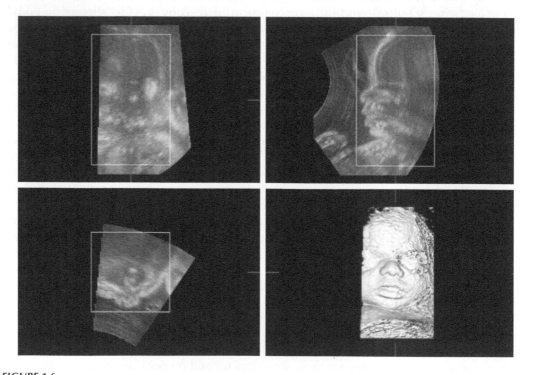

FIGURE 1.6
The two top images and the bottom left-hand-side image provide details of a fetus' skull using conventional medical ultrasound systems. The bottom right-hand-side 3D image is the result of image enhancement by means of postprocessing the original three ultrasound images. (Courtesy of Prof. G. Sakas, Fraunhofer IDG, Durmstadt, Germany. With permission.)

improvements in the image resolution and cognitive functionality that can be integrated in the next generation medical imaging systems.

Needless to say, the image postprocessing functionality of medical imaging systems is directly applicable in sonar and radar applications to reconstruct 2D and 3D image details of detected targets. This kind of image reconstruction postprocessing capability may improve the difficult classification tasks of sonar and radar systems.

At this point, it is also important to reemphasize the significant differences existing between the image content and system functionality of the various medical imaging systems mainly in terms of sensor array configuration and signal processing structures. Undoubtedly, a generic approach exploiting the conceptually similar processing functionalities among the various configurations of medical imaging systems will simplify OMI issues that would result in better interpretation of information of diagnostic importance. Moreover, the integration of data fusion functionality in the data manager of medical imaging systems, as discussed in Chapters 5, 6, and 19, will provide better diagnostic interpretation of the information inherent at the output of the medical imaging systems, by minimizing human errors in terms of interpretation.

Although these issues may appear as exercises of academic interest, it becomes apparent from the above discussion that system advances made in the field of sonar and radar systems may be applicable in medical imaging applications as well.

1.4.3 Signal and Target Tracking and Target Motion Analysis (TMA)

In sonar, radar and imaging system applications, single sensors or sensor networks are used to collect information on time-varying signal parameters of interest. The individual output data produced by the sensor systems result from complex estimation procedures carried out by the *signal processor*, introduced in Section 1.3 (sensor signal processing). Provided the quantities of interest are related to moving point-source objects or small extended objects (radar targets, for instance), relatively simple statistical models can often be derived from basic physical laws, which describe their temporal behavior and thus define the underlying dynamical system. The formulation of adequate dynamics models, however, may be a difficult task in certain applications. For an efficient exploitation of the sensor resources as well as to obtain information not directly provided by the individual sensor reports, appropriate data association and estimation algorithms are required (sensor data processing). These techniques result in tracks, i.e., estimates of state trajectories, which statistically represent the quantities or objects considered along with their temporal history. Tracks are initiated, confirmed, maintained, stored, evaluated, fused with other tracks, and displayed by the *tracking system* or *data manager*. The tracking system, however, should be carefully distinguished from the underlying sensor systems, though there may exist close interrelations, such as in the case of multiple target tracking with an agile-beam radar, increasing the complexity of sensor management.

In contrast to the target tracking via active sensors, Chapter 11 deals with a class of tracking problems that use passive sensors only. In solving tracking problems, active sensors certainly have an advantage over passive sensors. Nevertheless, passive sensors may be a prerequisite to some tracking solution concepts. This is the case, e.g., whenever active sensors are not feasible from a technical or tactical point of view, as in the case of passive sonar systems deployed by submarines and surveillance naval vessels. An important problem in passive target tracking is the TMA problem. The name TMA is normally used for the process of estimating the state of a radiating target from noisy measurements collected by a single passive observer. Typical applications can be found in passive sonar, IR, or radar tracking systems.

For signal followers, the parameter estimation process for tracking the bearing and frequency of detected signals consists of peak picking in a region of bearing and frequency space sketched by fixed gate sizes at the outputs of the conventional and nonconventional beamformers depicted in Figure 1.2. Figure 1.7 provides a schematic interpretation of

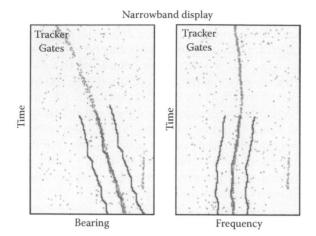

FIGURE 1.7

Signal following functionality in tracking the time-varying frequency and bearing of a detected signal (target) by a sonar or radar system. (Courtesy of William Cambell, Defence R&D Canada Atlantic, Dartmouth, Nova Scotia, Canada. With permission.)

the signal followers functionality in tracking the time-varying frequency and bearing estimates of detected signals in sonar and radar applications. Details about this estimation process can be found in Cambell et al. [34] and Chapter 11. Briefly, in Figure 1.7, the choice of the gate sizes was based on the observed bearing and frequency fluctuations of a detected signal of interest during the experiments. Parabolic interpolation was used to provide refined bearing estimates [35]. For this investigation, the bearings-only tracking process described in Campbell et al. [34] was used as a NB tracker, providing unsmoothed time evolution of the bearing estimates to the localization process [32,36].

Tracking of the time-varying bearing estimates of Figure 1.7 forms the basic processing step to localize a distant target associated with the bearing estimates. This process is called localization or TMA. The output results of a TMA process form the tactical display of a sonar or radar system, as shown in Figures 1.3 and 1.7. In addition, the temporal-spatial spectral analysis output results and the associated display (Figures 1.3 and 1.7), form the basis for classification and the target identification process for sonar and radar systems. In particular, data fusion of the TMA output results with those of a temporal-spatial spectral analysis output results outline an integration process to define the tactical picture for sonar and radar operations, as shown in Figure 1.8. For more details, the reader is referred to References 32–36, which provide detailed discussions of target tracking and TMA operations for sonar and radar systems.

It is apparent from the material presented in this section that for next-generation sonar and radar systems, emphasis should be placed on the degree of interaction between the operator and the system, through an OMI as shown schematically in Figure 1.1. Through this interface, the operator may selectively proceed with localization, tracking and classification tasks, as depicted in Figure 1.8.

In standard CT, image reconstruction is performed using projection data that are acquired in a time sequential manner [6,7]. Organ motion (cardiac motion, blood flow, lung motion due to respiration, patient's restlessness, etc.) during data acquisition produces artifacts, which appear as a blurring effect in the reconstructed image and may lead to inaccurate diagnosis [14]. The intuitive solution to this problem is to speed up the data acquisition process, so that the motion effects become negligible. However, faster CT scanners tend to

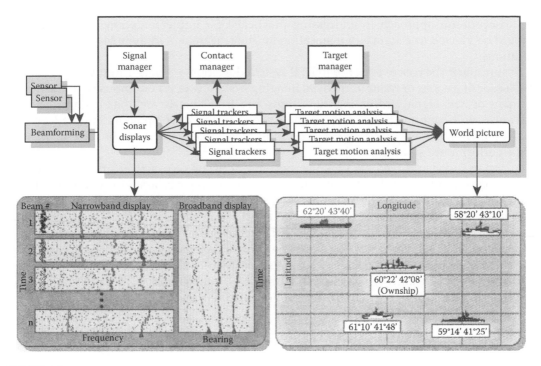

FIGURE 1.8
Formation of a tactical picture for sonar and radar systems. The basic operation is to integrate by means of data fusion the signal tracking and localization functionality with the temporal spatial spectral analysis output results of the generic signal processing structure of Figure 1.2. (Courtesy of Dr. William Roger, Defence R&D Canada Atlantic, Dartmouth, Nova Scotia, Canada. With permission.)

be significantly more costly and, with current X-ray tube technology, the scan times that are required are simply not realizable, as this is discussed in Chapters 14 and 15. Signal processing algorithms to account for organ motion artifacts are therefore needed. Several mathematical techniques have been proposed as a solution to this problem. These techniques usually assume a simplistic linear model for the motion, such as translational, rotational or linear expansion [14]. Some techniques model the motion as a periodic sequence and take projections at a particular point in the motion cycle to achieve the effect of scanning a stationary object. This is known as a retrospective electrocardiogram (ECG)-gating algorithm and projection data are acquired during 12–15 continuous 1-s source rotations while cardiac activity is recorded with an ECG. Thus, the integration of ECG devices with X-ray CT medical tomography imaging systems becomes a necessity in cardiac imaging applications using X-ray CT and MRI systems. However, the information provided by the ECG devices to select in phase segments of CT projection data can be available by signal trackers that can be applied on the sensor time series of the CT receiving array. This kind of application of signal trackers on CT sensor time series will identify the in-phase motion cycles of the heart under a similar configuration as the ECG-gating procedure. Moreover, the application of the signal trackers in cardiac CT imaging systems will eliminate the use of the ECG systems, thus making the medical imaging operations much simpler. These issues will be discussed in some detail in Chapters 14 and 15.

It is anticipated, however, that radar, sonar and medical imaging systems will exhibit fundamental differences in their requirements for information postprocessing functionality. Furthermore, bridging conceptually similar processing requirements may not always be an optimum approach in addressing practical DSP implementation issues, rather it should be viewed as a source of inspiration for the researchers in their search for creative solutions.

In summary, the past experience in DSP system development that "improving the signal processor of a sonar or radar or medical imaging system was synonymous with the development of new signal processing algorithms and faster hardware" has changed. While advances will continue to be made in these areas, future developments in data (contact) management represent one of the most exciting avenues of research in the development of high-performance systems.

In sonar, radar and medical imaging systems, an issue of practical importance is the operational requirement by the operator to be able to rapidly assess numerous images and detected signals in terms of localization, tracking, classification and diagnostic interpretation in order to pass the necessary information up through the chain of command to enable tactical or medical diagnostic decisions to be made in a timely manner. Thus, an assigned task for a *data manager* would be to provide the operator with quick and easy access to both the output of the *signal processor*, which is called processed data display, and the tactical display, which will show medical images and localization and tracking information through graphical interaction between the processed data and tactical displays.

1.4.4 Engineering Databases

The design and integration of engineering databases in the functionality of a *data manager* assists the identification and classification process, as shown schematically in Figure 1.3. To illustrate the concept of an engineering database, we will consider the identification process for landmines and other security surveillance systems discussed in Chapter 7, which is a highly essential functionality to minimize the false alarm rate. Although a lot of information on landmines exists, often organized in electronic databases, there is nothing like a CAD engineering database. Indeed, most databases serve either documentation purposes or are signatures of objects of interest related to a particular sensor technology. This wealth of information must be collected and organized in such a way so that it can be used online, through the necessary interfaces to the sensorial information, by each one of the future identification systems. Thus, an engineering database is intended to be the common core software applied to security surveillance detection systems [41] and to medical diagnostic applications, discussed in Chapter 16. It could be built around a specially engineered database storing all available information of interest to a specific application. The underlying idea is, using techniques of cognitive and perceptual sciences, to extract the particular features that characterize a particular medical diagnosis or a class of options, successively, define the sensorial information needed to identify these features in typical diagnostic applications. Such a diagnostic or security screening identification system would not only trigger an alarm for every suspect object but would also reconstruct a comprehensive model of the target. Successively, it would compare the model to an existing engineering database deciding or assisting the operator to make a decision as to the nature of the detected object.

A general approach of the engineering database concept and its applicability in the aforementioned DSP systems would assume that an effective engineering database will be a function of the available information on the subjects of interest, such as underwater targets, radar targets and medical diagnostic images. Most recently, the importance of engineering database concepts has been reinforced by the emerging field of biometrics for security screening applications. Chapter 8 discusses these issues within the scope of face and gait recognition for biometrics applications. Moreover, the functionality of an engineering database would be highly linked with the multisensor fusion process, which is the subject of discussion in the next section.

1.4.5 Multisensor Data Fusion

Data fusion refers to the acquisition, processing and synergistic combination of information from various knowledge sources and sensors to provide a better understanding of the situation under consideration [39]. Classification is an information processing task in which specific entities are mapped to general categories. For example, in the detection of land mines, the fusion of acoustic [41], electromagnetic (EM) and IR sensor data are in consideration to provide a better land-mine field picture and minimize the false alarm rates. Data and image fusion is of equal importance also for biometric recognition applications to optimize identification for security screening, as discussed in Chapter 8. The discussion of this section has been largely influenced by the work of Kundur and Hatzinakos on "blind image deconvolution" [39].

The process of multisensor data fusion addresses the issue of system integration of different type of sensors and the problems inherent in attempting to fuse and integrate the resulting data streams into a coherent picture of operational importance. The term integration is used here to describe operations wherein a sensor input may be used independently with respect to other sensor data in structuring an overall solution. *Fusion* is used to describe the result of joint analysis of two or more originally distinct data streams.

More specifically, while multisensors are more likely to correctly identify positive targets and eliminate false returns, using them effectively will require *fusing* the incoming data streams, each of which may have a different character. This task will require solutions to the following engineering problems:

- Correct combination of the multiple data streams in the same context
- Processing multiple-signals to eliminate false positives and further refine positive returns

For example, in humanitarian demining a positive return from a simple metal detector might be combined with a ground penetrating radar GPR evaluation resulting in the classification of the target as a spent shell casing, allowing the operator to safely pass by in confidence.

Given a design that can satisfy the above goals, it will then be possible to design and implement computer-assisted or automatic recognition in order to positively identify the nature, position and orientation of a target. Automatic recognition, however, will be pursued by the engineering database, as shown in Figure 1.3.

In data fusion, another issue of equal importance is the ability to deal with conflicting data, producing interim results that the algorithm can revise as more data become available. In general, the data interpretation process, as part of the functionality of data fusion, consists briefly of the following stages [39]:

- Low-level data manipulation
- Extraction of features from the data either using signal processing techniques or physical sensor models

- Classification of data using techniques such as Bayesian hypothesis testing, fuzzy logic, and neural networks
- Heuristic expert system rules to guide the previous levels, make high-level control decisions, provide operator guidance, and provide early warnings and diagnostics.

Current research and development (R&D) projects in this area include the processing of localization and identification of data from various sources, or type of sensors. The systems combine features of modern multihypothesis tracking methods and correlation. This approach, to process all available data regarding targets of interest allows the user to extract the maximum amount of information concerning target location from the complex "sea" of available data. Then a correlation algorithm is used to process large volumes of data containing localization and attribute information using multiple hypothesis methods.

In image classification and fusion strategies, many inaccuracies often result from attempting to fuse data that exhibit motion-induced blurring or defocusing effects and background noise [37,38]. Compensation for such distortions is inherently sensor-dependent and is nontrivial as the distortion is often time varying and unknown. In such cases, blind image processing, which relies on partial only information about the original data and the distorting process, is suitable [39].

In general, multisensor data fusion is an evolving subject, which is considered to be highly essential in resolving the sonar, radar detection/classification problem, the biometric recognition problem, and diagnostic problems in medical imaging systems, as discussed in Chapters 4, 5, 6, 8, and 16. Since a single sensor system with acceptable very low false alarm rate is rarely available, current developments in sonar, radar and medical diagnostic systems include multisensor configurations to minimize the false alarm rates. Then the multisensor data fusion process becomes highly essential. Although data fusion and databases have not been implemented yet in medical imaging systems, undoubtedly, their potential use in this area will be a rapidly evolving R&D subject in the near future. Then system experience in the areas of sonar and radar systems would be a valuable asset in that regard. For medical imaging applications the data and image fusion processes will be discussed in detail in Chapters 4 through 6.

Finally, Chapters 17 through 19 conclude the material of this handbook by providing clinical data and discussion on the role of arterial spin labeling, contrast agents, and computer aided diagnosis in magnetic resonance and mammography imaging diagnostic applications, respectively.

References

1. W.C. Knight, R.G. Pridham and S.M. Kay. 1981. Digital signal processing for sonar. *Proc. IEEE*, 69(11), 1451–1506.
2. B. Windrow, P.E. Manfey, L.J. Griffiths, and B.B. Goode. 1967. Adaptive antenna systems. *Proc. IEEE*, 55(12), 2143–2159.
3. B. Windrow and S.D. Stearns. 1985. *Adaptive Signal Processing*. Prentice Hall, Englewood Cliffs, NJ.
4. A.A. Winder. 1975. Sonar system technology. *IEEE Trans. Sonic Ultrasonics*, SU-22(5), 291–332.
5. A.B. Baggeroer. 1978. Sonar signal processing. In: *Applications of Digital Signal Processing*, A.V. Oppenheim, editor. Prentice Hall, Englewood Cliffs, NJ.

6. H.J. Scudder. 1978. Introduction to computer aided tomography. *Proc. IEEE*, 66(6), 628–637.

7. A.C. Kak and M. Slaney. 1992. *Principles of Computerized Tomography Imaging*. IEEE Press, NY.

8. D. Nahamoo and A.C. Kak. 1982. Ultrasonic Diffraction Imaging, TR-EE 82-80. Department of Electrical Engineering., Purdue Univesrity, West Lafayette, IN.

9. S.W. Flax and M. O'Donnell. 1988. Phase-aberration correction using signals from point reflectors and diffuse scatterers: Basic principles. *IEEE Trans. Ultrason., Ferroelectr. Freq. Control*, 35(6), 758–767.

10. G.C. Ng, S.S. Worrell, P.D. Freiburger and G.E. Trahey. 1994. A comparative evaluation of several algorithms for phase aberration correction. *IEEE Trans. Ultrason., Ferroelectr. Freq. Control*, 41(5), 631–643.

11. A.K. Jain. 1990. *Fundamentals of Digital Image Processing*. Prentice-Hall, Englewood Cliffs, NJ.

12. Q.S. Xiang and R.M. Henkelman. 1993. K-space description for the imaging of dynamic objects. *Magn., Reson. Med.*, 29, 422–428.

13. M.L. Lauzon, D.W. Holdsworth, R. Frayne and B.K. Rutt. 1994. Effects of physiologic waveform variability in triggered MR imaging: Theoretical analysis. *J. Magn. Reson. Imaging*, 4(6), 853–867.

14. C.J. Ritchie, C.R. Crawford, J.D. Godwin, K.F. King and Y. Kim. 1996. Correction of computed tomography motion artifacts using pixel-specific back-projection. *IEEE Trans. Med. Imaging*, 15(3), 333–342.

15. S. Stergiopoulos. 1998. Implementation of adaptive and synthetic-aperture processing schemes in integrated active–passive sonar systems. *Proc. IEEE*, 86(2), 358–396.

16. D. Stansfield. 1990. *Underwater Electroacoustic Transducers*. Bath University Press and Institute of Acoustics, Bath, England.

17. J.M. Powers. 1988. Long range hydrophones. In: *Applications of Ferroelectric Polymers*, T.T. Wang, J.M. Herbert, A.M. Glass, editors. Chapman and Hall, NY.

18. P.B. Boemer, W.A. Edelstein, C.E. Hayes, S.P. Souza and O.M. Mueller. 1990. The NMR phased array. *Magn., Reson. Med.*, 16, 192–225.

19. P.S. Melki, F.A. Jolesz and R.V. Mulkern. 1992. Partial RF echo planar imaging with the FAISE method. I. Experimental and theoretical assessment of artifact *Magn., Reson. Med.*, 26, 328–341.

20. N.L. Owsley. 1985. Sonar array processing. In: *Signal Processing Series*, S. Haykin, editor, A.V. Oppenheim, series editor. Prentice-Hall, Englewood Cliffs, NJ.

21. B. Van Veen and K. Buckley. 1988. Beamforming: a versatile approach to spatial filtering. *IEEE ASSP Mag.*, 4–24.

22. A.H. Sayed and T. Kailath. 1994. A state-space approach to adaptive RLS filtering. *IEEE Single Processing Mag.*, 18–60.

23. E.J. Sullivan, W.M. Carey and S. Stergiopoulos. 1992. Editorial, special issue on acoustic synthetic aperture processing. *IEEE J. Oceanic Eng.*, 17(1), 1–7.

24. C.L. Nikias and J.M. Mendel. 1992. Signal processing with higher-order spectra. *IEEE SP Mag.*, 10–37.

25. S. Stergiopoulos and A.T. Ashley. 1993. Guest editorial, special issue on sonar system technology. *IEEE J. Oceanic Eng.*, 18(4), 361–365.

26. A.B. Baggeroer, W.A. Kuperman and P.N. Mikhalevsky. 1993. An overview of matched field methods in ocean acoustics. *IEEE J. Oceanic Eng.*, 18(4), 401–424.

27. W. Miller, T. McKenna, and C. Lau. 1992. Office of Naval Research Contributions to Neural Networks and Signal Processing in Oceanic Engineering. *IEEEJ. Oceanic Eng.* 17(4), 299–307.

28. A. Kummert. 1993. Fuzzy technology implemented in sonar systems. *IEEE J. Oceanic Eng.*, 18(4), 483–490.

29. R.D. Doolitle, A. Tolstoy and E.J. Sullivan. 1993. Editorial, special issue on detection and estimation in matched field processing. *IEEE J. Oceanic Eng.*, 18, 153–155.

30. A. Antoniou. 1993. *Digital Filters: Analysis, Design, and Applications*, 2nd Ed. McGraw-Hill, NY.

31. Mercury Computer Systems Inc. 1997. Mercury news. Chelmsford, MA.

32. Y. Bar-Shalom and T.E. Fortman. 1988. *Tracking and Data Association*. Academic Press, Boston, MA.

33. S.S. Blackman. 1986. *Multiple-target Tracking with Radar Applications*. Artech House, Norwood, MA.

34. W. Cambell, S. Stergiopoulos and J. Riley. 1995. Effects of bearing estimation improvements of non-conventional beamformers on bearing-only tracking. *Proceedings of Oceans '95*, MTS/IEEE, San Diego, CA.

35. W.A. Roger and R.S. Walker. 1986. Accurate estimation of source bearing from line arrays. *Proceedings of Thirteen Biennial Symposium on Communications*, Kingston, Ont., Canada.

36. D. Peters. 1995. Long range towed array target analysis – principles and practice. *DREA Memorandum 95/217*, Defense Research Establishment Atlantic, Dartmouth, Nova Scotia, Canada.

37. A.H.S. Solberg, A.K. Jain and T.Taxt. 1994. A Markov random field model for classification of multisource satellite imagery. *IEEE Trans. Geosci. Remote Sensing*, 32, 768–778.

38. L.J. Chipman, T.M. Orr, and L.N. Graham. 1995. Wavelets and image fusion. *Proc. SPIE*, 2569, 208–219.

39. D. Kundur and D. Hatzinakos. 1996. Blind image deconvolution. *IEEE Signal Processing Mag.* 13, 43–64.

40. Mercury Computer Systems Inc. 1998. Mercury news. Chelmsford, MA.

41. Younis W., Stergiopoulos S., Havelock, D., Groski J. 2002. Non-distructive imaging of shallow buried objects using acoustic computed tomography. *J. Acoust. Soc. Am.*, 111(5), 2117–2127.

42. Stergiopoulos S. and Dhanantwari A. 2004. New DSPs aid non-invasive diagnostics. *EETimes Electr. Eng.*, 1334, 49–54.

Section I

Generic Topics
on Signal Processing

2

Adaptive Systems for Signal Process*

Simon Haykin

McMaster University

CONTENTS

---

* The material presented in this chapter is based on the author's two textbooks: (1) *Adaptive Filter Theory* (1996) and (2) *Neural Networks: A Comprehensive Foundation* (1999), Prentice-Hall, Englewood Cliffs, NJ.

2.1 The Filtering Problem

The term "filter" is often used to describe a device in the form of a piece of physical hardware or software that is applied to a set of noisy data in order to extract information about a prescribed quantity of interest. The noise may arise from a variety of sources. For example, the data may have been derived by means of noisy sensors or may represent a useful signal component that has been corrupted by transmission through a communication channel. In any event, we may use a filter to perform three basic information-processing tasks.

(1) *Filtering* means the extraction of information about a quantity of interest at time t by using data measured up to and including time t.

(2) *Smoothing* differs from filtering in that information about the quantity of interest need not be available at time t, and data measured later than time t can be used in obtaining this information. This means that in the case of smoothing there is a *delay* in producing the result of interest. Since in the smoothing process we are able to use data obtained not only up to time t, but also data obtained after time t, we would expect smoothing to be more accurate in some sense than filtering.

(3) *Prediction* is the forecasting side of information processing. The aim here is to derive information about what the quantity of interest will be like at some time $t+\tau$ in the future, for some $\tau > 0$, by using data measured up to and including time t.

We may classify filters into linear and nonlinear. A filter is said to be *linear* if the filtered, smoothed, or predicted quantity at the output of the device is a *linear function of the observations applied to the filter input*. Otherwise, the filter is *nonlinear*.

In the statistical approach to the solution of the *linear filtering problem* as classified above, we assume the availability of certain statistical parameters (i.e., *mean and correlation functions*) of the useful signal and unwanted additive noise, and the requirement is to design a linear filter with the noisy data as input so as to minimize the effects of noise at the filter output according to some statistical criterion. A useful approach to this filter-optimization problem is to minimize the mean-square value of the *error* signal that is defined as the difference between some desired response and the actual filter output. For stationary inputs, the resulting solution is *commonly* known as the *Wiener filter,* which is said to be *optimum in the mean-square sense.* A plot of the mean-square value of the error signal versus the adjustable parameters of a linear filter is referred to as the *error-performance surface.* The minimum point of this surface represents the Wiener solution.

The Wiener filter is inadequate for dealing with situations in which *nonstationarity* of the signal and/or noise is intrinsic to the problem. In such situations, the optimum filter has to assume a *time-varying form.* A highly successful solution to this more difficult problem is found in the *Kalman filter,* a powerful device with a wide variety of engineering applications.

Linear filter theory, encompassing both Wiener and Kalman filters, has been developed fully in the literature for *continuous-time* as well as *discrete-time* signals. However, for technical reasons influenced by the wide availability of digital computers and the ever-increasing use of digital signal-processing devices, we find in practice that the discrete-time representation is often the preferred method. Accordingly, in this chapter, we only consider the discrete-time version of Wiener and Kalman filters. In this method of representation, the input and output signals, as well as the characteristics of the filters themselves, are all defined at discrete instants of time. In any case, a continuous-time signal may always be represented by a *sequence of samples* that are derived by observing the signal at uniformly spaced instants of time. No loss of information is incurred during this conversion process provided, of course, we satisfy the well-known *sampling theorem*, according to which the sampling rate has to be greater than twice the highest frequency component of the continuous-time signal (assumed to be of a low-pass kind). We may thus represent a continuous-time signal $u(t)$ by the sequence $u(n)$, $n = 0, \pm 1, \pm 2, \ldots$, where for convenience we have normalized the sampling period to unity, a practice that we follow throughout this chapter.

2.2 Adaptive Filters

The design of a Wiener filter requires *a priori* information about the statistics of the data to be processed. The filter is optimum only when the statistical characteristics of the input data match the *a priori* information on which the design of the filter is based. When this information is not known completely, however, it may not be possible to design the Wiener filter or else the design may no longer be optimum. A straightforward approach that we may use in such situations is the "estimate and plug" procedure. This is a two-stage process whereby the filter first "estimates" the statistical parameters of the relevant signals and then "plugs" the results so obtained into a *nonrecursive* formula for computing the filter parameters. For a *real-time operation*, this procedure has the disadvantage of requiring excessively elaborate and costly hardware. A more efficient method is to use an adaptive filter. By such a device we mean one that is *self-designing* in that the adaptive filter relies on a *recursive algorithm* for its operation, which makes it possible for the filter to perform satisfactorily in an environment where complete knowledge of the relevant signal characteristics is not available. The algorithm starts from some predetermined set of *initial conditions*, representing whatever we know about the environment. Yet, in a stationary environment, we find that after successive iterations of the algorithm it *converges* to the optimum Wiener solution in some statistical sense. In a nonstationary environment, the algorithm offers a *tracking* capability, in that it can track time variations in the statistics of the input data, provided that the variations are sufficiently slow.

As a direct consequence of the application of a recursive algorithm whereby the parameters of an adaptive filter are updated from one iteration to the next, the parameters become *data dependent*. This, therefore, means that an adaptive filter is in reality a *nonlinear device, in the sense that it does not obey the principle of superposition*. Notwithstanding this property, adaptive filters are commonly classified as linear or nonlinear. An adaptive filter is said to be *linear* if the estimate of quantity of interest is computed adaptively (at the output of the filter) as a *linear combination of the available set of observations applied to the filter input*. Otherwise, the adaptive filter is said to be *nonlinear*.

A wide variety of recursive algorithms have been developed in the literature of the operation of linear adaptive filters. In the final analysis, the choice of one algorithm over another is determined by one or more of the following factors:

- *Rate of convergence* — This is defined as the number of iterations required for the algorithm, in response to stationary inputs, to converge "close enough" to the optimum Wiener solution in the mean-square sense. A fast rate of convergence allows the algorithm to adapt rapidly to a stationary environment of unknown statistics.

- *Misadjustment* — For an algorithm of interest, this parameter provides a quantitative measure of the amount by which the final value of the mean-squared error, averaged over an ensemble of adaptive filters, deviates from the minimum mean-squared error that is produced by the Wiener filter.

- *Tracking* — When an adaptive filtering algorithm operates in a nonstationary environment, the algorithm is required to *track* statistical variations in the environment. The tracking performance of the algorithm, however, is influenced by two contradictory features: (1) the rate of convergence and (2) the steady-state fluctuation due to algorithm noise.

- *Robustness* — For an adaptive filter to be *robust,* small disturbances (i.e., disturbances with small energy) can only result in small estimation errors. The disturbances may arise from a variety of factors internal or external to the filter.

- *Computational requirements* — Here, the issues of concern include (1) the number of operations (i.e., multiplications, divisions, and additions/subtractions) required to make one complete iteration of the algorithm, (2) the size of memory locations required to store the data and the program, and (3) the investment required to program the algorithm on a computer.

- *Structure* — This refers to the structure of information flow in the algorithm, determining the manner in which it is implemented in hardware form. For example, an algorithm whose structure exhibits high modularity, parallelism, or concurrency is well suited for implementation using very large-scale integration (VLSI).[†]

- *Numerical properties* — When an algorithm is implemented numerically, inaccuracies are produced due to *quantization errors.* The quantization errors are due to analog-to-digital conversion of the input data and digital representation of internal calculations. Ordinarily, it is the latter source of quantization errors that poses a serious design problem. In particular, there are two basic issues of concern: numerical stability and numerical accuracy. *Numerical stability* is an inherent characteristic of an adaptive filtering algorithm. *Numerical accuracy,* on the other hand, is determined by the number of *bits* (i.e., **bi**nary dig**i**ts used in the numerical representation of data samples and filter coefficients). An adaptive filtering algorithm is said to be numerically robust when it is insensitive to variations in the word length used in its digital implementation.

[†] VLSI technology favors the implementation of algorithms that possess high modularity, parallelism, or concurrency. We say that a structure is *modular* when it consists of similar stages connected in cascade. By *parallelism* we mean a large number of operations being performed side by side. By *concurrency,* we mean a large number of *similar* computations being performed at the same time. For a discussion of VLSI implementation of adaptive filters, see Shabhag and Parhi (1994). This book emphasizes the use of *pipelining,* an architectural technique used for increasing the throughput of an adaptive filtering algorithm.

These factors, in their own ways, also enter into the design of nonlinear adaptive filters, except for the fact that we now no longer have a well-defined frame of reference in the form of a Wiener filter. Rather, we speak of a nonlinear filtering algorithm that may converge to a local minimum or, hopefully, a global minimum on the error-performance surface.

In the sections that follow, we shall first discuss various aspects of linear adaptive filters. Discussion of nonlinear adaptive filters is deferred to Section 2.6.

2.3 Linear Filter Structures

The operation of a linear adaptive filtering algorithm involves two basic processes: (1) a *filtering* process designed to produce an output in response to a sequence of input data, and (2) an *adaptive* process, the purpose of which is to provide mechanism for the *adaptive control* of an *adjustable* set of parameters used in the filtering process. These two processes work interactively with each other. Naturally, the choice of a structure for the filtering process has a profound effect on the operation of the algorithm as a whole.

There are three types of filter structures that distinguish themselves in the context of an adaptive filter with *finite memory* or, equivalently, *finite-duration impulse response* (FIR). The three filter structures are transversal filter, lattice predictor, and systolic array.

2.3.1 Transversal Filter

The transversal filter[‡], also referred to as a *tapped-delay line filter,* consists of three basic elements, as depicted in Figure 2.1: (1) a *unit-delay element,* (2) a *multiplier,* and (3) an *adder.* The number of delay elements used in the filter determines the finite duration of its impulse response. The number of delay elements, shown as $M-1$ in Figure 2.1, is commonly referred to as the filter order. In Figure 2.1, the delay elements are each identified by the unit-delay operator z^{-1}. In particular, when z^{-1} operates on the input $u(n)$, the resulting output is $u(n-1)$. The role of each multiplier in the filter is to multiply the *tap input,* to which it is connected by a filter coefficient referred to as a *tap weight.* Thus, a multiplier connected to the kth tap input $u(n-k)$ produces the scalar version of the *inner product,* $W_k^* u(n-k)$, where W_k is the respective tap weight and $k=0, 1, \ldots, M-1$. The asterisk denotes *complex conjugation,* which assumes that the tap inputs and, therefore, the tap weights are all *complex valued.* The combined role of the adders in the filter is to sum the individual multiplier outputs and produce an overall filter output. For the transversal filter described in Figure 2.1, the filter output is given by:

$$y(n) = \sum_{k=0}^{m-1} W_k^* u(n-k) \tag{2.1}$$

Equation 2.1 is called a finite *convolution sum* in the sense that it convolves W_k^* the (FIR) of the filter, W_k^*, with the filter input $u(n)$ to produce the filter output $y(n)$.

[‡] The transversal filter was first described by Kallmann as a continuous-time device whose output is formed as a linear combination of voltages taken from uniformly spaced taps in a nondispersive delay line (Kallmann, 1940). In recent years, the transversal filter has been implemented using digital circuitry, charged-coupled devices, or surface acoustic wave devices. Owing to its versatility and ease of implementation, the transversal filter has emerged as an essential signal-processing structure in a wide variety of applications.

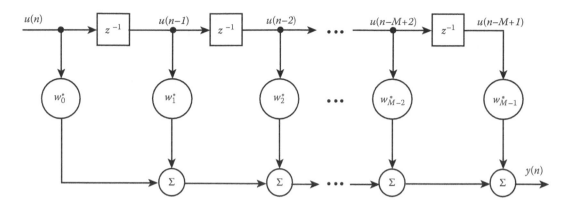

FIGURE 2.1
Transversal filter.

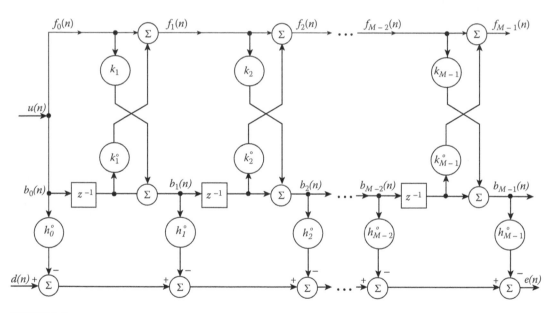

FIGURE 2.2
Multistage lattice filter.

2.3.2 Lattice Predictor

A *lattice predictor*[§] is *modular* in structure in that it consists of a number of individual stages, each of which has the appearance of a lattice, hence, the name "lattice" as a structural descriptor. Figure 2.2 depicts a lattice predictor consisting of $M-1$ stages; the number $M-1$ is referred to as the *predictor order*. The mth stage of the lattice predictor in Figure 2.2 is described by the pair of input–output relations (assuming the use of complex-valued, wide-sense stationary input data):

[§] The development of the lattice predictor is credited to Itakura and Saito (1972).

$$f_m(n) = f_{m-1}(n) + k_m^* b_{m-1}(n-1) \tag{2.2}$$

$$b_m(n) = b_{m-1}(n-1) + k_m f_{m-1}(n) \tag{2.3}$$

where $m = 1, 2, \ldots, M-1$, and $M-1$ is the final predictor order. The variable $f_m(n)$ is the mth *forward prediction error*, and $b_m(n)$ is the mth *backward prediction error*. The coefficient k_m is called the mth *reflection coefficient*. The forward prediction error $f_m(n)$ is defined as the difference between the input $u(n)$ and its *one-step predicted* value; the latter is based on the set of m past inputs $u(n-1), \ldots, u(n-m)$. Correspondingly, the backward prediction error $b_m(n)$ is defined as the difference between the input $u(n-m)$ and its "backward" prediction based on the set of m "future" inputs $u(n), \ldots, u(n-m+1)$. Considering the conditions at the input of stage 1 in Figure 2.2, we have:

$$f_0(n) = b_0(n) = u(n) \tag{2.4}$$

where $u(n)$ is the lattice predictor input at time n. Thus, starting with the *initial conditions* of Equation 2.4 and given the set of reflection coefficients $K_1, K_2, \ldots, K_{M-1}$, we may determine the final pair of outputs $F_{M-1}(n)$ and $B_{M-1}(n)$ by moving through the lattice predictor, stage by stage.

For a *correlated* input sequence $u(n), u(n-1), \ldots, u(n-M+1)$ drawn from a stationary process, the backward prediction errors $b_0, b_1(n), \ldots, b_{M-1}(n)$ form a sequence of *uncorrelated* random variables. Moreover, there is a one-to-one correspondence between these two sequences of random variables in the sense that if we are given one of them, we may uniquely determine the other and vice versa. Accordingly, a linear combination of the backward prediction errors $b_0, b_1(n), \ldots, b_{M-1}(n)$ may be used to provide an *estimate* of some desired response $d(n)$, as depicted in the lower half of Figure 2.2. The arithmetic difference between $d(n)$ and the estimate so produced represents the estimation error $e(n)$. The process described herein is referred to as a *joint-process estimation*. Naturally, we may use the original input sequence $u(n), u(n-1), \ldots, u(n-M+1)$ to produce an estimate of the desired response $d(n)$ directly. The indirect method depicted in Figure 2.2, however, has the advantage of simplifying the computation of the tap weights $h_0, h_1(n), \ldots, h_{M-1}$ by exploiting the uncorrelated nature of the corresponding backward prediction errors used in the estimation.

2.3.3 Systolic Array

A *systolic array*[1] represents a *parallel computing* network ideally suited for *mapping* a number of important linear algebra computations, such as *matrix multiplication, triangularization,* and *back substitution*. Two basic types of processing elements may be distinguished in a systolic array: *boundary cells* and *internal cells*. Their functions are depicted in Figure 2.3a and b, respectively. In each case, the parameter r represents a value *stored* within the cell. The function of the boundary cell is to produce an output equal to the input u divided by the number r stored in the cell. The function of the internal cell is twofold: (1) to multiply the input z (coming in from the top) by the number r stored in the cell, subtract the product rz from the second input (coming in from the left), and thereby produce the

---

[1] The systolic array was pioneered by Kung and Leiserson (1978). In particular, the use of systolic arrays has made it possible to achieve a high throughput, which is required for many advanced signal-processing algorithms to operate in *real time*.

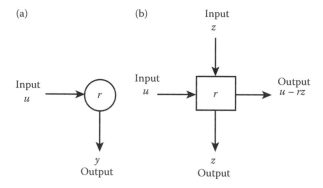

FIGURE 2.3
Two basic cells of a systolic array: (a) boundary cell and (b) internal cell.

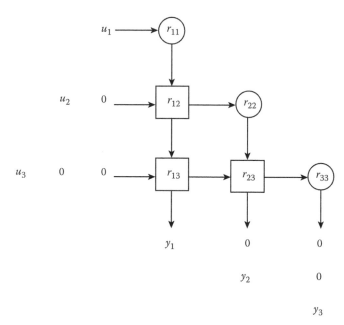

FIGURE 2.4
Triangular systolic array.

difference $u - rz$ as an output from the right-hand side of the cell; and (2) to transmit the first z downward without alteration.

Consider, for example, the 3×3 triangular array shown in Figure 2.4. This systolic array involves a combination of boundary and internal cells. In this case, the triangular array computes an output vector $\mathbf{y}$ related to the input vector $\mathbf{u}$ as follows:

$$Y = \mathbf{R}^{-T}\mathbf{u} \tag{2.5}$$

where the $\mathbf{R}^{-T}$ is the *inverse* of the transposed matrix $\mathbf{R}^{T}$. The elements of $\mathbf{R}^{T}$ are the respective cell contents of the triangular array. The zeros added to the inputs of the array in Figure 2.4 are intended to provide the delays necessary for pipelining the computation described in Equation 2.5.

A systolic array architecture, as described herein, offers the desirable features of *modularity*, *local interconnections*, and highly *pipelined* and *synchronized* parallel processing; the synchronization is achieved by means of a global *clock*.

We note that the transversal filter of Figure 2.1, the joint-process estimator of Figure 2.2 based on a lattice predictor, and the triangular systolic array of Figure 2.4 have a common property: all three of them are characterized by an impulse response of finite duration. In other words, they are examples of a *FIR filter*, whose structures contain *feedforward* paths only. On the other hand, the filter structure shown in Figure 2.5 is an example of an

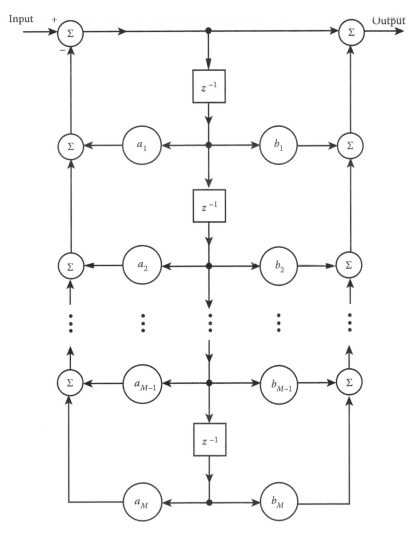

FIGURE 2.5
IIR filter.

infinite-duration impulse response (IIR) filter. The feature that distinguishes an IIR filter from an FIR filter is the inclusion of *feedback* paths. Indeed, it is the presence of feedback that makes the duration of the impulse response of an IIR filter infinitely long. Furthermore, the presence of feedback introduces a new problem, namely, that of *stability.* In particular, it is possible for an IIR filter to become unstable (i.e., break into oscillation), unless special precaution is taken in the choice of feedback coefficients. By contrast, an FIR filter in inherently *stable.* This explains the reason for the popular use of FIR filters, in one form or another, as the structural basis for the design of linear adaptive filters.

2.4 Approaches to the Development of Linear Adaptive Filtering Algorithms

There is no unique solution to the linear adaptive filtering problem. Rather, we have a "kit of tools" represented by a variety of recursive algorithms, each of which offers desirable features of its own. (For complete detailed treatment of linear adaptive filters, see the book by Haykin, 1996.) The challenge facing the user of adaptive filtering is (1) to understand the capabilities and limitations of various adaptive filtering algorithms and (2) to use this understanding in the selection of the appropriate algorithm for the application at hand.

Basically, we may identify two distinct approaches for deriving recursive algorithms for the operation of linear adaptive filters, as discussed next.

2.4.1 Stochastic Gradient Approach

Here, we may use a tapped-delay line or transversal filter as the structural basis for implementing the linear adaptive filter. For the case of stationary inputs, the *cost function***, also referred to as the *index of performance*, is defined as the *mean-squared error* (i.e., the mean-square value of the difference between the desired response and the transversal filter output). This cost function is precisely a second-order function of the tap weights in the transversal filter. The dependence of the mean-squared error on the unknown tap weights may be viewed to be in the form of a *multidimensional paraboloid* (i.e., punch bowl) with a uniquely defined bottom or *minimum point.* As mentioned previously, we refer to this paraboloid as the *error-performance surface;* the tap weights corresponding to the minimum point of the surface define the optimum Wiener solution.

To develop a recursive algorithm for updating the tap weights of the adaptive transversal filter, we proceed in two stages. We first modify the system of *Wiener–Hopf equations* (i.e., the matrix equation defining the optimum Wiener solution) through the use of the *method of steepest descent,* a well-known technique in optimization theory. This modification requires the use of a *gradient vector,* the value of which depends on two parameters: the *correlation matrix* of the tap inputs in the transversal filter and the *cross-correlation vector* between the desired response and the same tap inputs. Next, we use instantaneous values for these correlations so as to derive an *estimate* for the gradient vector, making it assume a *stochastic* character in general. The resulting algorithm is widely known as the *least-mean-*

** In the general definition of a function, we speak of a transformation from a vector space into the space of real (or complex) scalars (Luenberger, 1969; Dorny, 1975). A cost function provides a quantitative measure for assessing the quality of performance and, hence, the restriction of it to a real scalar.

square (LMS) algorithm, the essence of which may be described in words as follows for the case of a transversal filter operating on real-valued data:

$$
\begin{pmatrix} \text{updated value} \\ \text{of tap-weight} \\ \text{vector} \end{pmatrix} = \begin{pmatrix} \text{old value} \\ \text{of tap-weight} \\ \text{vector} \end{pmatrix} + \begin{pmatrix} \text{learning-} \\ \text{rate} \\ \text{parameter} \end{pmatrix} \begin{pmatrix} \text{tap-} \\ \text{input} \\ \text{vector} \end{pmatrix} \begin{pmatrix} \text{error} \\ \text{signal} \end{pmatrix}
$$

where the error signal is defined as the difference between some desired response and the actual response of the transversal filter produced by the tap-input vector.

The LMS algorithm, summarized in Table 2.1, is simple and yet capable of achieving satisfactory performance under the right conditions. Its major limitations are a relatively slow rate of convergence and a sensitivity to variations in the condition number of the

TABLE 2.1

Summary of the LMS Algorithm

Notations:

$\mathbf{u}(n)$	tap-input vector at time n
	$[u(n), u(n-1), ..., u(n-M+1)]^T$
M	number of tap inputs
$d(n)$	desired response at time n
$\hat{\mathbf{w}}(n)$	$[w_0(n), w_1(n), ..., w_{M-1}(n)]^T$
	tap-weight vector at time n
$y(n)$	actual response of the tapped-delay line filter
	$\hat{\mathbf{w}}^H(n)\mathbf{u}(n)$, where superscript H denotes Hermitian transposition
$e(n)$	error signal
	$d(n)-y(n)$

Parameters:

M	number of taps
m	step-size parameter

$$0 < \mu < \frac{2}{\text{tap-input power}}$$

$$\text{tap-input power} = \sum_{k=0}^{M-1} E[|u(n-k)|^2]$$

Initialization:

If prior knowledge on the tap-weight vector $\hat{\mathbf{w}}(n)$ is available, use it to select an appropriate value for $\hat{\mathbf{w}}(0)$. Otherwise, set $\hat{\mathbf{w}}(0)=0$.

Date:

Given:	$\mathbf{u}(n)=M\times 1$ tap-input vector at time n
	$d(n)=$ desired response at time n
To be computed:	$\hat{\mathbf{w}}(n+1)=$ estimate of tap-weight vector at time $n+1$
Computation:	For $n=0, 1, 2, ...,$ compute
	$e(n)=d(n)-\hat{\mathbf{w}}^H(n)\mathbf{u}(n)$
	$\hat{\mathbf{w}}(n+1)=\hat{\mathbf{w}}(n)=\mu\mathbf{u}(n)e^*(n)$

correlation matrix of the tap inputs; the *condition number* of a Hermitian matrix is defined as the ratio of its largest eigenvalue to its smallest eigenvalue. Nevertheless, the LMS algorithm is highly popular and widely used in a variety of applications.

In a nonstationary environment, the orientation of the error-performance surface varies continuously with time. In this case, the LMS algorithm has the added task of continually *tracking* the bottom of the error-performance surface. Indeed, tracking will occur provided that the input data vary slowly compared to the *learning rate* of the LMS algorithm.

The stochastic gradient approach may also be pursued in the context of a lattice structure. The resulting adaptive filtering algorithm is called the *gradient adaptive lattice (GAL) algorithm*. In their own individual ways, the LMS and GAL algorithms are just two members of the *stochastic gradient family* of linear adaptive filters, although it must be said that the LMS algorithm is by far the most popular member of this family.

2.4.2 Least-Squares Estimation

The second approach to the development of linear adaptive filtering algorithms is based on the *method of least squares*. According to this method, we minimize a cost function or index of performance that is defined as the *sum of weighted error squares*, where the *error* or *residual* is itself defined as the difference between some desired response and the actual filter output. The method of least squares may be formulated with *block estimation* or *recursive estimation* in mind. In block estimation, the input data stream is arranged in the form of blocks of equal length (duration), and the filtering of input data proceeds on a block-by-block basis. In recursive estimation, on the other hand, the estimates of interest (e.g., tap weights of a transversal filter) are *updated* on a sample-by-sample basis. Ordinarily, a recursive estimator requires less storage than a block estimator, which is the reason for its much wider use in practice.

Recursive least-squares (RLS) estimation may be viewed as a special case of Kalman filtering. A distinguishing feature of the Kalman filter is the notion of *state*, which provides a measure of all the inputs applied to the filter up to a specific instant of time. Thus, at the heart of the Kalman filtering algorithm we have a recursion that may be described in words as follows:

$$\begin{pmatrix} \text{updated value} \\ \text{of the} \\ \text{state} \end{pmatrix} = \begin{pmatrix} \text{old value} \\ \text{of the} \\ \text{state} \end{pmatrix} = \begin{pmatrix} \text{kalman} \\ \text{gain} \end{pmatrix} \begin{pmatrix} \text{innovation} \\ \text{vector} \end{pmatrix}$$

where the *innovation vector* represents new information put into the filtering process at the time of the computation. For the present, it suffices to say that there is indeed a one-to-one correspondence between the Kalman variables and RLS variables. This correspondence means that we can tap the vast literature on Kalman filters for the design of linear adaptive filters based on RLS estimation.

We may classify the *RLS family* of linear adaptive filtering algorithms into three distinct categories, depending on the approach taken:

(1) *Standard RLS algorithm* assumes the use of a transversal filter as the structural basis of the linear adaptive filter. Table 2.2 summarizes the standard RLS algorithm. Derivation of this algorithm relies on a basic result in linear algebra known as the *matrix inversion lemma*. Most importantly, it enjoys the same virtues and suffers

TABLE 2.2

Summary of the RLS Algorithm
_____

Notations:

$\mathbf{u}(n)$	tap-input vector at time n
	$[u(n), u(n-1), \ldots, u(n-M+1)]^T$
M	number of tap inputs
$d(n)$	desired response at time n
$\hat{\mathbf{w}}(n)$	$[\hat{w}_0(n), \hat{w}_1(n), \ldots, \hat{w}_{M-1}(n)]^T$
	tap-weight vector at time n
$\xi(n)$	innovation (i.e., _a priori_ error signal) at time n
l	exponential weighting factor
$\mathbf{k}(n)$	gain vector at time n
$\mathbf{P}(n)$	weight-error correlation matrix

Initialize the algorithm by setting

$\mathbf{P}(0) = \delta\Pi^1\mathbf{I}$, δ=small positive constant

$\mathbf{w}(0) = 0$

For each instant of time, $n-1, 2, \ldots$, compute

$$\mathbf{k}(n) = \frac{\lambda^{-1}\mathbf{P}(n-1)\mathbf{u}(n)}{1 + \lambda^{-1}u^H(n)\mathbf{P}(n-1)\mathbf{u}(n)}$$

$$\xi(n) = d(n) - \hat{\mathbf{w}}^H(n-1)\mathbf{u}(n)$$

$$\hat{\mathbf{w}}(n) = \hat{w}(n-1) + \mathbf{k}(n)\xi^*(n)$$

$$\lambda^{-1}\mathbf{P}(n-1) - \lambda^{-1}\mathbf{k}(n)\mathbf{u}(n)\mathbf{P}(n-1)$$
_____

from the same limitations as the standard Kalman filtering algorithm. The limitations include lack of numerical robustness and excessive computational complexity. Indeed, it is these two limitations that have prompted the development of the other two categories of RLS algorithms, described next.

(2) _Square-root RLS algorithms_ are based on _QR decomposition_ of the incoming data matrix. Two well-known techniques for performing this decomposition are the _Householder transformation_ and the _Givens rotation_, both of which are data adaptive transformations. At this point in the discussion, we need to merely say that RLS algorithms based on the Householder transformation or Givens rotation are numerically stable and robust. The resulting linear adaptive filters are referred to as _square-root adaptive filters_, because in a matrix sense they represent the square-root forms of the standard RLS algorithm.

(3) _Fast RLS algorithms_, which include the standard RLS algorithm and square-root RLS algorithms, have a computational complexity that increases as the square of M, where M is the number of adjustable weights (i.e., the number of degrees of freedom) in the algorithm. Such algorithms are often referred to as $O(M^2)$ algorithms, where $O(\cdot)$ denotes "order of". By contrast, the LMS algorithm is an $O(M)$ algorithm, in that its computational complexity increases linearly with M. When M is large, the computational complexity of $O(M^2)$ algorithms may become objectionable from a hardware implementation point of view. There is therefore a strong motivation to modify the formulation of the RLS algorithm in such a way that the computational complexity assumes an $O(M)$ form. This objective is indeed achievable, in the case of temporal processing, first by virtue of the inherent _redundancy_ in the _Toeplitz structure_ of the input data matrix and second by exploiting this redundancy through

the use of *linear least-squares prediction in both the forward and backward directions.* The resulting algorithms are known collectively as fast RLS algorithms; they combine the desirable characteristics of recursive linear least-squares estimation with an $O(M)$ computational complexity. Two types of fast RLS algorithms may be identified, depending on the filtering structure employed:

- *Order-recursive adaptive filters,* which are based on a lattice-like structure for making linear forward and backward predictions
- *Fast transversal filters,* in which the linear forward and backward predictions are performed using separate transversal filters

Certain (but not all) realizations of order-recursive adaptive filters are known to be numerically stable, whereas fast transversal filters suffer from a numerical stability problem and, therefore, require some form of stabilization for them to be of practical use.

An introductory discussion of linear adaptive filters would be incomplete without saying something about their tracking behavior. In this context, we note that stochastic gradient algorithms such as the LMS algorithm are *model independent;* generally speaking, we would expect them to exhibit good tracking behavior, which indeed they do. In contrast, RLS algorithms are *model dependent;* this, in turn, means that their tracking behavior may be inferior to that of a member of the stochastic gradient family, unless care is taken to minimize the mismatch between the mathematical model on which they are based and the underlying physical process responsible for generating the input data.

2.4.3 How to Choose an Adaptive Filter

Given the wide variety of adaptive filters available to a system designer, how can a choice be made for an application of interest? Clearly, whatever the choice, it has to be *cost effective.* With this goal in mind, we may identify three important issues that require attention: *computational cost, performance,* and *robustness.* The use of computer simulation provides a good first step in undertaking a detailed investigation of these issues. We may begin by using the LMS algorithm as an adaptive filtering tool for the study. The LMS algorithm is relatively simple to implement. Yet it is powerful enough to evaluate the practical benefits that may result from the application of adaptivity to the problem at hand. Moreover, it provides a practical frame of reference for assessing any further improvement that may be attained through the use of more sophisticated adaptive filtering algorithms. Finally, the study must include tests with real-life data, for which there is no substitute.

Practical applications of adaptive filtering are very diverse, with each application having peculiarities of its own. The solution for one application may not be suitable for another. Nevertheless, to be successful we have to develop a physical understanding of the environment in which the filter has to operate and thereby relate to the realities of the application of interest.

2.5 Real and Complex Forms of Adaptive Filters

In the development of adaptive filtering algorithms, regardless of their origin, it is customary to assume that the input data are in baseband form. The term "baseband" is used to designate the band of frequencies representing the original (message) signal as generated by the source of information.

In such applications as communications, radar, and sonar, the information-bearing signal component of the receiver input typically consists of a message signal *modulated* onto a carrier wave. The bandwidth of the message signal is usually small compared to the carrier frequency, which means that the modulated signal is a *narrowband signal*. To obtain the baseband representation of a narrowband signal, the signal is translated down in frequency in such a way that the effect of the carrier wave is completely removed, yet the information content of the message signal is fully preserved. In general, the baseband signal so obtained is *complex*. In other words, a sample $u(n)$ of the signal may be written as

$$u(n) = u_I(n) + j u_Q(n) \tag{2.6}$$

where $u_I(n)$ is the *in-phase* (real) *component*, and $u_Q(n)$ is the *quadrature* (imaginary) *component*. Equivalently, we may express $u(n)$ as

$$u(n) = |u(n)| e^{j\phi(n)} \tag{2.7}$$

where $|u(n)|$ is the *magnitude*, and $\phi(n)$ is the *phase angle*.

The LMS and RLS algorithms summarized in Tables 2.1 and 2.2 assume the use of complex signals. The adaptive filtering algorithm so described is said to be in *complex form*. The important virtue of complex adaptive filters is that they preserve the mathematical formulation and elegant structure of complex signals encountered in the aforementioned areas of application.

If the signals to be processed are *real*, we naturally use the *real form* of the adaptive filtering algorithm of interest. Given the complex form of an adaptive filtering algorithm, it is straightforward to deduce the corresponding real form of the algorithm. Specifically, we do two things:

(1) The operation of *complex conjugation*, wherever in the algorithm, is simply removed.

(2) The operation of *Hermitian transposition* (i.e., conjugate transposition) of a matrix, wherever in the algorithm, is replaced by ordinary transposition.

Simply put, complex adaptive filters include real adaptive filters as special cases.

2.6 Nonlinear Adaptive Systems: Neural Networks

The theory of linear optimum filters is based on the mean-square error criterion. The Wiener filter that results from the minimization of such a criterion, and which represents the goal of linear adaptive filtering for a stationary environment, can only relate to second-order statistics of the input data and no higher. This constraint limits the ability of a linear adaptive filter to extract information from input data that are nonGaussian. Despite its theoretical importance, the existence of Gaussian noise is open to question (Johnson and

Rao, 1990). Moreover, nonGaussian processes are quite common in many signal processing applications encountered in practice. The use of a Wiener filter or a linear adaptive filter to extract signals of interest in the presence of such nonGaussian processes will therefore yield suboptimal solutions. We may overcome this limitation by incorporating some form of *nonlinearity* in the structure of the adaptive filter to take care of higher order statistics. Although, by so doing, we no longer have the Wiener filter as a frame of reference and so complicate the mathematical analysis, we would expect to benefit in two significant ways: improving learning efficiency and a broadening of application areas.

In this section, we describe an important class of the nonlinear adaptive system commonly known as artificial neural networks or just simply *neural networks*. This terminology is derived from analogy with biological neural networks that make up the human brain.

A *neural network* is a massively parallel distributed processor that has a natural propensity for storing experiential knowledge and making it available for use. It resembles the brain in two respects:

1. Knowledge is acquired by the network through a learning process.
2. Interconnection strengths known as synaptic weights are used to store the knowledge.

Basically, learning is a process by which the free parameters (i.e., synaptic weights and bias levels) of a neural network are adapted through a continuing process of stimulation by the environment in which the network is embedded. The type of learning is determined by the manner in which the parameter changes take place. Specifically, learning machines may be classified as follows:

- Learning with a teacher, also referred to as supervised learning
- Learning without a teacher

This second class of learning machines may also be subdivided into:

- Reinforcement learning
- Unsupervised learning or self-organized learning

In the subsequent sections of this chapter, we will describe the important aspects of these learning machines and highlight the algorithms involved in their designs. For a detailed treatment of the subject, see Haykin (1999); this book has an up-to-date bibliography that occupies 41 pages of references.

In the context of adaptive signal-processing applications, neural networks offer the following advantages:

- *Nonlinearity,* which makes it possible to account for the nonlinear behavior of physical phenomena responsible for generating the input data.
- The ability to *approximate any prescribed input–output mapping* of a continuous nature.
- *Weak statistical assumptions* about the environment, in which the network is embedded.

- *Learning* capability, which is accomplished by undertaking a training session with input–output examples that are representative of the environment.
- *Generalization*, which refers to the ability of the neural network to provide a satisfactory performance in response to *test data* never seen by the network before.
- *Fault tolerance*, which means that the network continues to provide an acceptable performance despite the failure of some neurons in the network.
- *VLSI implementability*, which exploits the massive parallelism built into the design of a neural network.

This is indeed an impressive list of attributes, which accounts for the widespread interest in the use of neural networks to solve signal-processing tasks that are too difficult for conventional (linear) adaptive filters.

2.6.1 Supervised Learning

This form of learning assumes the availability of a labeled (i.e., ground truthed) set of training data made up of N input–output examples:

$$T = \{\mathbf{x}_i, d_i\}_{i=1}^{N} \tag{2.8}$$

where $\mathbf{x}_i$ = input vector of ith example; d_i = desired (target) response of ith example, assumed to be scalar for convenience of presentation; and N = sample size.

Given the training sample T, the requirement is to compute the free parameters of the neural network so that the actual output y_i of the neural network due to $\mathbf{x}_i$ is close enough to d_i for all i in a statistical sense. For example, we may use the mean-squared error

$$E(n) = \frac{1}{N} \sum_{i=1}^{N} (d_i - y_i)^2 \tag{2.9}$$

as the index of performance to be minimized.

2.6.1.1 Multilayer Perceptrons and Back-Propagation Learning

The back-propagation algorithm has emerged as the workhorse for the design of a special class of layered feedforward networks known as *multilayer perceptrons*. As shown in the block diagram of Figure 2.6, a multilayer perceptron consists of the following:

- *Input layer* of nodes, which provide the means for connecting the neural network to the source(s) of signals driving the network.
- One or more *hidden layers* of processing units, which act as "feature detectors".
- *Output layer* of processing units, which provide one final stage of computation and thereby produce the response of the network to the signals applied to the input layer.

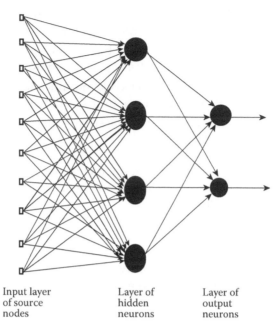

FIGURE 2.6
Fully connected feedforward of acyclic network with one hidden layer and one output layer.

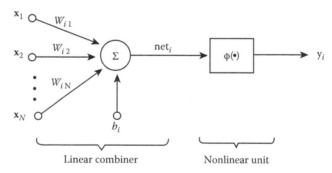

FIGURE 2.7
Simplified model of a neuron.

The processing units are commonly referred to as artificial neurons or just *neurons*. Typically, a neuron consists of a linear combiner with a set of adjustable synaptic weights, followed by a nonlinear activation function, as depicted in Figure 2.7.

Two commonly used forms of the activation function $\varphi(\cdot)$ are shown in Figure 2.8. The first one, shown in Figure 2.8a, is called the *hyperbolic function*, which is defined by

$$\varphi(V) = \tanh(V)$$
$$= \frac{1 - \exp(-2V)}{1 + \exp(-2V)} \tag{2.10}$$

The second one, shown in Figure 2.8b, is called the *logistic function*, which is defined by

$$\varphi(V) - \frac{1}{1 + \exp(-V)} \tag{2.11}$$

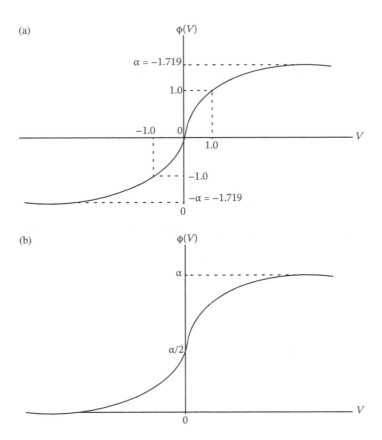

FIGURE 2.8
(a) Antisymmetric activation function; (b) nonsymmetric activation function.

From these definitions, we readily see that the logistic function is of a unipolar form that is nonsymmetric, whereas the hyperbolic function is bipolar that is antisymmetric.

The training of an MLP is usually accomplished by using the *back-propagation (BP) algorithm* that involves two phases (Werbos, 1974; Rumelhart et al., 1986):

- *Forward phase:* During this phase, the free parameters of the network are fixed, and the input signal is propagated through the network of Figure 2.6 layer by layer. The forward phase finishes with the computation of an error signal defined as the difference between a desired response and the actual output produced by the network in response to the signals applied to the input layer.

- *Backward phase:* During this second phase, the error signal e_i is propagated through the network of Figure 2.6 in the backward direction, hence the name of the algorithm. It is during this phase that adjustments are applied to the free parameters of the network so as to minimize the error e_i in a statistical sense.

BP learning may be implemented in one of two basic ways, as summarized here:

(1) *Sequential mode* (also referred to as the pattern mode, on-line mode, or stochastic mode): In this mode of BP learning, adjustments are made to the free parameters of

the network on an example-by-example basis. The sequential mode is best suited for pattern classification.

(2) *Batch mode:* In this second mode of BP learning, adjustments are made to the free parameters of the network on an epoch-by-epoch basis, where each epoch consists of the entire set of training examples. The batch mode is best suited for nonlinear regression.

The BP learning algorithm is simple to implement and computationally efficient in that its complexity is linear in the synaptic weights of the network. However, a major limitation of the algorithm is that it can be excruciatingly slow, particularly when we have to deal with a difficult learning task that requires the use of a large network.

Traditionally, the derivation of the BP algorithm is done for real-valued data. This derivation may be extended to complex-valued data by permitting the free parameters of the multilayer perceptron to assume complex values. However, in the latter case, care has to be exercised in how the activation function is handled for complex-valued inputs. For a detailed derivation of the complex BP algorithm, see Haykin (1996).

In any event, we may try to make BP learning perform better by invoking the following list of neuristics:

- Use neurons with antisymmetric activation functions (e.g., hyperbolic tangent function) in preference to nonsymmetric activation functions (e.g., logistic function). Figure 2.8 shows examples of these two forms of activation functions.

- Shuffle the training examples after the presentation of each epoch; an epoch involves the presentation of the entire set of training examples to the network.

- Follow an easy-to-learn example with a difficult one.

- Preprocess the input data so as to remove the mean and decorrelate the data.

- Arrange for the neurons in the different layers to learn at essentially the same rate. This may be attained by assigning a learning-rate parameter to neurons in the last layers that is smaller than those at the front end.

- Incorporate prior information into the network design whenever it is available.

One other heuristic that deserves to be mentioned relates to the size of the training set, N, for a pattern classification task. Given a multilayer perceptron with a total number of synaptic weights including bias levels, denoted by W, a rule of thumb for selecting N is

$$N = O\left(\frac{W}{\varepsilon}\right) \tag{2.12}$$

where O denotes "the order of", and ε denotes the fraction of classification errors permitted on test data. For example, with an error of 10%, the number of training examples needed should be about ten times the number of synaptic weights in the network.

Supposing that we have chosen a multilayer perceptron to be trained with the BP algorithm, how do we determine when it is "best" to stop the training session? How do we select the size of individual hidden layers of the MLP? The answers to these important questions may be obtained through the use of a statistical technique known as *cross-validation*, which proceeds as follows:

- The set of training examples is split into two parts:
    (1) Estimation subset used for training of the model.
    (2) Validation subset used for evaluating the model performance.
- The network is finally tuned by using the entire set of training examples and then tested on test data not seen before.

2.6.1.2 Radial-Basis Function (RBF) Networks

Another popular layered feedforward network is the radial-basis function (RBF) network, whose structure is shown in Figure 2.9. RBF networks use memory-based learning for their design. Specifically, learning is viewed as a curve-fitting problem in high-dimensional space (Broomhead and Lowe, 1988; Poggio and Girosi, 1990).

(1) Learning is equivalent to finding a surface in a multidimensional space that provides a best fit to the training data.

(2) Generalization (i.e., response of the network to input data not seen before) is equivalent to the use of this multidimensional surface to interpolate the test data.

A commonly used formulation of the RBFs, which constitute the hidden layer, is based on the *Gaussian function.* To be specific, let **u** denote the signal vector applied to the input

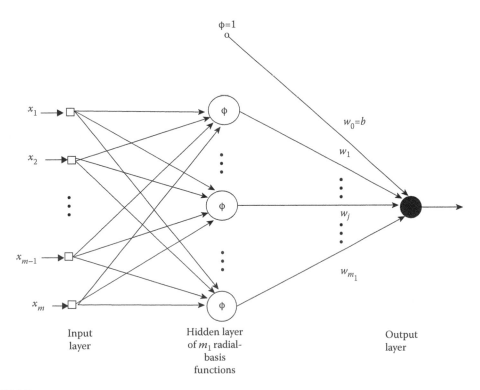

FIGURE 2.9
R1BF network.

layer and $\mathbf{u}_i$ denote the center of the Gaussian function assigned to hidden unit *i*. We may then define the corresponding RBF as

$$\varphi(\|\mathbf{u} - \mathbf{u}_i\|) = \exp\left(-\frac{\|\mathbf{u}-\mathbf{u}_i\|^2}{2\sigma^2}\right), \quad i = 1, 2, ..., K \tag{2.13}$$

where the symbol $\|\mathbf{u}-\mathbf{u}_i\|$ denotes the Euclidean distance between the vectors $\mathbf{u}$ and $\mathbf{u}_i$ and σ^2 is the width common to all K RBFs. (Each RBF may also be permitted to have a different width, but such a generalization results in increased complexity.) On this basis, we may define the input–output mapping realized by the RBF network (assuming a single output) to be

$$y = \sum_{i=1}^{K} W_i \varphi(\|\mathbf{u} - \mathbf{u}_i\|)$$

$$= \sum_{i=1}^{K} W_i \exp\left(-\frac{\|\mathbf{u} - \mathbf{u}_i\|^2}{2\sigma^2}\right) \tag{2.14}$$

where the set of weights $\{W_i\}_{i=1}^{K}$ constitutes the output layer. Equation 2.14 represents a *linear mixture* of Gaussian functions.

RBF networks differ from the multilayer perceptrons in some fundamental respects:

- RBF networks are local approximators, whereas multilayer perceptrons are global approximators.
- RBF networks have a single hidden layer, whereas multilayer perceptrons can have any number of hidden layers.
- The output layer of a RBF network is always linear, whereas in a multilayer perceptron it can be linear or nonlinear.
- The activation function of the hidden layer in an RBF network computes the Euclidean distance between the input signal vector and a parameter vector of the network, whereas the activation function of a multilayer perceptron computes the inner product between the input signal vector and the pertinent synaptic weight vector.

The use of a linear output layer in an RBF network may be justified in light of *Cover's theorem* on the separability of patterns. According to this theorem, provided that the transformation from the input space to the feature (hidden) space is nonlinear and the dimensionality of the feature space is high compared to that of the input (data) space, then there is a high likelihood that a nonseparable pattern classification task in the input space is transformed into a linearly separable one in the feature space.

Design methods for RBF networks include the following:

(1) Random selection of fixed centers (Broomhead and Lowe, 1988)
(2) Self-organized selection of centers (Moody and Darken, 1989)
(3) Supervised selection of centers (Poggio and Girosi, 1990)
(4) Regularized interpolation exploiting the connection between an RBF network and the Watson–Nadaraya regression kernel (Yee, 1998)

2.6.1.3 *Support Vector Machines*

Support vector machine (SVM) theory provides the most principled approach to the design of neural networks, eliminating the need for domain knowledge (Vapnik, 1998). SVM theory applies to pattern classification, regression, or density estimation using an RBF network (depicted in Figure 2.9) or an MLP with a single hidden layer (depicted in Figure 2.6).

Unlike BP learning, different cost functions are used for pattern classification and regression. Most importantly, the use of SVM learning eliminates the need for how to select the size of the hidden layer in an MLP or RBF network. In the latter case, it also eliminates the need for how to specify the centers of the RBF units in the hidden layer.

Simply stated, support vectors are those data points (for the linearly separable case) that are the most difficult to classify and are optimally separated from each other.

In an SVM, the selection of basis functions is required to satisty *Mercer's theorem*, that is, each basis function is in the form of a positive, definite, inner-product kernel (assuming real-valued data):

$$K(\mathbf{x}_i, \mathbf{x}_j) = \varphi^T(\mathbf{x}_i)\varphi(\mathbf{x}_j) \qquad (2.15)$$

where $\mathbf{x}_i$ and $\mathbf{x}_j$ are input vectors for examples i and j, and $\varphi(\mathbf{x}_i)$ is the vector of hidden-unit outputs for inputs $\mathbf{x}_i$. The hidden (feature) space is chosen to be of high dimensionality so as to transform a nonlinear, separable, pattern classification problem into a linearly separable one. Most importantly, however, in a pattern classification task, for example, the support vectors are selected by the SVM learning algorithm so as to maximize the margin of separation between classes.

The curse-of-dimensionality problem, which can plague the design of multilayer perceptrons and RBF networks, is avoided in SVMs through the use of quadratic programming. This technique, based directly on the input data, is used to solve for the linear weights of the output layer (Vapnik, 1998).

2.6.2 Unsupervised Learning

Turning next to unsupervised learning, adjustment of synaptic weights may be carried through the use of neurobiological principles such as Hebbian learning and competitive learning or information-theoretic principles. In this section we will describe specific applications of these three approaches.

2.6.2.1 *Principal Components Analysis*

According to *Hebb's postulate of learning*, the change in synaptic weight Δw_{ji} of a neural network is defined by (for real-valued data)

$$\Delta w_{ji} = \eta x_i y_j \qquad (2.16)$$

where η = learning-rate parameter; x_i = input (presynaptic) signal; and y_j = output (postsynaptic) signal.

Principal component analysis (PCA) networks use a modified form of this self-organized learning rule. To begin with, consider a linear neuron designed to operate as a maximum eigenfilter; such a neuron is referred to as *Oja's neuron* (Oja, 1982). It is characterized as follows:

$$\Delta w_{ji} = \eta y_j (x_i - y_j w_{ji}) \tag{2.17}$$

where the term $-\eta y_j^2 w_{ji}$ is added to stabilize the learning process. As the number of iterations approaches infinity, we find the following:

(1) The synaptic weight vector of neuron j approaches the eigenvector associated with the largest eigenvalue λ_{max} of the correlation matrix of the input vector (assumed to be of zero mean).

(2) The variance of the output of neuron j approaches the largest eigenvalue λ_{max}.

The generalized Hebbian algorithm (GHA), due to Sanger (1989), is a straightforward generalization of Oja's neuron for the extraction of any desired number of principal components.

2.6.2.2 Self-Organizing Maps

In a self-organizing map (SOM), due to Kohonen (1997), the neurons are placed at the nodes of a lattice, and they become selectively tuned to various input patterns (vectors) in the course of a competitive learning process. The process is characterized by the formation of a topographic map in which the spatial locations (i.e., coordinates) of the neurons in the lattice correspond to intrinsic features of the input patterns. Figure 2.10 illustrates the basic idea of an SOM, assuming the use of a two-dimensional lattice of neurons as the network structure.

In reality, the SOM belongs to the class of vector coding algorithms (Luttrell, 1989); that is, a fixed number of code words are placed into a higher dimensional input space, thereby facilitating data compression.

An integral feature of the SOM algorithm is the neighborhood function centered around a neuron that wins the competitive process. The neighborhood function starts by enclosing

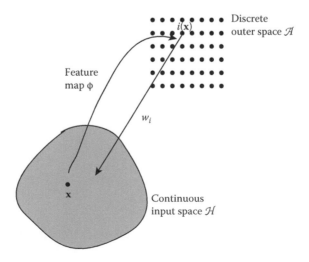

FIGURE 2.10
Illustration of the relationship between feature map ϕ and weight vector $\mathbf{w}_i$ of winning neuron i.

the entire lattice initially and is then allowed to shrink gradually until it encompasses the winning neuron.

The algorithm exhibits two distinct phases in its operation:

(1) *Ordering phase,* during which the topological ordering of the weight vectors takes place.

(2) *Convergence phase,* during which the computational map is fine tuned.

The SOM algorithm exhibits the following properties:

(1) Approximation of the continuous input space by the weight vectors of the discrete lattice.

(2) Topological ordering exemplified by the fact that the spatial location of a neuron in the lattice corresponds to a particular feature of the input pattern.

(3) The feature map computed by the algorithm reflects variations in the statistics of the input distribution.

(4) SOM may be viewed as a nonlinear form of principal components analysis.

2.6.3 Information-Theoretic Models

Mutual information, defined in accordance with Shannon's information theory, provides the basis of a powerful approach for self-organized learning. The theory is embodied in the maximum mutual information (Infomax) principle, due to Linsker (1988), which may be stated as follows:

> The transformation of a random vector **X** observed in the input layer of a neural network to a random vector **Y** produced in the output layer should be chosen so that the activities of the neurons in the output layer jointly maximize information about the activities in the input layer. The objective function to be maximized is the mutual information $I(\mathbf{Y};\mathbf{X})$ between **X** and **Y**.

The Infomax principle finds applications in the following areas:

- Design of self-organized models and feature maps (Linsker, 1989).
- Discovery of properties of a noisy sensory input exhibiting coherence across both space and time (first variant of Infomax due to Becker and Hinton, 1992).
- Dual-image processing designed to maximize the spatial differentiation between the corresponding regions of two separate images (views) of an environment of interest as in radar polarimetry (second variant of Infomax due to Ukrainec and Haykin, 1996).
- Independent components analysis (ICA) for blind source separation (due to Barlow, 1989); see also Comon (1994); ICA may be viewed as the third variant of Infomax (Haykin, 1999).

2.6.4 Temporal Processing Using Feedforward Networks

Time is an essential dimension of learning. We may incorporate time into the design of a neural network implicitly or explicitly. A straightforward method of implicit representation

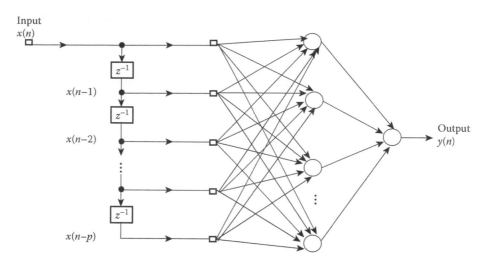

FIGURE 2.11
TLFN—the bias levels have been omitted for the convenience of presentation.

of time is to add a short-term memory structure at the input end of a static neural network (e.g., multilayer perceptron), as illustrated in Figure 2.11. This configuration is called a *focused time-lagged feedforward network* (TLFN). Focused TLFNs are limited to stationary dynamical processes.

To deal with nonstationary dynamical processes, we may use distributed TLFNs where the effect of time is distributed at the synaptic level throughout the network. One way in which this may be accomplished is to use FIR filters to implement the synaptic connections of an MLP. The training of a distributed TLFN is naturally a more difficult proposition than the training of a focused TLFN. Whereas we may use the ordinary BP algorithm to train a focused TLFN, we have to extend the BP algorithm to cope with the replacement of a synaptic weight in the ordinary MLP by a synaptic weight vector. This extension is referred to as the temporal BP algorithm due to Wan (1994).

2.6.5 Dynamically Driven Recurrent Networks

Another practical way of accounting for time in a neural network is to employ feedback at the local or global level. Neural networks so configured are referred to as recurrent networks.

We may identify two classes of recurrent networks:

1. Autonomous recurrent networks exemplified by the Hopfield network (Hopfield, 1982) and brain-state-in-a-box (BSB) model. These networks are well suited for building associative memories, each with its own domain of applications. Figure 2.12 shows an example of a Hopfield network involving the use of four neurons.
2. Dynamically driven recurrent networks are well suited for input–output mapping functions that are temporal in character.

A powerful approach for the design of dynamically driven recurrent networks with the goal of solving an input–output mapping task is to build on the state-space approach of modern control theory (Sontag, 1990). Such an approach is well suited for the two network

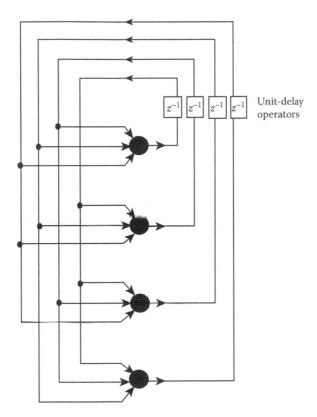

FIGURE 2.12
Recurrent network with no self-feedback loops and no hidden neurons.

configurations shown in Figures 2.13 and 2.14, which are respectively referred to as a *nonlinear autoregressive with exogeneous inputs (NARX) model* and a *recurrent multilayer perceptron (RMLP)*.

To design a dynamically driven recurrent network for input–output mapping, we may use any one of the following approaches:

- *Back-propagation through time* (BPTT) involves unfolding the temporal operation of the recurrent network into a layered feedforward network (Werbos, 1990). This unfolding facilitates the application of the ordinary BP algorithm.

- *Real-time recurrent learning* adjustments are made (using a gradient-descent method) to the synaptic weights of a fully connected recurrent network in real time (Williams and Zipser, 1989).

- An *extended Kalman filter* (EKF) builds on the classic Kalman filter theory to compute the synaptic weights of the recurrent network. Two versions of the algorithm are available (Feldkamp and Puskorius, 1998):

- Decoupled EKF.

- Global EKF.

The decoupled EKF algorithm is computationally less demanding but somewhat less accurate than the global EKF algorithm.

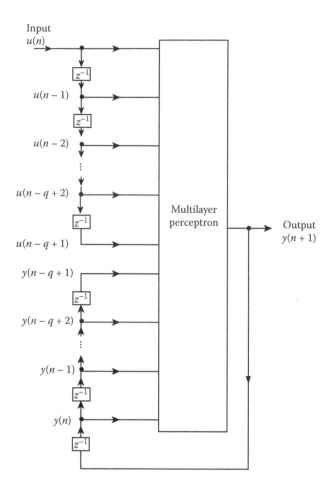

FIGURE 2.13
NARX model.

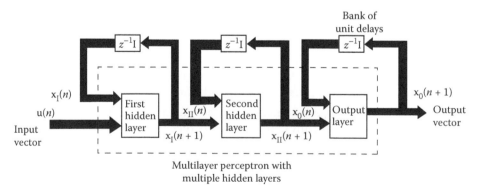

FIGURE 2.14
Recurrent multilayer perceptron.

A serious problem that can arise in the design of a dynamically driven recurrent network is the *vanishing gradients problem*. This problem pertains to the training of a recurrent network to produce a desired response at the current time that depends on input data in the distant past (Bengio et al., 1994). It makes the learning of long-term dependencies in gradient-based training algorithms difficult if not impossible in certain cases. To overcome the problem, we may use the following methods:

(1) EKF (encompassing second-order information) for training.

(2) Elaborate optimization methods such as pseudo-Newton and simulated annealing (Bengio et al., 1994).

(3) Use of long time delays in the network architecture (Giles et al., 1997).

(4) Hierarchically structuring of the network in multiple levels associated with different time scales (El Hihi and Bengio, 1996).

(5) Use of gating units to circumvent some of the nonlinearities (Hochreiter and Schmidhuber, 1997).

2.7 Applications

The ability of an adaptive filter to operate satisfactorily in an unknown environment and track time variations of input statistics make the adaptive filter a powerful device for signal-processing and control applications. Indeed, adaptive filters have been successfully applied in such diverse fields as communications, radar, sonar, seismology, and biomedical engineering. Although these applications are indeed quite different in nature, nevertheless, they have one basic common feature: an input vector and a desired response are used to compute an estimation error, which is in turn used to control the values of a set of adjustable filter coefficients. The adjustable coefficients may take the form of tap weights, reflection coefficients, rotation parameters, or synaptic weights, depending on the filter structure employed. However, the essential difference between the various applications of adaptive filtering arises in the manner in which the desired response is extracted. In this context, we may distinguish four basic classes of adaptive filtering applications, as depicted in Figure 2.15. For convenience of presentation, the following notations are used in Figure 2.15: u=input applied to the adaptive filter; y=output of the adaptive filter; d=desired response; and $e=d-y$=estimation error

The functions of the four basic classes of adaptive filtering applications depicted herein are as follows:

Class I.   *Identification* (Figure 2.15a): The notion of a *mathematical model* is fundamental to sciences and engineering. In the class of applications dealing with identification, an adaptive filter is used to provide a linear model that represents the best fit (in some sense) to an *unknown plant*. The plant and the adaptive filter are driven by the same input. The plant output supplies the desired response for the adaptive filter. If the plant is dynamic in nature, the model will be time varying.

Class II.   *Inverse modeling* (Figure 2.15b): In this second class of applications, the function of the adaptive filter is to provide an *inverse model* that represents the best fit (in some sense) to an *unknown noisy plant*. Ideally, in the case of a linear system, the inverse model has a transfer function equal to the *reciprocal* (*inverse*) of the plant's transfer function, such that the combination of the two constitutes an ideal transmission medium. A delayed version of the plant (system) input constitutes the desired response for the adaptive filter. In some applications, the plant input is used without delay as the desired response.

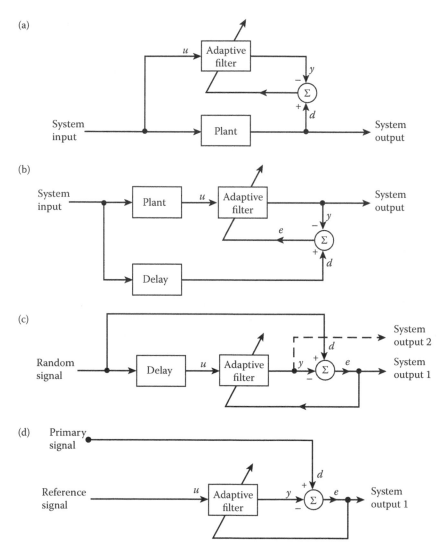

FIGURE 2.15
Four basic classes of adaptive filtering applications: (a) class I, identification; (b) class II, inverse modeling; (c) class III, prediction; (d) class IV, interference canceling.

Class III. *Prediction* (Figure 2.15c): Here, the function of the adaptive filter is to provide the best *prediction* (in some sense) of the present value of a random signal. The present value of the signal thus serves the purpose of a desired response for the adaptive filter. Past values of the signal supply the input applied to the adaptive filter. Depending on the application of interest, the adaptive filter output or the estimation (prediction) error may serve as the system output. In the first case, the system operates as a predictor; in the latter case, it operates as a *prediction-error filter.*

Class IV. *Interference canceling* (Figure 2.15d): In this final class of applications, the adaptive filter is used to cancel *unknown interference* contained (alongside an information-bearing signal component) in a *primary signal,* with the cancellation being opti-mized in some sense. The primary signal serves as the desired response for the adaptive filter. A *reference (auxiliary) signal* is employed as the input to the adap-tive filter. The reference signal is derived from a sensor or set of sensors located in relation to the sensor(s) supplying the primary signal in such a way that the information-bearing signal component is weak or essentially undesirable.

In Table 2.3, we have listed some applications that are illustrative of the four basic classes of adaptive filtering applications. These applications, totaling twelve, are drawn from the fields of control systems, seismology, electrocardiography, communications, and radar. A selected number of these applications are described individually in the remainder of this section.

2.7.1 System Identification

System identification is the experimental approach to the modeling of a process or a plant (Goodwin and Rayne, 1977; Ljung and Söderström, 1983; Åström and Eykhoff, 1971). It involves the following steps: experimental planning, the selection of a model structure, parameter estimation, and model validation. The procedure of system identification, as pursued in practice, is iterative in nature in that we may have to go back and forth between these steps until a satisfactory model is built. Here, we discuss briefly the idea of adap-tive filtering algorithms for estimating the parameters of an unknown plant modeled as a transversal filter.

Suppose we have an unknown dynamic plant that is linear and time varying. The plant is characterized by a *real-valued* set of discrete-time measurements that describe the vari-ation of the plant output in response to a known stationary input. The requirement is to develop an *on-line transferal filter model* for this plant, as illustrated in Figure 2.16. The model consists of a finite number of unit-delay elements and a corresponding set of adjust-able parameters (tap weights).

TABLE 2.3

Applications of Adaptive Filters

Class of Adaptive Filtering	Application
I. Identification	System identification
	Layered earth modeling
II. Inverse modeling	Predictive deconvolution
	Adaptive equalization
	Blind equalization
III. Prediction	Linear predictive coding
	Adaptive differential pulse-code modulation
	Autoregressive spectrum analysis
	Signal detection
IV. Interference canceling	Adaptive noise canceling
	Echo cancelation
	Adaptive beamforming

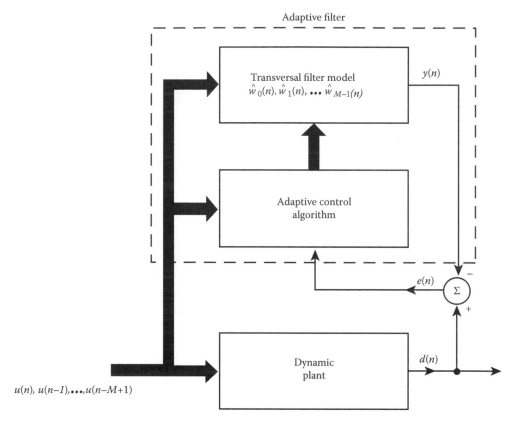

FIGURE 2.16
System identification.

Let the available input signal at time n be denoted by the set of samples: $u(n)$, $u(n-1)$, ..., $u(n-M+1)$, where M is the number of adjustable parameters in the model. This input signal is applied simultaneously to the plant and the model. Let their respective outputs be denoted by $d(n)$ and $y(n)$. The plant output $d(n)$ serves the purpose of a desired response for the adaptive filtering algorithm employed to adjust the model parameters. The model output is given by:

$$y(n) = \sum_{k=0}^{M-1} \hat{W}_k(n)u(n-k) \tag{2.18}$$

where $\hat{W}_0(n)$, $\hat{W}_1(n)$, ... and $\hat{W}_{M-1}(n)$ are the estimated model parameters. The model output $y(n)$ is compared with the plant output $d(n)$. The difference between them, $d(n)-y(n)$, defines the modeling (estimation) error. Let this error be denoted by $e(n)$.

Typically, at time n, the modeling error $e(n)$ is nonzero, implying that the model deviates from the plant. In an attempt to account for this deviation, the error $e(n)$ is applied to an *adaptive control algorithm*. The samples of the input signal, $u(n)$, $u(n-1)$, ..., $u(n-M+1)$, are also applied to the algorithm. The combination of the transversal filter and the adaptive control algorithm constitutes the adaptive filtering algorithm. The algorithm is designed to control the adjustments made in the values of the model parameters. As a result, the model parameters assume a new net of values for use on the next iteration. Thus, at time

$n+1$, a new model output is computed, and with it a new value for the modeling error. The operation described is then repeated. This process is continued for a sufficiently large number of iterations (starting from time $n=0$), until the deviation of the model from the plant, measured by the magnitude of the modeling error $e(n)$, becomes sufficiently small in some statistical sense.

When the plant is time varying, the plant output is *nonstationary* and so is the desired response presented to the adaptive filtering algorithm. In such a situation, the adaptive filtering algorithm has the task of not only keeping the modeling error small, but also continually tracking the time variations in the dynamics of the plant.

2.7.2 Spectrum Estimation

The *power spectrum* provides a quantitative measure of the second-order statistics of a discrete-time stochastic process as a function of frequency. In *parametric spectrum analysis*, we evaluate the power spectrum of the process by assuming a *model* for the process. In particular, the process is modeled as the output of a linear filter that is excited by a *white-noise process*, as in Figure 2.17. By definition, a white-noise process has a constant power spectrum. A model that is of practical utility is the *autoregressive (AR) model*, in which the transfer function of the filter is assumed to consist of poles only. Let this transfer function be denoted by

$$H(e^{j\omega}) = \frac{1}{1 + a_1 e^{j\omega} + \ldots + a_M e^{-jM\omega}}$$

$$= \frac{1}{1 + \sum_{k=1}^{M} a_k e^{-jk\omega}} \tag{2.19}$$

where the a_k are called the *AR parameters*, and M is the model order. Let σ_v^2 denote the constant power spectrum of the white-noise process $v(n)$ applied to the filter input. Accordingly, the power spectrum of the filter output $u(n)$ equals

$$S_{AR}(\omega) = \sigma_v^2 \left| H(e^{j\omega}) \right|^2 \tag{2.20}$$

We refer to $S_{AR}(\omega)$ as the *AR power spectrum*. Equation 2.19 assumes that the AR process $u(n)$ is real, in which case the AR parameters themselves assume real values.

When the AR model is time varying, the model parameters become time dependent, as shown by $a_1(n)$, $a_2(n)$, ..., $a_M(n)$. In this case, we express the power spectrum of the time-varying AR process as

$$S_{AR}(\omega, n) = \frac{\sigma_v^2}{\left| 1 + \sum_{k=1}^{M} a_k e^{-jk\omega} \right|} \tag{2.21}$$

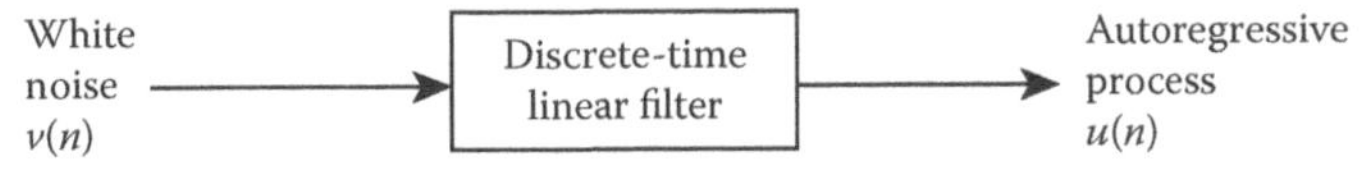

White noise $v(n)$ → Discrete-time linear filter → Autoregressive process $u(n)$

FIGURE 2.17
Black box representation of a stochastic model.

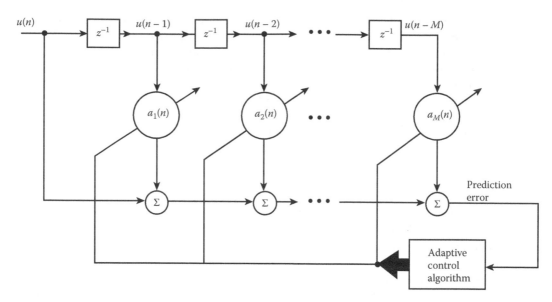

FIGURE 2.18
Adaptive prediction-error filter for real-valued data.

We may determine the AR parameters of the time-varying model by applying $u(n)$ to an *adaptive prediction-error filter*, as indicated in Figure 2.18. The filter consists of a transversal filter with adjustable tap weights. In the adaptive scheme of Figure 2.18, the prediction error produced at the output of the filter is used to control the adjustments applied to the tap weights of the filter. The *adaptive AR model* provides a practical means for measuring the *instantaneous frequency* of a frequency-modulated process. In particular, we may do this by measuring the frequency at which the AR power spectrum $S_{\mathrm{AR}}(\omega, n)$ attains its peak value for varying time n.

2.7.3 Signal Detection

The *detection problem*, that is, the problem of detecting an information-bearing signal in noise, may be viewed as one of *hypothesis testing* with deep roots in *statistical decision theory* (Van Trees, 1968). In the statistical formulation of hypothesis testing, there are two criteria of most interest: the *Bayes criterion* and the *Neyman–Pearson criterion*. In the Bayes test, we minimize the *average cost* or *risk* of the experiment of interest, which incorporates two sets of parameters: (1) *a priori probabilities* that represent the observer's information about the source of information before the experiment is conducted; and (2) a set of *costs* assigned to the various possible courses of action. As such, the Bayes criterion is directly applicable to digital communications. In the Neyman–Pearson test, on the other hand, we maximize the *probability of detection* subject to the constraint that the *probability of false alarm* does not exceed some preassigned value. Accordingly, the Neyman–Pearson criterion is directly applicable to radar or sonar. An idea of fundamental importance that emerges in hypothesis testing is that for a Bayes criterion or Neyman–Pearson criterion, the optimum test consists of two distinct operations: (1) processing the observed data to compute a test statistic called the *likelihood ratio* and (2) computing the likelihood ratio with a *threshold* to make a *decision* in favor of one of the two hypotheses. The choice of one criterion or the other merely affects the value assigned to the threshold. Let H_1 denote the hypothesis that the observed data consist of noise alone, and let H_2 denote the hypothesis that the data consist of signal plus noise. The likelihood ratio is defined as the ratio of two maximum likelihood functions, with the numerator assuming that hypothesis H_2 is true and the denominator assuming that hypothesis H_1 is true. If the likelihood ratio exceeds the threshold, the decision is made in favor of hypothesis H_2; otherwise, the decision is made in favor of hypothesis H_1.

In simple binary hypothesis testing, it is assumed that the signal is known and the noise is both white and Gaussian. In this case, the likelihood ratio test yields a *matched filter* (matched in the sense that its impulse response equals the time-reversed version of the known signal). When the additive noise is a *colored Gaussian noise* of known mean and correlation matrix, the likelihood ratio test yields a filter that consists of two sections: a *whitening filter* that transforms the colored noise component at the input into a white Gaussian noise process and a *matched filter* that is matched to the new version of the known signal as modified by the whitening filter.

However, in some important operational environments such as *communications, radar,* and *active sonar,* there may be inadequate information on the signal and noise statistics to design a fixed optimum detector. For example, in a sonar environment it may be difficult to develop a precise model for the received sonar signal, one that would account for the following factors completely:

- Loss in the signal strength of a *target echo* from an object of interest (e.g., enemy vessel), due to oceanic propagation effects and reflection loss at the target.
- Statistical variations in the additive *reverberation* component, produced by reflections of the transmitted signal from scatterers such as the ocean surface, ocean floor, biologies, and in homogeneities within the ocean volume.
- Potential sources of *noise* such as biological, shipping, oil drilling, and seismic and oceanographic phenomena.

In situations of this kind, the use of adaptivity offers a powerful approach to solve difficult signal detection problems. The particular application we have chosen for our present discussion is the detection of a small radar target in sea clutter (i.e., radar backscatter from the ocean surface). The radar target is a small piece of ice called a *growler;* the portion of which is visible above the sea surface is about the size of a grand piano. Recognizing that 90% of the volume of ice lies inside the water, a growler can indeed pose a threat to navigation in ice-infested waters as on the east coast of Canada.

The detection problem described herein is further compounded by the nonstationary nature of both sea clutter and the target echo from the growler. The strategy we have chosen to solve this difficult signal detection problem reformulates it into a pattern classification problem for which neural networks are well suited.

Figure 2.19 shows a block diagram of the detection strategy described in Haykin and Thomson (1998), and Haykin and Bhattacharya (1997). It consists of three functional units:

- *Time-frequency analyzer,* which converts the time-varying waveform of the input signal into a picture with two coordinates, namely, time and frequency.
- *Feature extractor,* the purpose of which is to compress the two-dimensional data produced by the time-frequency analyzer by extracting a set of features that retain the essential frequency content of the original signal.

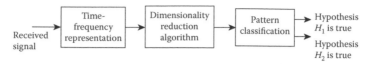

FIGURE 2.19
Block diagram of the detection strategy used in a nonstationary environment.

- *Pattern classifier,* which is trained to categorize the set of features applied to its input into two classes:
    1. No target present (i.e., the input signal consists of clutter only).
    2. Target present (i.e., the input signal consists of target echo plus clutter).

A signal-processing tool that is well suited for the application described herein is the *Wigner–Ville distribution* (WVD) (Cohen, 1995). The time-frequency map produced by this method is highly dependent on the nature of the input signal. If the input signal consists of clutter only, the resulting WVD picture is determined entirely by the time-frequency characteristics of sea clutter. Figure 2.20a shows a typical WVD picture due to sea clutter acting alone. On the other hand, if the input signal consists of a target echo plus sea clutter, the resulting WVD picture consists of three components: one due to the target echo, one due to clutter in the background, and one (commonly referred to as a "cross-product term") due to the interaction between these two components. Ordinarily, the cross-product terms are viewed as highly undesirable, as they tend to complicate the spectral interpretation of WVD pictures; indeed, much effort has been expended in the literature to reduce the effects of cross-product terms. However, in the application of WVD to target detection described herein, the cross-product terms perform a useful service by enhancing the detection power of the method. In particular, cross-product terms are there to be seen only when a target is present; they disappear when the input signal is target free. Figure 2.20b shows a typical WVD picture pertaining to the combined presence of sea clutter and radar echo from a small growler. The zebra-like pattern (consisting of an alternating set of dark and light stripes) is due to the cross-product terms. The point to note here is that the target echo in the original input signal is hardly visible; yet is shows up ever so clearly in the WVD picture of Figure 2.20b.

For the feature extraction, we may use PCA, which was briefly described in Section 2.6.2.1. Finally, for pattern classification, we may use a multilayer perceptron trained with the BP algorithm. Design details of these two functional units and those of the WVD are presented elsewhere (Haykin and Thomson, 1998; Haykin and Bhattacharya, 1997). For the present discussion, it suffices to compare the receiver operating characteristics of this new radar detection strategy against those of an ordinary constant false alarm rate (CFAR) receiver. Figure 2.21 presents the results of this comparison using real-life data, which were collected at a site on the east coast of Canada by means of an instrument-quality radar known as the IPIX radar (designed and built at McMaster University, Hamilton, Ontario). From Figure 2.21, we see that for probability of false alarm $P_{FA} = 10^{-3}$, we have the following values for probability of detection:

$$P_D = \begin{cases} 0.91 & \text{for adaptive receiver based on the detection strategy of Figure 2.19} \\ 0.71 & \text{for Doppler CFAR receiver} \end{cases}$$

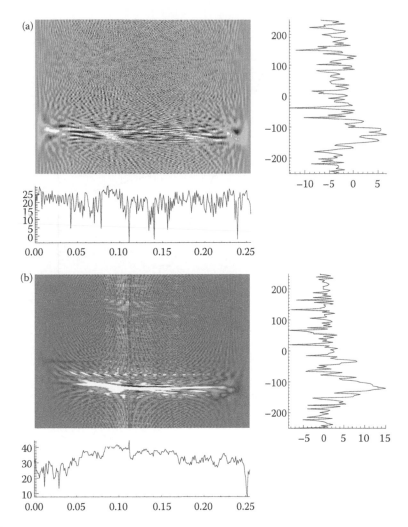

FIGURE 2.20

(a) WVD for sea clutter; (b) WVD for a barely visible growler. For the images, the horizontal axes are time in seconds and the vertical axes are frequency in Hertz. Horizontal axes of power spectra are in decibels.

2.7.4 Target Tracking

The objective of *tracking* is to estimate the *state* of a target of interest by processing measurements obtained from the target through the use of sensors and other means. The measurements are *noise-corrupted observables*, which are related to the current state of the target. Typically, the state consists of kinematic components such as the position, velocity, and acceleration of a moving target.

To state the tracking problem in mathematical terms, let the vector $\mathbf{x}(n)$ denote the state of a target, the vector $\mathbf{u}(n)$ denote the (known) input or control signal, and the vector $\mathbf{y}(n)$ denote the corresponding measurements obtained from the target. We may then express the state-space equations of the system in its most generic setting as follows:

$$\mathbf{x}(n+1) = \mathbf{f}(n, x(n), \mathbf{u}(n), \mathbf{v}_1(n)) \tag{2.22}$$

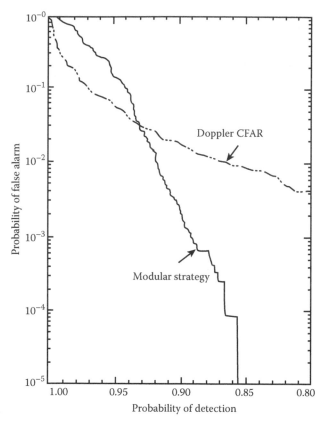

FIGURE 2.21
Composite receiver operating characteristics.

$$\mathbf{y}(n) = \mathbf{h}(n, \mathbf{x}(n), \mathbf{v}_2(n)) \tag{2.23}$$

where $\mathbf{f}(\cdot)$ and $\mathbf{h}(\cdot)$ are vector-valued functions, and $\mathbf{v}_1(n)$ and $\mathbf{v}_2(n)$ are noise vectors. The time argument n indicates that both $\mathbf{f}(\cdot)$ and $\mathbf{h}(\cdot)$ are time varying. Equations 2.22 and 2.23 are referred to as the process and measurement equations, respectively. The issue of interest is to estimate the state vector $\mathbf{x}(n)$, given the measurement vector $\mathbf{y}(n)$.

When the process and measurement equations are both linear, and the process noise vector $\mathbf{v}_1(n)$ and measurement noise vector $\mathbf{v}_2(n)$ are both modeled as zero-mean, white, Gaussian processes that are statistical independent, the Kalman filter provides the optimum estimate of the state $\mathbf{x}(n)$, given $\mathbf{y}(n)$ (Bar-Shalom and Fortmann, 1988). Optimality here refers to minimization of the mean-square error between the actual motion of the target and the *track* (i.e., state trajectory) estimated from the measurements associated with the target.

Unfortunately, in many of the radar and sonar target-tracking problems encountered in practice, the process and measurement equations are nonlinear, and the noise processes corrupting the state and measured data are nonGaussian. The traditional approach for dealing with nonlinear dynamics is to use the EKF, the derivation of which assumes knowledge of the nonlinear functions $\mathbf{f}(\cdot)$ and $\mathbf{h}(\cdot)$ and maintains the Gaussian assumptions about the process and noise vectors. The EKF closely resembles a Kalman filter except for the fact that each step of the standard Kalman filtering algorithm is replaced by its linearized equivalent (Bar-Shalom and Fortmann, 1988). However, the EKF approach to target tracking suffers from the following drawbacks when it is applied to an environment with nonlinear dynamics:

(1) Linearization of the vector-valued functions $\mathbf{f}(\cdot)$ and $\mathbf{h}(\cdot)$ can produce system instability if the time steps are not sufficiently short in duration.

(2) Linearization of the underlying dynamics requires the determination of two Jacobians (i.e., matrices of partial derivatives):

Jacobian of the vector-valued function $\mathbf{f}(\cdot)$, evaluated at the latest filtered estimate of the state at time n.

Jacobian of the vector-valued function $\mathbf{h}(\cdot)$, evaluated at the one-step predicted estimate of the state at time $n+1$.

The determination of these two Jacobians may lead to computational difficulties.

(3) The use of a short-time step to avoid system instability, combined with the determination of these two Jacobians, may impose a high computational overload on the system.

To overcome these shortcomings of the EKF, we may use the *unscented Kalman filter* (UKF) (Julier and Uhlmann, 1997; Wan et al., 1999), which is a generalization of the standard linear Kalman filter to systems whose process and measurement models are nonlinear. The UKF is preferable to the EKF for solving nonlinear filtering problems for two reasons:

(1) The UKF is accurate to the third order for Gaussian-distributed process and measurement errors. For nonGaussian distributions, the UKF is accurate to at least the second order. Accordingly, the UKF provides better performance than the traditional EKF.

(2) Unlike the EKF, the UKF does not require the computation of Jacobians pertaining to process and measurement equations. It is therefore simpler than the EKF in computational terms.

These are compelling reasons to reconsider the design of tracking systems for radar and sonar systems using the UKF.

2.7.5 Adaptive Noise Canceling

As the name implies, adaptive noise canceling relies on the use of *noise canceling* by subtracting noise from a received signal, an operation controlled in an *adaptive* manner for the purpose of improved signal-to-noise ratio. Ordinarily, it is inadvisable to subtract noise from a received signal, because such an operation could produce disastrous results by causing an increase in the average power of the output noise. However, when proper provisions are made, and filtering and subtraction are controlled by an adaptive process, it is possible to achieve a superior system performance compared to direct filtering of the received signal (Widrow et al., 1975; Widrow and Stearns, 1985).

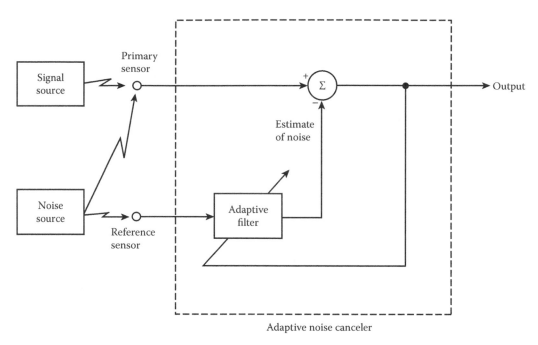

FIGURE 2.22
Adaptive noise cancelation.

Basically, an adaptive noise canceler is a *dual-input, closed-loop adaptive feedback system* as illustrated in Figure 2.22. The two inputs of the system are derived from a pair of sensors: a *primary sensor* and a *reference (auxiliary) sensor*. Specifically, we have the following:

(1) The primary sensor receives an information-bearing signal $s(n)$ corrupted by additive noise $v_0(n)$, as shown by

$$d(n) = s(n) + v_0(n) \tag{2.24}$$

The signal $s(n)$ and the noise $v_0(n)$ are uncorrelated with each other; that is,

$$E[s(n)v_1(n-k)] = 0 \text{ for all } k \tag{2.25}$$

where $s(n)$ and $v_0(n)$ are assumed to be real valued.

(2) The reference sensor receives a noise $v_1(n)$ that is uncorrelated with the signal $s(n)$, but correlated with the noise $v_0(n)$ in the primary sensor output in an unknown way; that is,

$$E[s(n)v_1(n-k)] = 0 \quad \text{for all } k \tag{2.26}$$

and

$$E[v_0(n)v_1(n-k)] = p(k) \tag{2.27}$$

where, as before, the signals are real valued, and $p(k)$ is an unknown cross-correlation for lag k.

The reference signal $v_1(n)$ is processed by an adaptive filter to produce the output signal

$$y(n) = \sum_{k=0}^{M-1} \hat{w}_k(n)v_1(n-k) \tag{2.28}$$

where $\hat{w}_k(n)$ is the adjustable (real) tap weights of the adaptive filter. The filter output $y(n)$ is subtracted from the primary signal $d(n)$, serving as the "desired" response for the adaptive filter. The error signal is defined by

$$e(n) = d(n) - y(n) \tag{2.29}$$

Thus, substituting Equation 2.22 into Equation 2.28, we get

$$e(n) = s(n) + v_0(n) - y(n) \tag{2.30}$$

The error signal is, in turn, used to adjust the tap weights of the adaptive filter, and the control loop around the operations of filtering and subtraction is thereby closed. Note that the information-bearing signal $s(n)$ is indeed part of the error signal $e(n)$, as indicated in Equation 2.30.

The error signal $e(n)$ constitutes the overall system output. From Equation 2.30, we see that the noise component in the system output is $v_0(n) - y(n)$. Now the adaptive filter attempts to minimize the mean-square value (i.e., average power) of the error signal $e(n)$. The information-bearing signal $s(n)$ is essentially unaffected by the adaptive noise canceler. Hence, minimizing the mean-square value of the error signal $e(n)$ is equivalent to minimizing the mean-square value of the output noise $v_0(n) - y(n)$. With the signal $s(n)$ remaining essentially constant, it follows that the *minimization of the mean-square value of the error signal is indeed the same as the maximization of the output signal-to-noise ratio of the system.*

The signal-processing operation described herein has two limiting cases that are noteworthy:

(1) The adaptive filtering operation is *perfect* in the sense that

$$y(n) = v_0(n)$$

In this case, the system output is *noise free,* and the noise cancelation is perfect. Correspondingly, the output signal-to-noise ratio is infinitely large.

(2) The reference signal $v_1(n)$ is completely uncorrelated with both the signal and noise components of the primary signal $d(n)$; that is,

$$E[d(n)v_1(n-k)] = 0 \quad \text{for all } k$$

In this case, the adaptive filter "switches itself off", resulting in a zero value for the output $y(n)$. Hence, the adaptive noise canceler has *no* effect on the primary signal $d(n)$, and the output signal-to-noise ratio remains unaltered.

The effective use of adaptive noise canceling therefore requires that we place the reference sensor in the noise field of the primary sensor with two specific objectives in mind. First, the information-bearing signal component of the primary sensor output is *undetectable* in the reference sensor output. Second, the reference sensor output is *highly correlated* with the noise component of the primary sensor output. Moreover, the adaptation of the adjustable filter coefficients must be near optimum.

In the remainder of this section, we described three useful applications of the adaptive noise canceling operation:

(1) *Canceling 60-Hz interference in electrocardiography:* In electrocardiography (ECG) commonly used to monitor heart patients, an *electrical discharge* radiates energy through human *tissue* and the resulting output is received by an *electrode.* The electrode is usually positioned in such a way that the received energy is maximized. Typically, however, the electrical discharge involves very low potentials. Correspondingly, the received energy is very small. Hence, extra care has to be exercised in minimizing signal degradation due to external *interference.* By far, the strongest form of interference is that of a 60-Hz periodic waveform picked up by the receiving electrode (acting like an antenna) from nearby electrical equipment (Huhta and Webster, 1973). Needless to say, this interference has undesirable effects in the interpretation of electrocardiograms. Widrow et al. (1975) have demonstrated the use of adaptive noise canceling (based on the LMS algorithm) as a method for reducing this form of interference. Specifically, the primary signal is taken from the ECG preamplifier, and the reference signal is taken from a wall outlet with proper attenuation. Figure 2.23 shows a block diagram of the adaptive noise canceler used by Widrow et al. (1975). The adaptive filter has two adjustable weights, $\hat{w}_0(n)$ and $\hat{w}_1(n)$. One weight, $\hat{w}_0(n)$, is fed directly from the reference point. The second weight, $\hat{w}_1(n)$, is fed from a 90° phase-shifted version of the reference input. The sum of the two weighted versions of the reference signal is then subtracted from the ECG output to produce an error signal. This error signal together with the weighted inputs are applied to the LMS algorithm, which, in turn, controls the adjustments applied to the

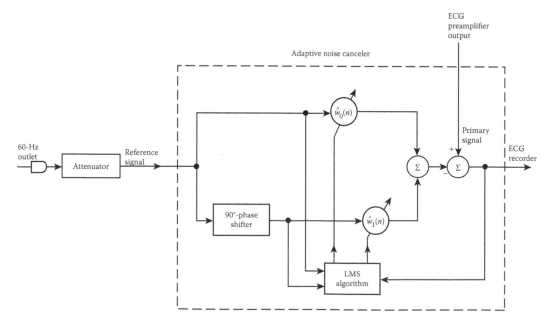

FIGURE 2.23
Adaptive noise canceler for suppressing 60 Hz interference in ECG. (After Widrow et al., 1975)

two weights. In this application, the adaptive noise canceler acts as a variable "notch filter". The frequency of the sinusoidal interference in the ECG output is presumably the same as that of the sinusoidal reference signal. However, the amplitude and phase of the sinusoidal interference in the ECG output are unknown. The two weights, $\hat{w}_0(n)$ and $\hat{w}_1(n)$, provide the *two degrees of freedom* required to control the amplitude and phase of the sinusoidal reference signal so as to cancel the 60-Hz interference contained in the ECG output.

(2) *Reduction of acoustic noise in speech:* At a noisy site (e.g., the cockpit of a military aircraft), voice communication is affected by the presence of *acoustic noise*. This effect is particularly serious when linear predictive coding (LPC) is used for the digital representation of voice signals at low bit rates; LPC was discussed earlier. To be specific, high-frequency acoustic noise severely affects the estimated LPC spectrum in both the low- and high-frequency regions. Consequently, the intelligibility of digitized speech using LPC often falls below the minimum acceptable level. Kang and Fransen (1987) describe the use of an adaptive noise canceler, based on the LMS algorithm, for reducing acoustic noise in speech. The noise-corrupted speech is used as the primary signal. To provide the reference signal (noise only), a reference microphone is placed in a location where there is sufficient isolation from the source of speech (i.e., the known location of the speaker's mouth). In the experiments described by Kang and Fransen, a reduction of 10–15 dB in the acoustic noise floor is achieved without degrading voice quality. Such a level of noise reduction is significant in improving voice quality, which may be unacceptable otherwise.

(3) *Adaptive speech enhancement:* Consider the situation depicted in Figure 2.24. The requirement is to listen to the voice of the desired speaker in the presence of background noise, which may be satisfied through the use of the adaptive noise canceling. Specifically, *reference microphones* are added at locations far enough away from the desired speaker such that their outputs contain *only* noise. As indicated

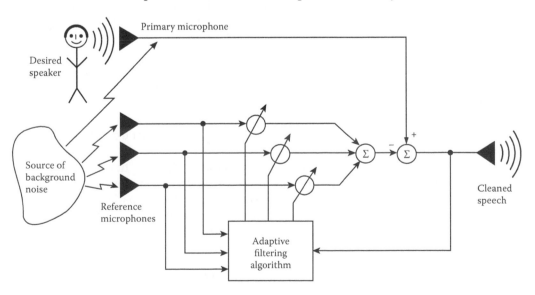

FIGURE 2.24
Block diagram of an adaptive noise canceler for speech.

in Figure 2.24, a weighted sum of the auxiliary microphone outputs is subtracted from the output of the desired speech-containing microphone, and an adaptive filtering algorithm (e.g., the LMS algorithm) is used to adjust the weights so as to minimize the average output power. A useful application of the idea described herein is in the adaptive noise cancelation for heating aids[tt] (Chazan et al., 1988). The so-called "cocktail party effect" severely limits the usefulness of hearing aids. The cocktail party phenomenon refers to the ability of a person with normal hearing to focus on a conversation taking place at a distant location in a crowded room. This ability is lacking in a person who wears hearing aids because of extreme sensitivity to the presence of *background noise*. This sensitivity is attributed to two factors: (1) the loss of directional cues and (2) the limited channel capacity of the ear caused by the reduction in both dynamic range and frequency response. Chazan et al. (1988) describe an adaptive noise-canceling technique aimed at overcoming this problem. The technique involves the use of an *array of microphones* that exploit the difference in spatial characteristics between the desired signal and the noise in a crowded room. The approach taken by Chazan et al. (1988) is based on the fact that each microphone output may be viewed as the sum of the signals produced by the individual speakers engaged in conversations in the room. Each signal contribution in a particular microphone output is essentially the result of a speaker's speech signal having passed through the *room filter*. In other words, each speaker (including the desired speaker) produces a signal at the microphone output that is the sum of the direct transmission of his/her speech signal and its reflections from the walls of the room. The requirement is to reconstruct the desired speaker signal, including its room reverberations, while canceling out the source of noise. In general, the transformation undergone by the speech signal from the desired speaker is not known. In addition, the characteristics of the background noise are variable. We thus have a signal-processing problem for which adaptive noise canceling offers a feasible solution.

2.7.6 Adaptive Beamforming

For our last application, we describe a *spatial* form of adaptive signal processing that finds practical use in radar, sonar, communications, geophysical exploration, astrophysical exploration, and biomedical signal processing.

In the particular type of spatial filtering of interest to us in this book, a number of independent *sensors* are placed at different points in space to "listen" to the received signal. In effect, the sensors provide a means of *sampling* the received signal in *space*. The set of sensor outputs collected at a particular instant of time constitutes a *snapshot*. Thus, a snapshot of data in spatial filtering (for the case when the sensors lie uniformly on a straight line) plays a role analogous to that of a set of consecutive tap inputs that exist in a transversal filter at a particular instant of time[tt].

In radar, the sensors consist of antenna elements (e.g., dipoles, horns, slotted waveguides) that respond to incident electromagnetic waves. In sonar, the sensors consist of hydrophones designed to respond to acoustic waves. In any event, spatial filtering, known as *beamforming*, is used in these systems to distinguish between the spatial properties of signal and noise. The device used to do the beamforming is called a *beamformer*. The term

---

[tt] This idea is similar to that of adaptive spatial filtering in the context of antennas, which is considered later in this section.

[tt] For a discussion of the analogies between time- and space-domain forms of signal processing, see Bracewell (1978) and Van Veen and Buckley (1988).

"beamformer" is derived from the fact that the early forms of antennas (spatial filters) were designed to form *pencil beams,* so as to receive a signal radiating from a specific direction and attenuate signals radiating from other directions of no interest (Van Veen and Buckley, 1988). Note that the beamforming applies to the radiation (transmission) or reception of energy.

In a primitive type of spatial filtering, known as the *delay-and-sum beamformer,* the various sensor outputs are delayed (by appropriate amounts to align spatial components coming from the direction of a target) and then summed, as in Figure 2.25. Thus, for a single target, the average power at the output of the delay-and-sum beamformer is maximized when it is steered toward the target. A major limitation of the delay-and-sum beamformer, however, is that it has no provisions for dealing with sources of *interference.*

In order to enable a beamformer to respond to an unknown interference environment, it has to be made *adaptive* in such a way that it places *nulls* in the direction(s) of the source(s) of interference automatically and in real time. By so doing, the output signal-to-noise ratio of the system is increased, and the *directional response* of the system is thereby improved. In the next section, we consider two examples of *adaptive beamformers* that are well suited for use with narrowband signals in radar and sonar systems.

2.7.6.1 Adaptive Beamformer with Minimum-Variance Distortionless Response

Consider an adaptive beamformer that uses a linear array of M identical sensors, as in Figure 2.26. The individual sensor outputs, assumed to be in *baseband* form, are weighted and then summed. The beamformer has to satisfy two requirements: (1) a *steering* capability whereby the target signal is always protected; and (2) the effects of sources of interference whereby the effects are minimized. One method of providing for these two requirements

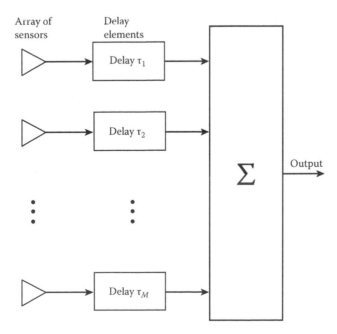

FIGURE 2.25
Delay-and-sum beamformer.

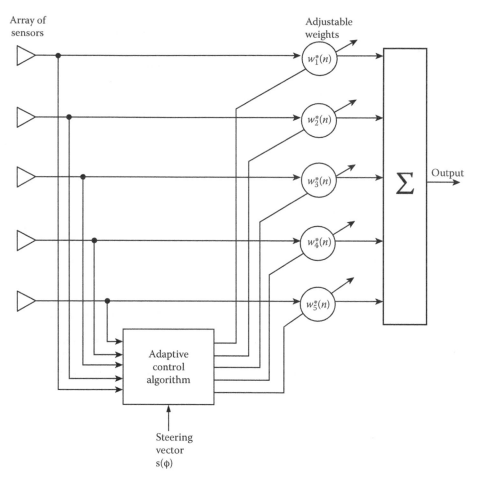

FIGURE 2.26
Adaptive beamformer for an array of five sensors. The sensor outputs (in baseband form) are complex valued; hence the weights are complex valued.

is to minimize the variance (i.e., average power) of the beamformer output, subject to the *constraint* that during the process of adaptation the weights satisfy the condition

$$\mathbf{w}^H(n)\mathbf{s}(\phi)=1 \quad \text{for all } n \quad \text{and} \quad \phi=\phi_t \tag{2.31}$$

where $\mathbf{w}(n)$ is the $M \times 1$ weight vector, and $\mathbf{s}(\phi)$ is an $M \times 1$ steering vector. The superscript H denotes Hermitian transposition (i.e., transposition combined with complex conjugation). In this application, the baseband data are complex valued, hence the need for complex conjugation. The value of the *electrical angle* $\phi=\phi_t$ is determined by the direction of the target. The angle ϕ is itself measured with sensor 1 (at the top end of the array) treated as the point of reference.

The dependence of vector $\mathbf{s}(\phi)$ on the angle ϕ is defined by

$$\mathbf{s}(\phi)=\left[1,\, e^{-j\phi},...,\, e^{-j(M-1)\phi}\right]^T$$

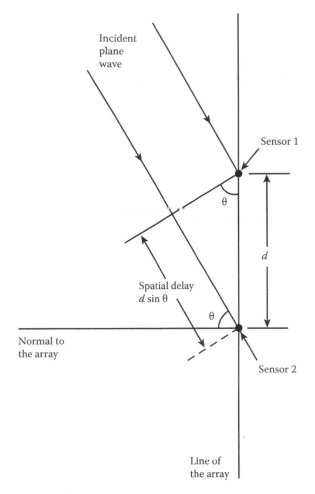

FIGURE 2.27
Spatial delay incurred when a plane wave impinges on a linear array.

The angle ϕ is itself related to incidence angle θ of a plane wave, measured with respect to the normal to the linear array, as follows[§§]:

$$\phi = \frac{2\pi d}{\lambda} \sin\theta, \tag{2.32}$$

where d is the spacing between adjacent sensors of the array and λ is the wavelength (see Figure 2.27). The incidence angle θ lies inside the range $-\pi/2$ to $\pi/2$. The permissible values that the angle ϕ may assume lie inside the range $-\pi$ to π. This means that we must choose the spacing $d < \lambda/2$, so that there is a one-to-one correspondence between the values of θ and ϕ without ambiguity. The condition $d < \lambda/2$ may be viewed as the spatial analog of the sampling theorem.

[§§] When a plane wave impinges on a linear array as in Figure 2.27, there is a spatial delay of $d \sin\theta$ between the signals received at any pair of adjacent sensors. With a wavelength of λ, this spatial delay is translated into an electrical angular difference defined by $\phi = 2\pi(d \sin\theta/\lambda)$.

The imposition of the *signal-protection constraint* in Equation 2.31 ensures that, for a pre-scribed look direction, the response of the array is maintained as constant (i.e., equal to 1), no matter what values are assigned to the weights. An algorithm that minimizes the variance of the beamformer output, subject to this constraint, is therefore referred to as the *minimum-variance distortionless response (MVDR) beamforming algorithm* (Capon, 1969; Owsley, 1985). The imposition of the constraint described in Equation 2.31 reduces the number of "degrees of freedom" available to the MVDR algorithm to $M-2$, where M is the number of sensors in the array. This means that the number of independent nulls pro-duced by the MVDR algorithm (i.e., the number of independent interferences that can be canceled) is $M-2$.

The MVDR beamforming is a special case of *linearly constrained minimum variance (LCMV) beamforming*. In the latter case, we minimize the variance of the beamformer output, sub-ject to the constraint

$$\mathbf{w}^{H}(n)\mathbf{s}(\phi) = g \quad \text{for all } n \quad \text{and} \quad \phi = \phi_t \tag{2.33}$$

where g is a complex constant. The LCMV beamformer linearly constraints the weights, such that any signal coming from electrical angle ϕ_t is passed to the output with response (gain) g. Comparing the constraint of Equation 2.31 with that of Equation 2.33, we see that the MVDR beamformer is indeed a special case of the LCMV beamformer for $g=1$.

2.7.6.2 Adaptation in Beam Space

The MVDR beamformer performs adaptation directly in the *data space*. The adaptation process for interference cancelation may also be performed in *beam space*. To do so, the input data (received by the array of sensors) are transformed into the beam space by means of an *orthogonal multiple beamforming network,* as illustrated in the block diagram of Figure 2.28. The resulting output is processed by a *multiple sidelobe canceler* so as to cancel interference(s) from unknown directions.

The beamforming network is designed to generate a set of *orthogonal* beams. The mul-tiple outputs of the beamforming network are referred to as *beam ports*. Assume that the sensor outputs are equally weighted and have a *uniform* phase. Under this condition, the response of the array produced by an incident plane wave arriving at the array along direction θ, measured with respect to the normal to the array, is given by:

$$A(\phi,\alpha) = \sum_{n=-N}^{N} e^{jn\varphi}e^{-jn\alpha} \tag{2.34}$$

where $M=(2N+1)$ is the total number of sensors in the array, with the sensor at the mid-point of the array treated as the point of reference. The electrical angle ϕ is related to θ by Equation 2.32, and α is a constant called the *uniform phase factor*. The quantity $A(\phi,\alpha)$ is called the array pattern. For $\phi=\lambda/2$, we find from Equation 2.31 that

$$\varphi = \pi \sin\theta$$

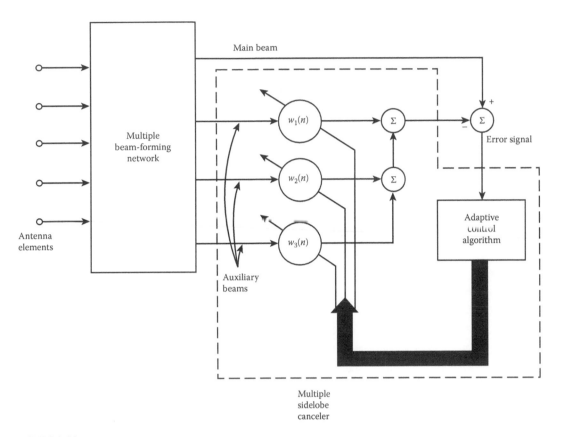

FIGURE 2.28
Block diagram of adaptive combiner with fixed beams; owing to the symmetric nature of the multiple beam-forming network, final values of the weights are real valued.

Summing the geometric series in Equation 2.34, we may express the array pattern as

$$A(\phi,\sigma) = \frac{\sin\left[\frac{1}{2}(2N+1)(\phi-\alpha)\right]}{\sin\left[\frac{1}{2}(\phi-\alpha)\right]} \qquad (2.35)$$

By assigning different values to α, the main beam of the antenna is thus scanned across the range $\pi < \phi \leq \pi$. To generate an orthogonal set of beams equal to $2N$ in number, we assign the following discrete values to the uniform phase factor

$$\alpha = \frac{\pi}{2N+1}k, \quad k = \pm1, \pm3, \ldots, \pm2N-1 \qquad (2.36)$$

Figure 2.29 illustrates the variations of the magnitude of the array pattern $A(\phi,\alpha)$ with ϕ for the case of $2N+1=5$ elements and $\alpha=\pm3\pi/5, \pm3\pi/5$. Note that owing to the symmetric nature of the beamformer, the final values of the weights are real valued.

The orthogonal beams generated by the beamforming network represent $2N$ independent *look directions,* one per beam. Depending on the target direction of interest, a particular beam in the set is identified as the *main beam* and the remainder are viewed as *auxiliary*

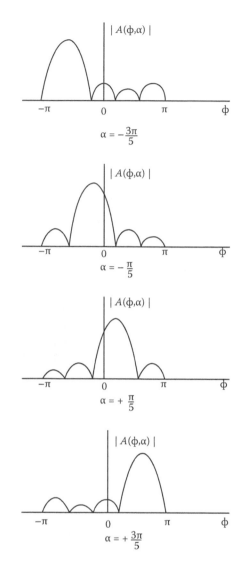

FIGURE 2.29
Variations of the magnitude of the array pattern, $A(\phi,\alpha)$ with ϕ and α.

beams. We note from Figure 2.29 that each of the auxiliary beams has a *null in the look direction of the main beam.* The auxiliary beams are adaptively weighted by the multiple sidelobe canceler so as to form a cancelation beam that is subtracted from the main beam. The resulting estimation error is fed back to the multiple sidelobe canceler so as to control the corrections applied to its adjustable weights.

Since all the auxiliary beams have nulls in the look direction of the main beam, and the main beam is excluded from the multiple sidelobe canceler, the overall output of the adaptive beamformer is constrained to have a constant response in the look direction of the main beam (i.e., along the direction of the target). Moreover, with $(2N-1)$ degrees of freedom (i.e., the number of available auxiliary beams), the system is capable of placing up to $(2N-1)$ nulls along the (unknown) directions of independent interferences.

Note that with any array of $(2N+1)$ sensors, we may produce a beamforming network with $(2N+1)$ orthogonal beam ports by assigning the uniform phase factor the following set of values:

$$\alpha = \frac{k\pi}{2N-1}, \quad k = 0, \pm 2, \ldots, \pm 2N \tag{2.37}$$

In this case, a small fraction of the main lobe of the beam port at either end lies in the nonvisible region. Nevertheless, with one of the beam ports providing the main beam and the remaining $2N$ ports providing the auxiliary beams, the adaptive beamformer is now capable of producing up to $2N$ independent nulls.

2.8 Concluding Remarks

Adaptive signal processing, be it in time, space, or space time, is essential to the design of modern radar, sonar, and biomedical imaging systems. We say so for the following reasons:

- The underlying statistics of the signals of interest may be unknown, which makes it difficult, if not impossible, to design optimum filters by classical methods. An adaptive signal processor overcomes this difficulty by learning the underlying statistics of the environment in an on-line fashion, off-line fashion, or combination thereof.

- Signals generated by radar, sonar, and biomedical systems are inherently *nonstationary*. Adaptive signal processing provides an elegant approach to deal with nonstationary phenomena by adjusting the free parameters of a filter in accordance with prescribed algorithms.

In this chapter, we have presented a guided tour of adaptive systems for signal processing by focusing attention on the following issues:

- Linear adaptive systems, exemplified by the LMS and RLS algorithms. The LMS algorithm is simple to design but slow to converge. The RLS algorithm, in contrast, is complex but fast to converge. When the adaptive filtering is of a temporal kind, the complexity of the RLS algorithm can be reduced significantly by exploiting the time-shifting property of the input signals. For details, see Haykin (1996).

- Nonlinear adaptive systems, exemplified by neural networks. This class of adaptive systems takes many different forms. The network can be of the feedforward type, which is exemplified by multilayer perceptrons and RBF networks. For the design of multilayer perceptrons, we can use the BP algorithm, which is a generalization of the LMS algorithm. A more principled approach for the design of multilayer perceptrons and RBF networks is to use SVM theory pioneered by Vapnik and coworkers (Vapnik, 1998). Another important type of neural network involves the abundant use of feedback, which is exemplified by Hopfield networks and dynamically driven recurrent networks. The state-space approach of the modern control theory provides a powerful basis for the design of dynamically driven recurrent networks, so as to synthesize input–output mappings of interest.

The choice of one of these adaptive systems over another can only be determined by the application of interest.

To motive the study of these two broadly defined classes of adaptive systems, we described five different applications:

- System identification
- Spectrum estimation
- Noise cancellation
- Signal detection
- Beamforming

The need for each and every one of these applications arises in radar, sonar, and medical imaging systems in a variety of different ways, as illustrated in the subsequent chapters of the book.

References

Åström, K.J. and P. Eykhoff. 1971. System identification—a survey. *Automatica*, 7, 123–162.

Barlow, H.B. 1989. Unsupervised learning. *Neural Computat.*, 1, 295–311.

Bar-Shalom, Y. and T.E. Fortmann. 1988. *Tracking and Data Association*. Academic Press, NY.

Becker, S. and G.E. Hinton. 1992. A self-organizing neural network that discovers surfaces in random-dot stereograms. *Nature (London)*, 355, 161–163.

Bengio, Y., P. Simard, and P. Frasconi. 1994. Learning long-term dependencies with gradient descent is difficult. *IEEE Trans. Neural Networks*, 5, 157–166.

Bracewell, R.N. 1978. *The Fourier Transform and its Applications*, 2nd ed. McGraw-Hill, NY.

Broomhead, D.S. and D. Lowe. 1988. Multivariable functional interpolation and adaptive networks. *Complex Syst.*, 2, 321–355.

Capon, J. 1969. High-resolution frequency-wavenumber spectrum analysis. *Proc. IEEE*, 57, 1408–1418.

Chazan, D., Y. Medan, and U. Shvadron. 1988. Noise cancellation for hearing aids. *IEEE Trans. Acoust. Speech Signal Process.*, ASSP-36, 1697–1705.

Cohen, L. 1995. *Time-Frequency Analysis*. Prentice Hall, Englewood Cliffs, NJ.

Comon, P. 1994. Independent component analysis: A new concept? *Signal Process.*, 36, 287–314.

Dorny, C.N. 1975. *A Vector Space Approach to Models and Optimization*. Wiley-Interscience, NY.

El Hihi, S. and Y. Bengio. 1996. Hierarchical recurrent neural networks for long-term dependencies. *Adv. Neural Inf. Process. Syst.*, 8, 493–499.

Feldkamp, L.A. and G.V. Puskorius. 1988. A signal processing framework based on dynamic neural networks with application to problems in adaptation, filtering and classification, Special issue. *Proc. IEEE Intelligent Signal Process*, 86.

Goodwin, G.C. and R.L. Rayne. 1977. *Dynamic System Identification: Experiment Design and Data Analysis*. Academic Press, NY.

Haykin, S. 1996. *Adaptive Filter Theory*, 3rd ed. Prentice Hall, Englewood Cliffs, NJ.

Haykin, S. 1999. *Neural Networks: A Comprehensive Foundation*, 2nd ed. Prentice Hall, Englewood Cliffs, NJ.

Haykin, S. and T. Bhattacharya. 1997. Modular learning strategy for signal detection in a non-stationary environment. *IEEE Trans. Signal Process.*, 45, 1619–1637.

Haykin, S. and D. Thomson. 1998. Signal detection in a nonstationary environment reformulated as an adaptive pattern classification problem. *Proc. IEEE,* 86(11), 2325–2344.

Hochreiter, S. and J. Schmidhuber. 1997. LSTM can solve hard long time lag problems. *Adv. Neural Inf. Process. Syst.,* 9, 473–479.

Itakura, F. and S. Saito. 1972. On the optimum quantization of feature parameters in the PARCOR speech synthesizer, In *IEEE 1972 Conf. Speech Commun. Process.,* New York, 434–437.

Johnson, D.H. and P.S. Rao. 1990. On the existence of Gaussian noise, In *The 1990 Digital Signal Processing Workshop,* New Paltz, NY, sponsored by IEEE Signal Processing Society, 8.13.1–8.14.2.

Julier, J.J. and J.K. Uhlmann. 1997. A new extension of the Kalman filter to nonlinear systems. Proceedings Eleventh International Symposium on Aerospace/Defence Sensing, Simulation, and Controls, Orlando, FL.

Kallmann, H.J. 1940. Transversal filters. *Proc. IRE,* 28, 302–310.

Kang, G.S. and L.J. Fransen. 1987. Experimentation with an adaptive noise-cancellation filter. *IEEE Trans. Circuits Syst.,* CAS-34, 753 758.

Kohonen, T. 1997. *Self-Organizing Maps,* 2nd ed. Springer-Verlag, Berlin.

Kung, H.T. and C.E. Leiserson. 1978. Systolic arrays (for VLSI). Sparse Matrix Proc. 1978, Soc. Ind. Appl. Math., 256–282.

Ljung, L. and T. Söderström. 1983. *Theory and Practice of Recursive Identification.* MIT Press, Cambridge, MA.

Linsker, R. 1988. Towards an organizing principle for a layered perceptual network. In *Neural Information Processing Systems,* D.Z. Anderson, Ed. American Institute of Physics, NY, pp. 485–494.

Linsker, R. 1989. How to generate ordered maps by maximizing the mutual information between input and output signals. *Neural Computat.,* 1, 402–411.

Luenberger, D.G. 1969. *Optimization by Vector Space Methods.* Wiley, NY.

Luttrell, S.P. 1989. Self-organization: a derivation from first principle of a class of learning algorithms. *IEEE Conf. Neural Networks,* 495–498.

Moody and Darken 1989. Fast learning in networks of locally-tuned processing units. *Neural Computat.,* 1, 281–294.

Oja, E. 1982. A simplified neuron model as a principal component analysis. *J. Math. Biol.,* 15, 267–273.

Owsley, N.L. 1985. Sonary array processing. In *Array Signal Processing,* S. Haykin, Ed. Prentice Hall, Englewood Cliffs, NJ, 115–193.

Poggio, T. and F. Girosi. 1990. Networks for approximation and learning. *Proc. IEEE,* 78, 1481–1497.

Rumelhart, D.E., G.E. Hinton, and R.J. Williams. 1986. *Learning Internal Representations by Error Propagation,* Vol. 1, D.E. Rumelhart and J.L. McCleland, Eds. MIT Press, Cambridge, MA.

Sanger, T.D. (1989). An optimality principle for unsupervised learning, *Adv. Neural Inf. Process. Syst.,* 1, 11–19.

Shabhag, N.R. and K.K. Parhi. 1994. *Pipelined Adaptive Digital Filters.* Kluwer, Boston, MA.

Sontag, E.D. (1990). *Mathematical Control Theory: Deterministic Finite Dimensional Systems,* Springer-Verlag, New York.

Ukrainec, A.M. and S. Haykin. 1996. A modular neural network for enhancement of cross-polar radar targets. *Neural Networks,* 9, 143–168.

Van Trees, H.L. 1968. *Detection, Estimation and Modulation Theory, Part I.* Wiley, NY.

Van Veen, B.D. and K.M. Buckley. 1988. Beamforming: a versatile approach to spatial filtering. *IEEE ASSP Mag.,* 5, 4–24.

Vapnik, V.N. 1998. *Statistical Learning Theory.* Wiley, New York.

Wan, E.A. 1994. Time series prediction by using a connectionist network with internal delay lines. In *Time Series Prediction: Forecasting the Future and Understanding the Past,* A.S. Weigend and N.A. Gershenfield, Eds. Addison-Wesley, Reading, MA, 195–217.

Wan, E.A., R. van der Merwe, and A.T. Nelson. 1999. Dual estimation and the unsented transformation. *Neural Inf. Process. Syst. (NIPS),* Denver, CO.

Werbos, P.J. 1974. Beyond regression: new tools for prediction and analysis in the behavioral sciences. Ph.D. thesis, Harvard University, Cambridge, MA.

Werbos, P.J. 1990. Backpropagation through time: what it does and how to do it. *Proc. IEEE*, 78, 1550–1560.

Widrow, B. et al . 1975. Adaptive noise cancelling: principles and applications. *Proc. IEEE*, 63, 1692–1716.

Widrow, B. and S.D. Stearns. 1985. *Adaptive Signal Processing*. Prentice Hall, Englewood Cliffs, NJ.

Williams, R.J. and D. Zipser. 1989. A learning algorithm for continually running fully recurrent neural networks, *Neural Computat.*, 1, 270–280.

Yee, P.V. 1998. Regularized radial basis function networks: theory and applications to probability estimation, classification, and time series prediction. Ph.D. thesis, McMaster University, Hamilton, Ontario.

3

Advanced Beamformers

Stergios Stergiopoulos

Defence R&D Canada Toronto

Unverisity of Toronto

Unviersity of Western Ontario

CONTENTS

3.1 Background

In general, the mainstream conventional signal processing of current sonar and ultra-sound systems consists of a selection of temporal and spatial processing algorithms [2–6]. These algorithms are designed to increase the signal-to-noise ratio (SNR) for improved signal delectability while simultaneously providing parameter estimates such as frequency, time-delay, Doppler and bearing for incorporation into localization, classification and signal tracking algorithms. Their implementation in real time systems had been directed at providing high-quality, artifact-free conventional beamformers, currently used in operational ultrasound and sonar systems. However, aberration effects associated with ultrasound system operations and the drastic changes in the threat acoustic signatures associated with sonars suggest that fundamentally new concepts need to be introduced into the signal processing structure of next-generation ultrasound and sonar systems.

To provide a context for the material contained in this chapter, it would seem appropriate to review briefly the basic requirements of high-performance sonar systems deploying multidimensional arrays of sensors. Figure 3.1 shows one possible high-level view of a generic warfare sonar system. The upper part of the figure presents typical sonar mine-hunting operations carried out by naval platforms (i.e., surface vessels). The lower left-hand side part of the figure provides a schematic representation of the coordinate system for a hull mounted cylindrical array of an active sonar [7,8]. The lower right-hand side part of Figure 3.1 provides a schematic representation of the coordinate system for a variable depth active sonar deploying a spherical array of sensors for mine warfare operations [9]. In particular, it is assumed that the sensors form a cylindrical or spherical array that allows for beam steering across 0–360° in azimuth and a 180° angular searching sector in elevation along the vertical axis of the coordinate system.

Thus, for effective sonar operations, the beam-width and the side-lobe structure of the beam steering patterns (shown in the lower part of Figure 3.1 for a given azimuth θ_s and

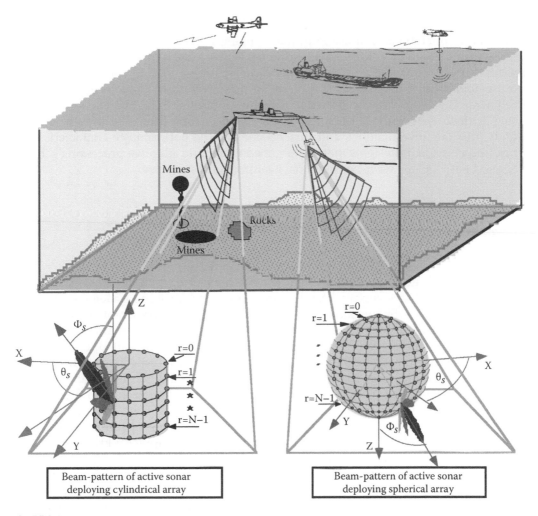

Beam-pattern of active sonar deploying cylindrical array

Beam-pattern of active sonar deploying spherical array

FIGURE 3.1
Upper part: Mine warfare sonar operations. Lower part (left): Schematic representation of the coordinate system for a hull mounted cylindrical array of an active sonar. Lower part (Right): Schematic representation of the coordinate system for a variable depth spherical array of an active sonar.

elevation Φ_s beam steering), should be very small to allow for high image and spatial resolution of detected mines that are in close proximity with other objects. More specifically, the beam steering pattern characteristics of a mine hunting sonar define its performance in terms of image and spatial resolution characteristics. For a given angular resolution in azimuth and elevation, a mine hunting sonar would not be able to distinguish detected objects and mines that are closer than the angular resolution performance limits. Moreover, the beam steering side-lobe structure would affect the image resolution performance of the system. Thus, for a high performance sonar it is desirable that the system should provide the highest possible angular resolution in azimuth and elevation as well as the lowest possible levels of side-lobe structures, properties that are defined by the aperture size of the receiving array. The above arguments are equally valid for ultrasound system operations since the beamforming process for ultrasound imaging assumes plane wave arrivals.

Because the increased angular resolution means longer sensor arrays with consequent technical and operational implications, many attempts have been made to increase the effective array length by synthesizing additional sensors (i.e., synthetic aperture processing) [1,6,11–16] or using adaptive beam processing techniques [1–5,17–24].

In previous studies, the impact and merits of these techniques have been assessed for towed array [1,4,5,10–20] and cylindrical array hull mounted [2,4,7,25,26] sonars and contrasted with those obtained using the conventional beamformer. The present material extends previous investigations and further assesses the performance characteristics of ultrasound and sonars systems that are assumed to include adaptive processing schemes integrated with a plane wave conventional beamforming structure.

3.2 Theoretical Remarks

Sonar operations can be carried out by a wide variety of naval platforms, as shown in Figure 3.1. This includes surface vessels, submarines and airborne systems such as airplanes and helicopters. Shown also in Figure 3.1 is a schematic representation of active and passive sonar operations in an underwater sea environment. Active sonar and ultrasound operations involve the transmission of well defined acoustic signals, called replicas, which illuminate targets in an underwater or human body medium, respectively. The reflected acoustic energy from a target or body organ provides the array receiver with a basis for detection and estimation. Passive sonar operations base their detection and estimation on acoustic sounds, which emanate from submarines and ships. Thus, in passive systems only the receiving sensor array is under the control of the sonar operators. In this case, major limitations in detection and classification result from imprecise knowledge of the characteristics of the target radiated acoustic sounds.

The passive sonar concept can be made clearer by comparing sonar systems with radars, which are always active. Another major difference between the two systems arises from the fact that sonar system performance is more affected than that of radar systems by the underwater medium propagation characteristics. All the above issues have been discussed in several review articles [1–6] that form a good basis for interested readers to become familiar with main stream sonar signal processing developments. Therefore, discussions of issues of conventional sonar signal processing, detection, estimation and influence of medium on sonar system performance are beyond the scope of this chapter. Only a very brief overview of the above issues will be highlighted in this section in order to define the basic terminology required for the presentation of the main theme of the present article. Let us start with a basic system model that reflects the interrelationships between the target, the underwater sea environment or the human body (medium) and the receiving sensor array of a sonar or an ultrasound system.

A schematic diagram of this basic system is shown in Figure 3.2, where array signal processing is shown to be 2-D [1,5,10,12,18] in the sense that it involves both temporal and spatial spectral analysis. The temporal processing provides spectral characteristics that are used for target classification and the spatial processing provides estimates of the directional characteristics (i.e., bearing and possibly range), of a detected signal. Thus, *space-time processing* is the fundamental processing concept in sonar and ultrasound systems and it will be the subject of our discussion in the next section.

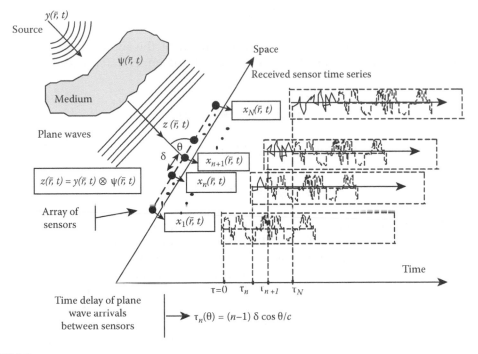

FIGURE 3.2
A model of space-time signal processing. It shows that ultrasound and sonar signal processing is two dimensional in the sense that it involves both temporal and spatial spectral analysis. The temporal processing provides characteristics for target classification and the spatial processing provides estimates of the directional characteristics (bearing, range-depth) of detected echoes (active case) or signals of interest (passive case). (Reprinted from IEEE ©1998: With permission.)

3.2.1 Space-Time Processing

For geometrical simplicity and without any loss of generality, we consider here a combination of N equally spaced acoustic transducers in a linear array, which may form a towed or hull mounted array system that can be used to estimate the directional properties of echoes and acoustic signals. As shown in Figure 3.2, a direct analogy between sampling in space and sampling in time is a natural extension of the sampling theory in space-time signal representation and this type of space-time sampling is the basis in array design that provides a description of an array system response. When the sensors are arbitrarily distributed, each element will have an added degree of freedom, which is its position along the axis of the array. This is analogous to nonuniform temporal sampling of a signal. In this chapter we extend our discussion to multidimensional array systems.

Sources of sound that are of interest in sonar and ultrasound system applications are harmonic narrowband, broadband and satisfy the wave equation [2,10]. Furthermore, their solutions have the property that their associated temporal-spatial characteristics are separable [10]. Therefore, measurements of the pressure field $z(\bar{r},t)$ which is excited by acoustic source signals, provide the spatial-temporal output response, designated by $x(\bar{r},t)$ of the measurement system. The vector $\bar{r}$ refers to the source-sensor relative position and t is the time. The output response $x(\bar{r},t)$ is the convolution of $z(\bar{r},t)$ with the line array system response $h(\bar{r},t)$ [10,30].

$$x(\overline{r}, t) = z(\overline{r}, t) \otimes h(\overline{r}, t) \tag{3.1}$$

where $\otimes$ refers to convolution. Since $z(\overline{r}, t)$ is defined at the input of the receiver, the convolution of the source's characteristics $y(\overline{r}, t)$ with the underwater medium's response $\psi(\overline{r}, t)$ is,

$$z(\overline{r}, t) = y(\overline{r}, t) \otimes \psi(\overline{r}, t) \tag{3.2}$$

Fourier transformation of Equation 3.1 provides:

$$X(\omega, \overline{k}) = \left\{ Y(\omega, \overline{k}) \cdot \Psi(\omega, \overline{k}) \right\} \cdot H(\omega, \overline{k}) \tag{3.3}$$

where, $\omega, \overline{k}$ are the frequency and wavenumber parameters of the temporal and spatial spectrums of the transform functions in Equations 3.1 and 3.2. Signal processing, in terms of beamforming operations, of the receiver's output $x(\overline{r}, t)$, provides estimates of the source bearing and possibly of the source range. This is a well-understood concept of the forward problem, which is concerned with determining the parameters of the received signal $x(\overline{r}, t)$ given that we have information about the other two functions $z(\overline{r}, t)$ and $h(\overline{r}, t)$ [5]. The inverse problem is concerned with determining the parameters of the impulse response of the medium $\psi(\overline{r}, t)$ by extracting information from the received signal $x(\overline{r}, t)$ assuming that the function $h(\overline{r}, t)$ is known [5]. The ultrasound and sonar problems, however, are quite complex and include both forward and inverse problem operations. In particular, detection, estimation and tracking-localization processes of sonar and ultrasound systems are typical examples of the forward problem, while target classification for passive–active sonars and diagnostic ultrasound imaging are typical examples of the inverse problem. In general, the inverse problem is a computationally very costly operation and typical examples in acoustic signal processing are seismic deconvolution and acoustic tomography.

3.2.2 Definition of Basic Parameters

This section outlines the context in which the sonar or the ultrasound problem can be viewed in terms of simple models of acoustic signals and noise fields. The signal processing concepts that are discussed in this chapter have been included in sonar and radar investigations with sensor arrays having circular, planar, cylindrical and spherical geometric configurations [7,25,26,28]. Thus, we consider a multidimensional array of equally spaced sensors with spacing δ. The output of the nth sensor is a time series denoted by $x_n(t_i)$, where $i = 1, ..., M_s$ are the time samples for each sensor time series.* denotes complex conjugate transposition so that $\overline{x}^*$ is the row vector of the received $\aleph$-sensor time series $\{x_n(t_i), n = 1, 2, ..., \aleph\}$.

Then $x_n(t_i) = s_n(ti) + \varepsilon_n(ti)$, where $s_n(ti)$, and $\varepsilon_n(ti)$ are the signal and noise components in the received sensor time series. $\overline{s}, \overline{\varepsilon}$ denote the column vectors of the signal and noise components of the vector $\overline{x}$ of the sensor outputs (i.e., $\overline{x} = \overline{s} + \overline{\varepsilon}$). $X_n(f) = \sum_{i=1}^{M_s} x_n(t_i) \exp(-j2\pi f t_i)$ is the Fourier transform of $x_n(t_i)$ at the signal with frequency f, $c = f\lambda$ is the speed of sound in the underwater, or human-body medium and λ is the wavelength of the frequency f. $S = E\{\overline{s} \quad \overline{s}^*\}$ is the spatial correlation matrix of the signal vector $\overline{s}$, whose n^{th} element is expressed by,

$$s_n(t_i) = s_n[t_i + \tau_n(\theta, \phi)] \tag{3.4}$$

$E\{.\}$ denotes expectation and $\tau_n(\theta, \phi)$ is the time delay between the $(n-1)$st and the nth sensor of the array for an incoming plane wave with direction of propagation of azimuth angle θ and an elevation angle ϕ, as depicted in Figure 3.2. In frequency domain, the spatial correlation matrix $\mathbf{S}$ for the plane wave signal $s_n(t_i)$ is defined by:

$$S(f_i, \theta, \phi) = A_s(f_i)\bar{D}(f_i, \theta, \phi)\bar{D}^*(f_i, \theta, \phi), \tag{3.5}$$

where $A_s(f_i)$ is the power spectral density of $s(t_i)$ for the ith frequency bin; and $\bar{D}(f, \theta, \phi)$ is the steering vector with the nth term being denoted by $d_n(f, \theta, \phi)$. Then matrix $\mathbf{S}(f, \theta, \phi)$ has as its nth row and nth column defined by, $S_{nm}(f_i, \theta, \phi) = A_s(f_i)d_n(f_i, \theta, \phi)d''_m(f_i, \theta, \psi)$. Moreover, $R(f_i)$ is the spatial correlation matrix of received sensor time series with elements, $R_{nm}(f, d_{nm})$. $R'_\varepsilon(f_i) = \sigma_n^2(f_i)R_\varepsilon(f_i)$ is the spatial correlation matrix of the noise for the ith frequency bin with $\sigma_n^2(f_i)$ being the power spectral density of the noise, $\varepsilon_n(t_i)$. In what is considered as an estimation procedure in this chapter, the associated problem of detection is defined in the classical sense as a hypothesis test that provides a detection probability and a probability of false alarm [31–33]. This choice of definition is based on the standard constant false alarm rate (CFAR) processor, which is based on the Neyman–Pearson (N–P) criterion [31]. The CFAR processor provides an estimate of the ambient noise or clutter level so that the threshold can be varied dynamically to stabilize the false alarm rate. Ambient noise estimates for the CFAR processor are provided mainly by noise normalization techniques [34] that account for the slowly varying changes in the background noise or clutter. The above estimates of the ambient noise are based upon the average value of the received signal, the desired probability of detection and probability of false alarms.

At this point, a brief discussion on the fundamentals of detection and estimation process is required in order to address implementation issues of signal processing schemes in sonar and ultrasound systems.

3.2.3 Detection and Estimation

In passive systems, in general, we do not have the *a priori* probabilities associated with the hypothesis H_1 that the signal is assumed present and the null hypothesis is H_0 that the received time series consists only of noise. As a result, costs can not be assigned to the possible outcomes of the experiment. In this case, the N–P criterion [31] is applied because it requires only a knowledge of the signal's and noise's probability density functions (pdf).

Let $x_{n=1}(t_i)$, $(i=1, ..., M)$ denote the received vector signal by a single sensor. Then for hypothesis H_1, which assumes that the signal is present, we have:

$$H_1: x_{n=1}(t_i) = s_{n=1}(t_i) + \varepsilon_{n=1}(t_i),$$

where $s_{n=1}(t_i)$ and $\varepsilon_{n=1}(t_i)$ are the signal and noise vector components in the received signal and $p_1(x)$ is the pdf of the received signal $x_{n=1}(t_i)$ given that H_1 is true. Similarly, for hypothesis H_0:

$$H_0: x_{n=1}(t_i) = \varepsilon_{n=1}(t_i)$$

and $p_0(x)$ is the pdf of the received signal given that H_0 is true. The N–P criterion requires maximization of probability of detection for a given probability of false alarm. So, there exists a nonnegative number η such that if hypothesis H_1 is chosen then

$$\lambda(x) = \frac{p_1(x)}{p_0(x)} \geq \eta \tag{3.6}$$

which is the likelihood ratio. By using the analytic expressions for $p_0(x)$ (the pdf for H_0) and $p_1(x)$ (the pdf for H_1) in Equation 3.6 and by taking the $\ln[\lambda(x)]$, we have [31],

$$\lambda_\tau = \ln[\lambda(x)] = \overline{s}^* R_\varepsilon' \overline{x} \tag{3.7}$$

where λ_τ is the log likelihood ratio and R_ε' is the covariance matrix of the noise vector, as defined in Section 3.2.2. For the case of white noise with $R_\varepsilon' = \sigma_n^2 I$ and I the unit matrix, the test statistic in Equation 3.7 is simplified into a simple correlation receiver (or replica correlator)

$$\lambda_\tau = \overline{s}^* \otimes \overline{x} \tag{3.8}$$

For the case of anisotropic noise, however, an optimum detector should include the correlation properties of the noise in the correlation receiver as this is defined in Equation 3.7.

For plane wave arrivals that are observed by a N-sensor array receiver the test statistics are [31]:

$$\lambda_\tau = \sum_{i=1}^{\frac{M_s}{2}-1} \overline{X^*(f_i)} \cdot R_\varepsilon'(f_i) \cdot S(f_i,\phi,\theta) \cdot [S(f_i,\phi,\theta)+R_\varepsilon'(f_i)]^{-1} \cdot \overline{X(f_i)}, \tag{3.9}$$

where the above statistics are for the frequency domain with parameters defined in Equations 3.4 and 3.5 in Section 3.2.2. Then, for the case of an array of sensors receiving plane wave signals, the log likelihood ratio λ_τ in Equation 3.9 is expressed by the following equation, which is the result of simple matrix manipulations based on the frequency domain expressions (Equations 3.4 and 3.5) and their parameter definitions presented in Section 3.2.2. Thus,

$$\lambda_\tau = \sum_{i=1}^{\frac{M}{2}-1} \left| \varphi(f_i)\overline{D}^*(f_i,\phi,\theta)R_\varepsilon'(f_i)^{-1}\overline{X(f_i)} \right|^2 \tag{3.10}$$

where [31]:

$$\varphi^2(f_i) = \frac{A_s(f_i)/\sigma_n^2(f_i)}{1+A_s(f_i)\overline{D}^*(f_i,\phi,\theta)R_\varepsilon'^{-1}(f_i)\overline{D}(f_i)/\sigma_n^2(f_i)} \tag{3.11}$$

Equation 3.10 can be written also as:

$$\lambda_\tau = \sum_{i=1}^{\frac{M}{2}-1} \left[\sum_{n=1}^{N} \zeta_n^*(f_i, \phi, \theta) X_n(f_i) \right]^2 \tag{3.12}$$

This last expression (Equation 3.12) of the log likelihood ratio indicates that an optimum detector in this case requires the filtering of each one of the N-sensor received time series $X_n(f_i)$ with a set of filters being the elements of the vector,

$$\overline{\zeta}(f_i, \phi, \theta) = \varphi(f_i) \overline{D}^*(f_i, \phi, \theta) R_\varepsilon'(f_i)^{-1} \tag{3.13}$$

Then, the summation of the filtered sensor outputs in frequency domain according to Equation 3.13 provides the test statistics for optimum detection. For the simple case of white noise $R_\varepsilon' = \sigma_n^2 I$ and for a line array receiver, the filtering operation in Equation 3.13 indicates plane wave conventional beamforming in frequency domain,

$$\lambda_\tau = \sum_{i=1}^{\frac{M}{2}-1} \left[\psi \sum_{n=1}^{N} d_n^*(f_i, \theta) X_n(f_i) \right]^2, \tag{3.14}$$

where, $\psi = \varsigma / (1 + N\varsigma)$, is a scalar, which is a function of SNR, $\varsigma = A_s^2 / \sigma_n^2$.

For the case of narrowband signals embedded in spatially and or temporaly correlated noise or interferes, it has been shown [13] that the deployment of very long arrays or application of acoustic synthetic aperture will provide sufficient array gain and will achieve optimum detection and estimation for the parameters of interest.

For the general case of broadband and narrowband signals embedded in a spatially anisotropic and temporally correlated noise field, Equation 3.14 indicates that the filtering operation for optimum detection and estimation requires adaptation of the sonar and ultrasound signal processing according to the ambient noise's and human body's noise characteristics, respectively. The family of algorithms for optimum beamforming that use the characteristics of the noise, are called *adaptive beamformers* [3,17–20,22,23]; and a detailed definition of an adaptation process requires knowledge of the correlated noise's covariance matrix $R_\varepsilon'(f_i)$. However, if the required knowledge of the noise's characteristics is inaccurate, the performance of the optimum beamformer will degrade dramatically [18,23]. As an example, the case of cancellation of the desired signal is often typical and significant in adaptive beamforming applications [18,24]. This suggests that the implementation of useful adaptive beamformers in real time operational systems is not a trivial task. The existence of numerous articles on adaptive beamforming suggests the dimensions of the difficulties associated with this kind of implementation. In order to minimize the generic nature of the problems associated with adaptive beamforming the concept of partially adaptive beamformer design was introduced. This concept reduces the degrees of freedom, which results in lowering the computational requirements and often improving the adaptive response time [17,18]. However, the penalty associated with the reduction of the degrees of freedom in partially adaptive beamformers is that they cannot converge to the same optimum solution as the fully adaptive beamformer.

Although a review of the various adaptive beamformers would seem relevant at this point, we believe that this is not necessary since there are excellent review articles [3,17,18,21] that summarize the points that have been considered for the material of this chapter. There are two main families of adaptive beamformers, the generalized side-lobe cancellers (GSC) [44,45] and the linearly constrained minimum variance beamformers (LCMV) [18]. A special case of the LCMV is Capon's maximum likelihood method [22], which is called minimum variance distortionless response (MVDR) [17,18,22,23,38,39]. This algorithm has proven to be one of the more robust of the adaptive array beamformers and it has been used by numerous researchers as a basis to derive other variants of MVDR [18]. In this chapter we will address implementation issues for various partially adaptive variants of the MVDR and a GSC adaptive beamformer [1], which are discussed in Section 3.4.2.

In summary, the classical estimation problem assumes that the *a priori* probability of the signal's presence $p(H_1)$ is unity [31–33]. However, if the signal's parameters are not known *a priori* and $p(H_1)$ is known to be less than unity, then a series of detection decisions over an exhaustive set of source parameters constitutes a detection procedure, where the results incidentally provide an estimation of source's parameters. As an example, we consider the case of a matched filter, which is used in a sequential manner by applying a series of matched filter detection statistics to estimate the range and speed of the target, which are not known *a priori*. This kind of estimation procedure is not optimal since it does not constitute an appropriate form of Bayesian minimum variance or minimum mean square error procedure.

Thus, the problem of detection [31–33] is much simpler than the problem of estimating one or more parameters of a detected signal. Classical decision theory [31–33,] treats signal detection and signal estimation as separate and distinct operations. A detection decision as to the presence or absence of the signal is regarded as taking place independently of any signal parameter or waveform estimation that may be indicated as the result of detection decision. However, interest in joint or simultaneous detection and estimation of signals arises frequently. Middleton and Esposito [46] have formulated the problem of simultaneous optimum detection and estimation of signals in noise by viewing *estimation* as a generalized detection process. Practical considerations, however, require different cost functions for each process [46]. As a result, it is more effective to retain the usual distinction between detection and estimation.

Estimation, in passive sonar and ultrasound systems, includes both the temporal and spatial structure of an observed signal field. For active systems, correlation processing and Doppler (for moving target indications) are major concerns that define the critical distinction between these two approaches (i.e., *passive, active*) to sonar and ultrasound processing. In this chapter, we restrict our discussion only to topics related to spatial signal processing for estimating signal parameters. However, spatial signal processing has a direct representation that is analogous to the frequency-domain representation of temporal signals. Therefore, the spatial signal processing concepts discussed here have direct applications to temporal spectral analysis.

3.2.4 Cramer–Rao Lower Bound (CRLB) Analysis

Typically, the performance of an estimator is represented as the variance in the estimated parameters. Theoretical bounds associated with this performance analysis are specified by the Cramer–Rao bound [31–33] and that has led to major research efforts by the sonar signal processing community in order to define the idea of an optimum processor for discrete sensor arrays [12,16,56–59]. If the *a priori* probability of detection is close to unity then the minimum variance achievable by any unbiased estimator is provided by the *Cramer–Rao lower bound* (CRLB) [31,32,46].

More specifically, let us consider that the received signal by the *n*th sensor of a receiving array is expressed by,

$$x_n(t_i) = s_n(t_i) + \varepsilon_n(t_i) \tag{3.15}$$

where, $s_n(t_i, \overline{\Theta}) = s_n[t_i + \tau_n(\theta, \phi)]$, defines the received signal model with $\tau_n(\theta, \phi)$ being the time delay between the $(n-1)$st and the *n*th sensor of the array for an incoming plane wave with direction of propagation of azimuth angle θ and an elevation angle ϕ, as depicted in Figure 3.2. The vector $\overline{\Theta}$, includes all the unknown parameters considered in relation (Equation 3.15). Let $\sigma_{\theta_i}^2$ denote the variance of an unbiased estimate of an unknown parameter θ_i in the vector $\overline{\Theta}$. The Cramer–Rao [31–33] bound states that the best unbiased estimate $\tilde{\Theta}$ of the parameter vector $\overline{\Theta}$ has the covariance matrix

$$\text{cov}\,\tilde{\Theta} \geq J(\overline{\Theta})^{-1}, \tag{3.16}$$

where *J* is the Fisher information matrix whose elements are:

$$J_{ij} = -E\left(\frac{\vartheta^2 \ln P\langle \overline{X}|\overline{\Theta}\rangle}{\vartheta\theta_i \vartheta\theta_j} \right). \tag{3.17}$$

In Equation 3.17, $P\langle \overline{X}|\overline{\Theta}\rangle$, is the probability density function (pdf) governing the observations:

$$\overline{X} = [x_1(t_i), x_2(t_i), x_3(t_i), ... x_N(t_i)]^*,$$

for each of the *N* and M_s independent spatial and temporal samples, respectively that are described by the model in Equation 3.15. The variance of the unbiased estimates $\tilde{\Theta}$ has a lower bound (called the CRLB), which is given by the diagonal elements of Equation 3.16. This CRLB is used as standard of performance and provides a good measure for the performance of a signal-processing algorithm which gives unbiased estimates $\tilde{\Theta}$ for the parameter vector $\overline{\Theta}$. In this case, if there exists a signal processor to achieve the CRLB, it will be the maximum-likelihood estimation (MLE) technique. The above requirement associated with the *a priori* probability of detection is very essential because if it is less than one, then the estimation is biased and the theoretical CRLBs do not apply. This general framework of optimality is very essential in order to account for Middleton's [32] warning that a system optimized for the one function (detection or estimation) may not be necessarily optimized for the other.

For a given model describing the received signal by a sonar or ultrasound system, the CRLB analysis can be used as a tool to define the information inherent in a sonar system. This is an important step related to the development of the signal processing concept for a sonar system as well as in defining the optimum sensor configuration arrangement under which we can achieve, in terms of system performance, the optimum estimation of signal parameters of our interest. This approach has been applied successfully to various studies related to the present development [12,15,56–59].

As an example, let us consider the simplest problem of one source with the bearing θ_1 being the unknown parameter. Following Equation 3.17, the results of the variance $\sigma_{\theta_1}^2$ in the bearing estimates are,

$$\sigma_{\theta_i}^2 = \frac{3}{2\psi N}\left(\frac{B_w}{\pi \sin\theta_1}\right)^2, \tag{3.18}$$

where, $\psi = M_s A_1^2 / \sigma_N^2$, the parameter $B_w = \lambda / (N-1)\delta$ gives the beamwidth of the physical aperture that defines the angular resolution associated with the estimates of θ_1. The SNR at the sensor level is $SNR = 10 \times \log_{10}(\psi)$ or

$$SNR = 20 \times \log_{10}(A_1 / \sigma_1) + 10 \times \log_{10}(M_s). \tag{3.19}$$

It is obvious from Equations 3.18 and 3.19 that the variance of the bearing $\sigma_{\theta_1}^2$ can get smaller when the observation period $T = M_s / f_s$ becomes long and the receiving array size, $L = (N-1)\lambda$ gets very long.

The next question needed to be addressed is about the unbiased estimator that can exploit this available information and provide results asymptotically reaching the CRLBs. For each estimator it is well known that there is a range of SNR in which the variance of the estimates rises very rapidly as SNR decreases. This effect, which is called the *threshold effect of the estimator*, determines the range of SNR of the received signals for which the parameter estimates can be accepted. In passive sonar systems the SNR of signals of interest are often quite low and probably below the threshold value of an estimator. In this case, high frequency resolution in both time and spatial domains for the parameter estimation of narrowband signals is required. In other words, the threshold effect of an estimator determines the frequency resolution for processing and the size of the array receivers required in order to detect and estimate signals of interest that have very low SNR [12,14,53]. The CRLB analysis has been used in many studies to evaluate and compare the performance of the various nonconventional processing schemes [17,18,55] that have been considered for implementation in the generic beamforming structure to be discussed in Section 3.4.1. In general, array signal processing includes a large number of algorithms for a variety of systems that are quite diverse in concept. There is a basic point that is common in all of them, however, and this is the beamforming process, which we are going to examine next.

3.3 Optimum Estimators for Array Signal Processing

Sonar signal processing includes mainly estimation (after detection) of the source's bearing, which is the main concern in sonar array systems because in most of the sonar applications the acoustic signal's wavefronts tend to be planar, which assumes distant sources. Passive ranging by measurement of wavefront curvature is not appropriate for the far-field problem. The range estimate of a distant source, in this case, must be determined by various target-motion analysis methods discussed elsewhere [1], which address the localization-tracking performance of nonconventional beamformers with real data.

More specifically, a 1-D device such as a line sensor array satisfies the basic requirements of a spatial filter. It provides direction discrimination, at least in a limited sense,
and an SNR improvement relative to an omni-directional sensor. Because of the simplified
mathematics and reduced number of the involved sensors, relative to multidimensional
arrays, most of the researchers have focused on the investigation of the line sensor arrays
in system applications [1–6]. Furthermore, implementation issues of synthetic aperture
and adaptive techniques in real time systems have been extensively investigated for line
arrays as well [1,5,6,12,17,19,20]. However, the configuration of the array depends on the
purpose for which it is to be designed. For example, if a wide range of horizontal angles is
to be observed, a circular configuration may be used, given rise to beam characteristic that
are independent of the direction of steering. Vertical direction may be added by moving
into cylindrical configuration [8]. In a more general case, where both vertical and horizontal steering is required and where a large range of angles is to be covered, a spherically
symmetric array would be desirable [9]. In modern ultrasound imaging systems planar
arrays are required to reconstruct real time 3-D images. However, the huge computational
load required for multidimensional conventional and adaptive beamformers makes the
applications of these 2-D and 3-D arrays in real time systems nonfeasible.

Furthermore, for modern sonar and radar systems, it has become a necessity these days
that all possible active and passive modes of operation should be exploited under an integrated processing structure that reduces redundancy and provides cost effective real
time system solutions [6]. Similarly, the implementation of computationally intensive data
adaptive techniques in real time systems is also an issue of equal practical importance.
However, when theses systems include multidimensional (2-D, 3-D) arrays with hundreds
of sensors, then the associated beamforming process requires very large memory and very
intensive throughput characteristics, something that makes its implementation in real time
systems a very expensive and difficult task.

To counter this implementation problem, this chapter introduces a generic approach of
implementing conventional beamforming processing schemes with integrated passive and
active modes of operations in systems that may include, planar, cylindrical or spherical
arrays [25–28]. This approach decomposes the 2-D and 3-D beamforming process into sets
of line and/or circular array beamformers. Because of the decomposition process, the fully
multidimensional beamformer can now be divided into subsets of coherent processes that
can be implemented in small size CPU's that can be integrated under the parallel configuration of existing computing architectures. Furthermore, application of spatial shading for
multidimensional beamformers to control side-lobe structures can now be easily incorporated. This is because the problem of spatial shading for line arrays has been investigated
thoroughly [36] and the associated results can be integrated into a circular and a multidimensional beamformer, which can be decomposed now into coherent subsets of line and
or circular beamformers of the proposed generic processing structure.

As a result of the decomposition process, provided by the generic processing structure,
the implementation effort for adaptive schemes is reduced to implementing adaptive processes in line and circular arrays. Thus, a multidimensional adaptive beamformer can
now be divided into two coherent modular steps which lead to efficient system oriented
implementations. In summary, the proposed approach demonstrates that the incorporation of adaptive schemes with near-instantaneous convergence in multidimensional arrays
is feasible [7,25–28].

At this point it is important to note that the proposed decomposition process of 2-D
and 3-D conventional beamformers into sets of line and/or circular array beamformers
is an old concept that has been exploited over the years by sonar system designers. Thus,

references on this subject may exist in Navy-Labs' and Industrial-Institutes' technical reports that are not always readily available and the author of this chapter is not aware of any kind of reports in this area. Previous efforts attempted to address practical implementation issues and had been focused on cylindrical arrays. As an example, a cylindrical array beamformer is decomposed into time-delay line array beamformers providing beams along elevation angles of the cylindrical array. These are called staves. Then, the beam time series associated with a particular elevation steering of interest are provided at the input of a circular array beamformer.

In this chapter the attempt is to provide a higher degree of development than the one discussed for cylindrical arrays. The task is to develop a generic processing structure that integrates the decomposition process of multidimensional planar, cylindrical and spherical array beamformers into line and or circular array beamformers. Furthermore, the proposed generic processing structure integrates passive and active modes of operation into a single signal processing scheme.

3.3.1 Generic Multidimensional Conventional Beamforming Structure

3.3.1.1 Line-Array Conventional Beamformer

Consider an N-sensor line array receiver with uniform sensor spacing δ, shown in Figure 3.3, receiving plane-wave arrivals with direction of propagation θ. Then, as a follow up of the parameter definition in Section 3.2,

$$\tau_n(\theta) = (n-1)\delta\cos\theta/c, \tag{3.20}$$

is the time delay between the 1st and the nth sensor of the line array for an incoming plane wave with direction θ, as illustrated in Figure 3.3.

$$d_n(f_i,\theta) = \exp\left[j2\pi \frac{(i-1)f_s}{M} \tau_n(\theta) \right], \tag{3.21}$$

is the nth term of the steering vector $\bar{D}(f,\theta)$. Moreover, because of Equations 3.13 and 3.14 the plane wave response of the N-sensor line array steered at a direction θ_s can be expressed by,

$$B(f,\theta_s) = \bar{D}^*(f,\theta_s)\bar{X}(f) \tag{3.22}$$

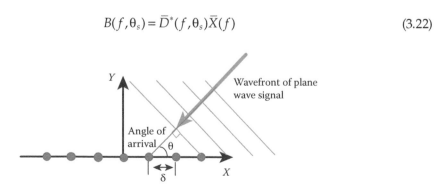

FIGURE 3.3
Geometric configuration and coordinate system for a line array of sensors.

Previous studies [1] have shown that for a single source this conventional beamformer without shading is an optimum processing scheme for bearing estimation. The side-lobe structure can be suppressed at the expense of a beam width increase by applying different weights (i.e., spatial shading window) [36]. The angular response of a line-array is ambiguous with respect to the angle θ_s, responding equally to targets at angle θ_s and $-\theta_s$ where θ_s varies over $[0, \pi]$.

Equation 3.22 is basically a mathematical interpretation of Figure 3.3 and shows that a line array is basically a spatial filter because by steering a beam in a particular direction we spatially filter the signal coming from that direction, as this is illustrated in Figure 3.3. On the other hand, Equation 3.22 is fundamentally a discrete Fourier transform relationship between the hydrophone weightings and the beam pattern of the line array and as such it is computationally a very efficient operation. However, Equation 3.22 can be generalized for nonlinear 2-D and 3-D arrays and this is discussed in the next section.

As an example, let us consider a distant monochromatic source. Then the plane wave signal arrival from the direction θ received by a N-hydrophone line array is expressed by Equation 3.21. The beam power pattern $P(f,\theta_s)$ is given by $P(f,\theta_s) = B(f,\theta_s) \times B^*(f,\theta_s)$ that takes the form

$$P(f,\theta_s) = \sum_{n=1}^{N} \sum_{m=1}^{N} X_n(f) X_m^*(f) \exp\left[\frac{j2\pi f \delta_{nm} \cos\theta_s}{c} \right], \tag{3.23}$$

where δ_{nm} is the spacing $\delta(n-m)$ between the nth and mth sensors. As a result of Equation 3.23, the expression for the power beam pattern $P(f,\theta_s)$, is reduced to:

$$P(f,\theta_s) = \left\{ \frac{\sin\left[N \frac{\pi\delta}{\lambda}(\sin\theta_s - \sin\theta) \right]}{\sin\left[\frac{\pi\delta}{\lambda}(\sin\theta_s - \sin\theta) \right]} \right\}^2 \tag{3.24}$$

Let us consider for simplicity the source bearing θ to be at array broadside, $\delta=\lambda/2$ and $L=(N-1)\delta$ is the array size. Then Equation 3.24 is modified as [4,10]:

$$P(f,\theta_s) = \frac{N^2 \sin^2\left[\frac{\pi L \sin\theta_s}{\lambda} \right]}{\left(\frac{\pi L \sin\theta_s}{\lambda} \right)^2}, \tag{3.25}$$

which is the farfield radiation or directivity pattern of the line array as opposed to near field regions. The results in Equations 3.24 and 3.25 are for a perfectly coherent incident acoustic signal and an increase in array size L results in additional power output and a reduction in beamwidth, which are similar arguments with those associated with the CRLB analysis expressed by Equation 3.18. The side-lobe structure of the directivity pattern of a line array, which is expressed by Equation 3.24, can be suppressed at the expense of a beamwidth increase by applying different weights. The selection of these weights will act as spatial filter coefficients with optimum performance [5,17,18]. There are two different approaches to select the above weights: *pattern optimization* and *gain optimization*. For pattern optimization the desired array response pattern $P(f,\theta_s)$ is selected first. A desired pattern is usually one with a narrow main lobe and low side-lobes. The weighting or shading coefficients in this case are real numbers from well known window functions that modify the array response pattern. Harris' review [36] on the use of windows in discrete Fourier transforms and temporal spectral analysis is directly applicable in this case to spatial spectral analysis for towed line array applications.

Using the approximation $\sin\theta \cong \theta$ for small θ at array broadside, the first null in Equation 3.22 occurs at $\pi L \sin\theta/\lambda = \pi$ or $\Delta\theta\, xL/\lambda \cong 1$. The major conclusion drawn here for line array applications is that [4,10]:

$$\Delta\theta \approx \lambda 4/L \text{ and } \Delta f \times T = 1 \tag{3.26}$$

where $T = M_s/F_s$ is the sensor time series length. Both the above relations in Equation 3.26 express the well known temporal and spatial resolution limitations in line array applications that form the driving force and motivation for adaptive and synthetic aperture signal processing that we will discuss later.

An additional constraint for sonar and ultrasound applications requires that the frequency resolution Δf of the hydrophone time series for spatial spectral analysis that is based on fast Fourier transform (FFT) beamforming processing must be such that

$$\Delta f \times \frac{L}{c} \ll 1 \tag{3.27}$$

in order to satisfy *frequency quantization* effects associated with discrete frequency domain beamforming following the FFT of sensor data [17,42]. This is because, in conventional beamforming Finite-duration impulse response (FIR) filters are used to provide realizations in designing digital phase shifters for beam steering. Since fast-convolution signal processing operations are part of the processing flow of a sonar signal processor, the effective beamforming filter length needs to be considered as the overlap size between successive snapshots. In this way, the overlap process will account for the wraparound errors that arise in the fast-convolution processing [1,40–42]. It has been shown [42] that an approximate estimate of the effective beamforming filter length is provided by Equations 3.25 and 3.27.

Because of the linearity of the conventional beamforming process, an exact equivalence of the frequency domain narrowband beamformer with that of the time-domain beamformer for broadband signals can be derived [42,43]. Based on the model of Figure 3.2, the time-domain beamformer is simply a time delaying [43] and summing process across the hydrophones of the line array, which is expressed by,

$$b(\theta_s, t_i) = \sum_{n=1}^{N} x_n(t_i - \tau_s) \tag{3.28}$$

Since,

$$b(\theta_s, t_i) = \text{IFFT}\{B(f, \theta_s)\}, \tag{3.29}$$

by using inverse FFTs (IFFT) and fast convolution procedures, continuous beam-time sequences can be obtained at the output of the frequency domain beamformer [42]. This is a very useful operation when the implementation of beamforming processors in sonar systems is considered.

The beamforming operation in Equation 3.28 is not restricted only for plane wave signals. More specifically, consider an acoustic source at the near field of a line array with r_s the source range and θ its bearing. Then the time delay for steering at θ is

$$\tau_s = \left(r_s^2 + d_{nm}^2 - 2r_s d_{nm}\cos\theta\right)^{1/2}/c \tag{3.30}$$

As a result of Equation 3.30, the steering vector $d_n(f,\theta_s)=\exp[j2\pi f\tau_s]$ will include two parameters of interest, the bearing θ and range r_s of the source. In this case the beamformer is called *focussed beamformer*, which is used mainly in ultrasound system applications There are, however, practical considerations restricting the application of the focused beamformer in passive sonar line array systems and these have to do with the fact that effective range focussing by a beamformer requires extremely long arrays.

3.3.1.2 Circular Array Conventional Beamformer

Consider M-sensors distributed uniformly on a ring of radius R receiving plane-wave arrivals at an azimuth angle θ and an elevation angle ϕ as shown in Figure 3.4. The plane-wave response of this circular array for azimuth steering θ_s and an elevation steering ϕ_s can be written as follows:

$$B(f,\theta_s,\phi_s) = \bar{D}^*(f,\theta_s,\phi_s)W(\theta_s)\bar{X}(f), \tag{3.31}$$

where $\bar{D}(f,\theta_s,\phi_s)$ is the steering vector with the mth term being expressed by $d_m(f,\theta_s,\phi_s)=\exp(j2\pi f R\sin\phi_s\cos(\theta_s-\theta_m)/c)$ and $\theta_m=2\pi m/M$ is the angular location of the mth sensor with $m=0,1,...,M-1$. $W(\theta_s)$ is a diagonal matrix with the off diagonal terms being zero and the diagonal terms being the weights of a spatial window to reduce the side-lobe structure [36]. This spatial window, in general, is not uniform and depends on the sensor location (θ_m) and the beam steering direction (θ_s). The beam power pattern $P(f,\theta_s,\phi_s)$ is given by $P(f,\theta_s,\phi_s)=B(f,\theta_s,\phi_s)\times B^*(f,\theta_s,\phi_s)$. The azimuth angular response of the circular array covers the range $[0,2\pi]$ and therefore there is no ambiguity with respect to the azimuth angle θ.

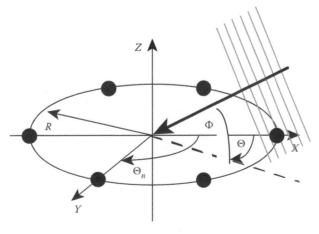

FIGURE 3.4
Geometric configuration and coordinate system for a circular array of sensors.

3.3.2 Multidimensional (3-D) Array Conventional Beamformer

Presented in this section is a generic approach to decompose the planar, cylindrical and spherical array beamformers into coherent subsets of line and/or circular array beamformers. In this chapter, we will restrict the discussion on 3-D arrays with cylindrical and planar geometric configuration. The details of the decomposition process for spherical arrays are similar and can be found elsewhere [7,25–28].

3.3.2.1 Decomposition Process for 2-D and 3-D Sensor Array Beamformers

3.3.2.1.1 Cylindrical Array Beamformer

Consider the cylindrical array shown in Figure 3.5 with $\aleph$ sensors and $\aleph = NM$, where N is the number of circular rings and M is the number of sensors on each ring. The angular response of this cylindrical array to a steered direction at (θ_s, ϕ_s) can be expressed as

$$B(f,\theta_s,\phi_s) = \sum_{r=0}^{N-1}\sum_{m=0}^{M-1} w_{r,m} X_{r,m}(f) d_{r,m}^*(f,\theta_s,\phi_s) \tag{3.32}$$

where $w_{r,m}$ is the $(r,m)^{th}$ term of a 3-D spatial window, $X_{r,m}(f)$ is the $(r,m)^{th}$ term of the matrix $\underline{X}(f)$, or $X_{r,m}(f)$ is the Fourier transform of the signal received by the mth sensor on the rth ring and $d_{r,m}(f,\theta_s,\phi_s) = \exp\{j2\pi f[(r\delta_z\cos\phi_s + R\sin\phi_s\cos(\theta_s - \theta_m)/c)]\}$ is the $(r,m)^{th}$ steering term of $\bar{D}(f,\theta_s,\phi_s)$. R is the radius of the ring, δ_z is the distance between each ring along z-axis, r is the index for the rth ring and $\theta_m = 2\pi m/M$, $m = 0, 1, ..., M-1$. Assuming $w_{r,m} = w_r \times w_m$, Equation 3.32 can be re-arranged as follows:

$$B(f,\theta_s,\phi_s) = \sum_{r=0}^{N-1} w_r d_r^*(f,\theta_s,\phi_s)\left[\sum_{m=0}^{M-1} X_{r,m}(f) w_m d_m^*(f,\theta_s,\phi_s)\right] \tag{3.33}$$

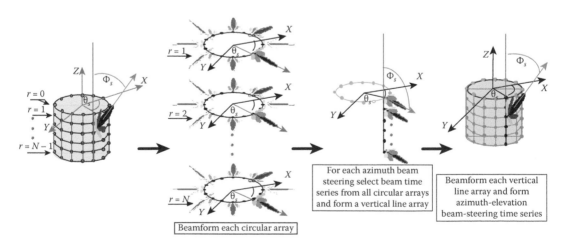

FIGURE 3.5

Coordinate system and geometric representation of the concept of decomposing a cylindrical array beamformer. The $\aleph = NM$ sensor cylindrical array beamformer consists of N circular arrays with M being the number of sensors in each circular array. Then, the beamforming structure for cylindrical arrays is reduced into coherent subsets of circular (for 0–360° azimuth bearing estimates) and line array (for 0–180° angular elevation bearing estimates) beamformers.

where $d_r(f, \theta_s, \phi_s) = \exp\{j2\pi f(r\delta_z \cos\phi_s/c)\}$ is the rth term of the steering vector for line-array beamforming, w_r is the rth term of a spatial window for line array spatial shading, $d_m(f, \theta_s, \phi_s) = \exp\{j2\pi f(R\sin\phi_s \cos(\theta_s - \theta_m)/c)\}$ is the mth term of the steering vector for a circular beamformer, discussed in Section 3.3.1, and w_m is the mth term of a spatial window for circular array shading.

Thus, Equation 3.33 suggests the decomposition of the cylindrical array beamformer into two steps, which is a well-known process in array theory. The first step is to perform circular array beamforming for each of the N rings with M sensors on each ring. The second step is to perform line array beamforming along z-axis on the N-beam time series outputs of the first step. This kind of implementation, which is based on the decomposition of the cylindrical beamformer into line and circular array beamformers is shown in Figure 3.5. The coordinate system is identical to that shown in Figure 3.4. The decomposition process of Equation 3.33 makes also the design and incorporation of 3-D spatial windows much simpler. Nonuniform shading windows can be applied to each circular beamformer to improve the angular response with respect to the azimuth angle, θ. A uniform shading window can then be applied to the line array beamformer to improve the angular response with respect to the elevation angle, ϕ. Moreover, the decomposition process, shown in Figure 3.5, leads to an efficient implementation in computing architectures based on the following two factors:

- The number of sensors for each of these circular and line array beamformers is much less than the total number of sensors, $\aleph$, of the cylindrical array. This kind of decomposition process for the 3-D beamformer eliminates the need for very large memory and CPU's with very high throughput requirements in one board for real time system applications.

- All the circular and line array beamformers can be executed in parallel, which allows their implementations in much simpler parallel architectures with simpler CPU's, which is a practical requirement for real time system applications.

Thus, under the restriction $w_{r,m} = w_r \times w_m$ for 3-D spatial shading, the decomposition process provides equivalent beam time series with those that would have been provided by a 3-D cylindrical beamformer, as this is shown by Equations 3.32 and 3.33.

3.3.2.1.2 Planar Array Beamformer

Consider the discrete planar array shown in Figure 3.6 with $\aleph$ sensors where $\aleph = NM$ and M, N are the number of sensors along x-axis and y-axis, respectively. The angular response of this planar array to a steered direction (θ_s, ϕ_s) can be expressed as

$$B(f, \theta_s, \phi_s) = \sum_{r=0}^{N-1}\sum_{m=0}^{M-1} w_{r,m} X_{r,m}(f) d_{r,m}^*(f, \theta_s, \phi_s), \tag{3.34a}$$

where $w_{r,m}$ is the (r,m)th term of matrix $W(\theta, \phi)$ including the weights of a 2-D spatial window, $X_{r,m}(f)$ is the (r,m)th term of the matrix $\underline{X}(f)$ including the Fourier transform of the received signal by the (m,r)th sensor along x-axis and y-axis, respectively. $\underline{D}(f, \theta_s, \phi_s)$ is the steering matrix having its (r,m)th term defined by

$$d_{r,m}(f, \theta_s, \phi_s) = \exp(j2\pi f(m\delta_x \sin\theta_s + r\delta_y \cos\theta_s \cos\phi_s)/c). \tag{3.34b}$$

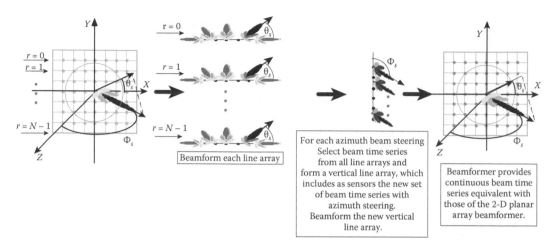

FIGURE 3.6

Coordinate system and geometric representation of the concept of decomposing a planar array beamformer. The $\aleph = NM$ sensor planar array beamformer consists of N line arrays with M being the number of sensors in each line array. Then, the beamforming structure for planar arrays is reduced into coherent subsets of line (for 0–180° azimuth bearing estimates) and line array (for 0–180° elevation bearing estimates) beamformers.

Assuming that the matrix of spatial shading (weighting) $W(\theta,\phi)$ is separable (i.e., $\underline{W}(\theta,\phi) = \underline{W}_1(\theta)\underline{W}_2(\phi)$), Equation 3.34 can be simplified as follows:

$$B(f,\theta_s,\phi_s) = \sum_{r=0}^{N-1} w_{1,r} d_r^*(f,\theta_s,\phi_s) \left[\sum_{m=0}^{M-1} w_{2,m} X_{r,m}(f) d_m^*(f,\theta_s,\phi_s) \right] \quad (3.35a)$$

where, $d_r(f,\theta_s,\phi_s) = \exp(j2\pi f r \delta_y \cos\theta_s \cos\phi_s / c)$, is the rth term of the steering vector, $\bar{D}_y(f,\theta_s,\phi_s)$ and $d_m(f,\theta_s,\phi_s) = \exp(j2\pi f m \delta_x \sin\theta_s / c)$ is the mth term of the steering vector, $\bar{D}_x(f,\theta_s,\phi_s)$. The summation term enclosed by parenthesis in Equation 3.35a is equivalent to the response of a line array beamformer along x-axis. Then all the steered beams from this summation term form a vector denoted by $\bar{B}_y(f,\theta_s)$. This vector defines a line array with directional sensors, which are the beams defined by the second summation process of Equation 3.35a. Therefore Equation 3.35a can be expressed as:

$$B(f,\theta_s,\phi_s) = \bar{D}_y^*(f,\theta_s,\phi_s) \underline{W}_1(\theta) \bar{B}_y(f,\theta_s) \quad (3.35b)$$

Equation 3.35b suggests that the 2-D planar array beamformer can be decomposed into two line array beamforming steps. The first step includes a line-array beamforming along x-axis and will be repeated N-times to get the vector $\bar{B}_y(f,\theta_s)$ that includes the beam times series $b_r(f,\theta_s)$, where the index $r = 0, 1, ..., N-1$ is along the y-axis. The second step includes line array beamforming along y-axis and will be done only once by treating the vector $\bar{B}_y(f,\theta_s)$ as the input signal for the line array beamformer to get the output $B(f,\theta_s,\phi_s)$. The separable spatial windows can now be applied separately on each line-array beamformer to suppress side-lobe structures. Figure 3.6 shows the involved steps of decomposing the 2-D planar array beamformer into two steps of line-array beamformers. The coordinate system is identical with that shown in Figure 3.3. The decomposition of the planar array beamformer into these two line-array beamforming steps leads to an efficient implementation based on the following two factors. First, the number of the involved sensors for each of these line array beamformers is much less than the total number of sensors, $\aleph$ of the planar array. This kind of decomposion process for the 2-D beamformer eliminates the need for very large memory and CPU's with very high throughput requirements in one board for real time system applications. Secondly, all these line array beamformers can be executed in parallel, which allows their implementation in much simpler parallel architectures with simpler CPU's, which is a practical requirement for real time system application. Besides the advantage of the efficient implementation, the proposed decomposition approach makes the application of the spatial window much simpler to be incorporated.

3.3.3 Influence of the Medium's Propagation Characteristics on the Performance of a Receiving Array

In ocean acoustics and medical ultrasound imaging, the wave propagation problem is highly complex due to the spatial properties of the nonhomogeneous underwater and human body mediums. For stationary source and receiving arrays, the space time properties of the acoustic pressure fields include a limiting resolution imposed by these mediums. This limitation is due either to the angular spread of the incident energy about a single arrival as a result of the scattering phenomena, or to the multipaths and their variation over the aperture of the receiving array.

More specifically, an acoustic signal that propagates through anisotropic mediums will interact with the transmitting medium microstructure and the rough boundaries, resulting in a net field that is characterized by irregular spatial and temporal variations. As a consequence of these interactions, a point source detected by a high-angular resolution receiver is perceived as a source of finite extent. It has been suggested [35,47] that due to the above spatial variations the sound field consists not of parallel, but of superimposed wavefronts of different directions of propagation. As a result, coherence measurements of this field by a receiving array give an estimate for the spatial coherence function. In the model for the spatial uncertainty of the above study [47], the width of the coherence function is defined as the coherence length of the medium and its reciprocal value is a measure of the angular uncertainty caused by the scattered field of the underwater environment.

By the *coherence* of acoustic signals in the sea or the human body, we mean the degree to which the acoustic pressures are the same at two points in the medium of interest located a given distance and direction apart. Pressure sensors placed at these two points will have phase coherent outputs if the received acoustic signals are perfectly coherent; if the two sensor outputs, as a function of space or time, are totally dissimilar, the signals are said to be incoherent. Thus, the loss of spatial coherence results in an upper limit on the useful aperture of a receiving array of sensors [10]. Consequently, knowledge of the angular uncertainty of the signal caused by the medium is considered essential in order to determine quantitatively the influence of the medium on the array gain, which is also influenced significantly by a partially directive anisotropic noise background. Therefore, for a given nonisotropic medium, it is desirable to estimate the optimum array size and achievable array gain for sonar and ultrasound array applications.

For geometrical simplicity and without any loss of generality we consider the case of a receiving line array. Quantitative estimates of the spatial coherence for a receiving line array are provided by the cross spectral density matrix in frequency domain between any set of two sensor time series of the line array. An estimate of the cross spectral density matrix $R(f)$ with its nm^{th} term defined by

$$R_{nm}(f,\delta_{nm}) = E[X_n(f)X_m^*(f)], \qquad (3.36)$$

The above space-frequency correlation function can be related to the angular power directivity pattern of the source, $\Psi_s(f,\theta)$, via a Fourier transformation by using a generalization of Bello's concept [48] of time-frequency correlation function $[t \Leftrightarrow 2\pi f]$ into space $[\delta_{nm} \Leftrightarrow 2\pi f \sin\theta/c]$, which gives

$$R_{nm}(f,\delta_{nm}) = \int_{-\pi/2}^{\pi/2} \Psi_s(f,\theta)\exp\left[\frac{-j2\pi f \delta_{nm}\theta}{c}\right]d\theta, \qquad (3.37)$$

or

$$\Psi_s(f,\theta) = \int_{-N\delta/2}^{N\delta/2} R_{nm}(f,\delta_{nm})\exp\left[\frac{j2\pi f \delta_{nm}\theta}{c}\right]d(\delta_{nm}), \qquad (3.38)$$

The above transformation can be converted into the following summation:

$$R_{nm}(f_o,\delta_{nm}) = \Delta\theta \sum_{g=-G/2}^{G/2} \psi_s(f_o,\theta_g)\exp\left[\frac{-j2\pi f_o \sin(g\Delta\theta)}{c}\right]\cos(g\Delta\theta), \qquad (3.39)$$

where $\Delta\theta$ is the angle increment for sampling the angular power directivity pattern, $\theta_g = g\Delta\theta$, g is the index for the samples and G is the total number of samples.

For line array applications, the power directivity pattern (calculated for a homogeneous free space) due to a distant source, which is treated as a point source, should be a delta function. Estimates, however, of the source's directivity from a line array operating in an anisotropic ocean are distorted by the underwater medium. In other words, the directivity pattern of the received signal is the convolution of the original pattern and the angular directivity of the medium (i.e., the angular scattering function of the underwater environment). As a result of the above, the angular pattern of the received signal, by a receiving line array system, is the scattering function of the medium.

In this chapter, the concept of spatial coherence is used to determine the statistical response of a line array to the acoustic field. This response is the result of the multipath and scattering phenomena discussed before, and there are models [10,35,47] to relate the spatial coherence with the physical parameters of an anisotropic medium for measurement interpretation. In these models, the interaction of the acoustic signal with the transmitting medium is considered to result in superimposed wavefronts of different directions of propagation. Then Equations 3.21 and 3.22, which define a received sensor signal from a distant source, are expressed by

$$x_n(t_i) = \sum_{l=1}^{J} A_l \exp\left[-j2\pi f_l(t_i - \frac{\delta(n-1)}{c}\theta_l)\right] + \varepsilon_{n,i}(0,\sigma_e), \qquad (3.40)$$

where $l = 1, 2, ..., j$, and J is the number of superimposed waves. As a result, a generalized form of the cross-correlation function between two sensors, which has been discussed by Carey and Moseley [10], is

$$R_{nm}(f,\delta_{nm}) = \tilde{X}^2(f)\exp\left[-\left(\frac{\delta_{nm}}{L_c}\right)^k\right], \quad k = 1, \text{ or } 1.5 \text{ or } 2, \qquad (3.41)$$

where L_c is the correlation length and $\tilde{X}^2(f)$ is the mean acoustic intensity of a received sensor time sequence at the frequency bin f. A more explicit expression for the Gaussian form of Equation 3.41 is given in [47],

$$R_{nm}(f,\delta_{nm}) \approx \tilde{X}^2(f)\exp\left[-\left(\frac{2\pi f\delta_{nm}\sigma_\theta}{c}\right)^2/2\right],$$ (3.42)

and the cross-correlation coefficients are given from

$$\rho_{nm}(f,\delta_{nm}) = R_{nm}(f,\delta_{nm})/\tilde{X}^2(f).$$ (3.43)

At the distance $L_c = c/(2\pi f\sigma_\theta)$, called *the coherence length*, the correlation function in Equation 3.43 will be 0.6. This critical length is determined from experimental coherence measurements plotted as a function of δ_{nm}. Then a connection between the medium's angular uncertainty and the measured coherence length is derived as

$$\sigma_\theta = 1/L_c, \quad \text{and} \quad L_c = 2\pi\delta_{1m}f/c,$$ (3.44)

here δ_{1m} is the critical distance between the first and the mth sensors at which the coherence measurements get smaller than 0.6. Using the above parameter definition, the effective aperture size and array gain of a deployed towed line array can be determined [10,47] for a specific underwater ocean environment.

Since the correlation function for a Gaussian acoustic field is given by Equation 3.42, the angular scattering function $\Phi(f,\theta)$ of the medium can be derived. Using Equation 3.38 and following a rather simple analytical integral evaluation, we have

$$\Phi(f,\theta) = \frac{1}{\sigma_\theta\sqrt{2\pi}}\exp\left[-\frac{\theta^2}{2\sigma_\theta^2}\right],$$ (3.45)

where $\sigma_\theta = c/(2\pi f\delta_{nm})$. This is an expression for the angular scattering function of a Gaussian underwater ocean acoustic field [10,47].

It is apparent from the above discussion that the estimates of the cross-correlation coefficients $\rho_{nm}(f_i,\delta_{nm})$ are necessary in order to define experimentally the coherence length of an underwater or human body medium. For details on experimental studies on coherence estimation for underwater sonar applications the reader may review the references [10,30].

3.3.4 Array Gain

The performance of a line array to an acoustic signal embodied in a noise field is characterized by the *array gain* parameter, AG. The mathematical relation of this parameter is defined by

$$AG = 10\log\frac{\displaystyle\sum_{n=1}^{N}\sum_{m=1}^{N}\tilde{\rho}_{nm}(f,\delta_{nm})}{\displaystyle\sum_{n=1}^{N}\sum_{m=1}^{N}\tilde{\rho}_{\varepsilon,nm}(f,\delta_{nm})}$$ (3.46)

where $\rho_{nm}(f_i,\delta_{nm})$ and $\rho_{\varepsilon,nm}(f,\delta_{nm})$ denote the normalized cross-correlation coefficients of the signal and noise field, respectively. Estimates of the correlation coefficients are given from Equation 3.43.

If the noise field is isotropic that it is not partially directive, then the denominator in Equation 3.46 is equal to N (i.e., $\sum_{n=1}^{N}\sum_{m=1}^{N}\tilde{\rho}_{\varepsilon,nm}(f,\delta_{nm})=N$), because the nondiagonal terms of the cross-correlation matrix for the noise field are negligible. Then Equation 3.46 simplifies to

$$AG = 10\log\frac{\sum_{n=1}^{N}\sum_{m=1}^{N}\tilde{\rho}_{nm}(f,\delta_{nm})}{N} \tag{3.47}$$

For perfect spatial coherence across the line array the normalized cross-correlation coefficients are $\rho_{nm}(f,\delta_{nm})\cong 1$ and the expected values of the array gain estimates are, $AG\times\log N$. For the general case of isotropic noise and for frequencies smaller than the towed array's design frequency the array gain term AG is reduced to the quantity called directivity index (DI),

$$DI = 10\times\log[(N-1)\delta/(\lambda/2)]. \tag{3.48}$$

When $\delta<<\lambda$ and the conventional beamforming processing is employed, Equation 3.26 indicates that the deployment of very long line arrays is required in order to achieve sufficient array gain and angular resolution for precise bearing estimates. Practical deployment considerations, however, usually limit the overall dimensions of a hull mounted line or towed array. In addition, the medium's spatial coherence [10,30] sets an upper limit on the effective towed array length. In general, the medium's spatial coherence length is of the order of $O(10^2)\lambda$ [10,30]. In addition to the above, for sonar systems very long towed arrays suffer degradation in the array gain due to array shape deformation and increased levels of self noise [49–53]. Although, towed line array shape estimation techniques [53] have solved the array deformation problem during course alterations of the vessels towing these arrays, the deployment issues of long towed arrays in littoral waters remains a prohibited factor for their effective use in sonar surveillance operations.

Alternatives to large aperture sonar arrays are signal processing schemes discussed elsewhere [1]. Theoretical and experimental investigations have shown that bearing resolution and detectability of weak signals in the presence of strong interferences can be improved by applying nonconventional beamformers such as adaptive beamforming [1–5,17–24], or acoustic synthetic aperture processing [1,11–16] to the sensor time series of deployed short sonar and ultrasound arrays, which are discussed in the next section.

3.4 Advanced Beamformers

3.4.1 Synthetic Aperture Processing

Various synthetic aperture techniques have been investigated to increase signal gain and improve angular resolution for line array systems. While these techniques have been successfully applied to aircraft and satellite-active radar systems, they have not been successful with sonar and ultrasound systems. In this section we will review synthetic aperture techniques that have been tested successfully with real data [11–16]. They are summarized in terms of their experimental implementation and the basic approach involved.

Let us start with the a few theoretical remarks. The plane wave response of a line array to a distant monochromatic signal, received by the nth element of the array, is expressed by Equations 3.20 through 3.22. In the above expressions, the frequency f includes the Doppler shift due to a combined movement of the receiving array and the source (or object reflecting the incoming acoustic wavefront) radiating signal. Let υ, denote the relative speed; it is assumed here that the component of the source's velocity along its bearing is negligible. If f_υ is the frequency of the stationary field, then the frequency of the received signal is expressed by

$$f = f_o\left(1 \pm \upsilon \sin\theta/c\right) \tag{3.49}$$

and an approximate expression for the received sensor time series, Equations 3.15 and 3.40 is given by

$$x_n(t_i) = A\exp\left[j2\pi f_o\left(t_i - \frac{\upsilon t_i + (n-1)\delta}{c}\sin\theta\right)\right] + \varepsilon_{n,i} \tag{3.50}$$

τ seconds later, the relative movement between the receiving array and the radiated source is $\upsilon\tau$. By proper choice of the parameters υ and τ, we have $\upsilon\tau = q\delta$, where q represents the number of sensor positions that the array has moved, and the received signal, $x_n(t_i+\tau)$ is expressed by,

$$x_n(t_i + \tau) = \exp(j2\pi f_o\tau)A\exp\left[j2\pi f_o\left(t_i - \frac{\upsilon t_i + (q+n-1)\delta}{c}\sin\theta\right)\right] + \varepsilon_{n,i}^\tau \tag{3.51}$$

As a result, we have the Fourier transform of $x_n(t_i+\tau)$, as

$$\tilde{X}_n(f)_\tau = \exp(j2\pi f_o\tau)\tilde{X}_n(f) \tag{3.52}$$

where, $\tilde{X}_n(f)_\tau$ and $\tilde{X}_n(f)$ are the discrete fourier transforms (DFTs) of $x_n(t_i+\tau)$, and $x_n(t_i)$, respectively. If the phase term $\exp(-j2\pi f_o\tau)$ is used to correct the line array measurements shown in Equation 3.52, then the spatial information included in the successive measurements at $t=t_i$ and $t=t_i+\tau$ is equivalent to that derived from a line array of $(q+N)$ sensors. When idealized conditions are assumed, the phase correction factor for Equation 3.49 in order to form a synthetic aperture, is $\exp(-j2\pi f_o\tau)$. However, this phase correction estimate requires *a priori* knowledge of the source receiver relative speed, υ and accurate estimates for the frequency f of the received signal. An additional restriction is that the synthetic aperture processing techniques have to compensate for the disturbed paths of the receiving array during the integration period that the synthetic aperture is formed. Moreover, the temporal coherence of the source signal should be greater or at least equal to the integration time of the synthetic aperture.

At this point it is important to review a few fundamental physical arguments associated with passive synthetic aperture processing. In the past [13] there was a conventional wisdom regarding synthetic aperture techniques, which held that practical limitations prevent them from being applicable to real-world systems. The issues were threshold.

(1) Since passive synthetic aperture can be viewed as a scheme that converts temporal gain to spatial gain, most signals of interest do not have sufficient temporal coherence to allow a long spatially coherent aperture to be synthesized.

(2) Since past algorithms required a *priori* knowledge of the source frequency in order to compute the phase correction factor, as shown by Equations 3.49 through 3.52, the method was essentially useless in any bearing estimation problem since Doppler would introduce an unknown bias on the frequency observed at the receiver.

(3) Since synthetic aperture processing essentially converts temporal gain to spatial gain, there was no new gain to be achieved, and therefore, no point to the method.

Recent work [12–16] has shown that there can be realistic conditions under which all of these objections are either not relevant or do not constitute serious impediments to practical applications of synthetic aperture processing in operational systems [1]. Theoretical discussions have shown [13] that the above three arguments are valid for cases that include the formation of synthetic aperture in mediums with isotropic noise characteristic. However, when the noise characteristics of the received signal are nonisotropic and the receiving array includes more than one sensor, then there is spatial gain available from passive synthetic aperture processing and this has been discussed analytically elsewhere [13]. Recently, there have been only two passive synthetic aperture techniques [11–16,54] and an MLE estimator [12] published in the open literature that deal successfully with the above restrictions. In this section, they are summarized in terms of their experimental implementation for sonar and ultrasound applications. For more details about these techniques the reader may review the references [11–16,54,55].

3.4.1.1 Fast Fourier Transform Based Synthetic Aperture Processing (FFTSA Method)

Shown in the upper part of Figure 3.7, under the title *Physical aperture*, are the basic processing steps of Equations 3.20 through 3.23 for conventional beamforming applications including line arrays. This processing includes the generation of the aperture function of the line array via FFT transformation (i.e., Equation 3.22), with the beamforming done in the frequency domain. The output (i.e., Equation 3.23) provides the directionality power pattern of the acoustic signal/noise field received by the N-sensors of the line array. As an example, the theoretical response of the power pattern for a 64-sensor line array is given in Figure 3.8. In the lower part of Figure 3.7, under the title *Synthetic aperture*, the concept of an FFT based synthetic aperture technique called FFTSA [55], is presented. The experimental realization of this method includes:

(1) The time series acquisition, using the N-sensor line array, of a number M of snapshots of the acoustic field under surveillance taken every τ seconds,

(2) the generation of the aperture function for each of the M snapshots,

(3) the beamforming in the frequency domain of each generated aperture function.

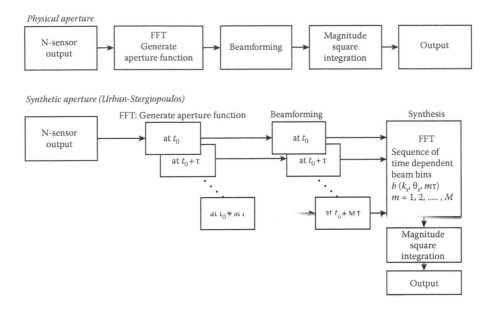

FIGURE 3.7

Shown in the upper part under the title *Physical aperture* is conventional beamforming processing in the frequency domain for a physical line array. Presented in the lower part under the title *Synthetic aperture* is the signal processing concept of the FFTSA method. (Reprinted from IEEE ©1992. With permission.)

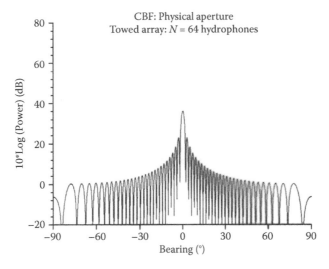

FIGURE 3.8

The azimuth power pattern from the beamforming output of the 64-sensor line array considered for the synthetic aperture processing in Figure 3.9. (Reprinted from IEEE ©1992. With permission.)

This beamforming processor provides M beam patterns with N beams each. For each beam of the beamforming output, there are M time-dependent samples with a τ seconds sampling interval.

The FFT transformation in the time domain of the M time-dependent samples of each beam provides the synthetic aperture output, which is expressed analytically by Equation 3.53. For more details please refer to Stergiopoulos and Urban [55].

$$P(f,\theta_s)_M = \left\{ \frac{\sin\left[N\frac{\pi\delta}{\lambda}(\sin\theta_s)\right]}{\sin\left[\frac{\pi\delta}{\lambda}(\sin\theta_s)\right]} \cdot \frac{\sin\left[M\frac{N}{2}\frac{\pi\delta}{\lambda}(\sin\theta_s)\right]}{\sin\left[\frac{N}{2}\frac{\pi\delta}{\lambda}(\sin\theta_s)\right]} \right\}^2 \tag{3.53}$$

The above expression assumes that $\upsilon\tau=(N\delta)/2$, which indicates that there is a 50% spatial overlap between two successive set of the M measurements and that the source bearing of θ is approximately at the boresight. The azimuthal power pattern of Equation 3.53 for the beamforming output of the FFTSA method is shown in Figure 3.9.

3.4.1.2 Yen and Carey's Synthetic Aperture Method

The concept of the experimental implementation of Yen and Carey's synthetic aperture method [54] is shown in Figure 3.10, which is also expressed by the following relation,

$$B(f_o,\theta_s)_M = \sum_{m=1}^{M} b(f_o,\theta_s)_{m\tau} \exp(-j\phi_m), \tag{3.54}$$

which assumes that estimates of the phase corrector ϕ_m require knowledge of the relative source receiver speed, υ or the velocity filter concept, introduced by Yen and Carey [54]. The basic difference of this method [54] with the FFTSA [55] technique is the need to estimate a phase correction factor ϕ_m in order to synthesize the M time-dependent beam patterns. Estimates of ϕ_m are given by

$$\phi_m = 2\pi f_o(1\pm\upsilon\sin\theta_s/c)m\tau, \tag{3.53}$$

and the application of a velocity filter concept for estimating the relative source receiver speed, υ. This method has been successfully applied to experimental sonar data including CW signals and the related application results have been reported [10,54].

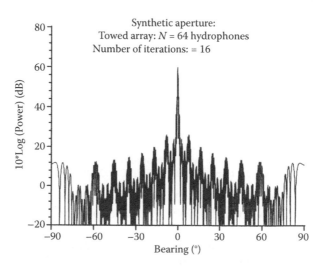

FIGURE 3.9
The azimuth power pattern from the beamforming output of the FFTSA method. (Reprinted from IEEE ©1992. With permission.)

Synthetic aperture (Yen–Carey)

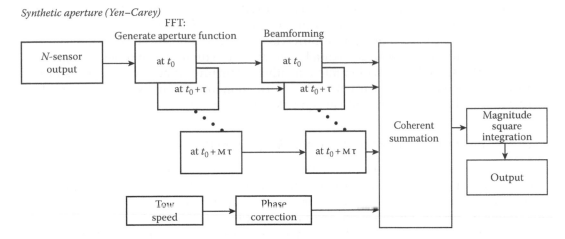

FIGURE 3.10
The concept of the experimental implementation of Yen–Carey's synthetic aperture method is shown under the same arrangement as the FFTSA method for comparison. (Reprinted from IEEE ©1992. With permission.)

3.4.1.3 Nuttall's MLE Method for Synthetic Aperture Processing

It is also important to mention here the development by Nuttall [12] of an MLE estimator for synthetic aperture processing. This MLE estimator requires the acquisition of very long sensor time series over an interval T, which corresponds to the desired length vT of the synthetic aperture. This estimator includes searching for the values of ϕ and ω that maximize the term,

$$\text{MLE}(\omega, \varphi) = \left| \Delta t \sum_{n=0}^{N-1} \exp(-jn\varphi) \left[\sum_{m=1}^{M} x_n(m\Delta t) \exp(-jm\Delta t\omega) \right] \right|, \tag{3.56}$$

where,

$$\omega = 2\pi f_o \left(1 - \frac{v \sin\theta}{c} \right), \quad \varphi = \frac{\delta}{c} 2\pi f_o \sin\theta. \tag{3.57}$$

The above relations indicate that the N complex vectors $X_n(\omega) = \sum_{m=1}^{M} x_n(t_m) \exp(-j\omega\Delta t)$, which give the spectra for the very long sensor time series $x_n(t)$ at ω, are phased together by searching ϕ over $(-p, p)$, until the largest vector length occurs in Equation 3.56. Estimates of (θ, f), are determined from Equation 3.57 using the values of ϕ and ω, that maximize Equation 3.56. The MLE estimator has been applied on real sonar data sets and the related application results have been reported [12].

A physical interpretation of the above synthetic aperture methods is that the realistic conditions for effective acoustic synthetic aperture processing can be viewed as schemes that convert temporal gain to spatial gain. Thus a synthetic aperture method requires that successive snapshots of the received acoustic signal have good cross-correlation properties in order to synthesize an extended aperture and the speed fluctuations are successfully compensated by means of processing. It has been also suggested [11–16] that the prospects for successfully extending the physical aperture of a line array require algorithms which are not based on the synthetic aperture concept used in active radars. The reported results [10–16,54,55] have shown that the problem of creating an acoustic synthetic aperture is centered on the estimation of a phase correction factor, which is used to compensate for the phase differences between sequential line-array measurements in order to coherently synthesize the spatial information into a synthetic aperture. When the estimates of this phase correction factor are correct, then the information inherent in the synthetic aperture is the same as that of an array with an equivalent physical aperture [11–16].

3.4.1.4 Spatial Overlap Correlator for Synthetic Aperture Processing (ETAM Method)

Recent theoretical and experimental studies have addressed the above concerns and indicated that the space and time coherence of the acoustic signal in the sea [10–16] appears to be sufficient to extend the physical aperture of a moving line array. In the above studies the fundamental question related to the angular resolution capabilities of a moving line array and the amount of information inherent in a received signal have been addressed. These investigations included the use of the CRLB analysis and showed that for long observation intervals of the order of 100 seconds the additional information provided by a moving line array over a stationary array is expressed as a large increase in angular resolution, which is due to the Doppler caused by the movement of the array (see Figure 3 in Stergiopoulos [12]). A summary of these research efforts has been reported in a special issue in the *IEEE J. Oceanic Eng.* [13]. The synthetic aperture processing scheme that has been used in broadband sonar applications [1] is based on the extended towed array measurements (ETAM) algorithm, which was invented by Stergiopoulos and Sullivan [11]. The basic concept of this algorithm is a phase-correction factor that is used to combine coherently successive measurements of the towed array to extend the effective towed array length.

Shown in Figure 3.11 is the experimental implementation of the ETAM algorithm in terms of the line array speed and sensor positions as a function of time and space. Between two successive positions of the N-sensor line array with sensor spacing δ, there are $(N-q)$ pairs of space samples of the acoustic field that have the same spatial information, their difference being a phase factor [11–12,55] related to the time delay these measurements were taken. By cross-correlating the $(N-q)$ pairs of the sensor time series that overlap, the

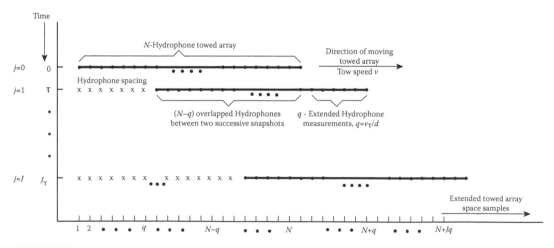

FIGURE 3.11
Concept of the experimental implementation of ETAM algorithm in terms of towed array positions and speed as a function of time and space. (Reprinted from IEEE ©1992. With permision.)

desired phase correction factor is derived, which compensates for the time delay between these measurements and the phase fluctuations caused by irregularities of the tow path of the physical array or relative speed between source and receiver; this is called the *overlap correlator*. Following the above, the key parameters in the ETAM algorithm are the time increment $\tau = q\delta/\upsilon$ between two successive sets of measurements, where υ is the tow speed and q represents the number of sensor positions that the towed array has moved during the τ seconds, or the number of sensors to which the physical aperture of the array is extended at each successive set of measurements. The optimum overlap size, $(N-q)$, which is related to the variance of the phase correction estimates, has been shown [13] to be $N/2$. The total number of sets of measurements required to achieve a desired extended aperture size is then defined by $J=(2/N)(T\upsilon/\delta)$, where T is the period taken by the towed array to travel a distance equivalent to the desired length of the synthetic aperture.

Then, for the frequency bin f_i and between two successive jth and $(j+1)$th snapshots, the phase-correction factor estimate is given by,

$$\tilde{\Psi}_j(f_i) = \arg\left\{ \frac{\displaystyle\sum_{n=1}^{N/2} X_{j,(\frac{n}{2}+n)}(f_i) \times X^*_{(j+1),n}(f_i) \times \rho_{j,n}(f_i)}{\displaystyle\sum_{n=1}^{N/2} \rho_{j,n}(f_i)} \right\} \tag{3.58}$$

where, for a frequency band with central frequency f_i and observation bandwidth Δf or $f_i-\Delta f/2 < f_i < f_i+\Delta f/2$, the coefficients

$$\rho_{j,n}(f_i) = \frac{\left| \displaystyle\sum_{i=-Q/2}^{Q/2} X_{j,(\frac{n}{2}+n)}(f_i) \times X^*_{(j+1),n}(f_i) \right|}{\sqrt{\displaystyle\sum_{i=-Q/2}^{Q/2} \left| X_{j,(\frac{n}{2}+n)} \right|^2 \times \displaystyle\sum_{i=-Q/2}^{Q/2} \left| X^*_{(j+1),n}(f_i) \right|^2}}, \tag{3.59}$$

are the normalized cross-correlation coefficients or the coherence estimates between the $N/2$ pairs of sensors that overlap in space. The above coefficients are used as weighting factors in Equation 3.58 in order to optimally weight the good against the bad pairs of sensors during the estimation process of the phase-correction factor.

The performance characteristics and expectations from the ETAM algorithm have been evaluated experimentally and the related results have been reported [1,12,55]. The main conclusion drawn from these experimental results is that for narrowband signals or for FM type of pulses from active sonar systems the overlap correlator in ETAM compensates successfully the speed fluctuations and effectively extends the physical aperture of a line array more than eight times. On the other hand, the threshold value of ETAM is −8 dB re 1-Hz band at the sensor. For values of SNR higher than this threshold, it has been shown that ETAM achieves the theoretical CRLB bounds and it has comparable performance to the maximum-likelihood estimator [12].

3.4.2 Adaptive Beamformers

Despite the geometric differences between the line and circular arrays, the underline beamforming processes for these arrays, as expressed by Equations 3.8 and 3.9

respectively, are time delay beamforming estimators, which are basically spatial filters. However, optimum beamforming requires the beamforming filter coefficients to be chosen based on the covariance matrix of the received data by the N-sensor array in order to optimize the array response [15,16], as discussed in Section 3.2. The family of algorithms for optimum beamforming that use the characteristics of the noise, are called *Adaptive beamformers* [3,17–20,22,23]. In this section we will address implementation issues for various partially adaptive variants of the MVDR method and a GSC adaptive beamformer [1,37].

Furthermore, the implementation of adaptive schemes in real time systems is not restricted into one method, such as the MVDR technique that is discussed next. In fact, the generic concept of the sup-aperture multidimensional array introduced in the chapter allows for the implementation of a wide variety of adaptive schemes in operational systems [7,25–28]. As for the implementation of adaptive processing schemes in active systems, the following issues need to be addressed.

For active applications that include matched filter processing, the outputs of the adaptive algorithms are required to provide coherent beam time series to facilitate the postprocessing. This means that these algorithms should exhibit near-instantaneous convergence and provide continuous beam time series that have sufficient temporal coherence to correlate with the reference signal in matched filter processing [1].

In a previous study [1], possible improvement in convergence periods of two algorithms in the subaperture configuration was investigated. The Griffiths–Jim generalized sidelobe canceller (GSC) [18,44] coupled with the normalized least mean square (NLMS) adaptive filter [45] has been shown to provide near-instantaneous convergence under certain conditions [1,37]. The GSC/NLMS in the subaperture configuration was tested under a variety of conditions to determine if it could yield performance advantages, and if its convergence properties could be exploited over a wider range of conditions [1,37]. The steered minimum variance beamformer (STMV) is a variant of the MVDR beamformer [38]. By applying narrowband adaptive processing on bands of frequencies, extra degrees of freedom are introduced. The number of degrees of freedom is equal to the number of frequency bins in the processed band. In other words, increasing the number of frequency bins processed decreases the convergence time by a corresponding factor. This is due to the fact that convergence now depends on the observation time bandwidth product, as opposed to observation time in the MVDR algorithm [38,39].

The STMV beamformer in its original form was a broadband processor. In order to satisfy the requirements for matched filter processing, it was modified to produce coherent beam time series [1]. The ability of the STMV narrowband beamformer to produce coherent beam time series has been investigated in another study [37]. Also, the STMV narrowband processor was implemented in the subaperture configuration to produce near-instantaneous convergence and to reduce the computational complexity required. The convergence properties of both the full aperture and subaperture implementations have been investigated for line arrays of sensors [1,37].

3.4.2.1 Minimum Variance Distortionless Response (MVDR)

The goal is to optimize the beamformer response so that the output contains minimal contributions due to noise and signals arriving from directions other than the desired signal direction. For this optimization procedure it is desired to find a linear filter vector $\overline{W}(f_i, \theta)$ which is a solution to the constrained minimization problem that allows signals from the look direction to pass with a specified gain [17,18],

$$\text{Minimize: } \sigma_{MV}^2 = \bar{W}^*(f_i,\theta)R(f_i)\bar{W}(f_i,\theta),$$

$$\text{subject to: } \bar{W}^*(f_i,\theta)\bar{D}(f_i,\theta) = 1 \tag{3.60}$$

where $\bar{D}(f_i,\theta)$ is the conventional steering vector based on Equation 3.21. The solution is given by,

$$\bar{W}(f_i,\theta) = \frac{R^{-1}(f_i)\bar{D}(f_i,\theta)}{\bar{D}^*(f_i,\theta)R^{-1}(f_i)\bar{D}(f_i,\theta)} \tag{3.61}$$

The above solution provides the adaptive steering vectors for beamforming the received signals by the N-hydrophone line array. Then in frequency domain, an adaptive beam at a steering θ_s is defined by

$$B(f_i,\theta_s) = \bar{W}^*(f_i,\theta_s)\bar{X}(f_i) \tag{3.62}$$

and the corresponding conventional beams are provided by Equation 3.22.

3.4.2.2 Generalized Sidelobe Canceller (GSC)

The GSC [44] is an alternative approach to the MVDR method. It reduces the adaptive problem to an unconstrained minimization process. The GSC formulation produces a much less computationally intensive implementation. In general GSC implementations have complexity $O(N^2)$, as compared to $O(N^3)$ for MVDR implementations, where N is the number of sensors used in the processing. The basis of the reformulation of the problem is the decomposition of the adaptive filter vector $\bar{W}(f_i,\theta)$ into two orthogonal components, $\bar{w}$ and $-\bar{v}$, where $\bar{w}$ and $\bar{v}$ lie in the range and the null space of the constraint of Equation 3.60, such that $W(f_i,\theta) = \bar{w}(f_i,\theta) - \bar{v}(f_i,\theta)$. A matrix C which is called signal blocking matrix, may be computed from $C\,\bar{I} = 0$ where $\bar{I}$ is a vector of ones. This matrix C whose columns form a basis for the null space of the constraint of Equation 3.60 will satisfy $\bar{v} = C\bar{u}$, where $\bar{u}$ is defined by Equation 3.64. The adaptive filter vector may now be defined as $\bar{W} = \bar{w} - C\bar{u}$ and yields the realization shown in Figure 3.12. Then the problem is reduced to:

$$\text{Minimize: } \sigma_u^2 = \{[\bar{w} - C\bar{u}]^* R[\bar{w} - C\bar{u}]\} \tag{3.63}$$

which is satisfied by:

$$\bar{u}_{opt} = (C^*RC)^{-1}C^*R\bar{w} \tag{3.64}$$

u_{opt} being the value of the weights at convergence.

The GSC in combination with the NLMS adaptive algorithm has been shown to yield near instantaneous convergence [44,45]. Figure 3.12 shows the basic structure of the so called memoryless generalized side-lobe canceller. The time delayed by $\tau_n(\theta_s)$ sensor time series defined by Equation 3.4, Equation 3.20 and Figure 3.2 form the presteered sensor time series, which are denoted by $x_n(t_i,\tau_n(\theta_s))$. In frequency domain these presteered sensor data are denoted by $X_n(f_i,\theta_s)$ and form the input data vector for the adaptive scheme in Figure 3.12. On the left-hand side branch of this figure the intermediate vector $\bar{Z}(f_i,\theta_s)$ is

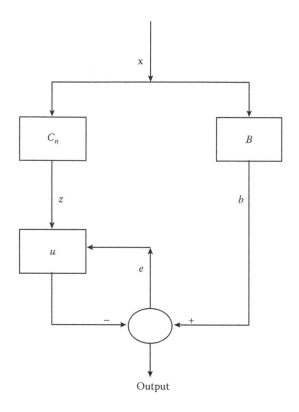

FIGURE 3.12
Basic processing structure for the memoryless GSC. The right-hand side branch is simply the shaded conventional beamformer. The left-hand side branch of this figure is the result of the signal blocking matrix (constraints) applied to presteered sensor time series. The output of the signal blocking matrix is the input to the NLMS adaptive filter. Then, the output of this processing scheme is the difference between the adaptive filter and the conventional output. (Reprinted from IEEE ©1998. With permision.)

the result of the signal blocking matrix C being applied to the input $\bar{X}(f_i, \theta_s)$. Next, the vector $\bar{Z}(f_i, \theta_s)$ is an input to the NLMS adaptive filter. The output of the right-hand branch is simply the shaded conventional output. Then, the output of this processing scheme is the difference between the adaptive filter output, and the conventional output:

$$e(f_i, \theta_s) = b(f_i, \theta_s) - \bar{u}^*(f_i, \theta_s)\bar{Z}(f_i, \theta_s) \tag{3.65}$$

The adaptive filter, at convergence, reflects the side-lobe structure of any interferers present, and it is removed from the conventional beamformer output. In the case of the NLMS this adaptation process can be represented by:

$$\bar{u}_{k+1}(f_i, \theta_s) = \bar{u}_k(f_i, \theta_s) + \frac{\mu \times e_k^*(f_i, \theta_s)}{\alpha + \bar{X}^*(f_i, \theta_s)\bar{X}(f_i, \theta_s)}\bar{Z}(f_i, \theta_s) \tag{3.66}$$

where, k is the iteration number, α is a small positive number designed to maintain stability. The parameter μ is the convergence controlling parameter or step size for the NLMS algorithm.

3.4.2.3 Steered Minimum Variance Broadband Adaptive (STMV)

Krolik and Swingler [38] have shown that the convergence time for broad-band source location can be reduced by using the space-time statistic called the steered covariance matrix (STCM). This method achieves significantly shorter convergence times than adaptive algorithms that are based on the narrowband cross spectral density matrix (CSDM) [17,18] without sacrificing spatial resolution. In fact, the number of statistical degrees of freedom available to estimate the STCM is approximately the time–bandwidth product ($T \times BW$) as opposed to the observation time, ($T = M/F_s$, F_s being the sampling frequency) in CSDM methods. This provides an improvement of approximately BW, the size of the broadband source bandwidth, in convergence time. The conventional beamformer's output in frequency domain is shown by Equation 3.22. The corresponding time domain conventional beamformer output $b(t_i, \theta_s)$ is the weighted sum of the steered sensor outputs, as expressed by Equation 3.29. Then, the expected broadband beam power, $B(\theta)$ is given by:

$$B(\theta_s) = E\{|b(\theta_s, t_i)|\} = \bar{h}^* E\{\bar{x}^*(t_i, \tau_n(\theta))\bar{x}(t_i, \tau_m(\theta))\}\bar{h} \tag{3.67}$$

where the vector $\bar{h}$ includes the weights for spatial shading [36].

$$\text{The term } \Phi(t_i, \theta_s) = E\{\bar{x}(t_i, \tau_n(\theta_s))\bar{x}^*(t_i, \tau_m(\theta_s))\} \tag{3.68}$$

is defined as the steered covariance matrix (STCM) in time domain and is assumed to be independent of t_i in stationary conditions. The name STCM is derived from the fact that the matrix is computed by taking the covariance of the presteered time domain sensor outputs. Suppose $X_n(f_i)$ is the fourier transform of the sensor outputs $x_n(t_i)$ and assuming that the sensor outputs are approximately band limited. Under these conditions the vector of steered (or time delayed) sensor outputs $x_n(t_i, \tau_n(\theta_s))$ can be expressed by

$$\bar{x}(t_i, \tau_n(\theta_s)) = \sum_{k=l}^{l+H} T_k(f_k, \theta_s)\bar{X}(f_k)\exp(j2\pi f_k t_i) \tag{3.69}$$

where $T(f_k, \theta)$ is the diagonal steering matrix in Equation 3.70 below with elements identical to the elements of the conventional steering vector, $\bar{D}(f_i, \theta)$

$$T(f_k, \theta) = \begin{bmatrix} 1 & & & \cdot & \cdot & \cdot & & 0 \\ 0 & d_1(f_k, \theta) & & & & & \\ \cdot & & & \cdot & & & \cdot \\ \cdot & & & & \cdot & & \cdot \\ \cdot & & & & & \cdot & \cdot \\ 0 & & & \cdot & \cdot & \cdot & d_N(f_k, \theta) \end{bmatrix} \tag{3.70}$$

Then it follows directly from the above equations that

$$\Phi(\Delta f, \theta_s) = \sum_{k=l}^{l+H} T(f_k, \theta_s)R(f_k)T^*(\theta_s) \tag{3.71}$$

where the index $k=l, l+1, ..., l+H$ refers to the frequency bins in a band of interest Δf, and $R(f_k)$ is the cross spectral density matrix (CSDM) for the frequency bin f_k. This suggests that $\Phi(\Delta f, \theta_s)$ in Equation 3.68 can be estimated from the CSDM, $R(f_k)$ and $T(f_k, \theta)$ expressed by Equation 3.70. In the steered minimum variance method (STMV), the broadband spatial power spectral estimate $B(\theta_s)$ is given by [38]:

$$B(\theta_s) = \left[\overline{I}^* \Phi(\Delta f, \theta_s)^{-1} \overline{I} \right]^{-1} \tag{3.72}$$

The STMV differs from the basic MVDR algorithm in that the STMV algorithm yields a STCM that is composed from a band of frequencies and the MVDR algorithm uses a CSDM that is derived from a single frequency bin. Thus, the additional degrees of freedom of STMV compared to those of CSDM provide a more robust adaptive process.

However, estimates of $B(\theta)$ according to Equation 3.72 do not provide coherent beam time series, since they represent the broadband beam power output of an adaptive process. In this investigation [1] we have modified the estimation process of the STMV matrix in order to get the complex coefficients of $\Phi(\Delta f, \theta_s)$ for all the frequency bins in the band of interest .

The STMV algorithm may be used in its original form to generate an estimate of $\Phi(\Delta f, \theta)$ for all the frequency bands Δf, across the band of the received signal. Assuming stationarity across the frequency bins of a band Δf, then the estimate of the STMV may be considered to be approximately the same with the narrowband estimate $\Phi(f_o, \theta)$ for the center frequency f_o of the band Δf. In this case, the narrowband adaptive coefficients may be derived from

$$\overline{w}(f_o, \theta) = \frac{\Phi(f_o, \Delta f, \theta)^{-1} \overline{D}(f_o, \theta)}{\overline{D}^*(f_o, \theta) \Phi(f_o, \Delta f, \theta)^{-1} \overline{D}(f_o, \theta)}, \tag{3.73}$$

The phase variations of $\overline{w}(f_o, \theta)$ across the frequency bins $i=l, l+1, ..., l+H$ (where H is the number of bins in the band Δf), are modeled by,

$$w_n(f_i, \theta) = \exp[2\pi f_i \Psi(\Delta f, \theta)], \quad i=l, l+1, ..., l+H \tag{3.74}$$

where, $\Psi_n(\Delta f, \theta)$ is a time delay term derived from,

$$\Psi_n(\Delta f, \theta) = F[w_n(\Delta f, \theta), 2\pi f_o]. \tag{3.75}$$

Then by using the adaptive steering weights $w_n(\Delta f, \theta)$, that are provided by Equation 3.74, the adaptive beams are formed as shown by Equation 3.62. Figure 3.13 shows the realization of the STMV beamformer and provides a schematic representation of the basic processing steps that include:

(1) Time series segmentation, overlap and FFT, shown by the group of blocks at the top-left part of the schematic diagram.

(2) Formation of steered covariance matrix (Equations 3.68 and 3.71) shown by the two blocks at the bottom left-hand side of Figure 3.13.

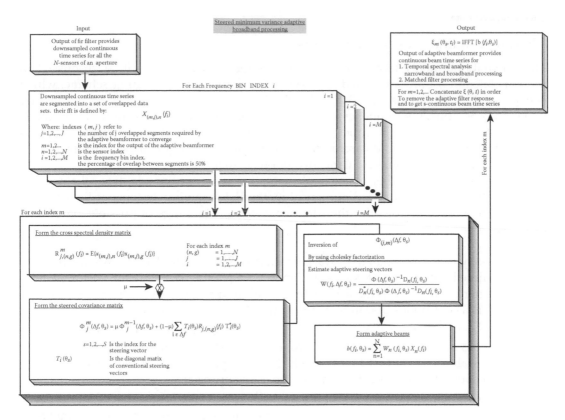

FIGURE 3.13

Realization of the steered covariance adaptive beamformer. The basic processing steps include: (1) time series segmentation, overlap and FFT, shown by the group of blocks at the top-left part of the schematic diagram, (2) formation of steered covariance matrix, shown by the two blocks at the bottom left-hand side, (3) inversion of covariance matrix using Cholesky factorization, estimation of adaptive steering vectors and formation of adaptive beams in frequency domain (middle and bottom blocks at the right-hand side), and finally (4) formation of adaptive beams in time domain through IFFT, discardation of overlap and concatenation of segments to form continuous beam time series (top right-hand side block). The various indexes provide details for the implementation of the STMV processing flow in a generic computing architecture. (Reprinted from IEEE ©1998. With permision.)

(3) Inversion of covariance matrix using Cholesky factorization, estimation of adaptive steering vectors and formation of adaptive beams in frequency domain, presented by the middle and bottom blocks at the right-hand side of Figure 3.13.

(4) Formation of adaptive beams in time domain through IFFT, discardation of overlap and concatenation of segments to form continuous beam time series, which is shown by the top right-hand side block.

The various indexes in Figure 3.13 provide details for the implementation of the STMV processing flow in a generic computing architecture. The same figure indicates that estimates of the STCM is based on an exponentially weighted time average of the current and previous STCM, which is discussed in the next section.

3.5 Implementation Considerations

The conventional and adaptive steering vectors for steering angles θ_s, ϕ_s discussed in Sections 3.3 and 3.4 are integrated in a frequency domain beamforming scheme, which is expressed by Equations 3.22, 3.25, 3.31 and 3.62. The beam time series are formed by Equation 3.29. Thus, the frequency domain adaptive and conventional outputs are made equivalent to the FFT of the time domain beamforming outputs with proper selection of beamforming weights and careful data partitioning. This equivalence corresponds to implementing FIR filters via circular convolution [40–42].

Matrix inversion is another major implementation issue for the adaptive schemes discussed in this chapter. Standard numerical methods for solving systems of linear equations can be applied to solve for the adaptive weights. The range of possible algorithms includes:

- Cholesky factorization of the covariance matrix $R(f_i)$, [17,29]. This allows the linear system to be solved by backsubstitution in terms of the received data vector. Note that there is no requirement to estimate the sample covariance matrix and that there is a continuous updating of an existing Cholesky factorization.

- QR decomposition of the received vector $\bar{X}(f_i)$, that includes the conversion of a matrix to upper triangular form via rotations. The QR decomposition method has better stability than the Cholesky factorization algorithm, but it requires twice as much computational efforts than the Cholesky approach.

- SVD (singular value decomposition) method. This is the most stable factorization technique. It requires, however, three times more computational requirements than the QR decomposition method.

In this implementation study we have applied the Cholesky factorization and the QR decomposition techniques in order to get solutions for the adaptive weights. Our experience suggests that there are no noticable differences in performance between the above two methods [1].

The main consideration, however, for implementing adaptive schemes in real time systems are associated with the requirements derived from Equations 3.61 and 3.62, which require knowledge of second order statistics for the noise field. Although these statistics are usually not known, they can be estimated from the received data [17,18,23] by averaging a large number of independent samples of the covariance matrixes $R(f_i)$ or by allowing the iteration process of the adaptive GSC schemes to converge [1,37]. Thus, if K is the effective number of statistically independent samples of $R(f_i)$, then the variance on the adaptive beam output power estimator detection statistic is inversely proportional to $(K-N+1)$ [17,18,22], where N is the number of sensors. Theoretical suggestions [23] and our empirical observations suggest that K needs to be three to four times the size of N in order to get coherent beam time series at the output of the above adaptive schemes. In other words, for arrays with a large number of sensors, the implementation of adaptive schemes as statistically optimum beamformers would require the averaging of a very large number of independent samples of $R(f_i)$ in order to derive an unbiased estimate of the adaptive weights [23]. In practice this is the most serious problem associated with the implementation of adaptive beamformers in real time systems.

Owsley [17,29] has addressed this problem with two important contributions. His first contribution is associated with the estimation procedure of $R(f_i)$. His argument is that in practice, the covariance matrix cannot be estimated exactly by time averaging because the received signal vector $\bar{X}(f_i)$ is never truly stationary and/or ergodic. As a result, the available averaging time is limited. Accordingly, one approach to the time-varying adaptive estimation of $R(f_i)$ at time t_k is to compute the exponentially time averaged estimator (geometric forgetting algorithm) at time t_k:

$$R^{t_k}(f_i) = \mu R^{t_{k-1}}(f_i) + (1-\mu)\bar{X}(f_i)\bar{X}^*(f_i) \tag{3.76}$$

where μ is a smoothing factor ($0<\mu<1$) that implements the exponentially weighted time averaging operation. The same principle has also been applied in the GSC scheme [1,37]. Use of this kind of exponential window to update the covariance matrix is a very important factor in the implementation of adaptive algorithms in real time systems.

Owsley's [29] second contribution deals with the dynamics of the data statistics during the convergence period of the adaptation process. As mentioned above, the implementation of an adaptive beamformer with a large number of adaptive weights in a large array sonar system, requires very long convergence periods that will eliminate the dynamical characteristics of the adaptive beamformer to detect the time varying characteristics of a received signal of interest. A natural way to avoid this kind of temporal stationarity limitation is to reduce the number of adaptive weights requirements. Owsley's [29] subaperture configuration for line array adaptive beamforming reduces significantly the number of degrees of freedom of an adaptation process. His concept has been applied to line arrays, as discussed elsewhere [1,37]. However, extension of the subaperture line array concept for multidimensional arrays is not a trivial task. In the following sections, the sup-aperture concept is generalized for circular, cylindrical, planar and spherical arrays.

3.5.1 Evaluation of Convergence Properties of Adaptive Schemes

To test the convergence properties of the various adaptive beamformers of this study, synthetic data were used that included one CW signal. The frequency of the monochromatic signal was selected to be 330 Hz, and the angle of arrival at 68.9° to directly coincide with the steering direction of a beam. The SNR of the received synthetic signal was very high, 10 dB at the sensor. By definition the adaptive beamformers allow signals in the look direction to pass undistorted, while minimizing the total output power of the beamformer. Therefore in the ideal case the main beam output of the adaptive beamformer should resemble the main beam output of the conventional beamformer, while the side beams outputs will be minimized to the noise level. To evaluate the convergence of the beamformers two measurements were made. From Equation 3.65, the mean square error (MSE) between the normalized main beam outputs of the adaptive beamformer and the conventional beamformer was measured, and the mean of the normalized output level of the side beam, which is the MSE when compared with zero, was measured. The averaging of the errors was done with a sliding window of four snapshots to provide a time varying average, and the outputs were normalized so that the maximum output of the conventional beamformer was unity.

3.5.1.1 Convergence Characteristics of GSC and GSC-SA Beamformers

The GSC/NLMS adaptive algorithm, which has been discussed in Section 3.4.2, and its subaperture configuration denoted by GSC-SA/NLMS were compared against each other to determine if the use of the subaperture configuration produced any improvement in the time required for convergence. The graph in the upper part of Figure 3.14 shows the comparison of the MSE of the main beams of both algorithms for the same step size μ, which is defined in Equation 3.65. The graphs show that the convergence rates of the main beams are approximately the same for both algorithms, reaching a steady state value of MSE within a few snapshots. The value of MSE that is achieved is dictated by the miss-adjustment, which depends on μ. The higher MSE produced by the GSC-SA algorithm indicates that the algorithm exhibits a higher misadjustment.

The graph in the lower part of Figure 3.14 shows the output level of an immediate side beam, again for the same step size μ. The side beam was selected as the beam right next to the main beam. The GSC-SA algorithm appears superior at minimizing the output of the side beam. It reaches its convergence level almost immediately, while the GSC algorithm requires approximately 30 snapshots to reach the same level. This indicates that the GSC-SA algorithm should be superior at canceling time varying interferers. By selecting a higher value for μ the time required for convergence will be reduced but the MSE of the main beam will be higher.

3.5.1.2 Convergence Characteristics of STMV and STMV-SA Beamformers

As with the GSC/NLMS and GSC-SA/NLMS beamformers, the STMV and the STMV subaperture (STMV-SA) beamformers were compared against each other to determine

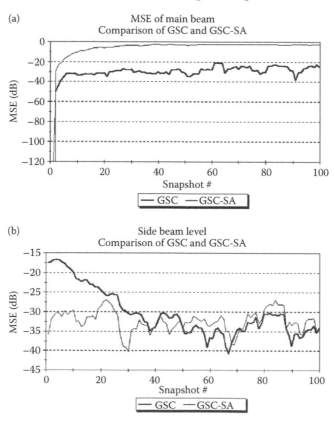

FIGURE 3.14
(a) MSE of the main beams of the GSC/NLMS and the GSC-SA/NLMS algorithms. (b) Side-beam levels of the above algorithms. (Reprinted from IEEE ©1998. With permission.)

if there was any improvement in the time required for convergence when using the subaperture configuration. The graph in the upper part of Figure 3.15 shows the comparison of the MSE of the main beams of both algorithms. The graph shows that the STMV-SA algorithm reaches a steady state value of MSE within the first few snapshots.

The STMV algorithm is incapable of producing any output for at least eight snapshots as tested. Before this time the matrices that are used to compute the adaptive steering vectors are not invertible. After this initial period the algorithm has already reached a steady state value of MSE. Unlike the case of the GSC algorithm the misadjustment from subaperture processing is smaller. The lower part of Figure 3.15 shows the output level of the side beam for both the STMV and the STMV-SA beamformers. Again the side beam was selected as the beam right next to the main beam. As before there is an initial period during which the STMV algorithm is computing an estimate of the STCM and is incapable of producing any output, after that period the algorithm has reached steady state, and produces lower side beams than the subaperture algorithm.

3.5.1.3 *Signal Cancellation Effects of the Adaptive Algorithms*

Testing of the adaptive algorithms of this study for signal cancellation effects was carried out with simulations that included two signals arriving from 64° and 69° [37]. All of the

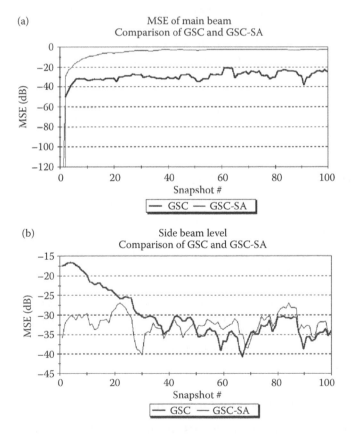

FIGURE 3.15
(a) MSE of the main beams of the STMV and the STMV-SA algorithms. (b) Side-beam levels of the above algorithms. (Reprinted from IEEE ©1998. With permission.)

parameters of the signals were set to the same values for all the beamformers, conventional, GSC/NLMS, GSC-SA/NLMS, STMV and STMV-SA. Details about the above simulated signal cancellation effects can be found elsewhere [37]. In the narrowband outputs of the conventional beamformer the signals appear at the frequency and beam at which they were expected. As anticipated, however, the side-lobes are visible in a number of other beams. The GRAM outputs of the GSC/STMV algorithm indicated that there is signal cancellation. In each case the algorithm failed to detect either of the two CWs. This suggests that there is a shortcoming in the GSC/NLMS algorithm, when there is strong correlation between two signal arrivals received by the line array. The narrowband outputs of the GSC-SA/NLMS algorithm showed that in this case the signal cancellation effects have been minimized and the two signals were detected only at the expected two beams with complete cancellation of the side-lobe structure. For the STMV beamformer, the GRAMs indicated a strong side-lobe structure in many other beams. However, the STMV-SA beamformer successfully suppresses the side-lobe structure that was present in the case of the STMV beamformer. From all these simulations [37], it was obvious that the STMV-SA beamformer, as a broadband beamformer, is not as robust for narrowband applications as the GSC-SA/NLMS.

3.5.2 Generic Multidimensional Subaperture Structure for Adaptive Schemes

The decomposition of the 2-D and 3-D beamformer into sets of line and/or circular array beamformers, which has been discussed in Section 3.3.2, provides a first-stage reduction of the numbers of degrees of freedom for an adaptation process. Furthermore, the subaperture configuration is considered in this study as a second stage reduction of the number of degrees of freedom for an adaptive beamformer. Then, the implementation effort for adaptive schemes in multidimensional arrays is reduced to implementing adaptive processes in line and circular arrays. Thus, a multidimensional adaptive beamformer can now be divided into two coherent modular steps which lead to efficient system oriented implementations.

3.5.2.1 Subaperture Configuration for Line Arrays

For a line array, a subaperture configuration includes a large percentage overlap between contiguous subapertures. More specifically, a line array is divided into a number of subarrays that overlap, as shown in Figure 3.16. These subarrays are beamformed using the conventional approach; and this is the first stage of beamforming. Then, we form a number of sets of beams with each set consisting of beams that are steered at the same direction but each one of them generated by a different subarray. A set of beams of this kind is equivalent to a line array that consists of directional sensors steered at the same direction, with sensor spacing equal to the space separation between two contiguous subarrays and with the number of sensors equal to the number of subarrays. The second stage of beamforming implements an adaptive scheme on the above kind of set of beams, as illustrated in Figure 3.16.

3.5.2.2 Subaperture Configuration for Circular Array

Consider a circular array with M-sensors as shown in Figure 3.17. The first circular subaperture consists of the first $M-G+1$ sensors with $n=1, 2, ..., M-G+1$, where n is the sensor index and G is the number of subapertures. The second circular subaperture array

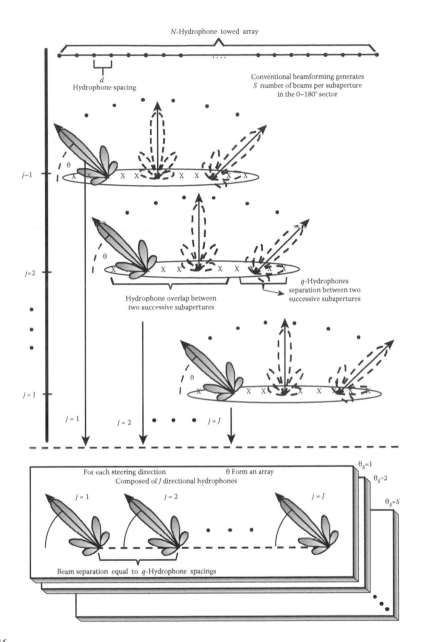

FIGURE 3.16
Concept of adaptive subaperture structure for line arrays. Schematic diagram shows the basic steps that include: (1) formation of J subapertures, (2) for each subaperture formation of S conventional beams, and (3) for a given beam direction, θ formation of line sensor arrays that consist of J number of directional sensors (beams). The number of line arrays with directional sensors (beams) are equal to the number S of steered conventional beams in each subaperture. For each line array, the directional sensor time series (beams) are provided at the input of an adaptive beamformer. (Reprinted from IEEE ©1998. With permision.)

consists of $M-G+1$ sensors with $n=2, 3, ..., M-G+2$. The subaperture formation goes on till the last subaperture consists of $M-G+1$ sensors with $n=G, G+1, ..., M$. In the first stage, each circular subaperture is beamformed as discussed in Section 3.3.1.2 and this first stage of beamforming generates G sets of beams.

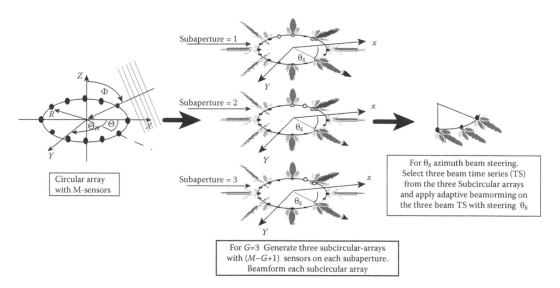

FIGURE 3.17
Concept of adaptive subaperture structure for circular arrays, which is similar to that for line arrays shown in Figure 3.16.

As in the previous section, we form a number of sets of beams with each set consisting of beams that are steered at the same direction but each one of them generated by a different subarray. For $G < 5$, a set of beams of this kind can be treated approximately as a line array that consists of directional sensors steered at the same direction, with sensor spacing equal to the space separation between two contiguous subarrays and with the number of sensors equal to the number of subarrays. The second stage of beamforming implements an adaptive scheme on the above kind of set of beams, as illustrated in Figure 3.17, for $G = 3$.

3.5.2.3 Subaperture Configuration for Cylindrical Array

Consider the cylindrical array shown in Figures 3.5 and 3.18 with the number of sensors $\aleph = NM$, where N is the number of circular rings and M is the number of sensors on each ring. Let n be the ring index, m be the sensor index for each ring and G be the number of subapertures. The formation of subapertures is as follows:

> The first subaperture consists of the first $(N - G + 1)$ rings, where $n = 1, 2, ..., N-G+1$. In each ring we select the first set of $(M - G + 1)$ sensors, where $m = 1, 2, ..., M-G+1$. However, each ring has M sensors, but only $(M-G+1)$ sensors are used to form the subaperture. These sensors form a cylindrical array cell, as shown in the upper right-hand side corner of Figure 3.18.

In other words, the subaperture includes the sensors of the full cylindrical array except for $G-1$ sensors from $G-1$ rings, which are denoted by small circles in Figure 3.18, that have been excluded in order to form the subaperture. Next, the generic decomposition concept of the conventional cylindrical array beamformer, presented in Section 3.3.2.1, is applied to the above subaperture cylindrical array cell. For a given pair of azimuth and elevation steering angles $\{\theta_s, \phi_s\}$, the output of the generic conventional multidimensional

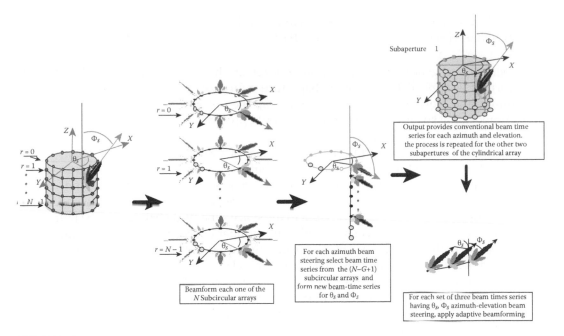

FIGURE 3.18

Coordinate system and geometric representation of the concept of adaptive subaperture structure for cylindrical arrays. In this particular example the number of subapertures was $G=3$. The $\aleph = NM$ sensor cylindrical array beamformer consists of N circular arrays with M being the number of sensors in each circular array. Then, the subaperture adaptive structure for cylindrical arrays is reduced to the basic steps of adaptive subaperture structures for circular and line arrays as defined in the schematic diagrams of Figure 3.17 and Figure 3.16, respectively. Thus, for a given azimuth θ_s and elevation ϕ_s beam steering and $G=3$, these steps include: (1) formation of a subaperture per circular array with $M-G+1$ sensors, (2) for each subaperture formation of S conventional beams, and (3) formation of $N-G+1$ vertical line sensor arrays that consist of directional sensors (beams). This arrangement defines a circulate subaperture. The process is repeated to generate two additional subaperture circular arrays. The beam output response of the $G=3$ subaperture circular arrays is provided at the input of a line array adaptive beamformer with $G=3$ number of directional sensors.

subaperture beamformer provides beam time series, $b_{g=1}(t_i,\theta_s,\phi_s)$, where the subscript $g=1$ is the subaperture index.

The second subaperture consists of the next set of $(N-G+1)$ rings, where $n=2, ...,$ $N-G+2$. In each ring we select the next set of $(M-G+1)$ sensors, where $m=2, ...,$ $M-G+2$. However, each ring has M sensors, but only $(M-G+1)$ sensors are used to form the subaperture. These sensors form the second subaperture cylindrical array cell.

Again, the generic decomposition concept of the conventional cylindrical array beamformer, presented in Section 3.3.2.1, is applied to the above subaperture cylindrical array cell. For a given pair of azimuth and elevation steering angles $\{\theta_s, \phi_s\}$, the output of the generic conventional multidimensional subaperture beamformer provides beam time series, $b_{g=2}(t_i,\theta_s,\phi_s)$ with subaperture index $g=2$.

This kind of subaperture formation continues untill the last subaperture which consists of a set of $(N-G+1)$ rings, where $n=G, G+1, ..., N$. In each ring we select the last set of $(M - G + 1)$ sensors, where $m = G, G+1, ..., M$. Please note also that each ring has M sensors, but only $(M-G+1)$ sensors are used to form the subaperture.

As before, the generic decomposition concept of the conventional cylindrical array beamformer is applied to the last subaperture cylindrical array cell. For a given pair of azimuth and elevation steering angles $\{\theta_s, \phi_s\}$, the output of the generic conventional multidimensional subaperture beamformer would provide beam time series, $b_{g=G}(t_i, \theta_s, \phi_s)$ with subaperture index $g = G$.

As in Section 3.4.2.2, we form a number of sets of beams with each set consisting of beams that are steered at the same direction but each one of them generated by a different subaperture cylindrical array cell. For $G < 5$, a set of beams of this kind can be treated approximately as a line array that consists of directional sensors steered at the same direction, with sensor spacing equal to the space separation between two contiguous subaperture cylindrical array cells and with the number of sensors equal to the number of subarrays. Then, the second stage of beamforming implements an adaptive scheme on the above kind of set of beams, as illustrated in Figure 3.18.

For the particular case, shown in Figure 3.18, the second stage of beamforming implements an adaptive beamformer on a line array that consists of the $G = 3$ beam time series $b_g(t_i, \theta_s, \phi_s)$, $g = 1, 2, ..., G$. Thus, for a given pair of azimuth and elevation steering angles $\{\theta_s, \phi_s\}$, the cylindrical adaptive beamforming process is reduced to an adaptive line array beamformer that includes as input only three beam time series $b_g(t_i, \theta_s, \phi_s)$, $g = 1, 2, 3$ with spacing $\delta = [(R2\pi / M)^2 + \delta_z^2]^{1/2}$, which is the spacing between two contiguous subaperture cylindrical cells, where $(R2\pi / M)$ is the sensor spacing in each ring and δ_z is the distance between each ring along z-axis of the cylindrical array. The output of the adaptive beamformer provides one or more adaptive beam time series with steering centered on the pair of azimuth and elevation steering angles $\{\theta_s, \phi_s\}$.

As expected, the adaptation process in this case will have near-instantaneous convergence because of the very small number of degrees of freedom. Furthermore, because of the generic characteristics, the proposed 3-D subaperture adaptive beamforming concept may include a wide variety of adaptive techniques such as MVDR, GSC and STMV that have been discussed elsewhere [1,37].

3.5.2.4 Subaperture Configuration for Planar and Spherical Arrays

The subaperture adaptive beamforming concepts for planar and spherical arrays are very similar to that of the cylindrical array. In particular, for planar arrays, the formation of subapertures is based on the subaperture concept of line arrays that has been discussed in Section 3.5.2.1. The different steps of subaperture formation for planar arrays as well as the implementation of adaptive schemes on the G beam time series $b_g(t_i, \theta_s, \phi_s)$, $g = 1, 2, ..., G$, that are provided by the G subapertures of the planar array, are similar with those in Figure 3.18 by considering the composition process for planar arrays shown in Figure 3.6. Similarly, the subaperture adaptive concept for spherical arrays is based on the subaperture concept of circular arrays, that has been discussed in Section 3.5.2.2.

3.5.3 Signal Processing Flow of a 3-D Generic Subaperture Structure

As stated earlier, the discussion in this chapter has been devoted in designing a generic subaperture beamforming structure that will decompose the computationally intensive multidimensional beamforming process into coherent subsets of line and/or circular subaperture array beamformers for ultrasound, radar and integrated active–passive sonar systems. In a sense the proposed generic processing structure is an extension of a previous effort [1].

The previous study [1] included the design of a generic beamforming structure that allows the implementation of adaptive, synthetic aperture and high-resolution temporal and spatial spectral analysis techniques in integrated active–passive line-array sonars. Figure 3.19, which is identical to Figure 9.4 in Chapter 9, shows the configuration of the signal processing flow of the previous generic structure that allows the implementation of finite impulse response (FIR) filters, conventional, adaptive and synthetic aperture beamformers [1,40–42]. A detailed discussion about the content of Figure 3.19 is provided in Chapter 9.

Shown in Figure 3.20 is the proposed configuration of the signal processing flow that includes the implementation of line and circular array beamformers as FIR filters [40–42]. The processing flow is for 3-D cylindrical arrays. The reconfiguration of the different processing blocks in Figures 3.19 and 3.20 allows the application of the proposed configuration to a variety of ultrasound, radar and integrated active–passive sonar systems with planar, cylindrical or spherical arrays of sensors. Chapters 9 and 11 present a set of real data results that were derived from the implementation of the signal processing flow of Figure 3.20 in integrated active–passive towed array sonars and 3-D/4-D ultrasound imaging systems, respectively.

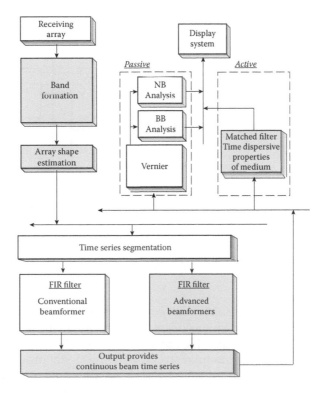

FIGURE 3.19
Schematic diagram of a generic signal processing flow that allows the implementation of conventional, adaptive and synthetic aperture beamformers in line-array sonar and ultrasound systems.

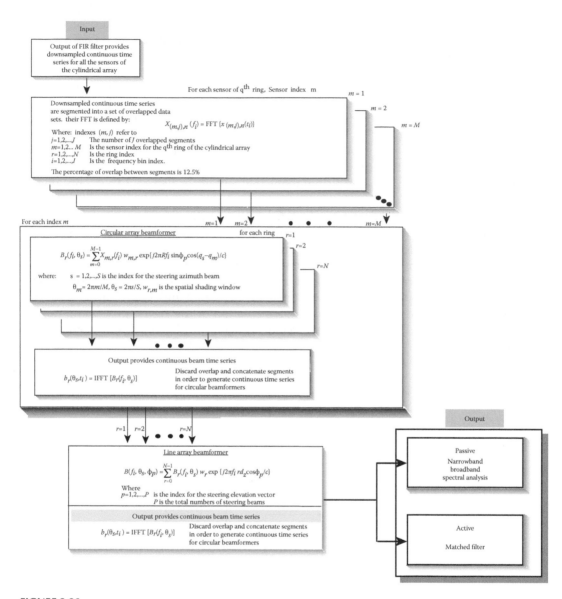

FIGURE 3.20
Signal processing flow of generic structure decomposing the 3-D beamformer for cylindrical arrays of sensors into coherent subsets of line and circular array beamformers.

As discussed at the beginning of this section, the output of the beamforming processing block in Figure 3.20 provides continuous beam time series. The beam time series are provided at the input of a vernier for passive narrowband/broadband analysis or a matched filter for active applications. This modular structure in the signal processing flow is a very essential processing arrangement in order to allow for the integration of a great variety of processing schemes such as the ones considered in this chapter. The details of the proposed generic processing flow, as shown in Figure 3.20, are very briefly the following:

- The first block in Figure 3.20 includes the partitioning of the time series from the receiving sensor array, the computation of their initial spectral FFT, the selection of the signal's frequency band of interest via band-pass FIR filters and downsampling. The output of this block provides continuous time series at reduced sampling rate [41,42].
- The second and third blocks titled *Circular array beamformer* and *Line array beamformer* provide continuous directional beam time series by using the FIR implementation scheme of the spatial filtering via circular convolution [40]. The segmentation and overlap of the time series at the input of each one of the above beamformers takes care of the wraparound errors that arise in fast-convolution signal processing operations. The overlap size is equal to the effective FIR filter's length [41,42].
- The block named *Active, Matched-filter* is for the processing of echos for active sonar and radar applications.
- The block *Passive, narrowband and broadband spectral analysis* includes the final processing steps of a temporal spectral analysis.

Finally, data normalization processing schemes are being used in order to map the output results into the dynamic range of the display devices in a manner which provides a CFAR capability [34].

In the passive unit, the use of verniers and the temporal spectral analysis (incorporating segment overlap, windowing and FFT coherent processing) provide the narrowband results for all the beam time series. Normalization and OR-ing are the final processing steps before displaying the output results. Since a beam time sequence can be treated as a signal from a directional sensor having the same array gain and directivity pattern as that of the beamformer, the display of the narrowband spectral estimates for all the beams follows the so-called GRAM presentation arrangements, as shown in Figures 3.25 through 3.28. This includes the display of the beam-power outputs as a function of time, steering beam (or bearing) and frequency [34].

Broadband outputs in the passive unit are derived from the narrowband spectral estimates of each beam by means of incoherent summation of all the frequency bins in a wideband of interest [34]. This kind of energy content of the broadband information is displayed as a function of bearing and time [1,34,43].

In the active unit, the application of a matched-filter (or replica correlator) on the beam time series provides coherent broadband processing. This allows detection of echoes as a function of range and bearing for reference waveforms transmitted by the active transducers of ultrasound, sonar or radar systems. The displaying arrangements of the correlator's output data are similar to the GRAM displays and include as parameters: range as a function of time and bearing [1].

Next, presented in Figure 3.21, is the signal processing flow of the generic adaptive subaperture structure for multidimensional arrays. The first processing block includes the formation of subapertures as discussed in Section 3.5.2. Then, the sensor time series from each subaperture are beamformed by the generic multidimensional beamforming structure that has been introduced in Section 3.3 and presented in Figure 3.20. Thus, for a given pair of azimuth and elevation steering angles $\{\theta_s, \phi_s\}$, the output of the generic conventional multidimensional beamformer would provide G beam time series, $b_g(t_i, \theta_s, \phi_s)$, $g=1, 2, ..., G$. The second stage of beamforming includes the implementation of an adaptive beamformer as discussed in Section 3.4.2.

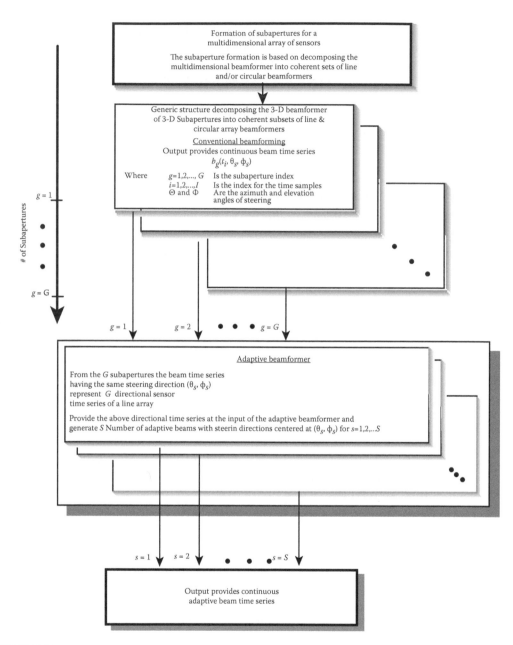

FIGURE 3.21
Signal processing flow of a generic adaptive subaperture structure for multidimensional arrays of sensors.

For the synthetic aperture processing scheme, however, there is an important detail regarding the segmentation and overlap of the sensor time series into sets of discontinuous segments. It is assumed here that the received sensor signals are stored as continuous time series. Therefore, the segmentation process of the sensor time series is associated with the tow speed and the size of the synthetic aperture as discussed in Section 3.4.1.4. So, in order to achieve continuous data flow at the output of the overlap correlator, the N-continuous time series are segmented into discontinuous data sets as shown in Figure 3.22. Our implementation scheme in Figure 3.22 considers five discontinuous segments

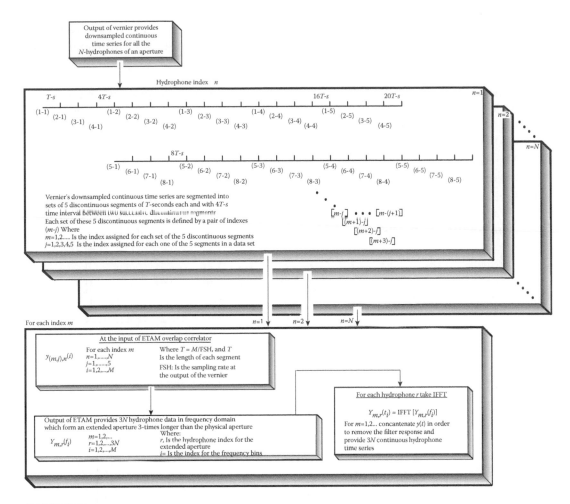

FIGURE 3.22

Schematic diagram of the data flow for the ETAM algorithm and the sensor time series segmentation into a set of five discontinuous segments for the overlap correlator. The basic processing steps include: time series segmentation, overlap and grouping of five discontinuous segments, which are provided at the input of the overlap correlator, (shown by group of blocks at the top part of schematic diagram). $T=M/f_s$, is the length in seconds of the discontinuous segmented time series and M defines the size of FFT. The rest of the blocks provide the indexing details for the formation of the synthetic aperture. These indexes provide details for the implementation of the segmentation process of the synthetic aperture flow in a generic computing architecture. The processing flow is shown in Figure 3.23. (Reprinted from IEEE ©1998. With permission.)

in each data set. This arrangement will provide at the output of the overlap correlator $3N$-continuous sensor time series, which are provided at the input of the conventional beamformer as if they were the sensor time series of an equivalent physical array. Thus the basic processing steps include: time series segmentation, overlap and grouping of five discontinuous segments, which are provided at the input of the overlap correlator, as shown by the group of blocks at the top part of Figure 3.22. $T=M/f_s$, is the length in seconds of the discontinuous segmented time series and M defines the size of FFT. The rest of the blocks provide the indexing details for the formation of the synthetic aperture. These indexes provide also details for the implementation of the segmentation process of the synthetic aperture flow in a generic computing architecture.

The processing arrangements and the indexes in Figure 3.23 provide the details needed for the mapping of this synthetic aperture processing scheme in sonar or ultrasound computing architectures. The basic processing steps include:

(1) Time series segmentation, overlap and grouping of five discontinuous segments, which are provided at the input of the overlap correlator, shown by the block at the top part of schematic diagram. Details of this segmentation process are shown also in Figure 3.22.

(2) The main block called ETAM: Overlap correlator provides processing details for the estimation of the phase correction factor to form the synthetic aperture.

(3) Formation of the continuous sensor time series of the synthetic aperture are obtained through IFFT, discardation of overlap and concatenation of segments to form continuous time series, which is shown by the left-hand side block.

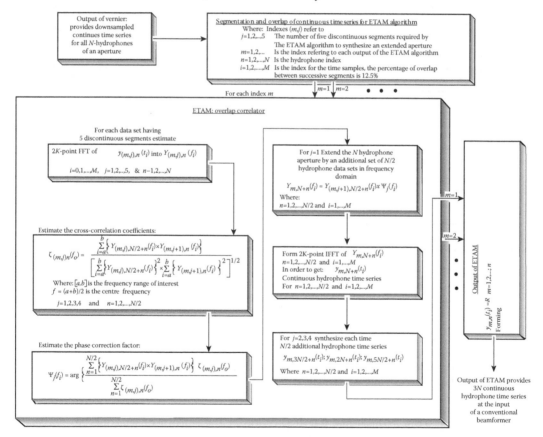

FIGURE 3.23
Schematic diagram for the processing arrangements of ETAM algorithm. The basic processing steps include: (1) time series segmentation, overlap and grouping of five discontinuous segments, which are provided at the input of the overlap correlator, shown by the block at the top part of schematic diagram. Details of the segmentation process are shown also by Figure 3.22. (2) The main block called ETAM: Overlap correlator provides processing details for the estimation of the phase correction factor to form the synthetic aperture, and finally (3) generation of the continuous sensor time series of the synthetic aperture are obtained through IFFT, discardation of overlap and concatenation of segments to form continuous time series (left-hand side block). The various indexes provide details for the implementation of the synthetic processing flow in a generic computing architecture. (Reprinted from IEEE ©1998. With permission.)

It is important to note here that the choice of five discontinuous segments was based on experimental observations [10,30] regarding the temporal and spatial coherence properties of the underwater medium. These issues of coherence are very critical for synthetic aperture processing and they have been addressed in Section 3.3.3.

3.6 Concept Demonstration: Simulations

Performance assessment and testing of the generic subaperture multidimensional adaptive beamforming structure has been carried out with synthetic and real data sets. Chapters 9 and 11 present a set of real data results that were derived from the implementation of the generic signal processing structure of this chapter in integrated active–passive towed array sonar and 3-D/4-D ultrasound imaging systems, respectively.

The synthetic data sets include narrowband and hyperbolic frequency modulated (HFM) signals for passive and active applications, respectively. For sonar applications, the frequencies of the passive narrowband signals are taken to be 330 Hz and the active signal consists of HFM pulses with pulse-width of 8 seconds, 100-Hz bandwidth centered at 330 Hz, with 120 seconds pulse repetition period, or pulses with pulse-width of 500 μs, 10-KHz bandwidth centered at 200 KHz, with arbitrary pulse repetition period. For ultrasound applications, the synthetic signals consist of FM pulses with 4 MHz bandwidth centered at 3 MHz. The scope here is to demonstrate that the implementation of adaptive schemes (i.e., GSC and STMV) in real time systems is feasible. Moreover, it is shown that the proposed generic configuration of adaptive schemes provides array gain improvements when compared with the performance characteristics of the multidimensional conventional beamformer. As for active applications, it is shown that the adaptive schemes of the proposed generic subaperture structure achieve near instantaneous convergence, which is essential for active ultrasound, sonar and radar applications.

The generic adaptive subaperture processing structure and the associated signal processing algorithms were implemented in a computer workstation. The memory of the workstation was sufficient to allow processing of long continuous sensor time series. However, the available memory restricts the number of sensors and the number of steered beams.

Nevertheless, the simulations of this chapter are sufficient to demonstrate that system oriented applications of the proposed generic subaperture adaptive structure for multidimensional arrays of sensors can be more effective than the relevant mainstream signal processing concepts. In fact, the conclusions derived from the present simulations are substantiated by the real data results reported in Chapters 9 and 11 for integrated active passive towed array sonar and 3-D/4-D ultrasound imaging applications.

3.6.1 Sonar Simulations: Cylindrical Array Beamformer

3.6.1.1 Synthetic Sonar Data: Passive

A cylindrical array with 160 sensors (16 rings with ten sensors on each ring) was considered where the distance between rings along z-axis is taken to be equal to the angular spacing between sensors of the rings (i.e., $\delta_z = 2\pi R / M = \delta = 2.09m$). Continuous sensor time series were provided at the inputs of the generic conventional and adaptive beamformers with the processing flows as shown in Figures 3.20 and 3.21, respectively. The total number of steering beams for both the adaptive and conventional beamformers, was 144. For the decomposition process of the generic beamformer, expressed by Equation 3.33, there were 16 beams steered in the angular sector of (0–360°) for azimuth bearing and nine beams formed in the angular sector of (0–180°) for elevation bearing. Thus, the generic beamformer provided 16 azimuth beams for each of the nine elevation steering angles, giving a total of 144 beams.

In the upper part of Figure 3.24, the left-hand side diagram shows the output power of the azimuth beams at the expected elevation bearing of the signal source for the generic 3-D cylindrical array conventional beamformer; the right-hand side of Figure 3.24 shows the output power of the elevation beams at the expected azimuth angle of the signal source. In both cases, no spatial window has been applied. The results at the left-hand side of the lower part of Figure 3.24 correspond to the azimuth beams for the conventional beamformer with Hamming as a spatial window (dotted line) and the adaptive (solid line) subaperture beamformer. In this case, the number of subapertures was $G=3$

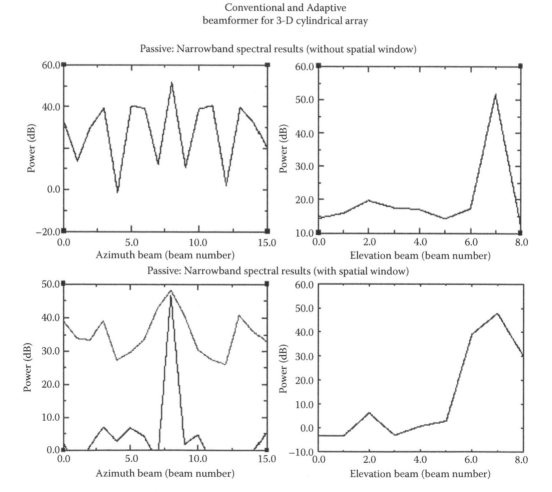

FIGURE 3.24
Passive beamforming results for a cylindrical array. Upper part shows azimuth and elevation bearing response for the proposed generic multidimensional beamformer. Lower left-hand side part shows beamforming results of the conventional with spatial window (dotted line) and adaptive (solid line) beamformers. Lower right-hand side part shows elevation bearing response for the conventional cylindrical beamformer with spatial window.

with Hamming as a spatial window applied on the subaperture conventional circular beamformer.

It is apparent by these results that the spatial shading has significantly suppressed the side-lobe structure of the conventional beamformer and has widen the beamwidth, as expected.

Moreover, the adaptive beamforming results demonstrate a significant improvement in suppressing the side-lobe structure as compared with the conventional results. The right-hand side of the lower part of Figure 3.24 includes elevation bearing response for the conventional beamformer with spatial shading. At this point it is important to note that the application of spatial shading on the fully coherent 3-D cylindrical beamformer would have been a much more elaborate process than the one that has been used for the generic multidimensional beamformer. This is because the decomposition process for the latter allows two much simpler and separate applications of spatial shading (i.e., one for circular arrays and the other for line arrays) discussed analytically in Section 3.3.2 [7,25–28].

Figure 3.25 shows the narrowband spectral estimates of the generic 3-D cylindrical array conventional beamformer with Hamming spatial shading for all the azimuth beams

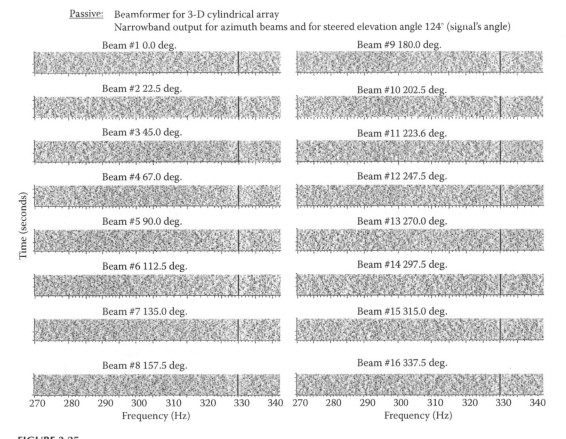

FIGURE 3.25

Narrowband spectral estimates of the generic 3-D cylindrical array conventional beamformer for all the azimuth beams steered at the signal's expected elevation angle. The 25 windows of this display correspond to the 25 steered beams equally spaced in [1, –1] cosine space. The acoustic field included two narrowband signals that the very poor angular resolution performance of the conventional beamformer has failed to resolve.

according to the so-called GRAM presentation arrangement, discussed in Section 3.4 [34]. The GRAMs in this figure represent the spectrograms of the output of the azimuth beams steered at the signal's expected elevation bearing. The GRAMs in Figure 3.26 show the corresponding results when the azimuth beams are steered at an elevation angle which is 55° off the expected elevation bearing of the signal. It is obvious from the results of Figure 3.26 that the array gain of the conventional beamformer with spatial shading is not very high.

For the same sensor time series, when the adaptive subaperture schemes are implemented in the generic multidimensional beamformer the corresponding results are shown in Figure 3.27. When the results of Figure 3.27 are compared with the corresponding conventional results of Figure 3.25, the directional array gain improvements of the generic multidimensional beamformer become apparent. In this case the adaptive technique was the subaperture GSC-SA. As expected and because of the array gain improvements provided by the adaptive beamformer, the signal of interest is not present in the GRAMs of Figure 3.28, which provides the azimuth beams steered at an elevation angle which is by 55° off the expected elevation bearing of the signal. The results of Figure 3.28 are in sharp contrast with those of Figure 3.26 for the conventional beamformer.

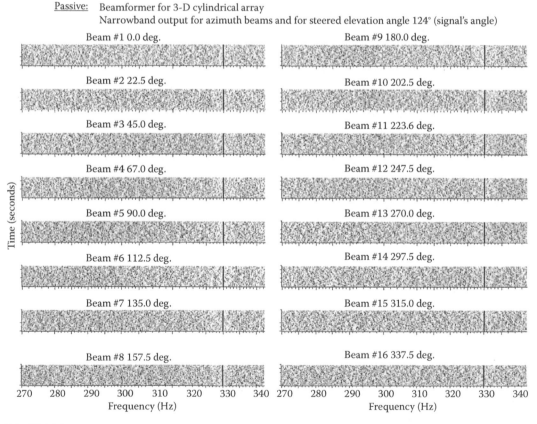

<u>Passive:</u> Beamformer for 3-D cylindrical array
Narrowband output for azimuth beams and for steered elevation angle 124° (signal's angle)

FIGURE 3.26
Narrowband spectral estimates of the generic 3-D cylindrical array conventional beamformer for all the azimuth beams steered at an elevation bearing, which is 55° off the expected signal's elevation angle. The 25 windows of this display correspond to the 25 steered beams equally spaced in [1, –1] cosine space. The acoustic field at this steering does not include signals. However, the very poor side-lobe suppression of the conventional beamformer reveals signals that do not exist at this steering.

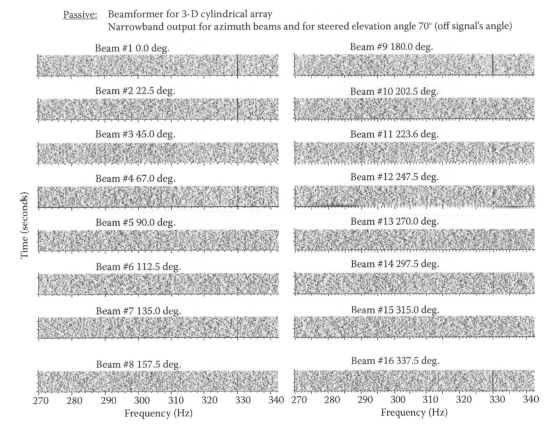

Passive: Beamformer for 3-D cylindrical array
Narrowband output for azimuth beams and for steered elevation angle 70° (off signal's angle)

FIGURE 3.27
Narrowband spectral estimates of the generic 3-D cylindrical array adaptive (GSC) beamformer for all the azimuth beams steered at the signal's expected elevation angle. Input data sets are the same as in Figure 3.24. The 25 windows of this display correspond to the 25 steered beams equally spaced in [1, –1] cosine space. The acoustic field included two narrowband signals that the very good angular resolution performance of the subaperture adaptive beamformer resolves the bearings of the two signals.

An explanation for the poor angular resolution performance of the conventional beamformer requires interpretation of the results of Figures 3.24 and 3.26. In particular, for the simulated cylindrical array, Figure 3.24 shows conventional beamformer with spatial shading has 13 dBs side-lobe suppression in azimuth beam steering and approximately 60° beam-width in elevation. Furthermore, to improve detection the power beam outputs shown in the GRAMs of Figures 3.25 and 3.27 have been normalized [34], since this is a typical processing arrangement for operational sonar displays. However, the detection improvements of the normalization process would enhance the detection of the side-lobe structure shown in Figure 3.24. Thus, the results of Figures 3.24 and 3.26 provide typical angular resolution performance characteristics for sonars deploying cylindrical array beamformers.

In summary, the results of Figures 3.24 through 3.28, with an appropriate scaling on the actual array dimensions and the frequency ranges of the signals that have been considered in the simulations, may project the performance characteristics for a variety of sonars deploying cylindrical arrays with conventional or adaptive beamformers.

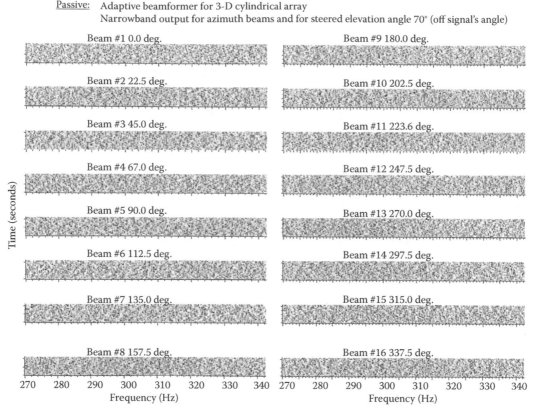

Passive: Adaptive beamformer for 3-D cylindrical array
Narrowband output for azimuth beams and for steered elevation angle 70° (off signal's angle)

FIGURE 3.28

Narrowband spectral estimates of the generic 3-D cylindrical array adaptive (GSC) beamformer for all the azimuth beams steered at an elevation bearing, which is 55° off the expected signal's elevation angle. Input data sets are the same as in Figure 3.25. The 25 windows of this display correspond to the 25 steered beams equally spaced in [1, –1] cosine space. The acoustic field at this steering does not include signals. Thus, the very good side-lobe suppression of the subaperture adaptive beamformer shows that there are no signals present at this steering.

3.6.1.2 Synthetic Sonar Data: Active

It was discussed before that the configuration of the generic beamforming structure providing continuous beam time series to the input of a matched filter or a temporal spectral analysis unit, forms the basis for integrated active or passive sonar applications. However, before the adaptive aperture processing schemes are integrated with a matched filter, it is essential to demonstrate that the beam time series from the outputs of the nonconventional beamformers have sufficient temporal coherence and correlate with the reference signal. For example, if the signal received by a sonar array consists of FM pulses with a pulse repetition period of a few minutes, then questions may be raised about the efficiency of an adaptive beamformer to achieve near-instantaneous convergence in order to provide beam time series with coherent content for the FM pulses. This is because partially adaptive processing schemes require at least a few iterations to converge to a suboptimum solution.

To address this question, the matched filter and the conventional and adaptive beamformers, shown in Figures 3.20 and 3.21, were tested with simulated data sets including

HFM pulses 8-seconds long with 100 Hz bandwidth. The pulse repetition period was 120 seconds. Although this may be considered as a configuration for bistatic active sonar applications, the findings from this experiment can be applied to monostatic active sonar systems as well.

In Figure 3.29, we will present some results from the output of the active unit of the generic signal processing structure. Figure 3.29 shows the output of the replica correlator for the conventional and adaptive beam time series of the subaperture GSC and STMV adaptive techniques [1,37]. In this case, the steering angles are the same with those of the data sets shown in Figure 3.25 through 3.28. The horizontal axis in this figure represents range or time delay ranging from 0 to 120-seconds, which is the pulse repetition period. While the three beamforming schemes provide artifact-free outputs, it is apparent from the values of the replica correlator-output that the conventional beam time series exhibit better temporal coherence properties than the beam time series of the subaperture GSC adaptive beamformer. The significance and a quantitative estimate of this difference can be assessed by comparing the amplitudes of the correlation outputs in Figure 3.29. The replica correlator amplitudes are 12.06, 11.81, 12.08 for the conventional, and the adaptive schemes: GSC-SA (subaperture), STMV-SA (subaperture), respectively. These results also show that the beam time series of the STMV subaperture scheme have temporal coherence properties equivalent to those of the conventional beamformer, which is the optimum case.

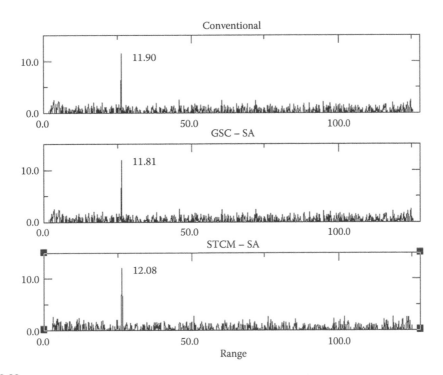

FIGURE 3.29
Output of replica correlator for the conventional and subaperture adaptive (GSC, STMV) beam time series of the generic cylindrical array beamformer. The azimuth and elevation beams are steered at the signal's expected bearings, which are the same with those in Figures 3.24 through 3.26.

3.6.1.3 Real Data: Active Cylindrical Sonar System

The proposed system configuration and the conventional and adaptive signal processing structures of this study was also tested with real data sets from an operational sonar system. Echoes of HFM pulses were received by a cylindrical array having comparable geometric configuration and sensor arrangements as discussed in the simulations. Continuous beam time series were provided at the input of the subaperture cylindrical adaptive beamformer with processing flow as defined in Figures 3.20 and 3.21. The total number of steering beams for both the adaptive and conventional beamformers was 144. Figure 3.30 provides the matched filter output results for a single pulse. The upper part of Figure 3.30 shows the matched filter output of the conventional beamformer and the lower part shows the output of the subaperture STMV adaptive algorithm [1,37]. The horizontal axis corresponds to the azimuth angular space ranging from 0 to 360°. The vertical axis corresponds to the time delay or range estimate, which is determined by the location of the pick of the matched filter output and the color shows the magnitude of the matched filter output. In a sense, each vertical color-coded line of Figure 3.30 represents the matched filter output (e.g., see Figure 3.29) for a giver azimuth steering angle.

Since these are unclassified results provided by an operational sonar, there were no real targets present during the experiments. In a sense, the results of Figure 3.30 present the scattering properties of the medium as they were defined by the received echoes. Although

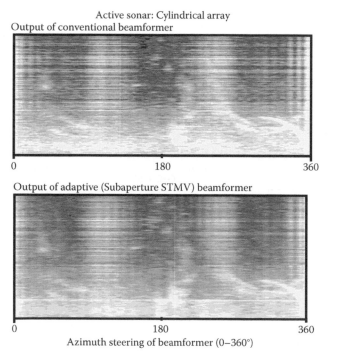

Active sonar: Cylindrical array
Output of conventional beamformer

0   180   360

Output of adaptive (Subaperture STMV) beamformer

0   180   360
Azimuth steering of beamformer (0–360°)

FIGURE 3.30
Upper part shows the matched filter output of the conventional and the lower part of the subaperture STMV adaptive algorithm for a cylindrical beamformer. The horizontal axis refers to the angular space covering the bearing range of 0–360°. The vertical axis refers to time delay or range estimates of the matched filter and the color refers to the correlation output. Each vertical color-coded line of Figure 3.30 represents a correlation output of Figure 3.27 for a given bearing angle. The basic difference between the conventional and adaptive matched filter output results is that the improved directionality (or array gain) of the adaptive beam time series localizes the detected HFM pulses and the associated echo returns in a smaller number of beams than the conventional beamformer.

the results of Figure 3.30 are normalized [34], the amplitudes of the output of the matched filter in Figure 3.30 for the conventional (upper figure) and adaptive (lower figure) beam time series were compared before the use of the normalization processing and they were found to be approximately the same. Again, these results show that the beam time series of the subaperture adaptive scheme have temporal coherence properties equivalent to those of the conventional beamformer, as this was also confirmed with simulated data, discussed in Section 3.6.1.2.

In summary, the basic difference between the conventional and adaptive matched filter output results is that the improved directionality (or array gain) of the adaptive beam time series localizes the detected HFM pulses and the associated echo returns more accurately than the conventional beamformer.

This kind of array gain improvement, provided by the adaptive beamformer, suppresses the reverberation effects during active sonar operations, as this is confirmed by the results of Figure 3.30. It is anticipated that the adaptive beamformers will enhance the performance of integrated active–passive and mine-hunting sonars by means of precise detection and localization of echoes that are embedded in reverberation noise fields.

3.6.2 Ultrasound Imaging Systems: Line and Planar Array Beamformers

Performance assessment and testing of the generic subaperture adaptive beamformers that have been discussed in this chapter, have been carried out with simulated ultrasound data. The parameters in these simulations were identical with those of an advanced 3-D/4-D experimental fully digital ultrasound imaging system that is discussed in Chapter 11.

The results presented in this section are divided into two parts. The first part discusses the simulations for linear phased arrays and the second part the results for planar phased array systems, respectively. The scope with these simulation is to evaluate the angular (azimuth) resolution performance of the 2-D, 3-D adaptive beamforming for ultrasound imaging.

The impact and merits of this technique will be contrasted with the angular resolution performance obtained using the 2-D, 3-D conventional phased array beamforming. The requirement for synthetic aperture processing for ultrasound imaging applications is discussed in detail in Chapter 11. Synthetic aperture processing in this case is required for the data acquisition and digitization process of the sensor channels of large size planar array ultrasound probes by A/DC peripherals that have smaller number of A/D channels than those in the probes. For details about the synthetic aperture processing, the reader is asked to see Reference [1]. The synthetic aperture scheme is called ETAM algorithm and has been tested only with line and planar arrays [11,12,14,30].

It was discussed in Section 3.5.3 that the configuration of the generic beamforming structure to provide continuous beam time series at the input of a matched filter and a temporal spectral analysis unit, forms the basis for ultrasound and integrated passive and active sonar applications. For ultrasound imaging applications, to address this question the matched filter and the subaperture adaptive processing scheme, shown in Figures 3.20 and 3.21, were tested with synthetic data sets including CW (for Doppler applications) and FM pulses.

Normalization of the output of a matched filter, such as the results of Figure 3.29, and the display of the beam time series as GRAMs provides a waterfall display of ranges (depth) as a function of beam-steering, which define the reconstructed images of an ultrasound system.

3.6.2.1 Ultrasound Imaging System with Linear Phased Array Probe

The first simulation considered a 32-elements linear phased array probe, having pitch equal to 0.4 mm. The sampling frequency was 33 MHz. The position and frequency characteristics of the received sensor time series relevant with the reference image are defined in Table 3.1. Figure 3.31 shows the normalized beam cross-sections obtained with adaptive beamforming and with conventional beamforming apodized in space. The adaptive beamformer beam width is noticeably smaller than the one obtained with the conventional apodized beamforming procedure, defined in Section 3.5.2.4.

Figure 3.32, shows the reconstructed images for the simulated point targets defined in Table 3.1. The first image at the left-hand side of Figure 3.32 shows the reconstructed image from the output of the beam time series of a conventional phased array beamformer applied on the synthetic 32-sensor probe time series. The middle image in Figure 3.32 shows the reconstructed image from the beam time series of the subaperture adaptive beamformer applied on the same synthetic 32-sensor probe time series, as before. The image at the right of Figure 3.32 shows the reconstructed image from the time series of a conventional phased array beamformer applied on 96-sensor phased array probe for the same structure of data defined in Table 3.1. It is apparent from these simulations that

TABLE 3.1

Parameters for simulated ultrasound time series for a linear phased array ultrasound probe

	Fc	BW	Bearing	Depth
Point target #1	4.0 MHz	2.0 MHz	80°	10 mm
Point target #2	4.0 MHz	2.0 MHz	92°	25 mm
Point target #3	2.0 MHz	1.0 MHz	84°	40 mm
Point target #4	2.0 MHz	1.0 MHz	96°	50 mm

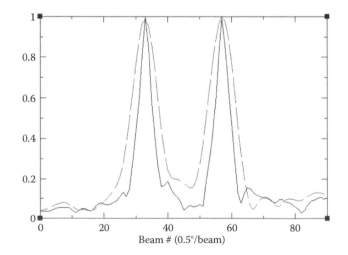

FIGURE 3.31

Beam cross-section for two sources at 65 mm depth: Left source −6° from broadside (at 2.1 MHz, BW 50%), right source +6° from broadside (at 2.0 MHz, BW 50%). Adaptive (solid line) and conventional beamforming (apodized in space, dashed line). (Reprinted from IEEE ©2003. With permission.)

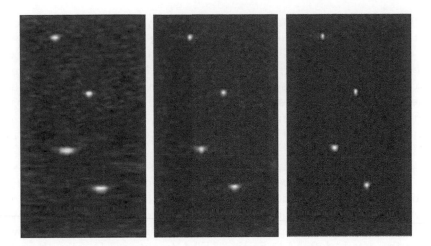

FFIGURE 3.32
First image from the left shows reconstructed image from beam time series of a conventional phased array beamformer applied on the synthetic data defined in Table 3.1, for a 32-sensor phased array probe. The central image in this figure, shows the output of the subaperture adaptive beamformer applied on the same 32-sensor data; and the right panel shows the reconstructed image from the time series of a conventional phased array beamformer applied on 96-sensor phased array probe for the same structure of data defined in Table 3.1. (Reprinted from IEEE ©2003. With permission.)

the subaperture adaptive beamforming provides better angular resolution with respect to conventional beamforming. Moreover, the 32-elements adaptive beamformer achieves almost the same performances as a 96-elements conventional beamformer.

3.6.2.2 *Ultrasound Imaging System with Planar Phased Array Probe*

Deployment of planar arrays by ultrasound medical imaging systems gains increasing popularity because of their advantage to provide 3-D images of organs under medical examination. However, if we consider that a state of the art line array ultrasound system consists of 256 sensors, then a planar array ultrasound system should include at least 4096 sensors (16×256) in order to achieve the angular resolution performance of a line array system and the additional 3-D image reconstruction capability provided by the elevation beam steering of a planar array. Thus, increased angular resolution in azimuth and elevation beam steering for ultrasound systems means larger sensor arrays with consequent technical and higher cost implications. As discussed in Section 3.4, the alternative is to use synthetic aperture and or subaperture adaptive beam processing.

In the simulations discussed in this section, a planar array with 121 (11×11) sensors was considered that provided continuous sensor time series at the input of the conventional and subaperture adaptive beamformers with processing flow similar to that shown in Figures 3.20 and 3.21 for a cylindrical array. As in the case of the cylindrical beamforming results, the power outputs of the beam time series of the conventional and the subaperture adaptive techniques implemented in a planar array, demonstrated the same performance characteristics with those of Figures 3.24 through 3.28 for the cylindrical array. Supporting evidence for this claim are the real data results, from an experimental 3-D/4-D ultrasound imaging system deploying a planar array with $16 \times 16 = 256$ sensors, that are presented in Chapter 11.

The synthetic data experiments for the planar array, discussed in this section, were carried out using the Field II ultrasound simulator program obtained from the Technical University of Denmark [60]. The Field II program simulates point sources, and was set up to simulate 5000 point sources arranged in a spherical shell, conforming to the specifications of an ultrasound imaging system with a planar phased array probe. More specifically, in the simulations the probe was assumed to have a 16×16 channel planar array as receiver and a 6x6 channel planar array as transmitter, with element spacing of 0.4 mm and sampling frequency of 33 MHz. The FM pulse was centred at 2.5 MHz, with a bandwidth of 4.0 MHz. The simulated illumination pattern was identical with that defined in Chapter 11 for an experimental planar array ultrasound system and included six illumination beams along azimuth and six beams along elevation spaced 10° apart covering 60° is each direction (azimuth and elevation). The result was a total of 36 angular sectors for illumination. A conventional beamformer was used to process the data received by the 16x16 planar array. The decomposition process of the 3-D planar array beamformer was carried according to the details defined in Section 3.5.2.4, to obtain 3-D azimuth and elevation beams. A complete image reconstruction of the beams into a 4-D (i.e., 3-D+time) volume was then performed.

Figure 3.33 shows the C-scans derived from the 3-D reconstructed volumes of the simulated spherical shell. In this image, which shows a slice (C-scan) of the spherical shell from the 3-D volumetric image, the expected ring that corresponds to the cross section of a shell is visible. The left and right images in Figure 3.33 correspond to the 3-D conventional and adaptive beamformers, respectively. The better performance of the adaptive beamformer is evident in this case as it provides better detection and image definition of the spherical shell than that of the corresponding conventional beamformer. The 3-D visualization software, which is discussed in Chapter 4, was provided by Prof. Sakas (i.e., Fraunhofer IGD, Germany), as part of our technical exchanges within the framework of the collaborative European–Canadian project ADUMS (EC-IST-2001-34088).

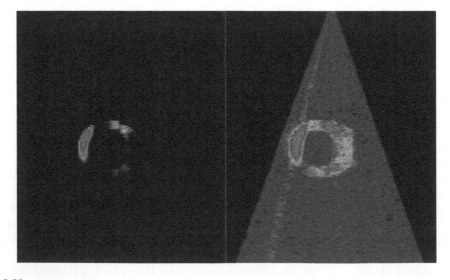

FIGURE 3.33
C-scans derived from the 3-D reconstructed images of the simulated spherical shell. Left image, shows image reconstruction from the beam time series of the 3-D conventional beamformer. Right image, shows reconstructed image from the beam time series of the 3-D adaptive beamformer.

Figure 3.34 shows the 3-D volume reconstruction of the spherical shell using the 3-D conventional (left image) and the 3-D adaptive (right image) ultrasound beamformers defined in this Chapter. As was the case with the C-scans (Figure 3.33), the results in Figure 3.34 show that the 3-D adaptive beamformer provides better image definition than the corresponding 3-D conventional beamforming results.

3.7 Conclusion

The synthetic data results of this chapter indicate that the generic multidimensional adaptive concept addresses practical concerns of near-instantaneous convergence, shown in Figures 3.24 through 3.33, for ultrasound imaging and integrated active–passive sonar systems. The performance characteristics of the subaperture adaptive beamformer compared with that of the conventional beamformer are reflected as improvements in directional estimates of azimuth and elevation angles and suppression of reverberation effects. This kind of improvement in azimuth and elevation bearing estimates is essential for 3-D ultrasound, sonar and radar operations. The conclusions of this Chapter are supported also by the real data results presented in Chapters 9 and 11 for integrated active passive towed array sonar and 3-D/4-D ultrasound imaging systems, respectively.

In summary, a generic beamforming structure has been developed for multidimensional sensor arrays that allows the implementation of conventional, synthetic aperture and adaptive signal processing techniques in integrated active–passive real time systems. The proposed implementation is based on decomposing the 2-D and 3-D beamforming process in subsets of coherent processes and creating subaperture configurations that allow the minimization of the number of degrees of freedom of the adaptive processing schemes. The proposed approach has been applied to line, planar, and cylindrical arrays of sensors where the multidimensional beamforming is decomposed into sets of line-array and/or circular array beamformers. Moreover, the application of spatial shading on the generic

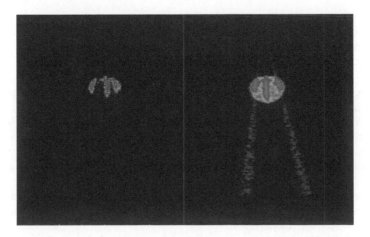

FIGURE 3.34
3-D volume reconstruction of the simulated spherical shell. Left image, reconstructed with the 3-D conventional beamformer. Right image, reconstructed with the 3-D adaptive beamformer. (Reprinted from IEEE ©2003. With permission.)

multidimensional beamformer is a much simpler process than that of the fully coherent 3-D beamformer. This is because the decomposition process allows two simple and separate applications of spatial shading (i.e., one for circular and the other for line arrays).

The fact that the subaperture adaptive beamformers provided array gain improvements for CW and HFM signals under a real time data flow as compared with the conventional beamformer demonstrates the merits of these advanced processing schemes for practical ultrasound, sonar and radar applications. In addition, the generic implementation scheme of this study suggests that the design approach to provide synergism between the conventional beamformer, the synthetic aperture and the adaptive processing schemes (e.g., see Figures 3.20 and 3.21) is an essential property for system applications.

Although the focus of the implementation effort included only a few adaptive processing schemes, the consideration of other types of spatial filters for real time ultrasound, sonar and radar applications should not be excluded. The objective here was to demonstrate that adaptive processing schemes can address some of the challenges that the next generation ultrasound and active–passive sonar systems will have to deal with in the near future. Once a generic signal processing structure is established, as suggested in the chapter, the implementation of a wide variety of processing schemes can be achieved with minimum efforts for real time systems deploying multidimensional arrays. Finally, the results presented in this chapter indicate that the subaperture STMV adaptive scheme address the practical concerns of near-instantaneous convergence associated with the implementation of adaptive beamformers in integrated active–passive sonar systems. It is the understanding of the investigators of this study that the CSA-SA (i.e., with near instantaneous convergence requirement for a single active transmission to generate a single image), is the most appropriate adaptive beamformer for cost efficient ultrasound imaging applications. However, the MVDR-SA adaptive beamformer, (i.e., with near instantaneous convergence requirement for three successive transmissions to generate a single image), may provide much better image resolution than the CSA-SA algorithm.

References

1. S. Stergiopoulos. 1998. Implementation of adaptive and synthetic aperture processing schemes in integrated active–passive sonar systems. *Proc. IEEE* , 86(2), 358–396.
2. W.C. Knight, R.G. Pridham and S.M. Kay. 1981. Digital signal processing for sonar. *Proc. IEEE*, 69(11), 1451–1506.
3. B. Windrow, P.E. Manfey, L.J. Griffiths, and B.B. Goode. 1967. Adaptive antenna systems. *Proc. IEEE*, 55(12), 2143–2159.
4. A.A. Winder. 1975. Sonar system technology. *IEEE Trans. Sonic Ultrasonics*, SU-22(5), 291–332.
5. A.B. Baggeroer. 1978. Sonar signal processing. In: *Applications of Digital Signal Processing*, A.V. Oppenheim, editor. Prentice Hall, Englewood cliffs, NJ.
6. S. Stergiopoulos and A.T. Ashley. 1993. Guest editorial, special issue on sonar system technology. *IEEE J. Oceanic Eng.*, 18(4), 361–366.
7. Dhanantwari A.C., S. Stergiopoulos and J. Grodski. 1998. Implementation of adaptive processing in integrated active–passive sonars deploying cylindrical arrays. Proceedings of Underwater Technology '98, UT'98, Tokyo, Japan.
8. W.C. Queen. 1970. The directivity of sonar receiving arrays. *J. Acoust. Soc. Am.*, 47, 711–720.
9. V. Anderson and J. Munson. 1963. Directivity of spherical receiving arrays. *J. Acoust. Soc. Am.*, 35, 1162–1168.

10. W.M. Carey and W.B. Moseley. 1991. Space-time processing, environmental-acoustic effects. *IEEE J. Oceanic Eng.*, 16, 285–301; also *Progress in Underwater Acoustics*, Plenum, NY, 743–758, 1987.

11. S. Stergiopoulos and E.J. Sullivan. 1989. Extended towed array processing by overlapped correlator. *J. Acoust. Soc. Am.*, 86(1), 158–171.

12. S. Stergiopoulos. 1990. Optimum bearing resolution for a moving towed array and extension of its physical aperture. *J. Acoust. Soc. Am.*, 87(5), 2128–2140.

13. E.J. Sullivan, W.M. Carey and S. Stergiopoulos. 1992. Editorial, special issue on acoustic synthetic aperture processing. *IEEE J. Oceanic Eng.*, 17(1), 1–7.

14. S. Stergiopoulos and H. Urban. 1992. An experimental study in forming a long synthetic aperture at sea. *IEEE J. Oceanic Eng.*, 17(1), 62–72.

15. G.S. Edelson and E.J. Sullivan. 1992. Limitations on the overlap-correlator method imposed by noise and signal characteristics. *IEEE J. Oceanic Eng.*, 17(1), 30–39.

16. G.S. Edelson and D.W. Tufts. 1992. On the ability to estimate narrow-band signal parameters using towed arrays. *IEEE J. Oceanic Eng.*, 17(1), 48–61.

17. N.L. Owsley. 1985. Sonar array processing. In: *Signal Processing Series*, S. Haykin, editor. A.V. Oppenheim, series editor. Prentice Hall, Englewood cliffs, NJ.

18. B. Van Veen and K. Buckley. 1988. Beamforming: a versatile approach to spatial filtering. *IEEE ASSP Mag.*, 4–24.

19. H. Cox, R.M. Zeskind and M.M. Owen. 1987. RobustAdaptive beamforming. *IEEE Trans. Acoustic Speech Signal Proc.*, ASSP-35(10), 1365–1376.

20. H. Cox. 1973. Resolving power and sensitivity to mismatch of optimum array processors. *J. Acoust. Soc. Am.*, 54(3), 771–785.

21. A.H. Sayed and T. Kailath. 1994. A state-space approach to adaptive RLS filtering. *IEEE SP Mag.*, July, 18–60.

22. J. Capon. 1969. High resolution frequency wavenumber spectral analysis. *Proc. IEEE*, 57, 1408–1418.

23. T.L. Marzetta. 1983. A new interpretation for Capon's maximum likelihood method of frequency-wavenumber spectra estimation. *IEEE Trans. Acoustic Speech Signal Proc.*, ASSP-31(2), 445–449.

24. S. Haykin. 1986. *Adaptive Filter Theory*. Prentice Hall, Englewood Cliffs, NJ.

25. A. Tawfik, A.C. Dhanantwari, and S. Stergiopoulos. 1997. A generic beamforming structure allowing the implementation of adaptive processing schemes into 2-D and 3-D arrays of sensors. Proceedings MTS/IEEE OCEANS'97, Halifax, Nova Scotia, Canada.

26. A. Tawfik, A.C. Dhanantwari, and S. Stergiopoulos. 1997. A generic beamforming structure for adaptive schemes implemented in 2-D and 3-D arrays of sensors. *J. Acoust. Soc. Am.*, 101(5), 3025.

27. S. Stergiopoulos, A. Tawfik and A.C. Dhanantwari. 1997. Adaptive microphone 2-D and 3-D arrays for enhancement of sound reception in coherent and incoherent noise environment. Proceedings of Inter-Noise'97, OPAKFI H-1027 Budapest, Hungary.

28. A. Tawfik and S. Stergiopoulos. 1997. A generic processing structure decomposing the beamforming process of 2-D and 3-D arrays of sensors into sub-sets of coherent processes. Proceedings CCECE'97, Canadian Conference on Electrical and Computer Engineering, St. John's, Newfoundland, Canada.

29. N.L. Owsley. 1987. Systolic array adaptive beamforming NUWC Report 7981.

30. S. Stergiopoulos. 1991. Limitations on towed-array gain imposed by a non isotropic ocean. *J. Acoust. Soc. Am.*, 90(6), 3161–3172.

31. A.D. Whalen. 1971. *Detection of Signals in Noise*. Academic Press, NY.

32. D. Middleton. 1960. *Introduction to Statistical Communication Theory*. McGraw-Hill, NY.

33. H.L. Van Trees. 1968. *Detection, Estimation and Modulation Theory*. Wiley, NY.

34. S. Stergiopoulos. 1995. Noise normalization technique for beamformed towed array data. *J. Acoust. Soc. Am.*, 97(4), 2334–2345.

35. S. Stergiopoulos. 1996. Influence of underwater environment's coherence properties on sonar signal processing. Proceedings of 3rd European Conference on Underwater Acoustics, FORTH-IACM, Heraklion-Crete, V-I.
36. F.J. Harris. 1978. On the use of windows for harmonic analysis with discrete Fourier transform. *Proc. IEEE*, 66, 51–83.
37. A.C. Dhanantwari. 1996. Adaptive beamforming with near-instantaneous convergence for matched filter processing. Master Thesis, Department of Electrical Engineering, Technical University of Nova Scotia, Halifax, Nova Scotia, Canada.
38. J. Krolik and D.N. Swingler. 1989. Bearing estimation of multiple broadband sources using steered covariance matrices. *IEEE Trans. Acoust. Speech, Signal Proc.*, ASSP-37, 1481–1494.
39. H. Wang and M. Kaveh. 1985. Coherent signal-subspace processing for the detection and estimation of angles of arrival of multiple wideband sources. *IEEE Trans. Acoust. Speech, Signal Proc.*, ASSP-33, 823–831.
40. A. Antoniou. 1993. *Digital Filters: Analysis, Design, and Applications*, 2nd Ed. McGraw-Hill, NY.
41. A. Mohammed. 1983. A high-resolution spectral analysis technique. DREA Memorandum 83/D, Defense Research Establishment Atlantic, Dartmouth, Nova Scotia, Canada.
42. A. Mohammed. 1985. Novel methods of digital phase shifting to achieve arbitrary values of time delays. DREA Report 85/106, Defense Research Establishment Atlantic, Dartmouth, Nova Scotia, Canada.
43. S. Stergiopoulos and A.T. Ashley. 1997. An experimental evaluation of split-beam processing as a broadband bearing estimator for line array sonar systems. *J. Acoust. Soc. Am.*, 102(6), 3556–3563.
44. L.J. Griffiths, C.W. Jim. 1982. An alternative approach to linearly constrained adaptive beamforming. *IEEE Trans. Antennas and Propagation*, AP-30, 27–34.
45. D.T.M. Slock. 1993. On the convergence behavior of the LMS and the normalized LMS algorithms. *IEEE Trans. Acoust. Speech, Signal Proc.*, ASSP-31, 2811–2825.
46. D. Middleton and R. Esposito. 1968. Simultaneous optimum detection and estimation of signals in noise. *IEEE Trans. Inform. Theory*, IT-14, 434–444.
47. P. Wille and R. Thiele. 1971. Transverse horizontal coherence of explosive signals in shallow water. *J. Acoust. Soc. Am.*, 50, 348–353.
48. P.A. Bello. 1963. Characterization of randomly time-variant linear channels. *IEEE Trans. Commun. Syst.*, 10, 360–393.
49. D.A. Gray, B.D.O. Anderson and R.R. Bitmead. 1993. Towed array shape estimation using Kalman filters—theoretical models. *IEEE J. Oceanic Eng.*, 18(4), 543–556.
50. B.G. Ferguson. 1993. Remedying the effects of array shape distortion on the spatial filtering of acoustic data from a line array of sensors. *IEEE J. Oceanic Eng.*, 18(4), 565–571.
51. J.L. Riley and D.A. Gray. 1993. Towed array shape estimation using Kalman filters—experimental investigation. *IEEE J. Oceanic Eng.*, 18(4), 572–581.
52. B.G. Ferguson. 1990. Sharpness applied to the adaptive beamforming of acoustic data from a towed array of unknown shape. *J. Acoust. Soc. Am.*, 88(6), 2695–2701.
53. Lu Feng, E. Milios, S. Stergiopoulos, and A. Dhanantwari. 2003. A new towed array shape estimation scheme for real time sonar systems. *IEEE J. Oceanic Eng.*, 28(3), 552–563.
54. N.C. Yen and W. Carey. 1989. Application of synthetic-aperture processing to towed-array data, *J. Acoust. Soc. Am.*, 86, 754–765.
55. S. Stergiopoulos and H. Urban. 1992. A new passive synthetic aperture technique for towed arrays. *IEEE J. Oceanic Eng.*, 17(1), 16–25.
56. V.H. MacDonald and P.M. Schulteiss. 1969. Optimum passive bearing estimation in a spatially incoherent noise environment. *J. Acoust. Soc. Am.*, 46(1), 37–43.
57. G.C. Carter. 1987. Coherence and time delay estimation. *Proc. IEEE*, 75(2), 236–255.
58. C.H. Knapp and G.C. Carter. 1976. The generalized correlation method for estimation of time delay. *IEEE Trans. Acoust. Speech Signal Processing*, ASSP-24, 320–327.
59. D.C. Rife and R.R. Boorstyn. 1974. Single-tone parameter estimation from discrete-time observations. *IEEE Trans. Inform. Theory*, 20, 591–598.
60. http://www.it.dtu.dk/~jaj/field

4

Advanced Applications of Volume Visualization Methods in Medicine

Georgios Sakas and Grigorios Karangelis

Fraunhofer Institute for Computer Graphics

Andreas Pommert

Institute of Mathematics and Computer Science in Medicine

CONTENTS

4.1 Volume Visualization Principles

4.1.1 Introduction

Medical imaging technology has experienced a dramatic change over the past 25 years. Previously, only X-ray radiographs were available, which showed the depicted organs as superimposed shadows on photographic film. With the advent of modern computers, new *tomographic* imaging modalities like CT, MRI, and positron emission tomography (PET) which deliver cross-sectional images of a patient's anatomy and physiology have been developed. These images show different organs free from overlays with unprecedented precision. Even the 3D structure of organs can be recorded if a sequence of parallel cross-sections is taken. For many clinical tasks like surgical planning, it is necessary to understand and communicate complex and often malformed 3D structures. Experience has shown that the "mental reconstruction" of objects from cross-sectional images is extremely difficult and strongly depends on the observer's training and imagination. For these cases, it is certainly desirable to present the human body as a surgeon or anatomist would see it. The aim of *volume visualization* (also known as *3D imaging*) in medicine is to create precise and realistic views of objects from medical volume data. The resulting images, even though they are of course two-dimensional (2D), are often called *3D images* or *3D reconstructions* to distinguish them from 2D cross sections or conventional radiographs. The first attempts date back to the late 1970s, with the first clinical applications reported on the visualization of bone from CT in craniofacial surgery and orthopedics. Methods and applications have since been extended to other subjects and imaging modalities. The same principles are also applied to sampled and simulated data from other domains, such as fluid dynamics, geology, and meteorology.[53]

4.1.2 Methods

An overview of the volume visualization pipeline is shown in Figure 4.1. After the acquisition of a series of tomographic images of a patient, the data usually undergo some

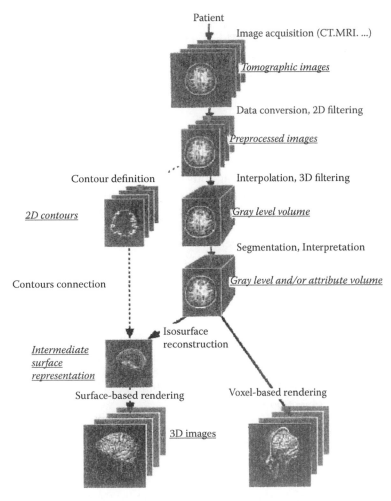

FIGURE 4.1
Overview of the volume visualization pipeline. Individual processing steps may be left out, combined, or reversed in order by a particular method.

preprocessing for data conversion and possibly image filtering. From this point, one of several paths may be followed.

The dotted line in Figure 4.1 represents an early approach where an object is reconstructed from its contours on the cross-sectional images. All other methods, represented by the solid line, start from a contiguous *data volume*. If required, equal spacing in all three directions can be achieved by interpolation. Like a 2D image, a 3D volume can be filtered to improve image quality. Corresponding to the *pixels* (picture elements) of a 2D image, volume elements are called *voxels* (volume elements).

The next step is to identify the different objects represented in the data volume so that they can be removed or selected for visualization. The simplest way is to binarize the data with an intensity threshold, e.g., to distinguish bone from other tissues in CT. For MRI data especially, however, more sophisticated *segmentation* methods are required.

After segmentation, there is a choice for which *rendering* technique is to be used. The more traditional surface-based methods first create an intermediate surface representation of the object to be shown. It may then be rendered with any standard computer graphics method. More recently, volume-based methods have been developed which create a 3D view directly from the volume data. These methods use the full grey level information to render surfaces, cuts, or transparent and semitransparent volumes. As a third way, transform-based rendering methods may be used.

Extensions to the volume visualization pipeline not shown in Figure 4.1, but also covered here, include volume registration, intelligent visualization, and intervention rehearsal.

4.1.2.1 Preprocessing

The data we consider usually come as a spatial sequence of 2D cross-sectional images. If they are put on top of each other, a contiguous *grey level volume* is obtained. The resulting data structure is an orthogonal 3D array of voxels, each representing an intensity value. This is called the *voxel model*. Many algorithms for volume visualization work on *isotropic* volumes, where the sampling density is equal in all three dimensions. In practice, however, only very few data sets have this property, especially for CT. In these cases, the missing information has to be reconstructed in an *interpolation* step. A quite simple method is linear interpolation of the intensities between adjacent images. Higher order functions such as splines usually give better results for fine details.[68] Shape-based methods are claimed to be superior in certain situations;[7] however, these are dependent on the results of a previous segmentation step.

With respect to later processing steps such as segmentation, it is often desirable to improve the signal-to-noise ratio of the data using image or volume filtering. Well-known *noise filters* are average, median, and Gaussian filters.[90] These methods, however, tend to smooth out small details as well; better results are obtained with *anisotropic diffusion* filters which largely preserve object boundaries.[34]

4.1.2.1.1 Data Structures for Volume Data

There are a number of different data structures for volume data. The most important are the following:

- *Binary voxel model:* Voxel values are either 1 (object) or 0 (no object). This very simple model is not in much use any more. In order to reduce storage requirements, binary volumes may be subdivided recursively into subvolumes of equal value; the resulting data structure is called an *octree*.

- *Grey level voxel model:* Each voxel holds an intensity information. Octree representations have also been developed for grey level volumes.[59]

- *Generalized voxel model:* In addition to an intensity information, each voxel contains *attributes*, describing its membership to various objects and/or data from other sources (e.g., MRI and PET).[47]

- *Intelligent volumes:* As an extension of the generalized voxel model, properties of anatomical objects (such as color, names in various languages, pointers to related information) and their relationships are modeled on a symbolic level.[49,84] This data structure is the basis for advanced applications such as medical atlases discussed later.

4.1.2.2 Segmentation

A grey level volume usually represents a large number of different structures obscuring each other. Thus, to display a particular one, we have to decide which parts of the data we want to use or ignore. Ideally, selection would be done with a command like "show only the brain". This, however, requires that the computer know which parts of the volume constitute the brain and which do not.

A first step toward object recognition is to partition the grey level volume into different regions which are homogeneous with respect to some formal criteria and correspond to real anatomical objects. This process is called *segmentation*. The generalized voxel model is a suitable data structure for representing the results. In a further *interpretation* step, the regions may be identified and labeled with meaningful terms such as "white matter" or "ventricle".

All segmentation methods can be characterized as being either "binary" or "fuzzy", corresponding to the principles of binary and fuzzy logic, respectively.[124] In *binary segmentation*, the question whether a voxel belongs to a certain region is always answered yes or no. This information is a prerequisite, e.g., for creating surface representations from volume data. As a drawback, uncertainty or cases where an object takes up only a fraction of a voxel (*partial volume effect*) cannot be handled properly. For example, a very thin bone would appear with false holes on a 3D image. Strict yes/no decisions are avoided in *fuzzy segmentation*, where a set of probabilities is assigned to every voxel, indicating the evidence for different materials. Fuzzy segmentation is closely related to the so-called volume rendering methods discussed later.

Currently, a large number of segmentation methods for 3D medical images are being developed, which may be roughly divided into three classes: point-, edge-, and region-based methods. The methods described often have been tested successfully on a number of cases; experience has shown, however, that the results should always be used with care.

4.1.2.2.1 Point-based Segmentation

In point-based segmentation, a voxel is *classified* depending only on its intensity, no matter where it is located. A very simple but nevertheless important example, which is very much used in practice, is *thresholding*: a certain intensity range is specified with lower and upper threshold values. A voxel belongs to the selected class if, and only if, its intensity level is within the specified range.

Thresholding is the method of choice for selecting bone or soft tissue in CT. In volume-based rendering, it is often performed during the rendering process itself so that no explicit segmentation step is required. In order to avoid the problems of binary segmentation, Drebin et al.[24] use a fuzzy maximum likelihood classifier which estimates the percentages of the different materials represented in a voxel, according to Bayes' rule. This method requires that the grey level distributions of different materials be different from each other and known *a priori*. This is approximately the case in musculoskeletal CT.

Unfortunately, these simple segmentation methods are not suitable if different structures have mostly overlapping or even identical grey level ranges. This situation frequently occurs, e.g., in the case of soft tissues from CT or MRI. The situation is somewhat simplified if multiparameter data are available, such as T_1- and T_2-weighted images in MRI, emphasizing fat and water, respectively. In this case, individual threshold values can be specified for every parameter. To somewhat generalize this concept, voxels in an n-parameter data set can be considered as n-dimensional vectors in an n-dimensional *feature space*. In *pattern recognition*, this feature space is partitioned into subspaces representing different tissue classes or organs. This is called the *training phase*. In supervised training, the partition is derived from feature vectors which are known to represent particular tissues.[19,35] In unsupervised training, the partition is automatically generated.[35] In the subsequent *test phase*, a voxel is classified according to the position of its feature vector in the partitioned feature space.

With especially adapted image acquisition procedures, pattern recognition methods have successfully been applied to considerable numbers of two- or three-parametric MRI data volumes.[19,35] Quite frequently, however, isolated voxels or small regions are incorrectly classified, such as subcutaneous fat in the same class as white matter. To eliminate these errors, a connected component analysis (see below) is often applied.

A closely related method, based on *neural network* methodology, was developed by Kohonen.[57] Instead of an n-dimensional feature space, a so-called *topological map* of $m \times m$ n-dimensional vectors is used. During the training phase, the map iteratively adapts itself to a set of training vectors which may either represent selected tissues (supervised learning) or the whole data volume (unsupervised learning).[40,115] Finally, the map develops several relatively homogeneous regions which correspond to different tissues or organs in the original data. The practical value of the topological map for 3D MRI data seems to be generally equivalent to that of pattern recognition methods.

4.1.2.2.2 Edge-based Segmentation

The aim of edge-based segmentation methods is to detect intensity discontinuities in a grey level volume. These edges (in 3D, they are actually surfaces; however, it is common to speak about edges) are assumed to represent the borders between different organs or tissues. Regions are subsequently defined as the enclosed areas. A common strategy for edge detection is to locate the maxima of the first derivative of the 3D intensity function. A method which very accurately locates the edges was developed by Canny.[17] All algorithms using the first derivative, however, share the drawback that the detected contours are usually not closed, i.e., they do not separate different regions properly. An alternative approach is to detect zero-crossings of the second derivative. The Marr–Hildreth operator convolves the input data with the Laplacian of a Gaussian; the resulting contour volume describes the locations of the edges. With a 3D extension of this operator, Bomans et al. segmented and visualized the complete human brain from MRI for the first time.[12] Occasionally, this operator creates erroneous "bridges" between different materials which have to be removed interactively. In addition, location accuracy of the surfaces is not always satisfactory.

Snakes[1,52] are 2D image curves which are adjusted from an initial approximation to image features by a movement of the curve caused by simulated forces (Figure 4.2). Image features produce the so-called external force. An internal tension of the curve resists against highly angled curvatures, which makes the Snakes movement robust against

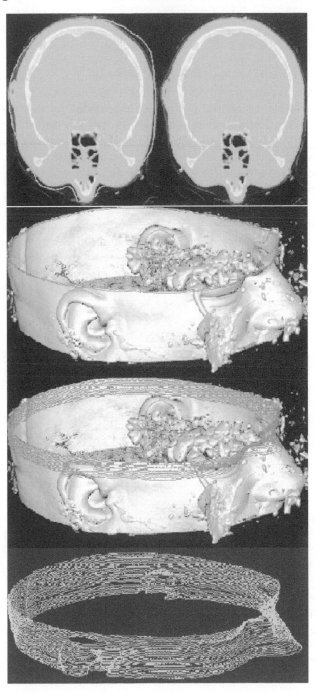

FIGURE 4.2
The principle of segmentation using Snakes.

noise. After a starting position is given, the Snake adapts itself to an image by relaxation to the equilibrium of the external force and internal tension. To calculate the forces, an external energy has to be defined. The gradient of this energy is proportional to the external force. The segmentation by Snakes is due to its 2D definition performed in a slice-by-slice manner, i.e., the resulting curves for a slice are copied into the neighboring slice and the minimization is started again. The user may control the segmentation process by stopping the automatic tracking if the curves run out of the contours and define a new initial curve.

For this reason, two methods have been applied to enter an initial curve for the Snake. The first is the interactive input of a polygon. Since the Snake contracts due to its internal energy, the contour to be segmented has to be surrounded by this polygon. The second one is a contour tracing method, using an *A* search tree to find the path with minimal costs between two interactively marked points.[76,116] The quality of the result depends on the similarity of two adjacent slices. Normally, this varies within a data set. Therefore, in regions with low similarity, the slices to be segmented by the interactive method must be selected rather tightly.

4.1.2.2.3 *Region-based Segmentation*

Region-based segmentation methods consider whole regions instead of individual voxels or edges. Since we are actually interested in regions, this approach appears to be the most natural. Properties of a region are, e.g., its size, shape, location, variance of grey levels, and its spatial relation to other regions.

A typical application of region-based methods is to postprocess the results of a previous point-based segmentation step. For example, a *connected component analysis* may be used to determine whether the voxels, which have been classified as belonging to the same class, are part of the same connected region. If not, some of the regions may be discarded.

A practical interactive segmentation system based on the methods of *mathematical morphology* was developed by Höhne and Hanson.[48] Regions are initially defined with thresholds; the user can subsequently apply simple but fast operations such as *erosion* (to remove small bridges between erroneously connected parts), *dilation* (to close small gaps), connected components analysis, region fill, or Boolean set operations. Segmentation results are immediately visualized on orthogonal cross sections and 3D images, so that they may be corrected or further refined in the next step (Figure 4.3). With this system, segmentation of gross structures is usually a matter of minutes. In Schiemann et al.,[101] this approach is extended to multiparameter data.

For automatic segmentation, the required knowledge about data and anatomy needs to be represented in a suitable model. A comparatively simply approach is presented by Brummer et al., who use a fixed sequence of morphological operations for the segmentation of brain from MRI.[16] For the same application, Raya and Udupa developed a rule-based system which successively generates a set of threshold values.[86] Rules are applied depending on measured properties of the resulting regions. Bomans generates a set of object hypotheses for every voxel, depending on its grey level.[11] Location, surface-to-volume ratio, etc. of the resulting regions are compared to some predefined values, and the regions are modified accordingly. Menhardt uses a rule-based system which models the anatomy with relations such as "brain is inside skull."[70] Regions are defined as fuzzy subsets of the volume, and the segmentation process is based on fuzzy logic and fuzzy topology.

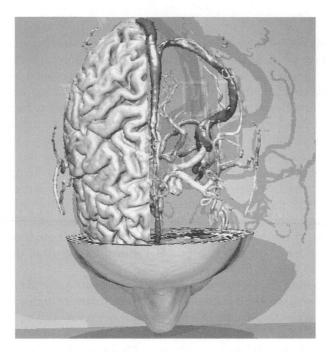

FIGURE 4.3
Results of interactive segmentation of MRI (skin, brain) and MRA (vessels) data.

One of the problems of these and similar methods for automatic segmentation is that the required anatomical knowledge is often represented in more or less ad hoc algorithms, rules, and parameters. A more promising approach is to use an explicit 3D organ model. For the brain, Atata et al. developed an atlas of the "normal" anatomy and its variation in terms of a probabilistic spatial distribution obtained from 22 MRI data sets of living persons.[3] The model was reported as suitable for the automatic segmentation of various brain structures, including white matter lesions. A similar approach is described by Kikinis et al.[56] Automatic segmentation of cortical structures based on a statistical atlas representing several hundred individuals is presented by Collins et al.[29]

Another interesting idea is to investigate object features in *scale-space*, i.e., at different levels of image resolution. This approach allows irrelevant image detail to be ignored. One such method developed by Pizer et al. considers the symmetry of previously determined shapes, described by medial axes.[129] The resulting ridge function in scale-space is called the *core* of an object. It may be used, e.g., for interactive segmentation, where the user can select, add, or subtract regions or move to larger "parent" or smaller "child" regions in the hierarchy. Other applications such as automatic segmentation or registration are currently being investigated.

In conclusion, automatic segmentation systems are not yet robust enough to be generally applicable to medical volume data. Interactive segmentation, which combines fast operations with the unsurpassed human recognition capabilities, is still the most practical approach.

4.1.2.3 Surface-Based Rendering

The key idea of *surface-based rendering* methods is to extract an intermediate surface description of the relevant objects from the volume data. Only this information is then used for rendering. If triangles are used as surface elements, this process is called *triangulation*. A clear advantage of surface-based methods is the possibly very high data reduction from volume to surface representations. Resulting computing times can be further reduced if standard data structures such as triangle meshes are used with common rendering hard- and software support. On the other hand, the *surface reconstruction* step throws away most of the valuable information on the cross-sectional images. Even simple cuts are meaningless because there is no information about the interior of an object. Furthermore, every change of surface definition criteria, such as thresholds, requires a recalculation of the whole data structure.

An early approach for the reconstruction of the polygonal mesh from a stack of contours is based on the Delauney interpolation developed by Boissinnat.[10] Using this heuristic method, the volume of the contours is computed by a 3D triangulation which allows an extraction of the surface of the object. Figure 4.4 shows the result of the triangle reduced surface of the Virtual Human Project patient.

A more recent method by Lorensen and Cline, called *marching cubes* algorithm, creates an *iso-surface*, representing the locations of a certain intensity value in the data.[65] This algorithm basically considers a cube of $2 \times 2 \times 2$ contiguous voxels. Depending on whether one or more of these voxels are inside the object (i.e., above a threshold value), a surface representation of up to four triangles is placed within the cube. The exact location of the

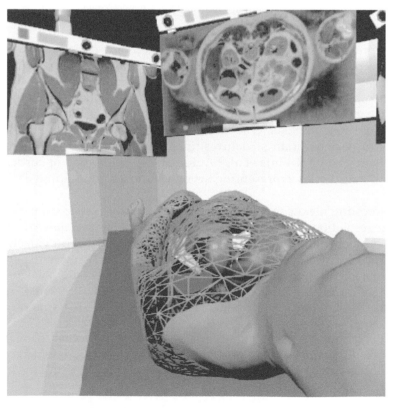

FIGURE 4.4
The virtual patient.

(a) (b)

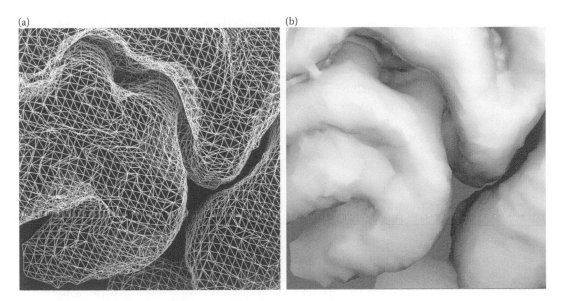

FIGURE 4.5
(a) Triangulated portion of the brain from MRI, created with the marching cubes algorithm. (b) Shaded portion of the brain from MRI, created with the marching cubes algorithm.

triangles is found by linear interpolation of the intensities at the voxel vertices. The result is a highly detailed surface representation with subvoxel resolution (Figure 4.5). Surface orientations are calculated from grey level gradients. Meanwhile, a whole family of similar algorithms has been developed.[80,118,122]

Applied to clinical data, the marching cubes algorithm typically creates hundreds of thousands of triangles. As has been shown, these numbers can be reduced considerably by a subsequent simplification of the triangle meshes, without much loss of information.[103,123] The reduction method can be parameterized and thus allows deriving models of different levels of detail.

4.1.2.3.1 Shading

In general, *shading* is the realistic display of an object based on the position, orientation, and characteristics of its surface and the light sources illuminating it.[32] The reflective properties of a surface are described with an *illumination model* such as the *Phong* model, which uses a combination of ambient light and diffuse (like chalk) and specular (like polished metal) reflections. A key input into these models is the local surface orientation, described by a *normal vector* perpendicular to the surface. The original marching cubes algorithm calculates the surface normal vectors from the grey level gradients in the data volume, described later.

4.1.2.4 Volume-Based Rendering

In *volume-based rendering*, images are created directly from the volume data. Compared to surface-based methods, the major advantage is that all grey level information, which has originally been acquired, is kept during the rendering process. As shown by Höhne et al.,[47] this makes it an ideal technique for interactive data exploration. Threshold values and other parameters, which are not clear from the beginning, can be changed interactively. Furthermore, volume-based rendering allows a combined display of different aspects such as opaque and semitransparent surfaces, cuts, and maximum intensity projections. A current drawback of volume-based techniques is that the large amount of data, which has to be handled, does not allow real-time applications on present-day computers.

4.1.2.4.1 *Scanning the Volume*

In volume-based rendering, we basically have the choice between two scanning strategies: pixel by pixel (image order) or voxel by voxel (volume order). These strategies correspond to the image and object order rasterization algorithms used in computer graphics.[32] In *image order* scanning, the data volume is sampled on rays along the view direction. This method is commonly known as *ray casting*:[47]

> FOR each pixel on image plane DO
> > FOR each sampling point on associated viewing ray DO
> > > compute contribution to pixel

The principle is illustrated in Figure 4.6. Along the ray, visibility of surfaces and objects is easily determined. The ray can stop when it meets an opaque surface. Yagel et al. extended this approach to a full *ray tracing* system which follows the viewing rays as they are reflected on various surfaces.[125] Multiple light reflections between specular objects can thus be handled.

Image order scanning can be used to render both voxel and polygon data at the same time.[63] Image quality can be adjusted by choosing smaller (oversampling) or wider (undersampling) sampling intervals.[62,82] As a drawback, the whole input volume must be available for random access to allow arbitrary view directions. Furthermore, interpolation of the intensities at the sampling points is required. A strategy to reduce computation times is based on the observation that most of the time is spent traversing empty space, far away

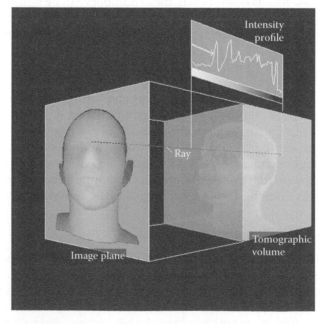

FIGURE 4.6
Principle of ray casting for volume visualization. In this case, the object surface is found using an intensity threshold.

from the objects to be shown. If the rays are limited to scan the data only within a pre-defined *bounding volume* around these objects, scanning times are greatly reduced.[4]

In *volume order* scanning, the input volume is sampled along the lines and columns of the 3D array, projecting a chosen aspect onto the image plane in the direction of view:

> FOR each sampling point in volume DO
> > FOR each pixel projected onto DO
> > > compute contribution to pixel

The volume can either be traversed in back-to-front (BTF) order from the voxel with maximal to the voxel with minimal distance to the image plane or vice versa in front-to-back (FTB) order. Scanning the input data as they are stored, these techniques are reasonably fast, even on computers with small main memory, and especially suitable for parallel processing. So far, ray casting algorithms still offer a higher flexibility in combining different display techniques. However, volume rendering techniques working in volume order are also available.[121]

4.1.2.4.2 Shaded Surfaces

Using one of the described scanning techniques, the visible surface of an object can be determined with a threshold and/or an object label. Unfortunately, using object labels will introduce a somewhat blocky appearance of the surface, especially when zooming into the scene. An algorithm which solves this problem, based on finding an iso-intensity surface, is presented by Tiede et al.[61]

As shown by Höhne and Bernstein,[45] a very realistic and detailed presentation is obtained if the grey level information present in the data is taken into account. Due to the partial volume effect, the grey levels in the 3D neighborhood of a surface voxel represent the relative proportions of different materials inside these voxels. The resulting *grey level gradients* can thus be used to calculate surface inclinations. The simplest variant is to calculate the components of a gradient G for a surface voxel at (i, j, k) from the grey level g of its six neighbors along the main axes as

$$Gx = g(i+1, j, k) - g(i-1, j, k)$$

$$Gy = g(i, j+1, k) - g(i, j-1, k)$$

$$Gz = g(i, j, k+1) - g(i, j, k-1)$$

Scaling G to unit length yields the normal surface. The grey level gradient may also be calculated from all 26 neighbors in a $3 \times 3 \times 3$ neighborhood, weighted according to their distance from the surface voxel.[109] Aliasing patterns are thus almost eliminated. A different approach is to use the first derivative of a higher order interpolation function, such as a cubic spline.[68]

4.1.2.4.3 Cut Planes

Once a surface view is available, a very simple and effective method to visualize interior structures is cutting. When the original intensity values are mapped onto the cut plane, they can be better understood in their anatomical context.[47] A special case is selective cutting, where certain objects are excluded (Figure 4.7).

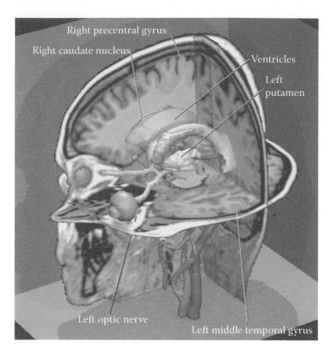

FIGURE 4.7
Brain from MRI. Original intensity values are mapped onto the cut planes.

4.1.2.4.4 *Maximum Intensity Projection*

For small, bright objects such as vessels from MRA, *maximum intensity projection* (MIP) is a suitable display technique. Along each ray through the data volume, the maximum grey level is determined and projected onto the image plane. The advantage of this method is that neither segmentation nor shading is needed, which may fail for very small vessels. But there are also some drawbacks: as light reflection is totally ignored, MIP does not give a realistic 3D impression. Spatial perception can be improved by rotating the object or by a combined presentation with other surfaces or cut planes.[47]

4.1.2.4.5 *Volume Rendering*

Volume rendering is the visualization equivalent to fuzzy segmentation. For medical applications, these methods were first described by Drebin et al.[24] and Levoy.[62] A commonly assumed underlying model is that of a colored, semitransparent gel with suspended low-albedo (low-reflectivity) particles.[9] Illumination rays are partly reflected and change color while traveling through the volume.

Each voxel is assigned a color and an opacity. This opacity is the product of an "object weighting function" and a "gradient weighting function". The object weighting function is usually dependent on the grey level, but it can also be the result of a more sophisticated fuzzy segmentation algorithm. The gradient weighting function emphasizes surfaces for 3D display. All voxels are shaded, using the grey level gradient method. The shaded values along a viewing ray are weighted and summed up. A somewhat simplified basic equation modeling frontal illumination with a ray casting system is given as follows:

$$\text{intensity}\,(p,\ l) = \alpha(p) \cdot l \cdot s(p) + (1.0 - \alpha(p)) \cdot \text{intensity}\,(p + 1 \cdot (1.0 - \alpha(p)) \cdot l)$$

Intensity = intensity of reflected light; p = index of sampling point on ray (0 … max. depth); l = fraction of incoming light (0.0 … 1.0); α = local opacity (0.0 … 1.0); s = local shading component.

The total reflected intensity as displayed on a pixel of the 3D image is given as *intensity* (0, 1.0). Since binary decisions are avoided in volume rendering, the resulting images do not show pronounced artifacts such as jagged edges and false holes. On the other hand, a number of superimposed, more or less transparent surfaces are often hard to understand. Spatial perception can be improved by rotating the object. Another problem is the large number of parameters, which have to be specified to define the weighting functions. Furthermore, volume rendering is comparably slow because weighting and shading operations are performed for many voxels on each ray.

4.1.2.5 Transform-Based Rendering

While both surface- and volume-based rendering are operated in a 3D space, 3D images may also be created from other data representations. One such method is *frequency domain rendering*, which creates 3D images in Fourier space, based on the Fourier projection-slice theorem.[112] This method is very fast, but the resulting images are somewhat similar to X-ray images, lacking real depth information.

A more promising approach is *wavelet transforms*. These methods provide a multiscale representation of 3D objects, with the size of the represented detail locally adjustable. Thus, the amount of data and rendering times may be dramatically reduced. Application to volume visualization is discussed by Muraki.[74,75]

4.1.2.6 Volume Registration

For many clinical applications, it is desirable to combine information from different imaging modalities. For example, for the interpretation of PET images, which show only physiological aspects, it is important to know the patient's morphology, as shown in MRI. In general, different data sets do not match geometrically. Therefore, it is required to transform one volume with respect to the other. This process is known as registration. The transformation may be defined by using corresponding *landmarks* in both data sets.[113] In a simple case, external markers attached to the patient are available which are visible on different modalities. Otherwise, arbitrary pairs of matching points may be defined. A more robust approach is to interactively match larger features such as surfaces (Figure 4.8), or selected internal features such as the AC-PC line (anterior/posterior commissure) in brain imaging.[100] All these techniques may also be applied in scale-space at different levels of resolution.[73]

In a fundamentally different approach, the results of a registration step are evaluated at every point of the combined volume using *voxel similarity measures*, based on intensity values.[107,120] Starting from a coarse match, registration is achieved by adjusting position and orientation until the mutual information between both data sets is maximized. Since these methods are fully automatic and do not rely on a possibly erroneous definition of landmarks, they are increasingly considered superior. A comparison of various approaches is found in West et al.[130]

4.1.2.7 Intelligent Visualization

Knowledge for the interpretation of the 3D images described so far still has to come from the observer. In contrast, the 3D brain atlas VOXEL-MAN/brain shown in Figure 4.9 is

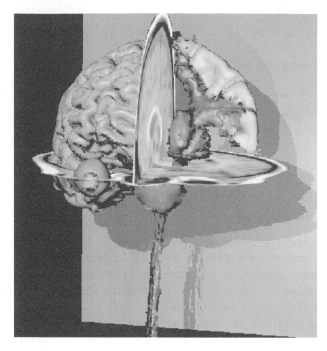

FIGURE 4.8
Volume registration of different imaging modalities MRI, showing morphology, is combined with an FDG-PET, showing the glucose metabolism of a volunteer. Since the entire volume is mapped, the activity can be explored at any location.

based on an *intelligent volume* (see Section 4.1.2.1.1), which has been prepared from an MRI data set.[49,84] It contains spatial and symbolic descriptions in terms of anatomical objects, their relationships, etc. of morphology, function, and blood supply. The brain may be explored on the computer screen in a style close to a real dissection and may be queried at any point. Besides, in education,[20] such atlases are also a powerful aid for the interpretation of clinical images.[100]

If high-resolution cryosections such as those created in the *Visible Human Project* of the National Library of Medicine[106] are used, even more detailed and realistic atlases can be prepared.[101,110] An example image from the VOXEL-MAN Junior/Inner Organs Atlas[67] is shown in Figure 4.10. The technical background of the VOXEL-MAN Junior Atlases,[20,67] based on precalculated QuickTime VR image matrices, is presented in Schroeder.[104]

4.1.2.8 Intervention Rehearsal

So far, we have focused on merely visualizing the data. A special case is to move the camera inside the patient for virtual endoscopy.[33] Besides, in education, potential applications are in noninvasive procedures, such as interventional radiology (Figure 4.11), gastrointestinal diagnosis, and virtual colonoscopy.

A step further is to manipulate the data at the computer screen for surgery simulation. These techniques are most advanced for craniofacial surgery, where a skull is dissected into small pieces and then rearranged to achieve a desirable shape (Figure 4.12). Several systems have been designed based on the binary voxel model[78,126] or polygon

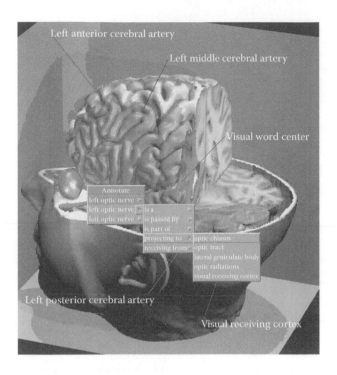

FIGURE 4.9

Exploration of a brain atlas. Arbitrary cutting reveals the interior. The user has accessed information available for the optic nerve concerning functional anatomy, which appears as a cascade of pop-up menus. He has asked the system to colormark some blood supply areas and cortical regions. He can derive from the colors that the visual word center is supplied by the left middle cerebral artery.

representations.[18] Pflesser et al. developed an algorithm which handles full grey level volumes.[81] Thus, all features of volume-based rendering, including cuts and semitransparent rendering of objects obscuring or penetrating each other, are available.

Another promising area is the simulation of soft tissue deformation. This may be due to applying force, e.g., using surgical tools,[98] or as a consequence of an osteotomy, modifying the underlying bone structures.[31]

4.1.2.9 Image Quality

For applications in the medical field, it is mandatory to assure that the 3D images show the true anatomical situation or to at least know about their limitations. A common approach for investigating *image fidelity* is to compare 3D images rendered by means of different algorithms. This method, however, is of limited value since the truth is usually not known. A more suitable approach is to apply volume visualization techniques to simulated data,[66,85,109] and to data acquired from corpses.[25,41,79,83,92] In both cases, the actual situation is available for comparison. In a more recent approach, the frequency behavior of the involved operations is investigated.[68,85]

Another aspect of *image quality* is image utility, which describes whether an image is really useful for a viewer with respect to a certain task. Investigations of 3D image utility in craniofacial surgery may be found elsewhere.[2,105,114]

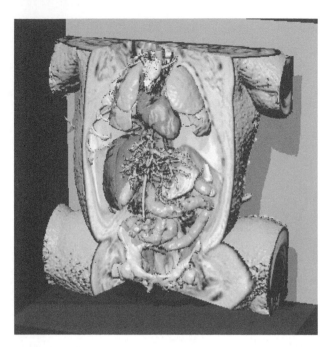

FIGURE 4.10
Dissection of the visible human. Used in a state-of-the-art visualization environment, these data represent a new quality of anatomical imaging.

4.2 Applications to Medical Data

4.2.1 Radiological Data

At first glance, one might expect diagnostic radiology to be the major field of application for volume visualization in medicine. In general, however, this is not the case. One of the reasons is clearly that radiologists are especially skilled in reading cross-sectional images. Another reason is that many diagnostic tasks such as tumor detection and classification can be done based on cross-sectional images. Furthermore, 3D visualization of these objects from MRI requires robust segmentation algorithms which are not yet available. A number of successful applications are presented in Bono and Sakas.[94]

The situation is generally different in all fields where therapeutical decisions have to be made by nonradiologists on the basis of radiological images.[46,127] A major field of application for volume visualization methods is *craniofacial surgery*.[2,22,64,128] Volume visualization not only facilitates understanding of pathological situations, but is also a helpful tool for planning optimal surgical access and cosmetic results of an intervention. A typical case is shown in Figure 4.12. Dedicated procedures for specific disorders have been developed,[105] which are now in routine application.

An application that is becoming more and more attractive with the increasing resolution and specificity of MRI is *neurosurgery planning*. Here, the problem is to choose a proper access path to a lesion (Figure 4.11). 3D visualization of brain tissue from MRI and blood vessels from MRA before surgical intervention allows the surgeon to find a path with minimal risk in advance.[21,82] In combination with an optical tracking system,

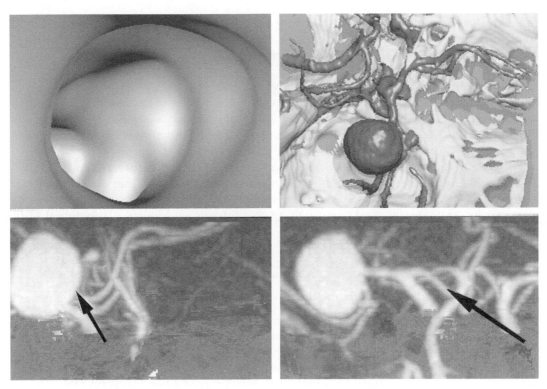

FIGURE 4.11
Virtual catheter examination of a large aneurysm of the right middle cerebral artery, based on CT angiography: (top left) inner structure of the blood vessels as seen from a virtual catheter camera; (top right) 3D overview image showing the blood vessels in relation to the cranial base; (bottom) corresponding MIP images in different orientations. The current position and view direction of the camera are indicated by arrows.

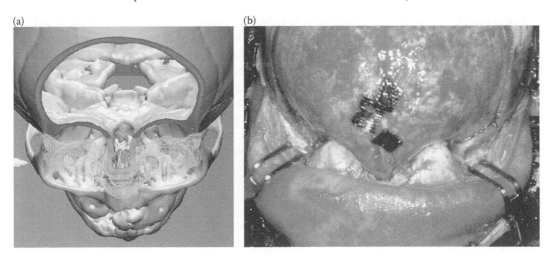

FIGURE 4.12
(a) Craniofacial surgery simulation of a complex congenital malformation of a seven-month-old patient. A CT data set was used to simulate a classic floating-forehead procedure. A frontobasal segment of the skull (shown as a wire mesh) was cut and can be moved in any direction. (Figure 4.12b) Corresponding intra-operative view. (b) Craniofacial surgery simulation of a complex congenital malformation of a seven-month-old patient. Corresponding intra-operative view of Figure 4.12a.

the acquired information can be used to guide the surgeon during the intervention.[6] In conjunction with functional information from PET images, localization of a lesion is facilitated (Figure 4.8). The state of the art in *computer-integrated surgery* is presented in Taylor et al.[108]

Another important application that reduces the risk of a therapeutical intervention is *radiotherapy planning*. Here, the objective is to focus the radiation as closely as possible to the target volume, while avoiding side effects in healthy tissue and radiosensitive organs at risk. 3D visualization of target volume, organs at risk, and simulated radiation dose allows an iterative optimization of treatment plans.[50,55,99,102]

Applications apart from clinical work include *medical research* (Figure 4.8) and *education* (Figure 4.10). In the current decade of brain research, exploring and mapping brain functions is a major issue (Figure 4.9). Volume visualization methods provide a framework to integrate information obtained from such diverse sources as dissection, functional MRI, or magnetoencephalography.[111]

4.2.2 3D Ultrasound

4.2.2.1 Introduction

3D ultrasound is a very new and interesting application in the area of "tomographic" medical imaging, able to become a fast, nonradiative, noninvasive, and inexpensive volumetric data acquisition technique with unique advantages for the localization of vessels and tumors in soft tissue (spleen, kidneys, liver, breast, etc.). In general, tomographic techniques (CT, MR, PET, etc.) allow for a high anatomical clarity when inspecting the interior of the human body (Figure 4.13).

In addition, they enable a 3D reconstruction and examination of regions of interest, offering obvious benefits (reviewing from any desired angle; isolation of crucial locations; visualization of internal structures; "fly-by"; accurate measurements of distances, angles, volumes, etc.).

The physical principle of ultrasound is as follows:[72] sound waves of high frequency (1–15 MHz) emanate from a row of sources that are located on the surface of a transducer which is in direct contact with the skin (Figure 4.14). The sound waves penetrate the human tissue, traveling with a speed of 1450–1580 m/s, depending upon the type of tissue. The sound waves are reflected partially if they hit an interface between two different types of tissue (e.g., muscle and bone). The reflected wavefronts are detected by sensors (microphones) located next to the sources on the transducer. The intensity of reflected energy is proportional to the sound impedance difference of the two corresponding types of tissue and depends on the difference of the sound impendances Z_1 and Z_2:

$$I_r = I_e \left(\frac{I - \dfrac{Z_2}{Z_1}}{I + \dfrac{Z_2}{Z_1}} \right) \tag{4.1}$$

An image of the interior structure can be reconstructed based upon the total traveling time, the (average) speed, and the energy intensity of the reflected waves. The resulting 3D images essentially represent hidden internal "surfaces". The principle is similar to radar, with the difference being that it uses mechanical instead of electromagnetic waves.

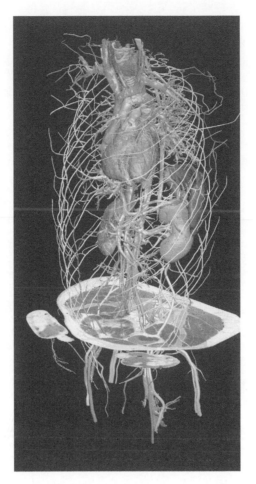

FIGURE 4.13
Gradual transition between surface and MIP visualization of a gamma camera data set of the pelvis. The heart, kidneys, liver, and spleen are visible. Three hemangiomas can be seen in the MIP mode.

4.2.2.2 Collecting 3D Ultrasound Data

In contrast to the common 2D case where a single image slice is acquired, 3D ultrasonic techniques cover a volume within the body with a series of subsequent image slices. The easiest way to collect 3D ultrasound data is to employ a Kretz Voluson 530 device. This is a commercially available device which allows direct acquisition of a whole volume area instead of a single slice. The principle of the Kretz device is based on a mechanical movement of the transducer during acquisition along a rotational or sweep path (see Figure 4.15)

The advantage of the Kretz system lies in its high decision and commercial availability. Its disadvantage is that the rather high system price makes it somehow difficult to purchase for physicians. The alternative is a free-hand scanning system which allows the upgrade of virtually any existing conventional (2D) ultrasound system to full 3D-capabilities. Such an update can be done exclusively by external components and hence does not require any manipulation of the existing hardware and software configuration. After the upgrade, the ultrasound equipment can be operated in both the 2D as well as the 3D mode almost simultaneously. Switching from the 2D to the 3D mode requires only a mouse click. As a

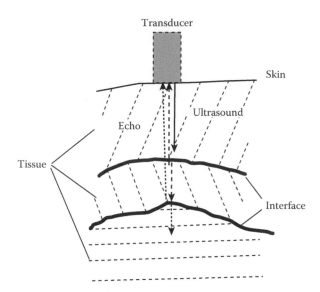

FIGURE 4.14
The principal function of ultrasound.

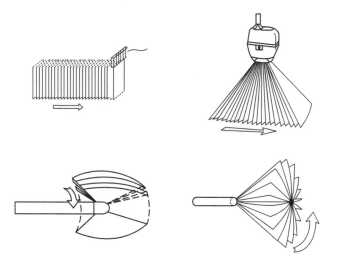

FIGURE 4.15
Different mechanical scanning methods.

result, the familiar 2D examination procedure remains unchanged, and the physician can switch on the 3D mode only when this is necessary. The system architecture is illustrated in Figure 4.16. The upgrade requires the employment of two external components:

(1) A 6-degrees-of-freedom (6DOF) tracking system for the transducer—Such a tracking system is mounted on the transducer and follows very precisely its position and orientation in 3D space. Thus, each 2D image is associated with corresponding position and orientation coordinates. The physician can now move the transducer free-hand over the region under examination. Commercially, several different types of 6DOF tracking systems exist: mechanical arms, electromagnetic trackers, and camera-based trackers (infrared or visible light).

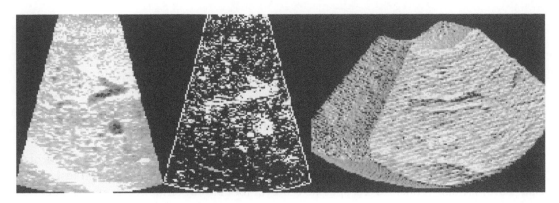

FIGURE 4.16
A grey image of the liver (left), the corresponding opacity values (middle), and a volume rendered data set (right). Note the high opacity values along the interface between data and empty space (middle) causing a solid "curtain" obscuring the volume interior (right).

(2) An image digitalization and volume rendering system—This component consists of a frame grabber, a workstation or PC with sufficient memory and processor power, a serial interface, and the usual peripheral devices (monitor, mouse, printer, etc.). The video output of the 2D ultrasound machine is connected to the frame grabber, and the 6DOF tracker is connected to the serial input. Every 2D image presented on the ultrasound screen is digitized in real time and stored together with its corresponding tracker coordinates in the memory. After finishing the scanning procedure, all acquired 2D slices are combined into a 3D volume sample of the examined area. This volume data set is then further processed.

4.2.2.3 Visualization of 3D Ultrasound

One of the major reasons for the limited acceptance of 3D ultrasound to date is the complete lack of an appropriate visualization technique able to display clear surfaces out of the acquired data. The very first approach was to use well-known techniques, such as those used for MRI and CT data, to extract surfaces. Such techniques, reported in more detail in the first part of this chapter, include binarization, iso-surfacing, contour connecting, marching cubes, and volume rendering either as a semitransparent cloud or as fuzzy gradient shading.[61] Manual contouring is too slow and impractical for real-life applications. Unfortunately, ultrasound images possess several features that cause all these techniques to fail totally. The general appearance of a volume rendered 3D ultrasound data set is that of a solid block covered with "noise snow" (Figure 4.16, right). The most important of these features,[95,97] are

(1) Significant amount of noise and speckle.

(2) Much lower dynamic range as compared to CT or MR.

(3) High variations in the intensity of neighboring voxels, even within homogeneous tissue areas.

(4) Boundaries with varying grey level caused by the variation of surface curvature and orientation to the sound source.

(5) Partially or completely shadowed surfaces from objects closer and within the direction of the sound source (e.g., a hand shadows the face).

(6) The regions representing boundaries are not sharp, but show a width of several pixels.

(7) Poor alignment between subsequent images (parallel-scan devices only).

(8) Pixels representing varying geometric resolutions, depending on the distance from the sound source (fan-scanning devices only).

The next idea in dealing with ultrasound data was to improve the quality of the data during a preprocessing step, i.e., prior to reconstruction, segmentation, and volume rendering. When filtering medical images, a trade-off between image quality and information loss must always be taken into account. Several different filters have been tested: 3D Gaussian for noise reduction, 2D speckle removal for contour smoothing, and 3D median for both noise reduction and closing of small gaps caused by differences in the average luminosity between subsequent images;[95] other filters such as mathematical topology and extended threshold-based segmentation have been tested as well. The best results have been achieved by combining Gaussian and median filters (see Figure 4.17).

However, preprocessing of large data sets (a typical 3D volume has a resolution of 256^3 voxels) requires several minutes of computing, reduces the flexibility to interactively adjust visualization parameters, and aliases the original data. For solving these problems, interactive filtering techniques based on multiresolution analysis and feature extraction have

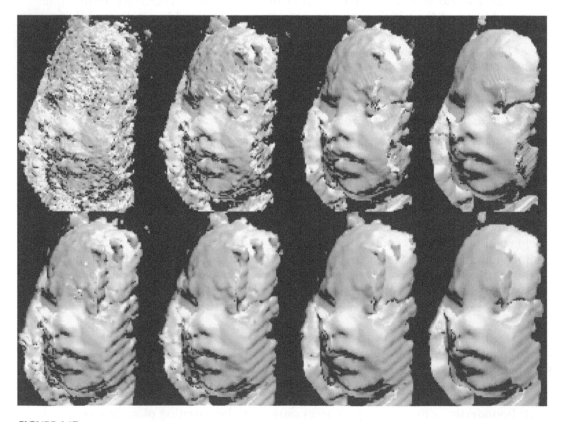

FIGURE 4.17
Volume rendering after off-line 3D median and 3D Gaussian filtering. From left to right: unfiltered and median with a width of 3^3, 5^3, and 7^3. In the lower row, the same data after additional Gaussian filtering with a width of 3^3.

been developed, allowing a user-adjustable, on-line filtering within a few seconds and providing an image quality comparable to the off-line methods[97] (see Figures 4.18 through 4.20).

In order to remove artifacts remaining in the image after filtering, semiautomatic segmentation has been applied because of the general lack of a reliable automatic technique. A segmentation can be provided by using the mouse to draw a few crude contours (see Schreyer et al.[95] for more details). The diagnostic value of surface reconstruction in prenatal diagnosis so far has been seen in the routine detection of small irregularities of the fetal surface, such as cheilo-gnatho-(palato)schisis or small (covered) vertebral defects, as well

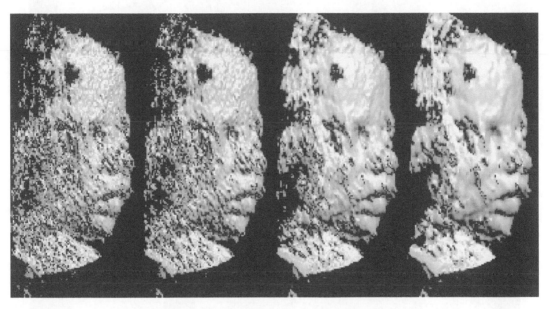

FIGURE 4.18
On-line filtering of the face of a fetus. This filtering is completed in less than 5 s.

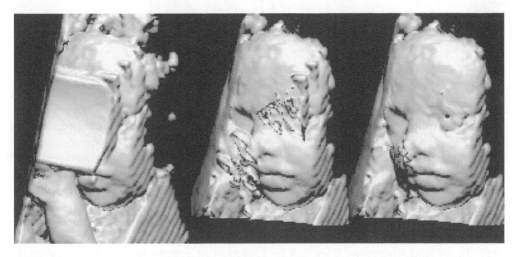

FIGURE 4.19
Fetal face before (left) and after (middle) removing the right hand and the remaining artifacts (right).

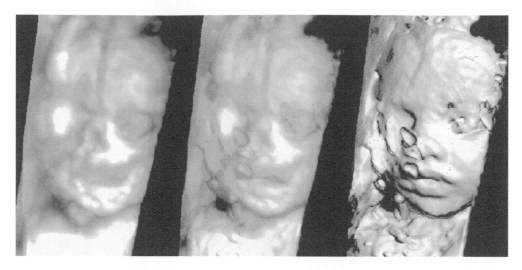

FIGURE 4.20
On-line mixing between surface and MIP models. This operation is performed in real time.

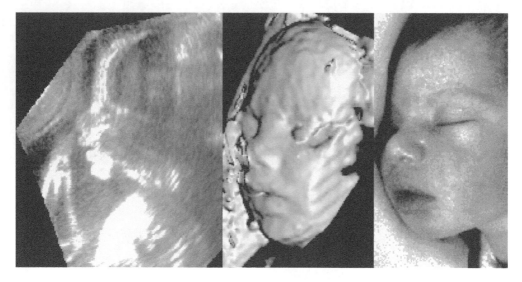

FIGURE 4.21
Comparison of a volume reconstructed from 3D ultrasound data acquired during the 25th pregnancy week (3.5 months before birth) with a photograph of the same baby taken 24 h after birth.

as in a better spatial impression of the fetus as compared to the 2D imaging. A useful side effect is a psychological one, the pregnant woman gets a plastic impression of the unborn.[5,44] Figure 4.21 compares an image reconstructed from data acquired in the 25th week of pregnancy with a photo of the baby 24 h after birth. The resolution of the data was $256 \times 256 \times 128$ (8 Mbytes); the time for volume rendering one image with a resolution of 300^2 pixels was about 1 s on a Pentium PC.

Figure 4.22 shows several other examples of fetal faces acquired at the Mannheim Clinic. It is important to note that these data sets have been acquired under routine clinical conditions, and, therefore, they can be regarded as representative. On average, 80% of the acquired volumes can be reconstructed within ca. 10 min with an image quality

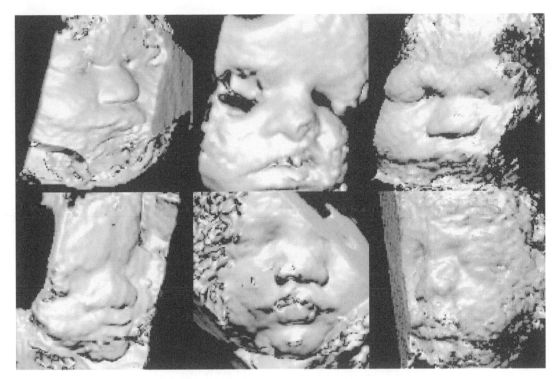

FIGURE 4.22
Six different examples of fetal faces acquired under daily clinical routine conditions.

comparable to that shown in Figure 4.22. All cases where the fetus was facing the abdominal wall could be reconstructed successfully.

Under clinical aspects, further work should be aimed toward a better distinction and automatic separation of surfaces lying close together and showing relatively small grey scale differences. The reconstruction of surfaces within the fetus, e.g., organs, is highly desirable. Surface properties of organs, but also of pathological structures (ovarian tumors, etc.), might give further information for the assessment of the dignity of tumors.

4.2.3 3D Cardiac Reconstruction from 2D Projections

Different imaging modalities are applied in order to acquire medical data. In terms of the human heart, 3D tomographic imaging techniques are not yet suitable for resolving either moving coronary arteries or the changing volume of the heart ventricles.

The golden standard for diagnosis of coronary artery disease or volumetry is X-ray angiography, recently combined with intra-vascular ultrasound (IVUS).[60] The main benefit of this technique is the high spatial and temporal resolution, as well as high image contrast.

For treatment planning of angioplasty or bypass surgery or for volumetry, sequences of X-ray images are traditionally acquired and evaluated. Despite the high quality of angiograms, an exact judgment of pathological changes (e.g., stenosis) requires a large amount of experience on the part of the cardiologist.

In order to improve the diagnostic accuracy, 3D reconstruction from 2D coronary angiograms appears desirable.[117] In general, two different approaches can be distinguished. The stereoscopic or multiscopic determination of ray intersections is a method which makes it necessary to identify correspondent features within different images. If this correspondence is impossible to establish, back-projection techniques[30] are more suitable. The choice of using either the stereoscopic or the back-projection approach mainly depends on the following criteria:

(1) **Number of images:** For the stereoscopic approach at least two images are necessary to perform the reconstruction. In order to achieve good results by using back-projection techniques, more than 20 images are necessary.

(2) **Relative orientation:** A small, relative orientation results in low accuracy for both stereoscopic and back-projection techniques. Nevertheless, the necessity of a large parallax angle is higher for back-projection techniques.

(3) **Morphology:** In order to reconstruct objects which are composed of a number of small, structured parts, stereoscopic techniques are more appropriated. On the other hand, large objects with low structure are easier to reconstruct by back-projection techniques.

(4) **Occluding objects:** Occluding objects cause problems when using stereoscopic methods. In contrast, back-projection techniques are able to separate different objects which lie on the same projection ray.

Since the choice of the right technique strongly depends on the current application, both approaches will be described briefly in the following sections.

4.2.3.1 Model-Based Restoration of Teeth

4.2.3.1.1 Introduction

An important goal in image processing of medical images, in addition to the analysis of the images, is the reconstruction and visualization of the scanned objects. The results assist the physician in making a diagnosis and, therefore, contribute to an improved treatment. By means of the reconstruction, it is possible to produce implants made to measure and adjusted to the individual patient's anatomy. Therefore, the object is scanned and afterward reconstructed. Then a high-quality implant is produced using the information of the 3D model. One field of application, where this method gained wide currency, is the restoration of teeth by range images.[13]

4.2.3.1.2 Related Work

The CEREC system, one of the most popular methods for the restoration of teeth, was introduced by Brandestini and colleagues.[13,14] It was developed at the University of Zürich in cooperation with the Brains Company (Brandestini Instruments of Switzerland). Today, the CEREC system is developed by Sirona, and in the latest version, it is possible to make crown restorations. Due to the complicated interactive construction process, the surface of the occlusal part of the inlay often has to be modeled manually after the inlay is inserted into the cavity of the prepared tooth.

The Minnesota system, developed from Rekow[87–89] at the University of Minnesota, uses affine transformations of a 3D model tooth to adapt it to the scanned tooth. Instead of covering all kinds of tooth restorations, they consider mainly the production of crowns. The system of Duret et al.,[26,27,28] developed in cooperation with the French company Hennson, works similar to the Minnesota system. They use affine transformations to produce crowns and small bridges.[69] In addition, they use the range image of an intact tooth to determine the occlusal surface of the inlay by 2D free-form deformations based on extracted feature points.

4.2.3.1.3 Occlusal Surface Restoration

Automatic occlusal surface reconstruction for all kinds of tooth restorations is an important ongoing research topic. It is undisputed that an automation of a restoration system is only possible if the typical geometry of teeth is known by the system. One realizable approach is the restoration of the occlusal surface by adapting an appropriate tooth model. Therefore, the starting point for automatic restorations is the theoretical tooth.[37] After positioning and scaling the tooth model, adjustments to the individual patient's anatomy are necessary to get a smooth join of the inlay with the undamaged parts of the real occlusal surface. A description of the method is given by Gürke.[37,38] The processing pipeline is shown in Figure 4.23.

4.2.3.1.4 Data Acquisition

For the 3D surface measurements of teeth, we use a new optical 3D laser scanner based on the principle of triangulation. Shaded areas due to steep inclines can be avoided by combining two scans from different directions.[77] The measurement time for 250,000 surface points is about 30 s. The distance between two scanned points is 27 m, and the scanner has a vertical resolution less than 1 m.

4.2.3.1.5 The Geometrically Deformable Model

For the purpose of shape modification, a 3D adaptation of a geometrically deformable model (GDM) is employed. GDMs were introduced by Miller et al.[15,71] for the segmentation and visualization of 2D- and 3D objects. Rückert[91] uses GDMs for the segmentation of 2D medical images. Gürke[39] extends the model by introducing free-function assignment in the control points in order to integrate *a priori* knowledge about the object.

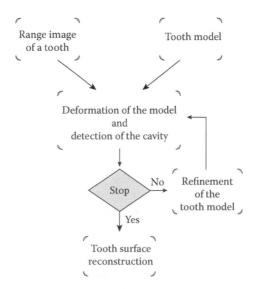

FIGURE 4.23
Occlusal surface restoration using a tooth model (heavy arrows indicate design loop).

4.2.3.1.6 *Optimization Methods*

It has been shown that the segmentation process of Miller et al.[15] is very susceptible to noise and artifacts. The reason lies in the usage of the Hillclimbing process. Rückert provided as an improvement the simulated annealing optimization. In theory, this method always finds the global minimum of a function. Due to the enormous complexity of our functions, we decided on the usage of simulated annealing as the optimization method.

4.2.3.1.7 *Cavity Detection*

In order to avoid a slip of the GDM into the cavity during the adaptation process, there has to be a mechanism to detect the control points of the GDM lying above a cavity. Later, we use this information to calculate the surface of the inlay. We decided on a criterion based on a distance measurement and an adaptive threshold. The threshold depends on the actual mean error calculated in the control points of the GDM.

The detected control points are labeled and removed from the deformation process. In the nonlabelled control points, we store the actual deformation vectors. After we pass all control points in the deformation step, we calculate the deformation vectors for the labeled control points by a weighted Shepard interpolation.

4.2.3.1.8 *Results*

The images we used in our experiments were captured with a 3D laser scanner at the dental school of Munich and registered with the Sculptor system 77 at the Fraunhofer Institute for Computer Graphics in Germany.

In Figure 4.24, you can see the range image of a prepared first upper molar (top). For better visibility of details, the rendered image is shown on the bottom.

Figure 4.25 shows the results of the deformations in the different resolutions starting with the initial model and terminating after four refinement steps.

The adaptation process terminates after a sufficient degree of refinement is reached. In our case, we finished after four refinement steps, respectively, five deformation steps. Figure 4.26 shows a 3D view of the reconstructed chewing surface.

Finally, we are able to calculate the inlay by using the information of the range image and the 3D model of the reconstructed occlusal surface. Figure 4.27 shows a volume representation of the inlay.

Figure 4.28 shows the result of a dental restoration of a first lower molar. Following the arrows, you can see, on the left side, the original prepared tooth; then different views of the reconstruction; and finally, the corresponding views of the resulting inlay. The whole process took 21 s on a SPARC 10 workstation, and we obtained an average error of 25 μm.

4.2.3.2 Reconstruction of Coronary Vessels

In this section, a method of reconstructing the 3D appearance of the coronary arteries, based on a sequence of angiograms, acquired by rotating a monoplane system around the heart, will be described. In order to determine the exact phase of the heart cycle for each image, an ECG is recorded simultaneously. In order to minimize user interaction and *a priori* knowledge introduced into the reconstruction process,[36] a new method has been developed and implemented. The technique requires a minimum of user interaction limited to the segmentation of vessels in the initial image of each angiographic sequence. The segmentation result is exploited in the entire series of angiograms to track each individual vessel. In contrast to the assumption for 3D reconstruction of objects from multiple

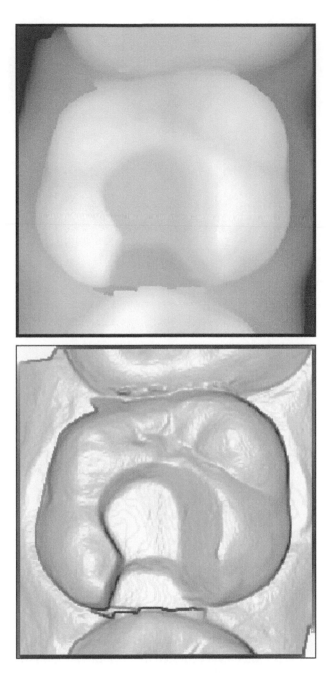

FIGURE 4.24
Range image and corresponding rendered image of a prepared tooth with a cavity.

projections, coronary arteries are not rigid. Due to the deterministic nature of the mobility of the heart with respect to the phase of the heart motion, distinct images are used, showing the heart at the same phase of the cardiac cycle.

The different processing steps used for reconstructing the 3D geometry of the vessel are shown in Figure 4.29 and are discussed later.[43] In order to separate the vessel tree to be reconstructed, the image has to be segmented. The major drawback of most of the existing

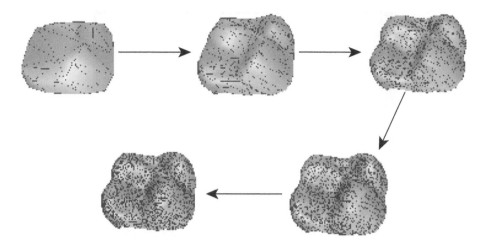

FIGURE 4.25
Result of the adaptation process in the form of triangular meshes.

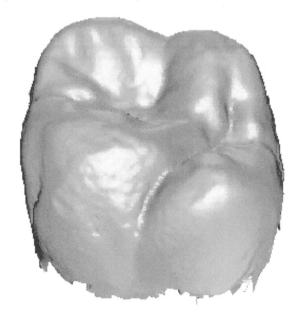

FIGURE 4.26
3D view of the reconstructed occlusal surface.

segmentation algorithms is either a very limited amount of variation in the input data amenable to processing by a fully automatic algorithm or the necessity of extensive user assistance. The approach leads to a compromise in which the user only identifies a very small number of points interactively. The segmentation process is separated into the detection of the vessel centerline and evaluation of the vessel contour. The algorithm works with a cost-minimizing A search tree,[76,116] which proves to be robust against noise and may be fully controlled by the user. Snakes track the obtained structure over the angiographic sequence.

Reconstruction is based on the extracted vessel tree structures, the known relative orientation (i.e., the angle) of the projections, and the imaging parameters of the X-ray system. The 3D reconstruction is performed from images of identical heart phases. It begins with the two projections of the same phase, defining the largest angle.

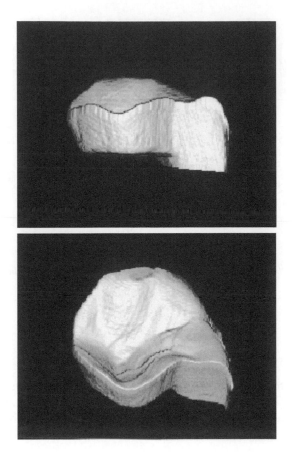

FIGURE 4.27
Volume representation of the calculated inlay.

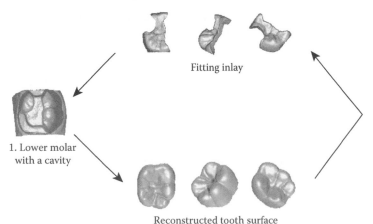

Fitting inlay

1. Lower molar
with a cavity

Reconstructed tooth surface

FIGURE 4.28
Dental restoration of a first lower molar.

The obtained result is improved afterward by introducing additional views. Applying a 3D optimization techniques the shape of a 3D Snake is adapted according to multiple 2D projections.[42] The obtained 3D structure can be either visualized by performing a volume rendering or, in order to be presented within VR systems, transferred into a polygonal representation.

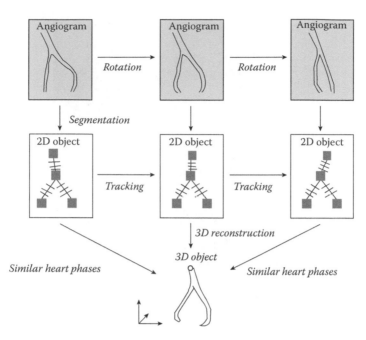

FIGURE 4.29
Processing steps used to reconstruct the 3D geometry of the coronary vessels.

Besides the 3D geometry of the coronary vessels, the trajectories of distinct points of the vessels are determined during the tracking process. As a result, these trajectories can be used to simulate the movement of the vessel caused during the heartbeat (Figure 4.30, bottom row).

4.2.3.3 Reconstruction of Ventricles

Beside the stereoscopic or multiscopic feature-based approach, the 3D structure can also be obtained using densitometric information. This technique, also known as the back-projection method, does not need any *a priori* knowledge or image segmentation. Similar to CT, the 3D information is obtained by determining the intensity of a volume element according to the density of the imaged structure. The intensity of each pixel within the angiogram correlates to the amount of X-ray energy, which is received at the image amplifier. This energy depends on the density and the absorption capabilities of the traversed material. As a result, a pixel represents the sum of the transmission coefficients of the different materials which are pierced by the X-ray. For homogeneous material and parallel mono-chromatic X-rays, the image intensity can be described by the rule of *Lambert–Beer*:[8]

$$I = I_0 e^{-\mu v d} \tag{4.2}$$

I = image intensity; I_0 = initial intensity; μ = absorption coefficient of the structure; v = density of the structure; and d = thickness of the structure.

If the X-ray travels through a material with varying densities, Equation 4.2 has to be split into parts with constant density. The total amount of transmitted intensity is the sum of these different parts.

$$I = I_0 e - \sum_i \mu_i v_i d_i \tag{4.3}$$

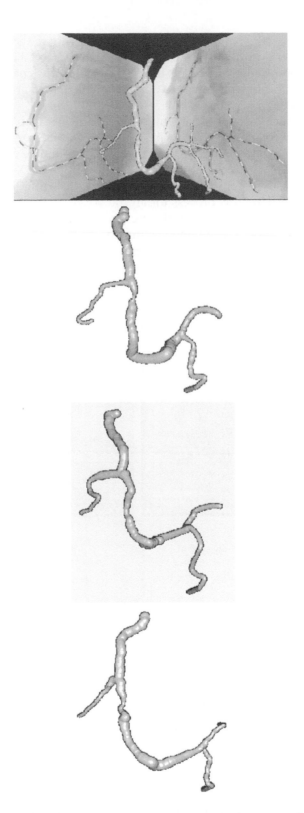

FIGURE 4.30
Reconstructed vessels rendered by InViVo (top: combined presentation of the volume rendered reconstruction result and angiograms; bottom: some frames of the 3D movement simulation).

To improve the image quality, contrast agent is injected during the acquisition process. For this purpose, a catheter is positioned in front of the ventricles (see Figure 4.31).

Applying the back-projection technique, the distribution of the coefficients can be determined. During the acquisition process, the X-ray system is rotated around the center of the heart (see Figure 4.32).

In order to reconstruct the appropriate intensities of the heart, all the images are translated into the center of rotation (see Figure 4.33); thus, according to the amount of images, a cylinder is defined by a number of sampling planes. The complete volume of the cylinder can now be determined. Therefore, all the rays starting from the X-ray source and intersecting a distinct voxel are accumulated and weighted according to the intensity of the different planes. Continuing this process for all the voxels of the cylinder, by taking the projection geometry into account by introducing a cone filter,[51] the intensity of each cylinder voxel can be determined. The obtained volume data can be visualized using a volume rendering technique and can be segmented by Snakes (Figure 4.34).

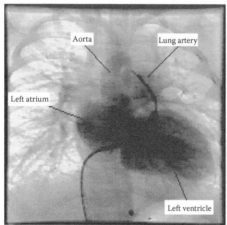
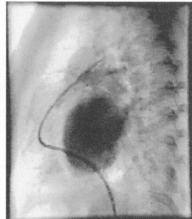

FIGURE 4.31
Angiograms acquired by the biplane X-ray system.

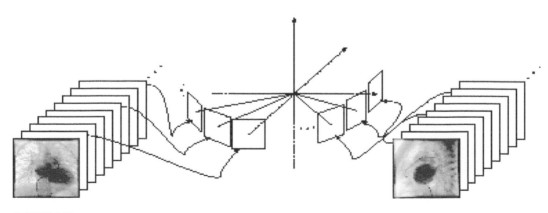

FIGURE 4.32
Acquisition of different angiograms by rotating a biplane X-ray system around the center of the heart.

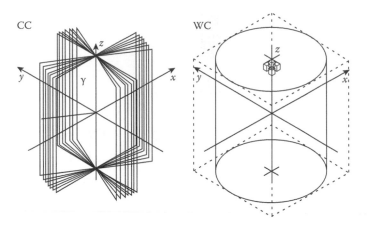

FIGURE 4.33
Translation of the angiograms in order to determine the voxel intensities.

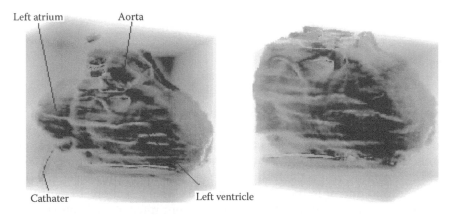

FIGURE 4.34
Volume rendering of the intensities obtained by the back-projection techique.

4.2.4 Visualization of Laser Confocal Microscopy Data Sets

Structures in the microscopic scale nerve cells, tissue and muscles, blood vessels, etc. show beautiful, complex, and still mostly unexplored patterns usually with higher complexity than those of organs. In order to understand the spatial relationship and internal structure of such microscopic probes, tomographic series of slices are required in analogy to the tomographies used for organs and other macroscopic structures.

Laser confocal microscopy is a relatively new method, allowing for a true tomographic inspection of microscopic probes. The method operates according to a simple, basic principle.[23]

A visible or ultraviolet laser emission is focused on the first confocal pinhole and then onto the specimen as a diffraction-limited light spot (see Figure 4.35). The primary incident light is then reflected from particular voxel elements or emitted from fluorescent molecules excited within it. Emissions from the object return along the primary laser light pathway and depart from it by lateral reflection from (or passage through, depending on the instrument) a dichroic mirror onto the second confocal pinhole. This aperture is confocal with

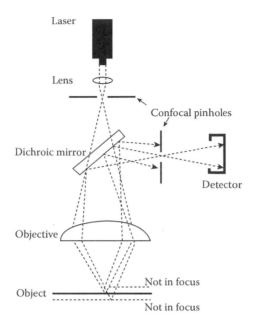

FIGURE 4.35
Principle of laser confocal microscopy.

the in-focus voxel elements in the specimen. The virtual elimination by defocusing of all distal and proximal flanking emissions at this physical point assures that the light passing onto the detector, a sensitive photodetector or camera, is specifically derived from in-focus object voxels with a resolution, e.g., in the Leica instrument, approaching 200–400 nm in the x/y and z directions, respectively. In order to image the entire object, the light spot is scanned by a second mirror in the x/y plane in successive z sections by means of a precision stage motor. Rapid scanning preserves fluorescent intensity, but must be reconciled with image quality. The storage, retrieval, and manipulation of light intensity information from the object make static and dynamic 3D imaging possible.

Although not perfect, the new method shows several significant benefits as compared to the traditional procedures. The most important of these benefits are the following: true tomographic method, significant freedom in choosing slice thickness and size, trivial registration of slices, very fast and easy in operation, capable of acquiring *in vivo* cells as well as static or dynamic structures, and nondestructive. Finally, by using different types of laser and fluorophore materials, different spatially overlapping structures can be visualized and superimposed within the same probe.

The data acquired with laser confocal microscopy (LCM) show several characteristics requiring specialized treatment in order to make the method applicable:

(1) Typical data sets have a large data size with a resolution of $512^2 \times 64$ pixels. These pixels are colored; thus, a typical RGB data set requires some 50 Mbytes of memory. Obviously, data sets of this size require efficient processing methods.

(2) These characteristics of low-contrast, low-intensity gradients and a bad signal-to-noise ratio make a straightforward segmentation between the structures of interest and the background (e.g., by using thresholding, region growing, homogeneity, color differences, etc.) impossible. All these methods listed apply more or less to

binary decision criteria whether a pixel/voxel belongs to the structure or not. Such criteria typically fail when used with signals showing the characteristics listed above.

(3) Due to unequal resolutions in the plane and the depth directions, a visualization method has to be able to perform with "blocks" or unequal size lengths instead of with cubic voxels. Resampling of the raw data to a regular cubic field will further reduce the signal quality, introduce interpolation artifacts, and generate an even larger data set, probably too large to be handled with conventional computers.

(4) Regarding the quality, artifacts have to be avoided as far as possible. Introducing artifacts in an unknown structure will often have fatal effects on their interpretation, since the human observer does not always have the experience for judging the correctness or the fidelity of the presented structures. As an example, an obvious artifact caused by bas parameter settings of the software during the visualization of human anatomy (e.g., of a head) is immediately detected by the observer, since the human anatomy is well known and such artifacts are trivially detected. This is not the case when inspecting an unknown data set.

(5) Choosing the "correct" illumination model (e.g., MIP, semitransparent, surface, etc.) has a significant impact on the clarity and information content of the visualization. Again, due to the lack of experience such a decision is typically much more difficult than in the case of anatomic tomographic data.

(6) The speed of visualization becomes the most crucial issue. The visualization parameters have to be adjusted in an interactive, trial-and-error procedure. This can take a very long time if, e.g., after an adjustment the user has to wait for several minutes to see the new result. Furthermore, inspection of new, unknown structures requires rapid changing of directions, illumination conditions, visualization models, etc. Looping and stereo images are of enormous importance for understanding unknown, complicated spatial structures.

The main requirement here is to employ a fast volumetric method which allows interactive and intuitive parameter settings during the visualization session. Detailed results of the employed volume visualization are reported in Vicker et al.[96] Figures 4.36 and 4.37 present a microscopic preparation of the tubular structure of a cat retina. The data set consist of $335 \times 306 \times 67$ voxels, each with a dimension of $0.16^2 \times 0.2$ μm. The first image presents the extracellular component of the blood vessel. The vessel diameter before the branch point is 19 μm. The second image shows the wire-like structure of the astrocyte cytoskeleton. Both data sets originate from the same probe. In all subsequent images, the difference of the visualization between slicing, MIP, surface, and semitransparent methods is shown.

Figure 4.38 shows the complicated structure of nerve cells networks. The resolution of the data set is 25 Mbytes ($512^2 \times 100$ voxels). As one can see on Figure 4.38, upper left, single slices are not able to provide full understanding of the complicated topology. The three other images of Figure 4.38 show in much better detail the internal structure of the cell network.

LCM plays a fundamental role for gathering *in vivo* data about not only static, but also dynamic structures, i.e., structures existing typically only within living cells and for a very short period of time (e.g., for a few seconds). Such structures are common in several biological applications. In the case referred to here, we present temporary structures formed by polymerized actin, a structure necessary for cell movements.

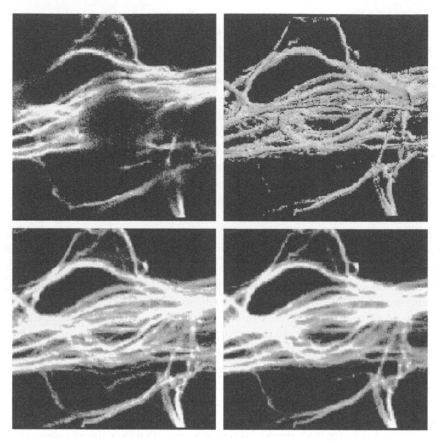

FIGURE 4.36
The wire-like structure of the astrocyte cytoskeleton of the same probe. Resolution of $335 \times 306 \times 67$ voxels with a size of $0.16^2 \times 0.2$ μm. Upper left, a single slice; upper right, surface reconstruction; lower left, MIP; lower right, transmission illumination models.

Figures 4.39 and 4.40 demonstrate the importance of LCM data visualization for detecting unknown structures. In this case, we studied actin filaments in *Dictyostelium* amoebae with time periods ranging from 10 to 100 s. The data resolution is $512 \times 484 \times 43$ voxels = 10 Mbytes. Note the structure of the surface visible in the "surface volume rendering" image. These structures are hardly visible and, therefore, are difficult to detect when regarding individual slices.

4.2.5 Virtual Simulation of Radiotherapy Treatment Planning

Radiation therapy (RT) is one of the most important techniques in cancer treatment. RT involves several steps which mainly take place in three equipment units: the CT unit, the simulator unit, and the treatment unit.

Before the patient goes to the treatment unit and the actual RT takes place, the treatment plan of the RT must be prepared (RTP). The RTP is performed on the simulator. The simulator machine can perform exactly the same movements and achieve the same position for the patient's RT as the treatment machine, but it uses conventional, diagnostic X-rays instead of high-energy treatment rays.

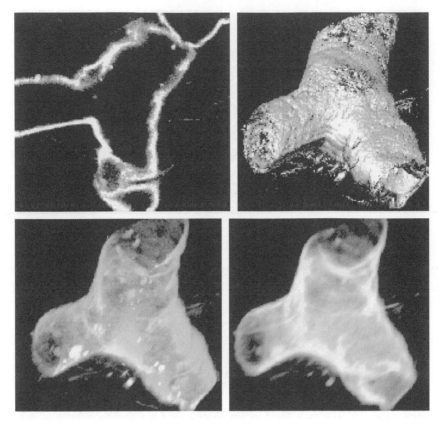

FIGURE 4.37
The extracellular component of a retina blood vessel of a cat. Resolution of 335×306×67 voxels with a size of 0.16²×0.2 μm. Upper left, a single slice; upper right, surface reconstruction; lower left, MIP; lower right, transmission illumination models.

The conventional simulation of the RT process has several limitations, mainly due to physical movement of the components of the simulator (e.g., gantry and table) and the long period of time the patient must remain in the simulator unit.

4.2.5.1 Current Clinical Routine of Radiation Therapy

Today the general procedure of an RT treatment is the following (see Figure 4.41):

(1) Move patient to the simulator. Physicians locate the region of interest (ROI, such as tumor) using the traditional X-ray fluoroscopy (diagnostic imaging).

(2) Move patient to the CT scanner. In both rooms, simulator unit and CT scanner, an identical laser system is installed defining the so-called "world coordinates". Physicians place the patient on the CT table in such a way that the CT laser coordinate system matches with the skin markers on the patient. This will assure to recover the patient position from Step 1.

(3) Physicians analyze the ROI and define the target volume(s) and the critical organ(s) on each CT slice. Then treatment parameters are selected.

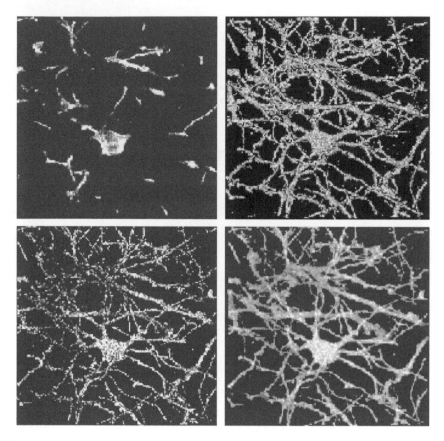

FIGURE 4.38
The complicated topology of nerve cell networks. Resolution 5122 × 100 voxels=25 Mbytes. Upper left, a single slice; upper right, surface reconstruction; lower left, MIP; lower right, transmission illumination models.

(4) Move patient to the simulator again. Physicians simulate the treatment plan to verify its effectiveness using X-rays instead of treatment rays. The treatment field is documented on X-ray films, and skin markers are placed on the patient's body. If the treatment plan is successfully verified, then the patient goes to the treatment unit; otherwise, Step 3 must be resumed.

(5) Move patient to the treatment unit. Physicians carry out the actual RT treatment according to the RTP derived during the previous steps.

4.2.5.2 Proposed "Virtual Simulation"

The general procedure of the virtual radiation treatment planning is the following (see Figure 4.42):

(1) Move patient to the CT scanner. Physicians digitize the patient using spiral CT.

(2) Transfer patient's CT data to the virtual simulator (VS). Physicians create the therapy plan on the VS using an interactive 3D planning and visualization interface. The VS system supports the aspects: VS interaction, digital reconstructed radiograph (DRR), and visualization, target volumes delineation beam shape design, and orientation determination in 3D space.

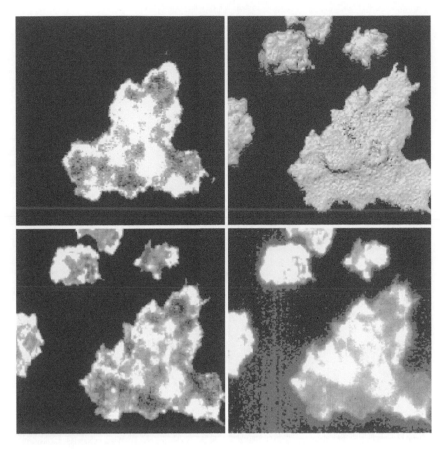

FIGURE 4.39

F-actin structures in *Dictyostelium* amoebae, resolution 512×484×43 voxels=10 Mbytes. Upper left, a single slice; upper right, surface reconstruction; lower left, MIP; lower right, transmission illumination models.

(3) Move patient to the irradiation machine, where physicians carry out the real treatment on the patient.

One significant feature of RTP systems (compared with other graphics applications) is that they support two rendering views: "beam's eye view" (BEV) and "observer's eye view" (OEV). In BEV, the patient's image is reconstructed as if the observer eye is placed at the location of the radiation source looking out along the axis of the radiation beam. The BEV window is used to detect the ROI and to define the target volume (tumor). However, BEV alone is insufficient for defining the ROI. Therefore, the OEV is also used as a second indicator to investigate the interaction among treatment beam, target volume (tumor), and its surrounding tissue (see Figure 4.43).

4.2.5.3 Digital Reconstruction Radiograph

In a VS, DRR (or X-ray) images, with which physicians are familiar, are required. In EXOMIO, two kinds of volume illumination methods are supported: DRR images and MIP. An MIP is physically impossible on a real simulator. In contrast to X-rays, the MIP makes the distinction between soft and hard tissues (e.g., bones) easier for the physicians (see Figure 4.44).

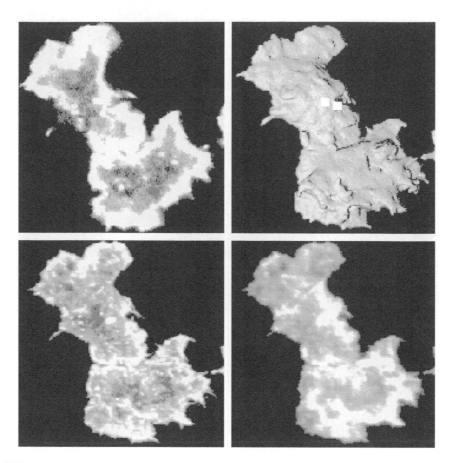

FIGURE 4.40
F-actin structures. Resolution 512×484×70 voxels=16.5 Mbytes. Upper left, a single slice; upper right, surface reconstruction; lower left, MIP; lower right, transmission illumination models.

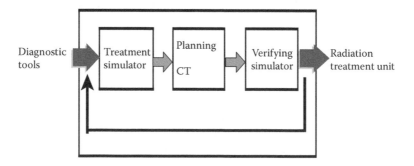

FIGURE 4.41
Clincal RT routine.

4.2.5.4 Registration

The patient's coordinates in different rooms, such as the CT room and the irradiation operation room, must be identical. The patient's position is labeled with several marks on his (or her) skin, at those points where the laser beams are projected onto their skin. These marks define the reference points of radiation iso-center. In EXOMIO, these marks

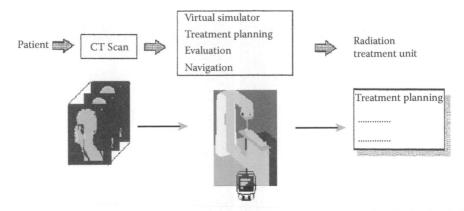

FIGURE 4.42
Virtual RT.

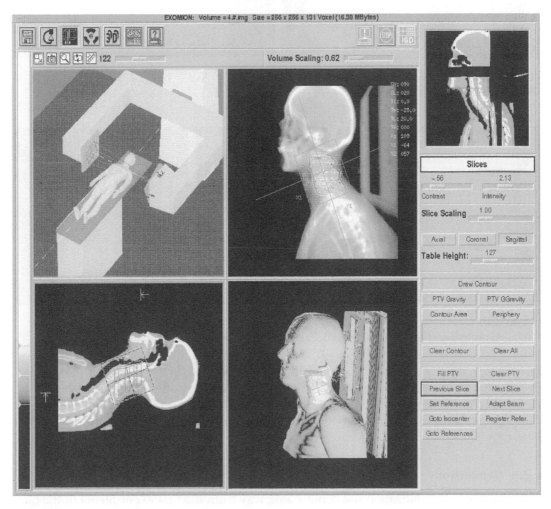

FIGURE 4.43
EXOMIO user interface.

(a)   (b)

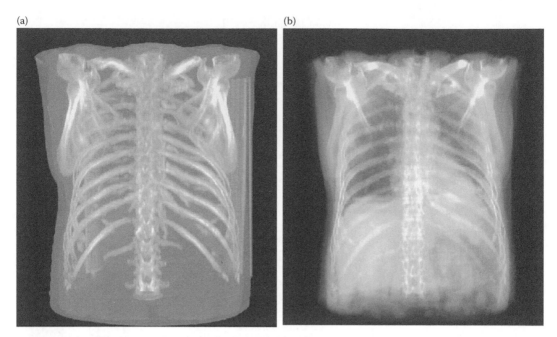

FIGURE 4.44
DRR images: (a) MIP image and (b) X-ray image.

(a)   (b)   (c)

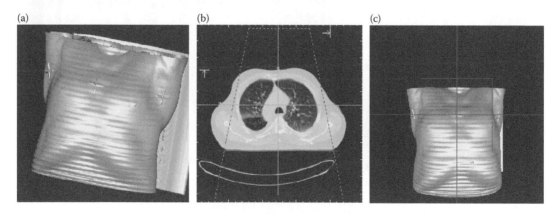

FIGURE 4.45
Marks and patient registration: (a) skin marks (OEV), (b) iso-center in slice, and (c) BEV after register.

can be seen on the axial slices and on the surface reconstructed model of the patient. In Figure 4.45, marks are displayed in both OEV and slices, allowing the patient to be identically positioned in the CT unit, the treatment unit, and the VS. The RTP parameters (gantry rotation, table position, beam size, etc.) are defined, based on these initial positions.

4.2.6 Conclusions

Medical volume visualization has come a long way from the first experiments to the current, highly detailed renderings. As the rendering algorithms are improved and the fidelity of the resulting images is investigated, 3D images are not just pretty pictures, but a powerful source of information for research, education, and patient care. In certain areas such as craniofacial surgery, volume visualization is increasingly becoming part of the standard preoperative procedures. New applications such as 3D cardiology, 3D ultrasound, and LCM are becoming more and more popular. Further rapid development of volume visualization methods is widely expected.[54]

A number of problems still hinder an even broader use of volume visualization in medicine. First, and most importantly, the segmentation problem is still unsolved. It is no coincidence that volume visualization is most accepted in all areas where clinicians are interested in bone from CT. Especially for MRI, however, automatic segmentation methods are still far from being generally applicable, while interactive procedures are too expensive. As has been shown, there is research in different directions going on; in many cases, methods have already proven valuable for specific applications.

The second major problem is the design of a user interface, which is suitable in a clinical environment. Currently, there is still a large number of rather technical parameters for controlling segmentation, registration, shading, and so on. Acceptance in the medical community will certainly depend heavily on progress in this field.

Third, current workstations are not yet able to deliver 3D images fast enough. For the future, it is certainly desirable to interact with the workstation in real time, instead of just looking at static images or precalculated movies. However, with computing power increasing further, this problem will be overcome in just a few years, even on low-cost platforms.

As has been shown, a number of applications based on volume visualization are becoming operational, such as surgical simulation systems and 3D atlases. Another intriguing idea is to combine volume visualization with virtual reality systems, which enable the clinician to walk around or fly through a virtual patient (Figure 4.46).[58,93] In *augmented reality*, images from the real and virtual world are merged to guide the surgeon during an intervention.[6] Integration of volume visualization with virtual reality and robotics toward *computer-integrated surgery* will certainly be a major topic in the coming decade.[93,108,119]

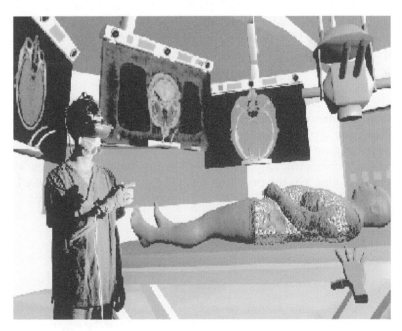

FIGURE 4.46
OP 2000 the operation theatre of the future.

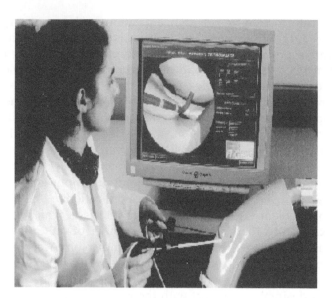

FIGURE 4.47
Virtual arthroscopy using VR and force feedback for training surgical procedures.

Acknowledgments

Georgios Sakas would like to thank Axel Hildebrand, Stefan Großkopf, Jürgen Jäger, Rainer Malkewitz, and Stefan Walter for their work in the different sections of the chapter, and Peter Plath and Mike Vicker from the University of Bremen for the laser confocal microscopy data. Data have been provided by Kretztechnik, Visible Human Project, Deutsche Klinik für Diagnostik Wiesbaden; H.T.M. van der Voort, Department of Molecular Biology, University of Amsterdam; Herbert Stüttler; Martin Hoppe, Leica Lasertechnik GmbH, State Hospital of Darmstadt; and Rolf Ziegler (Figure 4.47).

Many of the examples presented in this chapter are based on the work at the IMDM. Andreas Pommert is grateful to Karl Heinz Höhne (director); his colleagues Bernhard Pflesser, Martin Riemer, Thomas Schiemann, Rainer Schubert, and Ulf Tiede; and the students Sebastian Gehrmann, Stefan Noster, Markus Urban (Figure 4.3), and Frank Wilmer (Figure 4.5). Presented applications are in cooperation with Udo Schumacher, Department of Neuroanatomy (Figure 4.10); Christoph Koch, Department of Neuroradiology (Figure 4.11); and Andreas Fuhrmann, Department of Dental Radiology (Figure 4.12). The PET dataset (Figure 4.8) was kindly provided by Uwe Pietrzyk, Max Planck Institute of Neurological Research, Cologne. The Visible Human dataset is courtesy of the National Library of Medicine (Figure 4.10).

References

1. A. Blake and M. Isard. 1998. *Active Contours: The Application of Techniques from Graphics, Vision, Control Theory and Statistics to Visual Tracking of Shapes in Motion.* Springer-Verlag, London.
2. M. E. Alder, S. T. Deahl, and S. R. Matteson. 1995. Clinical usefulness of two-dimensional reformatted and three-dimensionally rendered computerized images: Literature review and a survey of surgeons' options. *J. Oral Maxillofac. Surg.*, 53(4), 375–386.

3. L. K. Atata, A. P. Dhawan, J. P. Broderick, M. F. Gaskil-Shipley, A. V. Levy, and N. D. Volkow. 1995. Three-dimensional anatomical model-based segmentation of MR brain images through principal axes registration. *IEEE Trans. Biomed. Eng.*, 42(11), 1069–1078.

4. R. S. Avila, L. M. Sobierajski, and Arie E. Kaufman. 1992. Towards a comprehensive volume visualization system. In *Proc. Visualization '92*. Boston, MA, 13–20.

5. K. Baba, K. Stach, and S. Sakamoto. 1989. Development of an ultra-sound system for 3d-reconstruction of the foetus. *J. Perinat. Med.*, 17, 19–24.

6. A. C. F. Colchester, J. Zhao, K. S. Holton-Tainter, C. J. Henri, N. Maitland, P. T. E. Roberts, C. G. Harris, and R. J. Evans. 1996. Development and preliminary evaluation of VISLAN, a surgical planning and guidance system using intra-operative video imaging. *Med. Image Anal.*, 1(1), 73–90.

7. W. Barrett and E. Bess. 1994. Interpolation by directed distance morphing. In R. A. Robb, editor. *Visualization in Biomedical Computing 1994*, Proc. SPIE 2359, Rochester, MN, 110–121.

8. J. Beier. 1993. *Automatische Quantifizierung von Koronarstenosen aus angiographischen Röntgenbildern*, Fortschr.-Ber. VDI Reihe 17 Nr.95, VDI-Verlag, Düsseldorf.

9. J. F. Blinn. 1982. Light reflection functions for simulation of clouds and dusty surfaces. *Comput. Graphics*, 16(3), 21–29.

10. J. D. Boissinat. 1985. Surface reconstruction from planar cross-sections. In *Proceedings of IEEE Conference on Computer Vision and Pattern Recognition*, 393–397.

11. M. Bomans. 1994. Segmentationsverfahren zur 3D-Visualisierung von Kernspintomogrammen des Kopfes: Evaluierung der Standardverfahren und Entwurf und Realisierung eines klinisch einsetzbaren Systems. Dissertation, Fachbereich Informatik, Universität Hamburg.

12. M. Bomans, K. H. Höhne, U. Tiede, and M. Riemer. 1990. 3D-segmentation of MR-images of the head for 3D-display. *IEEE Trans. Med. Imaging*, MI-9(2), 177–183.

13. M. Brandestini and W. Mörmann. 1989. Die CEREC Computer Rekonstruktion. Quint – essenz – Verlag.

14. M. Brandestini, W. Mörmann, A. Ferru, F. Lutz, and L. Kreijci. 1985. Computer machined ceramic inlays. In vitro marginal adaptation. *J. Dent. Res.*, 64, 208.

15. D. E. Breen, W. E. Lorensen, J. V. Miller, R. M. O'Bara, and M. J. Wozny. 1991. Geometrically deformed models: Method for extracting closed geometric models from volume data. *Comput. Graphics*, 25(4), 217–226.

16. M. E. Brummer, R. M. Mersereau, R. L. Eisner, and R. R. J. Lewine. 1993. Automatic detection of brain contours in MRI datasets. *IEEE Trans. Med. Imaging*, 12, 153–166.

17. J. Canny. 1985. A computational approach to edge detection. *IEEE Trans. Pattern Anal. Mach. Intell.*, PAMI-8(6), 679–698.

18. E. Keeve, S. Girod, and B. Girod. 1996. Craniofacial surgery simulation. In K. H. Höhne, editor. *Visualization in Biomedical Computing, Proc. VBC'96*. Springer-Verlag, Berlin, 541–546.

19. H. E. Cline, W. E. Lorensen, R. Kikinis, and F. Jolesz. 1990. Three-dimensional segmentation of MR images of the head using probability and connectivity. *J. Comput. Assist. Tomogr.*, 14(6), 1037–1045.

20. K. H. Höhne, editor. 1998. *VOXEL-MAN Junior: Interactive 3D Anatomy and Radiology in Virtual Reality Scenes, Part 1: Brain and Skull*. Springer-Verlag Electronic Media, Heidelberg (CD-ROM).

21. H. E. Cline, W. E. Lorensen, S. P. Souza, Ferenc A. Jolesz, R. Kikinis, G. Gerig, and T. E. Kennedy. 1991. 3D surface rendered MR images of the brain and its vasculature. *J. Comput. Assist. Tomogr.*, 15(2), 344–351.

22. D. J. David, D. C. Hemmy, and R. D. Cooter. 1990. *Craniofacial Deformities: Atlas of Three-Dimensional Reconstruction from Computed Tomography*. Springer-Verlag, New York, NY.

23. D. M. Shotton. 1989. Confocal scanning optical microscopy and its applications for biological specimens. 94, 175–206.

24. R. A. Drebin, L. Carpenter, and P. Hanrahan. 1988. Volume rendering. *Comput. Graphics*, 22(4), 65–74.

25. R. A. Drebin, D. Magid, D. D. Robertson, and E. K. Fishman. 1989. Fidelity of three-dimensional CT imaging for detecting fracture gaps. *J. Comput. Assist. Tomogr.*, 13(3), 487–489.

26. F. Duret. 1988. Method of making a prosthesis, especially a dental prosthesis. Technical report, United States Patent 4,742,464.

27. F. Duret, J. I. Blouin, and L. Nahami. 1985. Principes de fonctionnement et application techniques de l'empreine optique dans l'exercise des cabinet. *Cah Prothese*, 13, 73.

28. F. Duret, J. L. Blouin, and B. Duret. CAD/CAM in dentistry. 1988. *J. Am. Dent. Assoc.*, 117(11), 715–720.

29. D. L. Collins, A. P. Zijdenbos, W. F. C. Baare, and A. C. Evans. 1999. ANIMAL+INSECT: Improved cortical structure segmentation. In A. Kuba, M. Samal, and A. Todd-Pokropek, editors. *Information Processing in Medical Imaging, Proc. IPMI '99*. Vol. 1613, Lecture Notes in Computer Science. Springer-Verlag, Berlin, 210–223.

30. L. A. Feldkamp, L. C. Davis, and J. W. Kress. 1989. Practical cone-beam algorithm. *J. Opt. Soc. Am. A*, 1(6), 612–619.

31. R. M. Koch, M. H. Gross, F. R. Carls, D. F. von Büren, G. Fankhauser, and Y. I. H. Parish. 1996. Simulating facial surgery using finite element models. In *Computer Graphics Proceedings*. Annual Conference Series, ACM SIGGRAPH, New Orleans, 421–428.

32. J. D. Foley, A. van Dam, S. K. Feiner, and J. F. Hughes. 1996. *Computer Graphics: Principles and Practice. Second Edition in C*. Addison-Wesley Longman, Amsterdam, 1996.

33. B. Geiger and R. Kikinis. 1995. Simulation of endoscopy. In N. Ayache, editor. *Computer Vision, Virtual Reality and Robotics in Medicine, Proc. CVRMed '95*. Vol. 905, Lecture Notes in Computer Science. Springer-Verlag, Berlin, pp. 277–281.

34. G. Gerig, O. Kübler, R. Kikinis, and F. A. Jolesz. 1992. Nonlinear anisotropic filtering of MRI data. *IEEE Trans. Med. Imaging*, 11(2), 221–232.

35. G. Gerig, J. Martin, R. Kikinis, O. Kübler, M. Shenton, and F. A. Jolesz. 1991. Automating segmentation of dual-echo MR head data. In A. C. F. Colchester and D. J. Hawkes, editors. *Information Processing in Medical Imaging, Proc. IPMI '91*. Vol. 511, Lecture Notes in Computer Science. Springer-Verlag, Berlin, 175–187.

36. S. Großkopf and A. Hildebrand. Three-dimensional reconstruction of coronary arteries from X-ray projections. In P. Lanzer and M. Lipton, editors. *Vascular Diagnostics: Principles and Technology*. Springer-Verlag, Heidelberg, in press.

37. S. Gürke. 1994. Generation of tooth models for ceramic dental restorations. In The 4th International Conference on Computer Integrated Manufacturing, Singapore.

38. S. Gürke. 1994. Modellbasierte Rekonstruktion von Zähnen aus intraoralen Tiefenbildern. In *Digitale Bildverarbeitung in der Medizin*, Freiburger Workshop Digitale Bildverarbeitung in der Medizin. Freiburg, Germany, 231–237.

39. S. Gürke. 1998. Geometrically deformable models for model-based reconstruction of objects from range images. In *Computer Assisted Radiology and Surgery 98*. Tokyo, 824–829.

40. S. Haring, M. A. Viergever, and J. N. Kok. 1993. A multiscale approach to image segmentation using Kohonen networks. In H. H. Barrett and A. F. Gmitro, editors. *Information Processing in Medical Imaging, Proc. IPMI '93*. Vol. 687, Lecture Notes in Computer Science. Springer-Verlag, Berlin, 212–224.

41. D. C. Hemmy and P. L. Tessier. 1985. CT of dry skulls with craniofacial deformities: Accuracy of three-dimensional reconstruction. *Radiology*, 157(1), 113–116.

42. A. Hildebrand. 1996. Bestimmung Computer-Graphischer Beschreibungsattribute für reale 3D-Objekte mittels Analyse von 2D-Rasterbildern. Ph.D. thesis, TH Darmstadt, Darmstadt.

43. A. Hildebrand and S. Großkopf. 1995. 3D reconstruction of coronary arteries from X-ray projections. In *Proceedings of the Computer Assisted Radiology CAR'95 Conference*. Springer-Verlag, Berlin.

44. W. Hiltman. 1994. Die 3d-strukturrekonstruktion aus ultraschall-bildern.

45. K. H. Höhne and R. Bernstein. 1986. Shading 3D-images from CT using grey level gradients. *IEEE Trans. Med. Imaging*, MI-5(1), 45–47.

46. K. H. Höhne, M. Bomans, B. Pflesser, A. Pommert, M. Riemer, T. Schiemann, and U. Tiede. 1992. Anatomic realism comes to diagnostic imaging. *Diagn. Imaging*, 1, 115–121.

47. K. H. Höhne, M. Bomans, A. Pommert, M. Riemer, C. Schiers, U. Tiede, and G. Wiebecke. 1990. 3D-visualisation of tomographic volume data using the generalized voxel-model. *Visual Comput.*, 6(1), 28–36.

48. K. H. Höhne and W. A. Hanson. 1992. Interactive 3D-segmentation of MRI and CT volumes using morphological operations. *J. Comput. Assist. Tomogr.*, 16(2), 285–294.

49. K. H. Höhne, B. Pflesser, A. Pommert, M. Riemer, T. Schiemann, R. Schubert, and U. Tiede. 1995. A new representation of knowledge concerning human anatomy and function. *Nature Med.*, 1(6), 506–511.

50. A. Höss, J. Debus, R. Bendl, R. Engenhart-Cabillic, and W. Schlegel. 1995. Computerverfahren in der dreidimensionalen Strahlentherapieplanung. *Radiologe*, 35(9), 583–586.

51. J. Jäger. 1996. Volumetric reconstruction of heart ventricles from X-ray projections (in German). Dissertation in Technical University Darmstadt.

52. M. Kass, A. Witkin, and D. Terzopoulos. 1984. Snakes: active contour models. *IEEE First Int. Conf. Comput. Vision*, 259–268.

53. A. Kaufman, editor. 1991. *Volume Visualisation*. IEEE Computer Society Press, Los Alamitos, CA.

54. A. Kaufman, K. H. Höhne, W. Krüger, L. J. Rosenblum, and P. Schröder. 1994. Research issues in volume visualisation. *IEEE Comput. Graphics Appl.*, 14(2), 63–67.

55. M. L. Kessler and D. L. McShan. 1994. An application for design and simulation of conformal radiation therapy. In R. A. Robb, editor. *Visualisation in Biomedical Computing 1994*. Proc. SPIE 2359, Rochester, MN, 474–483.

56. R. Kikinis, M. E. Shenton, D. V. Iosifescu, et al. 1996. A digital brain atlas for surgical planning, model driven segmentation, and teaching. *IEEE Trans. Visualisation Comput. Graphics*, 2(3), 232–241.

57. T. Kohonen. 1988. *Self-Organisation and Associative Memory*. 2nd edition. Springer-Verlag, Berlin.

58. W. Krueger and B. Froehlich. 1994. The responsive workbench. *IEEE Comput. Graphics Appl.*, 14(3), 12–15.

59. D. Laur and P. Hanrahan. 1991 Hierarchical splatting: A progressive refinement algorithm for volume rendering. *Comput. Graphics*, 25(4), 285–288.

60. J. Leugyel, D. P. Greenberg, and R. Ç. Poop. 1995. Time-dependent three-dimensional intravascular ultrasound. In *Computer Graphics Proceedings*. SIGGRAPH, Los Angeles, CA, 457–464.

61. U. Tiede, T. Schiemann, and K. H. Höhne. 1988. High quality rendering of attributed volume data. In D. Ebert et al., editors. *Proc. IEEE Visualization '98*. IEEE Computer Society Press, Los Alamitos, CA, 255–262.

62. M. Levoy. 1988. Display of surfaces from volume data. *IEEE Comput. Graphics Appl.*, 8(3), 29–37.

63. M. Levoy. 1990 A hybrid ray tracer for rendering polygon and volume data. *IEEE Comput. Graphics Appl.*, 10(2), 33–40.

64. L.-J. Lo, J. L. Marsh, M. W. Vannier, and V. V. Patel. 1994. Craniofacial computer-assisted surgical planning and simulation. *Clin. Plast. Surg.*, 21(4), 501–516.

65. W. E. Lorensen and H. E. Cline. 1984. Marching cubes: A high resolution 3D surface construction algorithm. *Comput. Graphics*, 21(4), 163–169.

66. M. Magnusson, R. Lenz, and P.-E. Danielsson. 1991 Evaluation of methods for shaded surface display of CT volumes. *Comput. Med. Imaging Graphics*, 15(4), 247–256.

67. K. H. Höhne, editor. 2000. *VOXEL-MAN Junior: Interactive 3D Anatomy and Radiology in Virtual Reality Scenes, Part 2: Inner Organs*. Springer-Verlag Electronic Media, Heidelberg (CD-ROM).

68. S. R. Marschner and R. J. Lobb. 1994. An evaluation of reconstruction filters for volume rendering. In R. D. Bergeron and A. E. Kaufman, editors. *Proc. Visualisation '94*. IEEE Computer Society Press, Los Alamitos, CA, 100–107.

69. S. Meller, M. Wolf, D. Paulus, M. Pelka, P. Weierich, and H. Niemann. 1994. Automatic tooth restoration via image warping. In Proceedings of the Computer Assisted Radiology '97 Conference. Berlin.

70. W. Menhardt. 1992. Iconic fuzzy sets for MR image segmentation. In A. E. Todd-Pokropek and M. A. Viergever, editors. *Medical Images: Formation, Handling and Evaluation*. Vol. 98, NATO ASI Series F. Springer-Verlag, Berlin, 579–591.

71. J. V. Miller. 1990 On GDMs: Geometrically deformed models for the extraction of closed shapes from volume data. Master's thesis, Rensselaer Polytechnic Institute, Troy, New York, NY.

72. R. Millner. 1984. Ultraschalltechnik, grundlagen und anwendungen. Physik Verlag, Weinheim, ISBN 3-87664-106-3.

73. B. S. Morse, S. M. Pizer, and A. Liu. 1994. Multiscale medial analysis of medical images. *Image Vision Comput.*, 12(6), 327–338.

74. S. Muraki. 1993. Volume data and wavelet transforms. *IEEE Comput. Graphics Appl.*, 13(4), 50–56.

75. S. Muraki. 1995. Multiscale volume representation by a DOG wavelet. *IEEE Trans. Visualisation Comput. Graphics*, 1(2), 109–116.

76. P. J. Neugebauer. 1995. Interactive segmentation of dentistry range images in CIM systems for the construction of ceramic in-lays using edge tracing. In *Proceedings of the Computer Assisted Radiology CAR'95 Conference*. Springer-Verlag, Berlin.

77. P. J. Neugebauer. 1994. Geometrical cloning of 3d objects via simultaneous registration of multiple range images. In *International Conference on Shape Modeling and Applications 1994*. IEEE Computer Society Press, Los Alamitos, CA.

78. D. Ney and E. K. Fishman. 1991. Editing tools for 3D medical imaging. *IEEE Comput. Graphics Appl.*, 11(6), 63–70.

79. D. Ney, E. K. Fishman, D. Magid, D. D. Robinson, and A. Kawashima. 1991. Three-dimensional volumetric display of CT data: Effect of scan parameters upon image quality. *J. Comput. Assist. Tomogr.*, 15(5), 875–885.

80. P. Ning and J. Bloomenthal. 1993. An evaluation of implicit surface tilers. *IEEE Comput. Graphics Appl.*, 13(6), 33–41.

81. B. Pflesser, U. Tiede, and K. H. Höhne. 1998. Specification, modelling and visualization of arbitrarily shaped cut surfaces in the volume model. In W. M. Wells et al., editors. *Medical Image Computing and Computer-Assisted Intervention, Proc. MICCAI '98*. Vol. 1496, Lecture Notes in Computer Science. Springer-Verlag, Berlin, 853–860.

82. A. Pommert, M. Bomans, and K. H. Höhne. 1992. Volume visualisation in magnetic resonance angiography. *IEEE Comput. Graphics Appl.*, 12(5), 12–13.

83. A. Pommert, W.-J. Höltje, N. Holzknecht, U. Tiede, and K. H. Höhne. 1991. Accuracy of images and measurements in 3D bone imaging. In H. U. Lemke, M. L. Rhodes, C. C. Jaffe, and R. Felix, editors. *Computer Assisted Radiology, Proc CAR '91*. Springer-Verlag, Berlin, 209–215.

84. A. Pommert, R. Schubert, M. Riemer, T. Schiemann, U. Tiede, and K. H. Höhne. 1994. Symbolic modeling of human anatomy for visualisation and simulation. In R. A. Robb, editor. *Visualisation in Biomedical Computing 1994*. Proc. SPIE 2359, Rochester, MN, 412–423.

85. A. Pommert and K. H. Höhne. Towards an image quality index in medical volume visualization. In S. K. Mun, editor. *SPIE Medical Imaging 2000: Image Display and Visualization*. San Diego, CA, accepted for publication.

86. S. P. Raya and J. K. Udupa. 1990. Low-level segmentation of 3-D magnetic resonance brain images: A rule-based system. *IEEE Trans. Med. Imaging*, MI-9(3), 327–337.

87. D. E. Rekow. 1989. The Minnesota CAD/CAM System Denti-CAD. Technical report, University of Minnesota.

88. D. E. Rekow. 1991. CAD/CAM in dentistry: Critical analysis of systems. In *Computers in Clinical Dentistry*. Quintessence Publishing Co., 172–185.

89. D. E Rekow. 1993. Method and apparatus for modeling a dental prosthesis. Technical report, United States Patent 5,273,429.

90. B. Jahne. 1994. *Digital Image Processing: Concepts, Algorithms, and Scientific Applications*. Springer-Verlag, Berlin.

91. D. Rückert. 1993. Bildsegmentierung durch stochastisch optimierte Relaxation eines 'geometric deformable model.' Master's thesis, TU Berlin.

92. H. Rusinek, M. E. Noz, G. Q. Maguire, A. Kalvin, B. Haddad, D. Dean, and C. Cutting. 1991. Quantitative and qualitative comparison of volumetric and surface rendering techniques. *IEEE Trans. Nucl. Sci.*, 38(2), 659–662.

93. A. Hildebrand, R. Malkewitz, W. Mueller, R. Ziegler, G. Graschew, and S. Grosskopf. 1996. *Computer Aided Surgery – Vision and Feasibility of an Advanced Operation Theatre*. Vol. 20. Pergamon Press, Oxford, 825–835.

94. P. Bono and G. Sakas. 1996. *Special Issue on Medical Visualisation*. Vol. 20. Pergamon Press, Oxford, 759–838.

95. L. Schreyer, M. Grimm, and G. Sakas. 1994. *Case Study: Visualisation of 3D-Ultrasonic Data*. IEEE Computer Society Press, Los Alamitos, CA, 369–373.

96. M. Vicker, P. Plath, and G. Sakas. 1996. *Case Study: Visualisation of Laser Confocal Microscopy Data*. IEEE Computer Society Press, Los Alamitos, CA, 375–380.

97. S. Walter and G. Sakas. 1995. *Extracting Surfaces from Fuzzy 3D-Ultrasonic Data*. Addison-Wesley, Reading, MA, 465–474.

98. T. Schiemann and K. H. Höhne. 1994. Definition of volume transformations for volume interaction. In J. Duncan and G. Gindi, editors. *Information Processing in Medical Imaging, Proc. IPMI '94*. Vol. 1230, Lecture Notes in Computer Science 1230. Springer-Verlag, Berlin, 245–258.

99. T. Frenzel, D. Albers, K. H. Höhne, and R. Schmidt. 1994. Problems in medical imaging in radiation therapy. In H. U. Lemke et al., editors. *Computer Assisted Radiology and Surgery, Proc. CAR '94*. Excerpta Medica ICS 1134. Elsevier, Amsterdam, 381–387.

100. T. Schiemann, K. H. Höhne, C. Koch, A. Pommert, M. Riemer, R. Schubert, and U. Tiede. 1994. Interpretation of tomographic images using automatic atlas lookup. In R. A. Robb, editor. *Visualisation in Biomedical Computing 1994*. Proc. SPIE 2359, Rochester, MN, 457–465.

101. T. Schiemann, U. Tiede, and K. H. Höhne. 1994. Segmentation of the visible human for high quality volume based visualisation. *Med. Image Anal.*, 1(4), 263–271.

102. R. Schmidt, T. Schiemann, W. Schlegel, K. H. Höhne, and K.-H. Hübener. 1994. Consideration of time-dose-patterns in 3D treatment planning: An approach towards 4D treatment planning. *Strahlenther. Onkol.*, 170(5), 292–301.

103. W. J. Schroeder, J. A. Zarge, and W. E. Lorensen. 1992. Decimation of triangle meshes. *Comput. Graphics*, 26(2), 65–70.

104. R. Schubert, B. Pflesser, A. Pommert, K. Priesmeyer, M. Riemer, T. Schiemann, U. Tiede, P. Steiner, and K. H. Höhne. 1999. Interactive volume visualization using ``intelligent movies''. In J. D. Westwood, H. M. Hoffman, R. A. Robb, and D. Stredney, editors. *Medicine Meets Virtual Reality. The Convergence of Physical and Informational Technologies, Options for a New Era in Healthcare (Proc. MMVR '99)*. Vol. 62, Studies in Health Technology and Informatics. IOS Press, Amsterdam, 321–327.

105. R. Schubert, W.-J. Höltje, U. Tiede, and K. H. Höhne. 1991. 3D-Darstellungen für die Kiefer- und Gesichtschirurgie. *Radiologe*, 31, 467–473.

106. V. Spitzer, M. J. Ackerman, A. L. Scherzinger, and D. Whitlock. 1996. The visible human male: a technical report. *J. Am. Med. Inf. Assn.*, 3(2), 118–130.

107. C. Studholme, D. L. G. Hill, and D. J. Hawkes. 1996. Automated 3-D registration of MR and CT images of the head. *Med. Image Anal.*, 1(2), 163–175.

108. R. H. Taylor, S. Lavallée, G. C. Burdea, and R. Mösges. 1995. *Computer Integrated Surgery: Technology and Clinical Applications*. MIT Press, Cambridge, MA.

109. U. Tiede, K. H. Höhne, M. Bomans, A. Pommert, M. Riemer, and G. Wiebecke. 1990. Investigation of medical 3D-rendering algorithms. *IEEE Comput. Graphics Appl.*, 10(2), 41–53.

110. U. Tiede, T. Schiemann, and K. H. Höhne. 1996. Visualizing the visible human. *IEEE Comput. Graphics Appl.*, 16(1), 7–9.

111. A. W. Toga and J. C. Mazziotta. 1996. *Brain Mapping*. Academic Press, San Diego, CA.

112. T. Totsuka and M. Levoy. 1993. Frequency domain volume rendering. *Comput. Graphics*, 271–278.

113. P. A. van den Elsen, E.-J. D. Pol, and M. A. Viergever. 1993. Medical image matching: A review with classification. *IEEE Eng. Med. Biol. Mag.*, 12(1), 26–39.

114. M. W. Vannier, C. F. Hildebolt, J. L. Marsh, T. K. Pilgram, W. H. McAlister, G. D. Shackelford, C. J. Offutt, and R. H. Knapp. 1989. Craniosynostosis: Diagnostic value of three-dimensional CT reconstruction. *Radiology*, 173, 669–673.

115. E. Vaske. 1991. Segmentation von Kernspintomogrammen mit der topologischen Karte zur 3D-Visualisierung. IMDM Institutsbericht 91/1, Institut für Mathematics und Datenverarbeitung in der Medizin, Universität Hamburg.

116. G. Vosselman. 1992. *Relational Matching*. Springer-Verlag, New York, NY.

117. A. Wahle, E. Wellnhofer, I. Mugaragu, A. Trebeljahr, H. Oswald, and E. Fleck. 1995. Application of accurate 3D reconstruction from biplane angiograms in morphometric analyses and in assessment of diffuse coronary artery disease. In *CAR'95: Computer Assisted Radiology*. Springer Verlag, Berlin.

118. Å. Wallin. 1991. Constructing isosurfaces from CT data. *IEEE Comput. Graphics Appl.*, 11(6), 28–33.

119. J. D. Westwood, H. M. Hoffman, R. A. Robb, and D. Stredney, editors. *Medicine Meets Virtual Reality. The Convergence of Physical and Informational Technologies, Options for a New Era in Healthcare (Proc. MMVR '99)*. Vol. 62, Studies in Health Technology and Informatics. IOS Press, Amsterdam.

120. W. M. Wells III, P. Viola, H. Atsumi, S. Nakajima, and R. Kikinis. 1996. Multi-modal volume registration by maximization of mutual information. *Med. Image Anal.*, 1(1), 35–51.

121. L. Westover. 1990. Footprint evaluation for volume rendering. *Comput. Graphics*, 24(4), 367–376.

122. J. Wilhelms and A. van Gelder. 1990. Topological considerations in isosurface generation. *Comput. Graphics*, 24(5), 79–86.

123. I. J. Trotts, B. Hamann, and K. I. Joy. 1999. Simplification of tetrahedral meshes with error bounds. *IEEE Trans. Visualization Comput. Graphics*, 5(3), 224–237.

124. P. H. Winston. 1992. *Artificial Intelligence*. 3rd edition. Addison-Wesley, Reading, MA.

125. R. Yagel, D. Cohen, and A. Kaufman. 1992. Discrete ray tracing. *IEEE Comput. Graphics Appl.*, 12(5), 19–28.

126. T. Yasuda, Y. Hashimoto, S. Yokoi, and J.-I. Toriwaki. 1990. Computer system for craniofacial surgical planning based on CT images. *IEEE Trans. Med. Imaging*, MI-9(3), 270–280.

127. F. W. Zonneveld and K. Fukuta. 1994. A decade of clinical three-dimensional imaging: A review. Part 2: Clinical applications. *Invest. Radiol.*, 29, 574–589.

128. F. W. Zonneveld, S. Lobregt, J. C. H. van der Meulen, and J. M. Vaandrager. 1989. Three-dimensional imaging in craniofacial surgery. *World J. Surg.*, 13, 328–342.

129. S. M. Pizer, D. Eberly, D. S. Fritsch, and B. S. Morse. 1998. Zoom-invariant vision of figural shape: The mathematics of cores. *Comput. Vision Image Understanding*, 69(1), 55–71.

130. J. West, J. M. Fitzpatrick, M. Y. Wang, B. M. Dawant, C. R. Maurer, R. M. Kessler, R. J. Maciunas, C. Barillot, D. Lemoine, A. Collignon, F. Maes, P. Suetens, D. Vandermeulen, P. A. van den Elsen, S. Napel, T. Sumanaweera, B. Harkness, P. F. Hemler, D. L. G. Hill, D. J. Hawkes, C. Studholme, J. B. A. Maintz, M. A. Viergever, G. Malandain, X. Pennec, M. E. Noz, G. Q. Maguire, M. Pollack, C. A. Pelizzari, R. A. Robb, D. Hanson, and R. P. Woods 1994. Comparison and evaluation of retrospective intermodality brain image registration techniques. *J. Comput. Assist. Tomogr.*, 21(4), 554–566.

Appendix: Principles of Image Processing: Pixel Brightness Transformations, Image Filtering, and Image Restoration

A4.1 Introduction

What you see on an image is not the best of what you can get. In most cases, image quality can be further improved. An image "hides" information. To discover all this information and to make it visible, we need to process the image. Image processing is also an important and necessary process for medical imaging. Here, the term "medical imaging" is not limited only to the diagnostic imaging. There are several other applications, such as surgery, oncology, and simulation of medical systems, where image processing techniques are essential.

Image processing or (according to other authors[6,20]) image preprocessing can be separated into pixel brightness transformations (or image GSMT), image enhancement (or image filtering),[11,15] and image restoration.[12,18] In this section, we focus on the most commonly used techniques for medical image GSMT and enhancement. For the rest of the techniques, a number of references are given for further reading.

The main goal of image filtering is to enhance image quality and remove image artifacts such as noise or blurring (Figure A4.1). At the moment, several algorithms can perform these tasks. For smoothing operations filters like Gaussian, median or local averaging masks can be used. Most of the previous operations can be performed either on the 2D image basis or, in the case of volumetric data, in the 3D volume space.[20,21] These filters are applied on the image using multidimensional convolution, and they do not demand any further image transformation. Operations such as the Fourier,[7,14,16] wavelet,[6,22,25] and cosine[21] filter the image after it has been transformed.

The most popular and suitable transformation in medical application, including signal processing, used to be and probably still is the Fourier transform. When we transform an image or a signal using the Fourier transform, we mainly create a number of coefficients which represent the image or signal frequency components. One can realize that filtering in the frequency domain is ideal for removing periodic noise.

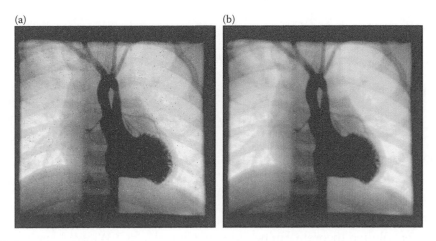

FIGURE A4.1
Random noise (a) and blurring (b) added on a digital angiography image.

Since 1990, the applications of wavelets started to increase rapidly in medicine. Except for image processing,[27,28] the wavelets have been applied also into applications like image compression,[1,3,26] a very important issue in medicine and tele-medicine, since today the digital data sets, especially from acquisition devices like CT, MRI, and PET, are very large.[5] The mathematical and application analysis of wavelets is beyond the scope of this book.

A4.2 2D Convolution Function

A very useful and important operation in image processing is convolution. In case of 2D functions f and h, the convolution is denoted by $(f*h)$ and is defined by the integral

$$G(x,y) = \int_{-\infty}^{\infty} \int_{-\infty}^{\infty} f(i,j)h(x-i,\ y-j)di\,dj$$

$$= \int_{-\infty}^{\infty} \int_{-\infty}^{\infty} f(x-i,\ y-j)h(i,j)di\,dj$$

$$= (f*h)(x,\ y)$$

In digital image processing, where images have a limited domain on the image plane, convolution is used locally. For example, assuming an image $I(x,\ y)$ and a filtering mask $h(j,\ k)$ of size $(M \times N)$, their 2D convolution is defined as

$$G(x,y) = \sum_{j=-\frac{M}{2}}^{\frac{M}{2}} \sum_{k=-\frac{N}{2}}^{\frac{N}{2}} I(x-j,\ y-k) \cdot h(j,\ k) \tag{A4.1}$$

There are a number of mathematical properties associated with convolution. Actually, they are the same as the mathematical operation of the multiplication.

1. Convolution is commutative

$$G = f*h = h*f$$

2. Convolution is associative

$$G = f*(e*h) = (f*e)*h = f*e*h$$

3. Convolution is distributive

$$G = f*(e+h) = (f*e)+(f*h)$$

where e, f, h, and G are all images, either continuous or discrete. An extensive description of the convolution theorem can be found in Gonzalez and Wintz,[7] Nussbaumer,[9] and Gonzalez and Woods.[13]

A4.3 The Fourier Transform

Using the Fourier transform, an image can be decomposed into harmonics.[10,14,16] To apply the Fourier transform in image processing, we have to make two assumptions:

1. The image under processing is periodic
2. The Fourier transform of the periodic function always exists

The 2D continuous Fourier transform of an image is defined as

$$Fr(u, v) = \int_{-\infty}^{\infty} \int_{-\infty}^{\infty} I(x, y) \exp^{-j2\pi(ux+vy)} dx dy$$

The inverse Fourier transform is

$$I(x, y) = \int_{-\infty}^{\infty} \int_{-\infty}^{\infty} Fr(u, v) \exp^{-j2\pi(ux+vy)} du dv$$

To be applicable in medical image processing, the continuous Fourier transform must be converted in a discrete form. Thus, the discrete Fourier transform is equal to

$$Fr(u, v) = \frac{1}{MN} \sum_{x=0}^{M-1} \sum_{y=0}^{N-1} I(x, y) e^{-2\pi j \left(\frac{xu}{M} + \frac{yv}{N} \right)}$$

$$u = 0, 1, \ldots, M-1, \ v = 0, 1, \ldots, N-1$$

The inverse Fourier transform is given by

$$I(x, y) = \sum_{u=0}^{M-1} \sum_{v=0}^{N-1} Fr(u, v) e^{2\pi j \left(\frac{xu}{M} + \frac{yv}{N} \right)}$$

$$x = 0, 1, \ldots, M-1, \ y = 0, 1, \ldots, N-1$$

For simplicity, the Fourier transform can be denoted by an operator *Fr*. Thus, the previous equation can be abbreviated to

$$Fr\{I(x, y)\} = Fr(u, v)$$

The result of the Fourier transform is a complex number, which is composed by a real and an imaginary part:

$$Fr(u, v) = Rl(u, v) + jI(u, v)$$

(a)   (b)

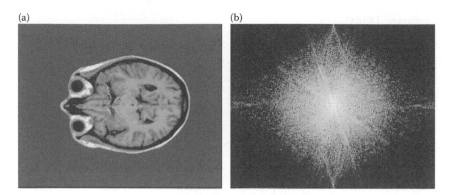

FIGURE A4.2
Head MRI image (a) and its frequency spectrum (b).

Using the real and the imaginary parts one can compute the values called frequency and the phase spectrum of the image $I(x, y)$. The frequency spectrum is defined as (Figure A4.2)

$$|Fr(u, v)| = \sqrt{Rl^2(u, v) * I^2(u, v)}$$

The phase spectrum is defined as

$$\phi(u, v)0 = \tan^{-1}\left[\frac{I(u, v)}{Rl(u, v)}\right]$$

Some interesting properties of the Fourier transform from the image processing point of view are the following:

1. Its relation with the convolution (convolution theorem):

$$Fr\{(I * h)(x, y)\} = Fr(u, v)H(u, v)$$

$$Fr\{I(x, y)h(x, y)\} = (Fr * H)(u, v)$$

2. Shift of the origin, e.g., to the center of the frequency range, in the image and in the frequency domain:

$$Fr\{I(x - u_0, y - y_0)\} = Fr(u, v)e^{-2\pi j \frac{(u_0 u + v_0 v)}{N}}$$

$$Fr\left\{I(x, y)e^{2\pi j\left(\frac{u_0 x + v_0 y}{N}\right)}\right\} = Fr(u - u_0, v - v_0)$$

3. Periodicity: The Fourier transform and its inverse are periodic. In practice, if one wants to specify $Fr(x, y)$ in the frequency domain, only one period is enough for that. In mathematical terms, periodicity can be expressed as

$$Fr(u, -v) = Fr(u, N-v), I(-x, y) = f(M-x, y)$$

$$Fr(-u, v) = Fr(M-u, v), I(x, -y) = f(x, N-y)$$

$$Fr(u_0M + u, v_0N + v) = Fr(u, v), I(u_0M + x, v_0N + y) = f(x, y)$$

4. Conjugate symmetry, when $I(x, y)$ contains real values:

$$Fr(u, v) = Fr'(-u, -v) \text{ or}$$

$$|Fr(u, v)| = |Fr'(-u, -v)|$$

with Fr' being a complex Fourier product.

Since the computational effort considering multiplication and addition is proportional to N^2. The fast Fourier transform (FFT) algorithm reduces this effort and makes it proportional to $N \log_2 N$. One way to do this is to compute the terms of $e^{-2\pi j(ux)/N}$ one time and store it in a table for all subsequent applications. Algorithms for FFT computation can be found in Nussbaumer,[9] Gonzalez and Woods,[13] and Pavlidis.[17]

A4.4 Grey Scale Manipulation Techniques

In most medical imaging diagnostic equipment, the obtained images are in grey scale. The grey level look-up table usually does not surpass the 256 values. Therefore, every depth value from the image matrix will be converted to an index to the current look-up table. The observer often must perform a grey scale transformation in order to make more clear specific organs and tissue types. The techniques used to perform this task are called GSMTs.

One can separate these techniques into two categories: (1) histogram techniques and (2) image windowing techniques.[8,13]

A4.4.1 Histogram Techniques

An image histogram provides a general description of the image appearance. It shows the corresponding number of pixels to every grey value of an image. A change of the histogram (histogram modification) shape has as a result a change to the image contrast. A very common histogram technique in medical image processing is the histogram equalization. The aim of this technique is to modify the current histogram so as to have an equal distribution of the pixel's grey level over the whole range of the grey scale values. The benefit of using this technique is the automatic contrast enhancement for grey levels near the maximum of the histogram. The result can be seen in Figure A4.3. The implementation details of the histogram equalization technique can be found in Gonzalez and Wintz[7] and Sonka et al.[8]

A4.4.2 Image Windowing Techniques

In windowing techniques, only a part of the image depth values is displayed with the available grey values. The term "window" refers to the range of the image depth values which are each time displayed. This window can be moved along the whole range of depth

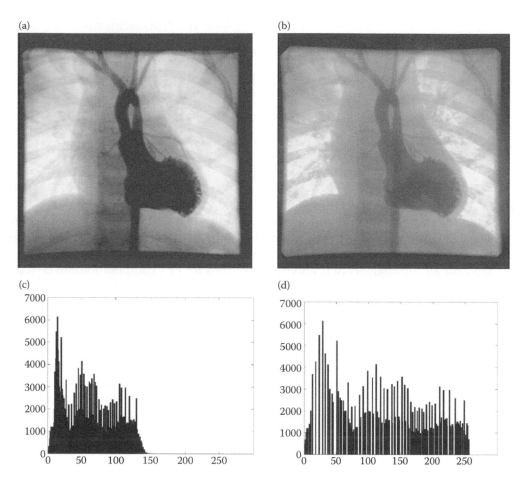

FIGURE A4.3
A digital angiographic image of the heart using injected contrast medium (a) and its histogram (c); the same image after histogram equalization (d).

values of the image, each time displaying different tissue types in the full range of the grey scale and achieving better image contrast. The new brightness value of the pixel Gv is given by the formula:

$$Gv = \left(\frac{Gv_{\max} - Gv_{\min}}{We - Ws}\right) \cdot (Wl - Ws) + Gv_{\min}$$

where $[Gv_{\max}, Gv_{\min}]$ is the grey level range, $[Ws, We]$ defines the window width, and Wl defines the window center. This is the simplest case of image windowing. Often, depending on the application, the window might have more complicated forms, such as double window, broken window, or nonlinear windows (exponential or sinusoid) (Figure A4.4).

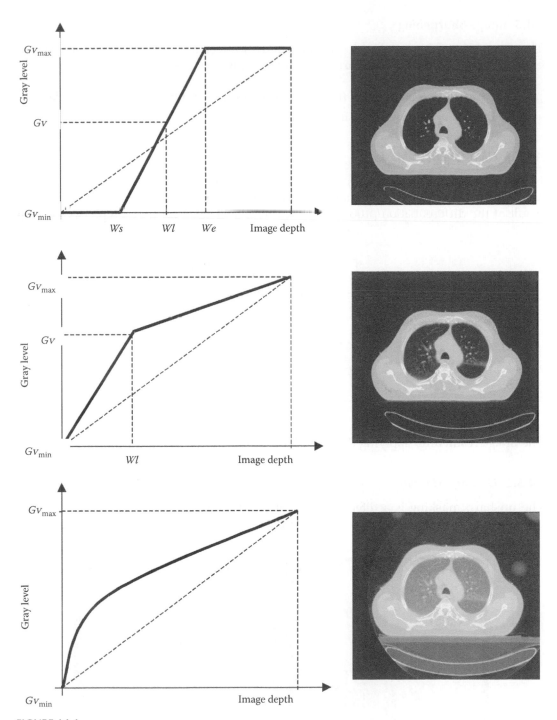

FIGURE A4.4
Examples of image windowing (from top to bottom): linear window, broken window, and nonlinear window.

A4.5 Image Sharpening

The sharpening when applied to an image aims to decrease the image blurring and to enhance image edges.[7,8,13] There are two ways to apply these filters on the image:

1. In the spatial domain using the convolution process and the appropriate masks
2. In the frequency domain using high-pass filters

A4.5.1 High-Emphasis Masks

High-emphasis masks are masks with dimensions 3×3, 5×5, 7×7, or even higher, which are applied via convolution (Equation A4.1) to the original image. These masks are the result of the differentiation process.[7,20] Common masks used for high-emphasis filtering are

$$Em_1(j,k) = \begin{bmatrix} 0 & -1 & 0 \\ -1 & 5 & -1 \\ 0 & -1 & 0 \end{bmatrix}, \quad Em_2(j,k) = \begin{bmatrix} -1 & -1 & -1 \\ -1 & 9 & -1 \\ -1 & -1 & -1 \end{bmatrix}$$

$$Em_3(j,k) = \begin{bmatrix} -1 & -2 & -1 \\ -2 & 13 & -2 \\ -1 & -2 & -1 \end{bmatrix}, \quad Em_4(j,k) = \begin{bmatrix} 1 & -2 & 1 \\ -2 & 5 & -2 \\ 1 & -2 & 1 \end{bmatrix}$$

The image result after applying these masks is demonstrated in Figure A4.5.

A4.5.2 Unsharp Masking

The unsharp masking is a filter which combines the original image with the result of the image if it is filtered using a Laplacian filter, which we will describe later. First, one

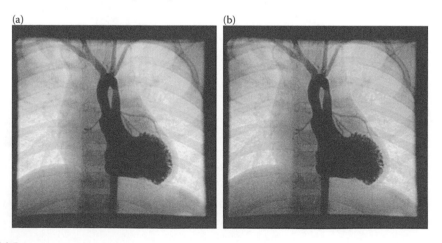

FIGURE A4.5
Removing blurring using high-emphasis masks Em_2 (a) and Em_3 (b). When using the masks Em_1 and Em_4, the image does not improve.

should enhance and isolate the edges, e.g., by using the Laplace operators, amplify them, and then add them back to the original image.[29] The results of this filter are similar to the high-emphasis filter.

A4.5.3 High-Pass Filtering

In the frequency spectrum of an image, one cannotice that the image edges and general high variations in grey levels result in high frequencies in the image spectrum. Using a high-pass filter in the frequency domain, we can attenuate the low frequencies without erasing the image edges. Some filter types are given here.

A4.5.3.1 Ideal Filter

The transfer function for a 2D ideal filter is given as

$$H(u, v) = \begin{cases} 0 & \text{if } T(u, v) \leq To \\ 1 & \text{if } T(u, v) > To \end{cases}$$

where *To* is the cut-off distance from the origin of the frequency plane and $T(u, v)$ is equal to

$$T(u, v) = \sqrt{(u^2 + v^2)} \tag{A4.2}$$

A4.5.3.2 Butterworth Filter

Having a cut-off frequency at a distance *To* from the origin, the Butterworth filter of *n* order is defined as

$$H(u, v) = \frac{1}{1 + [To/T(u, v)]^{2n}}$$

Note that when $T(u, v) = To$ it is down to half of the maximum value (Figure A4.6).

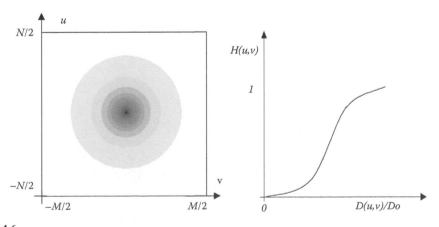

FIGURE A4.6
Butterworth high-pass filter. Shown are the 2D frequency spectrum and its cross section starting from the filter's center.

A4.5.3.3 Exponential Filter

Similar to the Butterworth filter, the high-pass exponential filter can be defined from the relation

$$H(u, v) = e^{-0.347(To/T(u, v))^n}$$

where *To* is the cut-off frequency and *n* is the filter order.

The above filters and general the filter in the frequency domain are applied using the equation

$$G(u, v) = H(u, v)Fi(u, v)$$

where *Fi(u, v)* it the Fourier transform of the image under processing, and *H(u, v)* is the function which describes the filter. The inverse Fourier transform of the function *G(u, v)* gives us the sharpened image (Figure A4.7).

A4.6 Image Smoothing

Image smoothing techniques are used in image processing to reduce noise. Usually, in medical imaging, the noise is distributed statistically, and it exists in high frequencies. Therefore, one can say that image smoothing filters are low-pass filters. The drawback of applying a smoothing filter is the simultaneous reduction of useful information and mainly detail features, which also exist in high frequencies.

A4.6.1 Local Averaging Masks

Most filters used in the spatial domain use matrices (or masks) which may have dimensions of 3×3, 5×5, 7×7, 9×9, or 11×11. These masks can be applied on the original image using the 2D convolution function. Assuming that $I(x, y)$ is the original image, $E(y, x)$ is the filtered image, and $La(j, k)$ is a mask of size $M=3, 5, 7, 9, 11$. If we choose $M=3$, then a 3×3 mask must be applied to each pixel of the image using (Equation A4.1). Different numeric values and different sizes of the masks will have different effects on the image.

(a)   (b)

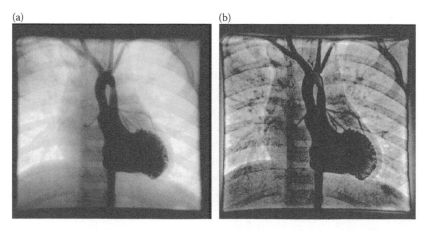

FIGURE A4.7
Filtering example of a digital angiographic image (a) using a high-pass Butterworth filter. In the filtered image (b) the vessels' branches are more distinct.

For example, if we increase the size of the matrix, we will have a more intensive smoothing effect. Usually, 3×3 masks are preferred. By using these masks, not only noise but also useful information will be removed (Figure A4.8). Some types of smoothing masks are

$$La_1(j, k) = \frac{1}{9} \begin{bmatrix} 1 & 1 & 1 \\ 1 & 1 & 1 \\ 1 & 1 & 1 \end{bmatrix}, \quad La_2(j, k) = \frac{1}{5} \begin{bmatrix} 0 & 1 & 0 \\ 1 & 1 & 1 \\ 0 & 1 & 0 \end{bmatrix}$$

$$La_3(j, k) = \frac{1}{10} \begin{bmatrix} 1 & 1 & 1 \\ 1 & 2 & 1 \\ 1 & 1 & 1 \end{bmatrix}, \quad La_4(j, k) = \frac{1}{16} \begin{bmatrix} 1 & 2 & 1 \\ 2 & 4 & 2 \\ 1 & 2 & 1 \end{bmatrix}$$

A4.6.2 Median Filter

The median filter is based upon applying an empty mask of size $M \times M$ on the image (as in convolution). While the mask moves around the image, each place P_i of this mask will copy the pixel with the same coordinates (x, y) and also that pixel which is about to be filtered (Figure A4.9).[15,30] Then these collected pixels, which are values of brightness, will be sorted from the lower to the higher value, and their median value will replace the pixel to be filtered, in our example P5 (Figure A4.10). A different approach for efficient median filtering can be found in Huang et al.[2] and Sonka et al.[8]

A4.6.3 Gaussian Filter

The Gaussian filter is a popular filter with several applications, including smoothing filtering. A detailed description of the Gaussian filter can be found in Sonka et al.,[8] Castleman,[15] and Young and Van Vliet.[19]

In general, the Gaussian filter is separable:

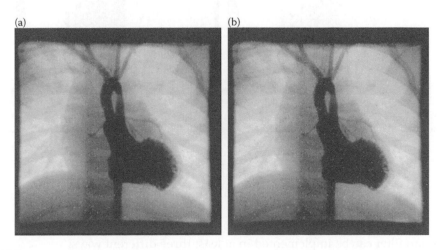

(a)   (b)

FIGURE A4.8
Filtering noise using the local averaging masks. The results are from (a) La_1 and (b) La_2. The masks La_3 and La_4 have very similar effects on the image.

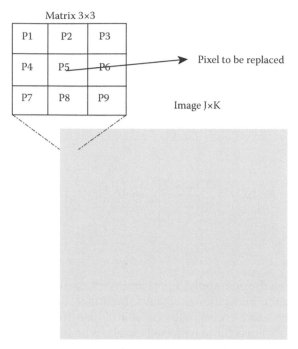

FIGURE A4.9
Applying a 3×3 median mask.

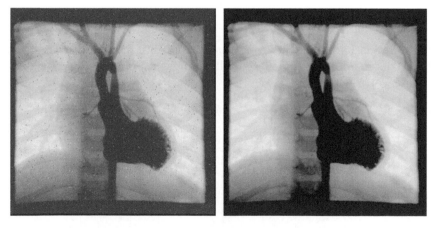

FIGURE A4.10
Filtering the random noise using the median filter. Observe how effective the filter is in this case.

$$h(x, y) = g_{2D}(x, y) = \left(\frac{1}{\sqrt{2\pi}\sigma} \exp\left(-\frac{x^2}{2\sigma^2} \right) \right) \cdot \left(\frac{1}{\sqrt{2\pi}\sigma} \exp\left(-\frac{y^2}{2\sigma^2} \right) \right)$$

$$= g_{1D}(x) \cdot g_{1D}(y)$$

The Gaussian filter can be implemented in at least three different ways:

1. Convolution: Using a finite number of samples M of the Gaussian as the convolution kernel—it is common to choose $M=3\sigma$ or 5σ, where σ is an integer number:

$$g_{1D(n)} = \begin{cases} \dfrac{1}{\sqrt{2\pi}\sigma} \cdot \exp\left(-\dfrac{n^2}{2\sigma^2}\right), & |n| \leq M \\ & |n| > M \\ 0 \end{cases}$$

2. Repetitive convolution: Using a uniform filter as a convolution kernel:

$$g_{1D} = u(n) * u(n) * u(n),$$

$$u(n) = \begin{cases} \dfrac{1}{(2M+1)}, & |n| \leq M \\ & |n| > M \\ 0 \end{cases}$$

where $M=\sigma$ and σ takes integer values. In each dimension, the filtering can be done as

$$E(n) = [[I(n) * u(n)] * u(n)] * u(n)$$

3. In the frequency domain: Similar to the Butterworth and exponential filters, one can create a filter using the Gaussian type and then multiply this filter with the image spectrum. The inverse Fourier transform will give the filtered image.

A4.6.4 Low-Pass Filtering

In low-pass filtering, we use same principles as in high-pass filtering. In this case, our aim is to cut the high frequencies, where the noise is usually classified. The benefit of low-pass filtering, compared to spatial domain filters, is that noise with a specific frequency can be isolated and completely cleared from the image. When filtering random noise the drawback is that the edge information will be suppressed. Common filter types are the following:

A4.6.4.1 Ideal Filter

The transfer function for a 2D low-pass ideal filter is given as

$$H(u, v) = \begin{cases} 1 & \text{if } T(u, v) \leq To \\ 0 & \text{if } T(u, v) > To \end{cases}$$

where To is the cut-off distance from the origin of the frequency plane and $T(u, v)$ is given from Equation A4.2.

A4.6.4.2 Butterworth Filter

Having a cut-off frequency at distance *To* from the origin, the Butterworth filter of *n* order is defined as

$$H(u, v) = \frac{1}{1 + [T(u, v)/To]^{2n}}$$

Note that when $T(u, v) = To$ is down to the half of the maximum value (Figure A4.11).

A4.6.4.3 Exponential Filter

Similar to the Butterworth filter, the low-pass exponential filter can be defined from the relation

$$H(u, v) = e^{-0.347(T(u, v)/To)^n}$$

where *To* is the cut-off frequency and *n* is the filter order.

Figure A4.12 is an example of image noise removal using the Butterworth low-pass filter.

A4.7 Edge Detection

Edge detection techniques aim to detect the image areas where we have a rapid change of the intensity.[4,8,13] In X-ray diagnostic imaging, these can be areas between bone and soft tissue or soft tissue and air. One can find similar examples in several types of diagnostic images. Here, we describe a number of gradient operators used in medical image edge

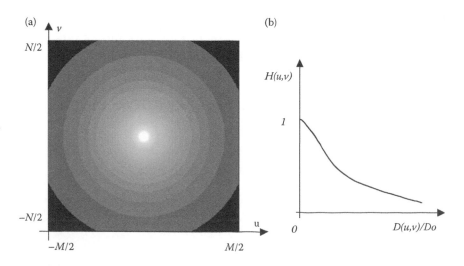

FIGURE A4.11
Butterworth low-pass filter, in a 2D representation of its frequency spectrum (a) and a cross-section of the same filter starting from the filter's center (b).

(a)　　　　　　　　　　　　　(b)

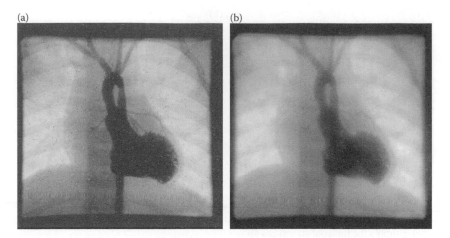

FIGURE A4.12
Removing random noise (a) from a digital angiographic image using a low-pass Butterworth filter. The smoothing effect in the filtered image is obvious.

detection as a mean of 3×3 masks. All of the masks can be applied to the original image via 2D convolution.

In general, the gradient magnitude $|grI(x, y)|$ of an image $I(x, y)$ is given as

$$|grI(x, y)| = \sqrt{\left(\frac{\partial I}{\partial x}\right)^2 + \left(\frac{\partial I}{\partial y}\right)^2}$$

A4.7.1 Laplacian Operator

A Laplacian or edge enhancement filter is used to isolate and amplify the edges of the image, but it completely destroys the image information at low frequencies (such as soft tissues). The Laplacian operator is invariant to image rotation and therefore has the same properties to all directions (Figure A4.13) and is calculated as

$$\nabla^2 I(x, y) = \frac{\partial^2 I(x, y)}{\partial x^2} + \frac{\partial^2 I(x, y)}{\partial y^2}$$

Common Laplacian masks are

$$Lp_1(j, k) = \begin{bmatrix} 0 & 1 & 0 \\ 1 & -4 & 1 \\ 0 & 1 & 0 \end{bmatrix} \quad Lp_2(j, k) = \begin{bmatrix} 1 & 1 & 1 \\ 1 & 1 & 1 \\ 1 & -8 & 1 \end{bmatrix}$$

$$Lp_3(j, k) = \begin{bmatrix} 1 & 2 & 1 \\ 2 & -12 & 2 \\ 1 & 2 & 1 \end{bmatrix} \quad Lp_4(j, k) = \begin{bmatrix} -2 & 2 & -1 \\ 2 & -4 & 2 \\ -1 & 2 & -1 \end{bmatrix}$$

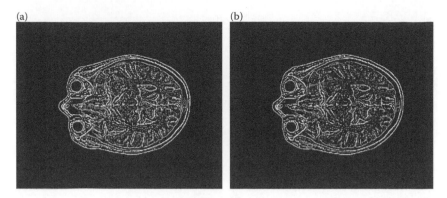

FIGURE A4.13

The Laplacian operator. The results are from (a) Lp_1 and (b) Lp_2. The mask Lp_3 gives a similar result to Lp_1, and the mask Lp_4 deteriorates the internal structures image.

A4.7.2 Prewitt, Sobel, and Robinson Operators

All these operators base their function on the first derivative. For a 3×3 mask, the gradient can be estimated for eight different directions. The gradient direction is indicated from the convolution result of greatest magnitude. In contrast to the Laplacian operator, these operators are related to the image orientation, and therefore, the image edges can be enhanced only at one direction for each time. We present here the first four masks from each operator. The rest of the masks are calculated considering the gradient direction we want to check (Figure A4.14).

Prewitt:

$$Pr_1(j, k) = \begin{bmatrix} 1 & 1 & 1 \\ 0 & 0 & 0 \\ -1 & -1 & -1 \end{bmatrix}, \quad Pr_2(j, k) = \begin{bmatrix} 0 & 1 & 1 \\ -1 & 0 & 1 \\ -1 & -1 & 0 \end{bmatrix}$$

$$Pr_3(j, k) = \begin{bmatrix} -1 & 0 & 1 \\ -1 & 0 & 1 \\ -1 & 0 & 1 \end{bmatrix}, \quad Pr_4(j, k) = \begin{bmatrix} 1 & 0 & -1 \\ 1 & 0 & -1 \\ 1 & 0 & -1 \end{bmatrix}$$

Sobel:

$$Sb_1(j, k) = \begin{bmatrix} 1 & 2 & 1 \\ 0 & 0 & 0 \\ -1 & -2 & -1 \end{bmatrix}, \quad Sb_2(j, k) = \begin{bmatrix} 0 & 1 & 2 \\ -1 & 0 & 1 \\ -2 & -1 & 0 \end{bmatrix}$$

$$Sb_3(j, k) = \begin{bmatrix} -1 & 0 & 1 \\ -2 & 0 & 2 \\ -1 & 0 & 1 \end{bmatrix}, \quad Sb_4(j, k) = \begin{bmatrix} 1 & 0 & -1 \\ 2 & 0 & -2 \\ 1 & 0 & -1 \end{bmatrix}$$

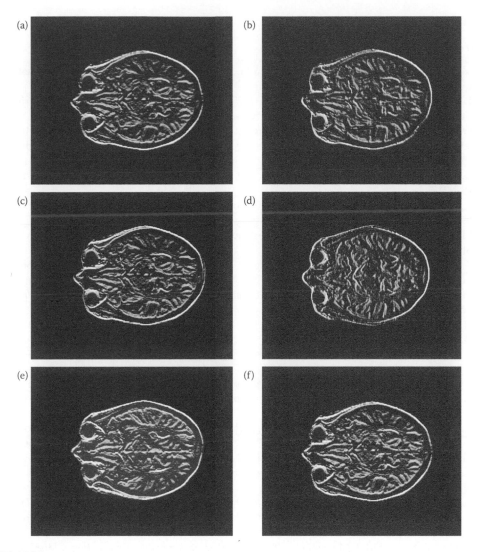

FIGURE A4.14
The Prewitt, Sobel, and Robinson operators applied on the MRI image. In (a) and (b), the results from Pr_1 and Pr_2 are equivalent. In (c) and (d), the results from Rb_1 and Rb_3 are equivalent. In (e) and (f), the results from Sb_0 and Sb_1 are equivalent.

Robinson:

$$Rb_1(j,k) = \begin{bmatrix} 1 & 1 & 1 \\ 1 & -2 & 1 \\ -1 & -1 & -1 \end{bmatrix}, \quad Rb_2(j,k) = \begin{bmatrix} 1 & 1 & 1 \\ -1 & -2 & 1 \\ -1 & -1 & 0 \end{bmatrix}$$

$$Rb_3(j,k) = \begin{bmatrix} -1 & 1 & 1 \\ -1 & -2 & 1 \\ -1 & 1 & 1 \end{bmatrix}, \quad Rb_4(j,k) = \begin{bmatrix} 1 & 1 & -1 \\ 1 & -2 & -1 \\ 1 & 1 & -1 \end{bmatrix}$$

References

1. M. L. Hilton, B. D. Jawerth, and A. Sengupta. 1994. Compressing still and moving images with wavelets. *J. Multimedia Syst.*, 2, 218–227.
2. T. S. Huang, G. J. Yang, and G. Y. Tang. 1979. A fast two-dimensional median filtering algorithm. *IEEE Trans. Acoust. Speech Signal Process.*, ASSP-27, 13–18.
3. W. R. Zettler, J. Huffman, and D. C. P. Linden. 1990. Application of compactly supported wavelets to image compression. In *Image Processing Algorithms and Techniques*. SPIE, Bellingham, WA, 150–160.
4. D. Marr and E. C. Hildreth. 1980. Theory of edge detection. *Proc. R. Soc. London Ser. B.*, 207, 187–217.
5. A. H. Delany and Y. Bresler. 1994. Multiresolution tomographic reconstruction using wavelets. *IEEE Int. Conf. Image Process.*, 1, 830–834.
6. M. D. Harpen. 1998. An introduction to wavelet theory and application for the radiological physicist. *Med. Physics*, 25, 1985–1993.
7. R. C. Gonzalez and P. Wintz. 1984. *Digital Image Processing*, 2nd edition. Addison-Wesley, Reading, MA.
8. M. Sonka, V. Hlavac, and R. Boyle. 1998. *Image Processing, Analysis and Machine Vision*, 2nd Edition. PWS Publishing.
9. H. J. Nussbaumer. 1982. *Fast Fourier Transform and Convolution Algorithms*, 2nd edition. Springer Verlag, Berlin.
10. A. C. Kak and M. Stanley. 1988. *Principles of Computerized Tomographic Imaging*. IEEE, Piscataway, NJ.
11. J. C. Russ. 1995. *The Image Processing Handbook*, 2nd edition, CRC Press, Boca Raton, FL.
12. H. C. Andrews and B. R. Hunt. 1974. *Digital Image Restoration*. Prentice Hall, Englewood Cliffs, NJ.
13. R. C. Gonzalez and R. E. Woods. 1992. *Digital Image Processing*. Addison-Wesley, Reading, MA.
14. J. W. Goodman. 1968. *Introduction to Fourier Optics*. McGraw-Hill Physical and Quantum Electronics Series, McGraw-Hill, New York, NY.
15. K. R. Castleman. 1996. *Digital Image Processing*, 2nd edition. Prentice Hall, Englewood Cliffs, NJ.
16. H. Stark. 1982. *Application of Optical Fourier Transforms*. Academic Press, New York, NY.
17. T. Pavlidis. 1982. *Algorithms for Graphics and Image Processing*. Computer Science Press, New York, NY.
18. R. H. T. Bates and M. J. McDonnell. 1986. *Image Restoration and Reconstruction*. Clarendon Press, Oxford.
19. I. T. Young and L. J. Van Vliet. 1995. Recursive implementation of the Gaussian filter. *Signal Process.*, 44(2), 139–151.
20. L. Schreyer, M. Grimm, and G. Sakas. 1994. *Case Study: Visualization of 3d-Ultrasonic Data*. IEEE, 369–373.
21. S. Walter and G. Sakas. 1995. *Extracting Surfaces from Fuzzy 3D-Ultrasonic Data*. Addison-Wesley, Reading, MA, 465–474.
22. A. Rosenfeld and A. C. Kak. 1982. *Digital Picture Processing*, 2nd edition. Academic Press, New York, NY.
23. K. R. Rao and P. Yip. 1990. *Discrete Cosine Transform, Algorithms, Advantages, Applications*. Academic Press, Boston, MA.
24. C. K. Chui. 1992. *An Introduction to Wavelets*. Academic Press, New York, NY.
25. G. Strang. 1993. Wavelet transforms versus Fourier transforms. *Bull. Am. Math. Soc.*, 28, 288–305.
26. R. Devore, B. Jaeverth, and B. J. Lucier. 1992. Image compression through wavelet transform coding. *IEEE Trans. Inf. Theory*, 38, 719–746.

27. D. Healy and J. Weaver. 1992. Two applications of wavelet transforms in MR imaging. *IEEE Trans. Inf. Theory*, 38, 840–860.
28. R. Devore, B. Jaeverth, and B. J. Lucier. 1996. Feature extraction in digital mammography. In A. Aldroubi and M. Unser, Editors. *Wavelets in Medicine and Biology*. CRC Press, Boca Raton, FL.
29. A. K. Jain. 1989. *Fundamentals of Digital Image Processing*. Prentice-Hall, Englewood Cliffs, NJ.
30. S. G. Tyan. 1981. Median filtering, deterministic properties. In T. S. Huang, Editor. *Two-Dimensional Digital Signal Processing*, Volume II. Springer-Verlag, Berlin.

5

Medical Image Registration and Fusion Techniques: A Review

George K. Matsopoulos

National Technical University of Athens

CONTENTS

5.1 Introduction

Registration between two (2-D) or three dimensional (3-D) images is a common problem encountered when more than one image of the same anatomical structure is obtained, either using different imagery or when performing dynamic studies, taken at different times. In all cases, the information present in the different images must be combined to produce fused or parametric images. For efficient fusion of information, the different images have to be registered. Several registration and fusion techniques have been proposed in the literature, with a varying degree of user intervention, and the most representative of which will be discussed in this chapter.

The chapter is divided in two main sections: medical image registration and medical image fusion. The medical image registration section concerns with the introduction of a generic medical image registration scheme based on global transformations. This scheme consists of the identification of common features in the images to be registered, the definition of geometrical transformation, the selection of an appropriate measure of match (MOM) between the reference and transformed image, and the application of various optimization techniques for the determination of the transformation parameters with respect to the MOM. In the application section, qualitative and quantitative results are presented for registering computed tomography (CT)-magnetic resonance (MR) human heads (3-D case) using four well-known registration algorithms and their performances in terms of registration accuracy are also compared.

The medical image fusion section concerns with the introduction of various techniques commonly used to combine information from different modalities, after the application of the medical image registration process. These techniques are divided into two main categories: fusion of information at the image level and at the object level. Finally, results of the information fusion are also qualitatively assessed from various modalities.

5.2 Medical Image Registration

5.2.1 Literature Review

The process of image registration can be formulated as a problem of optimizing a function that quantifies the match between the original (reference) and the transformed image. Several image features have been used for the matching process, depending on the modalities used, the specific application and the implementation of the transformation. Comprehensive surveys of medical image registration can be found elsewhere [1,2], in terms of image modalities and employed techniques. According to the matching features, the medical image registration process can be divided into three main categories: point-based, surface-based and volume-based methods.

Point-based registration involves the determination of the coordinates of corresponding points in different images and the estimation of geometrical transformation using these corresponding points. Such points can be defined using external markers placed on the patient's skin before the acquisition [3,4], stereotactic frames [5,6] and landmarks [7]. External markers have been used to register SPECT-MR images with the affine transformation method [3] and CT-MR data with 3-D global transformations [8]. Landmarks, as another point-based approach, were placed in the images by experts by means of software, in order to equivalently define and register anatomical areas [8–10]. A landmark-based method to register CT-MR images, using singular value decomposition, is described by Hill et al. [7]. A more advanced deformation based on the thin plate splines interpolation model is reported in Bookstein [11]. The low resolution along the transverse axis, the small number of corresponding markers, as well as possible inaccuracies in their placement during the acquisition from each modality, render these methods as *manual-based registration*, resulting in inaccuracies and inconsistencies.

Surface-based registration involves the determination of the surfaces of the images to be matched and the minimization of a distance measure between these corresponding surfaces. The surfaces are generally represented as set of large number of points obtained by segmenting contours in contiguous image slices. There may be a requirement for the points to be triangulated. The difference between point-based and surface-based registration algorithms is that point correspondence is defined by the user for the former, whereas is automatic for the later. Pelizzari et al. [12] introduced the idea of using surfaces to register brain images by obtaining a rigid body transformation, which when applied to "hat" coordinates (points that belong to the skin surface of the scan with the lower resolution) minimizes a residual that is the mean square distance between "hat" points and "head" surface (a stack of skin contours from the higher resolution scan) using an optimization technique described by Powell [13]. The Euclidean distance between a point of an image and the closest surface point is used in Kozinska et al. [14], as closest point projection rule, whereas the integer approximations of the Euclidean distance as well as its highly computation cost required were improved by using the well known chamfer distance transform [15]. This method was then applied to medical image registration [16,17]. Besl and McKay [18] presented a general purpose registration technique called "iterative closest point method" which was extended and implemented towards medical applications in Maurer et al. [19]. A method designed to register preoperative CT images to cerebral surface points acquired intra-operatively from ultrasound images was presented in Herring et al. [20] along with qualitative results and accuracy comparisons using marker based methods.

Volume-based registration involves the optimization of a quantity measuring the similarity of all geometrically corresponding voxel pairs, considering some predefined features. Multiple volume-based algorithms have been proposed [21,22], optimizing a measure of the absolute difference between image intensities of corresponding voxels within overlapping parts in a region of interest. These methods were based on the assumption that the two images are linearly correlated which is not the general case. Cross correlation of feature images derived from the original image data has been applied to CT-MR modeling using geometrical features, such as edges [23] and ridges [24,25] or using specially designed intensity transformations [26]. Misregistration was measured by the dispersion of the 2-D histogram of the image intensities of corresponding voxel pairs, which was assumed to be minimal in the registered position. Hills et al. [27] criterion required segmentation of the images or delineation of specific histogram regions, while Woods et al. [28] criterion was based on additional assumption concerning relationships between the gray values in the different modalities to reduce the complexity. In Collignon et al. [29], and in an extended work [30], the mutual information (MI) registration criterion was introduced, measuring the statistical dependence between two random variables of the amount of information that one variable contains about the other. The MI of the image intensity values of corresponding voxel pairs was maximal if the images were geometrically aligned. A comparative study between surface- and volume-based registration algorithms was published [31] which indicates that the volume-based techniques tented to give more accurate and reliable results when the CT-MR images are registered, and slightly more accurate results for the PET-MR images. The reason lies with the fact that surface-based registration methods require well-defined corresponding surfaces prior to registration.

Most of the aforementioned registration algorithms use rigid transformations to account for global misalignment. When merging brain image data or matching anatomical representations such as Atlases, elastic matching techniques seems most appropriate to record local shape differences [32,33]. Bookstein [34] used a user-defined set of landmarks and the thin plate splines interpolation model to interpolate the displacement function over the whole image. Approaches based on deformation models include similarity-based methods, polynomial transformations and contour matching. An elastic deformation method [35], based on the cross correlation coefficients (CCs), was applied iteratively to CT images until a match takes place with a predefined atlas model. Low degree polynomial transformations applied to register MR images causing a polynomial warping [36] whereas 2-D CT-MR chest images were registered using dynamic elastic method which matches corresponding contours and leads to a global translation vector for coarse alignment and local residual shifts along the contours [37]. Furthermore, a contour registration technique was presented, consisting of a boundary homothetic mapping together with an elastic shape deformation model based on active models [38]. The advantageous implementations of the elastic deformation models avoid the application of the geometrical distortion correction operation, due to the imaging modality, required prior to registration. Several imaging techniques, such as the MRI system, cause local geometrical distortions in their field of view. In West et al. [31], where surface- and volume-based registration methods were evaluated, MR images were corrected from static field inhomogeneities using various techniques [39].

In multimodal medical image registration, current trends are focused on the improvement of the registration accuracy as well as the implementation of various registration schemes involving approaches from different registration process categories (point-, surface- and volume-based methods). Fitzpatrick et al. [40], derived approximation expressions for the "target registration error" (TRE) and for the expected squared alignment error of an individual fiducial marker. They found out that the expected registration accuracy is worst when the fiducial points are closely aligned; thus providing the surgeons with the appropriate guidance in placing the fiducial markers before surgery and increasing the accuracy of point based guidance systems. Also, it was revealed in Maurer et al. [41], that fiducial alignment alone should not be trusted as an indicator of registration success. The efficiency of a weighted geometrical feature algorithm was demonstrated by registering CT and MR volume head images using fiducial points, surface and various weighted combinations of points and surfaces, according to Maurer et al. [19,41]. In Barillot et al. [42], a combined registration scheme based on the use of image features with the affine transformation and an elastic deformation model was applied on various 2-D and 3-D head data improving the registration performance by obtaining global and local brain deformations.

5.2.2 Generic Medical Image Registration Scheme

Before the presentation of a generic medical image registration scheme, the following notations should be made. A data set could be defined as a function $I(x_1, x_2, \ldots, x_n) : S \to Z$, where $(x_1, x_2, \ldots, x_n)$ are the co-ordinates of a point at the n-dimensional space $\Re^n$, S is a subset of $\Re^n$ and $I(x_1, x_2, \ldots, x_n)$ is the gray value of the point $(x_1, x_2, \ldots, x_n)$. In the case of tomography data, e.g., data from CT or MR, the dimensionality is $n = 3$; thus, $S \subset \Re^3$ and $I(x_1, x_2, x_3)$ is the gray value of the point (x_1, x_2, x_3) of the 3-D space $\Re^3$. In the next sections, all definitions will mainly concern 3-D data which can be generalized for data of any dimensionality.

Let two 3-D data sets: $I_R(x_1, x_2, x_3) : S_R \to Z$ and $I_F(x_1, x_2, x_3) : S_F \to Z$. The registration of these two data is defined as the determination of a geometrical transformation $T : S_F \to S_R$ such as:

$$I_F(x_1, x_2, x_3) = I_R(T(x_1, x_2, x_3)) \tag{5.1}$$

for every pixel $(x_1, x_2, x_3) \in S_F$. The data set $I_R(x_1, x_2, x_3)$ is called the *reference data set* while the $I_F(x_1, x_2, x_3)$ *the float data set*. Practically, due to the contrast and/or brightness differences between the two sets as well as the existence of noise, it is very difficult for Equation 5.1 to be satisfied for every $(x_1, x_2, x_3) \in S_F$. For this reason, the following equation could be used more generally for the definition of the registration between two data sets:

$$SM(I_F(x_1, x_2, x_3), I_R(T(x_1, x_2, x_3))) = \text{maximum} \tag{5.2}$$

SM is an MOM that quantifies the spatial matching between the two data sets. Consequently, the problem of the registration is equivalent to the determination of a geometrical transformation that maximizes a selected MOM of the two data sets. If $T^{-1} : S_R \rightarrow S_F$ is defined as the inverse transformation, then the set $I_{T_i} \cdot S_R \rightarrow Z$ such as $I_{Tr}(r_1, r_2, r_3) = I_r(T^{-1}(r_1, r_2, r_3))$ is called the *transformed data set*.

A generic medical image registration scheme based on local transformation may consist of the selection of the image features that will be used during the matching process, the definition of an MOM that quantifies the spatial matching between the reference and the float data sets, and the application of an optimization technique that determines the independent parameters of the transformation model employed, according to the MOM. The proposed generic medical image registration scheme is shown in Figure 5.1.

5.2.2.1 Preprocessing Techniques

The preprocessing is the first step towards registration that is often required and involves the application of various image processing techniques in order to: (a) enhance the image data or remove noise, (b) re-sample the data to approximate the same size of voxels of the two data sets, (c) determine features that can be used for the matching process, and (d) preregister the two data sets. The first two processes may be incorporated with any registration scheme whereas the latter two, mainly with any surface-based registration approach.

The acquired data often contained background noise and/or varying background due to the acquisition process. The noise should be removed without affecting the useful anatomical information within the data. Towards this direction, various approaches have

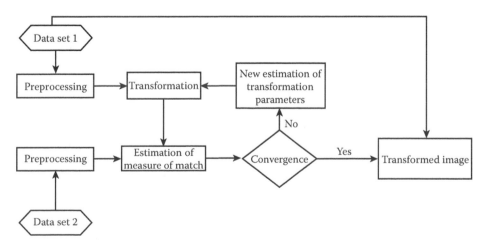

FIGURE 5.1
Generic medical image registration scheme.

been applied, mainly in 2-D data, including the mean averaging filter, the median filter, the γ-correction filter, the histogram-based enhancement filters, contrast and brightness correction filter [43], or an adaptive nonlinear filtering technique [44].

An important aspect towards image registration is the different resolution of the data sets in the case that they are acquired from different image modalities. For example, in CT, the common dimensions of a slice are $512 \times 512 \times 12$ bits per pixels and in MR are $256 \times 256 \times 8$ bits per pixel, respectively. Thus, re-sampling of the data is required before any processing in order to obtain a common voxel size for the data.

Let $I(x,y,z)$ is the gray level of the pixel (x,y) of a slice at z position. The re-sampling process defines the new grey level at a position (x',y',z') and it is implemented by the application of the correlation with an appropriate interpolation kernel as follows:

$$I(x',y',z') = \sum_x \sum_y \sum_z I(x,y,z)h(x'-x)h(y'-y)h(z'-z) \qquad (5.3)$$

where $h(u)$ is the interpolation kernel. The most representative interpolation kernels are the linear interpolation (according to Equation 5.4a) and the cubic spline interpolation (according to Equation 5.4b) [45], according to the following equations:

$$h_1(u) = \begin{cases} 1 - |u|, & |u| \leq 1 \\ 0, & |u| > 1 \end{cases} \qquad (5.4a)$$

$$h_3(u) = \begin{cases} \dfrac{1}{6}\left(4 - 6u^2 + 3|u|^3\right), & 0 \leq |u| < 1 \\ \dfrac{1}{6}(2 - |u|)^3, & 1 \leq |u| < 2 \\ 0, & |u| \geq 2 \end{cases} \qquad (5.4b)$$

An important issue related with the preprocessing step is the identification of common anatomical features to be used in the matching process. Several features have been proposed in the literature including, skin markers, landmarks (identified manually or automatically) and surfaces in the 3-D case, after the application of a segmentation process.

The use of external *skin markers* is a well-documented and relatively simple approach. They are placed physically on the patient prior to the acquisition of the image. The use of four skin markers for registering CT-MR data using 3-D global rigid transformation was reported in Clarysse et al. [8]. A similar application on CT images for stereotactic surgery was described in Schad et al. [46]. A combination of skin markers with affine transformation appears in Hawks et al. [47], considering SPECT and MR images whereas SPECT and PET images were registered with CT images using four skin markers and a least square fit method in Arun et al. [48].

Landmarks placed in the images by an expert to identify anatomically equivalent areas in the pair of images, is another approach. Usually the small number of the landmarks as well as possible inaccuracies in their placement renders these methods semi-automatic and able to estimate simple transforms. Landmarks and singular value decomposition methods were applied in CT-MR images [7], whereas registering 2-D myocardial perfusion images requires high order polynomial transformations [49]. Landmarks and a global affine transformation were employed for CT-SPECT registration in Kramer et al. [50], whereas more advanced transformation (thin plane splines) were used in Bookstein [11] for CT-MR data sets.

In cases of 3-D images, a match between *anatomical structures* is calculated (surface to surface or points to surface). For instance, in the case of registering CT with MR data sets, a match between the head–hat is estimated, with the hat extracted from the outer contours of the MR modality and the head being the surface generated from the outer contours of the CT image. The head–hat match, combined with rigid or affine transformation in Toennies et al. [51] was used in registering CT and MR data with PET data. Finally, an important process included in the preprocessing step in the surface-based registration could be the preregistration which involves the spatial realignment of the segmented surfaces based on their centers of mass.

5.2.2.2 Transformation Models

Several transformation models may be employed for the global medical image registration [1]. The most representatives include the affine, the bilinear, the rigid, and the projective transformations. Analytically,

1. *The affine transformation* can be decomposed into a linear transformation and a simple translation. It can be shown that this transformation maps straight lines into straight lines, whereas it preserves parallelism between lines. In the 3-D case, it can be mathematically expressed as follows:

$$
\begin{pmatrix} x' \\ y' \\ z' \end{pmatrix} = \begin{pmatrix} a_1 & a_2 & a_3 \\ b_1 & b_2 & b_3 \\ c_1 & c_2 & c_3 \end{pmatrix} \begin{pmatrix} x \\ y \\ z \end{pmatrix} + \begin{pmatrix} dx \\ dy \\ dz \end{pmatrix} \tag{5.5}
$$

   The affine transformation is completely defined by nine independent parameters (a_i, b_i, c_i) for $i = 1, 2, 3$, and dx, dy and dz, whereas, in the 2-D case, the affine transformation is defined by six independent parameters.

2. *The bilinear transformation* is the simplest polynomial transformation, which maps straight lines from the original image to curves. It can be expressed for the 3-D case as:

$$
x' = \alpha_0 + \alpha_1 x + \alpha_2 y + \alpha_3 z + \alpha_4 xy + \alpha_5 yz + \alpha_6 zx + \alpha_7 xyz
$$
$$
y' = b_0 + b_1 x + b_2 y + b_3 z + b_4 xy + b_5 yz + b_6 zx + b_7 xyz \tag{5.6}
$$
$$
z' = c_0 + c_1 x + c_2 y + c_3 z + c_4 xy + c_5 yz + c_6 zx + c_7 xyz
$$

   The bilinear transformation is completely defined by 24 independent parameters (a_i, b_i, c_i) for $i = 0, 1, 2, ..., 7$. For the 2-D case, eight independent parameters completely define the bilinear transformation.

3. *The rigid transformation* preserves the distance between any two points in the transformed image. The mathematical expression of the transformation for the 3-D case is given by:

$$\begin{pmatrix} x' \\ y' \\ z' \end{pmatrix} = \begin{pmatrix} 1 & 0 & 0 \\ 0 & \cos(t_x) & -\sin(t_x) \\ 0 & \sin(t_x) & \cos(t_x) \end{pmatrix} \begin{pmatrix} \cos(t_y) & 0 & \sin(t_y) \\ 0 & 1 & 0 \\ -\sin(t_y) & 0 & \cos(t_y) \end{pmatrix}$$

$$\times \begin{pmatrix} \cos(t_z) & -\sin(t_z) & 0 \\ \sin(t_z) & \cos(t_z) & 0 \\ 0 & 0 & 1 \end{pmatrix} \begin{pmatrix} x \\ y \\ z \end{pmatrix} + \begin{pmatrix} dx \\ dy \\ dz \end{pmatrix} \tag{5.7}$$

where t_x, t_y, and t_z represents the rotation angles and dx, dy, and dz the translation displacements along the x, y, and z axes, respectively.

4. *The projective transformation* maps any straight line in the original image onto a straight line in the transformed image. Parallelism is not preserved. The mathematical expression of the transformation for the 2-D case is given by:

$$\begin{pmatrix} u \\ v \\ w \end{pmatrix} = \begin{pmatrix} a_1 & a_2 & dx \\ a_3 & a_4 & dy \\ a_5 & a_6 & 1 \end{pmatrix} \begin{pmatrix} x \\ y \\ 1 \end{pmatrix}, \quad \begin{pmatrix} x' \\ y' \end{pmatrix} = \begin{pmatrix} u/w \\ v/w \end{pmatrix} \tag{5.8}$$

where w represents the extra homogeneous coordinate and u and v are dummy variables. The projective transformation is mainly employed in the 2-D case, is strongly resembles the bilinear and is completely defined by nine independent parameters $(a_{11}, ..., a_{33})$.

5.2.2.3 Measures of Match (MOM)

Several match measures have been employed, depending on the application as well as the implementation of the transform. The selection of an appropriate MOM should be based on two principals: (a) the MOM should contain less local maxima as possible, and (b) the MOM should be independent from any contrast/brightness differences of the data. The first principal reduces the possibility of the solution by the application of the optimization method employed to be trapped on local maxima whereas the second results in a registration without dependency on any contrast/brightness differences.

In cases of manual-based registration, the match is quantified by the *distance* between selected pairs of points or markers. In the case of matching the principal axes, the required translation is determined by the centroids of the two structures whereas scale and rotation is determined by the two systems of principal axes. For the head–hat match, the distance transform is often used [1,14].

The most representative measures of match are the following:

1. *Correlation coefficient (CC)* is suitable for both 2-D and 3-D image intramodality registration. Generally, the $CC(I_1, I_2)$ between two data sets $I_1 : S_1 \rightarrow Z$ and $I_2 : S_2 \rightarrow Z$ with $S = S_1 \cap S_2$, is defined for the 3-D case according to the following equation:

$$CC(I_1, I_2) = \frac{\displaystyle\sum_{(x_1, x_2, x_3) \in S} \left[I_1(x_1, x_2, x_3) - \bar{I}_1 \right] \left[I_2(x_1, x_2, x_3) - \bar{I}_2 \right]}{\sqrt{\displaystyle\sum_{(x_1, x_2, x_3) \in S} \left[I_1(x_1, x_2, x_3) - \bar{I}_1 \right]^2} \sqrt{\displaystyle\sum_{(x_1, x_2, x_3) \in S} \left[I_2(x_1, x_2, x_3) - \bar{I}_2 \right]^2}} \tag{5.9}$$

where $\bar{I_1}$ and $\bar{I_2}$ are the average values of the two data sets, respectively for the points $(x_1, x_2, x_3) \in S$.

Correlation techniques map regions of pixels on a best-fit basis and therefore require minimal user interaction. These methods have been combined with very simple transformations (translations only in Peli et al. [52] or translation and rotation in Junck et al. [53]). The use of global affine transform with cross correlation implemented in the spatial frequency domain was also reported in Aicella et al. [54], whereas curved transformations for digital subtraction angiography was described in Venot and Leclerc [55].

2. *Mutual information (MI)* as a similarity measure is also employed for registering multimodal medical images. MI is a concept measuring statistical dependence between two random variables, or the amount of information that one variable contains about the other. The MI registration criterion, proposed in Maes et al. [30], states that the MI of the image intensity values of corresponding voxel pairs is maximized if the images are geometrically aligned, according to the following equation:

$$MI = P(x,y)\log P((x,y)) - P(x)\log(P(x)) - P(y)\log(P(y)) \tag{5.10}$$

where $P(x)$ is the probability of an image pixel having value equal to x and $P(x,y)$ is the double histogram of the two registered images, defined as the probability of a pixel (i,j) having value of y in image B, given that the same pixel has a value of x in image A, according to the following equation: $P(x,y) = P(I_B(i,j) = y \mid I_A(i,j) = x)$.

3. *Average distance (AD)* as a criterion is employed to surface-based registration approaches where it represents the distance of segmented anatomical features. Let $I_1 : S_1 \to Z$ and $I_2 : S_2 \to Z$ be the two data sets and $U_1 = \{\mathbf{P}_i = (x_i, y_i, z_i), i = 1, 2, \ldots, n_1$ and $U_2 = \{\mathbf{Q}_j = (x_j, y_j, z_j), j = 1, 2, \ldots, n_2\}$ are the voxels sets of points corresponding to the segmented features of the two data sets, respectively. Then, the AD, of the set U_2 from the set U_1 is given by the following:

$$AD = \frac{1}{n_2}\sum_{j=1}^{n_2} d_j \tag{5.11}$$

where $d_j = \min_i \|\mathbf{Q}_j - \mathbf{P}_i\|$ and $\| \ \|$ is the Euclidean distance. The minimum the value of the AD, the more accurate the registration achieved. The estimation of the d_j is obtained by the application of the distance map [14,15]. According to the distance map, a table is constructed with the same dimensions for all data $I_1 : S_1 \to Z$ (e.g., for the tomography data, a 3-D table is formed). Each matrix element corresponds to the distance of each voxel belonging to S_1 from the closest point of U_1.

5.2.2.4 Determination of the Transformation Parameters Using Optimization Techniques

The determination of the transformation parameters strongly depends on the objective function, as well as on the medical data to be registered. Furthermore, the *search based methods*, provide an alternative, based on the optimization of a MOM between the reference and the float data sets, with respect to the transformation parameters. If the measure of match is well behaved (is continuous and has only one extreme), simple gradient based methods (steepest descent, conjugate gradient method [13]) or downhill simplex method [56] may suffice.

However, if the MOM has multiple extremes, presents discontinuities or can not be expressed analytically, as in the case of common clinical images, brute force based exhaustive search is a method that guarantees successful determination of the parameters. An exhaustive search in the case of an affine transformation model was proposed in order to register remote sensing images. It is assumed that some of the unknown parameters can be estimated one at a time, in a serial manner, thus converting the multidimensional search into a sequence of optimization problems of lower dimensionality [57]. Under this assumption, the authors were forced to narrow the range of the parameters in order to accelerate the execution of the program. However, in cases of nontrivial transformations with many independent parameters, exhaustive search is not possible.

The most attractive solution for search methods using nontrivial transformations is based on global optimization techniques. The most representatives are the simulated annealing (SA) and genetic algorithms (GAs).

A brief introduction of the most representative optimization techniques is then followed.

1. *The downhill simplex method (DSM)* is due to Nelder and Mead [58]. The method requires only function evaluations and not derivatives. The algorithm is supposed to make its own way downhill through the unimaginable complexity of an N-dimensional topography, until it encounters a local, at least, minimum (maximum). The DSM must be started not just with a single point, but with $N+1$ points, defining an initial simplex. If one of these points is an initial starting point, then the other N points could be expressed as $P_i = P_0 + e_i$, where e_i are N unit vectors. The DSM takes then a series of steps moving the point of simplex where the function is largest through the opposite face of the simplex to a lower point. Termination criteria can be delicate in any multidimensional minimization routine by identifying appropriate "steps". For example, it is possible to terminate the method wherever the vector distance moved in that step is fractionally smaller in magnitude than some tolerance.

2. *The genetic algorithms (GAs)* is a global optimization method, inspired by Darwinian evolution [59]. The method starts by creating a population of random solutions of the optimization problem. A solution to the problem usually consists of the values of the independent parameters of the function to be optimized (objective function). These values are often converted to binary and concatenated to a single string, called *individual*. In several cases it has been shown that this conversion does not offer substantial advantages and *real encoding* is used instead. The method treats each individual as an organism, assigning to it a measure of *fitness*. Each individual's fitness is estimated by the value of the objective function calculated over the values of the parameters that are stored in the individual. Using the principle of the *survival of the fittest*, pairs of fit individuals are selected to recombine their encoded parameters to produce offspring. The most basic genetic operators that act on the individuals are *crossover* and *mutation* [59]. In this way a new generation of solutions is produced which replaces the previous one. This process is formulated in pseudocode as follows:

> *initialize the first generation of n individuals randomly*
> *while (termination_condition is false)*

> { calculate the objective function of the n individuals
> select N/2 pairs of individuals
> apply crossover and mutation operator to produce offspring
> replace the current generation by the n offspring }

Because of the global search they provide, without the necessity for an optimal initial guess, GAs are a powerful tool in optimizing multidimensional, nonlinear objective functions, which present many local extremes. In such problems, the local search based methods will fail, unless an initial guess is given close to the required solution. Moreover, GAs have been successfully combined with local search techniques, such as the downhill simplex method or other gradient like methods (conjugate gradient), producing hybrid systems, in order to increase the convergence rate.

3. *The simulated annealing (SA)* presents an optimization technique that can process cost functions with arbitrary degrees of nonlinearities and arbitrary boundary conditions [60]. A function of a system's state, $f(\vec{x})$, described by the vector $\vec{x} = (x_i : i = 1,...,D)$, is minimized (respectively maximized) using a process called annealing. The system searches for optimal values of the function $f(\vec{x})$, while adding to the function a noise component, whose magnitude is a descending function of time. For every minimization problem, a quantity T, known as temperature, is defined as a descending function of time k and the function $T(k)$ is called annealing schedule. The very fast SA (VFSA) is then defined [61]:

$$T(k) = \frac{T_0}{k} \tag{5.12}$$

where k is a time index and T_0 is a starting value, large enough to cover the search space. In the case of a multidimensional cost function, a different temperature T could be assigned to each of the independent parameters of the transformation model used.

5.2.3 Application of 3-D Medical Image Registration

5.2.3.1 CT-MR Registration Schemes

In the specific application, four different well-known registration schemes based on global transformation were utilized: (a) a surface-based registration, (b) a volume-based registration based on the MI, (c) a volume-based registration based on the CC, and (d) a manual-based registration technique.

Analytically,

(a) Surface-based registration scheme

The main steps of the scheme are as follows:

Preprocessing step
- Noise removal.
- Extraction of outer surface of the CT and MR data sets using a threshold operation in conjunction with a filling algorithm to fill small holes or the 3-D seeded region growing algorithm to fill large holes [62].
- Preregistration by translating the points of the MR surface in order their center of mass to coincide with CT-surface's center of mass and scaling the MR surface points co-ordinates to compensate for the difference of voxel size between the two modalities according to following equation:

$$\begin{bmatrix} x' \\ y' \\ z' \end{bmatrix}_{\text{MRI}} = \begin{bmatrix} x_{\text{MRI}} + (\bar{x}_{CT} - \bar{x}_{\text{MRI}}) \\ y_{\text{MRI}} + (\bar{y}_{CT} - \bar{y}_{\text{MRI}}) \\ z_{\text{MRI}} + (\bar{z}_{CT} - \bar{z}_{\text{MRI}}) \end{bmatrix} \tag{5.13}$$

where $(x',y',z')_{\text{MRI}}$ and are $(x,y,z)_{\text{MRI}}$ the coarsely transformed and the original co-ordinates of the MR image data, $(\bar{x},\bar{y},\bar{z})_{\text{MRI}}$ and $(\bar{x},\bar{y},\bar{z})_{CT}$ are the mean values of the MR and CT data sets, respectively.

Transformation model	Affine transformation (according to Equation 5.5).
MOM	Average Euclidean distance between the CT and MR skin surfaces.
Optimization technique	GAs, with the followings implementation details: population=100, number of generations=100, total number of function evaluations=10,000, probability of crossover=1.0, probability of mutation/parameter=0.01–0.1 encoded into the individual, parameter encoding=real values, selection=tournament selection, type of crossover=linear and arithmetic, no hybridization, first generation=uniformly random, no speciation.

(b) Volume-based registration scheme based on the CC

The main steps of the scheme are as follows:

Preprocessing step	Noise removal.
Transformation model	Affine transformation (according to Equation 5.5).
MOM	Normalized CC (according to Equation 5.9).
Optimization technique	GAs, with the same implementation details.

(c) Volume-based registration scheme based on the MI

The main steps of the scheme are as follows:

Preprocessing step	Noise removal.
Transformation model	Affine transformation (according to Equation 5.5).
MOM	Normalized MI (according to Equation 5.10).
Optimization technique	GAs, with the same implementation details.

(d) Manual-based registration technique

In the case of the manual registration, the affine transformation parameters were obtained based on the co-ordinates of homologous anatomical pair points as they have been manually selected by an expert using the least squares method [63]. A total number of 25 pair points were selected by the expert at each execution corresponding to the most distinctive common anatomical points in the two data sets.

5.2.3.2 Acquired Medical Data Sets

X-ray CT and MR brain images were acquired from five patients (five pairs of data) at the Strahlenklinik of the Stadtische Kliniken Offenbach, in Germany. The CT images were acquired using a Siemens Somatom Plus scanner. Each CT slice contains 256×256 pixels,

TABLE 5.1

The Acquisition Parameters of the CT-MR Head Data Sets

Matching Scheme		Pixel Matrix	Acquisition Parameters				
			Slice Thickness (mm)	Slice Spacing (mm)	No Slices	Voxel Size (mm×mm×mm)	No Markers
Pair-1	CT	256×256	2	–	75	1.04×1.04×2	11
	MR	256×256	4	1	33	1.02×1.02×5	
Pair-2	CT	256×256	2	–	75	1.04×1.04×2	11
	MR	256×256	4	1	33	1.02×1.02×5	
Pair-3	CT	256×256	2	–	75	1.04×1.04×2	11
	MR	256×256	4	1	33	1.02×1.02×5	
Pair-4	CT	256×256	6	–	35	0.96×0.96×6	6
	MR	256×256	3	0.3	52	1.56×1.56×3.3	
Pair-5	CT	256×256	3	–	54	0.94×0.94×3	6
	MR	256×256	2	1	50	0.98×0.98×3	
	MR	256×256	2	1	48	0.94×0.94×3	

with 12 bits per pixel, with no interslice spacing or slice overlap for all CT data sets. The T1-weighted MR images were acquired using a Siemens SP 1.5-Tesla scanner. Each MR slice contains 256×256 pixels, using 8 bits per pixel, with constant interslice spacing for each MR data set. The acquisition parameters of the data used are listed in Table 5.1. The voxel size, for each data set in Table 5.1, represents the acquired size along the x- and y-axes, whereas the size along the z-axis is defined as the sum of the slice thickness and the slice spacing. External markers had been placed on the skull of each patient and were designed to be bright in CT and MR systems. They are constructed from plexus glass spheres with an external diameter of 5 mm and they were filled with a solution of gadolinium with water in portion 1/100 to enhance their visibility by the MR system. These skin markers were used for evaluating the performance of the four registration schemes.

5.2.3.3 Registration Results

The performance from the application of the aforementioned four medical registration schemes are visually demonstrated in Figures 5.2 and 5.3, where CT skull contours are superimposed on the corresponding MR transverse sections, for different CT-MR image pairs. The application of the surface-based and manual-based schemes result in slight inaccuracies in the placement of the CT skull contours, which often invade the brain and the zygomatic bone at the vicinity of the eye and they also fail to locate accurately the superior sagittal sinus. These inaccuracies were corrected by the combination of the two volume-based schemes which seems to produce similar qualitative registration results.

The performance of the four registration schemes was quantitatively assessed both under controlled environment as well as using the data sets listed in Table 5.1. Initially, the CT data set from Pair-1 of Table 5.1 was selected and it was subject to known transformations in terms of different displacements; thus, translations (dx, dy, dz) (in pixels) and rotations $(d\theta)$ around the perpendicular z-axis (in degrees) were made to the reference

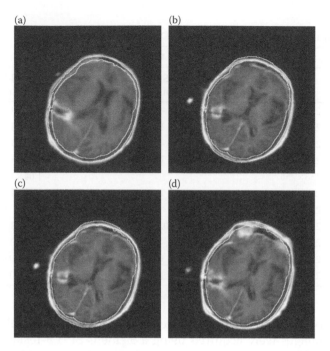

FIGURE 5.2
Visual assessment of the proposed registration schemes by superposing the CT skull contours (outer and inner) on the corresponding registered MR slice for the Pair-3, using: (a) The mutual information measure (MI). (b) The correlation coefficient measure (CC). (c) The surface-based registration. (d) The manual-based registration.

CT data set. Ten synthetic pairs were finally obtained. Then, the proposed four registration schemes were applied to each synthetic pair as follows: over ten executions for all synthetic pairs by the three automatic registration schemes and three trials for each pair by the manual-based registration. In Table 5.2, results on approximating the ideal translations and rotations are presented by the four registration schemes. It can be noticed that the two volume-based registrations produced better approximations against the surface-based and the manual-based registration schemes with the MI having an edge over the other three schemes.

Furthermore, the performance of the four registration schemes is quantitatively assessed on the five CT-MR data sets, as listed in Table 5.1. Since there is no golden standard criterion for comparing these registration schemes, it was decided to use the average distance of the centers of all skin markers before and after registration for each CT-MR data pair and for each registration scheme, as an independent criterion (MOM) to evaluate the performance of the registration schemes. The smaller the value of the specific criterion, the better the registration accuracy achieved. Yet again, the values of this criterion for all automatic registration schemes were averaged over ten independent executions for all image pairs to compensate for the stochastic (randomized) nature of the optimization method, whereas for the manual-based registration scheme, the values of this criterion were averaged over

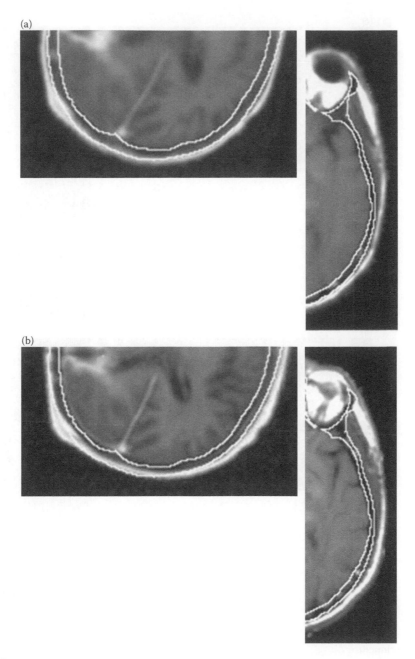

FIGURE 5.3
Magnified areas of the CT bone contours (outer and inner) superimposed on the corresponding MR slice for two pairs using: (a) The mutual information measure (MI). (b) The correlation coefficient measure (CC). (c) The surface-based registration. (d) The manual-based registration.

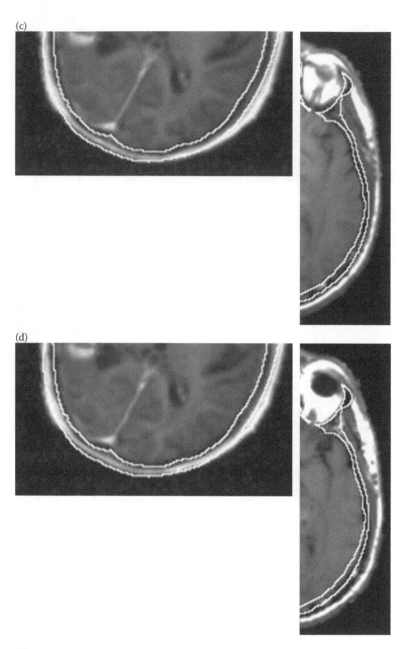

FIGURE 5.3   (continued)

three trials for each pair. Results from the application of these registration schemes are presented in Table 5.3 in terms of MOM (in mm). It can be observed that the volume-based registration scheme based on the MI is outperformed the other schemes for all image pairs. Also, the performance of the volume-based registration scheme based on the CC is slightly better than the surface-based registration scheme whereas the manual-based registration scheme performed significantly worse than all three automatic registration schemes.

TABLE 5.2

Performance of the Four Registration Schemes Based on the Mutual Information Measure (*MI*), the Correlation Coefficient Measure (*CC*), the Surface-based Registration, and the Manual-based Registration in Approximating Different Displacement Parameters (Translations (In Pixels) and Rotations (In Degrees)) for a CT Data Set

Synthetic Data Sets	Displacement Parameters	Registration Schemes			
		MI	CC	Surface-based	Manual-based
Pair-1	dx=5	5.04±0.14	5.31±0.16	5.28±0.17	6.08±0.73
	dy=0	−0.08±0.08	−0.21±0.16	−0.18±0.17	−0.74±1.16
	dz=0	−0.18±0.08	−0.15±0.18	−0.18±0.22	0.76±0.99
	θ=0	0.16±0.09	0.17±0.28	0.16±0.38	2.10±1.09
Pair-2	dx=10	9.89±0.07	10.27±0.12	9.78±0.07	8.98±1.26
	dy=0	−0.04±0.09	0.21±0.15	−0.24±0.15	−1.24±0.81
	dz=0	0.13±0.12	0.31±0.21	0.31±0.22	0.31±0.22
	θ=0	0.10±0.09	−0.25±0.26	0.27±0.25	2.37±1.35
Pair-3	dx=0	0.09±0.05	0.23±0.16	0.11±0.16	0.42±0.26
	dy=5	4.87±0.12	5.34±0.20	4.64±0.21	4.44±0.46
	dz=0	0.14±0.09	0.24±0.21	0.17±0.19	−0.79±0.32
	θ=0	0.26±0.15	0.29±0.26	0.31±0.24	0.55±0.29
Pair-4	dx=0	−0.14±0.09	0.22±0.19	−0.17±0.12	−0.23±0.11
	dy=10	10.09±0.06	10.26±0.12	9.72±0.09	9.74±0.14
	dz=0	0.07±0.07	−0.27±0.13	−0.17±0.13	−0.20±0.12
	θ=0	−0.22±0.25	−0.31±0.34	0.29±0.29	0.29±0.29
Pair-5	dx=0	0.11±0.12	−0.19±0.14	0.22±0.13	0.19±0.23
	dy=0	0.04±0.09	0.17±0.10	0.12±0.15	0.17±0.17
	dz=5	5.18±0.08	5.28±0.18	5.22±0.13	4.71±0.16
	θ=0	−0.06±0.04	−0.27±0.27	−0.17±0.24	0.17±0.24
Pair-6	dx=0	−0.05±0.09	−0.21±0.19	0.23±0.19	0.74±0.36
	dy=0	0.06±0.08	0.20±0.17	0.18±0.17	0.28±0.43
	dz=10	10.18±0.12	9.68±0.27	9.72±0.21	9.02±1.31
	θ=0	0.13±0.14	0.21±0.10	0.18±0.12	0.18±0.12
Pair-7	dx=0	0.10±0.04	−0.21±0.10	0.25±0.12	1.05±0.12
	dy=0	−0.11±0.09	−0.22±0.24	−0.18±0.14	−0.35±0.24
	dz=0	0.15±0.11	0.19±0.12	0.19±0.11	0.57±0.39
	θ=5	4.73±0.14	5.43±0.31	4.53±0.21	4.53±0.72
Pair-8	dx=0	0.07±0.08	−0.09±0.15	0.07±0.15	0.37±0.13
	dy=0	0.11±0.05	0.12±0.14	0.21±0.14	0.21±0.24
	dz=0	0.12±0.09	0.21±0.22	−0.19±0.17	−0.19±0.21
	θ=10	10.13±0.10	9.33±0.30	9.55±0.33	8.55±0.83
Pair-9	dx=5	4.98±0.07	5.38±0.14	5.42±0.21	5.82±0.49
	dy=5	5.02±0.10	5.12±0.19	5.12±0.26	4.70±0.25
	dz=5	5.11±0.08	5.36±0.21	5.23±0.14	4.23±0.24
	θ=5	5.08±0.07	4.58±0.19	4.63±0.27	4.33±0.41
Pair-10	dx=10	10.39±0.65	9.49±0.85	10.49±0.85	11.09±1.01
	dy=10	9.69±0.26	10.57±0.51	9.43±0.76	9.21±0.94
	dz=10	10.09±0.11	9.63±0.17	9.54±0.12	9.00±0.85
	θ=10	9.71±0.19	10.59±0.35	9.61±0.23	9.15±0.48

TABLE 5.3

Performance of the Four Registration Schemes Based on the Mutual Information Measure (*MI*), the Correlation Coefficient Measure (*CC*), the Surface-Based Registration, and the Manual-based Registration in Terms of the Average Distance of the Centers of all Skin Markers Before and After Registration (In mm) as a MOM for All Five CT-MR Data Sets (Continued)

Data Sets	Registration Schemes			
	MI	CC	Surface-based	Manual-based
Pair-1	1.06	1.52	1.61	2.32
Pair-2	1.08	1.78	1.92	2.27
Pair-3	0.95	1.34	1.29	2.86
Pair-4	1.23	2.07	2.25	3.01
Pair-5	1.59	2.31	2.76	3.45
MOM (Mean±S.D.)	1.182±0.249	1.804±0.395	1.966±0.569	2.782±0.495

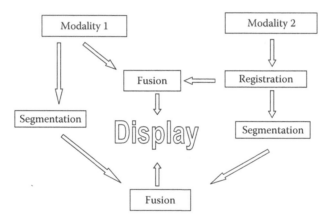

FIGURE 5.4
Generic Medical Image Fusion Scheme.

5.3 Medical Image Fusion

Imaging the same parts of human anatomy with the different modalities, or with the same modality at different times, provides the expert with a great amount of information, which must be combined in order to become diagnostically useful. Medical image fusion is a process that combines information from different images and displays it to the expert, so its diagnostic value is maximized.

Although much attention has been drown to the process of medical image registration, medical image fusion, as a prerequisite of the process of image registration, is not extensively explored in the literature, mainly because it is considered a straightforward step. A generic approach of the medical image fusion can be seen in Figure 5.4 where all the necessary steps are highlighted.

In this section, the concept of medical image fusion process, as well as the most representative techniques reported, will be revised and broadened to include several methods to combine diagnostic information.

5.3.1 Fusion at the Image Level

5.3.1.1 Fusion Using Logical Operators

This technique of fusing information from two images can be twofold. The reference image, which is not processed, accommodates a segmented region of interest from the second registered image. The simplest way to combine such information from the two images is by using a logical operator such as the XOR operator, according to the following equation:

$$I(x, y) = I_A(x, y)(1 - M(x, y)) + I_B(x, y)M(x, y) \tag{5.14}$$

where $M(x,y)$ is a Boolean mask that marks with 1s every pixel, which is copied from image B to the fused image $I(x,y)$.

Moreover, in certain cases, it is desirable to simply delineate the outline of the object of interest from the registered image and to position it in the coordinate system of the reference image. An example of this technique is the extraction of the boundary of an object of interest, such as a tumor from a MR image and overlay it on the coordinate system of the registered CT image of the same patient. Figure 5.5 demonstrates the fusion technique using the logical operators on the same patient data, where CT information (bone structure) is superimposed on two registered MR transverse slices.

5.3.1.2 Fusion Using Pseudocolor Map

According to this fusion technique, the registered image is rendered using a pseudo-color scale and transparently overlaid on the reference image [64]. There are a number of pseudocolor maps available, defined using psychophysiologic criteria or algorithmically. A pseudocolor map is defined as a correspondence of an *(R, G, B)* triplet to each distinct pixel value. Usually the pixel value ranges from 0 to 255 and each of the elements of the triplet varies in the same range, thus producing a true color effect. The color map is called pseudo because only one value is acquired for each pixel during the acquisition of the image data.

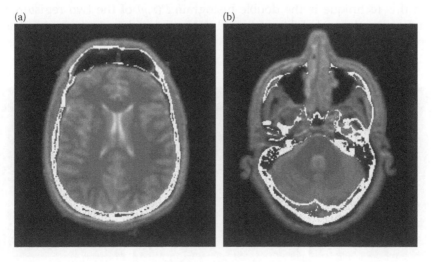

(a)   (b)

FIGURE 5.5
Fused images using the XOR operator: CT information, bone structures (a) and (b) superimposed on two registered MR transverse slices.

Two of the pseudocolor maps that are defined by psychophysiologic criteria are the *geographic color map* and the *hot body color map*. The *(R, G, B)* triplet values are defined as a function of the original pixel value, according to the following equation:

$$(R, G, B) = (R(pixel\_value), G(pixel\_value), B(pixel\_value)) \qquad (5.15)$$

The *RGB cube* color map is a positional color map, which maps the original values to a set of colors that are determined by traversing the following edges of a cube:

$$B(0,0,1) \rightarrow (1,0,1) \rightarrow R(1,0,0) \rightarrow (1,1,0) \rightarrow G(0,1,0) \rightarrow (0,1,1) \rightarrow B(0,0,1) \qquad (5.16)$$

This configuration sets up six color ranges each of which having $N/6$ steps, where N is the number of colors (distinct pixel values) of the initial image. If N is not a factor of 6, the difference is made up at the sixth range.

The *CIE diagram* color map follows the same philosophy, thus traversing a triangle on the *RGB cube*, in the following manner:

$$R(1,0,0) \rightarrow G(0,1,0) \rightarrow B(0,0,1) \qquad (5.17)$$

Another useful color map for biomedical images is the *gray code*. This color map is constructed by a three-bit gray code, where each successive value differs by a single bit. A number of additional color maps, such as *HSV (hue/saturation/value) rings* and *rainbow*, are constructed by manipulating the color in the HSV coordinate system. The conversion from HSV to RGB is straightforward and well documented in the literature. Figure 5.6 demonstrates the fusion using the RGB cube pseudocolor map of a SPECT image on the corresponding CT image of the same patient.

5.3.1.3 Clustering Algorithms for Unsupervised Fusion of Registered Images

Fusion of information on the image level can be achieved by processing both registered images in order to produce a fused image with appropriate pixel classification. The key quantity in this technique is the double histogram $P(x,y)$ of the two registered images, which is defined as the probability of a pixel (i,j) having value of y in image B, given that the same pixel has a value of x in image A:

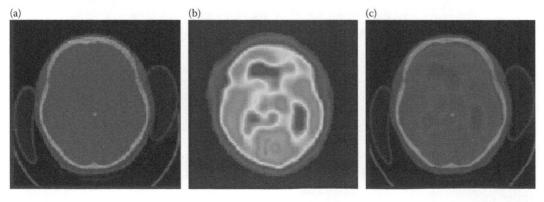

FIGURE 5.6
Fused image using the RGB cube pseudocolor map: (a) reference CT image, (b) registered SPECT image, in pseudocolor, and (c) fusion result.

$$P(x,y) = P(I_B(i,j) = y \mid I_A(i,j) = x) \qquad (5.18)$$

This quantity is very closely related to the entropy of the two images and can be very useful for effective tissue classification/segmentation, since it utilizes information from both (registered) images, rather than one. The concept of the double histogram can be generalized in the $\Re^n$ space fusing n registered images. Dynamic studies from SPECT are examples of such cases. In this general case, the n-dimensional histogram is defined as follows:

$$H^n(v_1, v_2, \ldots, v_n) = P(I_n(i,j) = y \mid I_1(i,j) = v_1; I_2(i,j) = v_2; \ldots; I_{n-1}(i,j) = v_{n-1}) \qquad (5.19)$$

Fusing the registered images to produced an enhanced or segmented/classified image becomes equivalent to partitioning the n-dimensional histogram to a desired number of classes, using a clustering algorithm.

The goal of clustering is to reduce the amount of data by categorizing similar data items together. Such grouping is pervasive in the way humans process information. Clustering is hoped to provide an automatic tool for constricting categories in data feature space. Clustering algorithms can be divided into two basic types: *hierarchical* and *partitional* [65]. Hierarchical algorithms are initialized by random definition of clusters and evolve by splitting large inhomogeneous clusters or merging small similar clusters. Partitional algorithms attempt to directly decompose the data into a set of disjoint clusters by minimizing a measure of dissimilarity between data points in the same clusters while maximizing dissimilarity between data points in different clusters.

5.3.1.3.1 The K-means Algorithm

The K-means algorithm is a partitional clustering algorithm, which is used to distribute points in feature space among a predefined number of classes [66]. An implementation of the algorithm applied, in pseudocode, can be summarized as follows:

Initialize the centroids of the classes, $i=1, \ldots, k$ at random $\vec{m}_i(0)$

$t=0$

  repeat

   for the centroids $\vec{m}_i(t)$ of all classes .

    locate the data points whose Euclidean distance from m_i is minimal

    set $\vec{m}_i(t+1)$ equal to the new center of mass of x_i, $\vec{m}_i(t+1) = \dfrac{\vec{x}_i}{n_i}$

   $t=t+1$

   until($\left|\vec{m}_i(t-1) - \vec{m}_i(t)\right| \le error$, for all classes i)

The above algorithm is applied to the n-dimensional histogram of the images to be fused, as defined in Equation 5.19. The centroids of the classes are selected randomly in the n-dimensional histogram and the algorithm evolves by moving the position of the centroids, so that the following quantity is minimized:

$$E = \sum_j \left(\vec{x}_j - \vec{m}_{c(\vec{x}_j)}\right)^2 \qquad (5.20)$$

where j is the number of data points index and $\vec{m}_{c(\vec{x}_j)}$ is the class centroid closest to data point $\vec{x}_j$. The above algorithm could be employed to utilize information from pairs of images, like CT, SPECT, MRI-T1, MRI-T2 and functional MRI to achieve tissue classification and

(a)   (b)   (c)

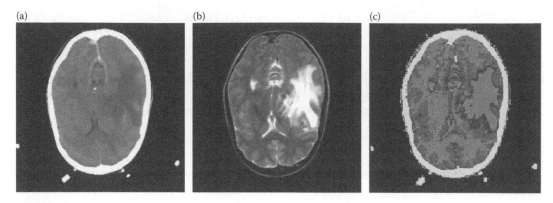

FIGURE 5.7
Fused image using the K-means algorithm for different classes of tissue: (a) reference CT image, (b) registered MRI image of the same patient, and (c) fusion result.

fusion. For tissue classification, nine different classes are usually sufficient, corresponding to background, cerebrospinal fluid, gray and white matter, bone material, abnormalities, etc. If fusion is required, then 64 or more classes could be defined, providing enhancement of the fine details of the original images. In Figure 5.7, a fused image using the K-means algorithm for different classes of tissue is produced by fusing CT and MR images of the same patient.

5.3.1.3.2 *The Fuzzy K-means Algorithm*

The fuzzy K-means algorithm (FKM) is a variation of the K-means algorithm, with the introduction of fuzziness in the form of a membership function. The membership function defines the probability with which each image pixel belongs to a specific class. It is also applied on the n-dimensional histogram of the images to be fused, as the previous method. The FKM and its variations have been employed in several types of pattern recognition problems [66,67].

If a number of n unlabelled data points is assumed, which have to be distributed over k clusters, a membership function u can be defined, such that:

$$u : (j, \vec{x}) \rightarrow [0, 1] \tag{5.21}$$

where $1 \leq j \leq k$ is an integer corresponding to a cluster. The membership function assigns to a data point x, a positive, less than or equal to one, probability u_{jx} which indicates that the point x belongs to class j. To meet computational requirements, the membership function is implemented as a 2-D matrix whose first index indicates the cluster whereas the second indicates the value of the data point. If the membership function is to express mathematical probability, the following constrain applies:

$$\sum_{i=1}^{n} u_{ij} = 1, \forall \ cluster \ j \tag{5.22}$$

The FKM algorithm evolves by minimizing the following quantity:

$$E = \sum_{j=1}^{k} \sum_{i=1}^{n} u_{ij}^{p} (\vec{x}_i - \vec{m}_j)^2 \tag{5.23}$$

where the exponential p is a real number greater than 1 and controls the fuzziness of the clustering process. The FKM algorithm can be described in pseudocode as follows:

$t=0$

initialize the matrix u randomly

repeat

calculate the cluster centroid using the formula: $m_j(t) = \dfrac{\sum\limits_{i=1}^{n} u_{ij}^{p} \vec{x}_i}{\sum\limits_{i=1}^{n} u_{ij}^{p}}$, $j = 1, \ldots, k$

calculate the new values for u: $u_{ij}(t) = \left(\sum \left(\dfrac{|\vec{x}_i - \vec{m}_j|}{|\vec{x}_i - \vec{m}_j|} \right)^{\frac{2}{p-1}} \right)^{-1}$

$t=t+1$

until $(\,|m_i(t-1) - m_i(t)| \le error$ or $|u_{uj\,i}(t-1) - u_{ij}(t)| \le error)$

5.3.1.3.3 Self Organizing Maps (SOMs)

The self-organizing map (SOM) is a neural network algorithm, which uses a competitive learning technique to train itself in an unsupervised manner. A comprehensive overview can be found in Section 5.2.3.1. The Kohonen model [68] comprises of a layer of neurons m, usually 1- or 2-D. The size of the network is defined by the purpose of the fusion operation. If it is desirable to simply fuse information to enhance fine detail visibility, a large number of neurons is required, typically 8×8 to 16×16. If it is *a priori* known that a small number of tissue classes exist within the images, then the number of neurons of the network is set equal to the number of classes. The resulting image is a segmented image, which classifies each pixel to one of these classes. A number up to nine is commonly the most meaningful choice in this case. Each neuron is connected to the input vector (data point) $\vec{x} \in \Re^n$ with a weight vector $\vec{w} \in \Re^n$. In the case of image fusion, the input signal has a dimensionality equal to the number of images to be fused, whereas the signal itself comprises of the values of the images at a random pixel (i,j):

$$\vec{x} = \{I_1(i,j), I_2(i,j), \ldots, I_n(i,j)\} \tag{5.24}$$

It becomes obvious that the proposed method can fuse an arbitrary number of images (assuming that registration has been performed for all of the images). However, the commonest used case is for $n=2$.

The above process achieves spatial coherence within the network of neurons as long as the input signal is concerned. This property is equivalent to clustering since after self-training, the neurons form clusters, which reflect to the input signal (data points). Figure 5.8 shows two fused images by applying the SOM algorithm on the same patient data imaged by a Gatholinium-enhanced MRI (Figure 5.8a) and MRI-T2 (Figure 5.8c). The upper right fused image (Figure 5.8b) is obtained using 256 classes whereas the lower right (Figure 5.8d) is classified using nine classes.

5.3.1.4 Fusion to Create Parametric Images

It is often necessary to fuse information from a series of images of a dynamic study, to classify tissues according to a specific parameter. The result of the fusion process is the

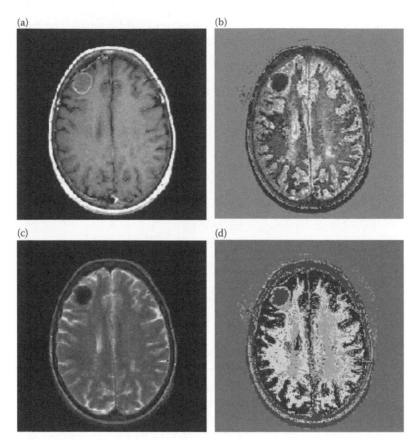

FIGURE 5.8
Fused images by applying the SOM algorithm on the same patient data imaged by: (a) Gatholinium-enhanced MRI and (c) MRI-T2. The user right fused image (b) is obtained using 256 classes whereas the lower right (d) is classified using nine classes.

creation of a *parametric* image, which visualizes pixel by pixel the value of the diagnostically useful parameter [69]. The required classification is performed by thresholding the parametric image at an appropriate level.

A typical example of this process is the gated blood pool imaging, performed for the clinical assessment of the ventricular function. A sequence of images, typically 16 per cardiac cycle, are acquired of the cardiac chamber of interest, which usually is the left ventricle. The assessment of ventricular function is performed by producing parametric images, visualizing several critical parameters:

- The *amplitude image* measures the change of volume as function of time, at a pixel by pixel basis, calculating the ejection fraction (EF) for every pixel in the sequence of images.
- The *phase image* visualizes the synchronization of cardiac contraction at a pixel by pixel basis.

The first parameter is valuable in assessing dyskinetic cardiac tissue, due to infarcts, whereas the second is a strong indication of fatal conditions such as fibrillation. Both images are calculated by means of Fourier analysis, using only the first few harmonics. In this way,

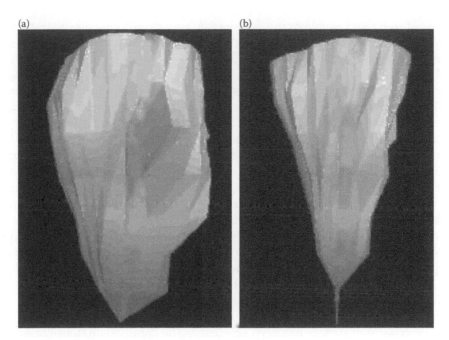

FIGURE 5.9
Fusion at the object level: the end-diastolic (a) and end-systolic (b) phases captured from the animated VRML files of a normal left ventricle.

information of a large number of images is fused into a single parametric image, thus enabling the expert to quantify clinical evaluation beyond any qualitative inaccuracies.

5.3.2 Fusion at the Object Level

Fusion at the object level involves the generation of either a spatio-temporal model or a 3-D textured object of the required object. Segmentation and triangulation algorithms must be performed prior to the fusion process.

Fusion at the object level is demonstrated by fusing temporal information on spatial domain. 4-D MR cardiac images, such as *gradient echo*, are obtained, providing both anatomical and functional information about the cardiac muscle [70]. The left ventricle from each 3-D image, covering the whole cardiac phase, is firstly segmented and the produced endocardiac surface is then triangulated. The radial displacement between the end-diastolic and end-systolic phase at each point of the triangulated surface is mapped on the surface, coded as pseudocolor, and thus producing a textured object (Figure 5.9) which visualizes information of myocardial wall motion during the entire cardiac cycle.

5.4 Conclusions

In this chapter, a comprehensive review of the most widely used techniques for automatic medical image registration and medical image fusion has been presented.

In terms of medical image registration, three main schemes have been reported: the point-based usually rendered as manual-based scheme, and the surface-based and the volume-based, rendered as automatic schemes. A generic automatic registration scheme based on global transformation was initially presented which consists of the selection of the image features that will be used during the matching process, the definition of an MOM that quantifies the spatial matching between the reference data set and the set to be transformed, and the application of an optimization technique that determines the independent parameters of the transformation model employed, according to the MOM. For each the aforementioned steps, a detailed description of alternative techniques that could be used towards image registration was presented. Based on the proposed generic registration scheme, four registration methods were utilized: a surface-based registration technique applied to common segmented anatomical features of the two data sets, two volume-based registration techniques using the CC and the MI as similarity measures, respectively, and finally a manual-based registration technique involving homologous anatomical points in both data sets as selected by an expert.

The aforementioned medical registration schemes have been successfully applied on 3-D CT-MR head registration. For all data sets, a global transformation has been used based on the implementation of the affine transformation method in conjunction with the GAs, as a global optimization process. The two volume-based schemes were operated on the image level without requiring any preprocessing while the surface-based registration scheme strongly depends on the initial segmentation step to obtain the CT-MR surfaces. The external skin surface has been selected as the common anatomical structure in both modalities for the specific application. Also, a sufficient number of anatomical points in two modalities have been selected by an experience radiologist and they were used towards manual-based registration. The qualitative and the quantitative results, for all CT-MR image pairs, have shown the advantageous performance of the automatic registration, in terms of average distance of the centers of all skin markers, against the manual-based registration.

Medical image fusion techniques have also been presented to increase diagnostic value from the combination of information from medical images of different modalities. The fusion process requires the prior application of a medical image registration scheme. Two main categories of the medical image fusion have been addressed: fusion at the image level and fusion at the object level. The most representative techniques of these categories have been revised and broadened throughout this chapter to include several methods, although more extensive research towards medical image fusion is required.

References

1. P.A. Van den Elsen, E.J. Pol, and M.A. Viergever. 1993. Medical image matching: A review with classification. *IEEE Eng. Med. Biol.*, 12(1), 26–39.
2. J.V. Hajnal, D.L.G. Hill, and D.J. Hawkes. 2001. *Medical Image Registration*. CRC Press, Florida, USA.
3. D.J. Hawks, D.L.G. Hill, and E.C.M.L. Bracey. 1992. Multimodal data fusion to combine anatomical and physiological information in the head and the heart. In *Cardiovascular Nuclear Medicine and MRI*, J.H.C. Reiber and E.E. Van der Wall (Eds.). Kluwer Academic Publishers, Dordrecht, The Netherlands, 113–130.
4. C.R. Maurer, J.M. Fitzpatrick, M.Y. Wang, R.L. Galloway, R.J. Maciunas, and G.G. Allen. 1997. Registration of head volume images using implantable fiducial markers. *IEEE Trans. Med. Imag.*, 16, 447–462.

5. W.E.L. Grimpson, G.J. Ettinger, S.J. White, T. Lozano-Perez, W.M. Wells, and R. Kikinis. 1996. An automatic registration method for frameless stereotaxy, image guided surgery, and enhanced reality visualization. *IEEE Trans. Med. Imag.*, 15, 129–140.

6. V. Morgioj, A. Brusa, G. Loi, E. Pignoli, A. Gramanglia, M. Scarcetti, E. Bomburdieri, and R. Marchesini. 1995. Accuracy evaluation of fusion of CT, MR, and SPECT images using commercially available software packages (SRS Proto with IFS). *Int. J. Rad. Oncol., Biol. Phy.*, 43(1), 227–234.

7. D.L.G. Hill, D.J. Hawkes, J.E. Crossman, M.J. Gleeson, T.C.S. Cox, et al. 1991. Registration of MR and CT images for skull base surgery using point-like anatomical features. *Brit. J. Radiol.*, 64, 1030–1035.

8. P. Clarysse, D. Gibon, J. Rousseau, S. Blond, C. Vasseur, et al. 1991. A computer-assisted system for 3-D frameless localization in stereotaxic MRI. *IEEE Trans. Med. Imag.*, 10, 523–529.

9. C. Evans, T.M. Peters, D.L. Collins, C.J. Henri, S. Murrett, G.S. Pike and W. Dai. 1992. 3-D correlative imaging and segmentation of cerebral anatomy, function and vasculature. *Automedica*, 14, 65–69.

10. R. Amdur, D. Gladstone, K. Leopold, and R.D. Hasis. 1999. Prostate seed implant quality assessment using MR and CT image fusion. *Int. J. Oncol., Biol., Phy.*, 43, 67–72.

11. F.L. Bookstein. 1989. Principal warps: thin-plate splines and the decomposition of deformation. *IEEE Trans. Patt. Anal. Mach. Intell.*, 11, 567–585.

12. C.A. Pelizzari, G.T.Y. Chen, D.R. Spelbring, R.R. Weichselbaum, and C.T. Chen. 1989. Accurate three-dimensional registration of CT, PET and/or MR images of the brain. *J. Comput. Assist. Tomogr.*, 13, 20–26.

13. W. Press, B. Flannery, S. Teukolsky, and W. Vetterling. 1992. *Numerical Recipes in C.* Cambridge University Press, Cambridge.

14. D. Kozinska, O.J. Tretiak, and J. Nissanov. 1997. Multidimensional alignment using Euclidean distance transform. *Graph. Models Image Proc.*, 59, 373–387.

15. G. Borgerfors. 1988. Multidimensional chamfer matching: A tree edge matching algorithm. *IEEE Trans. Patt. Anal. Mach. Intell.*, 10, 849–865.

16. M. Van Herk and H.M. Kooy. 1994. Automatic three-dimensional correlation of CT-CT, CT-MRI, and CT-SPECT using chamfer matching. *Med. Phy.*, 21, 1163–1178.

17. M. Jiang, R.A. Robb, and K.J. Molton. 1992. A new approach to 3-D registration of multimodality medical image by surface matching. *Visualization in Biomedical Computing*, Proc. SPIE-1808, 196–213.

18. P.J. Besl and N.D. McKay. 1992. A method for registration of 3-D shapes. *IEEE Trans. Patt. Anal. Mach. Intell.*, 14, 239–256.

19. A. Maurer, G.B. Aboutanos, B.M. Dawant, R.J. Maciunas, J.M. Fitzpatrick. 1996. Registration of 3-D images using weighted geometrical features. *IEEE Trans. Med. Imag.*, 15, 836–849.

20. J.L. Herring, B.M. Dawant, C.R. Maurer, D.M. Muratore, G.L. Galloway, and J.M. Fitzpatrick. 1998. Surface-based registration of CT images to physical space for image guided surgery of the spliene: A sensitivity study. *IEEE Trans. Med. Imag.*, 17, 743–752.

21. J.Y. Chianos and B.J. Sallivan. 1992. Coincident bit counting–A new criterion for image registration. *IEEE Trans. Med. Imag.*, 12, 30–38.

22. T. Radcliffe, R. Rajapekshe, and S. Shaler. 1994. Pseudocorrelation: A fast, robust, absolute, gray level image alignment algorithms. *Med. Phy.*, 41, 761–769.

23. J.B.A. Maintz, P.A. van den Elsen, and M.A. Viergever. 1995. Comparison of feature based matching of CT and MR brain images, In *Computer Vision, Virtual Reality, and Robotics in Medicine*, N. Ayache (Ed.). Springer-Verlag, Berlin 219–228.

24. J.B.A. Maintz, P.A. van den Elsen, and M.A. Viergever. 1996. Evaluation of ridge seeking operators for multimodality medical image matching. *IEEE Trans. Patt. Anal. Mach. Intell.*, 18, 353–365.

25. P.A. Van den Elsen, J.B.A. Maintz, E.J.D. Pol, and M.N. Viergever. 1995. Automatic registration of CT and MR brain images using correlation of geometrical features. *IEEE Trans. Med. Imag.*, 14, 384–396.

26. P.A. Van den Elsen, E.J.D. Pol, T.S. Samanaweera, P.F. Hemler, S. Napel, and S.R. Adler. 1994. Grey value correlation techniques used for automatic matching of CT and MRI brain and spine images. *Visual. on Biomed. Comput.*, 2359, 227–237.

27. C. Studholme, D.L.G. Hill, and D.J. Hawkes. 1995. Automated registration of truncated MR and CT datasets of the head. *Proc. Br. Mach. Vision, Assoc.*, 27–36.

28. R.P. Woods, J.C. Mazziotta, and S.R. Cherry. 1993. MRI-PET registration with automated algorithm. *J. Comput. Assist. Tomogr.*, 97, 536–546.

29. A. Collignon, F. Maes, D. Delaere, D. Vandermeulen, P. Suetens, and G. Marshal. 1995. Automated multimodality image registration based on information theory. In *Information Processing in Medical Imaging 1995*, Y. Bizais, C. Barillot, and R. Di Paola (Eds.). Kluwer, Dordrecht, The Netherlands, 263–274.

30. F. Maes, A. Collignon, D. Vandermeulen, G. Marchal, and P. Suetens. 1997. Multimodality image registration by maximization of mutual information. *IEEE Trans. Med. Imag.*, 16, 167–198.

31. J. West, J.M. Fitzpatrick, M.Y. Wang, B.M. Dawant, C.R. Maurer, R.M. Kassler, and R.J. Maciunas. 1999. Retrospective intermodality registration techniques for images of the head: Surface-based versus volume-based. *IEEE Trans. Med. Imag.*, 18, 147–150.

32. D. Lemoine, C. Barillot, C. Cibaud, and E. Pasqualinin. 1991. A 3-D CT stereoscopic deformation method to merge multimodality images and atlas datas. *Proc. Computer Assisted Radiology*, Springer-Verlag, Berlin, 663–668.

33. J.C. Gee, M. Reivich, and R. Bajscy. 1993. Elastically deforming a 3-D atlas to match anatomical brain images. *J. Comput. Assist. Tomogr.*, 17, 225–236.

34. L. Bookstein. 1991. Thin plate splines and the atlas problem for biomedical image. In *Lectures Notes in Computer Sciences, Information Processing in Medical Imaging*, A.C.F. Colchester, D.J. Hawkes (Eds.). Springer Verlag, Berlin, 26–342.

35. R. Bajcsy and S. Kovacic. 1989. Multiresolution elastic matching. *Comp. Vision, Graph, Image Processing*, 46, 1–29.

36. M. Singh, R. R. Brechner, and V. W. Henderson. 1984. Neuromagnetic localization using magnetic resonance images. *IEEE Trans. Med. Imag.*, 11, 125–134.

37. M. Moshfeghi. 1991. Elastic matching of multimodality medical images. *Graph. Models Image Proc.*, 53, 271–282.

38. C. Davatzikos, J.L. Prince, and R.N. Bryan. 1996. Image registration based on boundary mapping. *IEEE Trans. Med. Imag.*, 15, 112–115.

39. H. Chang and J.M. Fitzpatrick. 1992. A technique for accurate magnetic resonance imaging in the presence of field inhomogeneities. *IEEE Trans. Med. Imag.*, 11, 319–329.

40. J.M. Fitzpatrick, J.B. West, and C.R. Maurer. 1998. Predicting error in rigid-body point-based registration. *IEEE Trans. Med. Imag.*, 17, 694–702.

41. C.R. Maurer, R.J. Maciunas, and J.M. Fitzpatrick. 1998. Registration of head CT images to physical space using a weighted combination of points and surfaces. *IEEE Trans Med. Imag.*, 17, 753–761.

42. C. Barillot, B. Gibaud, J.C. Gee, and D. Lemoine. 1995. Segmentation and fusion of multimodality and multisubjects data for the preparation of neurosurgical procedures. In Medical Imaging: Analysis on Multimodality 2-D/3-D Image, L. Beolchi and M.H. Kuhn (Eds.). *Studies in Health Technology and Informatics*, 19, 70–82.

43. R.C. Gonzalez and P. Wintz. 1977. *Digital Image Processing*. Addison-Wesley, Reading, MA.

44. B. Fischl and E. Schwartz. 1995. Learning an integral equation aroximation to nonlinear anisotropic diffusion in image processing. *Tech Report CAS/CNS-TR-95-033*, Boston University, Department of Cognitive and Neural Systems.

45. J. Grevera and J.K. Udupa. 1996. Shape-based interpolation of multidimensional grey-level images. *IEEE Trans. Med. Imag.*, 15, 881–892.

46. L. Schad, R. Boesecke, W. Schlegel, G. Hartmann, V. Sturm, et al. 1987. Three-dimensional image correlation of CT, MR and PET studies in radiotherapy treatment planning of brain tumors. *J. Comput. Assisted Tomogr.*, 11(6), 948–954.

47. D. Hawks, D.L.G. Hill, and E.C.M.L. Bracey. 1992. Multimodal data fusion to combine anatomical and physiological information in the head and the heart. In: *Cardiovascular Nuclear Medicine and MRI*, J.H.C. Reiber, E.E. van der Wall (Eds.). Kluwer Academic Publishers, Dordrecht, The Netherlands, 113–130.

48. K.S. Arun, T.S. Huang, and S. Blostein. 1987. Least-squares fitting of two 3-D sets. *IEEE Trans. PAMI*, 9(5), 698–700.

49. M. Singh, W. Frei, T. Shibata, G. Huth, and N. Telfer. 1979. A digital technique for accurate change detection in nuclear medical images–with application to myocardial perfusion studies using thallium-201. *IEEE Trans. Nucl. Sci.*, 26(1), 565–575.

50. E. Kramer, M. Noz, J. Sanger, A. Megibaw, and G. Maguire. 1989, CT-SPECT fusion to correlate radiolabeled monoclonal antibody uptake with abdominal CT findings. *Radiology*, 172(3), 861–865.

51. K. Toennies, J. Udupa, G. Herman, I. Wornom, and S. Buchman. 1990. Registration of 3D objects and surfaces. *IEEE Comp Graph AI*, 10(3), 52–62.

52. E. Peli, R. Augliere, and G. Timberlake. 1987. Feature-based registration of retinal images. *IEEE Trans. Med. Imaging*, 6, 272–278.

53. L. Junck, J.G. Moen, G.D. Hutchins, M.B. Brown, and D.E. Kuhl. 1990. Correlation methods for the centering, rotation and alignment of functional brain images. *J. Nucl. Med.*, 31(7), 1220–1226.

54. A. Aicella, J.H. Nagel, and R. Duara. 1988. Fast multimodality image matching. *Ann. Int. Conf. IEEE Eng. Med. Biol. Soc.*, IEEE Comp Soc Press, Los Alamos, 10, 414–415.

55. A. Venot and V. Leclerc. 1984. Automated correction of patient motion and gray values prior to subtraction in digitized angiography. *IEEE Trans. Med. Im*, 3(4), 179–186.

56. S.L. Jacoby, J.S. Kowalik, and J.T. Pizzo. 1972. *Iterative Methods for Nonlinear Optimization Problems*. Prentice Hall, Englewood Cliffs, NJ.

57. C. Fuh and P. Maragos. 1991. Motion displacement estimation using an affine model for image matching. *Optical Eng.*, 30, 881–887.

58. J.A. Nelder and R. Mead. 1965. The downhill simplex method. *Comput. J.*, 7, 308–313.

59. D. Goldberg. 1989. *Genetic Algorithms in Optimization, Search and Machine Learning*. Addison-Wesley, Reading, MA.

60. E. Aarts and Van Laardhoven. 1987. *Simulated Annealing: Theory and Practice*. John Wiley and Sons, New York, NY.

61. L. Ingber. 1993. Simulated annealing practice versus theory. *J. Math. Comput. Modelling*, 18(11), 29–57.

62. R. Adams and L. Bischof. 1994. Seeded region growing. *IEEE Trans. Pattern Anal. Mach. Intell.*, 16(6), 641–647.

63. S. Umeyama. 1991. Least-squares estimation of transformation parameters between two point patterns. *IEEE Trans. Patt. Anal. Mach. Intell.*, 13, 376–380.

64. J. Gomes, L. Darsa, B. Costa, and L. Velho. 1998. Warping and morphing of graphical objects. In *Computer Graphics*. Morgan Kaufman Series, San Francisco, USA.

65. A.K. Jain and R.C. Dubes. 1988. *Algorithms for Clustering Data*. Prentice-Hall, Englewood Cliffs, NJ.

66. R. Rezaee, C. Nyqvist, P. van der Zwet, E. Jansen, and J. Reiber. 1995. Segmentation of MR images by a fuzzy C-means algorithm. *Comput. Cardiol.*, 21–24.

67. J.C. Bezdek. 1987. Partition structures: A tutorial. In *The Analysis of Fuzzy Information*, J.C. Bezdek (Ed.). CRC Press, Boca Raton, FL.

68. T. Kohonen. 1982. Self organized formation of topologically correct feature maps. *Biol. Cybernet.*, 43, 59–69.

69. P. Sharp, H. Gemmel, and F. Smith. 1989. *Practical Nuclear Medicine*. IRL Press, Oxford University, Oxford.

70. E. Wall. 1998. Magnetic resonance in cardiology: which clinical questions can be answered now and in the near future? In *What's New in Cardiovascular Imaging*, J.H.C. Reiber and E.E. Vander Wall (Eds.). Kluwer Academic Publishers, Dordrecht, 197–206.

6

Segmentation, Registration, and Fusion of Medical Images

Marius Erdt

Fraunhofer Institute for Computer Graphics

CONTENTS

6.1 Segmentation

In general, the term *segmentation* denotes the process of assigning sets of pixels to one or more distinct groups that are defined by the needs of the respective image processing task. Regarding medical imaging, volumetric segmentation is based on the classification of voxels to regions, which usually correspond to objects or organs in the data set. For example, the visualization of the human body's skeleton can be realized by dividing a computer tomography image into the two classes *bones* and *nonbones* and then pass the result to a renderer. Finding criteria to decide which voxels in the volume are similar or share a common property is therefore the essential part of every segmentation technique.

The result of segmentation is a classification that labels every voxel to be part of a certain region. This is referred to as *binary segmentation* since a voxel either shares a property with its neighbors or not. Because medical imaging techniques like *computed tomography* (CT) or *magnetic resonance imaging* (MRI) produce discrete volume grids, certain voxels may represent two different materials, e.g., on object boundaries. The so called *partial volume effect* leads to an uncertainty whether the voxel has to be assigned to the one object or the other. In contrast, *fuzzy segmentation* only computes a probability that a voxel belongs to a certain region. In the remainder of this chapter we will, however, focus on binary segmentation since most medical imaging classification techniques target at a clear distinction of the detected structures. For an overview of fuzzy segmentation techniques see Yoo [1].

Since segmentation often is a prerequisite for medical applications like therapy planning, visualization tasks or diagnosis, a large number of approaches have been proposed working either on 2D or 3D data. In the following section we will discuss some of the most important methods used in medical applications. In order to classify the different approaches from a medical point of view we will distinguish methods that are purely based on intensities from algorithms that incorporate prior knowledge of the structure to be segmented, e.g., through an organ model.

The methods to be described have proven to yield good results in the context of medical imaging. Some of them have been in practical use for over 20 years. However, the right choice of methods is critical for the success of the respective segmentation task. Algorithms that work well for bone segmentation may fail in the case of brain labeling and vice versa. Up to now, there is no universal method to deal with all medical segmentation tasks, so a decision which method to use has to be made from case to case.

6.1.1 Intensity-based Segmentation

6.1.1.1 Point-based Segmentation

The simplest approach to address the segmentation problem is to classify a voxel solely based on its intensity. Such methods can be denoted as *point-based segmentation* algorithms since they are not incorporating any local relationships between the voxels.

From that perspective, a segmentation can be made by determining a value range that assumingly uniquely contains the gray values of the structure to be segmented. This approach is called *thresholding* and is often used by more sophisticated methods as a preprocessing step to build a coarse-grained segmentation.

Medical acquisition techniques like CT or MRI have the advantage that it is roughly know *a priori* to which intensities different tissues will be mapped. Thresholds can therefore be directly determined. However, the value ranges of tissues slightly vary from scan to scan and between different patients, so a manual selection is generally not sufficient.

In order to automatically find suitable thresholds, usually the histogram of the image is inspected and searched for two or more local maxima. The appropriate thresholds are then determined by the minima between them. Unfortunately, most of the time the gray value ranges of different objects in images overlap. This has a direct impact on the appearance of the histogram, since the area of a local minimum is not necessarily the place with a minimum overlap of two corresponding distributions. For example, Figure 6.1 shows the histogram of an image that contains two different objects. One object has a very broad range of gray values (i.e., its histogram shows a wide distribution) while another is characterized by a high contrast to the background (and shows a narrow distribution). Here, the place in the cumulative histogram where both objects have an equal amount of misclassified pixels is not the local minimum between the two peaks.

An algorithm called *k-means* tries to solve this problem by dividing the histogram into k clusters such that a metric between the histogram's elements and the centroids of the clusters is minimized. For example, an image is divided into two clusters, one for an object and one for the background. First, the centroids of the clusters are, e.g., randomly, initialized. As a metric one may choose the minimum gray level distance. The intensity of each pixel is therefore compared to the mean intensity (centroid) of each cluster and is then assigned to the most similar one. Afterwards, the mean intensities are updated and the procedure starts anew until the algorithm converges. The result is a separation of the object from the background. One can think of generalizing this procedure to work with n-dimensional data vectors instead of just gray values. In addition, other metrics like the sum of the variance over all clusters or the total distance between every value and their centroids can be used.

The method of k-means clustering was also applied to fuzzy segmentation, here called *fuzzy k-means* or *fuzzy c-means*. Its idea is to assign weights to each element denoting the degree of membership to a cluster. Apart from that, the procedure is analogous to k-means.

While having the advantage of being simple and fast, the major drawback of these algorithms is that they do not necessarily always converge to the same result—a consequence of the random placing of cluster centroids.

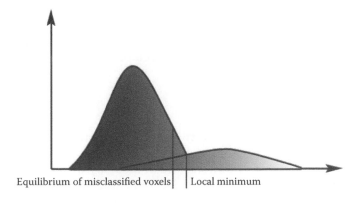

Equilibrium of misclassified voxels | Local minimum

FIGURE 6.1
The histogram of two different objects, which are characterized by a broad and a narrow distribution, respectively.

Another well known multimodal automatic thresholding-technique is the *otsu method* [2], which automatically divides the image into two or several (*multilevel otsu method*) classes. Talking about clusters, the idea is to find a threshold such that the gray value distribution in each cluster gets as narrow as possible. Since the distribution of one cluster gets wider when another gets tighter, the goal is to minimize the *combined* spread of all clusters.

Let us say we want to segment a single object from the image background. For a given threshold T we can compute the mean μ_{back} and μ_{object} for the two resulting clusters and weight those with the probabilities ω_{back} and ω_{object} of a pixel belonging to the respective cluster. The following term is called the *between-class* variance:

$$\sigma^2_{between}(T) = \omega_{back}(T) \cdot \omega_{object}(T) \cdot \left[\mu_{back}(T) - \mu_{object}(T) \right]^2 \tag{6.1}$$

You can think of the between-class variance as a measure of distance between the peaks of each distribution. In order to minimize the combined spread we therefore have to *maximize* the between-class variance. However, we practically have to test every possible threshold, calculate the means and then choose the maximal result, so this algorithm means a heavy computational burden. Therefore the above formula is further simplified. It can be shown that maximizing the between-class variance is equal to maximize a *modified between-class* variance denoted as:

$$\sigma^2_{betweenModified} = \sum_{j=0}^{k} \omega_j \mu_j^2 , \tag{6.2}$$

where k is the number of thresholds that shall be applied on the image. Liao et al. [3] showed that it is possible to precompute all sums of ω and μ in a look-up-table. The modified between-class variance can then be recursively computed by adding up the precomputed results of the table.

Using the described improvements allows for a noticeable speedup making the algorithm interesting for medical image processing. In Figure 6.2 the segmentation result of a cardiac CT data set using the otsu-method is shown.

Thresholds by their nature tend to produce either small islands that are not part of the object to be segmented (but share the same gray value) or result in segmentation holes (because the threshold was set too high). Therefore it is common to use the methods of *mathematical morphology* to close those holes and remove islands by applying *dilation* and *erosion* operators, respectively.

6.1.1.2 Edge-based Segmentation

In medical imaging, usually whole organs have to be segmented. It is therefore a natural approach to look at object boundaries which are often characterized by an intensity difference between the different tissues. Such intensity discontinuities or edges can be found by computing the derivatives of the local intensity function. Usually those derivatives are approximated by convolving the volume with filter masks. Several edge detection filters have been proposed, e.g., the popular Canny filter [4]. However, the filter results frequently need to be post processed, because object contours are often not closed. Furthermore noise can lead to the detection of erroneous edges which makes an

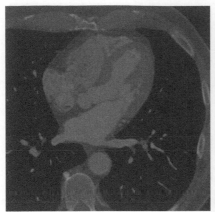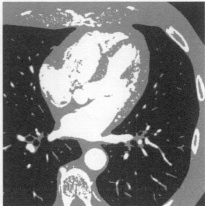

FIGURE 6.2
Cardiac CT data set and segmented result using the otsu-method. The different gray levels represent air (black), fat and muscle tissue (gray), bones, and cardiac structures which were enhanced by injection of a contrast agent (white).

automatic segmentation of structures problematic. Nevertheless edge detection plays an important role in the application pipeline of many more sophisticated methods like *snakes* covered in the following.

Snakes or *active contours* [5] are defined as 2D curves, which are placed (e.g., manually) in the image as an initial contour for the structure of interest. In order to adapt the contour to the real object boundaries while preventing high local deformations at the same time, two forces are defined. An *internal force* tries to preserve the form of the contour by providing a resistance against high curvatures. An *external force* opposes to the internal force by adapting the contour to the local gray values. This is often done by calculating the local gradient at the contour position and defining the external force to be antiproportional to the regularized gradient. The contour adaption is then iteratively performed by searching for the equilibrium between internal tension and external force. The principle of snake segmentation is shown in Figure 6.3.

The use of internal forces makes snakes robust against local noise. However, a good initial starting position is necessary in order to prevent the contour of being stuck in local maxima. In addition, the robustness against noise is achieved at the cost of an imprecise local adaption. Hence, fine structures with high curvature like vessels cannot be efficiently segmented.

The concept of snakes has also been applied to 3D using surfaces instead of curves. There exist numerous approaches, mostly differing in terms of proposed energy functions and optimization strategies. Usually all theses methods are summed up under the term *deformable models*. For a comprehensive introduction, see Yoo [1] and Singh et al. [1,5].

Extensions to snakes are the so called *active shape models* and their derivatives the *active appearance models* introduced by Cootes et al. [6,7]. They are characterized by incorporating statistical information about the shape of the structure to be found, i.e., the shape of the final contour is constrained to vary only between the results of previously labelled training images. Hereby, a *statistical shape model* is built by using principle components analysis. The training sets are first aligned such that the contour points build an *n*-dimensional cloud. Afterwards, the first few main axes of that cloud are extracted, which is usually sufficient to cover the typical modes of variation between all training

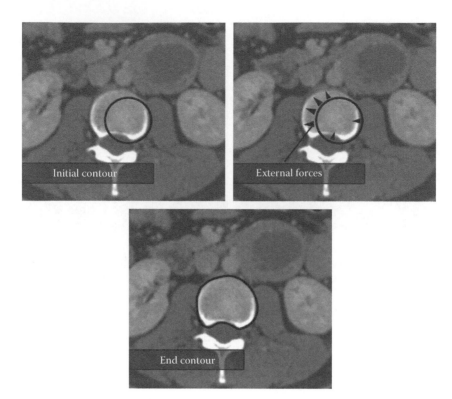

FIGURE 6.3
The snakes segmentation approach.

shapes. The adaptation to the current image is similar to the snake approach: a starting contour is iteratively refined by pulling it towards local features like edges. In every step the pose and shape parameters are updated according to the shape model in order to optimally fit the contour to the found points.

6.1.1.3 Region-based Segmentation

Instead of considering only single voxels as in point-based segmentation, several methods have been developed that incorporate local relationships between the voxels, i.e., target at building of regions.

One of the simplest yet very often used methods of 2D and 3D region based segmentation is the *region growing* algorithm. As the name suggests, a segmentation "grows" from initially placed points called *seeds* by aggregating neighboring pixels or regions according to some similarity criterion. Let us assume we want to segment a contrast enhanced vessel in a CT volume (see Figure 6.4a). We first choose a voxel as a seed point from which we know that it lies inside the vessel. In addition, we define a gray value range that should cover all intensities that occur within the vascular tree, i.e., we get a lower and an upper threshold, T_l and T_u, as a measure of similarity. The region growing is now iteratively going through all segmented voxels, comparing the current intensity with all neighbors and adding those voxels that are between T_l and T_u. As a result, a connected segmentation of the vessel is obtained (Figure 6.4b). In this example we approximately knew the intensity range of the vessel so we could set the thresholds accordingly. However, in many

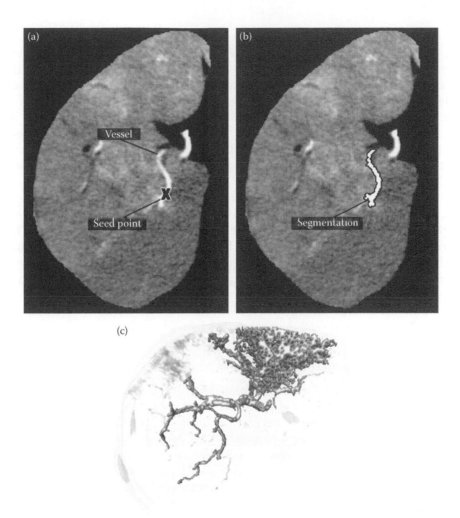

FIGURE 6.4
The region growing algorithm. Placing of a seed point inside a vessel (a) and resulting segmentation (b). (c): the segmentation "leaks" into the surrounding tissue.

image processing scenarios this value range cannot be precisely determined in advance. Wrong thresholds often result in a "leaking" of the region growing to the background as Figure 6.4c shows. Another problem arises if neighboring objects share a similar intensity to the structure to be segmented. The algorithm then leaks into those areas as well.

Nevertheless region growing is a widely used segmentation method due to its computational simplicity and the fact that the connectivity of all voxels grown from a seed point is ensured. This is a favourable property, because most image processing tasks require the segmentation of whole objects.

Another algorithm that iteratively groups pixels into regions is the so called *watershed transform*. Its idea is inspired by the observation of rain falling on a nonflat area with peaks and valleys. Typically the water will pool at the local minima of the region and build small lakes. Applied to image processing, the gray values of an image can be thought of as height differences on a rectangular surface. A segmentation is now created by picking the minimum gray value v_{min} as the sources of basic *catchment basins*. In the next step all neighboring pixels with an intensity of $v_{min}+1$ are added to the basins. If a pixel with the according intensity does not adjoin to a basin, a new one is created at that place. After some iterations the catchment basins will usually meet. The borders between them are called *watershed lines* and form the boundaries of the final segmentation. Because we are often interested in the segmentation of whole objects, the algorithm is usually applied to gradient images, where the edges denote the local maxima. Smoothing the image is also a widely used method to remove local minima that may occur due to noise. More sophisticated approaches to further improve the watershed segmentation were introduced by Meyer and Beucher [8,9] who developed methods to define a unique behavior in the presence of plateaus.

A fully automatic region-based segmentation method is known as the *split and merge algorithm*. It starts with considering the whole image as a single region. The region is then tested against a homogeneity criterion. If the test fails, the segmentation is split into four smaller regions (eight in 3D) of equal size. This procedure is recursively repeated until no further splitting is necessary. In a subsequent merging step, now adjacent regions are tested for similarity and merged accordingly. The result is an irregular segmentation of single homogeneous regions. Since the algorithm is splitting the image into equally sized blocks, it is suitable for segmenting local and contiguous structures instead of fine objects like vessels which may be spread over the entire data set.

6.1.2 Model-based Segmentation

So far we have been only working on the intensities without considering semantic, i.e., anatomical, knowledge. In *model-based segmentation* such knowledge is incorporated through a model of the object to be segmented. Recently, a variant of 3D deformable models based on polygonal organ meshes has become popular [10,11], because it allows a robust segmentation of anatomic structures with a minimum of user interaction. The approach, sometimes called *deformable surface model* or *deformable shape model*, starts by building a surface mesh of the organ to be segmented. To cover individual anatomical variations, often a mean shape from a database of different patients is generated. As it is known from deformable models, two energies are defined that strive towards local mesh adaptation and form preservation, respectively. Since the geometry of the model is not defined by a function (like 3D snakes), but by a surface mesh, the internal force is computed by comparing the original model to the deformed one, i.e., in every iteration the deviation from the initial surface is determined e.g., by rigidly registering both meshes and calculating the mean square error between the displacement of surface points. Such an approach is limiting the deformation in a global way, i.e., large deviations of the overall models shape are not possible, but (even large) local deformations may occur. The advantage of that approach in comparison to 3D snakes is that individual anatomical variations can be covered while the general shape of the organ is preserved.

Like 3D snakes, deformable surfaces need a good initial starting position. Therefore, in many medical imaging applications, the organ models are manually placed over the structure of interest.

6.1.3 Atlas-based Segmentation

Surface atlas: Instead of creating a single organ mesh to be used for every patient, there are approaches that build a statistical shape model to cover all possible anatomical variations learned from a training set of representative data sets. Lamecker et al. [12] create such an *atlas* for liver surface models out of 20 manually labeled CT volumes. A similar approach for MRI cardiac images has been proposed in Shang and Dössel [13]. Frangi et al. [14] present an automatic method to construct 3D statistical shape models that can be used as surface models in atlas-based segmentation.

Voxel atlas: In contrast to just model the organ contours, it is possible to incorporate voxel intensities themselves. Usually such an intensity based atlas is used together with *image registration techniques* to map labeled atlas structures to the target data set (the task of registration will be explained in detail in the upcoming section).

As in the case of surface models, there are several ways to create an intensity atlas. The simplest way is to manually label organ structures in a representative data set by an experienced physician. Furthermore, there are methods to build a mean atlas out of a training data base. Commowick and Malandain, and Rohling et al. [15, 16] give an overview of atlas construction techniques using databases in the context of CT imaging. In Guimond et al. [17] an automatic approach is proposed for building average brain models from MRI data sets. Shimizu et al. [18] build a probabilistic atlas for liver data sets by assigning to every voxel a probability of being part of the liver tissue. Segmentation is then performed using level sets.

So far we have described several methods for image segmentation using surface and voxel atlases. Moreover, it is possible to combine contour information as well as intensity structure in a combined model. This can be done by extending the active appearance model approach proposed by Cootes et al. [7] to 3D, i.e., a statistical model is built out of training images that incorporates both shape and structure. An adaption to the current data set can then be performed using the techniques of model-based segmentation described in the last section.

All atlas-based segmentation methods share the drawback that abnormal shapes are not covered by the training sets. This is especially problematic if the organ is affected by a disease like e.g., cancer, which is often the reason for image acquisition in the first place.

However, atlases can also be used to reduce the search space of an object. This means, segmentation algorithms does not have to be performed on the whole image content which drastically eases the segmentation task. For example, lymph nodes that are enlarged due to lymphoma cannot be modeled *directly* in an atlas, because the location of the cancer differs from patient to patient and only certain nodes are affected by the disease. Nevertheless, areas of lymph node clustering can be labeled in an atlas and mapped to the individual scan to support diagnosis. Som et al. [19] propose such a labeling for the cervical region.

6.2 Registration

In medical diagnosis and therapy often several images of the same patient from different image modalities are used. Having information from CT, MRI, PET (*positron emission tomography*) or ultrasound combined enables the physician to make more precise diagnosis, since all those acquisition techniques are sensible to different components of human anatomy like bones, soft tissue or, as in the case of PET, show functional information of anatomical structures. However, there are some problems that make a direct comparison of the resulting images difficult. Usually the position of patients changes when moving from one imaging modality to another. Figure 6.5 shows an example of a combination of a CT and MRT head scan which were taken under different angles. In addition, sometimes a significant amount of time elapses between two recordings, so the patient may lose or gain weight.

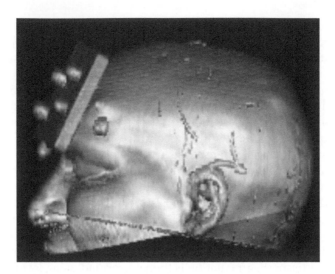

FIGURE 6.5
Registration of CT and MRI head images.

The goal of *registration* is therefore to provide a mapping of two different images that show comparable content. Finding the appropriate transformation between the images is the challenge of every registration algorithm. Usually one image is called the *reference image* and the other one the *target image*, whereby the target coordinates are mapped to the reference coordinates. There are several approaches that basically differ in the way what kind of transformations are used. For example, a so called *rigid registration* only allows translation and rotation to map the images while an *elastic registration* can deal with local deformations.

In the remainder of this section we will first discuss the different transformation models. Afterwards we will focus on how to get appropriate correspondences between the images that can be used in an optimization step to calculate the desired transformations. We then will present some medical registration application scenarios using the described methods. Finally, after registration, the task of a proper visualization of the result remains. This is referred to as *image fusion* and will be explained in detail in the last section.

6.2.1 Choice of Transformation

Before we are able to register images, i.e., to compute a one-to-one mapping between them, we have to define what kind of transformation model we want to use. Those models are characterized by the degrees of freedom (DOF) by which one of the images can be deformed to match another one. Typically, using only few free parameters limits the transformation to basic transformations like rotation or translation. By adding additional DOF it is possible to model more complex deformations to the price of higher computational costs. Furthermore, no closed form solutions exist to solve such problems in one step, but the result has to be computed iteratively.

Generally, all transformation models can be applied globally or locally, i.e., on the whole image content or on subsections, respectively. Using different local transformations for every subsection of an image is also possible.

6.2.1.1 Rigid

A *rigid transformation* only allows rotation and translation to register images. It is therefore suitable for mapping bones or other "rigid" structures. In 3D, an arbitrary rotation can be expressed by successive rotations, $R_z(\gamma) R_y(\beta) R_x(\alpha)$, around the three coordinate axes, where

$$R_x(\alpha) = \begin{pmatrix} 1 & 0 & 0 \\ 0 & \cos\alpha & \sin\alpha \\ 0 & -\sin\alpha & \cos\alpha \end{pmatrix}, R_y(\beta) = \begin{pmatrix} \cos\beta & 0 & -\sin\beta \\ 0 & 1 & 0 \\ \sin\beta & 0 & \cos\beta \end{pmatrix}, R_z(\gamma) = \begin{pmatrix} \cos\gamma & 0 & \sin\gamma \\ -\sin\gamma & \cos\gamma & 0 \\ 0 & 0 & 1 \end{pmatrix}. \tag{6.3}$$

Moreover, multiplying the described matrices yields a single 3×3 rotation matrix R. The translation part can be defined as a translation vector $t = (t_x, t_y, t_z)^T$. Combining both we get

$$T(x) = Rx + t, \tag{6.4}$$

with $x = (x, y, z)^T$ for the rigid transformation T. By using homogeneous coordinates a single matrix is sufficient to describe that transformation:

$$\begin{pmatrix} y_1 \\ y_2 \\ y_3 \\ 1 \end{pmatrix} = \begin{pmatrix} R_1 & R_2 & R_3 & t_1 \\ R_4 & R_5 & R_6 & t_2 \\ R_7 & R_8 & R_9 & t_3 \\ 0 & 0 & 0 & 1 \end{pmatrix} \begin{pmatrix} x_1 \\ x_2 \\ x_3 \\ 1 \end{pmatrix}, \tag{6.5}$$

where $y = (y_1, y_2, y_3)^T$ is a point in the reference image and x is a point in the target image. This matrix has six DOF: three for the rotations around the x-, y- and z-axis and three for the translation in x-, y- and z-direction.

6.2.1.2 Affine

Affine transformation is similar to rigid transformation but has additional DOF. Particularly, the transformation T is enhanced by scaling S_c and sheering S_h, with

$$S_c = \begin{pmatrix} s_x & 0 & 0 & 0 \\ 0 & s_y & 0 & 0 \\ 0 & 0 & s_z & 0 \\ 0 & 0 & 0 & 1 \end{pmatrix}, S_h = \begin{pmatrix} 1 & s_1 & s_2 & 0 \\ s_3 & 1 & s_4 & 0 \\ s_5 & s_6 & 1 & 0 \\ 0 & 0 & 0 & 1 \end{pmatrix}. \tag{6.6}$$

Depending on the order of multiplication, the transformation can be denoted as e.g.,

$$T(x) = RS_cS_hx + t.$$

Both S_c and S_h add three additional DOF since they can be applied in x-, y- and z-direction, respectively. Affine transformation therefore has 12 DOF in total. Again this can be expressed as a single 4×4 matrix:

$$\begin{pmatrix} y_1 \\ y_2 \\ y_3 \\ 1 \end{pmatrix} = \begin{pmatrix} a_1 & a_2 & a_3 & t_1 \\ a_4 & a_5 & a_6 & t_2 \\ a_7 & a_8 & a_9 & t_3 \\ 0 & 0 & 0 & 1 \end{pmatrix} \begin{pmatrix} x_1 \\ x_2 \\ x_3 \\ 1 \end{pmatrix}. \tag{6.7}$$

6.2.1.3 Projective

Projective transformations are somewhat a special case since those deformations by their nature only occur in projective surroundings. However, there are medical applications where it is necessary to model exactly this type of transformation. Figure 6.6 shows an example of a matching between 3D CT data and a 2D angiogram of the heart. Hereby, the latter can be seen as a perspective projection of the 3D volume onto a 2D plane.

The mapping of a 3D point $P^{3D} = (x, y, z, 1)^T$ in homogeneous coordinates to a point $P^{2D} = (u, v, 1)^T$ on a 2D image plane can be denoted by the following transformation

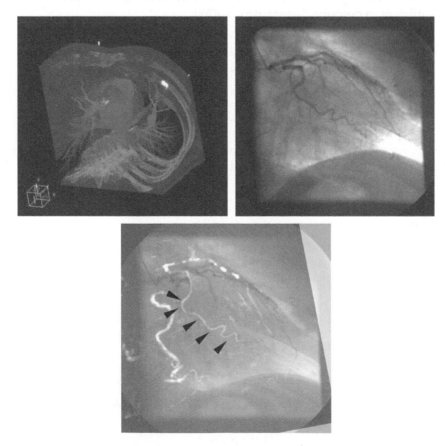

FIGURE 6.6
Registration of a 3D CT data set with a 2D angiogram via projective transformation.

$$p^{2D} = P_{\text{persp}}R \cdot (p^{3D} - t),\qquad\qquad(6.8)$$

with P_{persp} being the projection matrix of the pinhole camera model. R and t are a rotation and translation to transform the 3D point from world coordinates (i.e., a coordinate system placed in the 3D volume) to camera coordinates (a virtual camera that is placed somewhere around the 3D volume to perform the projection).

6.2.1.4 Elastic

The term *elastic transformation* describes a class of transformations that generally allows any kind of deformation. This, however, potentially results in an enormous large solution space, since there are usually countless combinations to map two images if any deformation is possible. Therefore in practice, the transformation will be constrained to ensure that the solution is physically meaningful.

Elastic deformations are generally too complex to be represented using constant matrices. A common way to model them is to use a vector displacement field that represents the deformations on every voxel. To reduce the computational complexity, grids can be placed over the images. The displacement between their control points is computed using interpolation, e.g., using spline-based methods. The application of grids in elastic deformation will be explained in more detail in the section "Integration".

As mentioned before, the deformation of the displacement field will be constrained. For example a "folding" of the field should be prevented, i.e., no points are allowed to be moved out of their local neighborhood. In addition, large distortions that map points over the whole scale of the image are usually prohibited. Such constraints limit the transformation in a way that global rotations, translations and scalings are difficult to model. Therefore, elastic registration is often accompanied by rigid or affine transformation as a pre computing step.

Up to now, several transformation methods have been described that can be used to map points or voxels between two different coordinate systems. The remaining sections will show what kind of correspondence and optimization strategies are needed to compute these transformations.

6.2.2 Finding Correspondences

In terms of registering two images, i.e., to find the transformation that maps points or voxels given in the coordinate system of one image space to the coordinate system of another, we generally have two options. Either we can search the images for points that correspond to each other and use them as input to compute the transformation matrices described in the last section. The alternative is to work directly on the intensities of the images. Here, a solution is found in an iterative fashion by constantly comparing both reference and the transformed target image using some kind of metric. Starting with an initial guess the appropriate transformation parameters are refined in every step as part of an optimization procedure to fit the images according to the metric.

For now, we focus on the former, i.e., we need to find at least some points which correspond to each other in order to solve a set of equations defined by the respective transformation. Depending on the choice of transformation, the registration contains several DOF, so more or less correspondences are needed. In certain cases like PET-CT combined devices are available that represent different acquisition modalities within the same machine.

Transformations between the images are therefore known *a priori*. However, usually those devices are not very common, so we need techniques that find correspondences in the images.

6.2.2.1 Extrinsic Landmarks

There is a variety of ways to manually set or compute correspondences. The simplest option is to use physical landmarks (also called *extrinsic landmarks* [20]) that are placed on the patient's body. For example, so called *stereotactic frames* can be attached to the head. On the frame, several markers are placed that can be seen both in CT and MRI scans. A unique assignment between the landmarks can therefore easily be defined. Often these frames are screwed into the skull to prevent shifting during patient movement—a procedure that is done in order to ensure a very high registration precision which is needed for example in brain surgery. Due to their massively invasive character those methods are rarely used in practice. Another and noninvasive approach is to place markers on the skin of the patient. However, apart from natural deformation, this is also an area where MRI shows a high degree of distortion which has a negative impact on the precision of the overall registration result.

The major drawback of external or internal physical landmarks is the fact that often the need for multimodal diagnostic arises *after* the first acquisition of images taken during patient inspection. In those images no landmarks are provided that can be taken to register new images from different modalities. Furthermore, because of their external placement, extrinsic landmarks do not allow any statement about local deformations e.g., caused by respiration. They are therefore limited to rigid transformations only. Fortunately, there are other ways to automatically find correspondences in medical image pairs without the need of external markers.

6.2.2.2 Intrinsic Landmarks

If there are no external markers available, we need to rely on the images only to extract the desired correspondences. Of course we can define the points manually, but this requires additional physician time, reduces reproducibility and often yields imprecise results. Therefore we are looking for an automatic detection of corresponding points or structures in the image. Generally this is the task of *feature* extraction algorithms, i.e., images are searched for edges, corners or local extrema etc. to identify certain recognizable anatomical points. Figure 6.7 gives an example of anatomical landmarks detected in both MRI and PET scans of the head.

Malsch et al. [21] use a block matching algorithm to automatically select point correspondences in the images. The correspondences are then used for an elastic registration. In most applications, however, only few intrinsic landmarks can be extracted, because the amount of clearly identifiable points is usually limited. Hence landmark based registration is mostly used to find rigid or affine transformations. In order to allow more complex deformations, i.e., transformations with a higher degree of freedom, we need more correspondences that are well spread over the area of interest.

6.2.2.3 Segmentations

In order to get a large amount of corresponding points, several methods rely on a previous segmentation of organs and structures. The idea is to segment an organ in both images and then transform the result to e.g., a surface representation like a polygonal mesh or a

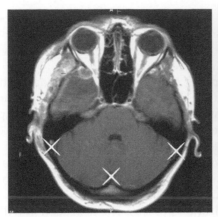 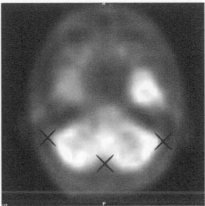

FIGURE 6.7
Intrinsic anatomical landmarks found in both MRI and PET data sets.

point cloud. Large point sets are therefore available in both datasets after segmentation. Certainly, there remain some problems: at first, we indeed get a lot of points from the segmentations, but how do we know which point in one data set correlates to what point in the other one? This issue will be addressed in the following section introducing some algorithms to automatically compute transformations between point clouds. Apart from that, the segmentation part itself is a non trivial procedure and there are many different approaches, all having advantages and drawbacks concerning precision, automation and speed as the previous section of this chapter shows. Among them, deformable surface models play a special role, because they only require a segmentation in one data set. Point correspondences are gained by placing the segmentation surface on the other data set, performing adaptation and comparing the result with the starting form. Recall that deformable surface models are meshes that iteratively adapt to a local structure by optimizing an energy function. This function contains internal and external forces which control preservation and adaptation of the form respectively. The surface is therefore constantly changing in every step. Hereby, the deviation of the new shape from the former is measured by registering both surfaces rigidly and then applying a metric to compare them. Although deformable models are used for segmentation, registration is an essential part of the procedure, i.e., the transformation gained from the optimization of the deformable model can be used to describe local deformations as well.

6.2.2.4 Grids and Voxel Properties

In the case of global elastic registration we are not focusing on local regions like organs but on the entire image itself. Localized point clouds are therefore not sufficient to cover all possible deformations. A common practice to deal with this problem is the use of *grids* that are placed on the images. The idea is that we cannot efficiently define correspondences which are spread over the whole image so we place an initial net on both data sets and let one of them iteratively adapt to the underlying voxel properties maximizing some similarity criterion. From initialization we know what grid points correspond to each other, so after reaching equilibrium we are able to calculate the elastic transformation between the unmodified grid in the one data set and the deformed one in the other image. Because the control points of the undeformed reference grid do not necessarily always lie in interesting areas (i.e., regions with a high degree of deformation), they can be placed irregular. However, this requires preprocessing to find appropriate positions in the data set. A simpler yet time-consuming way is to increase the number of grid points trusting that the net resolution is sufficient to cover all local deformations.

As we have seen, there are many ways of finding correspondences in images. The choice of method, however, depends on the class of the desired transformation, runtime, precision, the need of manual interaction and patient burden to name a few. The upcoming section "Integration" will give an overview of common approaches to combine transformations with appropriate correspondences and optimization methods.

Apart from points, surfaces and grids there is also the option to incorporate the full image content for registration as it were mentioned in the beginning of the section. In such a case, we do not use any explicit point correspondences between the images, because every voxel is directly involved in the optimization process. The next section will give an overview of strategies how to solve those *voxel based registration* problems.

6.2.3 Computing Transformations

6.2.3.1 Known Point to Point Correspondences

If we already know the correspondences of a sufficient amount of points, e.g., from extrinsic or intrinsic landmarks, rigid, affine and projective transformations can be directly computed. Recall that these transformations can be represented using a single constant matrix A:

$$y_i = A_{ij}x_j, \tag{6.9}$$

where x and y are the corresponding points in the images. Inserting the given point pairs will yield a set of equations with a number of independent unknowns according to the rank of A (i.e., according to the DOF of the transformation encoded in A). These equations can be summed up in the form

$$M \cdot s = 0, \tag{6.10}$$

with a matrix M encoding all point coordinates and s being a vector containing the unknown elements of A. Such a nullspace problem can be solved using the mathematical framework of singular value decomposition (SVD). Having more equations than unknowns (i.e., an over-determined system) using SVD is equivalent to minimizing the mean square error.

6.2.3.2 Point Clouds

In many registration applications we need to match surfaces to each other. As it was shown in the previous section, surfaces are often represented by point clouds derived from segmentations that were build in a preprocessing step. However, usually no information about correspondences is given, so we cannot directly compute the transformation between the surfaces. Instead we need algorithms that find both the optimal alignment of the two point clouds and the transformation to match them. A well known method to provide such a solution is the *iterative closest point algorithm* (ICP) proposed by Besl and McKay [22]. As its name implies, the approach is iteratively searching for an alignment of two surfaces such that the distance between them is minimized. Given two surfaces X and Y and a mapping T to match X to Y, the goal is to minimize the error function

$$d(T(X),Y) = \sqrt{\sum_{j=1}^{N} \|T(x_j) - y_j\|^2}, \qquad (6.11)$$

with x_j $(j = 1,..,N)$ are the points on X and

$$y_j = C(T(x_j),Y) \qquad (6.12)$$

are the points on Y corresponding to x_j. Starting with an initial guess of the alignment of X and Y the corresponding points for each iteration are found by projecting x_j perpendicularly to the triangulated surface Y. The transformation T can then be computed by minimizing the above formula using closed-form solutions. X and Y are afterwards aligned using T and the procedure starts anew. When reaching convergence T describes the final transformation minimizing the distance between both surfaces in a least squares manner.

There are several variants of the ICP algorithm using different approaches for selecting point correspondences and performing optimization. Rusinkievicz and Levoy give a detailed overview [23].

Another option for automatically matching surfaces contained in volumes is the use of *distance maps*. A distance map is a scalar volume that stores for every voxel the distance to the nearest surface point. Such a map has to be computed once for one of the data sets. The surface in the other data set is represented by a point cloud. By using the map, the distance between those points and the first surface can be directly determined. A transform that minimizes the sum of all distances is then computed as part of an iterative optimization using e.g., gradient descent.

Figure 6.8 shows an example of a distance map computed for a liver data set. Such a map is usually built by setting the surface contours in the volume to 1 followed by several filtering passes with different masks to count up the distance. Because filtering operations in 3D are still a computationally demanding task, several variations have been proposed to reduce the number of passes. Therefore usually not the Euclidian distance is computed but approximations like chessboard distance or chamfer distance.

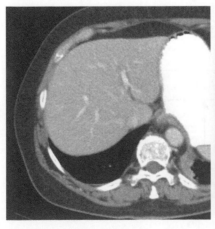 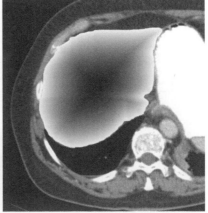

FIGURE 6.8
Liver data set and computed distance map.

6.2.3.3 Voxel Based Registration

Voxel based registration is probably the most challenging type of registration since it often incorporates the whole image content and does not need any point correspondences in advance. Therefore it is very well suited for modeling local deformations.

6.2.3.3.1 Principle Axes Registration

Most voxel based registration algorithms compute the desired transformation iteratively by optimizing some similarity criterion between the voxels of both data sets. However, there are methods to directly determine the transformation using *principle component analysis*: at first, the images are reduced to binary data sets which represent the objects to be transformed. Afterwards, the principle axes of the objects are extracted. A translation can then be computed between the objects centers of gravity while the orientation of the principle axes yields the rotation between the two coordinate systems. In addition a scaling is defined by the length of the principle axes.

As it was shown, this kind of registration is limited to rigid transform and scaling. Furthermore, the desired object has to be segmented and needs to be fully visible in both data sets. Principle axes registration is therefore only used in special cases like head registration [24].

6.2.3.3.2 Metrics

Apart from the principle axes approach, voxel based registration is usually done in an iterative manner using *metrics* and *optimizers*. Figure 6.9 shows the basic concept of this kind of registration. Reference and target image are compared based on their voxel intensities and by applying the metric. An optimizer now tries to maximize the similarity by varying the parameters of a given transform.

We will first focus on the different metrics that can be used to determine a similarity measure. Hereby, two types of metrics can be distinguished: monomodal and multimodal metrics. The former is used if two data sets were taken from the same image modality while the latter is applied otherwise.

In the case both data sets have identical intensity ranges the *sum of squared differences* (SSD) between the voxels can be used as a metric. Given the reference and target data sets f and g, the SSD is

$$M_{\text{SSD}} = \frac{1}{N} \sum_{x=1}^{N} (f(x) - g_T(x))^2 , \qquad (6.13)$$

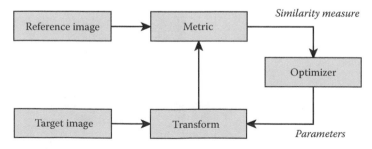

FIGURE 6.9
The basic concept of iterative image registration.

where T is the transformation that maps g to f and N is the number of voxels in f. Because gray value ranges are not comparable in multimodal imaging, SSD is limited to monomodal registration.

If there is a linear dependence between the intensities of both data sets, a metric called *cross-correlation* can be used. With μ_f and μ_g being the mean intensities of f and g_T, the normalized cross-correlation is denoted as

$$M_{CC} = \frac{\sum_{x=1}^{N}(f(x)-\mu_f)(g(x)-\mu_g)}{\sqrt{\left[\sum_{x=1}^{N}(f(x)-\mu_x)^2\right]\left[\sum_{x=1}^{N}(g(x)-\mu_g)^2\right]}}. \tag{6.14}$$

Cross-correlation is also a metric only used in monomodal imaging, because the assumption of linear dependence is usually not valid in multimodal acquisition.

A metric widely used in *multimodal registration* is the measure of *mutual information (MI)* introduced by Viola et al. [25] and Collignon et al. [26]. It regards images as the result of discrete random variables and therefore can apply statistical measures to compare them. MI of two random variables f and g_T is defined as

$$M_{MI} = \sum_{i} p(f(i), g_T(i))\log\frac{p(f(i), g_T(i))}{p(f(i))p(g_T(i))}, \tag{6.15}$$

where $p(f)$ and $p(g_T)$ are the probability density functions of the random variables and $p(f, g_T)$ is the joint probability density function of f and g_T respectively.

The idea of MI registration is that the voxels intensities of both input datasets are at least partially interdependent. The more f and g_T depend on each other, the more $p(f, g_T)$ and $p(f) \cdot p(g_T)$ are different leading to a higher value of M_{MI}. Maximization of MI regarding transformation T is therefore the goal to determine an optimal mapping between the data sets.

In order to compute the maximal MI between two volumes, it is necessary to estimate the probability density functions. This is usually done by histogram analysis or using the Parzen Window method [25, 26].

A reason why this algorithm has become very popular is that it only requires a dependence of data sets in a statistical way, i.e., as long as there is *some* correlation of intensities present, maximization of MI will find a solution to map the data sets.

6.2.3.3.3 *Optimization Strategies*

Having defined a set of metrics, we need strategies to optimize them by varying the parameters of the desired transformation T. Therefore, the goal is to find global maxima or minima of the functions that describe the metrics. There are several mathematical strategies for finding extrema of nonlinear functions. In terms of image registration we are looking for a fast method that is robust against converging to local extrema at the same time. However, the choice of optimization also depends on the transformation, the cost function and potential constraints to ensure a physically meaningful result. Popular methods to be used in registration are gradient descent, Levenberg–Marquardt optimization, genetic methods, downhill simplex [27], simulated annealing [28] or Powell's method to name a few. For an overview see Press et al. [29] and Crum et al. [30].

6.2.4 Integration

In this section the described transformations, correspondences and optimization strategies are combined to show some real world examples of monomodal and multimodal image registration.

6.2.4.1 Rigid Registration

Let us say we want to register two 3D data sets acquired from the same image modality (e.g., CT) and from the same patient in order to study the decrease in size of a tumor after chemotherapy. Because the tumor is located in an area that is not affected by respiration, we chose a rigid transformation to register both data sets. In addition the procedure should not require any user interaction, so the voxel intensities are used instead of landmarks. This means, we have to use a metric and an optimization strategy to iteratively determine the transformation parameters. Since both images are taken from CT, we chose normalized cross-correlation as a metric. Furthermore, gradient descent is chosen as an optimizer. Figure 6.10 shows an example of the registration result of two CT data sets.

In case the physician wants to find metastases of the tumor, usually PET is used to make them visible. Now, we have multimodal images, which need a different metric, e.g., MI, to register the data sets. An appropriate optimizer for this task is the downhill simplex method. The result of such a registration is shown in Figure 6.11.

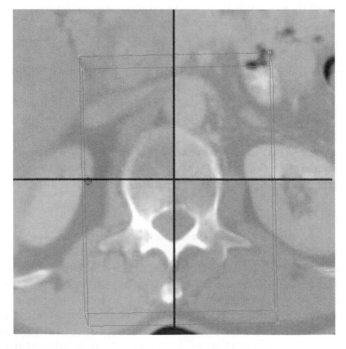

FIGURE 6.10
Registration result of two CT images using rigid transform.

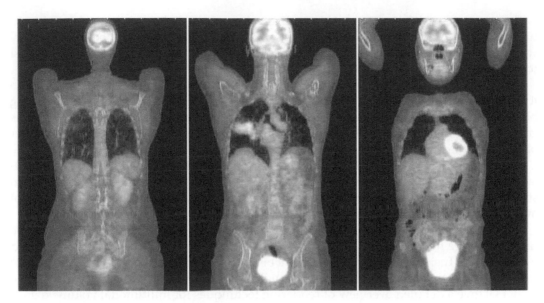

FIGURE 6.11
Visualization of registered CT and PET body scans.

6.2.4.2 Elastic Registration

In case we need elastic registration, e.g., if we have two images of the same patient with respiratory displacements, usually a reduction of the parameter space is necessary in order to reduce the computational complexity and to achieve physically meaningful results.

As mentioned before, grids are an opportunity to compute the displacement field by optimizing the voxel similiarities solely under the net points. That means, two grids are placed over the images and the control points of one data set are iteratively shifted such that a metric (e.g., MI) between the underlying intensities is optimized (using e.g., Levenberg-Marquardt optimization). In addition, splines are laid through the points to define the transformation in the space between them and to limit the overall deformation. Often B-Splines [31] are used as they provide only local "support". That means a shifting of one point does not affect all other points but only the ones in his neighborhood depending on the degree of the B-Spline functions. However, splines with global influence on control points are also used. Here, the so called *thin-plate splines* TPS [32] have become very popular.

The described methods are usually accompanied by an initial rigid registration since it is unfeasible to model large rotations or translations with these approaches.

6.3 Fusion

The term *fusion* denotes the combined visualization of registered data sets. Since the goal of registration is the comparison of images that show different information, methods are needed that present the clinically important structures from both data sets while hiding irrelevant details.

There are many different ways to visualize volume data and accordingly several techniques have been developed to adapt those methods to image fusion. Because 2D slice-wise imaging is still the method of choice for the majority of diagnostic tasks, we will first focus on this kind of imaging.

Probably the simplest opportunity to fuse two 2D images is to blend them with an opacity factor. Figure 6.12 shows an MRI and CT Image of the head and the corresponding blending result. As it can be easily seen, such a kind of fusion yields poor results in this case, because the strength of CT to clearly show bone structures is not preserved after blending the images.

Checkerboards are a very popular method to fuse 2D images while maintaining the familiar visualization of the respective imaging modality (Figure 6.13). However, the physician still has to mentally overlay the registered images since information of both data sets is not available at the same time.

In terms of 3D visualization, direct volume rendering is the dominating method to be used for image fusion. Several techniques have been proposed that can be distinguished by the kind of fusion (one property per point or multiple properties per point), the point of time of registration and fusion and how visualization parameters can be modified. Ferré et al. [33] give a detailed overview of rendering techniques for multimodal volume data.

As direct volume visualization makes use of transfer functions to map intensities to opacities and colors it is plausible to adapt this concept to image fusion. Because multiple

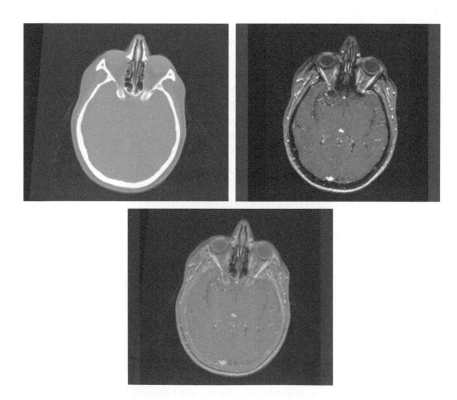

FIGURE 6.12
CT and MRI scans of the head. A blending (bottom) does not preserve all relevant details of both images.

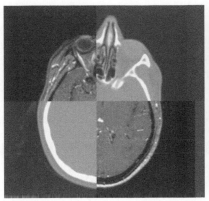
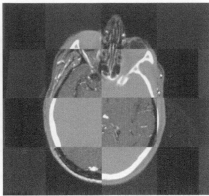

FIGURE 6.13
Fusion of CT and MRI scans in a checkerboard view.

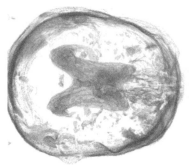

FIGURE 6.14
Fusion of CT and MRI head scans using spatialized transfer functions.

data sets are involved in the fusion process often *multidimensional* transfer functions are used. Among them the so called *spatialized transfer functions* [34, 35] have proven to ease the task of building appropriate mappings that can distinguish between different objects of the same tissue class. The method analyzes the combined 2D histogram (intensity + gradient) of both images and builds a colored classification that divides the histogram into clusters. The user can then choose a structure to be highlighted by picking the desired cluster of the histogram. Figure 6.14 shows the result of a fusion using spatialized transfer functions. The ventricles of the brain are clearly visible while the bone structures of the head are also shown.

Acknowledgments

The author would like to thank Evelyn Firle, Marion Jähne, and Christina Lacalli for their support and figures of the segmentation, registration and fusion part.

References

1. T. S. Yoo. 2004. *Insight into Images: Principles and Practice for Segmentation, Registration, and Image Analysis*. A. K. Peters, Ltd., Wellesley, Massachusetts, ISBN 1-56881-217-5.
2. N. Otsu. 1979. A threshold selection method from gray level histograms. *IEEE Trans. Systems, Man and Cybernetics*, 9:62–66.
3. P.S. Liao, T.S. Chen, P.C. Chung. 2001. A fast algorithm for multilevel thresholding. *J. Inform. Sci. Eng.*, 17:713–727.
4. J. Canny. 1985. A computational approach to edge detection. *IEEE Trans. Patt. Anal. Machine Intell.*, PAMI-8(6):679–698.
5. A. Singh, D. Terzopoulos, D. B. Goldgof. 1998. *Deformable Models in Medical Image Analysis*. Los Alamitos, CA: IEEE Computer Society Press. ISBN 0818685212.
6. T. F. Cootes, C. J. Taylor. 1992. Active shape models – smart 'snakes'. In: David Hogg u.a. (Hrsg.). *BMVC92. Proceedings of the British Machine Vision Conference*. Berlin: Springer-Verlag, S.266–275. ISBN 3-540-19777-X.
7. T. F. Cootes, G. J. Edwards, C. J. Taylor. 1998. Active appearance models. In: H. Burkhardt, B. Neumann (Hrsg.). *Proceedings of the European Conference on Computer Vision*, Vol. 2. Berlin: Springer, Heidelberg S.484–498.
8. S. Beucher, F. Meyer. 1993. The morphological approach to segmentation: the watershed transformation. In: *Mathematical Morphology in Image Processing*. Marcel Dekker Inc., New York 433–481.
9. F. Meyer. 1994. Topographic distance and watershed lines. *Signal Process*, 38:113–125.
10. J. v. Berg, C. Lorenz. 2005. *Multi-Surface Cardiac Modelling, Segmentation, and Tracking. Functional Imaging and Modelling of the Heart*. Berlin: Springer, Heidelberg 1–11. ISBN 978-3-540-26161-2.
11. C. Lorenz, J. v. Berg. 2006. A comprehensive shape model of the heart. Special issue on functional imaging and modelling of the heart. *Med. Image Anal.*, 10:657–670.
12. H. Lamecker, T. Lange, M. Seebaß. 2002. *A Statistical Shape Model for the Liver*. MICCAI 2002. Berlin: Springer, Heidelberg 421–427. ISBN 978-3-540-44225-7.
13. Y. Shang, O. Dössel. 2004. Statistical 3D shape-model guided segmentation of cardiac images. *Comput. Radiol.*, 553–556. ISBN: 0-7803-8927-1.
14. A. F. Frangi, D. Rueckert, J. A. Schnabel, W. J. Niessen. 2002. Automatic construction of multiple-object three-dimensional statistical shape models: application to cardiac modeling. *IEEE Trans. Med. Imag.*, 21(9):1151–1166.
15. O. Commowick, G. Malandain. 2006. Evaluation of atlas construction strategies in the context of radiotherapy planning. In: *Proceedings of the SA2PM Workshop (From Statistical Atlases to Personalized Models)*. Copenhagen, 2006. Held in conjunction with MICCAI 2006.
16. T. Rohlfing, R. Brandt, R. Menzel, C. R. Maurer, Jr. 2004. Evaluation of atlas selection strategies for atlas-based image segmentation with application to confocal microscopy images of bee brains. *NeuroImage*, 21(4):1428–1442.
17. A. Guimond, J. Meunier, J.-P. Thirion. 1999. Average brain models: a convergence study. *Comput. Vision Image Understand.*, 77:192–210.
18. A. Shimizu, R. Ohno, T. Ikegami, H. Kobatake, S. Nawano, D. Smutek. 2007. Segmentation of multiple organs in non-contrast abdominal CT images. *Int. J. Comput. Assist. Radiol. Surg.*, 2(3–4):135–142.
19. P. M. Som, H. D. Curtin, A. A. Mancuso. 1999. The new imaging-based classification for describing the location of lymph nodes in the neck with particular regard to cervical lymph nodes in relation to cancer of the larynx. *Arch Otolaryngol. Head Neck Surg.*, 125:388–396.
20. J. B. A. Maintz, M. A. Viergever. 1998. A survey of medical image registration. *Medical Image Analysis*, 2:1–16.
21. U. Malsch, C. Thieke, P. E. Huber, R. Bendl. 2006. An enhanced block matching algorithm for fast elastic registration in adaptive radiology. *Phy. Med. Biol.*, 51(19):4789–4806.

22. P. Besl, N. McKay. 1992. A method for registration of 3-D shapes. *IEEE Trans Patt. Anal. and Mach. Intell.*, 14:239–255.

23. S. Rusinkiewicz, M. Levoy. 2001. Efficient variants of the ICP algorithm. In: *Proceedings of the Third Intl. Conf. on 3D Digital Imaging and Modeling*, IEEE CS Press, Los Alamitos, Calif., 145–152.

24. L. K. Arata, A. P. Dhawan, J. P. Broderick, M. F. Gaskil-Shipley, A. V. Levy, N. D. Volkow. 1995. Three-dimensional anatomical model-based segmentation of MR brain images through principal axes registration. *IEEE Trans. Biomed. Eng.*, 42(11):1069–1077.

25. P. A. Viola and W. M. Wells. 1997. Alignment by maximization of mutual information. *Int. J. Comput. Vis.*, 24(2):137–154.

26. A. Collignon, F. Maes, D. Delaere, D. Vandermeulen, P. Suetens, G. Marchal. 1995. Automated multi-modality image registration based on information theory. In: *Information Processing in Medical Imaging (IPMI)*, Y. Bizais, Ed., Kluwer Academic, Dordrecht, 263–274.

27. J Nelder, R. Mead. 1965. A simplex method for function minimization. *Comput. J.*, 7:308 313.

28. S. Kirkpatrick, C. D. Gelatt, M. P. Vecchi. 1983. Optimization by simulated annealing. *Science*, 4598(220):671–680.

29. W. H. Press, S. A. Teukolsky, W. T. Vetterling, B. P. Flannery. 2007. *Numerical Recipes: The Art of Scientific Computing*. Cambridge University Press, Cambridge (MA). ISBN 0521880688.

30. W. R. Crum, T. Hartkens, D. L. G. Hill. 2004. Non-rigid image registration: theory and practice. *Brit. J. Radiol.*, 77(2):140–153.

31. D. Rueckert, L. Sonoda, C. Hayes, D. Hill, M. Leach, D. Hawkes. 1999. Nonrigid registration using free-form deformations: application to breast MR images. *IEEE Trans. Med. Imag.*, 18:712–721.

32. F. Bookstein. 1989. Principle warps – thin-plate splines and the decomposition of deformations. *IEEE Trans. Patt. Anal.*, 11:567–585.

33. M. Ferré, A. Puig, D. Tost. 2002. Rendering techniques for multimodal data. Technical Report LSI-64-R. *Proc. SIACG 2002 1st Jbero-American Symposium on Computer Graphics*, Grupo Portugues de Computa çao Gráfica, Guimaraes, Portugal, 305–313.

34. S. Roettger, M. Bauer, M. Stamminger. 2005. Spatialized transfer functions. In: *EUROGRAPHICS – IEEE VGTC Symposium on Visualization*. K. Brodlie, Ed., Eurographics Association, Leeds, United Kingdom, 271–278.

35. E. A. Firle, M. Keil. 2007. Multi-volume visualization using spatialized transfer functions. Gradient- vs. multi-intensity-based approach. In: H. U. Lemke, editor. *Computer Assisted Radiology and Surgery, Proc. CAR 07*. Elsevier, Amsterdam, 121–125.

7

Acoustic Diffraction Computed Tomography Imaging

Stergios Stergiopoulos

Defence R&D Canada Toronto
University of Toronto
University of Western Ontario

Waheed A. Younis and Julius Grodski

Defence R&D Canada Toronto

David Havelock

National Research Council

CONTENTS

7.1 Introduction

Acoustic techniques have been used to do subsurface image reconstruction for imaging and classifying buried objects for the past two decades. Most of these techniques use either well-to-well tomography [1] or surface-to-well tomography [2–4]. However, their methodology is not appropriate for detecting buried land mines or hazardous waste materials, since they require ground disturbances that could be hazardous for de-mining and destructive for archeological applications.

A recent study [5] investigates 2-D imaging of shallow buried objects by using ultrasound B-scan tomography [6]. The system concept includes a linear array of receivers and transmitters operating in the frequency range of 1–5 kHz. In this case, 3-D imaging can be accomplished by volume rendering of consecutive 2-D images [7]. If the linear sensor array is replaced by a planar array, then 3-D beamforming techniques [6] could provide the volume visualization of the underground area without combining multiple 2-D images.

Advancements in remote sensing have led to the development of nondestructive subsurface detection methods such as infra-red (IR) imaging [8,10], ground penetrating radar (GPR) [9,10], seismic refraction, electromagnetic sensing [10] and electrical conductivity. All these techniques use perturbations in seismic or electromagnetic waves to detect, locate, and identify the buried objects. Currently, GPR and IR imaging are widely used in de-mining applications. The performance of GPR for detecting landmines degrades considerably with increased moisture in the ground and, like the electromagnetic sensors, it fails to detect nonmetallic objects. The IR imaging methods can only detect recently buried (one month period) objects. The poor performance characteristics of the existing nondestructive imaging techniques can be improved by using data and image fusion techniques in a multisensor system that has shown to have the potential for reliable mine detection and classification [10].

Another challenge of international significance is the disposal of buried hazardous waste, which requires reliable estimation of the location and nature of the buried substances [11]. For archeologists, subsurface imaging techniques are useful for assessing the historical significance of a site before beginning costly digging [12].

The present chapter introduces an alternative approach to nondestructive 3-D imaging by using the concept of computed tomography (CT), which is successfully used for noninvasive medical imaging diagnostic applications [19]. Briefly, medical CT imaging uses x-rays to obtain cross-sectional images, or "slices", of the human body and image reconstruction algorithms that are based on the Radon theorem [13,14].

The techniques discussed in this chapter include subsurface image reconstruction of horizontal cross sections at different depths. Once such images of sufficient quality are available, a next step could include implementation of pattern recognition algorithms to identify and classify the mines. An advantage of this approach is the ability to detect mines that are buried beneath one another, as this is a common deployment procedure to deceive the conventional mine detection systems. The method, discussed in this chapter, is demonstrated with an experimental setup, which illustrates both the challenges and opportunities for imaging of shallow buried objects using acoustic CT.

7.2 Theoretical Remarks

7.2.1 Definition of Basic Parameters for CT Applications

The basic principles of a CT data acquisition process and the relevant image reconstruction algorithms are discussed in detail elsewhere [3,13–19]. The data acquisition process for fan or parallel beam CT applications is depicted in Figure 7.1.

The projection measurements, $p_n(r_n, \theta_i)$, $(n=1, ..., N)$, $(i=1, ...,M)$, shown schematically in this figure, are defined as the line integrals along lines passing through the object $f(x, y)$.

For a given detector n, projection i and projection angle θ_i, $p_n(r_n, \theta_i)$ is given by Equation 7.1,

$$p_n\left(r_n, \theta_i\right)=\iint f(x, y)\delta\{x\cos\theta_i + y\sin\theta_i - r_n\}dxdy \tag{7.1}$$

where, $\delta(\cdot)$ represents the Dirac delta function, θ_i, represents the projection angle for detector n, r_n is the shortest distance between the projection for detector n and the origin. The function $f(x, y)$ also represents the tomography image that we wish to reconstruct. Equation 7.1 is also applicable for parallel beam projection CT scanners. The angular step increment between two successive projections of a CT scanner is defined by: $\Delta\theta = 2\pi/M$, where M is

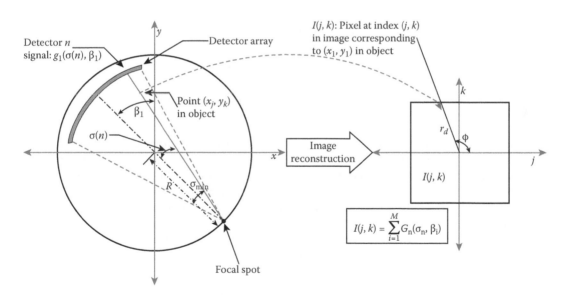

FIGURE 7.1
Schematic diagram of projection function for CT imaging applications. The projection measurements are denoted by g_n (σ_n, β_i), $(n=1, ..., N)$, $(i=1, ..., M)$, with the angular step increment between two successive projections of the x-ray scanner defined by: $\Delta\beta=2\pi/M$, and Δt is the time elapsed between these successive projections, where M is the number of projections taken during the period $T=M\Delta t$ that is required for one full rotation of the source and receiving N-detector array around the object $f(x, y)$ that is being imaged. (Reprinted by permission of IEEE © 2000.)

the number of projections taken during the period T required for one full rotation of the source and receiving array around the object $f(x, y)$ that is being imaged.

The process of reconstructing an image $I(x, y)$ from the projection data, using filtered back projection algorithm is given by

$$I(x, y) = \int_0^\pi G(x\cos\theta + y\sin\theta, \theta)d\theta$$

with $G(\cdot, \theta)$ being the filtered version of the projection $p(r_n, \theta)$. The filtering function used is the Ram–Lak filter [17], cascaded with the Parzen window. In the digital version of this process, the value of the pixel $I(j, k)$, corresponding to the Cartesian point (x_j, y_k) in the CT scan geometry is given by

$$I(j, k) = \sum_{i=1}^{M} G(r_n, \theta_i)$$

where M is the total number of projection measurements.

The transformations shown in Equation 7.2, are required to account for the geometry of the fan beam as shown in Figure 7.1.

$$r_n = R\sin\sigma_n, \quad \theta_i = \sigma_n + \beta_i \tag{7.2}$$

Then the fan beam projection function is now defined by:

$$g_n(\sigma_n, \beta_i) = p_n\{[r_n = R\sin\sigma_n], [\theta_i = \sigma_n + \beta_i]\} \tag{7.3}$$

The set of projection data $(n=1, \ldots, N)$, $(i=1, \ldots, M)$, acquired during a CT scan can be presented as a grey-scale image of the relative attenuation coefficient as a function of θ_i, the angular position of the source during the i-th projection, and σ_n, the angle of the n-th detector element within the fan beam. This representation is called a sinogram as illustrated in Figure 7.2. Each horizontal line displays one set of projection data acquired at a particular source angle. The projections of each point in the image plane trace out a quasi-sinusoidal curve when using fan-beam geometry. In parallel beam geometry, the curves are true sinusoids, hence the name sinogram.

7.2.2 Image Reconstruction in NonDiffracting CT Applications

Several methods of reconstructing CT images have been proposed over the years and described by many authors [3,13,16–19], including iterative algebraic techniques, the direct Fourier transform technique, and convolution backprojection. The direct Fourier method is perhaps the simplest method conceptually. It is based on the *Fourier slice theorem* [14], which states that the 1-D Fourier transform of a projection at an angle θ is the same as a slice through the 2-D Fourier transform of the image at the same angle [14], as depicted in Figure 7.3. When sufficient number of projections are acquired, the complete 2-D Fourier transform is interpolated from the samples and the image is obtained by taking the inverse 2-D Fourier transform. Although the technique is conceptually simple, it is computationally complex.

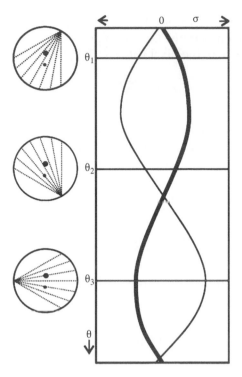

FIGURE 7.2
Sinogram data representation. Horizontal axis represents projections measured by the N-sensor array receiver for a source angular position θ. Vertical axis represents the M-angular positions of the N-sensor receiving array during a full rotation of the CT data acquisition process.

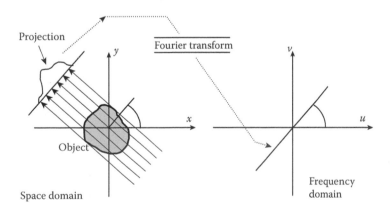

FIGURE 7.3
The Fourier slice theorem.

The most widely accepted reconstruction technique of greatest practical importance is known as the convolution backprojection (or filtered backprojection). Backprojection refers to the distribution of projections back across the image plane in the direction from which they were measured.

In the image reconstruction process the pixel $I(j, k)$ in the actual image, shown at the right of Figure 7.1, corresponds to the Cartesian point (x_j, y_k) in the CT scan plane. As discussed in the previous section, the pixel value $I(j, k)$ is given by:

$$I(j, k) = \sum_{i=1}^{M} G_n(\sigma_n, \beta_i), \tag{7.4}$$

where $G_n(\sigma_n, \beta_i)$ is the filtered version of the fan beam projection $g_n(\sigma_n, \beta_i)$.

The filtering function used is this investigation is the Ram–Lak filter [18], cascaded with the Parzen window. The angle σ_n, which defines the detector that samples the projection through a point (x_j, y_k), for a given projection angle β_i, is provided by Equation 7.5, where (r_d, φ) is the polar representation of (x_j, y_k),

$$\sigma_n = \tan^{-1} \left[\frac{r_d \sin(\phi - \beta_i)}{R + r_d \cos(\phi - \beta_i)} \right]. \tag{7.5}$$

7.2.3 Filtered Back Projection and Fourier Diffraction Theorem

There is a fundamental difference between tomography imaging with nondiffracting (e.g., x-rays) on one hand, and with acoustic waves on the other. X-rays, being nondiffracting, travel in straight lines, and therefore, the projection data measure the line integral of some object parameter along straight lines, as discussed in the previous section. This makes it possible to apply Fourier slice theorem as mentioned above. On the other hand, when acoustic waves are used for tomography imaging, the energy often does not propagate along straight lines. When object inhomogeneities are large compared to the wavelength, energy propagation is characterized by diffraction, refraction and multipath effects. These problems have been investigated and *filtered back propagation* algorithms have been developed [3,20] for diffracting tomography within either the first Born or Rytov approximations. These are based on the *Fourier diffraction projection theorem* [3,20] which states that 1-D Fourier transform of a projection of the object, yields the 2-D Fourier transform of the object over a semicircular arc, as depicted in Figure 7.4. This result is fundamental to diffraction tomography. It is shown in Pan and Kak [18] that filtered back propagation algorithm is computationally more expensive as compared to the equivalent, direct Fourier implementation for diffraction tomography.

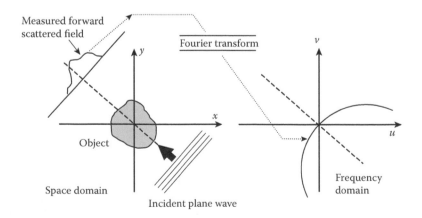

FIGURE 7.4
The Fourier diffraction projection theorem.

In our experimental investigation, we have implemented mainly the filtered back projection algorithm for the image reconstruction process. But since diffracting acoustic energy was used, we tested the image reconstruction process using Fourier diffraction projection theorem as well, even though there were prohibiting factors such as large fluctuations of the speed of sound in ground and the fact that we used broadband signal rather than continuous wave. When the Fourier diffraction projection theorem was considered in the image reconstruction process, it was implemented directly using bilinear interpolation [18].

In summary, when the energy used to illuminate the object is nondiffracting (e.g., x-rays), the image reconstruction algorithms are, *iterative algebraic techniques, Fourier slice theorem techniques* and *filtered back-projection* [13,14,17,19]. Whereas for diffracting energy (e.g., sound waves, ultrasound), these algorithms are *filtered back propagation* and *Fourier diffraction projection techniques* [3,16,18,20].

7.3 Implementation Issues

7.3.1 Acoustic CT Data Acquisition Process

In what follows, we present the results from an experimental investigation [27] that includes implementation of the CT concept using acoustic energy to reconstruct subsurface images of horizontal cross sections of the entire volume of a buried object at different depths. This approach assumes that the received signal carries information from different ground penetration depths that are represented as different time delays, indicated by the temporal index j below,

$$p\left(r_n, \theta_i, t_j\right); \quad \left(n=1,\dots,N\right), \quad \left(i=1,\dots,M\right), \quad \left(j=1,\dots,K\right). \tag{7.6}$$

As a result, for a given time delay t_o, image reconstruction of the projection data set $p(r_n, \theta_i, t_o)$ will provide horizontal tomography images of the buried object of interest corresponding to a specific depth in the ground. Then, repetition of the image reconstruction process for all the time delays ($j=1, \dots, K$) would define a 3-D volume consisting of horizontal underground tomography images. However, the correspondence between time delay t_o and specific depth in the ground requires stratification of the underground medium. For a nonstratified medium, such as the experimental set up discussed in Stergiopoulos [28], the above correspondence between time delay and depth is not valid due to the multipath effects of the acoustic waves in the homogenous underground.

Figure 7.5 depicts the experimental set-up of the acoustic tomography concept in Younis et al. [27] for nondestructive imaging of shallow buried objects. The acoustic sensors (microphone array) and source array, shown in Figure 7.5, do not have a contact with the surface of the ground. Thus, the whole system could be mounted on a platform that may be deployed by an armored vehicle or is remotely controlled. For each transmitted acoustic pulse, the receiving array collects the signal, which is reflected-refracted from the internal underground buried objects and the ground inhomogeneities. Therefore, the proposed approach addresses the inverse problem, which is the underground 2-D and 3-D imaging. Parameters related to the propagation problem (forward problem):

- the physics of acoustic/elastic wave propagating in a three dimensional structure,
- what waves are expected to generate in the geometry of the experimental set up,
- what are their expected arrivals, their travel time and their amplitude?

(a)   Horizontal arm, capable of rotating in a horizontal circle

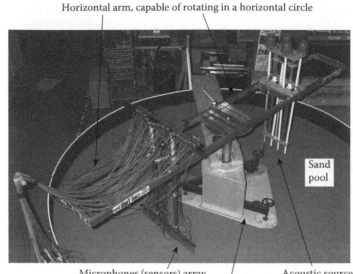

Microphones (sensors) array   Acoustic source
Triangular base

(b)   Top view
|←——Field of view of 31 detectors (90 cm)——→|

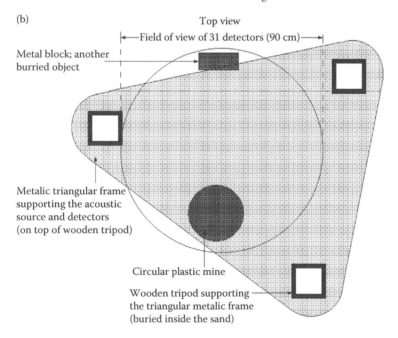

Metal block; another
buried object

Metalic triangular frame
supporting the acoustic
source and detectors
(on top of wooden tripod)

Circular plastic mine

Wooden tripod supporting
the triangular metalic frame
(buried inside the sand)

FIGURE 7.5
(a) Experimental set-up of the acoustic CT concept including the receiving 32-microphone array and the four-speaker source array. (b) Top schematic view of the experimental set-up. The dark features indicate buried objects. The triangular base was on the surface of the sand. The circle indicate the field of view of the 90-cm line array receiver during the CT data acquisition process. (Reprinted from Younis, W., Stergiopoulos, S., et al. *J. Acoust. Soc. Am.*, 111(5), 2117–2127, 2002. © AIP. With permission.)

are not required by the proposed approach to address the imaging problem. Although, solutions to the forward problem can provide the essential parameters for a better image reconstruction approach [26], they impose a practical implementation problem for a number of nondestructive imaging underground applications. In particular, to identify the 3-D propagation characteristics, it would require significant amount of information in terms of boundary conditions, density estimates, and stratification characteristics. This kind of information, although essential for an optimum approach, makes the acoustic tomography concept impractical for demining or archeological applications because it requires ground disturbances to collect the essential information. Our goal and objectives of this experimental study is the development of an acoustic CT technique for nondestructive imaging of underground features that is practically realizable. However, the theoretical and experimental aspects relevant with the propagation problem (forward problem) are discussed in Section 3 of this chapter, since they are essential to verify the experimental results reported in Younis et al. [27].

7.3.2 Propagation Characteristics

Experimental investigations such as the one in Younis et al. [27] are confronted with several challenges. First and foremost is the issue that the propagation characteristics of the acoustic signals in the setting shown in Figure 7.5 are not known. More specifically, what is received by the receiving microphone array, depicted in Figure 7.5, is a superposition of direct waves (direct arrivals through air), surface waves, subsurface waves, reflected-refracted energy from buried objects, reflected energy from various inhomogeneous ground layers and noise. Each received acoustic wavefront has a different time delay depending upon the propagation path it followed and the speed of sound in various sections of that path. Because of the poor air-to-ground-to-air coupling interface, another problem is to receive sufficient acoustic energy out of the soil by the receiving array [5] that is essential for reliable image reconstruction. The soil is a highly attenuating medium and it is difficult for acoustic waves in the frequency range of 1–5 kHz to penetrate deep in the ground [25]. Although high frequency acoustic waves (2–10 MHz) produce good image resolution in ultrasound imaging, they are not suitable for underground imaging applications due to their severe attenuation in the ground. Thus, in the experimental investigation [27], a lower frequency range of 0.2–3.0 kHz was selected due to its characteristics to penetrate in the ground, though it produces poor image resolution. Furthermore, the investigation [27] constructed a frequency-modulated FM pulse, linearly varying from 200 Hz to 2000 Hz, cascaded with a Parzen window. In order to generate a plane wave acoustic signal to illuminate the underground area of interest, a linear array of four synchronized acoustic point sources was used, as shown in Figure 7.5.

The poor signal-to-noise-ratio (SNR) in the received signal, as mentioned above, was mainly due to the poor air-to-ground-to-air coupling interface that induced a severe attenuation in the signal of interest. To minimize the impact of the very poor SNR in the image reconstruction process, the investigation [27] introduced an adaptive interference cancellation (AIC) [6,21,23] processing scheme. Its impact is demonstrated with real data in Section 7.4.4.

7.3.3 Sonar CT Concept for Detecting Buried Sea-Mines and Submarines

The diffraction CT concept that is discussed in this chapter can be used also in AntiSubmarine Warfare (ASW) surveillance applications to image buried sea mines and submarines. The diffraction CT concept and the deployment process of sources and receivers consists of sources and towed line array receivers as depicted in Figure 7.6.

The top part of this figure shows the diffraction CT sonar concept for imaging submarines. It requires the deployment of at least one naval vessel to tow an acoustic source and a

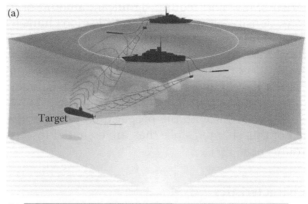

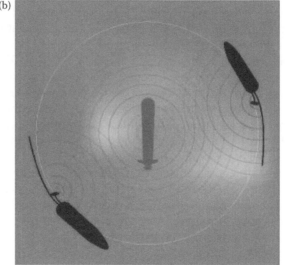

FIGURE 7.6
(Top) Diffraction CT sonar concept for imaging submarines. It requires the deployment of at least one naval vessel to tow an acoustic source and a receiving towed line array of hydrophones. (Bottom) Diffraction CT concept to detect underwater buried sea-mines. The deployment configuration of the source and the line array receiver should be similar to that in Figure 7.5 for land-mines and it can be accommodated either from two surface naval vessels or a submersible AUV.

receiving towed line array of hydrophones. The transmitted sonar signals and the deployment depths of the source and the towed line array receiver of hydrophones can be similar with those being used in sonar surveillance operations. For the detection of underwater buried sea-mines, the deployment configuration of the source and the line array receiver should be similar to that in Figure 7.5 for land-mines and it can be accommodated either from two surface naval vessels or submersible AUVs. However, for an effective implementation of the proposed sonar CT concept for surveillance applications it is essential to identify solutions in terms of deployment configurations of assets (i.e., source, receivers) to address a number of practical problems, which are the following:

First: the size (diameter) of the tomography area for CT sonar surveillance is determined by the size of the receiving line array, as depicted Figure 7.1. For practical applications this diameter needs to be as long as possible and in the order of a few kilometers. However, the size of sonar towed line arrays is very limited in length. A suggested solution to this problem is to use acoustic synthetic aperture processing [28] to synthesize a long receiving array to address this problem. Issues of SNR will not be of importance in this case because the ranges will be very short, which will result in low levels of signal propagation losses and with the radiated and reflected levels of the active signals being very high.

Second: it is highly desirable to differentiate the reconstructed images resulting from stationary (i.e., submarines at the bottom of a shallow sea) or moving targets. This requires the deployment of two towed sources combined with a single towed array and the implementation of a motion tracking CT algorithm as defined in Dhanantwari et al. [19]. The algorithm Dhanantwari et al. [19] was developed to track and correct for cardiac motion artifacts and to remove blurring imaging effects due to motion during the CT data acquisition process, which requires that the objects for imaging should be stationary. In the case of sonar CT applications, the requirement is to image and track only the moving target as this is defined in Dhanantwari et al. [19], in order to differentiate them from the reconstructed images of stationary objects.

Third: the time required for the acquisition of acoustic data for a complete image reconstruction sonar CT process can be very long, due to the very slow speed of the naval vessel towing the line arrays and the time required to travel the circumference of the tomography area of interest. This travel time requirement can be reduced by a factor of four, assuming that there are two vessels available to tow the two sources and receiving arrays. Their geometric deployment configuration can be derived from Younis et al. [27].

Furthermore, the results from the experimental investigation reported in this chapter and in Younis et al. [27], indicate that the acoustic coupling between the source-to-water-to-target-to-water-to-receiver for a sonar CT application will be much higher than the acoustic coupling between the source-to-air-to-ground-to-target-to-ground-to-air-to-receiver for demining CT applications. This better coupling for sonar CT applications will result into a significant improvement in SNR and in better quality of reconstructed images.

7.3.4 NonDestructive Imaging of Building Interiors

It is foreseen that in the near future antiterrorist missions would require remote monitoring of humans activities and tomography imaging of interior of buildings and other structures for surveillance applications. The relevant detection system should be mobile, easy to use and cost effective and this requirement can be addressed with the implementation of nondiffraction CT concepts at GPR frequency regimes below 2 GHz, as depicted in Figure 7.7.

The choice of this frequency regime combined with CT image processing will allow the radiated energy to penetrate building structures to achieve image reconstruction of the interior of building. Human activities, that may include movement, can be differentiated from the stationary structures by using the motion tracking methodology in Dhanantwari et al. [19]

The relevant system concept requires the deployment of two vehicles or two helicopters. The first vehicle/helicopter would deploy the GPR directional source to scan the building of interest and the second vehicle/helicopter in a diametrically opposite location should deploy the receiving array of GPR antennas to record the energy penetrating the building structure of interest. Both helicopters should circle the building structure by following a 180-degrees semi-circle movement. For small scale operations, the deployment procedure can be simplified by having two operators carrying the two sets of GPR sources and receivers respectively. Since the radiated GPR energy, being relatively

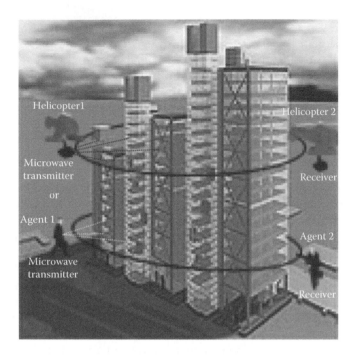

FIGURE 7.7
Concept of nondestructive tomography imaging of the interior of buildings using nondiffraction CT concepts at the frequency regime of 94 GHz. The source and receiver devices can be rotated around the building structure either from helicopters or from vehicles.

nondiffracting, will travel in straight lines, the signal processing requirements for the received data will be equivalent with that of the x-ray CT image reconstruction systems, discussed in Section 7.2.2.

7.4 Experiments

7.4.1 Experimental Set-up

The acoustic CT tomography imaging system was implemented as shown in Figure 7.5a. A minefield was simulated by filling a commercial swimming pool of 4 m diameter and 0.7 m depth with dry construction grade sand. A wooden tripod structure with three hollow wooden boxes buried in the sand (Figure 7.5b) was built to support the data acquisition system above the sand. The data acquisition system consisted of a 31-microphone receiving line array (Panasonic broad band capsule model WM063P) with 3-cm spacing and a four-speaker line array, mounted at the two sides of a horizontal metal bar (3.5 m long). Two objects (simulating landmines) were buried in the sand (Figure 7.5b), at different depths in the range of 10 to 50 cm: (1) an inert antivehicle plastic landmine 30 cm in diameter and (2) a metal flat box $5 \times 20 \times 30$ cm^3. A set of 32 preamplifiers (amplification=40 dB) was placed in the close vicinity of the microphones (Figure 7.5a). A 4-kHz, 32-channel antialiasing filter was used to condition the received signal for the 32-channel 12-bit A/DC (8 kHz sampling frequency). The digitized time series from the 32-channel were multiplexed and stored in a SCSI storage device. The CT data acquisition process included 360 set of measurements at 1-degree interval around a horizontal circle defined by the rotated metal bar with the mounted receiving and source arrays (Figure 7.5a). For each set of measurements, the four-speaker line array generated a plane wave pulse sent in the ground; and the data acquisition system was triggered simultaneously to acquire the reflected acoustic signals by the 31-microphone receiving array.

To align all the received time series from the 360 measurements, a reference microphone (32nd sensor) was placed near the four-speaker array. This procedure of temporal alignment, using the received signal from the reference microphone, will be discussed later in Section 7.4.3.

The transmitted acoustic signal was a FM pulse generated by a PC through a D/AC and the audio amplifiers of the four-speaker array (Figure 7.5). The FM pulse's characteristics were another major design issue, since they affect the image resolution. A very short pulse results in better image resolution and better differentiation of signal arrivals from different directions. However, the constraints imposed by the frequency range of 200–2000 Hz did not allow for the design of a very short FM pulse. A trial and error process provided a practical choice of a 0.1-ms FM pulse with 100 kHz sampling for the D/AC unit.

7.4.2 Transmitted Acoustic Signals

The acoustic diffraction CT process, as depicted in Figure 7.4, requires plane acoustic waves to scan the tomography area of interest. It is well known [6,21] that plane or cylindrical acoustic wavefronts can be generated by the simultaneous transmission from a number of point acoustic sources that form a planar or line array, respectively. This process is equivalent to broadside beamforming. In the present study, we used four acoustic sources, shown in Figure 7.5b, to simulate the generation of cylindrical wavefronts. Each source was rated at a maximum power of 100 Watt; and the measured acoustic level at the output of the source array was 135 dB.

The shape and frequency content of the acoustic pulse was a complex issue. For better resolution, we would like the spectrum of the transmitted acoustic pulse to include high frequencies. However, a high frequency acoustic pulse has a higher attenuation and does not penetrate deep into the soil. As a trade off, we selected a frequency-modulated FM pulse, linearly varying from 200 to 2000 Hz, cascaded with a Parzen window.

The advantages of using an FM type of acoustic pulse, would be similar with those in sonar and radar applications. In other words, correlation of the received time series with the replica of the FM acoustic pulse would allow for a better differentiation and definition of time delays of acoustic wavefront arrivals reflected and/or refracted from various underground structures. This process is called replica-correlator receiver [6,21].

The FM digital pulse waveform, defined above, was generated by a PC and was provided at the input of the PC's D/AC unit that was driving the audio-amplifiers of the four-speaker array. Another parameter in the pulse design, is the pulse length. The pulse length and the phase velocity of the propagating acoustic wave in the medium of interest define the image resolution capabilities of the system for the buried objects. However, a replica-correlator receiver would minimize the effects of this resolution problem, assuming that the SNR is sufficiently high to provide strong correlation peaks above the noise background. In the present experimental study, the replica-correlator receiver concept is not applicable because of the very poor SNR of the receiving microphone time series caused by the very poor air-to-ground coupling interface. This is an issue that is of practical importance for future attempts along the same experimental system concept.

In general, a very short acoustic pulse would allow for a better image resolution and better differentiation of acoustic wavefront arrivals from different directions at the receiving array. However, the design constraint imposed by the frequency range of 200–2000 Hz did not allow for the design of a very short FM pulse. A trial and error process provided a practical choice of a 0.1-ms FM pulse with 100 kHz sampling for the D/AC unit.

7.4.3 Propagation Characteristics of the Acoustic Signals in the Underground Medium

A set of experiments was conducted to determine the propagation speed, time of arrival, amplitude and SNR of the propagating acoustic signals in the air and in the ground, as these were important parameters to determine the sections of the received microphone time series that included the signals of interest.

The sections (a–f) of Figure 7.8 and (a–f) of Figure 7.9 show the various experimental arrangements in terms of microphone positions and the corresponding received signals, respectively. More specifically, in Figure 7.8, the microphone (a) is the reference microphone placed near the acoustic source; 2 m away is the microphone (b), which is one of the microphones of the 31-receiving array. The microphones (c) and (f), shown in Figure 7.8, are buried in the sand (5 cm deep), and just below the microphones (a) and (b), respectively. The time delay between the signal wavefront arrivals of microphones (a) and (b) and between (c) and (f) corresponds to the time taken by the signal to travel between the corresponding microphones through the air and in the sand, respectively. These time delays can be estimated from the results of Figure 7.9 that shows the corresponding time series of the microphones (a–f) of Figure 7.8.

To maximize the microphones receiving characteristics, a number of experiments were conducted to determine the optimum position configuration for the microphone array. For example, microphone (d), shown in Figure 7.8, was acoustically isolated from the

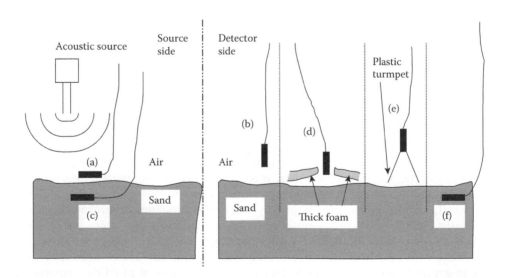

FIGURE 7.8
Experimental set-up of microphones to examine propagation effects for signals shown in Figure 7.9. (Reprinted from Younis, W., Stergiopoulos, S., et al. *J. Acoust. Soc. Am.*, 111(5), 2117–2127, 2002. © AIP. With permission.)

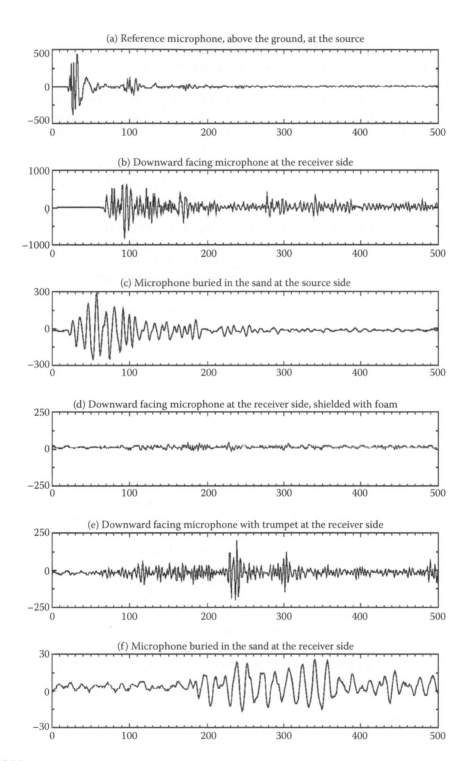

FIGURE 7.9
Received time series for microphone arrangements shown in Figure 7.8. (a), (b), (d) and (e) show received signals by microphone above the ground while (c) and (f) shows received signal by microphones buried in the sand. (a) and (c) are received signals by microphones located near the acoustic source while (b), (d), (e) and (f) are signals at the detector side. (Reprinted from Younis, W., Stergiopoulos, S., et al. *J. Acoust. Soc. Am.*, 111(5), 2117–2127, 2002. © AIP. With permission.)

surrounding acoustic signals propagating through the air by using thick foam to cover the sand surface. In another case, the microphone (e), shown in Figure 7.8, was put in a plastic trumpet to maximize its receiving directivity patter of the acoustic signals arriving from the ground.

From these experiments it became obvious that the amplitude of the received signals by microphone (d) (e.g., Figure 7.9d) were much smaller than the corresponding amplitudes of the signals of microphones (b) and (e), (e.g., Figure 7.9b and e). In particular, the differences in amplitude between the received signals (b) and (e) in Figure 7.9, show that a very significant portion of the energy collected by (b) is received directly from the air and hence the SNR of the component of the acoustic signal propagating through the ground is very small, as expected because of the very poor air-to-ground-to-air coupling. In another experimental set up, the amplitude of the acoustic signal received by microphone (e), (e.g., Figure 7.9e) has smaller amplitude than that of the signal of microphone (b) (e.g., Figure 7.9b). This confirms that the dominant signal components of the received acoustic signals by the microphone (e) (e.g., Figure 7.9e) are signals propagating in the ground. This observation is confirmed also by the signal of the shallow buried microphone (f), since the time arrival of the received signals for microphones (e) and (f) are nearly identical. Furthermore, it appears that the frequency spectrum for the times series (c) and (f) is different that those from the microphones in the air (d) and (e). This was expected since the ground acts as a low pass filter and the higher frequencies that do not penetrate the ground they are not included in (c) and (f).

Thus, the time delay of the signal wavefront arrivals between the reference microphone (a) (e.g., Figure 7.9a) and microphones (e) and (f), (e.g., Figure 7.9e and f), provides estimates of the speed of the acoustic signals propagating through the sand, which are in the range of 80–100 m/sec, which agrees with a similar experiment reported in Larson et al. [22].

In summary, the segments of the received microphone time series in the range of 180–500 samples for the microphones (b), (e) and (f) include the signal components of the pulses propagating through the air-to-ground-to-air. We will use this segment of the received signal to define the sinograms for the purpose of underground image reconstruction. It is worth noting that although the signal loss due to the poor air-to-ground-to-air coupling was great, this signal loss was tolerable in the image reconstruction process. This was confirmed also from the image results that were reconstructed using these very low SNR signals. Another important observation is the increased duration of the received signal by microphone (f), shown in Figure 7.9f as compared to that in Figure 7.9a. This is an indication that the received signal in Figure 7.9f includes multipath propagation effects in the ground. This, however, cannot be considered as justification of our claim that time delay t_j in the projection data $p(r_n, \theta_i, t_j)$ can be equated with depth, since the experimental set-up with the tank filled with sand is a nonstratified medium. Therefore, the reconstructed images of this investigation would represent near the surface 2-D horizontal tomography sections of the underground medium.

7.4.4 Experimental Results

Each experiment, which was based on the set-up of Figure 7.5, generated a three dimensional data set of projections:

$$p\left(r_n, \theta_i, t_j\right); \quad \left(n = 1, \ldots, N\right), \quad \left(i = 1, \ldots, M\right), \quad \left(j = 1, \ldots, K\right),$$

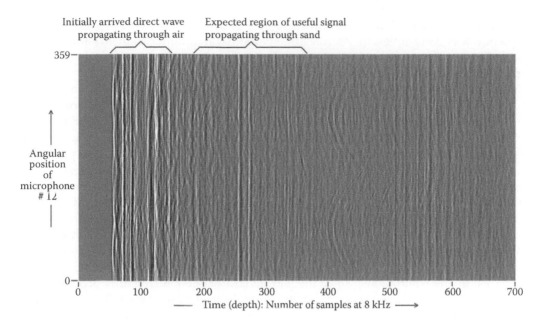

FIGURE 7.10

Temporal alignment of received signals for the 12th microphone of the receiving array and for all the 360 projections. This signal display is called "gram". (Reprinted from Younis, W., Stergiopoulos, S., et al. *J. Acoust. Soc. Am.*, 111(5), 2117–2127, 2002. © AIP. With permission.)

where $N=32$ microphones (31 sensors in the receiving array plus the reference microphone), $M=360$ projections around a circle, and $K=8192$ samples for each microphone time series. These 360 sets of time series had to be aligned according to their reference microphone's signal (Figure 7.9a). Figure 7.10, shows the aligned received signals for the 12th (a randomly selected) microphone. Such a signal display is called "gram". The alignment of the time series from the 360 different projections is clearly evident by the results of Figure 7.10.

7.4.4.1 *Interpolation for Improved Image Resolution and Simulations*

The image resolution characteristics of a CT reconstructed image are related to the number of sensors N, the sensor spacing and the number of projections M (Equation 7.6). A typical x-ray CT scanner deploys approximately 1400 sensors with 1 mm pitch and acquires 1100 projections to have sufficient image resolution for medical diagnostic purposes. In our case, however, due to limited resources, the present setup included only 31 sensors and the acquisition of 360 projections; hence the image resolution was anticipated to be very poor. Furthermore, because of our sensor spacing (e.g., pitch 3 cm) the spatial sampling frequency was 33 samples/m, which indicated that the maximum spatial frequency that could be handled by this pitch was 16.5 samples/m [18]. Another complication in terms of the image resolution capabilities of the current experimental set-up was due to the fact that our experimental observations in Figure 7.9 show that only acoustic frequencies below 1 KHz penetrated the sand medium {i.e., wavelength in the ground, $(100 \text{ m/s})/(1000\text{–}500 \text{ Hz})=10\text{–}20$ cm}. Thus, the image resolution was very limited in our diffraction tomography experiments since the wavelengths of the interrogating acoustic waves were in the range of 10–20 cm and the spatial sampling frequency was 33 samples/m. As a result, spatial details higher than 16 sample/m and object dimensions comparable to the wavelength of 10–20 cm cannot be detected with our current system configuration.

To minimize the impact of the poor image resolution inherent in our system, we used interpolation techniques, as defined in Pan and Kak [18]. To assess the performance of this interpolation technique we used diffraction tomography simulations for 800-Hz and the Shepp–Logan phantom [13,17,18] (left image in Figure 7.11). The upper-middle image results of Figure 7.11 have been generated from a sinogram that has the same parameters as in the experiment (i.e., $M=360$, $N=31$, 500–100-Hz). Image reconstruction results using the same sinogram and with interpolation parameters ($M=360$, $N=256$) are shown by the lower-middle image of Figure 7.11. The right upper and lower reconstructed images of Figure 7.11 indicate the anticipated improved image resolution performance of our experimental set-up for $N=62$ and with smaller sensor spacing of 1.5 cm (e.g., pitch 1.5 cm). It is apparent from these simulations that the experimental configuration of $N=31$ sensors would not be sufficient to provide reasonable image resolution even for high SNR (which is the case in the simulations), since the wavelength of the interrogating acoustic pulses is comparable to the dimensions of the objects and the sensor spacing is not sufficiently small. Thus, for poor SNR, image resolution would be even worse. For more details about the image resolution limits of diffraction tomography in terms of the wavelength of the interrogating energy, the size of the object and the spatial sampling of the receiving sensor array, the reader may review Kak and Slaney [29].

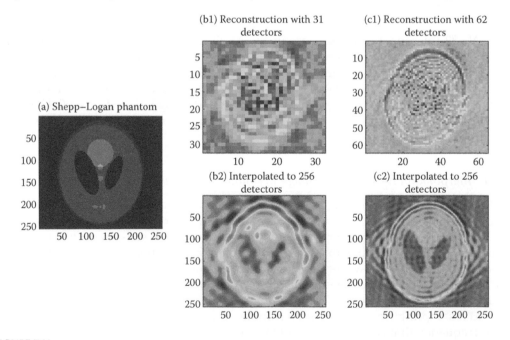

FIGURE 7.11
Simulation results using a Shepp–Logan phantom for diffraction image reconstruction integrated with interpolation techniques. (a) The simulated Shepp–Logan phantom. (b1) The 256×256 reconstructed image from a sinogram ($N=31$, $M=360$), without interpolation. (b2) The 256×256 reconstructed image from the sinogram ($N=31$, $M=360$) that has been interpolated to ($N=256$, $M=360$). (c1) The 256×256 reconstructed image from a sinogram ($N=62$, $M=360$), without interpolation. (c2) The 256×256 reconstructed image from the sinogram ($N=62$, $M=360$) that has been interpolated to ($N=256$, $M=360$). (Reprinted from Younis, W., Stergiopoulos, S., et al. *J. Acoust. Soc. Am.*, 111(5), 2117–2127, 2002. © AIP. With permission.)

7.4.4.2 Real Results

Based on Equation 7.6 and Figure 7.10, a 2-D sinogram for a predefined temporal sample is given by:

$$p(r_n, \theta_i, t_{j=\text{pre defined}}); \quad (n = 1, ..., N = 31), \quad (i = 1, ..., M = 360), \quad (j = \text{pre defined}) \quad (7.7)$$

Figure 7.12a displays a sinogram for the time delay $j=376$. The correspondence between the data sets of Figures 7.10 and 7.12a is that a vertical line for the sensor $n=12$ in Figure 7.12a represents the same data as the vertical line of Figure 7.10 for the time delay $j=376$.

The image in Figure 7.12b has been reconstructed from the sinogram of Figure 7.12a and for the temporal sample $j=376$. It shows a circular pattern (indicated by the arrow) that its location coincides with the location and time delay-depth of the buried plastic mine (Figure 7.5b). In this experiment the sand was wet in order to improve the acoustic propagation and compare results with dry sand. The image reconstruction algorithm for the results of Figure 7.12b was the Fourier diffraction algorithm [18] using interpolated data (from 31 to 256 sensors) for the same temporal samples. Thus, the image in Figure 7.12c has been derived from the same sinogram as the image in Figure 7.12b. Their differences represent better image resolution characteristics as those in the simulation results shown in Figure 7.11.

As discussed earlier, the temporal samples for $j=180, ..., 500$, are expected to provide information about buried objects. Although the images of the buried mine (Figure 7.12) have very poor image resolution, their spatial location coincides with that of the buried mine as in our previous experiments reported in Stergiopoulos et al. [24] and shown by Figure 7.12d. The image in Figure 7.12c was reconstructed using filtered back projection algorithm on noninterpolated data. The experiment was carried out in dry sand and for slightly different mine location and depth with those of Figure 7.12b and c. Figure 7.12e and f show reconstructed images for temporal sample $j=186$ using filter back projection and Fourier diffraction (interpolated data), respectively. The white spot in both figures corresponds to the expected location of the buried metallic mine, shown in Figure 7.5b.

7.4.5 Adaptive Interference Cancellation (AIC) in Acoustic CT

The results in Figure 7.13 demonstrate the advantages of using adaptive interference cancellation (AIC) processing to minimize the impact of the poor SNR in the image reconstruction process. The implementation of the AIC process in our acoustic CT concept has already been discussed briefly in Section 7.2 and for more delays the reader may review the papers [19,21,23]. The AIC implementation effort requires that the 31-microphone array should be re-arranged as follows: 16-microphones should be positioned along the receiving array downwards to receive the noise signal for the AIC process. 15-microphones should be positioned along the receiving array upwards to receive the interference, which is the propagating in the air component of the transmitted pulse. The spacing for both the upward and downward microphones was the same.

The time series in Figure 7.13a, refer to the reference microphone (a), (see Figure 7.8), located near the array of the acoustic sources. Figure 7.13g and b show the received time series for the upward and the downward facing microphones that represent the interference and noisy signals, respectively, that are essential in an AIC process. The time series in Figure 7.13i show the output of the AIC process and Figure 7.13f shows the signal of the buried microphone that defines the time position of the wavefront arrival in the ground. At this point it is important to note that a comparison between the time series (i) and (f) of

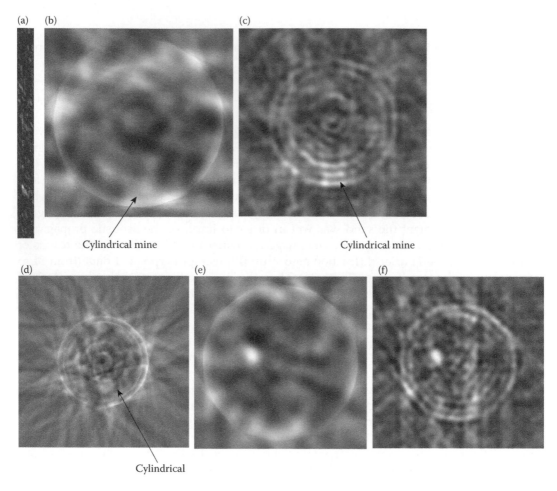

(a) (b)   (c)

Cylindrical mine   Cylindrical mine

(d)   (e)   (f)

Cylindrical

FIGURE 7.12
(a) Sinogram from the data set of Figure 7.10 corresponding to the temporal sample $j=376$, for the 31 microphones and 360 projections. (b) Image reconstruction based on filtered back projection and using the sinogram data set shown in (a) with no interpolation (wet sand). (c) Image reconstruction based on Fourier diffraction algorithm and using an interpolated version (31–256 sensors) of the same sinogram data set (a). The image of the buried mine is obvious at the bottom. This location coincides with the expected mine location in Figure 7.5b. (d) Image reconstruction based on filtered back projection of a data set from a previous experiment reported in Stergiopoulos et al. [24]. This image is for the same buried mine of Figure 7.5b. The mine location here is slightly different from (b) and (c). In this case the sand was dry. (e) Image reconstruction based on filtered back projection and using a sinogram data set with no interpolation (temporal sample $j=186$). In this experiment the sand was wet. (f) Image reconstruction based on Fourier diffraction algorithm and using an interpolated version of the sinogram data set from 31 to 256 sensors and for the same data set with those of image (e). The location of the white image spot in (e) and (f) coincides with that of the buried metallic object in Figure 7.5b. (Reprinted from Younis, W., Stergiopoulos, S., et al. *J. Acoust. Soc. Am.*, 111(5), 2117–2127, 2002. © AIP. With permission.)

Figure 7.13 indicates that these two time series are not similar since AIC filtering process (because of the very low SNR) has not been very effective to remove all the components of air contributions in the received signal. However, the AIC process has been sufficient to enhance the image resolution for this experimental set-up as shown by the image reconstruction results of Figure 7.14.

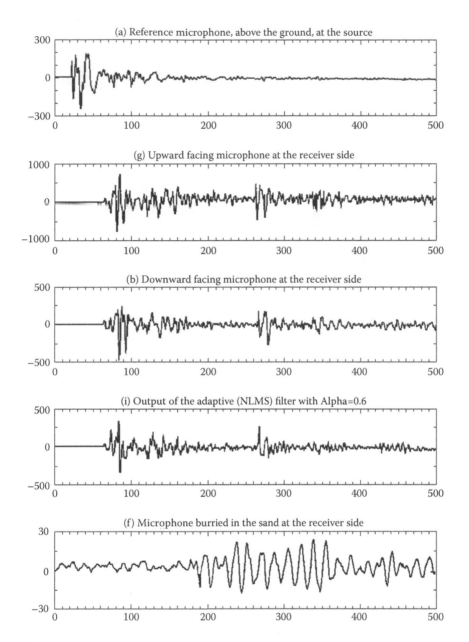

FIGURE 7.13
Time series from the adaptive acoustic CT experiment. (a) The signal from the reference microphone. (g) The signal from the upward microphone that is considered as the noise (interference) in the AIC process. (b) The signal from the downward microphone that is considered as the noisy signal in the AIC process. (i) The output signal of the AIC process. (f) The signal from the buried microphone. (Reprinted from Younis, W., Stergiopoulos, S., et al. *J. Acoust. Soc. Am.*, 111(5), 2117–2127, 2002. © AIP. With permission.)

Thus, for the temporal sample $j=55$ the Figure 7.14a and b are the sinograms for the 15-upward (interference) and the 16-downward (noisy signal) microphones, respectively. These sinograms define the inputs for the AIC processor. The output of the AIC processor is the sinogram shown in Figure 7.14c, which is supposed to have improved SNR as compared with the SNR of Figure 7.14b.

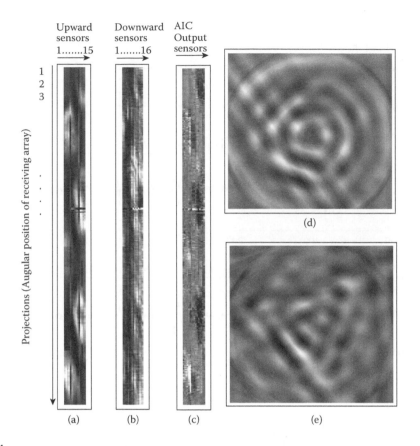

FIGURE 7.14
Sinograms from the acoustic CT experiment and for the temporal sample $j=55$. (a) For the 15-upward micro-phones. (b) For the 16-downward microphones. (c) Output of the AIC process. (d) Image reconstruction using the sinogram shown in (b). (e) Image reconstruction using the AIC sinogram shown in (c). The AIC image output shows the expected triangular base, shown in Figure 7.5. (Reprinted from Younis, W., Stergiopoulos, S., et al. *J. Acoust. Soc. Am.*, 111(5), 2117–2127, 2002. © AIP. With permission.)

The image reconstructed from the AIC output sinogram of Figure 7.14c, is shown in Figure 7.14e. This image result (Figure 7.14e) demonstrate the efficiency of the AIC processor in minimizing the impact of the air-reverberations in the downward microphones, when compared with the image of Figure 7.14d, which has been derived from the image reconstruction of the noisy sinogram of Figure 7.14b of the downward facing microphones.

The features in Figure 7.14e represent objects and structures above the ground (the triangular base shown in Figure 7.5), as this was expected since $j=55$, which indicates the early arrival of the signal wavefronts propagating through the air. It is interesting to note that these triangular base features were completely masked in the image results of Figure 7.14d, and were only revealed by the AIC process in Figure 7.14e. Although, the number of sensors in Figure 7.14e were halved from those in Figure 7.12b, the AIC process appeared to be effective in improving image quality by minimizing the impact of the surrounding interference noise. However, since the microphone spacing in the AIC process was 6 cm, the spatial sampling frequency was 16 samples/m, which indicates that the maximum spatial frequency that could be handled by this pitch would be insufficient to sample the dimensions of the buried mines, as indicated also by the simulation results of Figure 7.11.

Therefore, because of the limited number of microphones that were available in this experimental study, we were unable to assess the effectiveness of the AIC process in providing better image resolution for the mines shown in Figure 7.12.

7.5 Key Issues and Future Directions

The results of this experimental study and the discussions in the previous sections suggest that a successful implementation of the proposed acoustic tomography concept as a nondestructive underground imaging process should address the following issues:

- The receiving microphone array should have at least 64-downward-microphone (preferably 128-microphones) with smaller array pitch to improve the image resolution.

- The AIC process is essential to minimize the effects of the signal arrivals propagating through the air. Thus, the number of sensors should be increased to 128 (preferably 256) to include the 64 (preferably 128) upward microphones.

- The pulse design characteristics should guide the design efforts for the acoustic source array to avoid the nonlinearities discussed in Section 7.4.2 and to include a frequency range that would allow ground penetration with sufficiently small wavelengths for improved image resolution.

- Finally, if the AIC process would be effective to improve the SNR, the replica correlation process should be implemented on the microphone time series in order to differentiate the signal arrivals from various underground structures and from the air.

Although, the available resources were very limited and not sufficient to address the above issues, the results discussed in Section 7.4, suggest that the proposed acoustic tomography concept can be used successfully as a nondestructive underground imaging technique.

At this point it is important to note that the poor image results of our investigation are in agreement with the findings of another investigation by Crawford and Kak [26] that proposed a technique for breast tomography imaging using ultrasonic CT concepts. Although their physical parameters (forward problem) were more favorable than the poor air-to-ground-to-air coupling (and multipath) imposed in our experimental design, the image-quality of their experimental results was very poor and similar with those reported in our manuscript. In summary, investigators should anticipate that because of the very low frequencies (e.g., long wavelengths) used in acoustic CT imaging applications the associated image resolution would be poor and that they should not expect the high image resolution that is available by the x-ray CT scanners.

7.6 Conclusion

The present chapter describes a new approach to nondestructive subsurface imaging for detecting landmines and other buried objects. Subsurface imaging is performed by sending acoustic energy into the soil around a circular pattern and collected at the diametrically opposite end of it by a linear microphone array. From these collected signals, subsurface images are reconstructed employing image reconstruction algorithms, in the same manner as done in CT imaging. It is assumed that acoustic waves that propagate in the ground carry information about the subsurface buried objects to the sensors of the receiving array. It is also assumed that in the received signal, the signal amplitude at different time delays carries information from the different depths of the stratified soil. Consequently, we would be able to reconstruct the cross sectional images at different underground depths.

Among several challenges, the most important was the limited knowledge of the propagation characteristics. The very weak signals of interest, received by the microphones, resulted in poor SNR that was another major problem due to the poor air-to-ground-to-air acoustic coupling.

In summary, the results of this investigation provide supporting arguments that the implementation of the AIC process in combination with the deployment of large number of sensors in the receiving array could be a valuable method for 3-D subsurface imaging for the detection of shallow buried objects using acoustic diffraction CT.

References

1. Dines, K. A. and Lytle, J. 1979. Computerized geophysical tomography. *Proc. IEEE*, 67(7), 1065–1073.
2. King, W. C., Witten, A. J. and Reed, G. D. 1989. Detection and imaging of buried wastes using seismic wave propagation. *ASCE J. Environ. Eng.*, 115(3), 527–540.
3. Devaney, A. J. 1984. Geophysical diffraction tomography. *IEEE Trans. Geosci. Remote Sensing*, GE-22 (1), 3–13.
4. Witten, A. J. and Long, E. 1986. Shallow applications of geophysical diffraction tomography. *IEEE Trans. Geosci. Remote Sensing*, GE-24 (5), 654–662.
5. Reporting on Science and Technology, Archaeology—Underground Noises. *The Economist*, July, 83, 2000.
6. Stergiopoulos, S. 2000. Advanced beamformers. In: *Advanced Signal Processing Handbook: Theory and Implementation for Radar, Sonar and Medical Imaging Systems*. CRC Press, Boca Raton, FL.
7. Sakas, G., Karangelis, G. and Pommert, A. 2000. Advanced applications of volume visualization methods in medicine. In: *Advanced Signal Processing Handbook: Theory and Implementation for Radar, Sonar and Medical Imaging Systems*. CRC Press, Boca Raton, FL.
8. Haystead, J. 1997. Portable systems apply infrared technology to detect and classify mines. *Vision Systems Design*, 36–40.
9. Vaughan, C. J. 1986. Ground penetrating radar survey in archeological investigations. *Geophysics*, 51(3), 595–604.
10. McFee, J., et al. 1994. CRAD Countermine R&D Study, Final Report. DRES-SSP-174. Defense Research Establishment Suffield, Rlaston, Alberta, Canada.
11. Benson, R. C., Glaccum, R. P., and Noel, M. R. 1982. *Geophysical Techniques for Sensing Buried Wastes and Waste Migration*. US Environmental Protection Agency, Washington, D.C.
12. Frazier, C. H., Cadalli, N., Munson, D. C. and O'Brien, W. D. 2000. Acoustic imaging of objects buried in soil. *J. Acoust. Soc. Am.*, 108(1), 147–156.
13. Kak, A. C. 1979. Computerized tomography with x-ray, emission, and ultrasound sources. *Proc. IEEE*, 67, 1245–1272.

14. Radon, J. 1917. Uber die Bestimmung von Funktionen durch ihre Integral werte langs gewisser Mannigfaltigketiten. Berichte Saechsische Akademie der Wissenshcafien (Leipzig). *Mathematische-Physische Klasse*, 69, 262–279.
15. Healey, A. J. and Webber, W. T. 1995. *Sensors for the Detection of Land-Based Munitions*. Naval Postgraduate School, Monterey, CA.
16. Devaney, A. J. 1982. A filtered back projection algorithm for diffraction tomography. *Ultrasonic Imag.*, 4, 336–350.
17. Kak, A. C. 1984. *Digital Image Processing Techniques*. Academic Press, New York, NY.
18. Pan, S. X. and Kak, A. C. 1983. A computational study of reconstruction algorithms for diffraction tomography: interpolation versus filtered backpropagation. *IEEE Trans. Acoust., Speech, and Signal Process.*, ASSP-31(5), 1262–1275.
19. Dhanantwari, A., Stergiopoulos, S., and Iakovides, I. 2001. Correcting organ motion artifacts in x-ray CT medical imaging systems by adaptive processing (Part I: Theory). *Med. Phys.* 28(8), 1562–1576.
20. Mueller, R. K., Kaveh, M. and Wade, G. 1979. Reconstructive tomography and applications to ultrasonics. *Proc. IEEE*, 67, 567–587.
21. Stergiopoulos, S. 1988. Implementation of adaptive and synthetic aperture beamformers in sonar systems. *Proc. IEEE*, 86(2), 358–396.
22. Larson, G. D., Martin, J. S., et al. 2000. Air acoustic sensing of seismic waves. *J. Acoust. Soc. Am.*, 107(5), 2896.
23. Widrow, B. and Hoff, M. E. Jr. 1975. Adaptive noise cancellation: principles and applications. *Proc. IEEE*, 63, 1692–1716.
24. Stergiopoulos, S., Alterson, R., Havelock, D. and Grodski, J. 2000. Acoustic 3D computed tomography for demining and archeological applications. *J. Acoust. Soc. Am.*, 107(5), 2806.
25. Attenborough, K., et al. 1986. The acoustic transfer function at the surface of a layered poroelastic soil. *J. Acoust. Soc. Am.*, 79(5), 1353–1358.
26. Crawford, C. R. and Kak, A. C. 1982. Multipath artifact corrections in ultrasonic transmission tomography. *Ultrasonic Imag.*, 4, 234–266.
27. Younis, W., Stergiopoulos, S., Havelock, D., and Groski J. 2002. Non-distructive imaging of shallow buried objects using acoustic computed tomography. *J. Acoust. Soc. Am.*, 111(5), 2117–2127.
28. Stergiopoulos, S. 1998. Implementation of adaptive and synthetic aperture beamformers in sonar systems. *Proc. IEEE*, 86(2), 358–396.
29. Kak, A. C., and Slaney, M. 1998/2001. *Principles of Computerized Tomographic Imaging*. First published by IEEE Press, New York 1988 or Society of Industrial and Applied Mathematics (SIAM), Philadelphia, PA, 2001.

8

A Review on Face and Gait Recognition: System, Data, and Algorithms

Haiping Lu
University of Toronto

Jie Wang
Epson Edge

Konstantinos N. Plataniotis
University of Toronto

CONTENTS

8.1 Introduction

For thousands of years, humans have used visually perceived body characteristics such as face and gait to recognize each other. This remarkable ability of human visual system has inspired researchers to build automated systems to recognize individuals from digitally captured facial images and gait sequences [1]. Face and gait recognition belong to the field of biometrics, a very active area of research in the computer vision and pattern recognition society, and face and gait are two typical physiological and behavioral biometrics, respectively. Compared with other biometric traits, face and gait have the unique property that they facilitate human recognition at a distance, which is extremely important in surveillance applications. Moreover, their unintrusive nature leads to high collectability and acceptability, making them very promising technologies for wide deployments. The collectability refers to the ease of acquisition for measurement and the acceptability indicates the extent to which people are willing to accept the use of a particular biometric identifier in their daily lives [1].

8.1.1 Face Recognition

Face recognition has a wide range of applications, such as biometric authentication, surveillance, human–computer interaction and multimedia management. Table 8.1 lists some of these applications in detail. On the other hand, the rapid developments in technologies such as digital cameras, the Internet and mobile devices also increase the popularity of face recognition [2]. A number of commercial face recognition systems have been deployed, such as Cognitec [3], Eyematic [4], and L-1 Identity Solutions [5]. Face recognition is one of the three identification methods used in e-passports and it has an important advantage over other popular biometric technologies: it is nonintrusive and easy to use [2]. Among the six biometric attributes considered by Hietmeyer [6], facial features scored the highest compatibility in a machine readable travel documents (MRTD) system based on a number of evaluation factors, such as enrollment, renewal, machine requirements, and public perception [6]. In addition, in advanced human–computer interaction applications, it is very important for robots to be able to identify faces, expressions and emotions while interacting with humans.

8.1.2 Gait Recognition

Gait recognition [9,10], the identification of individuals in video sequences by the way they walk, is strongly motivated by the need for automated person identification system at a distance in visual surveillance and monitoring applications in security-sensitive environments, e.g., banks, parking lots, museums, malls, and transportation hubs such as airports and train stations [11], where other biometrics such as fingerprint, face or iris information are not available at high enough resolution for recognition [12,13]. Furthermore, night vision

TABLE 8.1

Applications of Face Recognition

Areas	Examples of Applications
Biometrics, mugshot identification	Drivers licenses, entitlement programs, smart cards, immigration, national ID, passports, voter registration, welfare fraud, airline industry, bank industry
Information	Desktop logon, secure trading terminals, application security, database security,
Security	File encryption, intranet security, internet access, medical records
Law enforcement and surveillance	Automated video surveillance (e.g., airport security checkpoints), portal control, postal-event analysis, face reconstruction from remain shoplifting and suspect tracking and investigation
Access control	Access to private buildings, facility, personal computers, PDA and cell phone
Others	Human–computer interaction, content-based image database management, information retrieval, multimedia communication (e.g., generation of synthetic faces)

Source: Adapted and expanded from W. Zhao, R. Chellappa, A. Rosenfeld, and P. Phillips., *ACM Computing Surveys*, 399–458, 2003, and R.-L. Hsu., Face detection and modeling for recognition. Ph.D. dissertation, Michigan State University. Available online: http://biometrics.cse.msu.edu/Publications/Thesis/VincentHsu_FaceDetection_PhD02.pdf, 2002.

capability (an important component in surveillance) is usually not possible with other biometrics due to the limited biometric details in an IR image at large distance [12,13].

Gait is a complex spatio-temporal biometric [10,12] that can address the problems above. In 1975 [14], Johansson used point light displays to demonstrate the ability of humans to rapidly distinguish human locomotion from other motion patterns. Similar experiments later showed the capability of identifying friends or the gender of a person [15,16], and Stevenage et al. show that humans can identify individuals based on their gait signature in the presence of lighting variations and under brief exposures [17].

Gait is a behavioral (habitual) biometric, in contrast with those physiological biometrics such as face and iris, and it is viewed as the only true remote biometric [18]. Capturing of gait is unobtrusive, which means that it can be captured without requiring the prior consent of the observed subject, and gait can be recognized at a distance (in low resolution video) [19]. In contrast, other biometrics either require physical contact (e.g., fingerprint) or sufficient proximity (e.g., iris). Furthermore, gait is harder to disguise than static appearance features such as face.

8.1.3 Organization

The problem of face and gait recognition is very challenging. The main challenge in vision-based face or gait recognition is the presence of a high degree of variability in human face images or gait sequences. For facial images [20], the intrasubject variations include pose (imaging angle), illumination, facial expression, occlusion, makeup, glasses, facial hair, time (aging) and imaging parameters such as aperture and exposure time. For gait, the intrasubject variations include pose (viewing angle), shoes, walking surface, carrying condition, clothing, time and also imaging device. Therefore, extracting the intrinsic information of a person's face or gait from their respective images or gait sequences, respectively, is a demanding task. Despite the difficulties, researchers have made significant advancement in this area. The rest of this chapter provides a survey on face and gait recognition. The next section is an overview of the face and gait recognition systems, where the key components are described and the two common approaches are introduced. The fusion of face and gait for recognition is also discussed in the next section. In Section 8.3, several commonly used face and gait databases are reviewed. Section 8.4 presents various feature extraction algorithms for face and gait recognition, ranging from linear, nonlinear to multilinear subspace learning algorithms. Finally, concluding remarks are drawn in Section 8.5.

8.2 Face and Gait Recognition System Overview

Figure 8.1 depicts a typical face or gait recognition system. By observing a subject in the view, a digital camera captures a digital raw facial image or a digital raw gait video. This image or video is then preprocessed (e.g., filtered to remove noise) to extract a facial image or a gait sequence for feature extraction. In the feature extraction, face or gait features are extracted from the input image or image sequence, and these features are passed to the recognition module, where classifiers are employed to match them with the stored features in the face or gait database and a person is recognized with his/her ID as the output.

Face localization and normalization (face detection and alignment) are preprocessing steps before face recognition (facial feature extraction and matching) is performed. Face detection segments the face area from the background and provides (coarse) information about the location and scale of each detected face. Face alignment aims to achieve more accurate localization and normalizes faces. Some specific facial components, such as eyes, nose, mouth and facial outline, are further located, and then the input face image is aligned and normalized in geometry (such as size and pose) and photometry (such as illumination and gray scale).

For gait recognition, binary gait silhouettes are usually extracted through background subtraction, where a background model is estimated from the input raw gait sequences and then it is subtracted to get the silhouettes [21]. The extracted silhouettes are then cropped and resized to a standard size.

After the preprocessing, features useful for discriminating different persons are extracted from the normalized face or gait sample, and in the recognition stage, the extracted features are matched against those of enrolled ones in the database. Finally, the system outputs the identity of the input when a match is found with sufficient confidence or indicates an unknown identity otherwise.

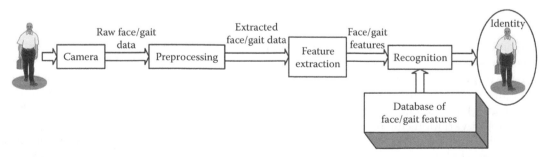

FIGURE 8.1
A typical face or gait recognition system.

There are three main recognition tasks: verification, identification and watch list [10]. Verification involves a one-to-one match that compares a query sample against a template of the claimed identity in the database. The claim is either accepted or rejected. The verification performance is usually measured by the receiver operating characteristic (ROC), which plots the false accept rates (FAR) versus the false rejection rate (FRR). Identification involves one-to-many matches that compares a query sample against all the templates in the database to output the identity or the possible identity list of the input query. In this scenario, it is often assumed that the query sample belongs to the persons who are in the database. The identification performance is usually measured by the cumulative match characteristic (CMC) [22,23], which plots the identification rate against the rank. The watch list scenario involves one-to-few matches that compares a query sample against a list of suspects. In this task, the size of database is usually very small compared to the possible queries and the identity of the probe may not be in the database. Therefore, the recognition system should first detect whether the query is on the list or not and if yes, correctly identify it. The performance of watch list tasks is usually measured by the detection rate, the identification rate and the false alarm rate.

Over the past two decades, a large number of face or gait recognition algorithms have been developed. These algorithms can be categorized into two approaches, the model-based approach and the appearance-based approach [12,24]. Although the model-based approach takes advantage of our prior knowledge on human face or gait, reliable recovery of model parameters from raw facial images or gait sequences is a very hard problem, especially in unconstrained conditions. Therefore, the appearance-based approach has been the most successful in reported literature. In the following, we briefly describe these two approaches, with more detailed discussions on the popular and more successful appearance-based approach in Section 8.4.

8.2.1 Model-Based Face Recognition Approach

The model-based face recognition approach aims to construct a model of the human face in order to capture the facial variations, based on prior knowledge of human faces. Then, the model is fitted to a given face image and parameters of the fitted model form the feature vector for recognition through similarity calculation between the query face and stored faces in the database [24].

In the so-called feature-based approach popular in the early days, local facial features such as eyes, nose, mouth, and chins are first accurately located. Properties of and relations between (such as areas, distances and angles) the features are used as descriptors for recognition. Typical examples include the hidden Markov model methods [25] and the elastic bunch graph matching algorithm [26]. Another model-based method is the active appearance models (AAM) [27] where face variations are learned through integrating shape and texture information with a 2-D morphable face model.

The main advantage of the model-based approach is that the model is based on intrinsic physical relationship with real faces and it is less sensitive to variations in illumination and viewpoint [24]. However, relatively high resolution and good quality face images are needed in this approach. Besides, the recognition performance relies heavily on the exact localization of facial features, but facial feature extraction techniques developed to date are not reliable enough for accurate recognition. For instance, most eye localization techniques assume some geometric and textural models of the eyes and they fail if the eyes are closed. Moreover, such geometric properties alone are inadequate for face recognition because rich information contained in the facial texture or appearance is discarded.

8.2.2 Model-Based Gait Recognition Approach

The model-based gait recognition approach considers a human subject as an articulated object represented by various body poses. Model-based gait recognition algorithms are usually based on 2-D fronto-parallel body models [19,28–30] and target to model human body structure explicitly, with support from the anthropometry and the biomechanics of human gait [31,32]. Body model parameters, such as joint angles, are searched in the solution space through matching edges and region-based information (e.g., silhouettes). The searching methods are either exhaustive [28,29] or in a Bayesian hypothesis-and-test fashion [19], where proper dealing with local extrema is an important problem. The estimated parameters are either used directly as features or fed into a feature extractor (e.g., frequency analyzer) to obtain gait features.

There are also works on coarser human body models. For instance, the work in Lee et al. [33] fits several ellipses to different parts (blobs) of the binary silhouettes and the parameters of these ellipses (e.g., location. and orientation) are used as gait features.

Recently, a full-body layered deformable model (LDM) is proposed in Lu et al. [34], inspired by the manually labeled body-part-level silhouettes [35]. The LDM has a layered structure to model selfocclusion between body parts and it is deformable so simple limb deformation is taken into consideration. In addition, it also models shoulder swing. The LDM parameters can be recovered from automatically extracted silhouettes and then used for recognition [36]. Figure 8.2 shows three examples of the LDM body poses (on the right) recovered from raw image frames (on the left) for illustration.

As in the case of face recognition, model-based gait recognition algorithms utilize our knowledge on gait and they are expected to be less sensitive to noise due to background clutters, cloths, shadows, etc. However, reliable estimation of model parameters is a very difficult task.

8.2.3 Appearance-Based Face Recognition Approach

In contrast with the model-based approach, the appearance-based face recognition approach operates directly on 2-D facial image and processes them as 2-D holistic patterns to avoid difficulties associated with 3-D modeling, and shape or landmark detection. Consequently, this class of methods tends to be easier to implement, more practical and reliable [37]. Furthermore, evidences have been shown in the studies in visual neuroscience that facial features are processed holistically and facial recognition is dependent on holistic processes involving an interdependency between featural and configural information [38].

(a)   (b)   (c)

FIGURE 8.2
Three examples of the human body poses recovered through the LDM. Left: the raw image frame. Right: the automatically reconstructed silhouette through the LDM recovery.

In this approach, the whole face region is the raw input to a recognition system and each face image is commonly represented by a high-dimensional vector consisting of the pixel intensity values in the image, i.e., a point in a high-dimensional vector space. Thus, face recognition is transformed to a multivariate statistical pattern recognition problem. Although the embedding is high-dimensional, the natural constraints of the face data indicate that the face vectors lie in a lower-dimensional subspace (manifold). The popular subspace learning is such a method to identify, represent, and parameterize this subspace with some optimality criteria [24].

8.2.4 Appearance-Based Gait Recognition Approach

The appearance-based gait recognition approach considers gait as a holistic pattern and uses a full-body representation of a human subject as silhouettes or contours. Most of the gait recognition algorithms proposed are appearance-based [11,33,35,39–49]. Some works use the silhouettes directly as the gait representation [23,50] and some others use the average silhouettes as the gait representation [43,51]. These gait recognition algorithms extract structural or shape statistics as features from silhouettes, e.g., width [41], contours [11,40], projections [44], and motion patterns [39,52–54]. There are also methods based on dense optical flow [55], which identify individuals by periodic variations (phase features) in the shape of their motion. In addition, as in face recognition, a gait subspace can be learnt for gait recognition as well [56].

8.2.5 Fusion of Face and Gait for Recognition

Besides using face or gait for recognition individually, these two biometric characteristics can also be fused to achieve multimodal recognition with enhanced results. The fusion can be done either at the feature (or sensor) level [57] or at the decision (or matching score) level [58,59]. In Zhou and Bhanu [60], face and gait features are integrated at the feature level, where the face and gait features are extracted separately and then they are normalized and concatenated to form a single feature vector for recognition. Improvement due to fusion over recognition with a single modality is observed in this work. In Shakhnarovich and Darrell [61], face and gait cues are derived from multiple simultaneous views and transformed to the canonical pose, frontal face and profile gait silhouettes. Then view-normalized face and gait sequences are used individually for recognition and the recognition results are combined using a number of conventional rules including the MAX, MIN, MEAN, PRODUCT and MAJORITY rules. The results presented in Shakhnarovich and Darrell [61] indicate that the MEAN and PRODUCT rules have better results. In Kale et al. [62], two ways of decision-level face and gait fusion are explored. The first uses the gait recognition algorithm as a filter to pass on a smaller set of candidates for face recognition. The second combines similarity scores obtained separately from the face and gait recognition algorithms, with the SUM, MIN and PRODUCT rules tested. Their results indicate that the second way is more effective.

8.3 Face and Gait Data Sets

The development of face or gait recognition algorithms largely depends on the availability of large and representative public databases of face images or gait sequences so that algorithms can be compared and advancements can be measured. In this section, we review three widely used face and gait databases: the Pose, Illumination, and Expression (PIE) database from the Carnegie Mellon University (CMU) [63], the Facial Recognition Technology (FERET) database [22] and the HumanID Gait Challenge data sets from the University of South Florida (USF) [23].

In typical pattern recognition problems of face and gait recognition, there are usually two types of data sets: the gallery and the probe [22,23]. The gallery set contains the set of data samples with known identities and it is used for training. The probe set is the testing set where data samples of unknown identity are to be identified and classified via matching with corresponding entries in the gallery set.

8.3.1 The PIE Database

Visually perceived human faces are significant affected by three factors: the pose, which is the angle at they are viewed from, the illumination/lighting conditions, and the facial expression such as happy, sad and anger. The collection of the PIE database is motivated by a need for a database with a fairly large number of subjects imaged a large number of times to cover these three significant factors, i.e., from a variety of different poses, under a wide range of illumination variation, and with several expressions [63].

This database was collected between October 2000 and December 2000 using the CMU 3D Room and it contains 41,368 face images from 68 individuals. Face images with 13 different poses are captured using 13 synchronized cameras. For the illumination variation, the 3D Room is augmented with a flash system having 21 flashes. Images are captured with and without background lighting, resulting in $21 \times 2 + 1$ different illumination conditions. In addition, the subjects were asked to pose with four different expressions. Figure 8.3 shows 144 sample face images for one subject in this database.

The PIE database can be used for a variety of purposes, including evaluating the robustness of face recognition systems against the three variations and three-dimensional modeling. In particular, this database has a very large number (around 600 on average) of facial images available for each subject, allowing us to study the effects of the number of training samples (per subject) on the recognition performance. In practice, a subset is usually selected with a specific range of pose, illumination and expression for experiments so that data sets with various difficulties can be obtained where a wider range of the three variations leads to a more difficult recognition task.

8.3.2 The FERET Database

The FERET database is a widely used database for face recognition performance evaluation. It was constructed through the FERET program, which aims to develop automatic face recognition systems to assist security, intelligence, and law enforcement personnel in the performance of their duties [22]. The face images in this database cover a wide range of variations in pose (viewpoint), illumination, facial expression, acquisition time, ethnicity and age.

The FERET database was collected in 15 sessions between August 1993 and July 1996 and it contains a total of 14,126 images from 1199 individuals with views ranging from frontal to left and right profiles. The face images were collected under relatively unconstrained conditions. The same physical setup and location was used in each session to maintain a degree of consistency throughout the database. However, since the equipment was reassembled for each session, images collected on different dates have some minor variation. Sometimes, a second set of images of an individual was captured on a later date, resulting

(a)

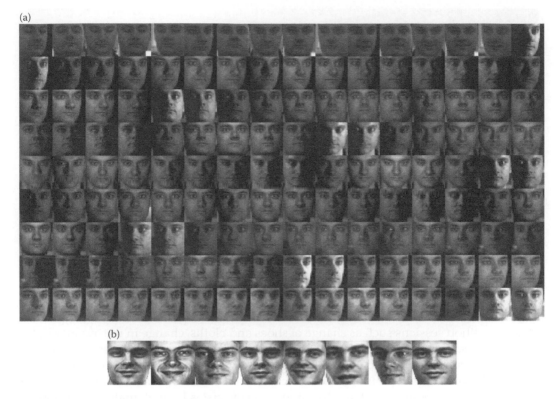

(b)

FIGURE 8.3
Examples of one subject from (a) the PIE face database and (b) the FERET database.

in variations in scale, pose, expression, and illumination of the face. Furthermore, for some people, over two years elapsed between their first and last capturing in order to study changes in a subject's facial appearance over a year. Figure 8.3b shows eight sample face images for one subject in this database. This database has a large number of subjects and it becomes the de facto standard for evaluating face recognition technologies [64], especially in the small sample size scenario, where a smaller number of training samples per subject and a larger number of total subjects lead to a more difficult recognition task [65].

8.3.3 The USF Gait Database

In the area of gait recognition, the Gait Challenge data sets from the USF captures the variations of a large number of covariates* for a large group of people and it has emerged as a standard testbed for new gait recognition algorithms. Other databases are limited in size, variations, capturing conditions, or of high resolutions [9,23].

The Gait Challenge data sets consists of 1870 sequences from 122 subjects and a set of 12 outdoor experiments are constructed to examine the effects of five covariates on performance. In the collection, the subjects are asked to walk in elliptical paths in front of the camera. The image frame is of size 720×480 and a subject's size in the back portion of the ellipse is on average 100 pixels in height. Two sample frames are shown in Figure 8.4.

---

* Covariates are random variables you treat as concomitants or as other influential variables that also affect the response.

FIGURE 8.4
Two sample frames from the USF Gait Challenge data sets.

The covariates either effect gait or effect the extraction of gait features from images. They are: change in viewing angle (left or right), change in shoe type (A or B), change in walking surface (Concrete or Grass), carrying or not carrying a briefcase and temporal (time) differences, where the time covariate implicitly includes other changes naturally occur between video acquisition sessions such as change of shoes and cloths, change in the outdoor lighting conditions, and inherent variation in gait over time. These covariates are selected (based on logistical issues and collection feasibility) from a larger list that was arrived at based on discussions with HumanID researchers at CMU, Maryland, MIT, Southampton and Georgia Tech about the potentially important covariates for gait analysis. It is shown in Sarkar et al. [23] that the shoe type has the least impact on the performance, next is the viewpoint, the third is briefcase, then surface type (flat concrete surface and typical grass lawn surface), and time (six months) difference has the greatest impact. The latter two are the most "difficult" covariates to deal with. In particular, it was found that the surface covariate impacts the gait period more than other covariates. Since its release, this database has made significant contributions to the advancement of the gait recognition technology.

8.4 Face and Gait Recognition Algorithms

In this section, we survey popular appearance-based face and gait recognition algorithms, and the focus is on subspace learning algorithms for feature extraction. We start from the classical linear subspace learning algorithms in Section 8.4.1. Then, we move on to the nonlinear kernel-based subspace learning algorithms that map the input to a high-dimensional space for (hopefully) better separation in Section 8.4.2. Next, Section 8.4.3 reviews recent multilinear subspace learning algorithms that operate directly on tensorial representations.

8.4.1 Linear Subspace Learning Algorithms

The linear algorithms reviewed here include the principal component analysis (PCA), the Bayesian method and the linear discriminant analysis (LDA). Linear subspace learning algorithms solve for a linear projection with some optimality criteria, given a set of training samples. The problem can be formulated mathematically as follows.

A set of M vectorial samples $\{x_1, x_2, ..., x_M\}$ is available for training, where each sample x_m is an $I \times 1$ vector in a vector space $\mathbb{R}^I$. The linear subspace learning objective is to find a linear transformation (projection) $\mathbf{U} \in \mathbb{R}^{I \times P}$ such that the projected samples (the extracted features) $\{\mathbf{y}_m = \mathbf{U}^T \mathbf{x}_m\}$ satisfy an optimality criterion, where $\mathbf{y}_m \in \mathbb{R}^{P \times 1}$ and $P < I$. In classification, these features are fed into a classifier, e.g., the nearest neighbor classifier, and the similarity is usually calculated based on some distance measure.

8.4.1.1 Principal Component Analysis

The PCA is one of the most influential linear subspace learning methods. The well-known eigenface method [66] for face recognition, which is built on PCA, started the era of the appearance-based approach to face recognition, and more generally to visual object recognition. The central idea behind PCA is to reduce the dimensionality of a data set consisting of a larger number of interrelated variables, while retaining as much as possible the variation present in the original data set [67]. This is achieved by transforming to a new set of variables, the so-called principal components (PCs), which are uncorrelated, and ordered so that the first few retain most of the original data variation. Thus, the PCA aims to derive the most descriptive features.

In practice, the variation to be maximized is measured by the total scatter through the total scatter matrix $\mathbf{S}_T$ defined as follows,

$$\mathbf{S}_T = \sum_{m=1}^{M} (\mathbf{x}_m - \bar{\mathbf{x}})(\mathbf{x}_m - \bar{\mathbf{x}})^T, \tag{8.1}$$

where $\bar{\mathbf{x}} = \frac{1}{M}\sum_{m=1}^{M} \mathbf{x}_m$ is the mean of all the training samples. The PCA projection matrix $\mathbf{U}_{PCA}$ is then composed of the eigenvectors corresponding to the largest P ($P < I$) eigenvalues of $\mathbf{S}_I$. The projection of a test sample $\mathbf{x}$ in the PCA space is obtained as:

$$\mathbf{y} = \mathbf{U}_{PCA}^T (\mathbf{x} - \bar{\mathbf{x}}). \tag{8.2}$$

The PCA is an unsupervised learning technique that does not take underlying class structure information into account, even when such information is available for use. In such cases, both intrasubject and intersubject variations are maximized in the PCA feature space. For classification purposes, however, large intrasubject (intraclass) variations have negative impact on classification performance. Figure 8.5a shows two classes of data samples, represented by "O" and "△", and the obtained PCA feature basis. It is not difficult to see that by projecting the data samples onto the PCA basis, these two classes are mixed together and become difficult to be separated. Therefore, it is generally believed that, for classification purposes, the PCA can not perform as well as those supervised learning techniques such as the LDA [68].

8.4.1.2 Bayesian Method

The Bayesian method [69] treats a multiclass recognition problem as a binary pattern classification problem. First, a feature space of Δ vectors is defined as the differences between two samples: $\Delta = \mathbf{x}_{m_1} - \mathbf{x}_{m_2}$. We can then define two class of variations: the intrasubject

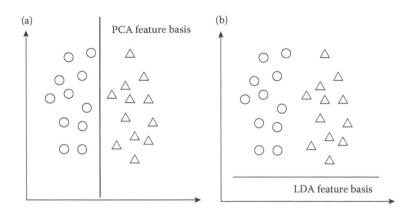

FIGURE 8.5
2-D training samples from two classes "O" and "△" and (a) the obtained PCA feature basis; (b) the obtained LDA feature basis.

variation Ω_I (corresponding to different appearances of the same individual) or an intersubject variation class Ω_E (corresponding to variations between different subjects). Let $\{\Delta_I\}$ be the intrasubject difference set consisting of difference vectors between samples from the same subject and $\{\Delta_E\}$ be the intersubject difference set consisting of difference vectors between samples from different individuals, two feature subspaces denoted as intrasubject space and intersubject space are then extracted by performing the PCA on $\{\Delta_I\}$ and $\{\Delta_E\}$, respectively. In the PCA subspace, the class conditional distributions of the two classes $p(\Delta|\Omega_I)$ and $p(\Delta|\Omega_E)$ are estimated using a Gaussian model. Thus, to determine if two samples $\mathbf{x}_i$ and $\mathbf{x}_j$ belong to the same individual, their difference $\Delta_{ij} = \mathbf{x}_i - \mathbf{x}_j$ is first calculated.

The decision is then made according to the maximum-likelihood classification rule, i.e.,

$$\Delta_{ij} \in \Omega_I \quad \text{if} \quad p(\Delta_{ij}|\Omega_I) > p(\Delta_{ij}|\Omega_E) \tag{8.3}$$

or the *maximum-a-posteriori* rule, i.e.,

$$\Delta_{ij} \in \Omega_I \quad \text{if} \quad P(\Omega_I)p(\Delta_{ij}|\Omega_I) > P(\Omega_E)p(\Delta_{ij}|\Omega_E), \tag{8.4}$$

where $P(\Omega_I)$ and $P(\Omega_E)$ are the *a priori* probabilities.

8.4.1.3 Linear Discriminant Analysis

The LDA is a classical supervised linear subspace learning method that has been very successful and applied widely in various applications. It aims to derive the most discriminative features and produces a class-specific feature space based on the maximization of the so-called Fisher's discriminant criterion (LDC), which is defined as the ratio of between-class scatter to within-class scatter:

$$\mathbf{U}_{\text{LDA}} = \arg\max_{\mathbf{U}} \frac{|\mathbf{U}^T \mathbf{S}_B \mathbf{U}|}{|\mathbf{U}^T \mathbf{S}_W \mathbf{U}|}, \tag{8.5}$$

where the $\mathbf{S}_B$ and $\mathbf{S}_W$ are the between-class and within-class scatter matrices, respectively, and they are defined as

$$\mathbf{S}_B = \sum_{c=1}^{C} M_c (\bar{\mathbf{x}}_c - \bar{\mathbf{x}})(\bar{\mathbf{x}}_c - \bar{\mathbf{x}})^T, \tag{8.6}$$

and

$$\mathbf{S}_W = \sum_{m=1}^{M} (\mathbf{x}_m - \bar{\mathbf{x}}_{c_m})(\mathbf{x}_m - \bar{\mathbf{x}}_{c_m})^T. \tag{8.7}$$

In the definitions above, C is the number of classes, c is the class index, and c_m is the class label for the mth training sample. M_C is the number of training samples in class c, and the mean for class c is

$$\bar{\mathbf{x}}_c = \frac{1}{M_c} \sum_{m, c_m = c} \mathbf{x}_m. \tag{8.8}$$

The maximization of Equation 8.5 leads to the following generalized eigenvalue problem:

$$\mathbf{S}_B \mathbf{u}_p = \lambda_p \mathbf{S}_W \mathbf{u}_p. \tag{8.9}$$

Thus, $\mathbf{U}_{LDA}$ consists of the generalized eigenvectors corresponding to the largest P generalized eigenvalues of Equation 8.9. When $\mathbf{S}_W$ is not singular, $\mathbf{U}_{LDA}$ can be obtained as the eigenvectors corresponding to the largest P eigenvalues of $\mathbf{S}_W^{-1}\mathbf{S}_B$. The projection of a test sample $\mathbf{x}$ in the LDA space is then obtained as:

$$\mathbf{y} = \mathbf{U}_{LDA}^T \mathbf{x}. \tag{8.10}$$

Figure 8.5b shows the LDA feature basis obtained from two classes of data samples (represented by "O" and "Δ", respectively). Obviously, compared to the PCA, the LDA provides a more ideal projection basis for classification purposes.

Although the LDA, as a class specific solution, is generally believed to be superior to the PCA for classification purposes, it is more susceptible to the small sample size problem, where the number of training samples per subject is much smaller than the dimensionality of the input sample space. In such cases, direct optimization of the ratio in Equation 8.5 becomes impossible as $\mathbf{S}_W$ is singular. In order to address this problem, the Fisherface method (FLDA) [68] proposes to apply the PCA as a preprocessing step to remove the null space of $\mathbf{S}_W$.

However, by removing the null space of $\mathbf{S}_W$, significant discriminatory information may be discarded since the maximum of Equation 8.5 can be reached when

$$\mathbf{u}_p^T \mathbf{S}_W \mathbf{u}_p = 0 \text{ and } \mathbf{u}_p^T \mathbf{S}_B \mathbf{u}_p \neq 0. \tag{8.11}$$

In other words, the null space of $\mathbf{S}_W$ spanned by $\mathbf{u}_p$ such that $\mathbf{u}_p^T \mathbf{S}_W \mathbf{u}_p = 0$, may contain significant discriminatory information. To avoid the possible loss of discriminatory information residing in the discarded null space of $\mathbf{S}_W$, a direct linear discriminant analysis (DLDA) [70] was proposed to solve the small sample size problem by diagonalizing $\mathbf{S}_B$ and $\mathbf{S}_W$ directly. The premise behind the DLDA solution is that the discriminatory information resides in the intersection of the null space of $\mathbf{S}_W$, denoted as A, and the complement space of the null space of $\mathbf{S}_B$, denoted as B′, i.e., A∩B′. At the same time, no significant information, in terms of maximization of Equation 8.5, will be lost if the null space of S_B (denoted as B) is discarded. The DLDA can be performed by diagonalizing either S_W first or S_B first. Given the fact that under the small sample size scenario, the rank of S_B, determined by rank(S_B)=min(I, $C-1$)is much smaller than I in most cases, the more attractive option is to diagonalize S_B first. In such cases, B' is first extracted which is spanned by the $C-1$ eigenvectors of S_B, denoted as $V_b = [v_{b1}, ..., v_{b(C-1)}]$, corresponding to the nonzero eigenvalues denoted as $[\lambda_{b1}, ..., \lambda_{b(C-1)}]$. Following that, S_W is projected to B' giving $U^T S_W U$, where $U = V_b \Lambda_b^{-1/2}$ is the transformation matrix from the $\mathbb{R}^I$ to B', $\Lambda_b = \text{diag}([\lambda_{b1}, ..., \lambda_{b(C-1)}])$ and diag($\cdot$) denotes the diagonalization operator. The null space of S_W thus can be easily found by solving an eigenvalue problem of $U^T S_W U$ in a low dimensional B', space.

In general, the LDA-based solutions are believed to be superior to the PCA-based solutions in the context of pattern classification. However, it should be noted that when the number of training samples per subject is small or the training samples are not representative to those in the test, the superiority of LDA technique can not be guaranteed and PCA may outperform LDA, as shown in Figure 8.6. There are two classes, each of which has a Gaussian-like class conditional distribution (represented by two ellipses in the figure). However, only two samples per class are provided for training. The PCA method calculates the projection basis such that the variance of the projected training samples is maximized, resulting in a horizontal projection basis as illustrated in the figure. The LDA method, however, extracts the projection basis such that the between-class variance of the projected samples is maximized while the within-class variance is minimized. It can be observed from the figure, in such cases, the projection basis obtained by the PCA solution is more desirable than that obtained by the LDA method for classification purposes.

8.4.2 Nonlinear Kernel-Based Subspace Learning Algorithms

The LDA-based methods have been shown to be successful and cost-effective techniques widely used in pattern recognition applications. However, as the complexity of patterns increases, the performance of the LDA-based solutions could deteriorate rapidly. This is

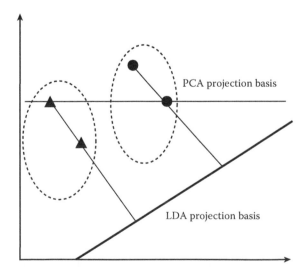

FIGURE 8.6
An example to illustrate why, in some cases, the PCA produces better features than the LDA. (Adapted from A. M. Martinez and A. C. Kak., *IEEE Trans. Pattern Anal. Machine Intell.*, 23(2), 228–233, 2001.)

due to the linear nature of the LDA-based methods assuming that the class conditional distribution of patterns is Gaussian with identical covariance structure. In such cases, the resulting class boundaries are restricted to linear hyper-planes. In appearance-based face and gait recognition, due to the large variations in the appearance, however, the distribution of patterns is far more complicated than Gaussian, usually multimodal and nonconvex [2]. To this end, nonlinear, kernel-based algorithms have been developed to handle complex distributed data.

Originating from the well-known support vector machine [72], the so-called "kernel machine" technique is considered an important tool in the design of nonlinear feature extraction techniques. The premise behind the kernel machine technique is to find a non-linear mapping from the original input space ($\mathbb{R}^I$) to a higher dimensional kernel feature space F^F by using a nonlinear function $\phi(\cdot)$, i.e.,

$$\phi : \mathbf{x} \in \mathbb{R}^I \to \phi(\mathbf{x}) \in \mathrm{F}^F \quad I < F \leq \infty. \tag{8.12}$$

In the kernel feature space F^F, the pattern distribution is expected to be simplified so that better classification performance can be achieved by applying traditional linear methodologies [73]. In general, the dimensionality of the kernel space is much larger than that of the original input space, sometimes even infinite. Therefore, an explicit determination of the nonlinear map ϕ is difficult or intractable. Fortunately, with the so-called "kernel trick", the nonlinear mapping can be performed implicitly in the original input space $\mathbb{R}^I$ by replacing dot products of the feature representations in F^F with a kernel function defined in $\mathbb{R}^I$ [73]. Thus, if $\mathbf{x}_i \in \mathbb{R}^I$ and $\mathbf{x}_j \in \mathbb{R}^I$ are two vectors in the original input space, the dot product of their feature representations $\phi(\mathbf{x}_i) \in \mathrm{F}^F$ and $\phi(\mathbf{x}_j) \in \mathrm{F}^F$ can be computed by a kernel function $k(\cdot)$ defined in $\mathbb{R}^I$, i.e.,

$$\phi(\mathbf{x}_i) \cdot \phi(\mathbf{x}_j) = k(\mathbf{x}_i, \mathbf{x}_j). \tag{8.13}$$

The function selected as the kernel function should satisfy the Mercer's condition [73]. Some commonly used kernel functions are summarized in Table 8.2. A toy example [20] is illustrated in Figure 8.7, where a second order polynomial kernel function is used, i.e.,

$$k(\mathbf{x}_i, \mathbf{x}_j) = (\mathbf{x}_i \cdot \mathbf{x}_j)^2, \ \phi(\mathbf{x}) = [\alpha_1^2, \sqrt{2}\alpha_1\alpha_2, \alpha_2^2], \tag{8.14}$$

where $\mathbf{x} = [\alpha_1, \alpha_2]^T$. It can be easily observed that by mapping the samples into the 3-D kernel space, use of a linear hyperplane to separate the two classes becomes more efficient than using a nonlinear ellipsoidal decision boundary in the original space.

TABLE 8.2

Commonly Used Kernel Functions

Kernel Name	Kernel Function
Gaussian	$k(\mathbf{x}_i, \mathbf{x}_j) = \exp\left(\frac{-\|\mathbf{x}_i - \mathbf{x}_j\|^2}{\sigma^2}\right), \sigma \in R$
Polynomial	$k(\mathbf{x}_i, \mathbf{x}_j) = (a(\mathbf{x}_i \cdot \mathbf{x}_j) + b)^d, a, b, d \in R$
Sigmoidal	$k(\mathbf{x}_i, \mathbf{x}_j) = \tanh(a(\mathbf{x}_i \cdot \mathbf{x}_j) + b), a, b \in R$

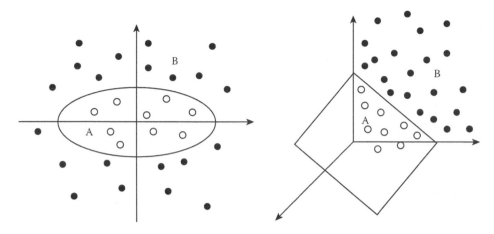

FIGURE 8.7
A two-class classification example showing nonlinear and linear decision boundaries. Left: samples in the 2-D input space; Right: samples in the 3-D kernel feature space.

The kernel-based solutions have been shown to be effective techniques to handle complicated pattern classification problems. However, the performance of the algorithm is significantly affected by the selected kernel functions and the corresponding kernel parameters. A toy example is given in Figure 8.8 which depicts the data distribution in different kernel spaces. Figure 8.8a shows a 2-D data set of 400 samples drawn from two Gaussian distributions (denoted as "+" and "o" respectively), i.e., $\mathbf{x} \sim N(\mu_i, \Sigma_i)$, $i=1,2$, where $\mathbf{x}$ denotes the 2-D data, $\mu_1 = [-2, 0]^T$, $\mu_2 = [2, 0]^T$, $\Sigma_1 = \Sigma_2 = \text{diag}([2.25, 1])$. Each class contains 200 samples. Figure 8.8b shows the projection of the data in a second order polynomial kernel space, i.e., $k(\mathbf{x}_1, \mathbf{x}_2) = (\mathbf{x}_1 \cdot \mathbf{x}_2)^2$. For better visual perception, the projection on the first two significant bases are depicted. Figure 8.8c and 8.8d illustrate the corresponding projection in a Gaussian kernel space when a Gaussian kernel function

$$k(\mathbf{x}_1, \mathbf{x}_2) = \exp\left(\frac{-\|\mathbf{x}_1 - \mathbf{x}_2\|^2}{\sigma^2}\right)$$

(8.15)

is employed with $\sigma^2 = 1000$ for Figure 8.8c and $\sigma^2 = 10$ for Figure 8.8d. It is not difficult to observe that different kernel functions and different kernel parameters significantly affect the geometric structure of the mapped data. If a polynomial kernel is used as shown in the Figure 8.8b, the class separability in the kernel space is even worse than that in the original input space. Therefore, for kernel-based solutions, selecting an appropriate kernel function is of vital importance to ensure good classification performance and methods have been proposed to tackle this problem [74]. The traditional approach is the cross validation framework and there are more systematic approaches for kernel parameter optimization [74].

In the context of the kernel trick, the key task of designing a kernelized feature extraction algorithm is to represent the linear feature extraction procedure using dot product forms. The corresponding kernel version of the algorithm can then be obtained by replacing the dot product with the kernel function. The kernel principal component analysis (KPCA) [75], the generalized discriminant analysis (GDA) [76] and the kernel direct discriminant analysis (KDDA) [77] are typical kernel-based learning algorithms commonly used for feature extraction and they are described below.

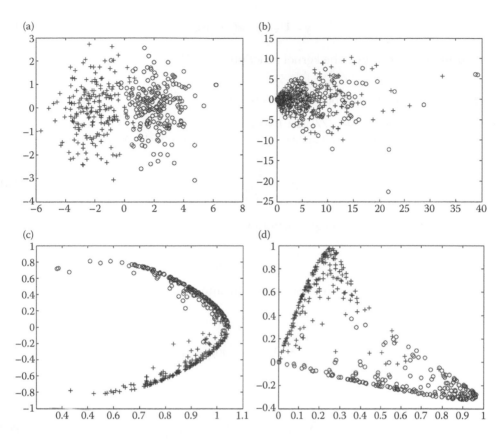

FIGURE 8.8
An example showing projections of a set of 2-D data points in different kernel spaces. (a) 2-D samples drawn from two Gaussian distributions; (b) projections in the polynomial kernel space; (c) projections in the Gaussian kernel space with $\sigma^2 = 1000$; (d) projections in the Gaussian kernel space with $\sigma^2 = 10$.

8.4.2.1 Kernel Principal Component Analysis

The KPCA [75] is actually an implementation of the traditional PCA algorithm in the kernel feature space. Let $\tilde{\mathbf{S}}_T$ be the total scatter matrix defined in $\mathbb{F}^F$, which could be expressed as follows,

$$\tilde{\mathbf{S}}_T = \sum_{m=1}^{M} (\phi(\mathbf{x}_m) - \bar{\phi})(\phi(\mathbf{x}_m) - \bar{\phi})^T \tag{8.16}$$

where

$$\bar{\phi} = \frac{1}{M} \sum_{m=1}^{M} \phi(\mathbf{x}_m) \tag{8.17}$$

is the mean of gallery samples in $\mathbb{F}^F$. Thus, the KPCA subspace is spanned by the first P significant eigenvectors of $\tilde{\mathbf{S}}_T$, denoted as $\tilde{\mathbf{U}}_{\text{KPCA}}$, corresponding to the P largest eigenvalues. The KPCA feature representation of a given input x is thus obtained by the dot product,

$$\tilde{\mathbf{y}} = \tilde{\mathbf{U}}_{\text{KPCA}} \cdot (\phi(\mathbf{x}) - \overline{\phi}), \tag{8.18}$$

computed implicitly through the kernel function $k(\cdot)$ [75].

8.4.2.2 Kernel-Based Discriminant Analysis (KDA)

The GDA [76] and the KDDA [77] are two kernel-based discriminant analysis solutions which produce corresponding subspaces by maximizing the Fisher's criterion defined in $\mathbf{F}^F$,

$$\tilde{\mathbf{U}}_{\text{KDA}} = \arg\max_{\tilde{\mathbf{U}}} \frac{\left| \tilde{\mathbf{U}}^T \tilde{\mathbf{S}}_B \tilde{\mathbf{U}} \right|}{\left| \tilde{\mathbf{U}}^T \tilde{\mathbf{S}}_W \tilde{\mathbf{U}} \right|} = [\tilde{\mathbf{u}}_1, ..., \tilde{\mathbf{u}}_P], \tag{8.19}$$

where $\tilde{\mathbf{u}}_p$ is the pth generalized eigenvector of

$$\tilde{\mathbf{S}}_B \tilde{\mathbf{u}}_p = \lambda_p \tilde{\mathbf{S}}_W \tilde{\mathbf{u}}_p, \tag{8.20}$$

and the corresponding between-class matrix $\tilde{\mathbf{S}}_B$ and within-class scatter matrix $\tilde{\mathbf{S}}_W$ are defined as

$$\tilde{\mathbf{S}}_B = \sum_{c=1}^{C} M_c (\overline{\phi}_c - \overline{\phi})(\overline{\phi}_c - \overline{\phi})^T \tag{8.21}$$

and

$$\tilde{\mathbf{S}}_W = \sum_{m=1}^{M} (\phi(\mathbf{x}_m) - \overline{\phi}_{c_m})(\phi(\mathbf{x}_m) - \overline{\phi}_{c_m})^T \tag{8.22}$$

where

$$\overline{\phi}_c = \frac{1}{M_c} \sum_{m, c_m = c} \phi(\mathbf{x}_m) \tag{8.23}$$

is the class mean in $\mathbf{F}^F$.

As discussed earlier, the LDA-based solutions often suffer from the small sample size problem. When applying the LDA techniques in the kernel feature space, the problem becomes even more severe due to the extremely high dimensionality of $\mathbf{F}^F$. Therefore, solving the small sample size problem is also a demanding task in the implementation of kernel-based discriminant analysis solutions. GDA attempts to solve the small sample size problem by removing the null space of $\tilde{\mathbf{S}}_W$ with a PCA routine as is done in the FLDA method, while the KDDA implements the DLDA solution in the kernel space.

8.4.3 Multilinear Subspace Learning Algorithms

The algorithms reviewed so far all take vectorial input. However, gray-level face images (row × column) and binary gait silhouette sequences (row × column × time) are naturally multidimensional objects, which are formally called tensor objects. Therefore, the linear and nonlinear algorithms above need to reshape these tensors into vectors in a very high-dimensional space, which not only results in high computation and memory demand, but also breaks the natural structure and correlation in the original data. This motivated the recent development of the multilinear subspace learning algorithms [78–84], which extract features directly from the tensorial representation rather than the vectorized representation, and it is believed that more compact and useful features can be obtained this way.

Tensors are conventionally denoted by calligraphic letters [85], e.g., $\mathcal{A}$. The elements of a tensor are to be addressed by N indices, where N (the number of indices used in the description) defines the order of the tensor object and each index defines one mode [85]. Thus, vectors are first-order tensors (with $N = 1$) and matrices are second-order tensors (with $N = 2$). Tensors with $N > 2$ can be viewed as a generalization of vectors and matrices to higher order.

The elements of a tensor are denoted with indices in brackets. Indices are denoted by lowercase letters and span the range from 1 to the uppercase letter of the index, e.g., $n = 1, 2, ..., N$. An Nth-order tensor is denoted as $\mathcal{A} \in \mathbb{R}^{I_1 \times I_2 \times ... \times I_N}$. It is addressed by N indices i_n, $n = 1, ..., N$, and each i_n addresses the n-mode of $\mathcal{A}$. The n-mode product of a tensor $\mathcal{A}$ by a matrix $\mathbf{U} \in R^{J_n \times I_n}$, denoted by $\mathcal{A} \times_n \mathbf{U}$, is a tensor with entries:

$$(\mathcal{A} \times_n \mathbf{U})(i_1, ..., i_{n-1}, j_n, i_{n+1}, ..., i_N) = \sum_{i_n} \mathcal{A}(i_1, ..., i_N) \cdot \mathbf{U}(j_n, i_n). \tag{8.24}$$

The scalar product of two tensors $\mathcal{A}, \mathcal{B} \in \mathbb{R}^{I_1 \times I_2 \times ... \times I_N}$ is defined as:

$$<\mathcal{A}, \mathcal{B}> = \sum_{i_1} \sum_{i_2} ... \sum_{i_N} \mathcal{A}(i_1, i_2, ..., i_N) \cdot \mathcal{B}(i_1, i_2, ..., i_N) \tag{8.25}$$

and the Frobenius norm of $\mathcal{A}$ is defined as

$$\|\mathcal{A}\|_F = \sqrt{<\mathcal{A}, \mathcal{A}>}. \tag{8.26}$$

The n-mode vectors of $\mathcal{A}$ are defined as the I_n-dimensional vectors obtained from $\mathcal{A}$ by varying the index i_n while keeping all the other indices fixed. A rank-1 tensor $\mathcal{A}$ equals to the outer product of N vectors:

$$\mathcal{A} = \mathbf{u}^{(1)} \circ \mathbf{u}^{(2)} \circ ... \circ \mathbf{u}^{(N)}, \tag{8.27}$$

which means that

$$\mathcal{A}(i_1, i_2, ..., i_N) = \mathbf{u}^{(1)}(i_1) \cdot \mathbf{u}^{(2)}(i_2) \cdot ... \cdot \mathbf{u}^{(N)}(i_N) \tag{8.28}$$

for all values of indices. Unfolding $\mathcal{A}$ along the n-mode is denoted as

$$\mathbf{A}_{(n)} \in \mathbb{R}^{I_n \times (I_1 \times ... \times I_{n-1} \times I_{n+1} \times ... \times I_N)}, \tag{8.29}$$

and the column vectors of $\mathbf{A}_{(n)}$ are the n-mode vectors of $\mathcal{A}$.

Based on the definitions above, a tensor can be projected to another tensor by N projection matrices $\mathbf{U}^{(1)}, \mathbf{U}^{(2)}, ..., \mathbf{U}^{(N)}$ as

$$\mathcal{Y} = \mathcal{X} \times_1 \mathbf{U}^{(1)^T} \times_2 \mathbf{U}^{(2)^T} ... \times_N \mathbf{U}^{(N)^T}. \tag{8.30}$$

The projection of an n-mode vector of $\mathcal{X}$ by an n-mode projection matrix $\mathbf{U}^{(n)^T}$ is computed as the inner product between the n-mode vector and the rows of $\mathbf{U}^{(n)^T}$. Figure 8.9 illustrate the 1-mode projection of a third-order tensor $\mathcal{X} \in \mathbb{R}^{10\times8\times6}$ in the 1-mode vector space by a projection matrix $\mathbf{U}^{(1)^T} \in \mathbb{R}^{5\times10}$, resulting in the projected tensor $\mathcal{X} \times_1 \mathbf{U}^{(1)^T} \in \mathbb{R}^{5\times8\times6}$. In the 1-mode projection, each 1-mode vector of $\mathcal{X}$ of length ten is projected by $\mathbf{U}^{(1)^T}$ to obtain a vector of length five, as indicated by the differently shaded vectors in Figure 8.9.

The problem of multilinear subspace learning based on the tensor-to-tensor projection above can be mathematically defined as follows:

A set of M Nth-order tensorial samples $\{\mathcal{X}_1, \mathcal{X}_2, ..., \mathcal{X}_M\}$ is available for training, where each sample $\mathcal{X}_m$ is an $I_1 \times I_2 \times ... \times I_N$ tensor in a tensor space $\mathbb{R}^{I_1\times I_2\times...\times I_N}$. The objective of multilinear subspace learning through tensor-to-tensor projection is to find a tensor-to-tensor projection $\{\tilde{\mathbf{U}}^{(n)} \in \mathbb{R}^{I_n\times P_n}, n=1,...,N\}$ mapping from the original tensor space $\mathbb{R}^{I_1} \otimes \mathbb{R}^{I_2} ... \otimes \mathbb{R}^{I_N}$ into a tensor subspace $\mathbb{R}^{P_1} \otimes \mathbb{R}^{P_2} ... \otimes \mathbb{R}^{P_N}$ (with $P_n < I_n$, for $n=1,...,N$):

$$\mathcal{Y}_m = \mathcal{X}_m \times_1 \tilde{\mathbf{U}}^{(1)^T} \times_2 \tilde{\mathbf{U}}^{(2)^T} ... \times_N \tilde{\mathbf{U}}^{(N)^T}, m=1,...,M, \qquad (8.31)$$

such that the projected samples (the extracted features) satisfy an optimality criterion, where the dimensionality of the projected space is much lower than the original tensor space. Here, $\otimes$ denotes the Kronecker product. This problem is usually solved in an iterative alternating projection manner. In classification, these tensorial features can be fed directly into a classifier, e.g., the nearest neighbor classifier, and the similarity is calculated according to some tensorial distance measure, e.g., based on the Frobenius norm in Equation 8.26, or they can be converted into vectors before the feeding so that more conventional distance measures and traditional classifiers can be used.

In the following, we describe two multilinear subspace learning algorithms. One is the multilinear principal component analysis (MPCA) and the other is a multilinear discriminant analysis algorithm.

8.4.3.1 Multilinear Principal Component Analysis

The recent MPCA algorithm is a multilinear extension of the PCA algorithm and it has been applied successfully to gait recognitin [56]. The MPCA maximizes the following tensor-based scatter measure:

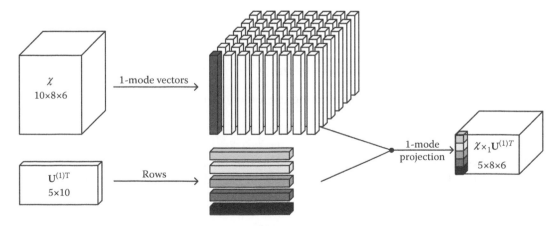

FIGURE 8.9
An illustration of the 1-mode projection: $\mathcal{X} \times_1 \mathbf{U}^{(1)^T}$.

$$\Psi_y = \sum_{m=1}^{M} \left\| \mathcal{Y}_m - \bar{\mathcal{Y}} \right\|_F^2, \tag{8.32}$$

named as the total tensor scatter, where

$$\bar{\mathcal{Y}} = \frac{1}{M} \sum_{m=1}^{M} \mathcal{Y}_m \tag{8.33}$$

is the mean sample.

This problem cannot be solved deterministically and it is decomposed into N simpler linear optimization problems, resulting in an iterative alternating projection procedure. In each subproblem, the following eigenvalue problem is solved: Given all the other projection matrices $\tilde{\mathbf{U}}^{(1)}, ..., \tilde{\mathbf{U}}^{(n-1)}, \tilde{\mathbf{U}}^{(n+1)}, ..., \tilde{\mathbf{U}}^{(N)}$, the matrix $\tilde{\mathbf{U}}^{(n)}$ that maximizes Ψ_y consists of the P_n eigenvectors corresponding to the largest P_n eigenvalues of the matrix

$$\Phi^{(n)} = \sum_{m=1}^{M} \left(\mathbf{X}_{m(n)} - \bar{\mathbf{X}}_{(n)} \right) \cdot \tilde{\mathbf{U}}_{\Phi^{(n)}} \cdot \tilde{\mathbf{U}}_{\Phi^{(n)}}^{T} \cdot \left(\mathbf{X}_{m(n)} - \bar{\mathbf{X}}_{(n)} \right)^{T}, \tag{8.34}$$

where

$$\tilde{\mathbf{U}}_{\Phi^{(n)}} = \left(\tilde{\mathbf{U}}^{(n+1)} \otimes \tilde{\mathbf{U}}^{(n+2)} \otimes ... \otimes \tilde{\mathbf{U}}^{(N)} \otimes \tilde{\mathbf{U}}^{(1)} \otimes \tilde{\mathbf{U}}^{(2)} \otimes ... \tilde{\mathbf{U}}^{(n-1)} \right). \tag{8.35}$$

The obtained projection matrices $\{\tilde{\mathbf{U}}^{(n)}, n = 1, ..., N\}$ can be viewed as $\prod_{n=1}^{N} P_n$ EigenTensors: $\tilde{U}_{p_1 p_2 ... p_N} = \tilde{\mathbf{u}}_{p_1}^{(1)} \circ \tilde{\mathbf{u}}_{p_2}^{(2)} \circ ... \circ \tilde{\mathbf{u}}_{p_N}^{(N)}$, where $\tilde{\mathbf{u}}_{p_n}^{(n)}$ is the p_n th column of $\tilde{\mathbf{U}}^{(n)}$. However, not all of them are useful for recognition and they can be selected according to their class discriminability $\Gamma_{p_1 p_2 ... p_N}$, where $\Gamma_{p_1 p_2 ... p_N}$ for the eigentensor $\tilde{U}_{p_1 p_2 ... p_N}$ is defined as

$$\Gamma_{p_1 p_2 ... p_N} = \frac{\sum_{c=1}^{C} M_c \cdot \left[\bar{\mathcal{Y}}_c(p_1, p_2, ..., p_N) - \bar{\mathcal{Y}}(p_1, p_2, ..., p_N) \right]^2}{\sum_{m=1}^{M} \left[\mathcal{Y}_m(p_1, p_2, ..., p_N) - \bar{\mathcal{Y}}_{c_m}(p_1, p_2, ..., p_N) \right]^2}. \tag{8.36}$$

$\mathcal{Y}_m$ is the feature tensor of $\mathcal{X}_m$ in the projected tensor subspace, and the class mean feature tensor

$$\bar{\mathcal{Y}}_c = \frac{1}{M_c} \sum_{m, c_m = c} \mathcal{Y}_m. \tag{8.37}$$

The entries in $\mathcal{Y}_m$ are then arranged into a feature vector $\mathbf{y}_m$ ordered according to $\Gamma_{p_1 p_2 ... p_N}$ in descending order and only the first P entries of $\mathbf{y}_m$ are kept. This feature vector can then be used directly as the input to conventional classifier or it can be further combined with the LDA (or other vector-based learning algorithms) to produce an MPCA+LDA algorithm [86].

8.4.3.2 Multilinear Discriminant Analysis

There are also multilinear extensions of the LDA algorithm [79–81]. The discriminant analysis with tensor representation (DATER)[†] was proposed to perform discriminant analysis

---

[†] Here, the name that was used when the algorithm was first proposed is adopted as it is more commonly refereed to in the literature.

directly on general tensor objects. Like in MPCA, the DATER algorithm solves for a tensor-to-tensor projection $\{\tilde{\mathbf{U}}^{(n)} \in \mathbb{R}^{I_n \times P_n}, P_n < I_n, n = 1, ..., N\}$ that project a tensor $\mathcal{X}_m \in \mathbb{R}^{I_1 \times ... I_N}$ to $\mathcal{Y}_m$. The DATER algorithm maximizes a tensor-based discrimination objective criterion: the tensor-based scatter ratio Ψ_{B_y} / Ψ_{W_y}, where the between-class scatter Ψ_{B_y} is defined as:

$$\Psi_{B_y} = \sum_{c=1}^{C} M_c \|\bar{\mathcal{Y}}_c - \bar{\mathcal{Y}}\|_F^2, \tag{8.38}$$

and the within-class scatter Ψ_{W_y} is defined as:

$$\Psi_{W_y} = \sum_{m=1}^{M} \|\mathcal{Y}_m - \bar{\mathcal{Y}}_{c_m}\|_F^2. \tag{8.39}$$

Similar to the case of the MPCA, a deterministic solution does not exist either and an iterative alternating projection procedure is followed. In addition, the method to obtain vectorial features $\{\mathbf{y}_m\}$ in MPCA above can also be applied to this algorithm. In its application to gait recognition [87], it is found that this algorithm does not converge and it appears to be very sensitive to parameter settings.

8.5 Concluding Remarks

In this section, we give a summary of this chapter and then discuss the current state and future directions of face and gait recognition.

8.5.1 Summary

This chapter presents a comprehensive review on the face and gait recognition technologies. The motivations and applications of face and gait recognition have been described in detail. Face and gait are both biometric characteristics that can be easily collected and well accepted by the public. Therefore, they have great potential in various applications. Face and gait recognition systems typically acquire images/videos of a subject through a digital camera. The acquired images/videos are then preprocessed to either detect facial images or extract gait silhouettes, from which features are extracted and classified. The two general approaches of model-based and appearance-based approaches are then reviewed, with their advantages and limitations pointed out. In addition, existing methods for the fusion of face and gait traits are also discussed. After the system-level overview, we move on to describe three widely used databases for research in face and gait recognition. Next, the important subspace learning algorithms for appearance-based recognition are surveyed in detail, including classical linear subspace learning algorithms such as the PCA and the LDA, and their nonlinear (kernel-based) and multilinear extensions. The intent here is not to discuss thoroughly all the face and gait recognition algorithms. Instead, we concentrate on the most representative algorithms in our view.

8.5.2 Current State and Future Directions of Face Recognition

Over the past two decades, face recognition has received substantial attention from both research communities and the market, and the recognition performance has improved significantly. Although progress in face recognition is encouraging, this task remains to be very challenging for real applications, especially for less constrained situations (e.g., outdoor) or even unconstrained situations where there are considerable variations in the viewpoint, illumination, expression and time [2]. Current face recognition technologies may work well for cooperative frontal faces without exaggerated expressions and under illumination without much shadow, but recognition in an unconstrained daily life environment without user cooperation, such as identifying persons in crowds and in surveillance videos, is still a very challenging problem that offers great opportunities for researchers in all related areas [2]. The following are some new developments and trends in face recognition.

Three-dimensional face recognition [88], where the 3-D geometry of the human face is used, is believed to have the potential to achieve better accuracy than its 2-D counterpart, which is sensitive to changes in variations like pose, illumination, and facial expressions. The 3-D face shape is usually extracted by a range camera and the 3-D model can also be used to improve accuracy of traditional image-based recognition by transforming the head into a preferred view. Furthermore, most range cameras acquire both a 3-D mesh and the corresponding texture, which allows the combination of the pure 3-D matcher output with traditional 2-D face recognition algorithms for better performance [89].

Some researchers believe that the use of video sequences, as opposed to a single image, will lead to much better recognition rates [12,90]. Their argument is that integrating the recognition performance over a sequence would give a better result than considering just one single image from that sequence.

The Face Recognition Grand Challenge (FRGC) [91] is recently designed to advance face recognition by presenting a six-experiment challenge problem to researchers, along with data corpus of 50,000 images. The data consists of 3-D scans and high resolution still imagery taken under controlled and uncontrolled conditions. The experiments can be used to measure performance on controlled images, uncontrolled images, 3-D images, multistill images and the matching between 3-D and still images. Having seen the contributions made by the public databases such as PIE and FERET, it is expected that this FRGC database will further advance the state-of-the-art for face recognition.

8.5.3 Current State and Future Directions of Gait Recognition

Gait is a new biometric characteristics receiving increasing attention. It has the unique advantage over the other biometrics in surveillance applications, where the recognition needs to be performed at a distance and only low-resolutions videos can be captured. On one hand, the state-of-the-art gait recognition algorithms have achieved high recognition rates on gait sequences captured indoors with controlled environment (such as controlled lighting, background clutter) or sequences captured outdoor under the same surface, with variation in viewing angle and shoe type. On the other hand, the recognition on sequences captured under uncontrolled environment, such as outdoor with different surfaces, different carrying conditions, and different time, is still very challenging, thus offering great opportunities to researchers. In particular, the recent development in multilinear subspace learning algorithms is encouraging since they can handle natural gait sequences directly.

The uniqueness of gait is lower than the other biometrics such as face and fingerprint. Thus, in practice, a gait recognition system may not be used alone. Instead, it can be deployed as part of a multimodal biometric recognition system. Besides security surveillance applications, studies on gait, especially the model-based approach, can benefit the entertainment and communication industries, and also the medical field, such as clinical rehabilitation of patients of stroke or spinal cord injuries and diagnosis of disorders [12,92].

References

1. A. K. Jain, A. Ross, and S. Prabhakar. 2004. An introduction to biometric recognition. *IEEE Trans. Circuits Syst. Video Technol.*, 14(1), 4–20.
2. S. Z. Li and A. K. Jain. 2004. Introduction. In: *Handbook of Face Recognition*, S. Z. Li and A. K. Jain, Eds. Springer-Verlag, Berlin, Germany, 1–11.
3. Cognitec Systems GmbH | The Face Recognition Company. [Online]. Available: http://www.cognitec-systems.de/(Accessed 2008, November 1).
4. Eyematic Interfaces Inc. [Online]. Available: http://www.siggraph.org/s2002/exhibition/detail/225.html(Accessed 2008, November 1).
5. L-1 Identity Solutions. [Online]. Available: http://www.l1id.com/(Accessed 2008, November 1).
6. R. Hietmeyer. 2000. Biometric identification promises fast and secure processing of airline passengers. *Int. Civil Aviat. Org. J.*, 55(9), 10–11.
7. W. Zhao, R. Chellappa, A. Rosenfeld, and P. Phillips. 2003. Face recognition: A literature survey. *ACM Computing Surveys*, 399–458.
8. R-L. Hsu. 2002. Face detection and modeling for recognition. Ph.D. dissertation, Michigan State University. [Online]. Available: http://biometrics.cse.msu.edu/Publications/Thesis/VincentHsu_FaceDetection_PhD02.pdf
9. M. S. Nixon and J. N. Carter. 2006. Automatic recognition by gait. *Proc. IEEE*, 94(11), 2013–2024.
10. A. K. Jain, A. Ross, and S. Prabhakar. 2004. An introduction to biometric recognition. *IEEE Trans. Circuits Syst. Video Technol.*, 14(1), 4–20.
11. L. Wang, T. Tan, H. Ning, and W. Hu. 2003. Silhouette analysis-based gait recognition for human identification. *IEEE Trans. Pattern Anal. Machine Intell.*, 25(12), 1505–1518.
12. R. Chellappa, A. Roy-Chowdhury, and S. Zhou. 2005. *Recognition of Humans and Their Activities Using Video*. Morgan and Claypool Publishers, California.
13. A. Kale. 2003. Algorithms for gait-based human identification from a monocular video sequences. Ph.D. dissertation, Department of Electrical and Computer Engineering, University of Maryland College Park. [Online]. Available: http://www.cs.uky.edu/amit/thesis.pdf
14. G. Johansson. 1975. Visual motion perception. *Sci. Am.*, 232(6), 76–88.
15. J. Cutting and L. Kozlowski. 1977. Recognizing friends by their walk: Gait perception without familiarity cues. *Bull. Psych. Soc.*, 9(5), 353–356.
16. C. Barclay, J. Cutting, and L. Kozlowski. 1978. Temporal and spatial factors in gait perception that influence gender recognition. *Percep. Psychophy.*, 23(2), 145–152.
17. S. V. Stevenage, M. S. Nixon, and K. Vince. 1999. Visual analysis of gait as a cue to identity. *Appl. Cog. Psychol.*, 13(6), 513–526.
18. A. K. Jain, R. Chellappa, S. C. Draper, N. Memon, P. J. Phillips, and A. Vetro. 2007. Signal processing for biometric systems. *IEEE Signal Processing Mag.*, 24(6), 146–152.
19. L. Wang, H. Ning, T. Tan, and W. Hu. 2004. Fusion of static and dynamic body biometrics for gait recognition. *IEEE Trans. Circuits Syst. Video Technol.*, 14(2), 149–158.
20. J. Lu. 2004. Discriminant learning for face recognition. Ph.D. dissertation, University of Toronto. [Online]. Available: http://www.dsp.utoronto.ca/juwei/Publication/JuweiThesisUT04.pdf (Accessed 2008, April 8).
21. H. Lu, K. N. Plataniotis, and A. N. Venetsanopoulos. 2006. Coarse-to-fine pedestrian localization and silhouette extraction for the gait challenge data sets. In: *Proc. IEEE Conf. on Multimedia and Expo*, The Institute of Electrical and Electronics Engineers (IEEE), New Jersey, 1009–1012.

22. P. J. Phillips, H. Moon, S. A. Rizvi, and P. Rauss. 2000. The FERET evaluation method for face recognition algorithms. *IEEE Trans. Pattern Anal. Machine Intell.*, 22(10), 1090–1104.

23. S. Sarkar, P. J. Phillips, Z. Liu, I. Robledo, P. Grother, and K. W. Bowyer. 2005. The human ID gait challenge problem: Data sets, performance, and analysis. *IEEE Trans. Pattern Anal. Machine Intell.*, 27(2), 162–177.

24. X. Lu. 2003. Image analysis for face recognition. Personal notes. [Online]. Available: http://www.face-rec.org/interesting-papers/General/ImAna4FacRcg_lu.pdf (Accessed 2008, November 1).

25. F. Samaria and S. Young. 1994. HMM based architecture for face identification. *Image and Vision Comput.*, 12, 537–583.

26. L. Wiskott, J. M. Fellous, N. Kruger, and C. von der Malsburg. 1997. Face recognition by elastic bunch graph matching. *IEEE Trans. Pattern Anal. Machine Intell.*, 19(7), 775–779.

27. T. F. Cootes, G. J. Edwards, and C. J. Taylor. 2001. Active appearance models, *IEEE Trans. Pattern Anal. Machine Intell.*, 23(6), 681–685.

28. C. Y. Yam, M. S. Nixon, and J. N. Carter. 2004. Automated person recognition by walking and running via model-based approaches. *Patt. Recog.*, 37(5), 1057–1072.

29. D. Cunado, M. S. Nixon, and J. N. Carter. 2003. Automatic extraction and description of human gait models for recognition purposes. *Comput. Vision Image Understand.*, 90(1), 1–41.

30. D. K. Wagg and M. S. Nixon. 2004. On automated model-based extraction and analysis of gait. In: *Proc. IEEE Int. Conf. on Automatic Face and Gesture Recognition*, The Institute of Electrical and Electronics Engineers (IEEE), New Jersey, 11–16.

31. D. A. Winter. 1991. *The Biomechanics and Motor Control of Human Gait: Normal, Elderly and Pathological.* University of Waterloo Press, Waterloo, Canada.

32. D. A. Winter. 2005. *The Biomechanics and Motor Control of Human Movement.* John Wiley & Sons, New Jersey.

33. L. Lee, G. Dalley, and K. Tieu. 2003. Learning pedestrian models for silhouette refinement. In: *Proc. IEEE Conf. on Computer Vision*, The Institute of Electrical and Electronics Engineers (IEEE), New Jersey, 663–670.

34. H. Lu, K. N. Plataniotis, and A. N. Venetsanopoulos. 2006. A layered deformable model for gait analysis. In: *Proc. IEEE Int. Conf. on Automatic Face and Gesture Recognition*, 249–254.

35. Z. Liu and S. Sarkar. 2005. Effect of silhouette quality on hard problems in gait recognition. *IEEE Trans. Syst., Man, Cybern. B*, 35(2), 170–178.

36. I I. Lu, K. N. Plataniotis, and A. N. Venetsanopoulos. 2008. A full-body layered deformable model for automatic model-based gait recognition. *EURASIP J. Adv. Signal Process. Special Issue on Advanced Signal Processing and Pattern Recognition Methods for Biometrics*, article ID 261317, doi:10.1155/2008/261317.

37. R. Brunelli and T. Poggio. 1993. Face recognition: Features versus templates. *IEEE Trans. Pattern Anal. Machine Intell.*, 15(10), 1042–1052.

38. P. Sinha, B. Balas, Y. Ostrovsky, and R. Russell. 2006. Face recognition by humans: 19 results all computer vision researchers should know about. *Proc. IEEE*, 94(11), 1948–1962.

39. R. Cutler, C. Benabdelkader, and L. Davis. 2002. Motion-based recognition of people in eigen-gait space. In: *Proc. IEEE Int. Conf. on Automatic Face and Gesture Recognition*, The Institute of Electrical and Electronics Engineers (IEEE), New Jersey, 254–259.

40. L. Wang, T. Tan, W. Hu, and H. Ning. 2003. Automatic gait recognition based on statistical shape analysis. *IEEE Trans. Image Process.*, 12(9), 1120–1131.

41. A. Kale, A. N. Rajagopalan, A. Sunderesan, N. Cuntoor, A. Roy-Chowdhury, V. Krueger, and R. Chellappa. 2004. Identification of humans using gait. *IEEE Trans. Image Processing*, 13(9), 1163–1173.

42. D. Tolliver and R. T. Collins. 2003. Gait shape estimation for identification. In: *Proc. Int. Conf. on Audio and Video-Based Biometric Person Authentication*, Springer-Verlag, Berlin, Germany, 734–742.

43. Z. Liu and S. Sarkar. 2004. Simplest representation yet for gait recognition: averaged silhouette. In: *Proc. Int. Conf. on Pattern Recognition*, Vol. 4, The Institute of Electrical and Electronics Engineers (IEEE), New Jersey, 211–214.

44. J. P. Foster, M. S. Nixon, and A. Prügel-Bennett. 2003. Automatic gait recognition using area-based metrics. *Patt. Recog. Lett.*, 24(14), 2489–2497.

45. M. Soriano, A. Araullo, and C. Saloma. 2004. Curve spreads—a biometric from front-view gait video. *Patt. Recog. Lett.*, 25(14), 1595–1602.

46. J. Han and B. Bhanu. 2005. Performance prediction for individual recognition by gait. *Patt. Recog. Lett.*, 26, 615–624.

47. J. E. Boyd. 2001. Video phase-locked loops in gait recognition. In: *Proc. IEEE Conf. on Computer Vision*, Vol. 1, The Institute of Electrical and Electronics Engineers (IEEE), New Jersey, 696–703.

48. N. Cuntoor, A. Kale, and R. Chellappa. 2003. Combining multiple evidences for gait recognition. In: *Proc. IEEE Conf. on Multimedia and Expo*, Vol. 3, 113–116.

49. M. G. Grant, J. D. Shutler, M. S. Nixon, and J. N. Carter. 2004. Analysis of a human extraction system for deploying gait biometrics. In: *Proc. IEEE Southwest Symposium on Image Analysis and Interpretation*, The Institute of Electrical and Electronics Engineers (IEEE), New Jersey, 46–50.

50. N. V. Boulgouris, K. N. Plataniotis, and D. Hatzinakos. 2006. Gait recognition using linear time normalization. *Patt. Recog.*, 39(5), 969–979.

51. J. Han and B. Bhanu. 2006. Individual recognition using gait energy image. *IEEE Trans. Pattern Anal. Machine Intell.*, 28(2), 316–322.

52. R. Cutler and L. S. Davis. 2000. Robust real-time periodic motion detection, analysis, and applications. *IEEE Trans. Pattern Anal. Machine Intell.*, 22(8), 781–796.

53. Y. Liu, R. T. Collins, and Y. Tsin. 2004. A computational model for periodic pattern perception based on frieze and wallpaper groups. *IEEE Trans. Pattern Anal. Machine Intell.*, 26(3), 354–371.

54. I. R. Vega and S. Sarkar. 2003. Statistical motion model based on the change of feature relationships: human gait-based recognition. *IEEE Trans. Pattern Anal. Machine Intell.*, 25(10), 1323–1328.

55. J. J. Little and J. E. Boyd. 1998. Recognizing people by their gait: the shape of motion. *Videre*, 1(2), 1–32.

56. H. Lu, K. N. Plataniotis, and A. N. Venetsanopoulos. 2008. MPCA: Multilinear principal component analysis of tensor objects. *IEEE Trans. Neural Networks*, 19(1), 18–39.

57. A. Ross and R. Govindarajan. 2005. Feature level fusion of hand and face biometrics. In: *Proc. SPIE Conf. on Biometric Technology for Human Identification II*, Berllingham, Washington, 196–204.

58. A. K. J. A. Ross and J. Z. Qian. 2003. Information fusion in biometrics. *Patt. Recog. Lett.*, 24, 2115–2125.

59. J. Kittler, M. Hatef, R. P. W. Duin, and J. Matas. 1998. On combining classifiers, *IEEE Trans. Pattern Anal. Machine Intell.*, 20(3), 226–239.

60. X. Zhou and B. Bhanu. 2006. Feature fusion of face and gait for human recognition at a distance in video. In: *Proc. Int. Conf. on Pattern Recognition*, Vol. 4, The Institute of Electrical and Electronics Engineers (IEEE), New Jersey, 529–532.

61. G. Shakhnarovich and T. Darrell. 2002. On probabilistic combination of face and gait cues foridentification. In: *Proc. IEEE Int. Conf. on Automatic Face and Gesture Recognition*, Vol. 5, The Institute of Electrical and Electronics Engineers (IEEE), New Jersey, 169–174.

62. A. Kale, A. K. Roychowdhury, and R. Chellappa. 2004. Fusion of gait and face for human identification. In *Proc. IEEE Int. Conf. on Acoustics, Speech, and Signal Processing*, Vol. 5, The Institute of Electrical and Electronics Engineers (IEEE), New Jersey, 901–904.

63. T. Sim, S. Baker, and M. Bsat. 2003. The CMU pose, illumination, and expression database. *IEEE Trans. Pattern Anal. Machine Intell.*, 25(12), 1615–1618.

64. C. Liu. 2006. Capitalize on dimensionality increasing techniques for improving face recognition grand challenge performance. *IEEE Trans. Pattern Anal. Machine Intell.*, 28(5), 725–737.

65. J. Lu, K. N. Plataniotis, A. N. Venetsanopoulos, and S. Z. Li. 2006. Ensemble-based discriminant learning with boosting for face recognition. *IEEE Trans. Neural Networks*, 17(1), 166–178.

66. M. Turk and A. Pentland. 1991. Eigenfaces for recognition. *J. Cog. Neurosci.*, 3(1), The MIT Press, Boston, Massachusetts, 71–86.

67. I. T. Jolliffe. 2002. *Principal Component Analysis,* second edition. Springer Series in Statistics, Springer, Heidelberg, Germany.

68. P. N. Belhumeur, J. P. Hespanha, and D. J. Kriegman. 1997. Eigenfaces vs. fisherfaces: Recognition using class specific linear projection. *IEEE Trans. Pattern Anal. Machine Intell.,* 19(7), 711–720.

69. B. Moghaddam, T. Jebara, and A. Pentland. 2000. Bayesian face recognition. *Patt. Recog.,* 33(11), 1771–1782.

70. H. Yu and J. Yang. 2001. A direct LDA algorithm for high-dimensional data with application to face recognition. *Patt. Recog.,* 34, 2067–2070.

71. A. M. Martinez and A. C. Kak. 2001. PCA versus LDA. *IEEE Trans. Pattern Anal. Machine Intell.,* 23(2), 228–233.

72. B. Schölkopf. 1997. *Support Vector Learning.* Munich: R. Oldenbourg Verlag. [Online]. Available: http://www.kernel machines org/papers/book_ref.ps.gz (Accessed 2008, November 1).

73. K.-R. Müller, S. Mika, G. Rätsch, K. Tsuda, and B. Schölkopf. 2001. An introduction to kernel-based learning algorithms. *IEEE Trans. Neural Networks,* 12(2), 181–201.

74. J. Wang, H. Lu, K. N. Plataniotis, and J. Lu. 2008. Gaussian kernel optimization for pattern classification. *Patt. Recog.,* Accepted pending minor revisions.

75. B. Schölkopf, A. Smola, and K. R. Müller. 1998. Nonlinear component analysis as a kernel eigenvalue problem. *Neural Comput.,* 10(5), 1299–1319.

76. G. Baudat and F. Anouar. 2000. Generalized discriminant analysis using a kernel approach. *Neural Comput.,* 12, 2385–2404.

77. J. Lu, K. N. Plataniotis, and A. N. Venetsanopoulos. 2003. Face recognition using kernel direct discriminant analysis algorithms. *IEEE Trans. Neural Networks,* 14(1), 117–126.

78. H. Lu, K. N. Plataniotis, and A. N. Venetsanopoulos. 2008. Uncorrelated multilinear principal component analysis through successive variance maximization. In: *Proc. International Conference on Machine Learning* (ICML 2008), Helsinki, Finland, 616–623.

79. S. Yan, D. Xu, Q. Yang, L. Zhang, X. Tang, and H. Zhang. 2007. Multilinear discriminant analysis for face recognition. *IEEE Trans. Image Processing,* 16(1), 212–220.

80. H. Lu, K. N. Plataniotis, and A. N. Venetsanopoulos. 2007. Uncorrelated multilinear discriminant analysis with regularization for gait recognition. In: *Proc. Biometrics Symposium 2006,* The Institute of Electrical and Electronics Engineers (IEEE), New Jersey.

81. J. Ye, R. Janardan, and Q. Li. 2004. Two-dimensional linear discriminant analysis. In: *Advances in Neural Information Processing Systems (NIPS),* The MIT Press, Boston, Massachusetts, 1569–1576.

82. H. Lu, K. N. Plataniotis, and A. N. Venetsanopoulos. 2006. Multilinear principal component analysis of tensor objects for recognition. In: *Proc. Int. Conf. on Pattern Recognition,* Vol. 2, The Institute of Electrical and Electronics Engineers (IEEE), New Jersey, 776–779.

83. J. Ye, R. Janardan, and Q. Li. 2004. GPCA: An efficient dimension reduction scheme for image compression and retrieval. In: *The Tenth ACM SIGKDD International Conference on Knowledge Discovery and Data Mining,* 354–363.

84. H. Lu,, K. N. Plataniotis, and A. N. Venetsanopoulos. 2007. Boosting LDA with regularization on MPCA features for gait recognition. In: *Proc. Biometrics Symposium 2006,* The Institute of Electrical and Electronics Engineers (IEEE), New Jersey.

85. L. D. Lathauwer, B. D. Moor, and J. Vandewalle. 2000. On the best rank-1 and rank-$(R_1, R_2, ..., R_N)$ approximation of higher-order tensors. *SIAM J. Matrix Anal. Appl.,* 21(4), 1324–1342.

86. H. Lu, K. N. Plataniotis, and A. N. Venetsanopoulos. 2006. Gait recognition through MPCA plus LDAI. In: *Proc. Biometrics Symposium 2006,* The Institute of Electrical and Electronics Engineers (IEEE), New Jersey, 1–6.

87. D. Xu, S. Yan, D. Tao, L. Zhang, X. Li, and H-J. Zhang. 2006. Human gait recognition with matrix representation. *IEEE Transactions on Circuits and Systems for Video Technology* 16(7), 896–903.

88. A. M. Bronstein, M. M. Bronstein, and R. Kimmel. 2005. Three-dimensional face recognition. *Int. J. Comput. Vis.,* 64(1), 5–30.

89. Three-dimensional face recognition. [Online]. Available: http://en.wikipedia.org/wiki/Three-dimensional_face_recognition (Accessed 2008, November 1).

90. S. Zhou, R. Chellappa, and B. Moghaddam. 2004. Visual tracking and recognition using appearance-adaptive models in particle filters. *IEEE Trans. Image Processing*, The Institute of Electrical and Electronics Engineers (IEEE), New Jersey, 13(11), 1491–1506.

91. P. J. Phillips, P. Flynn, T. Scruggs, K. Bowyer, J. Chang, K. Hoffman, J. Marques, J. Min, and W. Worek. 2005. Overview of the face recognition grand challenge. In: *Proc. IEEE Computer Society Conf. on Computer Vision and Pattern Recognition*, The Institute of Electrical and Electronics Engineers (IEEE), New Jersey, 947–954.

92. N. V. Boulgouris, D. Hatzinakos, and K. N. Plataniotis. 2005. Gait recognition: a challenging signal processing technology for biometrics. *IEEE Signal Processing Mag.*, 22(6), 78–90.

Section II

Sonar and Radar System Applications

9

Detection Paradigms for Radar

Bhashyam Balaji, Michael K. McDonald, and Anthony Damini

Defence R&D Canada Ottawa

CONTENTS

9.1 Introduction

The radar sensor has proven to be an invaluable tool to both civilian agencies and military forces for the undertaking of wide area surveillance. The general requirement for any all weather, wide area surveillance radar, is that it should be capable of detections and tracking airborne or surface targets from a stand-off position. Targets to be detected include aircraft and missiles in the airspace environment, surface vessels and submarines in the maritime environment, and both wheeled and tracked vehicles in the land environment. Despite its widespread application, radar technology should not be considered a solved problem. The continued evolution of supporting technologies and signal processing capabilities promises the expansion of radar surveillance capabilities to previously undetectable target types, such as current work by the authors to achieve wide area airborne surveillance of moving persons on the ground.

Conventional radar signal processing is performed in the single dimension of time, that is, a time series of data received from a single channel (i.e., aperture or antenna) is analysed for the presence of targets. In general terms, traditional noncoherent detection of a signal in an interference background requires matched filtering, envelope detection, video integration and thresholding. Coherent detection of a complex, coherent signal in an interference background introduces the additional step of coherent integration prior to envelope detection.

The radar echoes received by an air or space-borne surveillance radar (hereafter referred to as "radar returns" or "returns") include many forms of interference, both electromagnetic and natural. A natural and significant source of interference or "clutter" arises due to reflection of signals from the earth's surface. The source of the clutter can be conveniently split into two broad groups, referred to as ground clutter or sea clutter depending on the surface type being surveyed. Each class of clutter present its own unique challenges. A significant portion of the useable signal bandwidth from the antenna's mainbeam and sidelobes is often occupied by clutter which can posses a significant Doppler spread due to motion of distributed scatterers as might occur from a moving sea surface or the movement of foliage in the wind. Alternatively, motion of the actual radar itself can induce substantial Doppler spread when the systems are mounted on fast moving platforms such as aircraft or satellites.

The presence of Doppler spread clutter complicates or eliminates the possibility of target detection when a single channel radar is used. Low velocity targets are often masked by clutter occupying the same Doppler region in the main lobe and strong sidelobes (endoclutter). In these cases a weak target must possess a minimum detectable velocity high enough for the target Doppler to be clear of the mainbeam and strong sidelobe clutter spectrum (i.e., it must be exoclutter) so as to ensure robust target detection.

Over the past two decades, there has been a proliferation of research within the international community to extend the detection capabilities of radar beyond that achieved by the traditional noncoherent and simple coherent processing strategies. In particular, surveillance with air and space-borne multiple aperture radar systems utilizing space-time adaptive processing (STAP) has resulted in a decreased minimum velocity at which a target can be detected in a clutter background, thus increasing the robustness of the radar system. Furthermore, these multiple channel radar systems offer the opportunity for not only detection, but also accurate location of slow-moving targets within the clutter band. STAP signal processing techniques operate by not only analysing the time series of data received from each aperture, as is done in traditional coherent processing, but also by further capitalizing on the spatial diversity afforded through the use of multiple apertures which allows the targets and clutter to be viewed from a diversity of aspect angles. This added degree of freedom provides the capability to simultaneously suppress both clutter and jammer interference, thus increasing the probability of target detection. Similarly, in the case of fixed surface platforms, one and two-dimensional array processing techniques introduce great gains in the area of target detection in the presence of spatially localized interference.

In this chapter, we present a brief overview of the radar detection problem as it applies to the wide area surveillance of targets, and discuss a series of detection strategies of increasing complexity for addressing increasingly difficult target detection scenarios. The layout is as follows. In Section 9.2, the nature of the observations (or measurement data) is presented and organized through the concept of the radar data cube. Section 9.3 then presents the basic aspects of detection theory through stochastic models of land and sea clutter, and the Swerling models. Section 9.4 describes detector design for single channel systems, both noncoherent and coherent. This is followed in Section 9.5 by a review of STAP, for moving target detection. Finally, Section 9.6 discusses the use of detect-before-track (DBTk) and track-before-detect (TkBD) techniques to further refine the detection process through knowledge of the constraints on a target's possible motion. A brief introduction is given into the solution of the Bayesian filtering problem through the application of finite-difference based numerical solutions of partial differential equations and particle filter techniques.

Though the detection techniques discussed herein are in the context of radar due to the authors' familiarity with the radar application, it is noted that they are equally applicable to other types of sensors, particularly single channel imaging sensors in the domains of electro-optic/infrared, multi and hyper-spectral. Each of these types of sensors produces an image whose distribution in space and amplitude is analogous to the reflectivity field of radar, upon which the following derivations are based.

9.2 Input to Detector: the Data Cube and Related Products

In this section, the intermediate data products (i.e., prior to application of the detection algorithms) from the radar are described [1].

Radars are used for both civilian and military applications. Typically, military radars are high-resolution and/or large volume/wide area surveillance radars, while civilian radars are low resolution radars with large coverage area. Another point of distinction in system parameters occurs between airborne and spaceborne platforms. An airborne

TABLE 9.1

Airborne and Spaceborne Radars: a General Comparison

Parameter	Airborne	Spaceborne
Motion compensation	Significant	Less
Constraints	Flight path	Orbit
Swath	10–200 km	30–500 km
Altitude	3–12 km	600–800 km
Speed	0.1–0.2 km/sec	7.5 km/sec
Incident angle	5–30 degrees	37–40 degrees

platform is usually at a few kilometres in altitude, and tens of kilometers in range while moving on the order of 100 m/sec, while a spaceborne platform orbits at altitudes on the order of 800 km and moves at 5–10 km/sec. See Table 9.1 for some key differences. These differences present unique challenges with resulting intermediate products which have dissimilar properties. Hence, the outcome is different, leading to different preferable solutions. Nevertheless, the general detection concepts presented in this chapter are applicable and relevant in all cases.

Throughout this chapter, the transmitter and receiver are assumed to be co-located, i.e., the monostatic case. In recent times, there has been considerable interest in the bistatic (and multistatic) case, where the transmitter and receiver(s) are not co-located [2]. This leads to significant differences from the monostatic case as the bistatic return depends on the bistatic angle (the angle between the transmit-to-target and receive-to-target lines). However, many of the underlying signal processing techniques in regards to detection are still relevant since they are based on general statistical principles.

9.2.1 Data Cube

Pulsed radars transmit electromagnetic pulses and record reflected responses from the target and clutter environment. A portion of the transmitted pulse is reflected back (the "backscatter") and is measured as an amplitude and a phase, or an in-phase and quadrature phase response. The amplitude depends on various quantities such as range, polarization, and antenna gain, and is summarized in the radar equation, while the phase depends only on the wavelength, target composition, modulation and the data acquisition geometry between the target and the radar [1].

There are two time scales implicit in radar measurements. The time interval between pulses, termed the pulse repetition interval (T_{PRI}) is referred to as the "slow time" dimension. In contrast, the round-trip time for the pulse is very small (relative to the T_{PRI}) and is referred to as the "fast time" dimension. Radar motion can be considered negligible during the round-trip time as the velocity of the platform (airborne or spaceborne) is much smaller than the speed of light. It is assumed that the pulse repetition frequency (PRF) and range to the target are such that there are no returns beyond the unambiguous range, $R_u = 0.5cT_{PRI}$, where T_{PRI} is the pulse repetition interval and c is the speed of light. When data is collected from a series of pulses a two-index tensor is formed. That is, the received data is naturally binned along the slow-time, or azimuth, dimension. Similarly, when digitized it is also binned along the fast-time, or range, dimension. If there is more than one antenna on receive, the result is a three-index tensor of voltages comprised of returns along range or fast time, pulse or slow time and aperture dimension. This is referred to as the data cube.

The simplest technique to achieve range resolution is to transmit a sinusoidal pulse of duration T, i.e., no modulation. The return signal is a time-shifted (and attenuated) version of the transmitted signal. The received signal is then match filtered to maximize the signal-to-noise ratio (SNR). The matched filter result is proportional to the autocorrelation, $r_c(t)$. The autocorrelation of rectangular impulse is given by a triangular envelope.

To successfully to resolve the returns from two targets, the respective triangular envelopes must be separated by at least T so that the maxima of both pulses can be separated. Conceptually, the simplest way to increase resolution is to transmit a shorter pulse. This implies a higher bandwidth, B, as pulse length is inversely proportional to bandwidth of the transmitted signal as per $cT/2=c/2B$. However, in order to maximize SNR, and hence detection, it is generally desirable that the pulse length be as long as possible to maximize transmitted energy.

A method to accommodate these two seemingly conflicting requirements is to impose a frequency modulation on a relatively long transmitted pulse to increase the effective bandwidth, and use matched filtering to "compress" the resulting measurements to obtain a finer resolution. This approach is commonly referred to as pulse compression. An explanation of the pulse compression process can be found, for instance, in Skolnik [1]. Commonly, a linear frequency modulated (FM) pulse of the following form is transmitted:

$$s(t) = A(t)e^{i(2\pi(ft+\beta t^2))}, \tag{9.1}$$

where $-(T/2) \le t \le (T/2)$, f is the carrier frequency, β is the so called chirp rate or phase modulation constant, A is the amplitude of the transmitted signal and the signal bandwidth is $2\beta T$. The returned echo is

$$s_0(t) = \sigma A(t-t_0)e^{i2\pi(f(t-t_0)+\beta(t-t_0)^2)}, \tag{9.2}$$

where σ is the composite reflectivity from multiple scatterers and $t_0 = 2R_0/c$ is the round-trip propagation time to the target at range R_0. After down-conversion to an intermediate frequency (IF), digitization, digital demodulation to baseband, and pulse compression according to the chirp rate, the sampled return from one pulse is of the form

$$z_0(t_k) = A(t_k)e^{i2\pi R_0(t_k)}, \quad k = 1, 2, 3..., \tag{9.3}$$

where t_k is discretized time. A signal of bandwidth $B=2BT$ has a time-domain resolution of $1/B$, which, in the units of range, after two-way propagation, is $c/2B$.

Finally, the following is important from a statistical perspective. The measured signal, $z(t)$, can usually be modelled as a complex zero-mean stationary stochastic process. Some of the commonly used stochastic models are discussed later.

9.3 The Detection Problem—Detecting Targets in Clutter

9.3.1 Preliminary Remarks

The fundamental challenge inherent in any detection problem is the reliable, unambiguous detection of a target signal that is itself obscured by interference from other sources.

Broadly speaking the interference in the radar problem can be broken into two categories (the impact of jammers will be discussed in Section 9.5):

- White noise
- Clutter

The dominant source of white noise is typically receiver noise from the radar itself. The treatment of signals in white noise is well documented in the literature and current attempts to improve signal detection in white noise commonly revolve around the minimization of receiver noise. This is a system engineering design issue and is outside the scope of the current chapter.

As discussed in the introduction, clutter arises due to signal being reflected from other features or surfaces in the antenna field of view. This means that the source of the clutter signal is external to the radar receiver itself and, in effect, beyond the radar design engineer's control. This statement is not strictly true as the design engineer can make design choices to minimize clutter return, e.g., antenna beamwidth, resolution, nevertheless these parameters are often constrained by other considerations and the fact remains that the designer can't redesign the area that is being sensed.

Given the complexity of a sea or land surface, it is clear that the behavior of the radar returns from these surfaces will be nondeterministic and complex. As such, when formulating a detection approach, one is forced to rely on stochastic characterizations to describe clutter and target returns. Understandably these characterisations provide varying degrees of success and fidelity and typically represent a trade-off between accuracy and mathematical tractability. The following section summarizes some commonly used stochastic models for land and sea clutter and the Swerling or chi square target models [1,2]. The basic formulation of the statistical detection theory is also presented.

Note that the detection problem is different over land than over water. Over land the clutter is usually stationary (i.e., small spectral width) but has a large backscatter (except in areas like smooth roads). As such, detection over land is practically limited to the detection of movers. In addition, the coherence is quite high over land, i.e., target phase remains correlated from measurement to measurement. This coherence is quantified in terms of clutter rank in the discussion of STAP in Section 9.5.

In contrast, while the clutter returns from the ocean are typically of lower average power they are the result of a distributed surface which is undergoing complex dynamic motion. Sea clutter returns tend to have a broader and more time-varying spectrum then land clutter. The lower average power of sea clutter permits the detection of stationary and moving targets on the basis of signal strength alone, but the dynamic nature of the sea surface leads to a broad Doppler spectrum and the creation of so called sea spikes, extremely strong, localized and relatively long lived clutter returns which can be extremely difficult to distinguish from moderate sized real targets.

It should be emphasized that the measured returns from real targets frequently exhibit quite complex behavior due to the rapidly varying radar cross-section of a complex target with viewing angle. The mathematical formulations of the commonly employed target models as discussed in Section 9.3.3 represent a trade off between mathematical tractability and actual behavior, and frequently offer only a weak approximation of actual observed behavior of real target returns.

9.3.2 Sea Clutter

9.3.2.1 Stochastic Models

The simplest model for sea clutter assumes that the amplitude of the clutter is Rayleigh-distributed (or equivalently the in-phase and quadrature components of the received signals are jointly Gaussian processes). The Rayleigh assumption is valid at coarse resolutions, i.e., when the surface is rough on the scale of the radar wavelength and homogeneous with constant mean radar cross section (RCS). For sea surfaces, the Rayleigh assumption is frequently useful for measurements with range resolutions on the order of tens of metres. The fundamental reason for this is the central limit theorem (CLT).

As the resolution is reduced to below 10 m, the Rayleigh distribution tends to become less accurate due to the increased occurrence of high amplitude events. The observed breakdown of the Gaussian behavior is the result of bunching of scatterers due to correlations in the underlying sea surface (so that the i.i.d. assumption for the CLT is not met). This structure presents itself as an extended tail on a histogram of the observations and is clearly evident in Figure 9.1 where a histogram of real, submetre resolution data is compared with a distribution of a Rayleigh probability distribution function (pdf).

To better describe the observed clutter behavior at high resolution a number of alternate pdfs such as the Weibull and K-distribution have been proposed [3]. The K-distribution is a popular choice and is an example of a compound Gaussian distribution in which the mean power of a Gaussian distribution is modulated by a specified power distribution [4–11]. Figure 9.1 also displays an example of the K-distribution fitted to real data, note the better fit at high amplitudes. Specifically the K-distribution is given by

$$p(x) = \int_{-\infty}^{\infty} p(x \mid y) p(y) dy \tag{9.4}$$

where $p(y)$ is the gamma pdf corresponding to the distribution of the local mean power, y, and $p(x \mid y)$ is given by

$$p(x \mid y) = \frac{2x^{N-1}}{(2y)^{N/2} \Gamma(N/2)} \exp\left(-\frac{x^2}{2y}\right), \tag{9.5}$$

which is simply the generalized Rayleigh distribution resulting from the square root of the sum of the squares of N i.i.d. Gaussian scalar random variables with zero mean and variance y. The texture is gamma distributed with paramaters σ, α, i.e.,

$$p(y) = \frac{\sigma^\alpha y^{\alpha-1}}{\Gamma(\alpha)} \exp(-\sigma y). \tag{9.6}$$

Thus, the K-distribution is given by

$$p(x) = \frac{x^{\alpha-1+N/2} \, 2^{2-N/4-\alpha/2}}{\Gamma(N/2)\Gamma(\alpha)} K_{\alpha-N/2}(\sqrt{2\sigma} x), \tag{9.7}$$

where $K_v(\cdot)$ is a modified Bessel function of the second kind of order v.

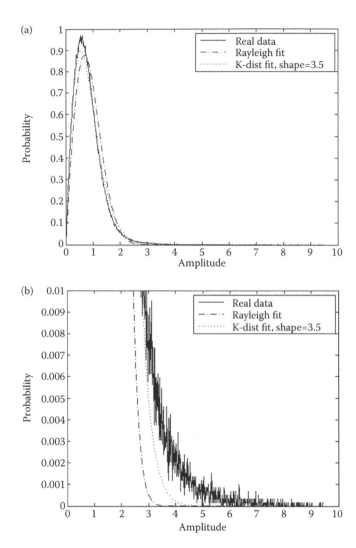

FIGURE 9.1
Comparison of real, submetre resolution data histogram versus fitted Rayleigh and K-distributions.

Note that for the shape parameter $v \to \infty$, the K-distribution becomes the Rayleigh distribution. In practice, even $v \approx 10$ can often be considered to be Rayleigh to a good approximation. The key point is that the relationship is multiplicative, and that the mean level of the Rayleigh distribution is modulated by the underlying texture.

The K-distribution represents a useful choice due to its tractability and its ability to provide a reasonably good fit to a broad range of sea clutter. The multiplicative structure also lends itself to a convenient, if not entirely accurate, physical model to justify its application as discussed below.

An examination of sea clutter returns reveals that the observed time and spatial variability is actually a function of two parameters, conveniently identified with the texture and speckle component above, each parameter possessing a different scale of correlation [7]. Texture (local mean power) measurements tend to decorrelate over ranges on the order of tens of meters or alternatively over periods on the order of seconds. The texture is interpreted as the average reflected power due to gross features of the sea surface such as swell length. Speckle tends to decorrelate on scales of milliseconds or between contiguous resolution cells. Physically, this variability is considered to arise as a result of variations in reflectivity within a single resolution cell due to the fine scale structure on the sea surface e.g., capillary waves. The implications of the different correlation lengths of the texture and speckle can have a significant impact on the detection performance and the detection strategies. This will be discussed in further detail in Section 9.4.

9.3.2.2 Chaotic Models of Sea Clutter

In recent years, there was considerable interest in the idea that the dynamic behavior of radar returns from sea clutter might be modelled as a chaotic dynamical system [12]. The ability to model sea clutter as a deterministic system is an appealing one as a chaotic system would posses a relatively small number of degrees of freedom with correspondingly greater predictability. The net result would be greater detection sensitivity for a given false alarm rate. Early results seemed to suggest that real returns were in fact chaotic as one could show that traditional characteristic measures of chaotic behavior such as fractal dimension and positive Lyapunov dimension were also observed with sea clutter returns. Unfortunately, further study revealed that the standard techniques that were employed to calculate the chaotic invariants were not a reliable method of distinguishing between chaotic and stochastic processes. In particular, tests against simulated stochastic data of a colored noise process were shown to produce identical measures as similar chaotic processes [13]. In other words, while the invariant measures would correctly characterize a known chaotic process they were incapable of distinguishing between chaotic and stochastic processes. The chaotic paradigm was also investigated for the high-resolution radar case where it was postulated that only the underlying texture component was chaotic in nature. Once again no convincing evidence of chaotic behavior was revealed. Even if the issue of identifying discriminating test parameters is ignored, practical attempts to develop nonlinear prediction techniques based on radial basis function neural networks (designed to model chaotic systems) failed to produce any performance improvement over linear prediction techniques used with stochastic approaches [14–16]. Put simply, nonlinear and more particularly, chaotic modeling has not been shown to date to offer any improvement over linear stochastic approaches.

9.3.3 Land Clutter

As with the sea clutter case, the Rayleigh assumption is valid for land clutter at coarse resolutions under similar conditions as sea clutter, i.e., when the surface is rough on the scale of one wavelength and homogeneous with constant mean RCS (e.g., agricultural areas, or deserts). For finer resolution, nonGaussian distributions provide a better fit. Most nonGaussian clutter distributions arise from a compound (or multiplicative) model for the clutter, i.e., product of a Gaussian distribution and an independently modulating random variable A, or $z_i = Ax_i \ \forall i = 1, 2, \ldots, N$. While physical justification for the multiplicative noise model can be provided in some cases, the fundamental reason for the effectiveness of such models is that this introduces an additional parameter in the probability density function which can be used to "tune" the pdf to the data. Three such cases are [17]:

- K-distribution: A is chi-square distributed;
- Weibull distribution: density of A in terms of Meijers G functions;
- G_c^0 distribution: A is the square root of the generalized inverse Gaussian distribution.

The third case is found to be useful in modelling extremely heterogeneous clutter (like urban areas with different landscape types), especially the important tail region, for determining detector thresholds where the K-distribution fails. Also, unlike the K-distribution, there are no modified Bessel functions involved; so calculations of tests and thresholds are easier to evaluate for large arguments, as in along-track-interferometry with high coherence and high backscatter RCS.

9.3.4 Common Target Models

In principle, the radar cross-sections of a known target can be precisely evaluated using electromagnetic theory. However, the cross-section depends on many factors, such as electromagnetic properties of the scatterers, aspect angle, etc. Generally, the net result is highly complex cross-section behavior versus viewing angle. It is impractical to incorporate this level of complexity into a generalized detection or modeling approach and, as such, it is typically more convenient to study the effects of fluctuating cross-section by postulating a reasonable, simple and analytically tractable stochastic model for the target return. The most commonly used radar target models are the Swerling target models which are based on the chi-square pdf with specific degrees of freedom [2]. Despite their widespread use the reader should bear in mind that these models are, in fact, highly idealized and actual observed performance against a real target could vary greatly from that predicted by the idealized target models.

The Swerling models are all special cases of the chi-square distribution of degree $2m$ of the following pdf:

$$p(\sigma) = \frac{m}{(m-1)!\sigma_{avg}} \left(\frac{m\sigma}{\sigma_{avg}}\right)^{m-1} \exp\left(-\frac{m\sigma}{\sigma_{avg}}\right), \quad \sigma > 0, \tag{9.8}$$

where σ_{avg} is the average RCS over all target fluctuations. The choice of parameter m determines the pdf corresponding to the amplitude returns and partly distinguishes the Swerling models. The final characteristic to fully delineate between the different Swerling models is the timescale over which the target is presumed to fully decorrelate. To understand this latter distinction one must consider the usual operating mode of a radar.

A typical surveillance radar operates in a scanning fashion whereby the antenna look direction is repeatedly swept across a defined surveillance area. Therefore, an individual target will only be under surveillance during the small portion of the scan in which the target is within the antenna beam pattern extent. The time required to complete each scan and return to the starting position is referred to as the scan-to-scan revisit time or scan rate. Typical scan rates can vary from a few rpm up to rates on the order of 300 rpm. During the course of a scan the radar will transmit pulses at rates approaching thousands of pulses per second. With these definitions in mind, we can complete the Swerling model definition by specifying whether target returns decorrelate on a pulse-to-pulse or scan-to-scan basis. The importance of this distinction becomes clear when pulse-to-pulse integration is employed to improve signal-to-interference plus noise ratio (SINR). If only a single pulse is considered or integration is purely on a scan-to-scan basis the distinction becomes irrelevant.

With these definitions in mind, we can explicitly state the Swerling target models as follows. When $m=1$, the equation above reduces to the well known exponential or Rayleigh power pdf:

$$p(\sigma) = \frac{1}{\sigma_{avg}} \exp\left(-\frac{\sigma}{\sigma_{avg}}\right), \quad \sigma \geq 0. \tag{9.9}$$

This is the prescribed pdf for the Swerling models 1 and 2. The Swerling 1 model amplitude is presumed constant over the extent of a single scan but will completely decorrelate on a scan-to-scan basis. In contrast, the Swerling 2 target amplitude is presumed to decorrelate on a pulse-to-pulse basis.

The $m=2$ case corresponds to the pdf for the Swerling 3 and 4 models. In analogy with the Swerling 1 and 2 cases, the Swerling 3 is presumed to decorrelate on a scan-to-scan basis while Swerling 4 decorrelates on a pulse-to-pulse. The final case, $m=\infty$, corresponds to the Swerling 0 model and represents the condition in which the general Swerling target model collapses to a probability of 1, i.e., a constant cross-section. Swerling 1 and 2 models exhibit a larger variance than Swerling 3 and 4 models.

In practical terms, it is empirically observed that the Swerling 1 and 2 models are appropriate when the complex target is composed of many independent scatterers of approximately equal cross-section whereas the Swerling 3 and 4 models are found to be most appropriate when the target is composed of one dominant scatterer section and a number of smaller cross-section scatterers [1].

9.3.5 Basics of Detection Theory and Performance Metrics

The detection problem can be viewed as one of choosing between two hypotheses for a given measurement:

H_0: the hypothesis that no target is present;

H_1: the hypothesis that a target is present.

The most common detection approach is to define a simple threshold of detection; if a measured return has a greater amplitude than the threshold the H_1 hypothesis is presumed, if it is less than the H_0 hypothesis is selected. The assignment of a hypothesis permits four possible outcomes:

(a) H_1 is chosen but no target is actually present. This is called a false alarm.
(b) H_1 is chosen and a target is present. This is called a detection.
(c) H_0 is chosen but a target is present. This is called a missed detection.
(d) H_0 is chosen and no target was present.

The detection process can be mathematically formulated as follows. Let the probability distribution under the null hypothesis be denoted by $p(x|H_0)$. Then, the probability of false alarm (P_{fa}) is given by

$$P_{fa} = \int_{\eta}^{\infty} p(x|H_0)dx. \tag{9.10}$$

The quantity η is the threshold of detection discussed above. Similarly, let the probability density for the H_1 hypothesis be denoted by $p(x|H_1)$. Then, the probability of detection for threshold η is given by

$$P_d = \int_{-\infty}^{\eta} p(x|H_1)dx. \tag{9.11}$$

This later equation is dependent on the SNR of the target and noise model.

From the above, two optimal detectors follow. The first is the Bayesian approach that minimizes the probability of error criterion. The Bayesian approach requires that the *a priori* probability of each hypothesis be known, and hence is not suitable for most radar detection applications. An alternative is the Neyman–Pearson criterion that seeks to maximize the probability of detection, for a false alarm rate not exceeding a certain value. It can be shown that the optimal decision rule under the Neyman–Pearson criterion is the likelihood ratio test [18].

The canonical example is detection of target in Rayleigh amplitude clutter. The probability of false alarm for Rayleigh clutter is:

$$P_{fa} = \int_{\eta}^{\infty} \frac{R}{\psi_0} \exp\left(-\frac{R^2}{2\psi_0}\right) dR,$$

$$= \exp\left(-\frac{\eta^2}{2\psi_0}\right). \tag{9.12}$$

Here ψ_0 is the variance, or mean square value of each of the zero-mean I or Q Gaussian noise processes.

The output of the envelope detector for a Swerling 0 target of amplitude A is given by the Rice probability distribution

$$p_s(R) = \frac{R}{\psi_0} \exp\left(-\frac{R^2 + A^2}{2\psi_0}\right) I_0\left(\frac{RA}{\psi_0}\right), \tag{9.13}$$

$I_0(\cdot)$ being the modified Bessel function of zero order. Therefore, the probability of detection is given by:

$$P_d = \int_{\eta}^{\infty} \frac{R}{\psi_0} \exp\left(-\frac{R + A^2}{2\psi_0}\right) I_0\left(\frac{RA}{\psi_0}\right) dR. \tag{9.14}$$

The results for the other Swerling models follow in similar, if slightly more complicated fashion, and the interested reader is referred to the following references [1,18].

The above analysis can, in theory, be carried out for any clutter probability distribution. Similarly, the formulation can be expanded to include the effects of integration prior to detection. However, it should be noted that, depending on the complexity of the clutter model, target models and detection schemes, it may not be possible to develop a closed form solution for the P_{fa} and/or P_d. In these cases, the thresholds must be evaluated numerically. In some cases $p(x|H_0)$ may be known analytically, while $p(x|H_1)$ has to be evaluated numerically. This is an important distinction as the requirement to calculate P_{fa} numerically can be a practical barrier (due to the computational load) to the implementation of a real time detection scheme since the detection threshold must be calculated in real time (whereas an accurate estimate of $p(x|H_1)$ is not typically required during real time operation). Frequently, a trade-off must be made between the fidelity of the pdf model and the available computational capacity of the signal processing system.

9.4 Detector Design

The following sections discuss several detector designs that are commonly employed to achieve detection of maritime and land targets. All the approaches can be considered "threshold based" as they ultimately rely on the final strength of postprocessed measurements exceeding a specified threshold to determine if a target is present. A number of different approaches and techniques of increasing complexity (both from a signal processing and system design perspective) are presented.

9.4.1 Constant False Alarm Rate (CFAR) Detections

The most common design goal for a threshold based detection scheme is to achieve a constant false alarm rate (CFAR), i.e., Neyman–Pearson criterion. Essentially this equates to designing an algorithm which adapts the threshold so as to achieve a CFAR across varying clutter conditions. The practical reason for adopting this design goal is easily understood in terms of operator loading. If the false alarm rate (FAR) from a detector becomes too high, the display presented to the radar operator will become swamped with false alarms. The larger the area that is being surveyed the more false alarms that are presented to the operator for a given FAR and the more difficult it becomes for the operator to distinguish between real and false targets. The problem remains even if the operator is removed from the loop; a typical automatic detection scheme is followed by a tracking stage where excessive FAR can cause problems in two ways; the available maximum track capacity may be exceeded or real tracks may be seduced off course by false alarms. In a CFAR detector, the FAR is typically fixed at the largest value that can be tolerated by the operator or tracking algorithms so as to maximize surveillance effectiveness. To get a sense of the scale of the problem consider a 1 m resolution radar with a scan rate of 60 rpm and a PRF of 1000 which is tasked with the surveillance of a 200 km deep by 180 degree wide sector. Over the course of each scan the system will record on the order of 10^8 measurements. Hence, on average, even with a seemingly very small specified P_{fa} of 10^{-6}, false measurements will occur on each scan.

The CFAR condition is not a particularly restrictive design constraint and a wide range of detector and system CFAR design options are available to the designer to tailor the system to different target and clutter environments. The following sections provide a brief introduction to a sampling of approaches which represent increasing levels of complexity (both from a signal processing and system design perspective) and, hopefully improved surveillance capabilities.

9.4.2 Single Channel Noncoherent Detection

Single channel noncoherent detection represents the most straightforward detection concept in which amplitude (or power) returns are measured using a single aperture antenna. No phase information is preserved so spectral knowledge such as target Doppler is lost. Since a single antenna phase centre is used, the azimuth (or bearing to target) measurements typically have uncertainties on the order of one-half of the antenna azimuth beamwidth. The advantages of this approach are that it imposes the least computational burden and the antenna and receiver design are the simplest of all the approaches presented herein.

Traditionally the single channel noncoherent approach has been utilized for maritime surveillance where target scattering cross-sections, and therefore target returns, are commonly large with respect to surrounding clutter cross-sections. That said, it is typically still necessary to integrate several measurements to achieve a practical signal-to-interference ratio (SIR) for reliable detection. The simplest integration approach is to sum the detected amplitudes (square law or linear detected values can be used with little difference in performance, Skolnik [1]) across a specified dwell and/or scan interval. The detection statistic, z, is then

$$z = \sum_{i=1}^{mN_s} |z_i|, \qquad (9.15)$$

where m is the number of measurements of the target location within one beamwidth (or CPI), z_i is the amplitude (or power) measured for each pulse and N_s is the number of scans over which the integration is carried out. The required threshold value to achieve a given P_{fa} for *a priori* specified underlying Rayleigh interference is easily determined (see Skolnik [1], for instance). In practice the underlying parameters for the appropriate clutter pdf are not know *a priori* and they must be adaptively estimated from the received signal. This is further discussed below.

9.4.2.1 *Practical Implementations of Adaptive Detection in Single Channel Noncoherent Detection*

For the Rayleigh clutter model, the only parameter required to fully characterize the pdf is the average clutter power (or amplitude in the linear case). The most straightforward approach to adaptively determining this value is the cell averaging (CA) CFAR [9]. The CA configuration is illustrated in Figure 9.2. It consists of a series of sampling cells on either side of the cell under test (CUT) over which the background clutter power is measured and averaged. These background cells are referred to as secondary data. The calculated average provides the required estimate of background clutter power. The approach introduces additional performance loss with respect to *a priori* or clairvoyant knowledge of clutter power as the act of estimating the background power from a limited size sample set introduces additional uncertainty.

Various elaborations on this basic approach exist to deal with a variety of complicating factors such as the presence of other targets or land in the secondary and correlated clutter conditions. Examples include "least of CA-CFAR" in which a separate average is formed from secondary data collected on each side of the CUT and only the side with the smaller average power (i.e., least of) is used. This approach is useful for excluding secondary data containing strong anomalous clutter such as from land clutter returns. In contrast a "greater of CA-CFAR" is useful for suppressing false alarms from leading edges of abrupt clutter regions [1]. Another approach is the ordered statistics (OS-CFAR) detector in which secondary data is ranked according to amplitude and the value occupying a set rank position is used as the estimate of background power [20]. This approach is frequently advocated for situations in which contamination of secondary data by other nearby targets is anticipated. If, for instance, it is anticipated that in the worst case scenario three contaminating targets may be present, then the fourth highest value in the data ranking may be used to determine the required threshold. This has the effect of excluding the measurements that may have been derived from cells containing targets.

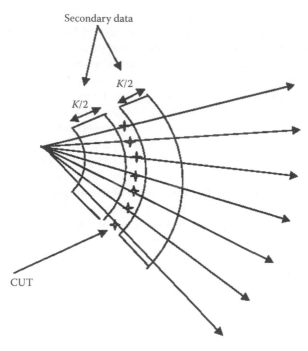

FIGURE 9.2
The cell averaging CFAR (CA-CFAR) concept. The mean power of the clutter in the cell under test (CUT) is estimated using background samples on either side of the CUT. Guard cells are used to prevent contamination of the background estimate by distributed targets.

The basic consideration for all configurations is that the more data that is excised from the secondary data set, the less stochastically accurate the estimate derived from it will be. This performance loss must be balanced against the protection that the various schemes offer against anomalous conditions that can often lead to catastrophic collapse of detection performance. In general, the structures described above have analogous structures for use with the more sophisticated detection approaches outlined in the following sections. We will refer back to them as applicable section.

To this point we have only considered the simple Gaussian clutter cause. It is clear to the thoughtful reader that the use of more complicated clutter pdfs will require the estimation of additional parameters. The K-distribution, for instance, requires the specification of the shape parameter which is understood as a measure of the spikieness of the data. Various approaches can be used to empirically estimate the shape parameter from the measured data, but all rely on the calculation of higher order moments (in comparison with the first order moment for average power) and, as such, require significantly larger sample size to achieve an acceptable statistical accuracy. This requirement can, in practice, be difficult to achieve as clutter characteristics may not remain homogenous over the required sample size. This rather inconvenient truth reflects the real world nature of clutter which can change over short distances due to differences in winds, prevailing currents etc.

One may ask the question why a designer would want to use a more complicated pdf when the Gaussian in Figure 9.1 would appear, at first glance, to provide a reasonably good fit? The answer relates back to the earlier discussion of Section 9.4.1 in which the requirement for a very low P_{fa} was illustrated by example. It turns out that the small addition of the tail of the real data histogram that was observed in Figure 9.1 leads to a large number of false alarms when a large area is surveyed, i.e., use of Gaussian model for spiky data will lead to unacceptably high false alarm rates. One potential solution to this problem is to simply provide the radar operator with the capability to tweak the required threshold so as to reduce false alarms to a manageable level. Unfortunately, this technique becomes impractical if clutter characteristics vary widely over the surveillance area covered on each scan and, as will be discussed below, can lead to a decrease in detection performance.

Another important consideration for the size of the secondary data sets is revealed by considering the effect of correlation in the underlying texture (or mean clutter power) that was discussed in Section 9.3.2. This effect is illustrated in Figure 9.3 where simulated K-distributed clutter returns are plotted. The dotted line in the Figure 9.3 illustrates the variation of the underlying texture. If a histogram of the amplitude returns across the full data set is formed the expected K-distribution is observed. On the other hand if individual histograms are formed using localized sections of data that are short in comparison with the scale of the underlying texture variation, each histogram will be found to be well represented by a Rayleigh distribution. This observation suggests a useful approach to maximize detection performance under conditions of correlated texture. ... CUT, the calculation of the local required threshold can be calculated using Rayleigh statistics [19]. The dotted curve in the Figure 9.3 illustrates this approach in which the threshold is locally tailored to the underlying texture conditions. The global approach of simply specifying a threshold based on the global characteristics of the K-distributed data set results in the fixed threshold indicated by the dot-dash line. The drawbacks to the local CFAR approach is that fewer samples are available to calculate the estimate of mean and the secondary data set extent must be somehow tailored to the existing texture correlation lengths. A particularly perceptive reader might, at this point, ask themselves how it is possible to get better performance with less measurements? The answer to this apparent paradox is that we are actually utilizing additional information. By utilizing a local mean we are

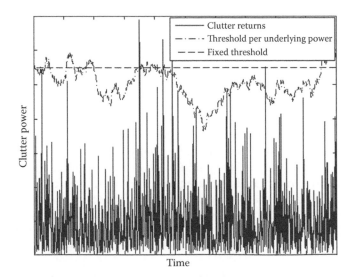

FIGURE 9.3
Simulated K-distributed clutter returns with underlying correlated texture. The dashed line indicates the fixed threshold level required to achieve a specified P_{fa}. The dash-dot line indicates the equivalent adaptive threshold that would achieve an identical P_{fa}.

exploiting the fact that we know (or at least suspect) the texture is correlated over these spatial scales. If this condition is not met then the local mean approach will provide inferior results to a global mean.

To get around some of the sensitivities noted above, detector designers have utilized approaches which are not dependent on a strict definition of clutter pdf; the so called distribution free approaches. One example of a distribution free approach, presented by Dillard and Antoniak [21], utilizes a transformation to convert measurements to a sufficient statistic in a distribution free space. However, as just discussed, you rarely get something for nothing and the approach is only applicable if a number of conditions are met, namely the pdf of all observations on a single scan must be the same and the transformed statistics must be independent. This approach will generally not work for sea clutter due to the underlying correlation of the clutter texture discussed above.

Alternately, a radar designer may chose to determine required thresholds using an empirical distribution free approach in which a range of thresholds is applied against the real data and the resulting P_{fa} calculated. A curve is then fitted to these points and extrapolated to determine the threshold required for a desired P_{fa}. This is the automated equivalent to having an operator tweak the threshold. The primary difficulties of this approach are due to two factors: are the very small P_{fa} that are required in an operational system, and the limited regions over which clutter remains homogenous. To calculate the threshold corresponding to a practical P_{fa} one must either use unrealistically large sample sets so as to obtain a statistically meaningful results or the extrapolation of the curve must be performed over such a large range that significant inaccuracies may result.

The moral to take out of the above discussion is that the design of a sensitive, robust detection scheme is by no means a trivial task. In fact this superficially simple task (i.e., threshold and detect) often proves to be one of the more difficult tasks of the overall signal processor design.

9.4.3 Single Channel Coherent Detection for Fast Movers

The next level of design complexity is implementation of a coherent single channel receiver. In this context the term coherent implies that both an amplitude and a phase is measured for each resolution cell (or alternately in-phase and quadrature-phase).

The measurement of phase is a powerful discriminant due to the Doppler effect that is associated with measured targets. By taking the Fourier transform of a series of measurements of a moving target (or alternately passing the signal though a filter bank of plausible Doppler filters) the target's Doppler can be isolated. This form of processing, often referred to as pulse Doppler processing, is particularly effective at separating stationary clutter returns from the moving target signal, thereby significantly improving the SINR. Air surveillance radar makes extensive use of Doppler processing due to readily distinguishable Doppler associated with fast moving air targets. Unfortunately, the approach is less effective for maritime surveillance as the dynamic nature of the sea surface imposes a complex fluctuating spectrum on the sea clutter returns that often overlaps with the Doppler signature of the slower moving maritime targets. Approaches for dealing with this issue are discussed in the following section on adaptive coherent processing. A further complication with maritime targets arises due to the rough motion of the sea surface which can cause the target's Doppler signature to be smeared and less distinctive, especially during high sea states.

Many of the approaches discussed in Section 9.4.2 for implementing detection schemes are applicable to pulse Doppler systems. After the application of the Fourier transform one is left with a two dimensional detection plane corresponding to range and Doppler (as opposed to a 1D coordinate of range for noncoherent data). One can now apply CA-CFAR as described above where the secondary data is distributed in range or, alternately, one can collect secondary data cells across the Doppler dimension. This latter approach can provide some gain in detecting a target signal in a smoothly structured clutter spectrum but is generally limited by the small number of independent Doppler background cells (i.e., small sample size for secondary data) and the unpredictable variation of the spectrum with Doppler. One approach to help mitigate the problem of limited sample size is to collect secondary data across a region distributed in both range and Doppler around the CUT. Analogous formulations of OS and empirical distribution free detectors can also be developed. In the case of land clutter, an additional approach to CFAR involves capitalizing on the phase information and including in the secondary data only the range samples whose phase corresponds to the phase predicted from the data collection geometry.

9.4.4 Adaptive Single Channel Coherent Detection

As briefly discussed above, when performing coherent processing on single channel data, two situations can arise: the Doppler signature of the target can be completely outside of the clutter spectrum or the Doppler signature can overlap with the clutter spectrum. These conditions are referred to as exoclutter and endoclutter, respectively. As discussed, an endoclutter situation can arise during maritime surveillance due to motion of the sea surface. In addition, it can also arise over land (or be exacerbated over water) due to spreading of the clutter spectrum as a result of movement of the sensor platform as might occur with an airborne radar. The latter mechanism is most easily understood by considering a side-looking radar mounted on an aircraft.

Stationary clutter located at positions along a boresight line at precisely 90 degrees to the platform velocity vector will have no velocity component parallel to the range coordinate and, therefore, no associated Doppler. This happy situation represents only a small portion of the area that is illuminated by a typical sidelooking antenna as any finite aperture antenna will have an associated azimuth spread associated with its receive pattern. The end result is that any clutter within the azimuth beam spread that is located in front of, or behind, the boresight line will have an apparent component of motion along its corresponding range vector and, therefore, a corresponding Doppler. Since a single channel system has no way of further refining the direction of arrival, these clutter returns will be indistinguishable from a moving target located elsewhere in the antenna beam pattern that have an equivalent Doppler velocity. This effect is referred to as Doppler beam spreading.

It is difficult to apply standard thresholding techniques to the endoclutter conditions due to the nonflat or coloured character of the clutter spectrum. This section describes the application of adaptive filtering techniques that act to whiten the clutter spectrum prior to application of detection thresholding, thereby simplifying the threshold calculation. This approach also relies on the use of secondary samples, but in this case it uses the secondary data to form an estimate of the clutter covariance matrix. This can be viewed as the coherent analogy of the CA-CFAR techniques discussed in Section 9.4.2. The end result after filtering is the desired signal in white noise detection problem.

The adaptive coherent detection techniques are based on the optimum Neyman–Pearson LRT. Recall that the object of the Neyman–Pearson test is to maximize the P_d for a given P_{fa}. Let z be the M-dimensional complex variable vector and $p(z|H_1)$ and $p(z|H_0)$ be the clutter plus target and clutter alone M-dimensional pdfs, respectively.

The optimum formulation of the LRT is determined by optimizing with respect to any unknown variables.

There are several coherent detectors that can be deduced using the LRT formulation. The Kelly detector [16], perhaps the best known of the coherent detectors, is described below to illustrate the general approach. The Kelly detector uses both CUT data and secondary data drawn from the surrounding range cells. The following conditions are assumed to apply during the Kelly detector formulation. Secondary data is presumed to be free of targets and the secondary clutter is presumed to have statistical properties identical to the CUT, specifically the same covariance matrix. The Kelly detector further assumes that the in-phase and quadrature components of the received signals are jointly Gaussian processes, i.e., the amplitude is Rayleigh distributed. This implies that

$$p(z \mid H_o) = \frac{1}{\pi^m \det R} \exp(-z^{T*} R^{-1} z),$$

$$p(z \mid H_1) = \frac{1}{\pi^m \det R} \exp(-(z - bs)^{*T} R^{-1} (z - bs)),$$

(9.16)

where z is an m-dimensional vector of values at a given location over a specified period of time commonly referred to as the coherent processing interval (CPI), b is the (unknown) target amplitude and R is the covariance matrix. s is an m-dimensional steering vector of complex values which describes the phase progression of the target returns across the CPI. Typically the phase is presumed to correspond to a constant velocity or, equivalently, constant Doppler target signature but can, in theory, be tailored to any phase progression. The steering vector is typically normalized to unity. The combined pdf of the CUT and secondary data is given by

$$p(z, z_1, ..., z_k \mid H_0) = \frac{1}{\pi^m \det R} \exp(-tr(R^{-1} T_0)),$$

$$p(z, z_1, ..., z_k \mid H_1) = \frac{1}{\pi^m \det R} \exp(-tr(R^{-1} T_1)),$$

(9.17)

where

$$T_0 = \frac{1}{K+1} \left(zz^{*T} + \sum_{k=1}^{K} z(k) z^{*T}(k) \right),$$

$$T_1 = \frac{1}{K+1} \left((z - bs)(z - bs)^{*T} + \sum_{k=1}^{K} z(k) z^{*T}(k) \right)$$

(9.18)

Here z is an m-dimensional vector corresponding to measurements taken within the CUT while $z(k)$ corresponds to a secondary data vector measured at each range cell, k, in the background with K secondary vectors in total. When maximization of the pdfs for H_0 and H_1 is carried out individually, it is found that the maximizing positive-definite matrices are T_0 and T_1, respectively. The maximum-likelihood (ML) estimate of the unknown signal amplitude is obtained by extremizing T_0 with respect to b so that the resulting ML estimate of b is given by

$$b_{ML} = \frac{s^{*T} R^{-1} z}{s^{*T} R^{-1} s}.$$ (9.19)

The ML estimate of R is the sample covariance matrix (SCM), and it follows from considering

$$p(z_1, \ldots, z_k) = \frac{1}{\pi^m \det R} \exp(-tr(R^{-1} T_0)),$$

$$= \frac{1}{\pi^m} \exp(-tr(\ln R + R^{-1} T_0)),$$ (9.20)

where the identity $\det R = \exp(tr \ln R)$ has been used. Extremizing with respect to R (easiest in the diagonal basis of R) leads to the ML estimate of the covariance matrix $\hat{R}$ defined as

$$\hat{R} = \frac{1}{K} \sum_{k=1}^{K} z(k) z^{*T}(k).$$ (9.21)

When these ML estimates are substituted into the LRT, the following detector results:

$$\frac{|s^{*T} \hat{R} z|^2}{(s^{*T} \hat{R} s)\left[1 + \frac{1}{K} z^{*T} \hat{R}^{-1} z\right]} > \eta.$$ (9.22)

This is referred to as the Kelly GLRT (G for generalized) and the left-hand side is termed the Kelly GLRT test statistic [22]. If we exclude the background samples from the above definition, i.e.,

$$p(z \mid H_0) = \frac{1}{\pi^m \det R} \exp(-tr(R^{-1} T_0)),$$

$$p(z \mid H_1) = \frac{1}{\pi^m \det R} \exp(-tr(R^{-1} T_1)),$$ (9.23)

a simpler detector structure arises

$$\frac{|s^{*T} \hat{R}^{-1} z|^2}{(s^{*T} \hat{R}^{-1} s)} > \eta.$$ (9.24)

This is referred to simply as the GLRT in contrast with the Kelly GLRT.

Both detector structures presented so far presume that the in-phase and quadrature components of the clutter are jointly Gaussian processes. Often, clutter may be better modelled using compound Gaussian distribution as described in Section 9.3. To extend this concept to the CPI measurement vector, we introduce the concept of the spherically invariant random process (SIRP) where all pulses within a CPI have the same underlying power [23]. An analysis similar to that carried out above for the Kelly detector is unfeasible as it requires multidimensional integrations over the underlying power or texture. However by presuming the GLRT structure and substituting a ML estimate of the underlying power, the following suboptimal adaptive linear quadratic (ALQ) detector can be shown to result [23]

$$\frac{|s^{*T}\hat{R}_{NSCM}^{-1}z|}{(s^{*T}\hat{R}_{NSCM}^{-1}s)(z^{*T}\hat{R}_{NSCM}^{-1}z)} \overset{H_1}{\underset{H_0}{\gtrless}} \eta, \tag{9.25}$$

where $\hat{R}_{NSCM}$ is the normalized sample covariance matrix (NSCM)

$$\hat{R}_{NSCM} = \frac{m}{K}\sum_{k=1}^{K}\frac{z_k z_k^{*T}}{z_k^{*T}z_k}. \tag{9.26}$$

Note that the NSCM is insensitive to changes in the underlying power of the background cells due to the normalization with respect to the ML estimate of the underlying power of z. This property may be useful when large background sample sizes are required to help compensate for changes in underlying clutter power.

A variety of other formulations and techniques exist to address additional violations of the conditions assumed during the derivation of the Kelly detector. Most can be shown to produce superior results against simulated clutter possessing appropriately matched conditions, but often produce inferior or varying performance when applied to real data. As with the noncoherent design process, the radar design engineer must be careful when choosing a coherent detection approach to ensure that it is properly matched to the clutter and target conditions of the actual data. In addition, in cases where conditions are quite variable, it is often more effective to use a simpler but more robust detector structure.

9.5 Space-Time Adaptive Processing

To this point we have examined increasingly complex detector structures which exploit increasing levels of information in the radar return measurements. In Section 9.4.2, we introduced single channel noncoherent detectors which relied solely on the amplitude of the measured returns as the discriminator of a target's presence. Essentially, these approaches employ no form of clutter suppression and, as such, the target cross-section must be relatively large with respect to clutter cross-sections to ensure detection. In Sections 9.4.3 and 9.4.4 we further refined the detection process by considering the motion of the target and the corresponding Doppler shift it imposes on the radar returns. By preserving the signal phase in a measurement we are able to discriminate between the phase histories of the target and clutter over time. When the target signature is exoclutter this is a particularly effective way of suppressing the clutter contribution. Unfortunately, as discussed earlier for the case of radars mounted on moving platforms, the effect of Doppler spreading can cause the clutter spectrum to occupy a significant portion of the signal bandwidth and many targets will not possess Doppler velocities that cause their signals to fall within the exoclutter region.

STAP represents a further increase in dimensionality whereby an array of spatially distributed antenna phase centres are used to capture multiple measurements of the return signal for each pulse. The addition of this third dimension corresponds to construction of the so called data cube, introduced in Section 9.1.

The minimum requirement for STAP is two antenna phase centres separated in space although, in principle, the STAP formulation allows for any number of phase centers in any desired configuration. It is the combination of multiple measurements in space (i.e., antenna phase centres) and time (measurements within a CPI) which accounts for the "space-time" designation in the name. The adaptive portion of the name label arises due to the clutter cancellation and whitening techniques which are employed. This is strongly analogous to the whitening approach discussed in Section 9.2. They will be discussed further in the sections below.

STAP represents a generalized approach to the problem of exploiting spatial and temporal diversity of measurements to allow clutter cancellation and subsequent target detection. It includes within its general framework the interferometric approaches that were historically exploited to detect moving targets. While the following sections will primarily focus on the STAP approach, it is conceptually helpful to briefly consider the physically intuitive concepts of interferometric processing before moving onto the more mathematical, but more general, formulation of the STAP problem.

9.5.1 Interferometric Methods

Interferometric synthetic aperture radar systems employ more than one antenna that may be displaced along the platform velocity direction (along-track interferometry), or along the direction perpendicular to the platform motion (across-track interferometry). The across-track configuration is used for accurate reconstruction of the height profile on the earth's surface, while the along-track configuration enables detection of ground moving targets and features as well as velocity estimation. This is possible because the interferometric phase (rather the wrapped phase in ($[-\pi,\pi]$) is related to the height values on the ground and radial velocity for the across-track interferometric and the along-track interferometric (ATI) cases, respectively [24,25].

The basic idea of along-track interferometry is as follows. Let the measurement be repeated by two identical radars Δt apart in time given by $\Delta t = d/v_p$, where v_p is the platform speed and d the distance between the antennas. Then, the signals measured by the two identical antennas at the same point in space are given by

$$z_1(t) = A(t)\exp\left(-i\frac{2\pi}{\lambda}2R(t)\right),$$

$$z_2(t) = A(t+\Delta t)\exp\left(-i\frac{2\pi}{\lambda}2R(t+\Delta t)\right).$$

(9.27)

The magnitude and phase of the quantity $z_1(t)z_2*(t+\Delta t)$, termed the interferogram, are called the ATI magnitude and the ATI phase, respectively.

Now, for a stationary target or stationary clutter, the ATI phase is zero since $R(t) = R(t+\Delta t)$. Thus, a nonzero ATI phase indicates a moving target. More precisely, the ATI phase can be modelled as a random variable with zero mean for stationary clutter. Following this, the variance of the phase defines a limit on the minimum detectable velocity since the ATI phase is proportional to the target across-track speed. Thus, a CFAR ATI phase detector follows from computing an appropriate phase threshold for a given false alarm rate. This is analytically done using the expression for the ATI phase statistics. The second-order statistics, or the covariance matrix R, are found to be adequate to define CFAR detection rules for moving targets.

In the original formulation, ATI means co-registration (via interpolation), slow-time SAR filtering and then formation of an interferogram. A natural extension is the multilooked interferogram to reduce speckle, or phase noise [26].

A general view is shaped by first noting that the ATI phase is the off-diagonal element of the two-dimensional covariance matrix (for a two-channel system). This naturally leads one to consider information on movers in other elements of the covariance matrix. Such an investigation naturally leads to other GMTI metrics such as the normalized magnitude of the off-diagonal element, the displaced phase centre antenna (DPCA) metric and recently developed combinations of quantities arising from the eigendecomposition of the covariance matrix. For a recent review of these topics, the reader is referred to Sikaneta [27], where the ATI processing was done in the image domain. However, it can also be performed in the raw data domain as well. While the subsequent sections on STAP focus on the detection of moving targets, it should be noted that general interferometric approach is also applicable to other areas such as polarimetry (target discrimination/classification [24]) and across-track interferometry (for DEM generation).

Traditionally many of the interferometric approaches evolved prior to and independently of the formulation of STAP and, as such, the terminology and formulations of the approaches are often somewhat different to that encountered in the literature on STAP. Nevertheless the interferometric approaches can be framed within the context of STAP. The interested reader is encouraged to explore the traditional interferometric development as it can provide useful insights on both approaches.

9.5.2 General Theory: Optimal Filters

STAP [28] refers to techniques that adaptively combine signals from multiple channels (N) and pulses (M) to suppress interference due to clutter in airborne (or spaceborne) radar. These techniques have been intensively studied when applied to adaptive interference suppression and are reviewed elsewhere [29–31] in the context of STAP. They are based on the filtering techniques common in the domains of spatial adaptive processing, or array processing. The main distinguishing feature of STAP is the exploitation of both spatial and temporal degrees of freedom to cancel clutter more efficiently.

Recall that an optimal filter is defined to be the filter that provides the best performance under a particular criterion, usually specified in terms of extremization of a cost function. Thus, there is a different optimal filter for a different choice of cost function. The performances of the optimum filters in practical applications are not the same. For instance, they may have different sample size requirements for the same level of performance. Another issue in their practical implementation is the computational complexity and numerical stability of the algorithms. Finally, the interference cancellation performance may depend on the type and nature of the interference. It is thus, important to study the different optimal solutions.

The optimum filters may be broadly classified into two categories: inversion and projection. However, there are several solutions in each of the categories depending on the choice of the cost function that has to be extremized.

The maximum SINR filter maximizes the signal-to-interference-plus-noise ratio (ρ), i.e.,

$$\rho = \frac{|\alpha_t|^2 |w^H s|^2}{w^H R w} \leq \frac{|\alpha_t|^2 |w^H R^{1/2}|^2 |R^{-1/2} s|}{w^H R w}, \tag{9.28}$$

where $\alpha_t, w, s,$ and R are target signal amplitude, weight vector, steering vector, and $N_R \times N_R$ interference covariance matrix, respectively. The Cauchy–Schwarz inequality has been employed to obtain the inequality. This implies that w_{SINR}, the optimal filter under maximum SINR criterion, is given by

$$R^{1/2}w_{SINR} = \kappa R^{-1/2}s \Rightarrow w_{SINR} = \kappa R^{-1}s, \ \rho_{max} = |\alpha_t|^2 \, s^H R^{-1}s, \tag{9.29}$$

where κ is a constant. Note that no assumption regarding the statistics of interference (e.g., Gaussian) has been made, so w_{SINR} is optimal under this criterion irrespective of the statistics of the interference. However, when the interference is Gaussian, w_{SINR} also maximizes the probability of detection (P_d) for a given probability of false alarm (P_{fa}).

Alternately, the Wiener filter is defined to be the optimum filter under the least mean square (LMS) error criterion. The data vector, x, is weighted with a weight vector w such that the mean-squared difference between the output and the known desired signal (d) is minimized, i.e.,

$$\text{Minimize } E\left\{ \left| \sum_{k=1}^{N_R} w_k x_k - d \right|^2 \right\} \Rightarrow w_{Wiener} = Q^{-1}v, \quad \text{where } Q_{kl} = E\{x_k x_l\}, \ v_k = E\{x_k d\}. \tag{9.30}$$

Here w_k, x_k are the components of the weight vector and data vector, Q the covariance matrix of the data vector, w_{Wiener} is the optimal Wiener weight vector and $E\{\bullet\}$ denotes expectation operator. Note that the Wiener filter also does not make any assumptions about the statistics. Using the matrix inversion lemma, it can be shown that the Wiener filter is proportional to the maximum SINR filter, and hence they may be considered to be equivalent.

The constant κ in the expression for w_{SINR} above can be specified by imposing the condition that $w_{SINR}^H s = 1$, the resulting optimal weight vector is termed the minimum variance distortionless response (MVDR):

$$w_{MVDR} = \frac{R^{-1}s}{s^H R^{-1}s}. \tag{9.31}$$

This optimal solution also results from minimizing the output power subject to a unity gain constraint in the target steering vector direction, i.e.,

$$\text{Minimizing } E\left\{ |w^H x|^2 \right\} + \zeta\left(w^H s - 1 \right) = w^H R w + \zeta\left(w^H s - 1 \right) \text{ implies } w_{opt} = w_{MVDR},$$

where ζ is a Lagrange multiplier introduced to solve the constrained minimization problem. This constant $\kappa = 1/(s^H R^{-1}s)$ is also referred to as the minimum variance or Capon superresolution spectral estimator.

An important inversion type filter is derived from $\min\limits_{w} w^H R w$, subject to $w^H w = \delta^2$ and

$$|w^H s|^2 = 1 \Rightarrow w_{LSMI} = \frac{(R + \zeta I)^{-1}s}{s^H (R + \zeta I)^{-1}s}, \tag{9.32}$$

where δ^2, ξ are constants and the optimum solution can be easily obtained using the method of Lagrange multipliers. This solution is termed the loaded sample matrix inversion (LSMI) solution and is very important in practical applications. It is shown below that the ad-hoc solution of adding a small amount to each diagonal element is in fact the optimum solution for this problem, not merely introduced for better numerical stability. The LSMI solution can also be viewed as the optimum solution of the constrained optimization problem of maximization of SINR subject to constant norm of w and desired signal response. Typical guidelines for the diagonal loading are the addition of three times the measured noise variance. It also has the added benefit of avoiding the instability of inversion of the SCM when the number of samples is small.

The eigenvector projection (EVP) filter results when the criterion to be optimized is to find the s_{sig} dimensional subspace S that best approximates the data vectors in the mean square error (MSE) sense:

$$E\|X - P_S X\| = \min \Rightarrow E\|P_S X\|^2 = trace\{P_S R\} = trace\{S^H R S\} = \max \quad \text{with } S^H S = I. \quad (9.33)$$

Here X is the data vector and P_S the projector onto the subspace S spanned by the clutter. More specifically, the solution is the subspace spanned by the eigenvectors corresponding to the s_{sig} dominant eigenvalues (or unitarily related to it) and the projection matrix is written as $P_S^\perp = I - SS^H$, with the filter vector $w_{proj} = \alpha P_S^\perp s$. Unlike the inversion based optimum filters, the dimension of the interference subspace to be projected out needs to be known in advance. The number of dominant eigenvectors (by thresholding against the largest eigenvalues) can be ascertained from the eigenspectrum of the interference covariance matrix.

Rewriting the inverse of the covariance matrix in terms of its eigenvectors and eigenvalues, it is seen that the inversion techniques reduce to the projection class, e.g., $w_{MVDR} = w_{proj}$ when the interference to noise ratio is infinite. More generally, the inversion class of weight vectors are a superposition of vectors in the noise and interference subspaces, while the projection class of weight vectors lie entirely in the noise subspace.

9.5.3 Maximum-Likelihood Estimate of Covariance Matrix and Finite Sample Size

Since the asymptotic (or exact) covariance matrix is not known it must be estimated from a given number of samples, K. Typically, it is estimated by taking the average of dyads of the snapshots $X_k, k = 1, ..., K$:

$$R_{est} = \frac{1}{K} \sum_{k=1}^{K} X_k X_k^H \quad (9.34)$$

If the statistics of the samples is that of a multivariate Gaussian distribution, then R_{est} may be interpreted as the ML estimator of the covariance matrix which is complex Wishart-distributed with K-degrees of freedom and parameter matrix R of size N_R.

The finite sample size version of the optimum weight vectors can be shown to be

$$\max_w \frac{|w^H s|^2}{w^H R w} \Rightarrow w^H R_{est} w = \min \text{ and } w^H s = 1 \Rightarrow \hat{w}_{SMI} = \alpha R_{est}^{-1} s \quad \text{(SMI)}$$

$$\min_w w^H R_{est} w \text{ with constraints } w^H s = 1, w^H w = c \Rightarrow \hat{w}_{LSMI} = \alpha (R_{est} + \zeta I)^{-1} s \quad \text{(LSMI)}$$

$$\max_S \sum_{k=1}^{K} \|P_S x_k\|^2 = \max_S K trace\{S^H R_{est} S\}, S^H S = I \Rightarrow \hat{w}_{EVP} = (I - \hat{S}\hat{S}^H) s \quad \text{(EVP)}$$

Where $\hat{S}$ is the finite-sample size quantity corresponding to S.

It has been shown that normalized SINR for SMI is beta-distributed with parameters $(2(N_R - 1), 2(K - N_R + 2))$, i.e.,

$$f_\rho(\rho) = \frac{\Gamma(K+1)}{\Gamma(N_R - 1)\Gamma(K - N_R + 2)}\rho^{N_R}(1-\rho)^{K-N_R+2}, \quad 0 < \rho \le 1. \tag{9.35}$$

The expectation value is $E\{\rho\} = (K - N_R + 2)/(K + 1)$ which leads to the famous result of Brennan, Mallett and Reed that SMI requires $2 \times N_R$ samples for a SINR loss of 3dB (i.e., for $E\{\rho\} = 0.5$) relative to the optimal [32]. This is loosely referred to as "3dB performance" (see Figure 9.4a). Thus, for large N and M, the sample requirement is not practically achievable.

A less well-known result is that the probability density function of the finite sample normalized SINR for LSMI is also beta-distributed with parameters $(2(K - r + 1), 2r)$, where r is the rank of the (asymptotic) interference covariance matrix and it is assumed that $K > r$ [33]. Thus, its pdf is given by

$$f_\rho(\rho) = \frac{\Gamma(K+1)}{\Gamma(K - r + 1)\Gamma(r)}\rho^{K-r}(1-\rho)^r, \quad 0 < \rho \le 1, \tag{9.36}$$

and the expectation is $E\{\rho\} = (K - r + 2)/(K + 1)$. The derivation of this result also assumes that the data vectors are samples drawn from the independent, identically distributed (i.i.d.) multivariate Gaussian random variate, and the load is of the order of the noise power. In addition, it also assumes that the interference level is well-separated from the noise level, an assumption not strictly valid for some interferences like clutter.

The remarkable feature of LSMI is that it requires only $2r$ samples for 3dB performance (see Figure 9.4b). This is the reason why LSMI should be used in practical applications, and not SMI. In addition, it obviously has better numerical stability than SMI.

The probability distribution of the finite sample normalized SINR has also been derived for EVP when the signal level is well separated from the noise level and it has been shown to be identical to that of LSMI [34]. Hence, EVP also requires only $2r$ samples for 3dB performance. However, for some types of interferences, this assumption is not valid. In such cases, it may actually be better to project out fewer interference eigenvectors. Intuitively, this makes sense as the eigenvectors corresponding to the smaller eigenvalues will be estimated poorly. A plausible threshold is presented elsewhere [35–37] for determining the number of eigenvectors to project out.

There also exist projection methods that do not require eigenanalysis. The Hung–Turner projection (HTP) takes the data vectors directly as a basis for the clutter subspace. The statistical analysis of HTP has shown that it converges more slowly than EVP or LSMI. A better solution is the matrix transformation based projection (MTP) which estimates the interference subspace to be $\hat{S} = \hat{R}T = XX^H T$, where T is a random matrix of size $N_R \times s_{est}$. The corresponding projection matrix is

$$\left(I - \hat{S}\left(\hat{S}^H\hat{S}\right)^{-1}\hat{S}^H\right). \tag{9.37}$$

Under assumptions similar to that for the analysis of EVP, it has been shown that the finite sample normalized SINR has the same probability distribution function as EVP or LSMI. Since, MTP also requires only $2r$ samples for 3dB performance and requires fewer computations than EVP, it is an excellent choice of a projection class of algorithm [38,39].

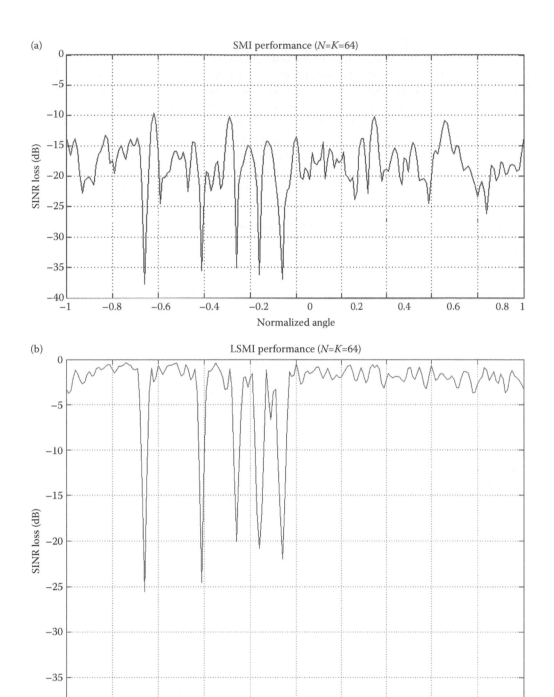

FIGURE 9.4
An example of (a) finite sample size SMI performance, (b) finite size LSMI performance (likewise for EVP or MTP). This is for a 64-element array with five jammers.

Finally, observe that the sample requirements vary for the different optimal processors. The probability distribution functions of the normalized SINR are derived assuming that the data vectors are drawn from an i.i.d. Gaussian random vector, an assumption unlikely to be valid in practise. However, there are no closed-form solutions for the more complicated multi-variate distributions and the sample size properties in those cases will have to be obtained from simulations. Thus, the sample requirements may not be valid for the real-world case of nongaussian and heterogeneous clutter environments. In addition, the steering vector com-ponents are assumed from the terms in a geometric series in exponentials (and perhaps ten-sor product of such vectors in the multidimensional case). Nevertheless, it is useful to note the sample size requirements when comparing the various optimal algorithms.

9.5.4 Interference Types and their Rank Properties

9.5.4.1 Sensor Noise

The white noise at the receiver has the covariance matrix $R_n = \sigma_n^2 I$, which is of full rank NM. Note that there is no need to estimate the covariance matrix since the optimal weight vector is simply the steering vector.

9.5.4.2 Jammer

Next, consider the case of narrowband sidelobe noise jamming which is correlated in angle. This may be suppressed by placing a spatial null at the jammer. Of course, if a mainlobe jammer is suppressed in this way, the spatial null may null, or partially null, the mainlobe target as well. If J is the number of jammers, then the rank of the jammer covariance matrix is JM.

Spatial adaptive processing is adequate for suppressing sidelobe jammers and also for locating exoclutter targets, i.e., targets not buried in clutter. The use of adaptive beam-forming techniques requires special measures to avoid the inclusion of mainlobe clutter during the adaptation process, like using a special listening interval to sample the jam-ming. But that takes away from the radar timeline and creates a vulnerability to a nonsta-tionary jamming environment. Furthermore, adaptive beamforming changes the receive antenna patterns and so affects the clutter that must be subsequently suppressed.

9.5.4.3 Clutter

Ground clutter in a monostatic radar is spread in Doppler due to platform motion. Ground clutter exhibits correlation in both spatial and temporal dimensions or equivalently angle and Doppler.

The most important property of the clutter covariance matrix is its rank. In particu-lar, the rank of the clutter covariance matrix indicates the severity of the clutter scenario and also the number of degrees of freedom required to produce effective cancellation. For the side-looking case, the rank of the asymptotic clutter covariance matrix is given by Brennan's rule:

$$N + \beta(M-1), \quad \beta = \frac{2v_p T}{d}, \tag{9.38}$$

where v_p is the platform velocity, d the phase centre separation, so that β can be viewed as twice (due to two-way propagation) the number of interelement spacings traversed by the platform during one PRI (β is the number of times the clutter Doppler spectrum aliases into the unambiguous Doppler space).

Brennan's rule is easy to understand for integral β. For instance, we see that clutter observations are effectively repeated by different elements on different pulses as the platform moves during a CPI; in fact, there are only independent distinct observations, as predicted by Brennan's rule. See Figure 9.5 for an example. Note Brennan's rule is valid for side-looking uniform linear array (SLULA) even for the range-ambiguous case. Strictly speaking, when β is not an integer, the clutter covariance matrix is no longer singular although many of the eigenvalues are quite small. Then, Brennan's rule merely provides an estimate of the knee of the eigenspectrum. Clearly, the more Doppler ambiguous the clutter, the greater is the number of targets which qualify as slow movers and hence the more difficult is the detection problem, as the clutter then occupies a larger fraction of the Doppler spectrum. The clutter rank can also been estimated in the case of subarrays [40].

In the presence of both cold clutter and narrowband sidelobe jammers, the interference rank for the SLULA case has been shown to be given by $N + (\beta + J)(M-1)$. This is different from the naive reasoning which overcounts the number of independent interference measurements [41].

Note that the estimated clutter covariance matrix rank is always the number of snapshots K (which are never linearly dependent due to their random nature) if the number of snapshots is less than the spatio-temporal dimension. The sample clutter covariance matrix is always full rank, even though the asymptotic covariance is not full rank. Thus, from a practical point of view, it is more important to distinguish the eigenvalues corresponding to the dominant eigenvectors (or the desired signal subspace) from those that correspond to eigenvalues close to the noise floor.

Brennan's rule is extremely important and useful in understanding clutter as seen from a uniform linear array on an airborne platform. The clutter rank characterizes the severity

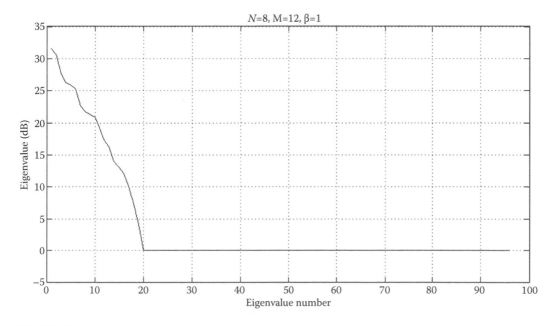

FIGURE 9.5
An example of the eigenspectrum of the asymptotic clutter covariance matrix (N=8, M=12, β=1. Note the agreement with Brennan's rule, i.e., $N+(M-1)=19$ clutter eigenvalues.

of the clutter—the larger the rank, the more severe is the clutter. It is very useful for the projection class of algorithms, although, it often over-estimates the subspace dimension that should be projected out when the number of samples is small. Brennan's rule also motivates the design of the nearly optimal processing algorithms discussed in the next subsection. Finally, note that the general idea that the interference is strong and occupies a small fraction of the total space can be exploited by using compression techniques, such as wavelets [42].

9.5.5 Partially Adaptive STAP Algorithms

Fully adaptive STAP refers to the case when all the available spatial and temporal degrees of freedom are used. However, fully adaptive STAP is typically impractical due to computational load and the sample size requirements discussed above. Near-optimal solutions can still be obtained by exploiting the low clutter rank property. In this subsection, we briefly discuss such partially adaptive STAP algorithms. Essentially, a partially adaptive processor linearly transforms the full processor so that the reduced-dimension adaptive filtering problem can be solved.

From a practical point of view, the partially adaptive STAP algorithms may be broadly classified into the following two categories:

- Predoppler STAP: adaptation performed prior to Doppler filtering.
- Postdoppler STAP: adaptation performed after Doppler filtering.

One may further distinguish between element-space and beamspace depending on whether a beamformer is also used. The advantages of partially adaptive STAP are:

- Sample size requirements are much smaller. This is extremely important since it is often not possible to have a very large number of clutter samples in real world operation due to violation of the homogeneity condition across the background extent.
- Computational load is clearly lower. For instance, a much smaller matrix to invert or upon which to perform eigendecomposition.

Several pre and postdoppler algorithms have been designed and shown to yield near optimal performance but with different strengths and weaknesses. Here we restrict ourselves to widely-used algorithms, including PRI-staggered STAP [43] and joint domain localized (JDL) [44].

For each antenna element in PRI-staggered STAP, there are subCPIs (subcoherent processing intervals) each of length M (i.e., pulses 1 to M, 2 to $M+1, ..., N$ to $N+K'-1$, where K' is the number of subCPIs or temporal taps). Doppler processing is performed on each subCPI for each aperture, producing processed outputs per aperture. The N outputs for each Doppler bin, N being the number of apertures, are then used for adaptive processing. The covariance matrix is of the order $K'N$. A $K'N$ predoppler STAP architecture , constructed similarly, also shows excellent performance. For both configurations, the reason is that the clutter rank is now $N + \beta(K' - 1)$, which is often less than $K'N$. The adjacent-bin postdoppler STAP algorithm adaptively combines the spatial samples from a cluster of adjacent Doppler bins. It has a rank behavior similar to PRI-staggered STAP, but (unlike PRI-staggered STAP) only if no windowing is applied.

The JDL algorithm is a postdoppler, beamspace approach. The received data is transformed into the angle-Doppler domain using a two-dimensional DFT, and the adaptive processing is done by restricting it to a localized processing region (LPR) in the transform domain. In the original formulation, it was assumed that the receiving antenna was an equi-spaced linear array of ideal, isotropic, point sensors, so that the targets are ideally localized in a single angle-Doppler bin. This was extended to the real antenna case by replacing the DFT with the measured steering vector, though the target information is now spread in the angle domain. The spread in target information is accounted for by JDL, resulting in significantly improved performance [45,46].

The postdoppler STAP algorithms are usually preferred. Some of the reasons for considering the postdoppler class of algorithms (over predoppler) are:

- It is possible to detect multiple targets in the test cell (same range), which differ in Doppler, even if they are close in angle. This is because a separate adaptive problem is solved for each Doppler bin.

- The clutter eigenvalues fall off more sharply in the Doppler domain than in the time domain, leading to better signal preservation after clutter cancellation.

- The clutter suppression need only be performed for those Doppler bins that are within the antenna's clutter bandwidth (endo-clutter), thus limiting the computational requirements of STAP. This leads to significant savings especially since the exo-clutter region can be quite large for the narrow beamwidth case.

- In the presence of many targets, postdoppler STAP is more suitable as it leads to less likelihood of targets being in the guard bins used to estimate the SCM [47].

The general STAP approach to GMTI may thus be summarized as follows:

- Algorithm architecture: choice of CPI and aperture combination;
- Domain of adaptation: predoppler or postdoppler and element-space or beam-space;
- Weight training: estimation of covariance matrix from guard and reference bins (secondary data set);
- Weight computation approach: application of weight (e.g., LSMI or EVP) to data vector;
- CFAR normalization and detection.

9.5.6 Some Practical Implementation Issues

9.5.6.1 Data Selection

The ML estimate of the covariance matrix relies on the successful estimation of the covariance matrix in the CUT using independent identically distributed (i.i.d.) secondary data samples that have similar stochastic and spectral properties. In the discussion so far, it has been assumed that the clutter is homogeneous and Gaussian across the CUT and background samples. In practice, this is typically not the case due to due to the presence of anomalous clutter events such as spikes or due to fluctuations of the underlying power (or texture) clutter returns as discussed in Section 9.3. The methodology for dealing with these inhomogeneities varies depending on which of the above two sources is dominant [48].

Despite the fact that background cells are typically selected from nearby range cells so as to ensure the maximum degree of homogeneity with the CUT, clutter spikes or other unwanted outliers can still be present in the clutter background and will distort the covariance matrix estimates resulting in degraded detection performance. Clearly it is desirable to have a method to suppress the effect of these outliers during computation of the SCM [49,50].

Past approaches to outlier suppression have focussed on the pruning (i.e., removal) of suspect background samples from the total background set available. The remaining samples are then used to form the final desired SCM. Theoretically one could evaluate the performance of all possible background sets to determine which provide the best performance, but this approach is computationally expensive. One suboptimal simplification that is often employed is to evaluate the sample test factors (STF), $z_k^H \hat{R}_K^{-1} z_k$, for each background sample using the SCM formed using all available background samples (i.e., prior to pruning) and then pruning the background samples associated with the largest STFs. One drawback with the pruning approach is that it requires that a decision on the number of samples to prune be made prior to processing; furthermore it treats all pruned samples as being equally undesirable.

An alternate approach is the de-emphasis weighting approach in which the contribution of each background sample to the SCM is controlled through application of a sample weight during the averaging process. The applied weights are derived as a function of the STFs with background samples with high STFs being assigned a low weight (i.e., weaker effect on SMI) then background samples with low STFs [51].

As discussed above, the counterpart to outlier contamination occurs when the background samples and CUT are presumed to have similar spectrums and be i.i.d. up to a scaling factor. In this case a situation can occur whereby clutter discretes may be undernulled due to the inclusion of lower power background samples in the covariance matrix calculations. This may be remedied by power selected training (PST), which selectively includes only the higher power snapshots. While this approach can lead to over-nulling of weaker clutter the degradation in detection performance is usually minimal. A more serious problem arises due to the inclusion of targets in the background sample set as this can lead to target self-nulling and overall degradation in clutter cancellation performance. Phase-selected training seeks to address this problem by excluding target contributions from the adaptive weight training based on angle or phase. The phase difference between antenna phase centers across the array can be used to estimate phase of a return, the quality of the estimate improves with the strength of the incoming signal (snapshots selected on basis of power). If the measured phase differs significantly from the expected phase (angle) of clutter for the given Doppler bin, it should not be included in STAP training. Thus, strong targets can be excised from the STAP adaptive weight training.

9.5.6.2 Knowledge-Aided STAP

Real-world ground clutter consists of quite complex man-made objects (like buildings, power lines, roadways). The presence of large discretes and abrupt discontinuities leads to significant loss when using STAP, if uncompensated. In addition, the appropriate STAP algorithm to choose might depend on the clutter characteristics. The idea behind knowledge-aided STAP (KA-STAP) is conceptually straightforward. Databases from several sources (like digital terrain and elevation data (DTED)) and SAR have information about the environment useful in the STAP adaptation process. For example, resolution cells corresponding to potentially competing ground traffic can be excised from the training data resulting in improvement in detection performance of targets with Doppler frequencies similar to the competing background traffic. From a statistical perspective, note that this is equivalent to incorporation of priors in the statistical estimation process, as in Bayesian statistics.

Knowledge-aided STAP incorporates additional information that partially characterize aspects of the clutter environment. This includes databases, GPS data, and information from other sensors to accomplish some of the following: enhance training data selection, preadaptively null discretes, minimize the number of unknowns requiring estimation, and condition the space-time data through prefiltering. Knowledge of site-specific clutter will further enhance STAP capability. For a recent review on the subject, see [52].

9.5.7 Minimal STAP: $\Sigma - \Delta$ and Multimode STAP

9.5.7.1 $\Sigma - \Delta$ STAP

Typically, STAP has been studied for large planar arrays; studies in GMTI using reflector antennas to effect clutter suppression are rare. A notable exception is $\Sigma - \Delta$ STAP that employs the sum and difference beams in monopulse systems [53]. A monopulse system is very affordable since very low sidelobe $\Sigma - \Delta$ beams can be designed via analog beamforming; there is no need for expensive digital beamforming. Note that $N = 2$ is the minimum number of apertures required for clutter cancellation.

The advantages of $\Sigma - \Delta$ STAP include:

- Use of a real difference beam eliminates the need to calibrate and store the steering vector information as for subarray STAP.
- Even if both subarrays are designed to have low sidelobes, combining the two does not ensure a sum beam of the same sidelobe level; the gain and sidelobe levels of the transmit beam have a direct impact on the overall system performance.
- Applicability to existing systems, affordability (analog beamforming), data efficiency (sample size requirements), easier channel calibration, response pattern, and much lower computational load.

However, $\Sigma - \Delta$ STAP has limitations, since it cannot include spatial ambiguities (low PRF and/or high clutter Doppler due to fast platform $\beta > 1$), it can suffer from beam pattern mismatch, and jammer cancellation is impossible.

9.5.7.2 Multimode STAP

Reflector antenna feedhorns are normally fed by rectangular waveguides that can propagate TE and TM modes. Conventional feedhorns use the dominant TE11 mode. The higher order modes are typically used for beam shaping or tracking. To demonstrate our concept, we use a circular horn, fed by a circular waveguide. The main idea behind these multimode techniques is to implement secondary modes to control the radiation characteristics of the feedhorn.

Here, we use them to displace the reflector phase centre. By manipulating the amplitudes and phases of the dominant and secondary modes of the circular waveguide and combining them, two physically separated phase centres are realized. In our example, we use TE11 and TE21 beams, which correspond to the Σ and Δ beams, respectively [54]. For horizontal polarization, the TM01 mode is used instead of TE21. The horizontally

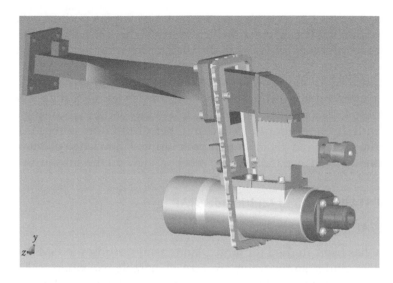

FIGURE 9.6
A horizontally polarized multimode feedhorn with circularly symmetric TM01 port (bottom right) and TE11 (top left).

polarized feedhorn is shown in Figure 9.6; it has been successfully employed in several flight trials. The electromagnetic and signal processing aspects of multimode STAP are discussed in [55–57].

Finally, note that a natural extension of the above follows if more than two modes can also be controlled on receive at a single physical antenna, be it a reflector antenna or a phased array. Unlike in conventional antenna arrays, the receive beams need not be parallel. The important point is the availability of independent measurements of voltages.

9.6 Detect-Before-Track (DBTk) and Track-Before-Detect (TkBD)

9.6.1 Preliminary Comments

To this point we have only considered the basic problem of detecting a target in clutter. While important, this aspect typically comprises only one component of an overall processing chain to obtain the final which (in most instances) is the association of a track with the target. The most general expression of the track problem is as follows. Consider an object moving along a trajectory where we are given a sequence of snapshots of the "mostly stationary" region of space in which it evolves. We would like to extract from the measurement sequence the coordinates of the object of interest.

The traditional approach to associating a track with a target relies on a multiphase strategy in which target detections, obtained using the approaches outlined in the proceeding sections, act as the input to various target tracking techniques, such as the Kalman filter, probabilistic data association (PDA) and multiple hypothesis testing (MHT). These follow-on trackers output a target trajectory or track for display to the operator. While the detection and tracking stages must be designed to complement each other, they are nevertheless distinct processing steps. This approach is hereafter referred to as detect before track (DBTk).

The DBTk methodology has been a design mainstay for many years and has a proven record of performance provided certain conditions are met. The first requirement is that the target cross-section produces a sufficiently high SINR to ensure that detections occur with sufficient reliability to provide adequate updates to the tracker. This requirement for sufficiently high probability of detection is of course constrained by the need to keep false alarm rates at low enough level to ensure that tracker performance is not compromised. For a target with an inherently high SINR these conditions are easily met on each measurement. However, as the target cross-section decreases it becomes necessary to employ some form of integration to improve the SINR prior to detection thresholding. The application of integration becomes problematic for a moving target as the target migrates through resolution cells over the course of the integration period thereby causing measurements without a target signal to be collapsed into the integrated value.

A common approach to addressing the target migration problem is to apply a linear track before detect scheme in which a range of potential constant velocities are postulated. For each postulated velocity the resolution cells that align with the resulting trajectory are integrated. While this approach leads to a slight increase in false alarms due the multiplicity of integrated outcomes, it can work well provided the constant velocity requirement is met.

The DBTk and linear track before detect approaches will fail when applied to highly manoeuvrable and dim targets such as a small manoeuvrable speedboat due to the constraints noted above. Interestingly, if an operator were tasked with closely monitoring the PPI display of raw radar return amplitudes corresponding to such a target, they might, in fact, be able to easily detect and visually track the target (although they might not be able to achieve this feat if they were forced to spread their concentration over a large surveillance area) while the automatic detection and tracking would fail. Why should the operator succeed where the DBTk will not? The answer again comes down to a question of available information. In DBTk the detection stage makes a hard decision on the presence or absence of a target (i.e., H0 versus H1). When a detection is declared, the location of the detection is passed to the tracker input. In simple DBTk processors other available information, such as details of the target strength and its spatial distribution, are then discarded. In truth, many trackers try to recover some of this lost information by accepting additional parameters such as estimated detection SINR, but in general terms the information that is passed on to the tracker is less rich in detail than that which is presented to the operator's eye.

It is these observations which motivate the use of track-before-detect (TkBD) techniques. The most obvious distinction between DBTk and TkBD relates to the output products that each approach produces. The DBTk output corresponds to an estimate of the current state of a target with error bounds, e.g., position of the target. If one target is present then one estimate of its current state will be output with error bounds on each state variable. In contrast, the output of TkBD is a distributed probability across the full state space dimensionality which specifies the probability that the target could be occupying any of the available state positions within the available state space. For example, consider a simple 2D state vector in which only the x and y position of the target is specified, the output of the TkBD processor would correspond to a probability distribution of the target occupying each of the x-y positions in the field of view.

A similar distinction exists when examining the inputs to the TkBD and the tracker stage of the DBTk. For DBTk the measurements supplied to the tracker are local in nature, they typically represent a direct measurement of at least a subset of the state variables, for example the x and y location of the detection, or a measurement that is linearly related to a state variable such as target Doppler is a function of velocity. In contrast the measurements used in track before detect are nonlocal is a function of represent measurements from across the entire field of view or the "swath" in the case of a radar. This implies that the tracker input measurements in DBTk are linearly related to the state variables while the measurements input to TkBD have a nonlinear relationship to the state. This concept is illustrated in Figure 9.7. The measurement that is supplied to the input of the tracker stage in the DBTk is the location of the detection while the measurement that is supplied to the TkBD is the record of signal return strengths that were measured as the radar swept across the field of view. The latter measurement contains information on the strength of the returns and distribution of the returns across the entire field of view while the former provides only a single detected location which has a certain probability of being a false alarm and associated error bounds.

Both DBTk and TkBD are Bayesian filtering problems; however the DBTK is usually a linear filtering problem whereas the TkBD is typically nonlinear due to the relationship between the state and measurement. A reader familiar with traditional tracking problems will note that the underlying dynamic state equation for some situations such as a ballistic missile experiencing drag on re-entry may, in fact, also be nonlinear. This does not alter the usefulness of the distinction outlined above as the typical approach for handling nonlinearities in DBTk is to linearize the dynamic state equation around a specified state point. The key point for the reader to bear in mind is that TkBD, as defined in this chapter almost always implies a nonlinear measurement equation. The distinction should become clearer to the reader in the mathematical formulation presented in the following sections.

Two approaches which are commonly applied to the solution of nonlinear filtering problems are grid-based methods based on: solving the Fokker–Planck–Kolmogorv

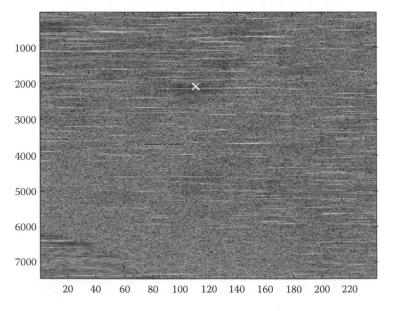

FIGURE 9.7
Comparison of measurement information passed on to tracker linear DBTk approach and nonlinear TkBD approaches. The white "x" in the Figure represents the x-y position of the target in the frame which is the only information passed by the DBTk detection stage. In contrast the full underlying amplitude distribution is passed to the TkBD process in which the spread amplitude signature of the target is clearly visible.

forward equation for the continuous-discrete model (see, for example [58,59]) and the Duncan–Mortensen-Zakai equation for the continuous–continuous model; and particle filters (PF) [60–63]. The distinctions between the models are clarified in the following discussion on the nonlinear Bayesian filtering problem. Other previously developed techniques for TkBD include direct ML, the Hough transform and dynamic programming [64]. These latter methods, while effective, are batch methods with some limitations (e.g., linear trajectories for Hough transform) which generally require discretisation of the state space and are very computationally intensive. As such they are not discussed further in this chapter.

9.6.2 Bayesian Filtering Theory: A Brief Overview

Bayesian filtering is a general approach to solving a broad class of problems that seeks to estimate an indirectly observable state process from the measurement history of a related stochastic process [58]. TkBD is a specific example of one type of problem for which Bayesian techniques can be employed. The general Bayesian approach is summarized briefly in the following sections and its application to TkBD is clarified.

The filtering problem is as follows. The time evolution of the state, or signal of interest, is assumed to be well-described by a discrete-time (continuous-time) stochastic process. However, the state process is not directly observable, i.e., the state process is a hidden continuous-time (or discrete-time) Markov process. Instead, what is measured is a related discrete-time (or continuous-time) stochastic process termed the measurement process. The filtering problem is to estimate the state of the system given the measurements.

The conditional probability density function, or the probability density of the state given the measurement history, provides a complete solution in a probabilistic sense. From the conditional probability density various state estimators can be obtained. Frequently, the conditional mean is of interest since it is the optimal estimator in the least squares sense.

Consider first the discete-time nonlinear filtering problem where the state and measurement processes are described by

$$x(k+1) = f(x(k), v(k)),$$
$$y(k) = h(x(k), w(k)). \tag{9.39}$$

where f is the function describing the evolution of the state and h is the function describing the relationship between the state and measurement. v and w correspond to the state and measurement noise, respectively, and reflect our inability to fully specify the above functions in a deterministic manner. An example of a state vector x is the location and velocity of a target. In contrast the measurement y can take very different forms in the linear and nonlinear cases. For the linear case it could correspond to an x-y location of a detection. In the nonlinear track before detect problem it may correspond to the distribution of measured amplitudes across the surveillance swath.

Assume that the state transition probability density function, $p(x(k+1)|x(k))$, and $p(y(k)|h(x(k))$ are known, and let $Y(k)$ denote the measurement history till time k. Then, the formal solution of the problem consists of the prediction step

$$p(x(k+1)|Y(k)) = \int p(x(k+1)|x(k))p(x(k)|Y(k)), \tag{9.40}$$

and a correction step that incorporates the measurements:

$$p(x(k+1)|Y(k+1)) \propto p(y(k+1)|x(k+1))p(x(k+1)|Y(k)), \quad (9.41)$$

where the constant of proportionality follows from the requirement that the conditional probability density be normalized. It is important that the reader fully understand the implications of the above two equations. Practically it means that one can first calculate the predicted evolution of the state from one time step to the next based on the state model alone and then filter this predicted state based on the measurement and its relative uncertainty.

In continuous-discrete filtering theory (see, for example [58]), the state model is given by the Itô stochastic differential equation of the form

$$dx(t) = f(x(t), t)dt + e(x(t), t)dv(t). \quad (9.42)$$

Here $x(t)$ is a n-dimensional real-valued process, $f \in R^n, e \in R^{n \times p}$ and v is a R^p-valued Brownian process with covariance $Q(t)$. The forward diffusion operator, L, of the state process generated by the state process is given by

$$L(\bullet) = -\sum_{i=1}^{n} \frac{\partial(\bullet f_i)}{\partial x_i} + \frac{1}{2} \sum_{i,j=1}^{n} \frac{\partial^2[\bullet(eQe^T)_{ij}]}{\partial x_i \partial x_j}. \quad (9.43)$$

The measurement process is a discrete-time process such that $p(y(t_k)|x)$ is assumed known.

The continuous-discrete filtering problem is solved as follows. Let the initial distribution be $\sigma_0(x)$ and let the measurements be collected at time instants $t_1, t_2, \ldots, t_k, \ldots$. We use the notation $Y(\tau)$ to denote all measurements prior to time τ. Then, at observation at time t_k, the conditional density is given by

$$p(t_k, x|Y(t_k)) \propto p(y(t_k)|x)p(t_k, x|Y(t_{k-1})), \quad (9.44)$$

and $p(t_k, x|Y(t_{k-1}))$ is given by the solution of the Fokker–Planck–Kolmogorov forward equation (FPKfe)

$$\frac{\partial p}{\partial t}(t, x) = L(p(t, x)), \quad t_{k-1} \le t \le t_k, \quad (9.45)$$

with initial condition $p(t_{k-1}, x|Y(t_{k-1}))$ i.e., we must integrate the above partial differential equation over time to solve for the evolved probability.

9.6.3 Signal and Measurement Models in the Track-Before-Detect Problem

While in principle Bayesian formalism is powerful enough to handle nonlinear state models, it is frequently the case that the best overall performance for the application of TkBD against real targets is obtained by using linear constant velocity (CV) or constant acceleration (CA) state model. In the following discussion we restrict our development to the planar CV model so as to limit the complexity of the discussion and more clearly illustrate the necessary conceptual points. The extension to additional dimensions and/or the CA model is reasonably straightforward.

In the continuous formulation, the four-dimensional CV state model is given by

$$
\begin{bmatrix} dx_1(t) \\ dx_2(t) \\ dx_3(t) \\ dx_4(t) \end{bmatrix} = \begin{bmatrix} 0 & 1 & 0 & 0 \\ 0 & 0 & 0 & 0 \\ 0 & 0 & 0 & 1 \\ 0 & 0 & 0 & 0 \end{bmatrix} \begin{bmatrix} x_1(t) \\ x_2(t) \\ x_3(t) \\ x_4(t) \end{bmatrix} dt + \begin{bmatrix} 0 \\ \sigma_2 dv_2(t) \\ 0 \\ \sigma_4 dv_4(t) \end{bmatrix}. \tag{9.46}
$$

Since the CV state model is linear in the state variables, it is straightforward to integrate over the time period T to obtain the resulting state model for the discrete-time case,

$$
\begin{bmatrix} x_1(k+1) \\ x_2(k+1) \\ x_3(k+1) \\ x_4(k+1) \end{bmatrix} = \begin{bmatrix} 1 & T & 0 & 0 \\ 0 & 1 & 0 & 0 \\ 0 & 0 & 1 & T \\ 0 & 0 & 0 & 1 \end{bmatrix} \begin{bmatrix} x_1(k) \\ x_2(k) \\ x_3(k) \\ x_4(k) \end{bmatrix} + V(k), \tag{9.47}
$$

where the covariance of the noise, Q, is given by

$$
Q = E\{V(k)V(K)^T\} = \begin{bmatrix} T^3/3 & T^2/2 & 0 & 0 \\ T/2 & T & 0 & 0 \\ 0 & 0 & T^3/3 & T^2/2 \\ 0 & 0 & T^2/2 & T \end{bmatrix}. \tag{9.48}
$$

In the above we have assumed that σ is one with no loss of generality.

The CV model FPKfe to be solved is then given by

$$
\frac{\partial u}{\partial t}(t,x) = \left(\frac{\sigma_2^2}{2} \frac{\partial^2}{\partial x_2^2} + \frac{\sigma_4^2}{2} \frac{\partial^2}{\partial x_4^2} - x_2 \frac{\partial}{\partial x_1} - x_4 \frac{\partial}{\partial x_3} \right) u(t,x), \tag{9.49}
$$

where u is the unnormalized state prediction probability distribution, also know as the prior.

To perform the measurement update it is necessary to calculate the measurement probability distribution, $p(y(t_k)|x)$, for all points within the state space. This function corresponds to the probability that the observed measurement $y(t_k)$ could have arisen at any point in the state space. We obtain the final filtered state probability distribution, also known as the posterior, after updating the prior by multiplication with $p(y(t_k)|x)$. In general the solution for the prior and posterior cannot be obtained as a closed form solution and some sort of point wise approximation is employed. The following sections illustrate two of the more common approaches to solving the state evolution portion (i.e., FPKfe) of the nonlinear filtering problem.

9.6.4 Computational Techniques

9.6.4.1 *Finite-Difference Method with Multiplicative Operator Splitting*

As indicated above, except for some very specific problems, the FPKfe cannot generally be solved in closed form. As a result one is generally forced to resort to numerical approaches which provide an approximation (hopefully a very accurate one) to the desired solution.

A naïve approach to the problem might utilize a straightforward discretization of the available state space into a multidimensional grid followed by direct solution of the FPKfe using finite difference or finite element techniques. In practice, this is generally not feasible since for any reasonable dimensionality of the state space and grid size the required computational effort becomes exponentially large. Practically, some approximation method is required. However, one must ensure that whatever approximation method is utilized the requirements for consistency, stability, positivity and convergence are maintained. A thorough discussion of these issues is beyond the scope of this chapter, but it is useful to examine the concept of multiplicative operator splitting as it provides an easily understandable and commonly used technique which greatly decrease the required computational power while meeting the above criteria.

Consider the FPKfe for the CV model given above. It requires a solution of a PDE of the following form:

$$\frac{\partial u}{\partial t}(t, x) = \sum_{i-1}^{s} L_i u(t, x). \tag{9.50}$$

Two flavours exist. In the forward Euler explicit scheme (see, for instance, [65]), Equation 9.13 is numerically solved using the following approximation:

$$\frac{u(t + \Delta t, x) - u(t, x)}{\Delta t} = \sum_{i=1}^{s} L_i u(t, x), \tag{9.51}$$

so that

$$u(t + \Delta t, x) = (1 + \Delta t \sum_{i=1}^{s} L_i) u(t, x),$$

$$= \prod_{i=1}^{s} (1 + \Delta t L_i) \, u(t, x) + O(\Delta t^2). \tag{9.52}$$

Thus, the summation structure can be well approximated by a product formulation which allows independent, sequential evaluation along each of the state space axis.

Note that if the time step is too large, the explicit scheme is unstable. This problem is evaded by splitting up the time interval between measurements into NT time steps prior to applying the forward Euler scheme, i.e.,

$$\left(1 + \Delta t \sum_{i=1}^{s} L_i\right) \approx \left(1 + \frac{\Delta t}{N_T} L_i\right). \tag{9.53}$$

Furthermore, stability of the discretization of the convection operators requires that "upwind differencing" be used for the first order derivative operator. This also ensures that the probability remains positive.

In the backward Euler (or Laasonen) implicit scheme, the following approximation is made:

$$u(t + \Delta t, x) = \prod \left(1 - \Delta t \sum_{i=1}^{s} L_i\right)^{-1} u(t, x). \tag{9.54}$$

The backward scheme is superior in terms of stability, but it is not as accurate as the forward scheme.

9.6.4.2 Particle Filtering

In the grid based approaches the computational issues discussed above are exacerbated by the fact that the state space grid points are predefined and, as such, cannot be partitioned according to the probability mass in any region. This can lead to significant waste of computational resources as a significant amount of computational effort is expended computing the updated state probability at points where it has become vanishingly small. One approach to mitigate this effect is to use sparse tensors, i.e., multidimensional arrays with most elements zero, so that there is a large savings, in memory and computational requirements. Another approach that has generated significant interest in recent years is to utilize PF.

The particle filter produces a Monte Carlo estimate of the resulting posterior probability distribution based on a state sample or particle distribution. The idea is analogous to drawing samples or particles from a probability distribution and plotting the particles at their corresponding locations within the state space grid. Areas in which the plotted points or particles are densely clustered will correspond to areas of high state probability while areas in which few or no particles are present will indicate low state probabilities. Mathematically the Monte Carlo estimate of the distribution can be expressed as follows:

$$p(X_k \mid Y_k) \approx \sum_{i=1}^{N} w_k^i \delta(X_k - X_k^i), \qquad (9.55)$$

where X_k represents the sequence of all target states up to time k, X_k^i represents a sample, i, of the corresponding state space that has be drawn from a known distribution and Y_k represent the sequence of measurements up to time k. w_k^i represents a currently undefined weighting factor which modifies the importance of a given sample. If the actual distribution corresponding to $p(X_k \mid Y_k)$ is used to draw the samples then the above weights are all equal to one. In many instances, however, the distribution $P(X_k \mid Y_k)$ is too complex to allow the drawing or direct generation of samples corresponding to it. In this case one is forced to utilize a simpler, more tractable distribution, $q(X_k \mid Y_k)$, from which to draw samples. q is often referred to as the proposal or importance distribution and this approach is commonly know as importance sampling.

For a given proposal distribution the corresponding weights can be easily shown to obey the following relationship

$$w_k^i \propto \frac{p(X_k^i \mid Y_k)}{q(X_k^i \mid Y_k)}. \qquad (9.56)$$

Typically the weights are normalized so that they sum to one.

Importance sampling is also helpful when dealing with highly peaked distributions where it would be necessary to draw a very large numbers of samples if any representation of the low probability regions is to be achieved. By utilizing a flatter proposal distribution, these regions of the state space will be more frequently sampled. However, it should be emphasized the over probability representation is maintained as the weights associated with these low probability samples will be correspondingly lower.

In effect, a PF is a form of adaptive grid approach in which most particles will be clustered in regions of higher state probability thereby maximizing the use of computational resources. PF represents a form of sequential Monte Carlo (SMC). The inclusion of the term sequential is important as it indicates that the PF can be used to generate the distribution at step k through an update of the particle and weights that were calculated at step $k-1$. This relationship is illustrated by factoring the proposal density as follows

$$q(X_k \mid Y_k) \stackrel{\Delta}{=} q(x_k \mid X_{k-1}, Y_k)q(X_{k-1} \mid Y_{k-1}), \tag{9.57}$$

where x_k represents the state at time k. This factorisation illustrates the relationship between the proposal distribution at time $k-1$ and k. To generate the samples at time step k one simply draws new samples based on $q(x_k \mid X_{k-1}, Y_k)$. As an example, consider a case where $q(x_k \mid X_{k-1}, Y_k)$ is chosen to correspond to a simple normal distribution which has no dependence on Y_k. To generate the particles at time step k we draw one sample x_k^i from each of N normal distributions centred at X_{k-1}^i. The weights corresponding to each of these new samples can be shown to be given by

$$w^i(k) \propto w^i(k-1)\frac{p(y(k) \mid x^i(k))p(x^i(k) \mid x^i(k-1))}{q(x^i(k) \mid x^i(k-1), y(k))}, \tag{9.58}$$

where filtering via the measurements is incorporated per Bayes formula. To clarify, the sequential estimation and filtering process involves drawing new updated samples at each step per $q(x_k \mid X_{k-1}^i, Y_k)$ and then calculating updated weights per the above relationship. The resulting distribution represents the filtered posterior.

A particularly convenient choice is to choose the proposal distribution to be the state transition probability density, i.e.,

$$q(x(k) \mid x^i(k-1), y(k)) = q(x(k) \mid x^i(k-1)) = p(x(k) \mid x^i(k-1)), \tag{9.59}$$

in which case

$$w^i(k) \propto w^i(k-1)p(y(k) \mid x^i(k)). \tag{9.60}$$

Note that in this case, the proposal distribution actually has no dependence on the current measurement. The inclusion of the filtering per the measurements in the calculation of the new weight is readily evident by the presence of the $p(y_k \mid x_k^i)$ term on the right hand side of the weight equation. While the state transition probability density is a convenient choice of proposal distribution, it is not necessarily the best or even a good choice in many situations. Maskell et al. provide a useful discussion on the choice of proposal distribution [66].

While computationally efficient, the implementation of PFs is not without challenge. One particularly acute problem is degeneracy in which the weights of all but one particle tend to approach zero after a number of recursive steps. To counter this problem a particle resampling step is almost always employed to eliminate samples with excessively low probability and multiply samples with high probability. This step significantly improves performance but can itself lead to the problem of sample impoverishment in which the resampled particles all collapse to the same state point. A number of approaches have evolved to deal with these issues with varying degrees of effectiveness. As with the grid based approaches, a comprehensive review of the PF is beyond the scope of this chapter and the interested reader is referred to the existing literature for a more thorough treatment (see for instance [61–63]).

9.6.5 Track Before Detect for Targets In Gaussian Noise

To provide a simple illustration of the potential benefits of utilising TkBD in comparison with simple detection approaches we consider a simple case of a constant amplitude point target buried in Rayleigh clutter. Two examples of target motion are illustrated, one relatively straight-line and the other, highly manoeuvrable. For more details, including performance on real data, the reader is referred to [67–69].

In Figure 9.8a the results of applying a peak detector to the raw amplitudes returns (i.e., choose the highest amplitude measurement) from each frame are plotted (stars) along with the actual path followed by the target over the series of scans (solid). While the linear track of the target is relatively easy to identify in the record of detected peak amplitudes there is, nevertheless, a large number of false detections which lie well removed from the actual target track. In contrast, Figure 9.8b displays the results of choosing the peak of the posterior density function that is computed for each frame. It is readily apparent that these estimates of target location that are derived using the TkBD approach are much more tightly constrained in the region of the actual target.

Figure 9.9a and b again display the detection records for the peak amplitude and the peak posterior, respectively, but in this case for a target displaying a very high level of manoeuvrability. The superiority of the TkBD approaches is readily evident. Clearly, an operator would have a much more difficult time identifying a target in the highly scattered results of the peak amplitude detections in comparison with the peak posterior results which are tightly clustered around the target path.

It should be remarked that the improvement noted above is due primarily to the inclusion of information contained within the dynamic constraints of the state model. No spread function, such as was observed for the real measurement data shown in Figures 9.8 and 9.9, was imposed on the target signature for this simulation. The implication is that the presence of a unique target spread function (as is almost always the case with scanning surveillance radar data) would serve to further distinguish the target from surrounding uncorrelated clutter and would lead to additional detection gains in comparison with simpler detection methods. Unfortunately, this latter condition is not always met, i.e., the surrounding clutter discretes sometimes posseses a spread function to the target thereby limiting the actual performance gains. An example of this situation is sea clutter spikes. Sea clutter spike events are often localized in extent, persist over periods of several seconds and their corresponding amplitude signature frequently displays spreading similar to the desired maritime targets. This should come as no surprise to the reader since they are, in fact, just another "target" in the field of view.

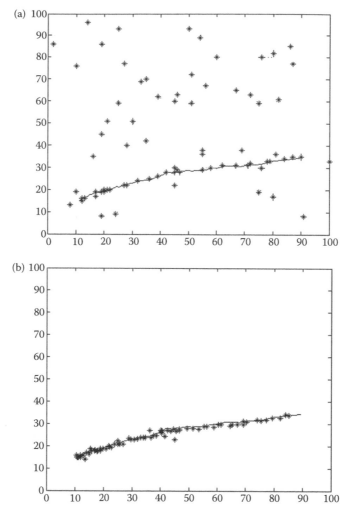

FIGURE 9.8
Comparison of peak amplitude detections and TkBD results for nonmanoeuvring target. (a) Peak amplitude detections for all scans in measurement set. Actual path of point target (solid line) across all scans. (b) Peaks of posterior distribution (stars) for all scans in measurement set. Actual path of point target (solid line) across all scans.

9.7 Conclusion

Given the complexity of the rapidly changing electromagnetic environment, it is obvious that the requirement for radar sensors to perform optimally in any environment will only increase to maintain safety and effectiveness. The development of sophisticated radars which can detect aircraft and missiles with small radar cross-sections, as well as slow-moving ground targets, presents significant technical challenges. Furthermore, for such radars to be produced in a cost-controlled and timely manner, it will require that all of these challenges be met using existing technologies. The preceding discussion did not address the bottlenecks faced by the real-time implementation issues in terms of memory, bandwidth and processor throughput. Radar surveillance and detection strategies are

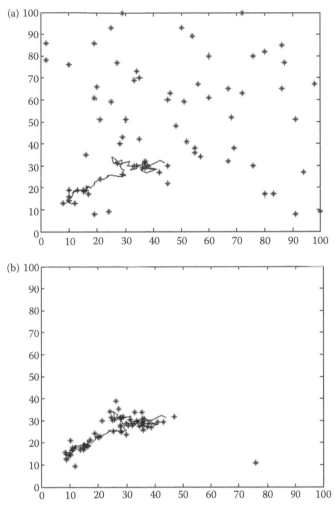

FIGURE 9.9
Comparison of peak amplitude detection and TkBD results for highly manoeuvrable target. (a) Peak amplitude detections for all scans in measurement set. Actual path of point target (solid line) across all scans. (b) Peaks of posterior distribution (stars) for all scans in measurement set. Actual path of point target (solid line) across all scans.

commonly required to accomodate 8000 or more digitized samples to be per pulse and pulse repetition frequencies on the order of 0.5–2 kHz.

The data rates implied by the above parameters further increased by the need to employ multiple apertures for the application of STAP techniques. Signal processors must reduce these high data rates to manageable bandwidths, which will allow the radar operator to make a timely and intelligent decision. One of the key issues with multichannel radars is the impact that the number of required apertures has on the complexity of the multiprocessor implementation. Moreover, operators' tools to allow for autonomous system operations have become critical in environments which are target rich and prone to interference. Although this chapter has not addressed all these issues, the material in the preceding sections has been written for the purpose of providing the scientists and engineers with an appreciation of the fundamental tools needed to design state-of-the-art radar detectors.

References

1. M. Skolnik. 2002. *Introduction to Radar Systems*. McGraw-Hill, New York.
2. N. Willis. 2007. *Bistatic Radar: Principles and Practice*. SciTech Publishing, Raleigh, NC.
3. P. Swerling. 1960. Probability of detection for fluctuating targets. *IEEE Trans., Inform. Theory*, 6(2), 269–308.
4. E. Jakeman and P. N. Pusey. 1978. Significance of K distributions in scattering experiments. *Phys. Rev. Lett.*, 40, 546–550.
5. K. Ward. 1981. Compound representation of high resolution sea clutter. *Elec. Lett.*, 17(16), 561–563.
6. S. Watts. 1985. Radar detection and prediction in sea clutter using the compound K-distribution model. *IEE Proc.*, F-132, 613–620.
7. S. Watts and K. Ward. 1987. Spatial correlation in K-distributed sea clutter. *IEE Proc.*, F-134(6), 526–532.
8. M. Skolnik. 2002. *Radar Handbook*. McGraw-Hill, New York.
9. K. Sangston, F. Gini, M. Greco, and A. Farina. 1999. Structures for radar detection in compound gaussian clutter. *IEEE Trans., Aero. Elec. Syst.*, I-35(2), 445–458.
10. K. D. Ward, C. J. Baker, and S. Watts. 1990. Maritime surveillance radar part 1: Radar scattering from the ocean surface. *IEE Proc., Radar, Sonar, Navig.*, 137(2), 51–62.
11. S. Watts, R. Tough, and K. Ward. 2006. Sea Clutter. *Radar, Sonar, Navigation and Avionics*. Institution of Engineering and Technology.
12. H. Leung and S. Haykin. 1990. Is there a radar clutter attractor? *Appl. Phys. Lett.*, 56(6), 593–595.
13. C. P. Unsworth, M. R. Cooper, S. McLaughlin, and B. Mulgrew. 1999. Re-examining the nature of sea clutter. *IEE Proc., Radar, Sonar, Navig.*, 149(3), 104–114.
14. J. C. Schouten, F. Takens, and C. M. van den Bleek. 1994. Estimation of the dimension of a noisy attractor. *Phys. Rev.*, E-50, 1851–1861.
15. H. Kantz. 1994. A robust method to estimate the maximal Lyapunov exponent of a time series. *Phys. Lett.*, A-185, 77–87.
16. M. McDonald and A. Damini. 2004. Limitations of nonlinear chaotic dynamics in predicting sea clutter returns. *IEE Proc., Radar, Sonar, Navig.*, 151(2), 105–113.
17. A. C.Frery, H.-J. Muller, C. F. Yanasse, and S. J. S. Sant'Anna. 1997. A model for extremely heterogeneous clutter. *IEEE Trans., Geosci. Remote Sensing*, 35(3), 648–659.
18. H. L. Van Trees. 2001. *Detection, Estimation, and Modulation Theory, Part I.* John Wiley and Sons, New York, reprint paperback edition.
19. S. Watts. 1996. Cell-averaging CFAR gain in spatially correlated K-distributed clutter. *IEE Proc., Radar, Sonar, Navig*, 143 (1996) 321–327.
20. H. Rohling. 1983. Radar CFAR thresholding in clutter and multiple target situations. *IEEE Trans., Aero. Elec. Sys.*, 19, 608–621.
21. G. M. Dillard and C. E. Antoniak. 1970. A practical distribution-free detection detection procedure for multiple-range-bin radar. *IEEE Trans., Aero. Elec. Sys.*, 6, 629–635.
22. E. J. Kelly. 1986. An adaptive detection algorithm. *IEEE Trans., Aero. Elec. Sys.*, AES-22, 115–127.
23. E. Conte, M. Lops, and G. Ricci. 1996. Adaptive matched lter detection in spherically invariant noise. *IEEE Signal Proc. Lett.*, 3(8), 248–250.
24. R. Bamler. 1992. A comparison of range-doppler and wavenumber domain SAR focusing algorithms. *IEEE Trans., Geosci. Remote Sens.*, 30, 706–713.
25. A. Budillon, G. Ferraiuolo, V. Pascazio, and G. Schirinzi. 2005. Multichannel SAR interferometry via classical and Bayesian estimation techniques. *EURASIP J. Appl. Signal Process.*, 20, 3180–3193.

26. C. Gierull. 2004. Statistical analysis of multilook SAR interferograms for CFAR detection of ground moving targets. *IEEE Trans., Geosci. Remote Sens.*, 42, 691–701.

27. I. C. Sikaneta. 2004. Detection of ground moving objects with synthetic aperture radar. PhD thesis, University of Ottawa.

28. L. E. Brennan, J. D. Mallett, and I. S. Reed. 1976. Adaptive arrays in airborne MTI. *IEEE Trans., Antennas Propag.*, 74(5), 607–615.

29. J. Ward. 1994. Space-time adaptive processing for airborne radar. Technical report 1015, Lincoln Laboratory, MIT.

30. R. Klemm. 1998. *Space-Time Adaptive Processing*. IEE Press, Stevenage, UK.

31. J. Guerci. 2003. *Space-Time Adaptive Processing for Radar*. Artech House Publishers, Norewood, MA.

32. I. Reed, J. Mallett, and L. Brennan. 1974. Rapid convergence rate in adaptive arrays. *IEEE Trans., Aero. Elec. Sys.*, AES-10(6), 853–863.

33. O. P. Cheremisin. 1982. Efficiency of adaptive algorithms with regularised sample covariance matrix (in Russian). *Radiotechnik und Elektronik*, 2(10), 1933–1941.

34. C. H. Gierull. 1997. Statistical analysis of the eigenvector projection method for adaptive spatial ltering of interference. *IEE Proc. Radar, Sonar, Navig.*, 144(2), 57-63.

35. C. H. Gierull and B. Balaji. 2002. Minimal sample support space-time adaptive processing with fast subspace techniques. *IEE Proc. Radar, Sonar Navig.*, 149, 209–220.

36. C. H. Gierull and B. Balaji. 2002. Application of fast projection technique without eigenanalysis to space-time adaptive processing. In: *IEEE Radar Conference*. Long Beach, CA, April 22–25, 379–385.

37. B. Balaji and C. Gierull. 2002. Theoretical analysis of small sample size behaviour of eigenvector projection technique applied to STAP. In: *IEEE Radar Conference*. Long Beach, CA, April 22–25, 373–378.

38. C. H. Gierull. 1997. A fast subspace estimation method for adaptive beamforming based on covariance matrix transformation. *AEU Int. J. Electr. Commun.*, 51(4), 196–205.

39. C. H. Gierull. 1998. Fast and effective method for low-rank interference suppression in presence of channel errors. *Elec. Lett.*, El-34, 518–520.

40. Q. Zhang and W. B. Mikhael. 1997. Estimation of the clutter rank in the case of subarraying for space-time adaptive processing. *Elec. Lett.*, 33, 419–420.

41. P. G. Richardson. 2001. STAP covariance matrix structure and its impact on clutter plus jamming suppression solutions. *IEE Elec. Lett.*, 37(2), 118–119.

42. B. Balaji. 2001. A wavelet based subspace technique for space-time adaptive processing. In: *International Symposium on Spectral Sensing Research*. Quebec City, Quebec, Canada.

43. J. Ward. 1994. Multiwindow post-Doppler space-time adaptive processing. In: *Proceedings of the Seventh SP Workshop on Statistical Signal and Array Processing*. Quebec City, Canada, June 26–29.

44. H. Wang and L. Cai. 1994. On adaptive spatial-temporal processing for airborne surveillance radar systems. *IEEE Trans., Aero. Elec. Sys.*, 30, 660–669.

45. R. Adve, T. Hale, and M. Wicks. 2000. Practical joint domain localised adaptive processing in homogeneous and nonhomogeneous environments. part 1: Homogeneous environments. *IEE Proc. Radar, Sonar Navig.*, 147(2), 57–65.

46. R. Adve, T. Hale, and M. Wicks. 2000. Practical joint domain localised adaptive processing in homogeneous and nonhomogeneous environments. Part 2: Nonhomogeneous environments. *IEE Proc. Radar, Sonar Navig.*, 147(2), 66–74.

47. B. MacEachern. 2006. Performance evaluation of pre-Doppler and post-Doppler space-time adaptive processing (STAP) techniques for a multimode ground moving target indication system. Master's thesis, Royal Military College, Kingston, ON, Canada.

48. S. Kogon. 2004. *Adaptive Weight Training for Post-Doppler STAP Algorithms in Non-Homogeneous Clutter*. Scitech Publishing, Inc, Raleigh, NC.

49. K. Gerlach, Outlier resistant adaptive matched ltering. *IEE Proc. Radar, Sonar Navig.*, 38(3), 885–901.

50. E. Conte, A. De Maio, A. Farina, and G. Foglia. 2004. Data-adaptive training selection for radar applications. Signal Processing and Information Technology, 2004. In: *Proceedings of the Fourth IEEE International Symposium*, Rome, Italy, 18–21 December, 179–182.

51. M. McDonald and B. Balaji. 2007. Outlier suppression in adaptive filtering through de-emphasis weighting. *IET Radar, Sonar, Navig.*, 1(1), 38–49.

52. M. Wicks, M. Rangaswamy, R. Adve, and T. Hale. 2006. Space-time adaptive processing: a knowledge-based perspective for airborne radar. *Signal Process. Mag., IEEE,* 23(1), 51–65.

53. H. Wang, R. Schneible, and R. D. Brown. 2004. Sigma-Delta-STAP: an efficient, affordable approach for clutter suppression. In: *Applications of Space-Time Adaptive Processing*, R. Klemm, ed. IEE, London, UK, 123–147.

54. B. Balaji and A. Damini. 2006. Multimode adaptive signal processing: a new approach to GMTI radar. *IEEE Trans., Aero. Elec. Sys.*, 42(3), 1121–1126.

55. A. Damini, B. Balaji, L. Shafai, and G. Haslam. 2004. Novel multiple phase centre re ector antenna for GMTI radar. *IEE Proc., Micro., Antennas, Propag.*, 151, 199–204.

56. A. Damini, M. McDonald, and G. E. Haslam. 2005. X-band wideband experimental airborne radar for SAR, GMTI and maritime surveillance. *IEEE Trans., Aero. Elec. Sys.*, 150, 305–312.

57. A. Damini, B. Balaji, G. Haslam, and M. Goulding. 2006. X-band experimental airborne radar phase II: synthetic aperture radar and ground moving target indication. *IEE Proc. Radar, Sonar Navig.*, 153 (2006), no. 2, 144–151.

58. A. H. Jazwinski. 2007. *Stochastic Processes and Filtering Theory*. Dover Publications, Mineola, NY.

59. A. G. Tartakovsky, S. Kligys, and A. Petrov. 1999. Adaptive sequential algorithms for detecting targets in a heavy IR clutter. In: *Signal and Data Processing of Small Targets*, O. E. Drummond, ed., Vol. 3809. SPIE, Orlando, FL, 119–130.

60. Z. S. Haddad and S. R. Simanaca. 1995. Filtering image records using wavelets and the Zakai equation. *IEEE Trans., Patt. Anal. Machine Intell.*, 17, 1069–1078.

61. B. Ristic, S. Arulampalam, and N. Gordon. 2004. *Beyond the Kalman Filter: Particle Filters for Tracking Applications*. Artech House, Norwood, MA.

62. M. Arulampalam, S. Maskell, N. Gordon, and T. Clapp. 2002. A tutorial on particle lters for online nonlinear/non-gaussian bayesian tracking. *IEEE Trans., Signal Process.* [see also *IEEE Trans., Acoustics, Speech, Signal Proc.*], 50(2), 174–188.

63. A. Doucet, S. Godsill, and C. Andrieu. 2000 On sequential Monte Carlo sampling methods for Bayesian ltering. *Stat. Comput.*, 10(3), 197–208.

64. V. Krishnamurthy and L. A. Johnston. 2002. Performance analysis of a dynamic programming track before detect algorithm. *IEEE Trans., Aero. Elec. Sys.* 38, 228–242.

65. W. F. Ames. 1977. *Numerical Methods for Partial Differential Equations*. 2nd edition. Academic Press, New York, NY.

66. S. Maskell, M. Briers, R. Wright, and P. Horridge. 2005. Tracking using a radar and a problem specic proposal distribution in a particle filter. *IEE Proc. Radar, Sonar Navig.*, 152(5), 315–322.

67. M. K. McDonald and B. Balaji. 2007. Continuous-discrete filtering for dim manoeuvring maritime targets. In: *Proccedings of the Tenth International Conference on Information Fusion*. ICIF, Quebec City, Quebec, Canada.

68. M. K. McDonald and B. Balaji. 2008. Track-before-detect using Swerling 0, 1 and 3 target models for small manoeuvring targets. *EURASIP Journal on Advances in Signal Processing Volume* 2008, 9 p., doi: 10.1155/2008/326259.

69. M. K. McDonald and B. Balaji. 2008. Impact of measurement model mismatch on nonlinear track-before-detect performance. In: *IEEE Radar Conference*, 26–30 May 2008, pp. 351–356. Rome, Italy.

10

Sonar Systems*

G. Clifford Carter, Sanjay K. Mehta, and Bernard E. McTaggart[†]

Naval Undersea Warfare Center

CONTENTS

---

* This work represents a revised version from the CRC Press *Electrical Engineering Handbook*, R. Dorf, Ed., 1993
and from the CRC Press *Electronics Handbook*, J. C. Whitaker, Ed., 1996.
† Retired.

10.1 Introduction

10.1.1 What is a Sonar System?

A system that uses acoustic signals propagated through the water to detect, classify, and localize underwater objects is referred to as a sonar system.[‡] Sonars are typically on surface ships (including fishing vessels), submarines, autonomous underwater vehicles (including torpedoes), and aircraft (typically helicopters). A sonar system generally consists of four major components. The first component is a transmitter that (radiates or) transmits a signal through the water. For active sonars, the system transmits energy to be reflected off objects. In contrast, for passive sonar, the object itself is the radiator of acoustic energy. The second component is a receiving array of hydrophones that receives the transmitted (or radiated) signal which has been degraded due to underwater propagation effects, ambient noise, or interference from other signal sources such as surface war ships and fishing vessels. A signal processing subsystem which then processes the received signals to minimize the degradation effects and to maximize the detection and classification capability of the signal is the third component. The fourth component consists of the various displays that aid machine or human operators to detect, classify, and localize sonar signals.

10.1.2 Why Exploit Sound for Underwater Applications?

Acoustic signals propagate better underwater than do other types of energy. For example, both light and radio waves (used for satellite or in-air communications) are attenuated to a far greater degree underwater than are sound waves.[§] For this reason, sound waves have generally been used to extract information about underwater objects. A typical sonar signal processing scenario is shown in Figure 10.1.

10.1.3 Background

In underwater acoustics, the metric system has seldom been universally applied and a number of nonmetric units are still used: distances of nautical miles (1852 m), yards (0.9144 m), and kiloyards; speeds of knots (nautical miles per hour); depths of fathoms (6 ft or 1.8288 m); and bearing in degrees (0.1745 radian). However, in the past decade, there has been an effort to become totally metric, i.e., to use MKS or standard international units.

Underwater sound signals that are processed electronically for detection, classification, and localization can be characterized from a statistical point of view. When time averages

---

‡ Also known as sonar.
§ There has been some limited success propagating blue–green laser energy in clear water.

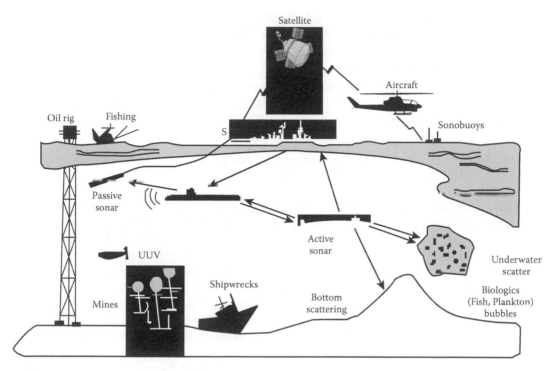

FIGURE 10.1
Active and passive underwater acoustical signal processing.

of each signal are the same as the ensemble average of signals, the signals are said to be *ergodic*. When the statistics do not change with time, the signals are said to be *stationary*. The spatial equivalent to stationary is *homogeneous*. For many introductory problems, only stationary signals and homogeneous noise are assumed; more complex problems involve nonstationary, inhomogeneous environments of the type experienced in actual underwater acoustic environments.

Received sound signals have a first-order probability density function (PDF). For example, the PDF may be Gaussian, or in the case of clicking, sharp noise spikes, or crackling ice noise, the PDF may be nonGaussian. In addition to being characterized by a PDF, signals can be characterized in the frequency domain by their power spectral density functions, which are Fourier transforms of the autocorrelation functions. White signals, which are uncorrelated from sample to sample, have a delta function autocorrelation or a flat (constant) power spectral density. Ocean signals, in general, are much more colorful, and they are neither stationary nor homogeneous.

10.1.4 Sonar

SONAR (sound navigation and ranging), the acronym adapted in the 1940s, is similar to the popular RADAR (radio detection and ranging) and involves the use of sound to explore the ocean and underwater objects.

Passive sonar uses sound radiated from the underwater object itself. The duration of the radiated sound may be short or long in time and narrow or broad in frequency. Transmission through the ocean, from the source to a receiving sensor, is one way.

Active sonar involves transmitting an acoustical signal from a source and receiving reflected echoes from the object of interest. Here, the transmissions from a transmitter to an object and back to a receiving sensor are two way. There are three types of active sonar systems:

Monostatic: In this most common form, the source and receiver can be identical or distinct, but are located on the same platform (e.g., a surface ship or torpedo).

Bistatic: In this form, the transmitter and receiver are on different platforms (e.g., ships).

Multistatic: Here, multiple transmitters and multiple receivers are located on different platforms (e.g., multiple ships).

Passive sonar signals are primarily modeled as random signals. Their first-order PDFs are typically Gaussian; one exception is a stable sinusoidal signal that is nonGaussian and has a power spectral density function that is a Dirac delta function in the frequency domain. Such sinusoidal signals can be detected by measuring the energy output of narrowband filters. This can be done with fast Fourier transform (FFT) electronics and long integration times. However, in actual ocean environments, an arbitrarily narrow frequency width is never observed, and signals have some finite bandwidth. Indeed, the full spectrum of most underwater signals is quite "colorful". In fact, the signals of interest are not unlike speech signals, except that the signal-to-noise (SNR) ratio is much higher for speech applications than for practical sonar applications.

Received active sonar signals can be viewed as consisting of the results of (1) a deterministic component (known transmit signal) convolved with the medium and reflector transfer functions and (2) a random (noise) component. The Doppler imparted (frequency shift) to the reflected signal makes the total system effect nonlinear, thereby complicating analysis and processing of these signals. In addition, in active systems the noise (or reverberation) is typically correlated with the signal, making detection of signals more difficult.

10.2 Underwater Propagation

10.2.1 Speed/Velocity of Sound

Sound speed, c, in the ocean, in general, lies between 1450 and 1540 m/s and varies as a function of several physical parameters, such as temperature, salinity, and pressure (depth). Variations in sound speed can significantly affect the propagation (range or quality) of sound in the ocean. Table 10.1 gives approximate expressions for sound speed as a function of these physical parameters.

10.2.2 Sound Velocity Profiles

Sound rays that are normal (perpendicular) to the signal acoustic wavefront can be traced from the source to the receiver by a process called ray tracing.[1] In general, the acoustic ray

---

[1] Ray tracing models are used for high-frequency signals and in deep water. Generally, if the depth-to-wavelength ratio is 100 or more, ray tracing models are accurate. Below that, corrections must be made to these models. In shallow water or low frequencies, i.e., when the depth-to-wavelength is about 30 or less, "mode theory" models are used.

TABLE 10.1

Expressions for Sound Speed in Meters Per Second

Expression	Limits	Reference
$c = 1492.9 + 3(T-10) - 6 \times 10^{-3}(T-10)^2$	$-2 \leq T \leq 24.5°$	1[a]
$\quad - 4 \times 10^{-3}(T-18)^2 + 1.2(S-35)$	$30 \leq S \leq 42$	
$\quad - 10^{-2}(T-18)(S-35) + D/61$	$0 \leq D \leq 1000$	
$c = 1449.2 + 4.6T - 5.5 \times 10^{-2}T^2$	$0 \leq T \leq 35°$	2[b]
$\quad + 2.9 \times 10 - 4T^3 + (1.34 - 10^{-2}T)(S-35)$	$0 \leq S \leq 45$	
$\quad + 1.6 \times 10^{-2}D$	$0 \leq D \leq 1000$	
$c = 1448.96 + 4.591\,T - 5.304 \times 10^{-2}T^2$	$0 \leq T \leq 30°$	3[c]
$\quad + 2.374 \times 10^{-4}T^3 + 1.340(S-35)$	$30 \leq S \leq 40$	
$\quad + 1.630 \times 10^{-2}D + 1.675 \times 10^{-7}D^2$	$0 \leq D \leq 8000$	

Note: D=depth, in meters; S=salinity, in parts per thousand; and T=temperature, in degrees Celsius.

[a] Leroy, C. C. 1969. Development of simple equations for accurate and more realistic calculation of the speed of sound in sea water. *J. Acoust. Soc. Am.*, 46, 216.

[b] Medwin, H. 1975. Speed of sound in water for realistic parameters. *J. Acoust. Soc. Am.*, 58, 1318.

[c] Mackenzie, K. V. 1981. Nine-term equation for sound speed in the oceans. *J. Acoust. Soc. Am.*, 70, 807.

Source: Urick, R. J., *Principles of Underwater Sound*. McGraw-Hill, New York, NY, 1983. With permission.

paths are not straight, but bend in a manner analogous to optical rays focused by a lens. In underwater sound, the ray paths are determined by the *sound velocity profile* (SVP) or the *sound speed profile* (SSP): i.e., the speed of sound in water as a function of water depth. The sound speed not only varies with depth, but also varies in different regions of the ocean and with time as well. In deep water, the SVP fluctuates the most in the upper ocean due to variations of temperature and weather. Just below the sea surface is the *surface layer,* where the sound speed is greatly affected by temperature and wind action. Below this layer lies the *seasonal thermocline,* where the temperature and speed decrease with depth and the variations are seasonal. In the next layer, the *main thermocline,* the temperature and speed decrease with depth, and surface conditions or seasons have little effect. Finally, there is the *deep isothermal layer,* where the temperature is nearly constant at 39°F and the sound velocity increases almost linearly with depth. A typical deep water SVP as a function of depth is shown in Figure 10.2.

If the sound speed is a minimum at a certain depth below the surface, then this depth is called the axis of the underwater sound channel.** The sound velocity increases both above and below this axis. When the sound wave travels through a medium with a sound speed gradient, the direction of travel of the sound wave is bent toward the area of lower sound speed.

Although the definition of shallow water can be signal dependent, in terms of depth-to-wavelength ratio, a water depth of less than 100 m is generally referred to as shallow water. In shallow water, the SVP is irregular and difficult to predict because of large surface temperature and salinity variations, wind effects, and multiple reflections of sound from the ocean bottom.

** Often called the SOFAR, SOund Fixing And Ranging, channel.

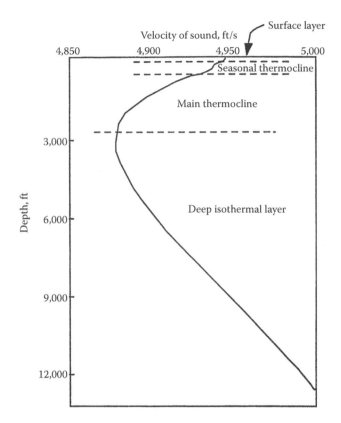

FIGURE 10.2
A typical SVP. (From Urick, R. J., *Principles of Underwater Sound*, McGraw-Hill, New York, 1983. With permission.)

10.2.3 Three Propagation Modes

In general, there are three dominant propagation modes that depend on the distance or range between the sound source and the receiver.

1. *Direct path*: Sound energy travels in a (nominal) straight-line path between the source and receiver, usually at short ranges.
2. *Bottom bounce path*: Sound energy is reflected from the ocean bottom (at intermediate ranges, see Figure 10.3).
3. *Convergence zone* (CZ) *path*: Sound energy converges at longer ranges where multiple acoustic ray paths add or recombine coherently to reinforce the presence of signal energy from the radiating/reflecting source.

10.2.4 Multipaths

The ocean splits signal energy into multiple acoustic paths. When the receiving system can resolve these multiple paths (or multipaths), then they should be coherently recombined by optimal signal processing to fully exploit the available signal energy for detection.[4] It is also theoretically possible to exploit the geometrical properties of multipaths present in the bottom bounce path by investigation of a virtual aperture that is created by the different path arrivals to localize the energy source. In the case of a first-order bottom bounce transmission (i.e., only one bottom interaction), there are four paths (from source to receiver):

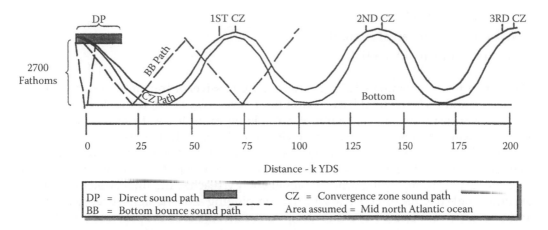

FIGURE 10.3
Typical sound paths between source and receiver. A fathom is a unit of length or depth generally used for underwater measurements. 1 fathom = 6 feet. (From Cox, A.W., *Sonar and Underwater Sound*, Lexington Books, D.C. Health and Company, Lexington, MA, 1974. With permission.)

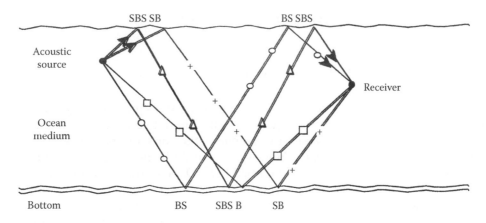

FIGURE 10.4
Multipaths for a first-order bottom bounce propagation model.

1. A bottom bounce ray path (B)
2. A surface interaction followed by a bottom interaction (SB)
3. A bottom bounce followed by a surface interaction (BS)
4. A path that first hits the surface, then the bottom, and finally the surface (SBS)

Typical first-order bottom bounce ocean propagation paths are depicted in Figure 10.4.

10.2.5 Sonar System Performance

The performance of sonar systems, at least to first order, is often assessed by the passive and active sonar equations. The major parameters in the sonar equation, measured in decibels (dB), are as follows:

l_{s} = source level
L_{N} = noise level

N_{DI}=directivity index

N_{TS}=echo level or target strength

N_{RD}=recognition differential

Here, L_S is the target-radiated signal strength (for passive) or transmitted signal strength (for active), and L_N is the total background noise level. N_{DI}, or DI, is the directivity index, which is a measure of the capability of a receiving array to electronically discriminate against unwanted noise. N_{TS} is the received echo level or target strength. Underwater objects with large values of N_{TS} are more easily detectable with active sonar than those with small values of N_{TS}. In general, N_{TS} varies as a function of object size, aspect angle (i.e., the direction at which impinging acoustic signal energy reaches the underwater object), and reflection angle (i.e., the direction at which the impinging acoustic signal energy is reflected off the underwater object). N_{RD} is the recognition differential of the processing system.

The *figure of merit* (FOM), a basic performance measure involving parameters of the sonar system, ocean, and target, is computed for active and passive sonar systems (in dB) as follows:

For passive sonar,

$$\text{FOM}_p = L_S - (L_N - N_{DI}) - N_{RD}$$

For active sonar,

$$\text{FOM}_A = (L_S + N_{TS}) - (L_N - N_{DI}) - N_{RD}$$

Sonar systems are designed so that the FOM exceeds the signal propagation loss for a given set of parameters of the sonar equations. The amount above the FOM is called the *signal excess*. When two sonar systems are compared, the one with the largest signal excess is said to hold the *acoustic advantage*. However, it should be noted that the set of parameters in the above FOM equations is simplified here. Depending on the design or parameter measurability conditions, parameters can be combined or expanded in terms of such quantities as the frequency dependency of the sonar system in particular ocean conditions, the speed and bearing of the receiving or transmitting platforms, reverberation loss, and so forth. Furthermore, due to multipaths, differences in sonar system equipment and operation, and the constantly changing nature of the ocean medium, the FOM parameters fluctuate with time. Thus, the FOM is not an absolute measure of performance, but rather an average measure of performance over time.

10.2.6 Sonar System Performance Limitations

In a typical reception of a signal wavefront, noise and interference can degrade the performance of the sonar system and limit the system's capability to detect signals in the underwater environment. The effects of these degradations must be considered when any sonar system is designed. The noise or interference could be from a school of fish, shipping (surface or subsurface), active transmission operations (e.g., jammers), or the use of multiple receivers or sonar systems simultaneously. Also, the ambient noise may have unusual vertical or horizontal directivity, and in some environments, such as the Arctic, the noise due to ice motion may produce unusual interference. Unwanted backscatterers, similar to the headlights of a car driving in fog, can cause a signal-induced and signal-correlated noise that degrades processing gain. Some other performance-limiting factors are the loss of signal level and acoustic coherence due to boundary interaction as a function of grazing angle; the radiated pattern (signal level) of the object and its spatial coherence; the presence of surface, bottom, and volume reverberation (in active sonar); signal spreading (in time, frequency, or bearing) owing to the modulating effect of surface motion; biologic noise as a function of time (both time of day and time of year); and statistics of the noise in the medium (e.g., does the noise arrive in the same ray path angles as the signal?).

10.2.7 Imaging and Tomography Systems

Underwater sound and signal processing can be used for bottom imaging and underwater oceanic tomography.[10] Signals are transmitted in succession, and the time delay measurements between signals and measured multipaths are then used to determine the speed of sound in the ocean. This information, along with bathymetry data, is used to map depth and temperature variations of the ocean. In addition to mapping ocean bottoms, such information can aid in quantifying global climate and warming trends.

10.3 Underwater Sound Systems: Components and Processes

In this section, we describe a generic sonar system and provide a brief summary for some of its components. In Section 10.4, we describe some of the signal processing functions. A detailed description of the sonar components and various signal processing functions can be found in Knight et al.,[7] Hueter,[6] Oppenheim,[8] and Winder.[13] Figures 10.5 and 10.6 show block diagrams of the major components of a typical active and passive sonar system, respectively. Except for the signal generator, which is present only in active sonar, there are many similarities in the basic components and functions for the active and passive sonar system.

In an active sonar system, an electronic signal generator generates a signal. The signal is then inverse beamformed by delaying it in time by various amounts. A separate projector is used to transmit each of the delayed signals by transforming the electrical signal into an acoustic pressure wave that propagates through water. Thus, an array of projectors is used to transmit the signal and focus it in the desired direction. Depending on the desired range and Doppler resolution, different signal waveforms can be generated and transmitted.

At the receiver (an array of hydrophones), the acoustic or pressure waveform is converted back to an electrical signal. The received signal consists of the source signal (usually the transmitted signal in the active sonar case) embedded in ambient noise and interference from other sources present in water. The signal then goes through a number of signal processing functions. In general, each channel of the analog signal is first filtered in a signal conditioner. It is then amplified or attenuated within a specified dynamic range using an automatic gain control (AGC). For active sonar, we can also use a time-varied gain (TVG) to amplify or attenuate the signal. The signal, which is analog until this point, is then sampled and digitized by analog-to-digital (A/D) converters. The individual digital sensor outputs are next combined by a digital beamformer to form a set of beams. Each beam represents a different search direction of the sonar. The beam output is further processed (bandshifted, filtered, normalized, downsampled, etc.) to obtain detection, classification, and localization (DCL) estimates, which are displayed to the operator on single or multiple displays. Based on the display output (acoustic data) and other nonacoustic data (environmental, contact, navigation, and radar/satellite), the operators make their final decision.

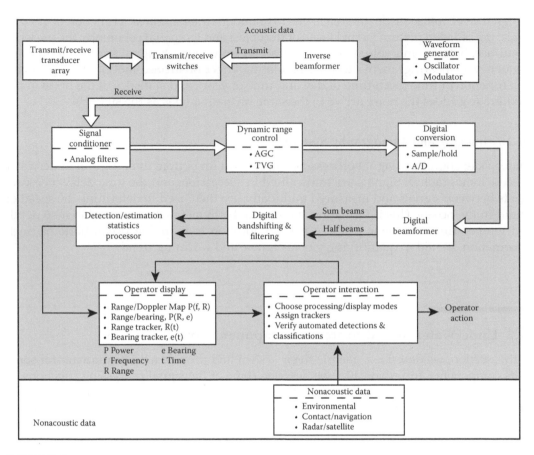

FIGURE 10.5
Generic active sonar system. (Modified from Knight, W. C. et al., *Proc. IEEE*, 69(11), 1451–1506, 1981.)

10.3.1 Signal Waveforms

The transmitted signal is an essential part of an active sonar system. The properties of the transmitted signal will strongly affect the quality of the received signal and the information derived from it. The main objective in active sonar is to detect a target and estimate its range and velocity. The range and velocity information is obtained from the reflected signal. To show how the range and velocity information is contained in the received signal, we consider a simple case where $s(t)$, $0 \leq t \leq T$ is transmitted. Neglecting medium, noise, and other interference effects, we let the transmitted signal be reflected back from a moving target located at range $R = R_0 + vt$, where v is the velocity of the target. The received signal is then given by

$$r(t) = a(R)\, s\, [(1-b)t - \tau]$$

$$\tau = 2R_0/c$$

$$b = 2v/c$$

Where $a(R)$ is the propagation loss attenuation factor[††] and c is the speed of sound. Measuring the delay τ gives the propagation time to and from the target. We can then calculate the range from the time delay. The received signal is also time compressed or

---

[††] More generally, $a(R)$ is also a function of frequency.

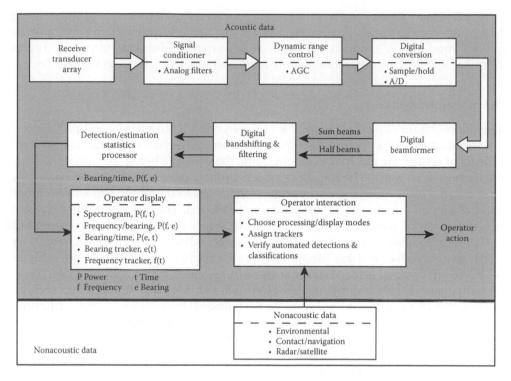

FIGURE 10.6

Generic Passive sonar system. (Modified from Knight, W. C. et al., *Proc. IEEE*, 69(11), 1451–1506, 1981.)

expanded by *b*. The velocity of the target can be estimated by determining the compression or expansion factor of the received signal. Each signal has different range and Doppler (velocity) resolution properties. Some signals are good for range resolution, but not for Doppler; some for Doppler, but not for range; some for reverberation-limited environments; and some for noise-limited environments.[13] Commonly used signals in active sonar are continuous wave (CW), linear frequency modulation (LFM), hyperbolic frequency modulation (HFM), and pseudo-random noise (PRN) signals. CW signals have been used in sonar for decades, whereas signals like frequency hop codes (FHC) and Newhall waveforms are recently "rediscovered" signals[9] that work well in high-reverberation, shallow water environments.

Some of the most commonly used signals are described below. So far, we have made generalized statements about the effectiveness of signals, which only provide a broad overview. More specifically, the signal has properties that depend on a number of factors not discussed here, such as time duration and frequency bandwidth. Rihaczek[9] provides a detailed analysis of the properties and effectiveness of signals. The simplest signal is a rectangular CW pulse, which is a single frequency sinusoid. The CW signal may have high resolution in range (short CW) or Doppler (long CW), but not in both simultaneously. LFM signals are waveforms whose instantaneous frequency varies linearly with time; in HFM signals, the instantaneous frequency sweeps monotonically as a hyperbola. Both these signals are good for detecting low Doppler targets in reverberation-limited conditions. PRN signals, which are generated by superimposing binary data on sinusoid carriers, provide simultaneous resolution in range and Doppler. However, such range resolution may not be as good as LFM alone, and Doppler resolution is not as good as CW alone. An FHC signal is a waveform that consists of subpulses of equal duration. Each subpulse has a distinct frequency, and these frequencies jump or hop in a defined manner. Similar to PRN, FHC also provides simultaneous resolution in range and Doppler. Newhall waveforms (also known as "coherent pulse trains" or "saw-tooth frequency modulation"), which are trains of repeated modulated subpulses (typically HFM or LFM), allow reverberation suppression and low Doppler target detection.

10.3.2 Sonar Transducers

A transducer is a fundamental element of both receiving hydrophones and projectors. It is a device that converts one form of energy into another. In the case of a sonar transducer, the two forms of energy are electricity and pressure, the pressure being that associated with a signal wavefront in water. The transducer is a reciprocal device, such that when electricity is applied to the transducer, a pressure wave is generated in the water, and when a pressure wave impinges on the transducer, electricity is developed. The heart of a sonar transducer is its active material, which makes the transducer respond to electrical or pressure excitations. These active materials produce electrical charges when subjected to mechanical stress and conversely produce a stress proportional to the applied electrical field strength when subjected to an electrical field. Most sonar transducers employ piezoelectric materials, such as lead zirconate titanate ceramic, as the active material. Magnetostictive materials such as nickel can also be used. Figure 10.7 shows a flextensional transducer. In this configuration, the ceramic stack is mounted on the major axis of a metallic elliptical cylinder. Stress is applied by compressing the ellipse along its minor axis, thereby extending the major axis. The ceramic stacks are then inserted into the cylinder, and the stress is released, which places the ceramic stacks in compression. This design allows a small change imparted at the ends of the ceramic stack to be converted into a larger change at the major faces of the ellipse.

10.3.3 Hydrophone Receivers

Hydrophone sensors are microphones capable of operating in water under hydrostatic pressure. These sensors receive radiated and reflected sound energy that arrives through the multiple paths of the ocean medium from a variety of sources and reflectors. As with a microphone, hydrophones convert acoustic pressure to electrical voltages or optical signals. Typical hydrophones are hollow piezoelectric ceramic cylinders with end caps. The cylinders are covered with rubber or polyethylene as water proofing, and an electrical cable exits from one end. These hydrophones are isolation mounted so that they do not

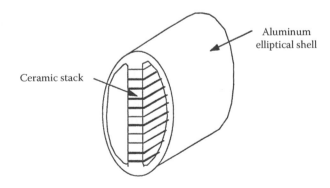

FIGURE 10.7
Flextensional transducer. (From Burdic, W. S., *Underwater Acoustic System Analysis*, Prentice-Hall, Inc. Englewood Cliffs, NJ, 1984. With permission.)

pick up stray vibrations from the ship. Most hydrophones are designed to operate below their resonance frequency, thus resulting in a flat or broadband receiving response.

10.3.4 Sonar Projectors (Transmitters)

A sonar projector is a transducer designed principally for transmission, i.e., to convert electrical voltage or energy into sound energy. Although sonar projectors are also good receivers (hydrophones), they are too expensive and complicated (designed for a specific frequency band of operation) for just receiving signals. Most sonar projectors are of a tonpilz (sound mushroom) design with the piezoelectric material sandwiched between a head and tail mass. The head mass is usually very light and stiff (aluminum) and the tail mass is very heavy (steel). The combination results in a projector that is basically a half-wavelength long at its resonance frequency. The tonpilz resonator is housed in a watertight case that is designed so that the tonpilz is free to vibrate when excited. A pressure release material like corprene (cork neoprene) is attached to the tail mass so that the housing does not vibrate, and all the sound is transmitted from the front face. The piezoelectric material is a dielectric and, as such, acts like a capacitor. To ensure an efficient electrical match to driver amplifiers, a tuning coil or even a transformer is usually contained in the projector housing. A tonpilz projector is shown in Figure 10.8.

10.3.5 Active Sources

One type of sonar transducer primarily used in the surveillance community is a low-frequency active source. The tonpilz design is commonly used for such projectors at frequencies down to about 2 kHz (a tonpilz at this frequency is almost 3/4 m long). For frequencies below 2 kHz, other types of transducer technology are employed, including

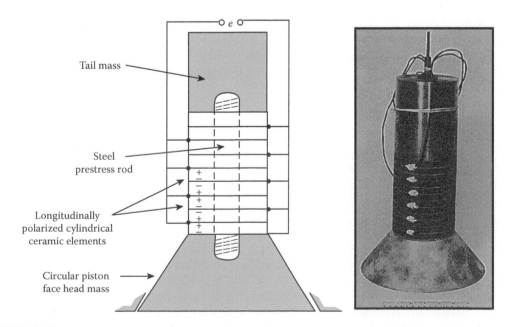

FIGURE 10.8
Tonpilz projector. (Modified from Burdic, W. S., *Underwater Acoustic System Analysis*, Prentice-Hall, Englewood Cliffs, NJ, 1984; and Hueter, T. F., *J. Acoust. Soc. Am.*, 51 (3, Part 2), 1032, 1972.)

mechanical transformers such as flexing shells, moving coil (loud speaker) devices, hydraulic sources, and even impulse sources such as spark gap and air. Explosives are a common source for surveillance, and when used with towed arrays, they make a very sophisticated system.

10.3.6 Receiving Arrays

Most individual hydrophones have an omnidirectional response: that is, sound emanates almost uniformly in all directions. Focusing sound in a particular direction requires an array of hydrophones or projectors. The larger the array, the narrower and more focused is the beam of sound energy, and hence, the more the signal is isolated from the interfering noise. An array of hydrophones allows discrimination against unwanted background noise by focusing its main response axis (MRA) on the desired signal direction. Arrays of projectors and hydrophones are usually designed with elements spaced one-half wavelength apart. This provides optimum beam structure at the design frequency. As the frequency of operation increases and exceeds three-quarter wavelength spacing, the beam structure, although narrowing, deteriorates. As the frequency decreases, the beam increases in width to the point that focusing diminishes. Some common array configurations are linear (omnidirectional), planar (fan shaped), cylindrical (searchlight), and spherical (steered beams). Arrays can be less than 3.5 in. in diameter and can have on the order of 100 hydrophones or acoustic channels. Some newer arrays have even more channels. Typically, these channels are nested, subgrouped, and combined in different configurations to form the low-frequency (LF), mid-frequency (MF), and high-frequency (HF) apertures of the array. Depending on the frequency of interest, one can use any one of these three apertures to process the data. The data are then prewhitened, amplified, and lowpass filtered before being routed to A/D converters. The A/D converters typically operate or sample the data at about three times the lowpass cutoff frequency.

A common array, shown in Figure 10.9a, is a single linear line of hydrophones that makes up a device called a towed array.[‡‡] The line is towed behind the ship and is effective for searching for low level and LF signals without interference from the ship's self-noise. Figure 10.9b shows a more sophisticated bow array (sphere) assembly for the latest Seawolf submarine.

10.3.7 Sonobuoys

These small expendable sonar devices contain a transmitter to transmit signals and a single hydrophone to receive the signal. Sonobuoys are generally dropped by fixed-wing or rotary-wing aircraft for underwater signal detection.

10.3.8 Dynamic Range Control

Today, most of the signal processing and displays involve the use of electronic digital computers. Analog signals received by the receiving array are converted into a digital format, while ensuring that the dynamic range of the data is within acceptable limits. The receiving array must have sufficient dynamic range so that it can detect the weakest signal, but also not saturate upon receiving the largest signal in the presence of noise and interference. To be able to convert the data into digital form and display it, large fluctuations in the data

---

[‡‡] In the oil exploration business, they are called streamers.

(a)

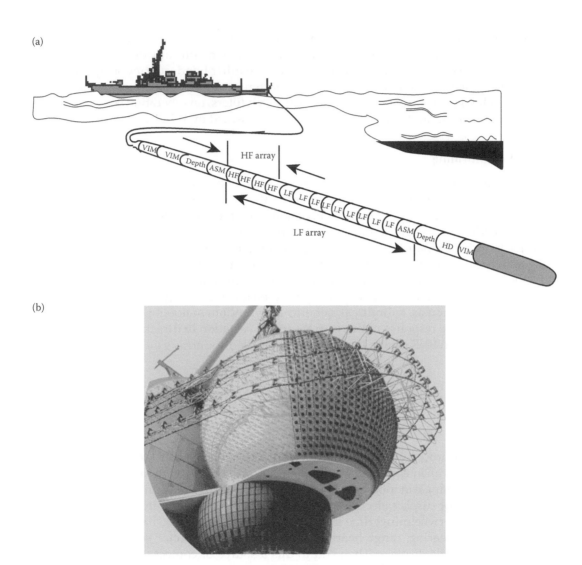

(b)

FIGURE 10.9
(a) Schematic of a towed line array. (From Urick, R. J., *Principles of Underwater Sound*, McGraw-Hill, New York, 1983. With permission.). (b) Bow array assembly for Seawolf submarine (SSN-21). (Taken from the Naval Undersea Warfare Center, Division Newport, BRAC 1991 presentation.)

must be eliminated. Not only do these fluctuations overload the computer digital range capacity, they affect the background quality and contrast of the displays as well. It has been shown that the optimum display background, a background with uniform fluctuations as a function of range, time, and bearing, for detection is one that has constant temporal variance at a given bearing and a constant spatial variance at a given range. Since the fluctuations (noise, interference, propagation conditions, etc.) are time varying, it is necessary to have a threshold level that is independent of fluctuations. The concept is to use techniques that can adapt the high dynamic range of the received signal to the limited dynamic range of the computers and displays. TVG for active sonar and AGC are two popular techniques to control the dynamic range. TVG controls the receiver gain so that it follows a prescribed variation with time, independent of the background conditions. The disadvantage of TVG is that the variations of gain with time do not follow the variations in reverberation. TVG is sufficient if the reverberation is uniform in bearing and monotonically decreasing in range (which is not the case in shallow water). AGC, on the other hand, continuously varies the gain according to the current reverberation or interference conditions. Details of how TVG, AGC, and other gain control techniques such as notch filters, reverberation controlled gain, logarithmic receivers, and hard clippers work are presented by Winder.[13]

10.3.9 Beamforming

Beamforming is a process in which outputs from the hydrophone sensors of an array are coherently combined by delaying and summing the outputs to provide enhanced detection and estimation. In underwater applications, we are trying to detect a directional (single direction) signal in the presence of normalized background noise that is ideally isotropic (nondirectional). By arranging the hydrophone (array) sensors in different physical geometries and electronically steering them in a particular direction, we can increase the SNR in a given direction by rejecting or canceling the noise in other directions. There are many different kinds of arrays that can be beamformed (e.g., equally spaced line, continuous line, circular, cylindrical, spherical, or random sonobuoy arrays). The beam pattern specifies the response of these arrays to the variation in direction. In the simplest case, the increase in SNR due to the beamformer, called the *array gain* (in dB), is given by

$$AG = 10\log\frac{SNR_{array\ (output)}}{SNR_{single\ sensor\ (input)}}$$

10.3.10 Displays

Advancements in processing power and display technologies over the last two decades have made displays an integral and essential part of any sonar system today. Displays have progressed from a single monochrome terminal to very complicated, interactive, real-time, multiterminal, color display electronics. The amount of data that can be provided to an operator can be overwhelming; time series, power spectrum, narrowband and broadband lofargrams, time bearing, range bearing, time Doppler, and sector scans are just some of the many available displays. Then add to this a source or contact tracking display for single or multiple sources over multiple (50–100) beams. The most recent displays provide these data in an interactive mode to make it easier for the operator to make a decision.

For passive systems, the three main parameters of interest are time, frequency, and bearing. Since three-dimensional data are difficult to visualize and analyze, they are usually displayed in the following formats:

Bearing time: Obtained by integrating over frequency; useful for targets with significant broadband characteristics; also called the BTR for bearing time recorder.

Bearing frequency: Obtained at particular intervals of time or by integrating over time; effective for targets with strong stable spectral lines; also called FRAZ for frequency azimuth.

Time frequency: Usually effective for targets with weak or unstable spectral lines in a particular beam; also called lofargram or sonogram.

In general, all three formats are required for the operator to make an informed decision. The operator must sort the outputs from all the displays before classifying them into targets. For example, in passive narrowband sonar, classification is usually performed on the

DCL Analysis display

DCL Search display

Command and control

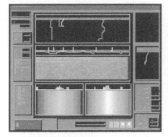 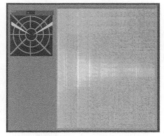 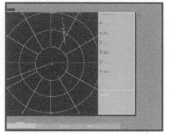

- Torpedo alert window
- Displays of raw/normalized data
- Acoustic LOFARGRAMS
- Specialized acoustic displays
- Beam cursors Hot-linked for quick contact association
- Time bearing window
- Nonacoustic data window
- Recorder control window
- DCL algorithm status window
- Algorithm alert history
- Operator alert verification

- Provides a full suite of search beams of presenting 360° of azimuthal detection coverage to the operator
- Provides tactical surface picture to support classification

- Threat evaluation and tactical advice (TETA) data
- Display with evasion maneuver and Launched Expendable Acoustic Device (LEAD) countermeasure deployment recommendations
- Evasion order block with step by step procedures to execute selected tactic
- Automatic torpedo Target Motion Analysis (TMA) solution

FIGURE 10.10
DCL displays for a passive sonar system. (From Personal communication, E. Marvin, MSTRAP Project, Naval Undersea Warfare Center, Division Newport, Detachment New London, CT.)

outputs from spectral/tonal contents of the targets. The operator uses the different tonal content and its harmonic relations of each target for classification. In addition to the acoustic data information and displays, nonacoustic data such as environmental, contact and navigation information, and radar/satellite photographs are also available to the operator. Figure 10.10 illustrates the displays of a recently developed passive sonar system.

Digital electronics have also had a major impact in sonar in the last two decades and will have an even greater impact in the next decade. As sonar arrays become larger and algorithms become more complex, even more data and displays will be available to the operator. This trend, which requires more data to be processed than before, is going to continue in the future. Due to advancement in technologies, computer processing power has increased, permitting additional signal and data processing. Figure 10.11a shows approximate electronic sonar loads of the past, present, and future sonar systems. Figure 10.11b shows the locations of thecontrols and displays in relation to the different components of a typical sonar system on a submarine.

10.4 Signal Processing Functions

10.4.1 Detection

Detection of signals in the presence of noise, using classical Bayes or Neyman–Pearson decision criteria, is based on hypothesis testing. In the simplest binary hypothesis case, the detection problem is posed as two hypotheses:

H_0 : Signal is not present (referred to as the null hypothesis).

H_1 : Signal is present.

(a)

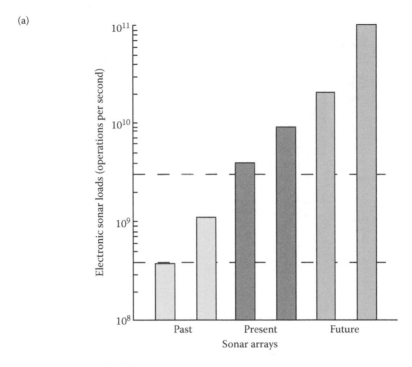

(b)

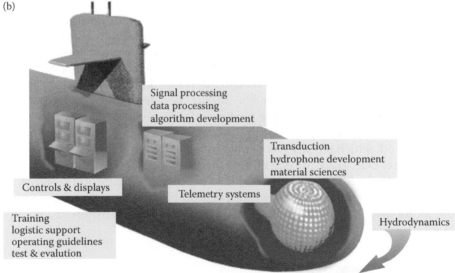

FIGURE 10.11
(a) Electronic sonar load. (From Personal communication, J. Law and N. Owsley, Naval Undersea Warfare Center, Division Newport, Detachment New London, CT.) (b) Control, display, and other components of a submarine sonar system. (From the Naval Undersea Warfare Center, Division Newport, BRAC 1991 presentation.)

For a received wavefront, H_0 relates to the noise-only case and H_1 to the signal-plus-noise case. Complex hypotheses (M-hypotheses) can also be formed if detection of a signal among a variety of sources is required.

Probability is a measure, between zero and unity, of how likely an event is to occur. For a received wavefront, the likelihood ratio, L, is the ratio of P_{H_1} (probability that hypothesis

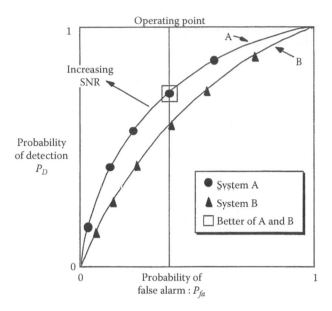

FIGURE 10.12

Typical ROC curves. Note that points (0,0) and (1,1) are on all ROC curves; upper curve represents higher P_D for fixed P_{fa} and hence better performance by having higher SNR or processing time.

H_1 is true) to P_{H_0} (probability that hypothesis H_0 is true). A decision (detection) is made by comparing the likelihood[§§] to a predetermined threshold h. That is, if $L = P_{H_1}/P_{H_0} > h$, a decision is made that the signal is present.

Probability of detection, P_D, measures the likelihood of detecting an event or object when the event does occur. *Probability of false alarm*, P_{fa}, is a measure of the likelihood of saying something happened when the event did NOT occur. Receiver operating characteristics (ROC) curves plot P_D vs. P_{fa} for a particular (sonar signal) processing system. A single plot of P_D vs. P_{fa} for one system must fix the SNR and processing time. The threshold h is varied to sweep out the ROC curve. The curve is often plotted on either log-log scale or "probability" scale. In comparing a variety of processing systems, we would like to select the system (or develop one) that maximizes the P_D for every given P_{fa} Processing systems must operate on their ROC curves, but most processing systems allow the operator to select where on the ROC curve the system is operated by adjusting a threshold; low thresholds ensure a high probability of detection at the expense of a high false alarm rate. A sketch of two monotonically increasing ROC curves is given in Figure 10.12. By proper adjustment of the decision threshold, we can trade off detection performance for false alarm performance. Since the points (0,0) and (1,1) are on all ROC curves, we can always guarantee 100% P_D with an arbitrarily low threshold (albeit at the expense of 100% P_{fa}) or 0% P_{fa} with an arbitrarily high threshold (albeit at the expense of 0% P_D). The (log) *likelihood detector* is a detector that achieves the maximum P_D for fixed P_{fa}; it is shown in Figure 10.13 for detecting Gaussian signals reflected or radiated from a stationary object. For spiky nonGaussian noise, clipping prior to filtering improves detection performance.

In active sonar, the filters are matched to the known transmitted signals. If the object (acoustic reflector) has motion, it will induce Doppler on the reflected signal, and the receiver will be complicated by the addition of a bank of Doppler compensators. Returns

---

[§§] Sometimes the logarithm of the likelihood ratio, called the log-likelihood ratio, is used.

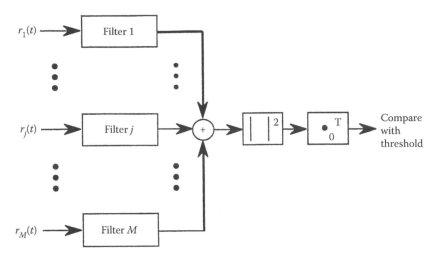

FIGURE 10.13
Log-likelihood detector structure for uncorrelated Gaussian noise in the received signal $r_j(t)$, $j=1, ..., M$.

from a moving object are shifted in frequency by $\Delta f = (2v/c)f$, where v is the relative velocity (range rate) between the source and object, c is the speed of sound in water, and f is the operating frequency of the source transmitter.

In passive sonar, at low SNR, the optimal filters in Figure 10.13 (so-called Eckart filters) are functions of $G_{ss}^{1/2}(f)/G_{nn}(f)$, where f is frequency in Hertz, $G_{ss}(f)$ is the signal power spectrum, and $G_{nn}(f)$ is the noise power spectrum.

10.4.2 Estimation/Localization

The second function of underwater signal processing estimates the parameters that localize the position of the detected object. The source position is estimated in range, bearing, and depth, typically from the underlying parameter of *time delay* associated with the acoustic signal wavefront. The statistical uncertainty of the positional estimates is important. Knowledge of the first-order PDF or its first- and second-order moments, the mean (expected value), and the variance are vital to understanding the expected performance of the processing system. In the passive case, the ability to estimate range is extremely limited by the geometry of the measurements; indeed, the variance of passive range estimates can be extremely large, especially when the true range to the signal source is long when compared with the aperture length of the receiving array. Figure 10.14 depicts direct path passive ranging uncertainty from a collinear array with sensors clustered so as to minimize the bearing and uncertainty region. Beyond the direct path, multipath signals can be processed to estimate source depth passively. Range estimation accuracy is not difficult with the active sonar, but active sonar is not covert, which for some applications can be important.

10.4.3 Classification

The third function of sonar signal processing is classification. This function determines the type of object that has radiated or reflected signal energy. For example, was the sonar signal return from a school of fish or a reflection from the ocean bottom? The action taken by the operator is highly dependent upon this important function. The amount of radiated or reflected signal power relative to the background noise (that is, SNR) necessary to achieve good classification may be many decibels higher than for detection. Also, the type of signal processing required for classification may be different than the type of processing

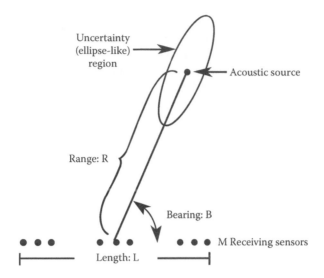

FIGURE 10.14
Array geometry used to estimate source position. (From Carter, G. C., *Proc. IEEE*, 75(2), 251, 1987. With permission.)

for detection. Processing methods that are developed on the basis of detection might not have the requisite SNR to adequately perform the classification function. Classifiers are, in general, divided into feature (or clue) extractors followed by a classifier decision box. A key to successful classification is feature extraction. Performance of classifiers is plotted (as in ROC detection curves) as the probability of deciding on class A, given A was actually present, or $P(A|A)$, vs. the probability of deciding on class B, given that A was present, or $P(B|A)$, for two different classes of objects, A and B. Of course, for the same class of objects, the operator could also plot $P(B|B)$ vs. $P(A|B)$.

10.4.4 Target Motion Analysis

The fourth function of underwater signal processing is to perform contact or target motion analysis (TMA): that is, to estimate parameters of bearing and speed. Generally, nonlinear filtering methods, including Kalman–Bucy filters, are applied; typically, these methods rely on a state space model for the motion of the contact. For example, the underlying model of motion could assume a straight-line course and a constant speed for the contact of interest. When the signal source of interest behaves like the model, then results consistent with the basic theory can be expected. It is also possible to incorporate motion compensation into the signal processing detection function. For example, in the active sonar case, proper signal selection and processing can reduce the degradation of detector performance caused by uncompensated Doppler. Moreover, joint detection and estimation can provide clues to the TMA and classification processes. For example, if the processor simultaneously estimates depth in the process of performing detection, then a submerged object would not be classified as a surface object. Also, joint detection and estimation using Doppler for detection can directly improve contact motion estimates.

10.4.5 Normalization

Another important signal processing function for the detection of weak signals in the presence of unknown and (temporal and spatial) varying noise is normalization. The statistics of noise or reverberation for oceans typically varies in time, frequency, and/or bearing from measurement to measurement and location to location. To detect a weak signal in a broadband, nonstationary, inhomogeneous background, it is usually desirable to make the noise background statistics as uniform as possible for the variations in time, frequency, and/or bearing. The noise background estimates are first obtained from a window of resolution cells (which usually surrounds the test data cell). These estimates are then used to normalize the test cell, thus reducing the effects of the background noise on detection. Window length and distance from the test cell are two of the parameters that can be adjusted to obtain accurate estimates of the different types of stationary or nonstationary noise.

10.5 Advanced Signal Processing

10.5.1 Adaptive Beamforming

Beamforming was discussed earlier in Section 10.3. The cancellation of noise through beamforming can also be done adaptively, which can improve array gain further. Some of the various adaptive beamforming techniques[7] use Dicanne, sidelobe cancellers, maximum entropy array processing, and maximum-likelihood (ML) array processing.

10.5.2 Coherence Processing

Coherence is a normalized,[11] cross-spectral density function that is a measure of the similarity of received signals and noise between any sensors of the array. The complex coherence functions between two wide-sense stationary processes x and y are defined by

$$\gamma_{Xy}(f) = \frac{G_{Xy}(f)}{\sqrt{G_{XX}(f)G_{yy}(f)}}$$

where, as before, f is the frequency in Hertz and G is the power spectrum function. Array gain depends on the coherence of the signal and noise between the sensors of the array. To increase the array gain, it is necessary to have high signal coherence, but low noise coherence. Coherence of the signal between sensors improves with decreasing separation between the sensors, frequency of the received signal, total bandwidth, and integration time. Loss of coherence of the signal could be due to ocean motion, object motion, multipaths, reverberation, or scattering. The coherence function has many uses, including measurement of SNR or array gain, system identification, and determination of time delays.[2,3]

10.5.3 Acoustic Data Fusion

Acoustic data fusion is a technique that combines information from multiple receivers or receiving platforms about a common object or channel. Instead of each receiver making a decision, relevant information from the different receivers is sent to a common control unit where the acoustic data are combined and processed (hence the name *data fusion*). After fusion, a decision can be relayed or "fed" back to each of the receivers. If data transmission is a concern, due to time constraints, cost, or security, other techniques can be used in which each receiver makes a decision and transmits only the decision. The control unit

---

[11] So that it lies between zero and unity.

makes a global decision based on the decisions of all the receivers and relays this global decision back to the receivers. This is called "distributed detection." The receivers can then be asked to re-evaluate their individual decisions based on the new global decision. This process could continue until all the receivers are in agreement or could be terminated whenever an acceptable level of consensus is attained.

An advantage of data fusion is that the receivers can be located at different ranges (e.g., on two different ships), in different mediums (e.g., shallow or deep water, or even at the surface), and at different bearings from the object, thus giving comprehensive information about the object or the underwater acoustic channel.

10.6 Application

Since World War II, in addition to military applications, there has been an expansion in commercial and industrial underwater acoustic applications. Table 10.2 lists the

TABLE 10.2

Underwater Acoustic Applications

Function	Description
	Military
Detection	Deciding if a target is present or not.
Classification	Deciding if a detected target does or does not belong to a specific class.
Localization	Measuring at least one of the instantaneous positions and velocity components of a target (either relative or absolute), such as range, bearing, range rate, or bearing rate.
Navigation	Determining, controlling, and/or steering a course through a medium (includes avoidance of obstacles and the boundaries of the medium).
Communications	Instead of a wire link, transmitting and receiving acoustic power and information.
Control	Using a sound-activated release mechanism.
Position marking	Transmitting a sound signal continuously (beacons) or transmitting only when suitably interrogated (transponders).
Depth sounding	Sending short pulses downward and timing the bottom return.
Acoustic-speedometers	Using pairs of transducers pointing obliquely downward to obtain speed over the bottom from the Doppler shift of the bottom return.
	Commercial Applications
Industrial	Oceanographic.
Fish finders/fish herding	Subbottom geological mapping.
Fish population estimation	Environmental monitoring.
Oil and mineral explorations	Ocean topography.
River flow meter	Bathyvelocimeter.
Acoustic holography	Emergency telephone.
Viscosimeter	Seismic simulation and measurement.
Acoustic ship docking system	Biological signal and noise measurement.
Ultrasonic grinding/drilling	Sonar calibration.
Biomedical ultrasound	

Source: Modified from Urick, R. J., *Principles of Underwater Sound*. McGraw-Hill, New York, NY, 1983; and Cox, A. W., *Sonar and Underwater Sound*, Lexington Books, D. C. Health and Company, Lexington, MA, 1974.

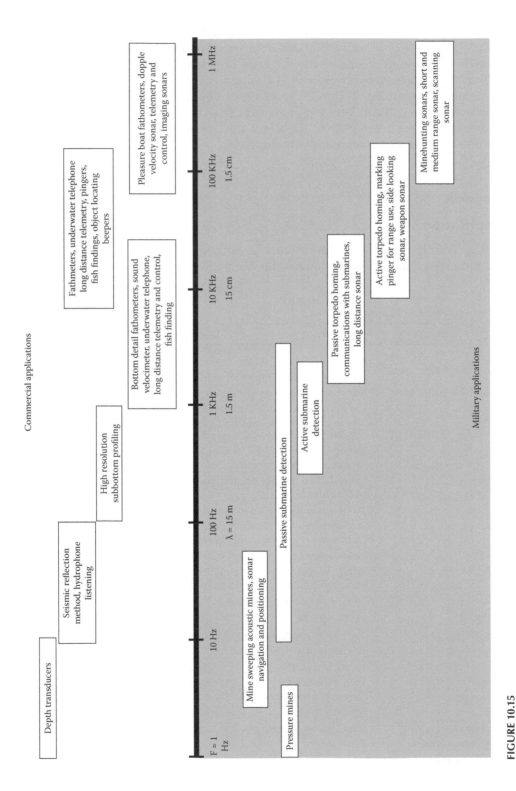

FIGURE 10.15
Sonar frequency allocation. (Modified from Neitzel, E. B., Civil uses of underwater acoustics, Lectures on Marine Acoustics, AD 156-052, 1973.)

military and nonmilitary functions of sonar along with some of the current applications. Figure 10.15 shows the sonar frequency allocations for military and commercial applications.

Acknowledgment

The authors thank Louise Miller for her assistance in preparing this chapter.

References

1. Burdic, W. S. 1984. *Underwater Acoustic System Analysis*, Prentice Hall, Englewood Cliffs, NJ.
2. Carter, G. C. 1987. Coherence and time delay estimation, *Proc. IEEE*, 75(2), 236–255.
3. Carter, G. C., Ed. 1993. *Coherence and Time Delay Estimation*, IEEE Press, Piscataway, NJ.
4. Chan, Y. T., Ed. 1989. *Digital Signal Processing for Sonar Underwater Acoustic Signal Processing*, NATO ASI Series, Series E: Applied Sciences, Vol. 161. Kluwer Academic Publishers, Dordrecht.
5. Cox, A. W. 1974. *Sonar and Underwater Sound*. Lexington Books, D.C. Health and Company, Lexington, MA.
6. Hueter, T. F. 1972. Twenty years in underwater acoustics: generation and reception. *J. Acoust. Soc. Am.*, 51(3), 1025–1040.
7. Knight, W. C., Pridham, R. G., and Kay, S. M. 1981. Digital signal processing for Sonar. *Proc. IEEE*, 69(11), 1451–1506.
8. Oppenheim, A. V. Ed. 1980. *Applications of Digital Signal Processing*. Prentice Hall, Englewood Cliffs, NJ.
9. Rihaczek, A. W., Ed. 1985. *Principles of High Resolution Radar*. Peninsula Publishing, Los Altos, CA.
10. Spindel, R. C. 1985. Signal processing in ocean tomography. In: *Adaptive Methods in Underwater Acoustics*, H.G. Urban, Ed. D. Reidel Publishing Company, Dordrecht, 687–710.
11. Urick, R. J. 1983. *Principles of Underwater Sound*. McGraw-Hill, New York, NY.
12. Van Trees, H. L. 1968. *Detection, Estimation, and Modulation Theory*. John Wiley & Sons, New York, NY.
13. Winder, A. A. 1975. Sonar system technology. *IEEE Trans. Sonics and Ultrasonics*, su-22(5), 291–332.

Further Information

Journal of Acoustical Society of America (JASA); *IEEE Transactions on Signal Processing* (formerly the *IEEE Transactions on Acoustics, Speech and Signal Processing*), and *IEEE Journal of Oceanic Engineering* are professional journals providing current information on underwater acoustical signal processing.

The annual meetings of the International Conference on Acoustics, Speech and Signal Processing, spon-sored by the IEEE, and the biannual meetings of the Acoustical Society of America are good sources for current trends and technologies.

Digital Signal Processing for Sonar (Knight, W.C., Pridham, R.G., and Kay, S.M. 1981 *Proc. IEEE*, 69(11), 1451–1506) and *Sonar System Technology* (Winder, A.A. 1975. *IEEE Trans. Sonics and Ultrasonics*, su-22(5), 291–332.) are informative and detailed tutorials on underwater sound systems. Also, the March 1972 issue of *Journal of Acoustical Society of America* (Hueter) has historical and review papers on underwater acoustics related topics.

11

Theory and Implementation of Advanced Signal Processing for Active and Passive Sonar Systems

Stergios Stergiopoulos
Defence R&D Canada Toronto
University of Toronto
University of Western Ontario

Geoffrey Edelson
Advanced Systems & Technology

CONTENTS

11.1 Introduction

Several review articles[1-4] on sonar system technology have provided a detailed description of the mainstream sonar signal processing functions along with the associated implementation considerations. The attempt with this chapter is to extend the scope of these articles[1-4] by introducing an implementation effort of nonmainstream processing schemes in real-time sonar systems. The organization of the chapter is as follows.

Section 11.1 provides a historical overview of sonar systems and introduces the concept of the signal processor unit and its general capabilities. This section also outlines the practical importance of the topics to be discussed in subsequent sections, defines the sonar problem, and provides an introduction into the organization of the chapter.

Section 11.2 introduces the development of a realizable generic processing scheme that allows the implementation and testing of *nonlinear processing techniques* in a wide spectrum of real-time active and passive sonar systems. Finally, a concept demonstration of the above developments is presented in Section 11.3, which provides real data outputs from an advanced beamforming structure incorporating adaptive and synthetic aperture beamformers.

11.1.1 Overview of a Sonar System

To provide a context for the material contained in this chapter, it would seem appropriate to briefly review the basic requirements of a high-performance sonar system. A *sonar* (SOund, NAvigation, and Ranging) system is defined as a "method or equipment for determining by underwater sound the presence, location, or nature of objects in the sea."[5] This is equivalent to detection, localization, and classification as discussed in Chapter 6.

The main focus of the assigned tasks of a modern sonar system will vary from the detection of signals of interest in the open ocean to very quiet signals in very cluttered underwater environments, which could be shallow coastal sea areas. These varying degrees of complexity of the above tasks, however, can be grouped together quantitatively, and this will be the topic of discussion in the following section.

11.1.2 The Sonar Problem

A convenient and accurate integration of the wide variety of effects of the underwater environment, the target's characteristics, and the sonar system's designing parameters is provided by the sonar equation.[8] Since World War II, the sonar equation has been used extensively to predict the detection performance and to assist in the design of a sonar system.

11.1.2.1 The Passive Sonar Problem

The passive sonar equation combines, in logarithmic units (i.e., units of decibels [dB] relative to the standard reference of energy flux density of rms pressure of 1 μPa integrated over a period of 1 s), the following terms:

$$(S - TL) - (N_e - AG) - DT \geq 0, \tag{11.1}$$

which define signal excess where S is the source energy flux density at a range of 1 m from the source; TL is the propagation loss for the range separating the source and the sonar array receiver. Thus, the term $(S - TL)$ expresses the recorded signal energy flux density at the receiving array; N_e is the noise energy flux density at the receiving array; AG is the array gain that provides a quantitative measure of the coherence of the signal of interest with respect to the coherence of the noise across the line array; DT is the detection threshold associated with the decision process that defines the SNR at the receiver input required for a specified probability of detection and false alarm.

A detailed discussion of the DT term and the associated statistics is given in References 8 and 28–30. Very briefly, the parameters that define the detection threshold values for a passive sonar system are the following:

- The time-bandwidth product defines the integration time of signal processing. This product consists of the term T, which is the time series length for coherent processing such as the FFT, and the incoherent averaging of the power spectra over K successive blocks. The reciprocal, $1/T$, of the FFT length defines the bandwidth of a single frequency cell. An optimum signal processing scheme should match the acoustic signal's bandwidth with that of the FFT length T in order to achieve the predicted DT values
- The probabilities of detection, P_D, and false-alarm, P_{FA}, define the confidence that the correct decision has been made.

Improved processing gain can be achieved by incorporating segment overlap, windowing, and FFT zeroes extension as discussed by Welch[31] and Harris.[32] The definition of DT for the narrow-band passive detection problem is given by[8]

$$DT = 10\log\frac{S}{N_e} = 5\log\left(\frac{d \cdot BW}{t}\right), \tag{11.2}$$

where N_e is the noise power in a 1-Hz band, S is the signal power in bandwidth BW, t is the integration period in displays during which the signal is present, and $d=2t(S/N_e)$ is the detection index of the receiver operating characteristic (ROC) curves defined for specific values of P_D and P_{FA}.[8,28] Typical values for the above parameters in the term DT that are considered in real-time narrowband sonar systems are $BW=O(10^{-2})$ Hz, $d=20$, for $P_D=50\%$, $P_{FA}=0.1\%$, and $t=O(10^2)$ seconds.

The value of TL that makes Equation 11.1 become an equality leads to the equation

$$FOM=S-N_e-AG-DT, \tag{11.3}$$

where the new term figure of merit (FOM) equals the transmission loss TL and gives an indication of the range at which a sonar can detect a target.

The noise term N_e in Equation 11.1 includes the total or composite noise received at the array input of a sonar system and is the linear sum of all the components of the noise processes, which are assumed independent. However, detailed discussions of the noise processes related to sonar systems are beyond the scope of this chapter and readers interested in these noise processes can refer to other publications on the topic.[8,33–39]

When taking the sonar equation as the common guide as to whether the processing concepts of a passive sonar system will give improved performance against very quiet targets, the following issues become very important and appropriate:

- During passive sonar operations, the terms S and TL are beyond the sonar operators' control because S and TL are given as parameters of the sonar problem. DT is associated mainly with the design of the array receiver and the signal processing parameters. The signal processing parameters in Equation 11.2 that influence DT are adjusted by the sonar operators so that DT will have the maximum positive impact in improving the FOM of a passive sonar system. The discussion in Section 11.1.2.2 on the active sonar problem provides details for the influence of DT by an active sonar's signal processing parameters.

- The quantity (N_e-AG) in Equations 11.1 and 11.3, however, provides opportunities for sonar performance improvements by *increasing the term AG* (e.g., deploying large size array receivers or using new signal processing schemes) and by *minimizing the term* N_e (e.g., using adaptive processing by taking into consideration the directional characteristics of the noise field and by reducing the impact of the sensor array's self noise levels).

Our emphasis in the sections of this chapter that deal with passive sonar will be focused on the minimization of the quantity (N_e-AG). This will result in new signal processing schemes in order to achieve a desired level of performance improvement for the specific case of a line array sonar system.

11.1.2.2 The Active Sonar Problem

The criterion for sonar system detection requires the signal power collected by the receiver system to exceed the background level by some threshold. The minimum SNR needed to achieve the design false alarm and detection probabilities is called the detection threshold as discussed above. Detection generally occurs when the signal excess is nonnegative, i.e., $SE = \text{SNR} - DT \geq 0$. The signal excess for passive sonar is given by Equation 11.1.

A very general active sonar equation for signal excess in decibels is

$$SE = EL - IL - DT, \tag{11.4}$$

in which EL and IL denote the echo level and interference level, respectively. For noise-limited environments with little to no reverberation, the echo and interference level terms in Equation 11.4 become

$$EL = S - TL_1 + TS - TL_2 + AGS - L_{sp}$$
$$IL = NL + AGN, \tag{11.5}$$

in which TL_1 is the transmission loss from the source to the target, TS is the target strength, TL_2 is the transmission loss from the target to the receiver, L_{sp} denotes the signal processing losses, AGS is the gain of the receiver array on the target echo signal, and AGN is the gain of the receiver on the noise. Array gain (AG), as used in Chapter 6, is defined as the difference between AGS and AGN. All of these terms are expressed in decibels.

In noise-limited active sonar, the SNR, defined as the ratio of signal energy (S) to the noise power spectral density at the processor input (NL) and expressed in decibels, is the fundamental indicator of system performance. Appropriately, the detection threshold is defined as $DT = 10 \log(S/NL)$. From the active sonar equation for noise-limited cases, we see that one simple method of increasing the signal excess is to increase the transmitted energy.

If the interference is dominated by distributed reverberation, the echo level term does not change, but the interference level term becomes

$$IL = S - TL_1' + 10\log(\Omega_s) + S_X - TL_2' + AGS' - L_{sp}', \tag{11.6}$$

in which the transmission loss parameters for the out and back reverberation paths are represented by the primed TL quantities and S_x is the scattering strength of the bottom (dB re m²), surface (dB re m²), or volume (dB re m³). The terms for the gain of the receive array on the reverberation signal and for the signal processing losses are required because the reverberation is different in size from the target and they are not co-located. Ω_s is the scattering area in square meters for the bottom (or surface) or the scattering volume in cubic meters. The scattering area for distributed bottom and surface reverberation at range R is $R\phi((c\tau)/2)$, in which ϕ is the receiver beamwidth in azimuth, c is the speed of sound, and τ is the effective pulse length after matched filter processing. For a receiver with a vertical beamwidth of θ, the scattering volume for volume reverberation is $(R\phi((c\tau)/2))R\theta$.

The resulting active sonar equation for signal excess in distributed reverberation is

$$SE = (TL'_1) + (TS - 10\log(\Omega_s) + (TL'_2 - TL_2)$$

$$+(AGS - AGS') - (L_{sp} - L'_{sp}) - DT \tag{11.7}$$

Of particular interest is the absence of the signal strength from Equation 11.7. Therefore, unlike the noise-limited case, increasing the transmitted energy does not increase the received signal-to-reverberation ratio.

In noise-limited active sonar, the formula for DT depends on the amount known about the received signal.[111] In the case of a completely known signal with the detection index as defined in Section 11.1.2.1, the detection threshold becomes $DT = 10 \log(d/2\omega t)$, where ω is the signal bandwidth. In the case of a completely unknown signal in a background of Gaussian noise when the SNR is small and the time-bandwidth product is large, the detection threshold becomes $DT = 5 \log(d/\omega t)$, provided that the detection index is defined as $d = \omega t \cdot (S/NL)^2$.[111] Thus, the noise-limited detection threshold for these cases improves with increasing pulse length and bandwidth.

In reverberation-limited active sonar, if the reverberation power is defined at the input to the receiver as $R = U_R t$ in which U_R is the reverberation power per second of pulse duration, then S/U_R becomes the measure of receiver performance.[112] For the cases of completely known and unknown signals, the detection thresholds are $DT = 10 \log(d/2\omega_R)$ and $DT = 5 \log(dt/2\omega_R)$, respectively, with ω_R defined as the effective reverberation bandwidth. Therefore, the reverberation-limited detection threshold improves with increasing ω_R.

Thus, a long-duration, wideband active waveform is capable of providing effective performance in both the noise-limited and reverberation-limited environments defined in this section.

11.2 Theoretical Remarks

Sonar operations can be carried out by a wide variety of naval platforms, as shown in Figure 11.1. This includes surface vessels, submarines, and airborne systems such as airplanes and helicopters. Shown also in Figure 11.1A is a schematic representation of active and passive sonar operations in an underwater sea environment. Active sonar operations involve the transmission of well-defined acoustic signals, which illuminate targets in an underwater sea area. The reflected acoustic energy from a target provides the sonar array receiver with a basis for detection and estimation. The major limitations to robust detection and classification result from the energy that returns to the receiver from scattering bodies also illuminated by the transmitted pulses.

Passive sonar operations base their detection and estimation on acoustic sounds that emanate from submarines and ships. Thus, in passive systems only, the receiving sensor array is under the control of the sonar operators. In this case, major limitations in detection and classification result from imprecise knowledge of the characteristics of the target radiated acoustic sounds.

The depiction of the combined active and passive acoustic systems shown in Figure 11.1 includes towed line arrays, hull-mounted arrays, a towed source, a dipping sonar, and vertical line arrays. Examples of some active systems that operate in different frequency regimes are shown in Figures 11.1B through 11.3C. The low-frequency (LF) sources in Figure 11.1B are used for detection and tracking at long ranges, while the hull-mounted spherical and cylindrical mid-frequency (MF) sonars shown in Figure 11.2A and B are designed to provide the platform with a tactical capability.

The shorter wavelengths and higher bandwidth attributable to high-frequency (HF) active sonar systems like those shown in Figure 11.3A and C yield greater range and bearing resolution compared to lower frequency systems. This enables better spatial

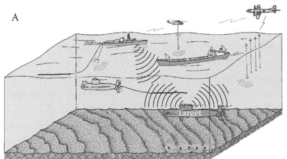

FIGURE 11.1
(A) Schematic representation for active and passive sonar operations for a wide variety of naval platforms in an underwater sea environment. (Reprinted by permission of IEEE © 1998.) (B) Low-frequency sonar projectors inside a surface ship. (Photo provided courtesy of Sanders, A Lockheed Martin Company.)

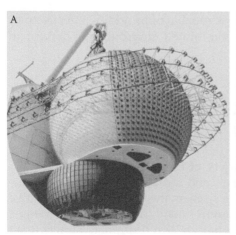

FIGURE 11.2
(A) A bow-installed, mid-frequency spherical array. (Photo provided courtesy of the Naval Undersea Warfare Center.) (B) A mid-frequency cylindrical array on the bow of a surface ship. (Photo provided courtesy of the Naval Undersea Warfare Center.)

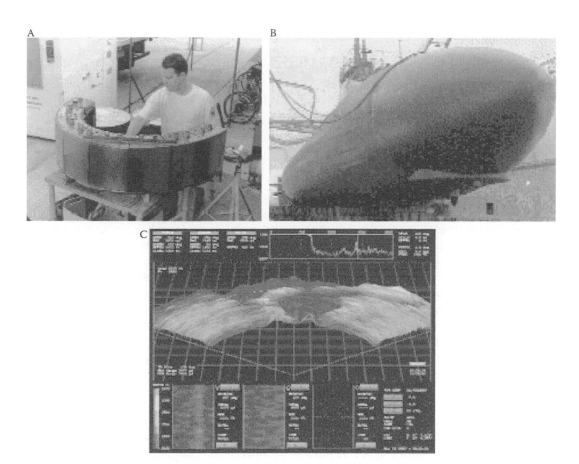

FIGURE 11.3
(A) Preparation of a high-frequency cylindrical array for installation in a submarine. (Photo provided courtesy of Undersea Warfare Magazine.) (B) High-frequency receiver and projector arrays visible beneath the bow dome of a submarine. (Photo provided courtesy of Undersea Warfare Magazine.) (C) Output display of a high-frequency sonar system showing the geological features of an undersea volcano. (Photo provided courtesy of Undersea Warfare Magazine.)

discrimination, which can be broadly applied, from the geological mapping of the seafloor to the detection and classification of man-made objects. Figure 11.3C shows the geological features of an undersea volcano defined by an HF active sonar. These spatial gains are especially useful in shallow water for differentiating undersea objects from surface and bottom reverberation. HF arrays have also been used successfully as passive receivers.[113]

The passive sonar concept, in general, can be made clearer by comparing sonar systems with radars, which are always active. Another major difference between the two systems arises from the fact that sonar system performance is more affected than that of radar systems by the underwater medium propagation characteristics. All the above issues have been discussed in several review articles[1–4] that form a good basis for interested readers to become familiar with "main stream" sonar signal processing developments. Therefore, discussions of issues of conventional sonar signal processing, detection, and estimation and the influence of the medium on sonar system performance are briefly highlighted in this section in order to define the basic terminology required for the presentation of the main theme of this chapter.

Let us start with a basic system model that reflects the interrelationships between the target, the underwater sea environment (medium), and the receiving sensor array of a sonar system. A schematic diagram of this basic system is shown in Figure 6.3 of Chapter 6, where sonar signal processing is shown to be two-dimensional (2-D)[1,12,40] in the sense that it involves both temporal and spatial spectral analysis. The temporal processing provides spectral characteristics that are used for target classification, and the spatial processing provides estimates of the directional characteristics (i.e., bearing and possibly range) of a detected signal. Thus, *space-time processing* is the fundamental processing concept in sonar systems, and it has already been discussed in Chapter 6.

11.2.1 Definition of Basic Parameters

This section outlines the context in which the sonar problem can be viewed in terms of models of acoustic signals and noise fields. The signal processing concepts that are discussed in Chapter 6 have been included in sonar and radar investigations with sensor arrays having circular, planar, cylindrical, and spherical geometric configurations. Therefore, the objective of our discussion in this section is to integrate the advanced signal processing developments of Chapter 6 with the sonar problem. For geometrical simplicity and without any loss of generality, we consider here an N hydrophone line array receiver with sensor spacing δ. The output of the nth sensor is a time series denoted by $x_n(t_i)$, where ($i=1, \ldots, M$) are the time samples for each sensor time series. An $*$ denotes complex conjugate transposition so that $\bar{X}^*$ is the row vector of the received N hydrophone time series $\{x_n(t_i), n=1, 2, \ldots, N\}$. Then $x_n(t_i)=s_n(t_i)+\varepsilon_n(t_i)$, where $s_n(t_i)$, $\varepsilon_n(t_i)$ are the signal and noise components in the received sensor time series. $\bar{S}$, $\bar{\varepsilon}$ denote the column vectors of the signal and noise components of the vector $\bar{X}$ of the sensor outputs (i.e., $\bar{X}=\bar{S}+\bar{\varepsilon}$). $X_n(f)=\sum_{f=1}^{M}X_n(t_i)\exp(-j2\pi ft_i)$ is the Fourier transform of $x_n(t_i)$ at the signal with frequency f, $c=f\lambda$ is the speed of sound in the underwater medium, and λ is the wavelength of the frequency f. $S=E\{\bar{S}\,\bar{S}^*\}$ is the spatial correlation matrix of the signal vector $\bar{S}$, whose nth element is expressed by

$$s_n(t_i) = s_n[t_i + \tau_n(\theta)] , \tag{11.8}$$

E{...} denotes expectation, and

$$\tau_n(\theta) = (n-1)\delta\cos\theta/c \tag{11.9}$$

is the time delay between the first and the nth hydrophone of the line array for an incoming plane wave with direction of propagation θ, as illustrated in Figure 6.3 of Chapter 6.

In this chapter, the problem of detection is defined in the classical sense as a hypothesis test that provides a detection probability and a probability of false alarm, as discussed in Chapter 6. This choice of definition is based on the standard CFAR processor, which is based on the Neyman–Pearson criterion.[28] The CFAR processor provides an estimate of the ambient noise or clutter level so that the threshold can be varied dynamically to stabilize the false alarm rate. Ambient noise estimates for the CFAR processor are provided mainly by noise normalization techniques[42–45] that account for the slowly varying changes in the background noise or clutter. The above estimates of the ambient noise are based upon the average value of the received signal, the desired probability of detection, and the probability of false alarms.

Furthermore, optimum beamforming, which has been discussed in Chapter 6, requires the beamforming filter coefficients to be chosen based on the covariance matrix of the received data by the N sensor array in order to optimize the array response.[46,47] The family of algorithms for optimum beamforming that use the characteristics of the noise are called *adaptive beamformers*,[2,11,12,46–49] and a detailed definition of an adaptation process requires knowledge of the correlated noise's covariance matrix $R(f_i)$. For adaptive beamformers, estimates of $R(f_i)$ are provided by the spatial correlation matrix of received hydrophone time series with the nmth term, $R_{nm}(f,d_{nm})$, defined by

$$R_{nm}(f,\delta_{nm}) = E[X_n(f)X_m^*(f)] \tag{11.10}$$

$R_{\varepsilon}'(f_i) = \sigma_n^2(f)R_a(f_i)$ is the spatial correlation matrix of the noise for the ith frequency bin with $\sigma_n^2(f_i)$ being the power spectral density of the noise $\varepsilon_n(t_i)$. The discussion in Chapter 6 shows that if the statistical properties of an underwater environment are equivalent with those of a white noise field, then the *conventional beamformer* (CBF) without shading is the optimum beamformer for bearing estimation, and the variance of its estimates achieve the CRLB bounds. For the narrowband CBF, the plane wave response of an N hydrophone line array steered at direction θ_s is defined by[12]

$$B(f,\theta_s) = \sum_{n=1}^{N} X_n(f)d_n(f,\theta_s), \tag{11.11}$$

where $d_n(f, \theta_s)$ is the nth term of the steering vector $\bar{D}(f,\theta_s)$ for the beam steering direction θ_s, as expressed by

$$d_n(f_i,\theta) = \exp\left[j2\pi \frac{(i-1)f_s}{M}\tau_n(\theta) \right], \tag{11.12}$$

where f_s is the sampling frequency.

The beam power pattern $P(f, \theta_s)$ is given by $P(f, \theta_s)=B(f, \theta_s)B^*(f, \theta_s)$. Then, the power beam pattern $P(f, \theta_s)$ takes the form

$$P(f,\theta_s) = \sum_{n=1}^{N}\sum_{m=1}^{N} X_n(f)X_m^*(f)\exp\left[\frac{j2\pi f\delta_{nm}\cos\theta_s}{c} \right], \tag{11.13}$$

where δ_{nm} is the spacing $\delta(n-m)$ between the nth and mth hydrophones. Let us consider for simplicity the source bearing θ to be at array broadside, $\delta=\lambda/2$, and $L=(N-1)\delta$ to be the array size. Then Equation 11.13 is modified as[3,40]

$$P(f,\theta_s) = \frac{N^2 \sin^2\left[\dfrac{\pi L\sin\theta_s}{\lambda} \right]}{\left(\dfrac{\pi L\sin\theta_s}{\lambda} \right)^2}, \tag{11.14}$$

which is the far-field radiation or directivity pattern of the line array as opposed to near-field regions.

Equation 11.11 can be generalized for nonlinear 2-D and 3-D arrays, and this is dis-cussed in Chapter 6. The results in Equation 11.14 are for a perfectly coherent incident acoustic signal, and an increase in array size $L=\delta(N-1)$ results in additional power out-put and a reduction in beamwidth. The sidelobe structure of the directivity pattern of a receiving array can be suppressed at the expense of a beamwidth increase by applying different weights. The selection of these weights will act as spatial filter coefficients with optimum performance.[4,11,12] There are two different approaches to select these weights: *pattern optimization* and *gain optimization*. For pattern optimization, the desired array response pattern $B(f, \theta_s)$ is selected first. A desired pattern is usually one with a narrow main lobe and low sidelobes. The weighting or shading coefficients in this case are real numbers from well-known window functions that modify the array response pattern. Harris' review[32] on the use of windows in discrete Fourier transforms and temporal spec-tral analysis is directly applicable in this case to spatial spectral analysis for towed line array applications.

Using the approximation $\sin\theta \cong \theta$ for small θ at array broadside, the first null in Equation 11.14 occurs at $\pi L\sin\theta/\lambda=\pi$ or $\Delta\theta \times L/\lambda \cong 1$. The major conclusion drawn here for line array applications is that[3,40]

$$\Delta\theta \approx \lambda/L \quad \text{and} \quad \Delta f \times T = 1, \tag{11.15}$$

where $T=M/F_s$ is the hydrophone time series length. Both relations in Equation 11.15 express the well-known temporal and spatial resolution limitations in line array applica-tions that form the driving force and motivation for adaptive and synthetic aperture signal processing techniques that have been discussed in Chapter 6.

An additional constraint for sonar applications requires that the frequency resolution Δf of the hydrophone time series for spatial spectral analysis, which is based on FFT beam-forming processing, must be

$$\Delta f \times \frac{L}{c} \ll 1 \tag{11.16}$$

to satisfy *frequency quantization* effects associated with the implementation of the beam-forming process as Finite-Duration Impulse Response (FIR) filters that have been dis-cussed in Chapter 6. Because of the linearity of the conventional beamforming process, an exact equivalence of the frequency domain narrowband beamformer with that of the time domain beamformer for broadband signals can be derived.[64,68,69] The time domain beamformer is simply a time delaying[69] and summing process across the hydrophones of the line array, which is expressed by

$$b(\theta_s, t_i) = \sum_{n=1}^{N} X_n(t_i - \tau_s). \tag{11.17}$$

Since $b(\theta_s, t_i)=\text{IFFT}\{B(f, \theta_s)\}$, by using FFTs and fast-convolution procedures, continuous beam time sequences can be obtained at the output of the frequency domain beamform-er.[64] This is a very useful operation when the implementation of adaptive beamforming processors in sonar systems is considered.

When gain optimization is considered as the approach to select the beamforming weights, then the beamforming response is optimized so that the output contains minimal contributions due to noise and signals arriving from directions other than the desired signal direction. For this optimization procedure, it is desired to find a linear filter vector $\overline{W}(f_i,\theta)$, which is a solution to the constrained minimization problem that allows signals from the look direction to pass with a specified gain,[11,12] as discussed in Chapter 6. Then in the frequency domain, an adaptive beam at a steering θ_s is defined by

$$B(f_i,\theta_s) = \overline{W}^*(f_i,\theta_s)\overline{X}(f_i), \tag{11.18}$$

and the corresponding conventional beams are provided by Equation 11.11. Estimates of the adaptive beamforming weights $\overline{W}(f_i,\theta)$ are provided by various adaptive processing techniques that have been discussed in detail in Chapter 6.

11.2.2 System Implementation Aspects

The major development effort discussed in Chapter 6 has been devoted to designing a generic beamforming structure that will allow the implementation of adaptive, synthetic aperture, and spatial spectral analysis techniques in integrated active-passive sonar system. The practical implementation of the numerous adaptive and synthetic aperture processing techniques, however, requires the consideration of the characteristics of the signal and noise, the complexity of the ocean environment, as well as the computational difficulty. The discussion in Chapter 6 addresses these concerns and prepares the ground for the development of the above generic beamforming structure.

The major goal here is to provide a *concept demonstration* of both the sonar technology and advanced signal processing concepts that are proving invaluable in the reduction risk and in ensuing significant innovations occur during the formal development process.

Shown in Figure 11.4 is the proposed configuration of the signal processing flow that includes the implementation of FIR filters and conventional, adaptive, and synthetic aperture beamformers. The reconfiguration of the different processing blocks in Figure 11.4 allows the application of the proposed configuration into a variety of active and/or passive sonar systems. The shaded blocks in Figure 11.4 represent advanced signal processing concepts of next-generation sonar systems, and this basically differentiates their functionality from the current operational sonars. In a sense, Figure 11.4 summarizes the signal processing flow of the advanced signal processing schemes shown in Figure 6.14 and Figures 6.20 through 6.24 of Chapter 6.

The first point of the generic processing flow configuration in Figure 11.4 is that its implementation is in the frequency domain. The second point is that the frequency domain beamforming (or spatial filtering) outputs can be made equivalent to the FFT of the broadband beamformers outputs with proper selection of beamforming weights and careful data partitioning. This equivalence corresponds to implementing FIR filters via circular convolution. It also allows spatial-temporal processing of narrowband and broadband types of signals as well. As a result, the output of each one of the processing blocks in Figure 11.4 provides continuous time series. This modular structure in the signal processing flow is a very essential processing arrangement, allowing the integration of a great variety of processing schemes such as the ones considered in this study. The details of the proposed generic processing flow, as shown in Figure 11.4, are very briefly the following:

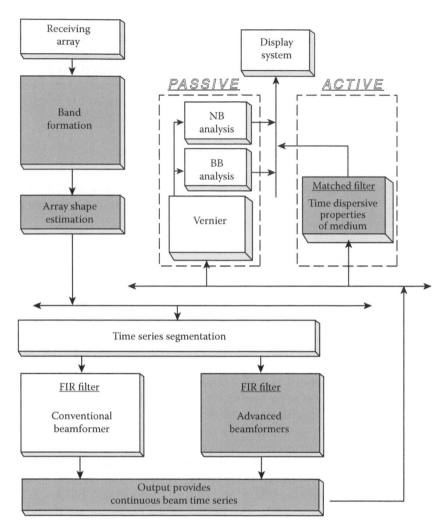

FIGURE 11.4
Schematic diagram of a generic signal processing flow that allows the implementation of nonconventional processing schemes in sonar systems. (Reprinted by permission of IEEE © 1998.)

- The block named as *initial spectral FFT—and formation* includes the partitioning of the time series from the receiving sensor array, their initial spectral FFT, the selection of the signal's frequency band of interest via bandpass FIR filters, and downsampling.[65–67] The output of this block provides continuous time series at a reduced sampling rate.

- The major blocks including *conventional spatial FIR filtering and adaptive and synthetic aperture FIR filtering* provide continuous directional beam time series by using the FIR implementation scheme of the spatial filtering via circular convolution.[64–67] The segmentation and overlap of the time series at the input of the beamformers takes care of the wraparound errors that arise in fast-convolution signal processing operations. The overlap size is equal to the effective FIR filter's length.

- The block named *matched filter* is for the processing of echoes for active sonar applications. The intention here is to compensate also for the time dispersive properties of the medium by having as an option the inclusion of the medium's propagation characteristics in the replica of the active signal considered in the matched filter in order to improve detection and gain.

- The blocks *vernier, NB analysis,* and *BB analyisis*[67] include the final processing steps of a temporal spectral analysis. The inclusion of the vernier here is to allow the option for improved frequency resolution capabilities depending on the application.

- Finally, the block *display system* includes the data normalization[42,44] in order to map the output results into the dynamic range of the display devices in a manner which provides a CFAR capability.

The strength of this generic implementation scheme is that it permits, under a parallel configuration, the inclusion of nonlinear signal processing methods such adaptive and synthetic aperture, as well as the equivalent conventional approach. This permits a very cost-effective evaluation of any type of improvements during the concept demonstration phase.

All the variations of adaptive processing techniques, while providing good bearing/frequency resolution, are sensitive to the presence of system errors. Thus, the deformation of a towed array, especially during course alterations, can be the source of serious performance degradation for the adaptive beamformers. This performance degradation is worse than it is for the CBF. So, our concept of the generic beamforming structure requires the integration of towed array shape estimation techniques[73-78] in order to minimize the influence of system errors on the adaptive beamformers. Furthermore, the fact that the advanced beamforming blocks of this generic processing structure provide continuous beam time series allows the integration of passive and active sonar application in one signal processor. Although this kind of integration may exist in conventional systems, the integration of adaptive and synthetic aperture beamformers in one signal processor for active and passive applications has not been reported yet, except for the experimental system discussed in Reference 1. Thus, the beam time series from the output of the conventional and non-CBFs are provided at the input of two different processing blocks, the passive and active processing units, as shown in Figure 11.4.

In the passive unit, the use of verniers and the temporal spectral analysis (incorporating segment overlap, windowing, and FFT coherent processing[31,32]) provide the narrowband results for all the beam time series. Normalization and OR-ing[42,44] are the final processing steps before displaying the output results. Since a beam time sequence can be treated as a signal from a directional hydrophone having the same AG and directivity pattern as that of the above beamforming processing schemes, the display of the narrowband spectral estimates for all the beams follows the so-called LOFAR presentation arrangements, as shown in Figures 11.10 through 11.19. This includes the display of the beam-power outputs as a function of time, steering beam (or bearing), and frequency. LOFAR displays are used mainly by sonar operators to detect and classify the narrowband characteristics of a received signal.

Broadband outputs in the passive unit are derived from the narrowband spectral estimates of each beam by means of incoherent summation of all the frequency bins in a wideband of interest. This kind of energy content of the broadband information is displayed as a function of bearing and time, as shown by the real data results of Section 11.3.

In the active unit, the application of a matched filter (or replica correlator) on the beam time series provides coherent broadband processing. This allows detection of echoes as a function of range and bearing for reference waveforms transmitted by the active transducers of a sonar system. The displaying arrangements of the correlator's output data are similar to the LOFAR displays and include, as parameters, range as a function of time and bearing, as discussed in Section 11.2.

At this point, it is important to note that for active sonar applications, waveform design and matched filter processing must not only take into account the type of background interference encountered in the medium, but should also consider the propagation characteristics (multipath and time dispersion) of the medium and the features of the target to be encountered in a particular underwater environment. Multipath and time dispersion in either deep or shallow water cause energy spreading that distorts the transmitted signals of an active sonar, and this results in a loss of matched filter processing gain if the replica has the properties of the original pulse.[1–4,8,54,102,114,115] Results from a study by Hermand and Roderick[103] have shown that the performance of a conventional matched filter can be improved if the reference signal (replica) compensates for the multipath and the time dispersion of the medium. This compensation is a model-based matched filter operation, including the correlation of the received signal with the reference signal (replica) that consists of the transmitted signal convolved with the impulse response of the medium. Experimental results for a one-way propagation problem have shown also that the model-based matched filter approach has improved performance with respect to the conventional matched filter approach by as much as 3.6 dB. The above remarks should be considered as supporting arguments for the inclusion of model-based matched filter processing in the generic signal processing structure shown in Figure 11.4.

11.2.3 Active Sonar Systems

Emphasis in the discussion so far has been centered on the development of a generic signal processing structure for integrated active-passive sonar systems. The active sonar problem, however, is slightly different than the passive sonar problem. The fact that the advanced beamforming blocks of the generic processing structure provide continuous beam time series allows for the integration of passive and active sonar application into one signal processor. Thus, the beam time series from the output of the conventional and non-CBFs are provided at the input of two different processing blocks, the passive and active processing units, as shown in Figure 11.4. In what follows, the active sonar problem analysis is presented with an emphasis on long-range, LF active towed array sonars. The parameters and deployment procedures associated with the short-range active problem are conceptually identical with those of the LF towed array sonars. Their differences include mainly the frequency range of the related sonar signals and the deployment of these sonars, as illustrated schematically in Figure 6.1 of Chapter 6.

11.2.3.1 *Low-Frequency Active Sonars*

Active sonar operations can be found in two forms. These are referred to as monostatic and bistatic. Monostatic sonar operations require that the source and array receivers be deployed by the same naval vessel, while bistatic or multistatic sonar operations require the deployment of the active source and the receiving arrays by different naval vessels, respectively. In addition, both monostatic and bistatic systems can be air deployed. In bistatic or multistatic sonar operations, coordination between the active source and the receiving arrays is essential. For more details on the principles and operational deployment procedures of multistatic sonars, the reader is referred to References 4, 8, and 54. The signal processing schemes that will be discussed in this section are applicable to both bistatic and monostatic LF active operations. Moreover, it is assumed that the reader is familiar with the basic principles of active sonar systems which can be found in References 4, 28, and 54.

11.2.3.1.1 Signal Ambiguity Function and Pulse Selection

It has been shown in Chapter 6 that for active sonars the optimum detector for a known signal in white Gaussian noise is the correlation receiver.[28] Moreover, the performance of the system can be expressed by means of the ambiguity function, which is the output of the quadrature detector as a function of time delay and frequency. The width of the ambiguity function along the time-delay axis is a measure of the capacity of the system to resolve the range of the target and is approximately equal to

- The duration of the pulse for a continuous wave (CW) signal
- The inverse of the bandwidth of broadband pulses such as linear frequency modulation (LFM), hyperbolic frequency modulation (HFM), and pseudo-random noise (PRN) waveforms

On the other hand, the width of the function along the frequency axis (which expresses the Doppler-shift or velocity tolerance) is approximately equal to

- The inverse of the pulse duration for CW signals
- The inverse of the time-bandwidth product of frequency modulated (FM) types of signals

Whalen[28] has shown that in this case there is an uncertainty relation, which is produced by the fact that the time-bandwidth product of a broadband pulse has a theoretical bound. Thus, one cannot achieve arbitrarily good range and Doppler resolution with a single pulse. Therefore, the pulse duration and the signal waveform, whether this is a mono-chromatic or broadband type of pulse, is an important design parameter. It is suggested that a sequence of CW and FM types of pulses, such as those shown in Figure 11.5, could address issues associated with the resolution capabilities of an active sonar in terms of a

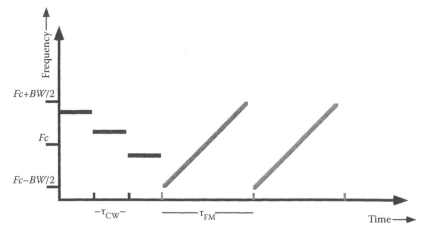

FIGURE 11.5
Sequence of CW and FM types of pulses for an LF active towed array system.

detected target's range and velocity. Details regarding the behavior (in terms of the effects of Doppler) of the various types of pulses, such as CW, LFM, HFP, and PRN, can be found in References 8, 28, 54, and 110.

11.2.3.1.2 *Effects of Medium*

The effects of the underwater environment on active and passive sonar operations have been discussed in numerous papers[1,4,8,41,54,110] and in Chapter 6. Briefly, these effects for active sonars include

- Time, frequency, and angle spreading
- Surface, volume, and bottom scattering
- Ambient and self receiving array noise

Ongoing investigations deal with the development of algorithms for model-based matched filter processing that will compensate for the distortion effects and the loss of matched filter processing gain imposed by the time dispersive properties of the medium on the transmitted signals of active sonars. This kind of model-based processing is identified by the block, *matched filter: time dispersive properties of medium*, which is part of the generic signal processing structure shown in Figure 11.4. It is anticipated that the effects of angle spreading, which are associated with the spatial coherence properties of the medium, will have a minimum impact on LF active towed array operations in blue (deep) waters. However, for littoral water (shallow coastal areas) operations, the medium's spatial coherence properties would impose an upper limit on the aperture size of the deployed towed array, as discussed in Chapter 6.

Furthermore, the medium's time and frequency spreading properties would impose an upper limit on the transmitted pulse's duration τ and bandwidth Bw. Previous research efforts in this area suggest that the pulse duration of CW signals in blue waters should be in the range of 2 to 8 s, and in shallow littoral waters in the range of 1–2 s long. On the other hand, broadband pulses, such as LFM, HFM, and PRN, when used with active towed array sonars should have upper limits

- For their bandwidth in the range of 300 Hz
- For their pulse duration in the range of 4–24 s

Thus, it is apparent by the suggested numbers of pulse duration and the sequence of pulses, shown in Figure 11.5, that the anticipated maximum detection range coverage of LF active towed array sonars should be beyond ranges of the order of $O(10^2)$ km. This assumes, however, that the intermediate range coverage will be carried out by the MF hull-mounted active sonars.

Finally, the effects of scattering play the most important role on the selection of the type of transmitted pulses (whether they will be CW or FM) and the duration of the pulses. In addition, the performance of the matched filter processing will also be affected.

11.2.3.2 *Effects of Bandwidth in Active Sonar Operations*

If an FM signal is processed by a matched filter, which is an optimum estimator according to the Neyman-Pearson detection criteria, theory predicts[28,110] that a larger bandwidth FM signal will result in improved detection for an extended target in reverberation. For extended targets in white noise, however, the detection performance depends on the SNR of the received echo at the input of the replica correlator.

In general, the performance of a matched filter depends on the temporal coherence of the received signal and the time-bandwidth product of the FM signal in relation to the relative target speed. Therefore, the signal processor of an active sonar may require a variety of matched filter processing schemes that will not have degraded performance when the coherence degrades or the target velocity increases. At this point, a brief overview of some of the theoretical results will be given in order to define the basic parameters characterizing the active signal processing schemes of interest.

It is well known[28] that for a linear FM signal with bandwidth, Bw, the matched filter provides pulse compression and the temporal resolution of the compressed signal is $1/Bw$. Moreover, for extended targets with virtual target length, $T\tau$ (in seconds), the temporal resolution at the output of the matched filter should be matched to the target length, $T\tau$. However, if the length of the reverberation effects is greater than that of the extended target, the reverberation component of bandwidth will be independent in frequency increments, $\Delta Bw > 1/T\tau$.[30,110] Therefore, for an active LF sonar, if the transmitted broadband signal $f(t)$ with bandwidth Bw is chosen such that it can be decomposed into n signals, each with bandwidth $\Delta Bw = Bw/n > 1/T\tau$, then the matched filter outputs for each one of the n signal segments are independent random variables. In this case, called reverberation limited, the SNR at the output of the matched filter is equal for each frequency band ΔBw, and independent of the transmitted signal's bandwidth Bw as long as $Bw/n > 1/T\tau$. This processing arrangement, including segmentation of the transmitted broadband pulse, is called *segmented replica correlator* (SRC).

To summarize the considerations needed to be made for reverberation-limited environments, the area (volume) of scatterers decreases as the signal bandwidth increases, resulting in less reverberation at the receiver. However, large enough bandwidths will provide range resolution narrower than the effective duration of the target echoes, thereby requiring an approach to recombine the energy from time-spread signals. For CW waveforms, the potential increase in reverberation suppression at low Doppler provided by long-duration signals is in direct competition with the potential increase in reverberation returned near the transmit frequency caused by the illumination of a larger area (volume) of scatterers. Piecewise coherent (PC) and geometric comb waveforms have been developed to provide good simultaneous range and Doppler resolution in these reverberation-limited environments. Table 11.1 provides a summary for waveform selection based on the reverberation environment and the motion of the target.

In contrast to the reverberation-limited case, the SNR in the noise-limited case is inversely proportional to the transmitted signal's bandwidth, Bw, and this case requires long replica correlation. Therefore, for the characterization of a moving target, simultaneous estimation of time delay and Doppler speed is needed. But for broadband signals, such as HFM,

TABLE 11.1

Waveform Considerations in Reverberation

Doppler	Background Reverberation		
	Low	Medium	High
Low	FM	FM	FM
Moderate	FM (CW)	PC (CW, HFM)	CW (PC)
High	CW (HFM)	CW (HFM)	CW

LFM, and PRN, the Doppler effects can no longer be approximated simply as a frequency shift. In addition, the bandwidth limitations, due to the medium and/or the target characteristics, require further processing considerations whether or not a long or SRC will be the optimum processing scheme in this case.

It is suggested that a sequence of CW and broadband transmitted pulses, such as those shown in Figure 11.5, and the signal processing scheme, presented in Figure 11.6, could address the above complicated effects that are part of the operational requirements of LF active sonar systems. In particular, the CW and broadband pulses would simultaneously provide sufficient information to estimate the Doppler and time-delay parameters characterizing a detected target. As for the signal processing schemes, the signal processor of an LF active towed array system should allow simultaneous processing of CW pulses as well as bandwidth-limited processing for broadband pulses by means of replica correlation integration (RCI) and/or segmented replica correlation.

11.2.3.2.1 *Likelihood Ratio Test Detectors*

This section deals with processing to address the effects of both the bandwidth and the medium on the received waveform. As stated above, the RC function is used to calculate the likelihood ratio test (LRT) statistic for the detection of high time-bandwidth waveforms.[28] These waveforms can be expected to behave well in the presence of reverberation due to the $1/B$ effective pulse length. Because the received echo undergoes distortion during its two-way propagation and reflection from the target, the theoretical RC gain of $10 \log BT$ relative to a zero-Doppler CW echo is seldom achievable, especially in shallow water where multipath effects are significant. The standard RC matched

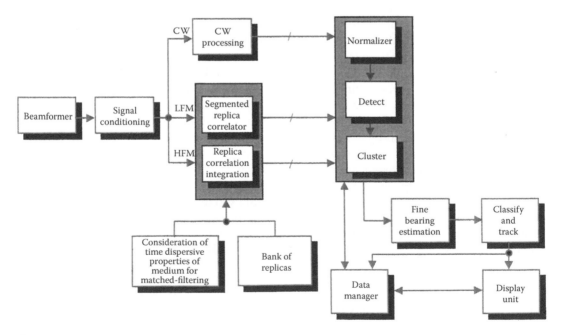

FIGURE 11.6
Active waveform processing block diagram. Inputs to the processing flow of this schematic diagram are the beam time series outputs of the advanced beamformers of Figure 11.4. The various processing blocks indicate the integration of the medium's time dispersive properties in the matched filter and the long or segmented replica correlations for FM type of signals to improve detection performance for noise-limited or reverberation-limited cases discussed in Section 11.2.3.2.

filter assumes an ideal channel and performs a single coherent match of the replica to the received signal at each point in time. This nth correlation output is calculated as the inner product of the complex conjugate of the transmitted waveform with the received data so that

$$y(n) = \left| \sqrt{2/N} \sum_{i=0}^{N-1} s^*(i) r(i+n) \right|^2 . \tag{11.19}$$

One modification to the standard RC approach of creating the test statistic is designed to recover the distortion losses caused by time spreading.[114,115] This statistic is formed by effectively placing an energy detector at the output of the matched filter and is termed RCI, or long replica correlator (LRC). The RCI test statistic is calculated as

$$y(n) = \sum_{k=0}^{M-1} \left| \sqrt{2/N} \sum_{i=0}^{N-1} s^*(i-k) r(i+n) \right| . \tag{11.20}$$

The implementation of RCI requires a minimal increase in complexity, consisting only of an integration of the RC statistic over a number of samples (M) matched to the spreading of the signal. Sample RCI recovery gains with respect to standard RC matched filtering have been shown to exceed 3 dB.

A second modification to the matched filter LRT statistic, called SRC and introduced in the previous section, is designed to recover the losses caused by fast-fading channel distortion, where the ocean dynamics permit the signal coherence to be maintained only over some period T_c that is shorter than the pulse length.[114,115] This constraint forces separate correlation over each segment of length T_c so that the receiver waveform gets divided into $M_s = T/T_c$ segments, where T is the length of the transmitted pulse. For implementation purposes, M_s should be an integer so that the correlation with the replica is divided into M_s evenly sized segments. The SRC test statistic is calculated as

$$y(n) = \sum_{k=0}^{M_S-1} \left| \sqrt{2M_S/N} \sum_{i=0}^{N/M_S-1} s^* \left(i + \frac{kn}{M_S} \right) r \left(i + n \frac{kN}{M_S} \right) \right|^2 . \tag{11.21}$$

One disadvantage of SRC in comparison to RCI is that SRC does not support multihypothesis testing when the amount of distortion is not known *a priori*.[114]

11.2.3.2.2 Normalization and Threshold Detection for Active Sonar Systems

Figure 11.6 presents a processing scheme for active sonars that addresses the concerns about processing waveforms like the one presented in Figure 11.5 and about the difficulties in providing robust detection capabilities. At this point, it is important to note that the block named *normalizer*[42,44] in Figure 11.6 does not include simple normalization schemes such as those assigned for the LOFAR-grams of a passive sonar, shown in Figure 11.4.

The ultimate goal of any normalizer in combination with a threshold detector is to provide a system-prescribed and constant rate of detections in the absence of a target, while maintaining an acceptable probability of detection when a target is present. The detection statistic processing output (or FFT output for CW waveforms) is normalized and threshold detected prior to any additional processing. The normalizer estimates the power (and frequency) distribution of the mean background (reverberation plus noise) level at the output of the detection statistic processing.

The background estimate for a particular test bin that may contain a target echo is formed by processing a set of data that is assumed to contain no residual target echo components. The decision statistic output of the test bin gets compared to the threshold that is calculated as a function of the background estimate. A threshold detection occurs when the threshold is exceeded. Therefore, effective normalization is paramount to the performance of the active processing flow.

Normalization and detection are often performed using a split window mean estimator.[42,44,45] Two especially important parameters in the design of this estimator are the guard window and the estimation window sizes placed on both sides (in range delay) of the test bin (and also along the frequency axis for CW). The detection statistic values of the bins in the estimation windows are used to calculate the background estimate, whereas the bins in the guard windows provide a gap between the test bin of interest and the estimation bins. This gap is designed to protect the estimation bins from containing target energy if a target is indeed present. The estimate of the background level is calculated as

$$\hat{\sigma}^2 = \frac{1}{K} \sum_{\{K\}} y(k),$$
(11.22)

in which $\{y(k)\}$ are the detection statistic outputs in the K estimation window bins. If the background reverberation plus noise is Gaussian, the detection threshold becomes[28]

$$\lambda_T = -\hat{\sigma}^2 \ln P_{fa}.$$
(11.23)

The split window mean estimator is a form of CFAR processing because the false alarm probability is fixed, providing there are no target echo components in the estimation bins. If the test bin contains a target echo and some of the estimation bins contain target returns, then the background estimate will likely be biased high, yielding a threshold that exceeds the test bin value so that the target does not get detected. Variations of the split window mean estimator have been developed to deal with this problem. These include (1) the simple removal of the largest estimation bin value prior to the mean estimate calculation and (2) clipping and replacement of large estimation bin values to remove outliers from the calculation of the mean estimate.

Most CFAR algorithms also rely on the stationarity of the underlying distribution of the background data. If the distribution of the data used to calculate the mean background level meets the stationarity assumptions, then the algorithm can indeed provide CFAR performance. Unfortunately, the real ocean environment, especially in shallow water, yields highly nonstationary reverberation environments and target returns with significant multipath components. Because the data are stochastic, the background estimates made by the normalizer have a mean and a variance. In nonstationary reverberation environments, these measures may depart from the design mean and variance for a stationary background. As the nonstationarity of the samples used to compute the background estimate increases, the performance of the CFAR algorithm degrades accordingly, causing

1. Departure from the design false alarm probability
2. A potential reduction in detectability

For example, if the mean estimate is biased low, the probability of false alarm increases. And, if the mean estimate is biased high, the reduction in signal-to-reverberation-plus-noise ratio causes a detection loss.

Performance of the split window mean estimator is heavily dependent upon the guard and estimation window sizes. Optimum performance can be realized when both the guard window size is well matched to the time (and frequency for CW) extent of the target return and the estimation window size contains the maximum number of independent, identically distributed, reverberation-plus-noise bins. The time extent for the guard window can be determined from the expected multipath spread in conjunction with the aspect-dependent target response. The frequency spread of the CW signal is caused by the dispersion properties of the environment and the potential differential Doppler between the multipath components. The estimation window size should be small when the background is highly nonstationary and large when it is stationary.

Under certain circumstances, it may be advantageous to adaptively alter the detection thresholds based on the processing of previous pings. If high-priority detections have already been confirmed by the operator (or by postprocessing), the threshold can be lowered near these locations to ensure a higher probability of detection on the current ping. Conversely, the threshold can be raised near locations of low-priority detections to drop the probability of detection. This functionality simplifies the postprocessing and relieves the operator from the potential confusion of tracking a large number of contacts.

The normalization requirements for an LF active sonar are complicated and are a topic of ongoing research. More specifically, the bandwidth effects, discussed in Section 11.2.3.2, need to be considered also in the normalization process by using several specific normalizers. This is because an active sonar display requires normalized data that retain bandwidth information, have reduced dynamic range, and have constant false alarm rate capabilities which can be obtained by suitable normalization.

11.2.3.3 Display Arrangements for Active Sonar Systems

The next issue of interest is the display arrangement of the output results of an LF active sonar system. There are two main concerns here. The first is that the display format should provide sufficient information to allow for an unbiased decision that a detection has been achieved when the received echoes include sufficient information for detection. The second concern is that the repetition rate of the transmitted sequence of pulses, such as the one shown in Figure 11.5, should be in the range of 10–15 min. These two concerns, which may be viewed also as design restrictions, have formed the basis for the display formats of CW and FM signals, which are discussed in the following sections.

11.2.3.3.1 Display Format for CW Signals

The processing of the beam time series, containing information about the CW transmitted pulses, should include temporal spectral analysis of heavily overlapped segments. The display format of the spectral results associated with the heavily overlapped segments should be the same with that of a LOFAR-gram presentation arrangement for passive sonars.

Moreover, these spectral estimates should include the so-called *ownship Doppler nullification*, which removes the component of Doppler shift due to ownship motion. The left part of Figure 11.7A shows the details of the CW display format for an active sonar as well as the mathematical relation for the ownship Doppler nullification.

Accordingly, the display of active CW output results of an active sonar should include LOFAR-grams that contain all the number of beams provided by the associated beam-former. The content of output results for each beam will be included in one window, as shown at the left-hand side of Figure 11.7B. Frequencies will be shown by the horizontal axis. The temporal spectral estimates of each heavily overlapped segment will be plotted as a series of gray-scale pixels along the frequency axis. Mapping of the power levels of the temporal spectral estimates along a sequence of gray-scale pixels will be derived according to normalization processing schemes for the passive LOFAR-gram sonar displays.

If three CW pulses are transmitted, as shown in Figure 11.5, then the temporal spectral estimates will include a frequency shift that would allow the vertical alignment of the spectral estimates of the three CW pulses in one beam window. Clustering across frequencies and across beams would provide summary displays for rapid assessment of the operational environment.

11.2.3.3.2 *Display Format for FM Type of Signals*

For FM type of signals, the concept of processing heavily overlapped segments should also be considered. In this case, the segments will be defined as heavily overlapped replicas derived from a long broadband transmitted signal, as discussed in Section 11.2.3.2. However, appropriate time shifting would be required to align the corresponding time-delay estimates from each segmented replica in one beam window. The display format of the output results will be the same as those of the CW signals. Shown at the right-hand side of Figure 11.7A are typical examples of FM types of display outputs. One real data example of an FM output display for a single beam is given in Figure 11.7B. This figure shows replica correlated data from 30 pings separated by a repetition interval of approximately 15 min.

At this point, it is important to note that for a given transmitted FM signal a number of Doppler shifted replicas might be considered to allow for multidimensional search and estimation of range and velocity of a moving target of interest.

Thus, it should be expected that during active LF towed array operations the FM display outputs will be complicated and multidimensional. However, a significant downswing of the number of displays can be achieved by applying clustering across time delays, beams, and Doppler shift. This kind of clustering will provide summary displays for rapid assessment of the operational environment, as well as critical information and data reduction for classification and tracking.

In summary, the multidimensional active sonar signal processing, as expressed by Figures 11.5 through 11.7, is anticipated to define active sonar operations for the next-generation sonar systems. However, the implementation in real-time active sonars of the concepts that have been discussed in the previous sections will not be a trivial task. As an example, Figures 11.8 and 11.9 present the multidimensionality of the processing flow associated with the SRCs and LRCs shown in Figure 11.6. Briefly, the schematic interpretation of the signal processing details in Figures 11.8 and 11.9 reflects the implementation and mapping in sonar computing architectures of the multidimensionality requirements of next-generation active sonars. If operational requirements would demand large number of beams and Doppler shifted replicas, then the anticipated multidimensional processing, shown in Figures 11.8 and 11.9, may lead to prohibited computational requirements.

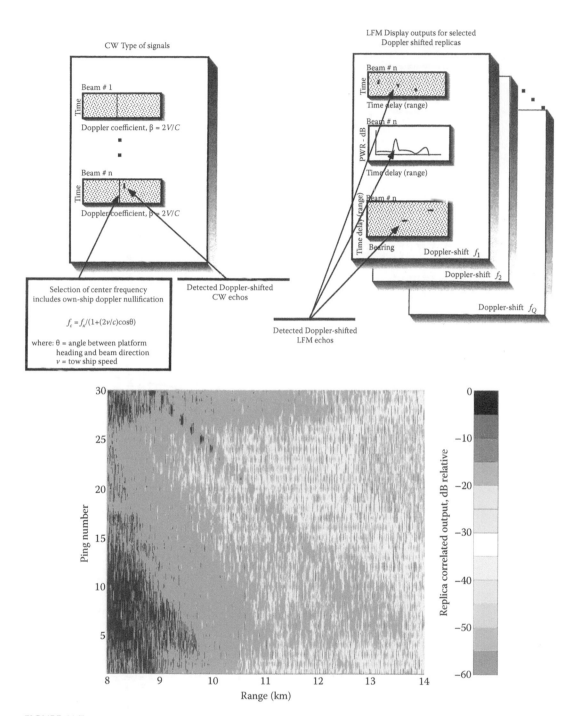

FIGURE 11.7
(A) Display arrangements for CW and FM pulses. The left part shows the details of the CW display format that includes the ownship Doppler nullification. The right part shows the details of the FM type display format for various combinations of Doppler shifted replicas. (B) Replica correlated FM data displayed with time on the vertical axis and range along the horizontal axis for one beam. The detected target is shown as a function of range by the received echoes forming a diagonal line on the upper left corner of the display output.

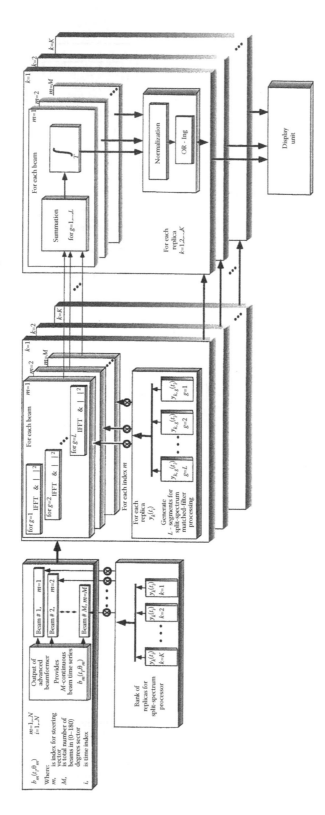

FIGURE 11.8

Signal processing flow of an SRC. The various layers in the schematic diagram represent the combinations that are required between the segments of the replica correlators and the steering beams generated by the advanced beamformers of the active sonar system. The last set of layers (at the right-hand side) represent the corresponding combinations to display the results of the SRC according to the display formats of Figure 11.7.

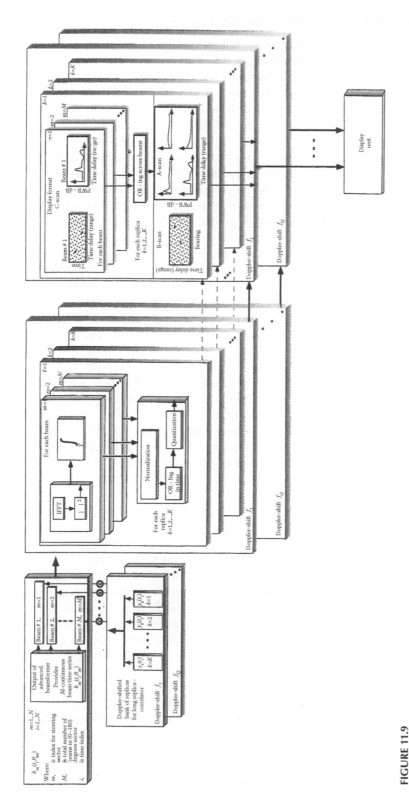

FIGURE 11.9

Processing flow of an LRC. The various layers in the schematic diagram represent the combinations that are required between the Doppler shifted replicas of the LRC and the steering beams generated by the advanced beamformers of the active sonar system. The last set of layers (at the right-hand side) represent the corresponding combinations to display the results of the LRC according to the display formats of Figure 11.7.

11.2.4 Comments on Computing Architecture Requirements

The implementation of this investigation's nonconventional processing schemes in sonar systems is a nontrivial issue. In addition to the selection of the appropriate algorithms, success is heavily dependent on the availability of suitable computing architectures.

Past attempts to implement matrix-based signal processing methods, such as adaptive beamformers reported in this chapter, were based on the development of systolic array hardware, because systolic arrays allow large amounts of parallel computation to be performed efficiently since communications occur locally. None of these ideas are new. Unfortunately, systolic arrays have been much less successful in practice than in theory. The fixed-size problem for which it makes sense to build a specific array is rare. Systolic arrays big enough for real problems cannot fit on one board, much less one chip, and interconnects have problems. A 2-D systolic array implementation will be even more difficult. So, any new computing architecture development should provide high throughput for vector- as well as matrix-based processing schemes.

A fundamental question, however, that must be addressed at this point is whether it is worthwhile to attempt to develop a system architecture that can compete with a multiprocessor using stock microprocessors. Although recent microprocessors use advanced architectures, improvements of their performance include a heavy cost in design complexity, which grows dramatically with the number of instructions that can be executed concurrently. Moreover, the recent microprocessors that claim high performance for peak MFLOP rates have their net throughput usually much lower, and their memory architectures are targeted toward general purpose code.

These issues establish the requirement for dedicated architectures, such as in the area of operational sonar systems. Sonar applications are computationally intensive, as shown in Chapter 6, and they require high throughput on large data sets. It is our understanding that the Canadian DND recently supported work for a new sonar computing architecture called the next-generation signal processor (NGSP).[10] We believe that the NGSP has established the hardware configuration to provide the required processing power for the implementation and real-time testing of the non-CBFs such as those reported in Chapter 6.

A detailed discussion, however, about the NGSP is beyond the scope of this chapter, and a brief overview about this new signal processor can be found in Reference 10. Other advanced computing architectures that can cover the throughput requirements of computationally intensive signal processing applications, such as those discussed in this chapter, have been developed by Mercury Computer Systems, Inc.[104] Based on the experience of the authors of this chapter, the suggestion is that implementation efforts of advanced signal processing concepts should be directed more on the development of generic signal processing structures as in Figure 11.4, rather than the development of very expensive computing architectures. Moreover, the signal processing flow of advanced processing schemes that include both scalar and vector operations should be very well defined in order to address practical implementation issues.

In this chapter, we address the issue of computing architecture requirements by defining generic concepts of the signal processing flow for integrated active-passive sonar systems, including adaptive and synthetic aperture signal processing schemes. The schematic diagrams in Figure 6.14 and Figures 6.20 through 6.24 of Chapter 6 show that the implementation of advanced sonar processing concepts in sonar systems can be carried out in existing computer architectures[10,104] as well as in a network of general purpose computer workstations that support both scalar and vector operations.

11.3 Real Results from Experimental Sonar Systems

The real data sets that have been used to test the implementation configuration of the above nonconventional processing schemes come from two kinds of experimental setups. The first one includes sets of experimental data representing an acoustic field consisting of the tow ship's self noise and the reference narrowband CWs, as well as broadband signals such as HFM and pseudo-random transmitted waveforms from a deployed source. The absence of other noise sources as well as noise from distant shipping during these experiments make this set of experimental data very appropriate for concept demonstration. This is because there are only a few known signals in the received hydrophone time series, and this allows an effective testing of the performance of the above generic signal processing structure by examining various possibilities of artifacts that could be generated by the non-CBFs.

In the second experimental setup, the received hydrophone data represent an acoustic field consisting of the reference CW, HFM, and broadband signals from the deployed source that are embodied in a highly correlated acoustic noise field including narrowband and broadband noise from heavy shipping traffic. During the experiments, signal conditioning and continuous recording on a high-performance digital recorder were provided by a real-time data system.

The generic signal processing structure, presented in Figure 11.4, and the associated signal processing algorithms (minimum variance distortionless response [MVDR], generalized sidelobe cancellers [GSC], steered minimum variance [STMV], extended towed array measuremnts [ETAM], matched filter), discussed in Chapter 6, were implemented in a workstation supporting a UNIX operating system and FORTRAN and C compilers, respectively.

Although the CPU power of the workstation was not sufficient for real-time signal processing response, the memory of the workstation supporting the signal processing structure of Figure 11.4 was sufficient to allow above of continuous hydrophone time series up to 3 h long. Thus, the output results of the above generic signal processing structure were equivalent to those that would have been provided by a real-time system, including the implementation of the signal processing schemes discussed in this chapter.

The results presented in this section are divided into two parts. The first part discusses passive narrowband and broadband towed array sonar applications. The scope here is to evaluate the performance of the adaptive and synthetic aperture beamforming techniques and to assess their ability to track and localize narrowband and broadband signals of interest while suppressing strong interferers. The impact and merits of these techniques will be contrasted with the localization and tracking performance obtained using the CBF.

The second part of this section presents results from active towed array sonar applications. The aim here is to evaluate the performance of the adaptive and synthetic aperture beamformers in a matched filter processing environment.

11.3.1 Passive Towed Array Sonar Applications

11.3.1.1 Narrowband Acoustic Signals

The display of narrowband bearing estimates, according to a LOFAR presentation arrangement, are shown in Figures 11.10 through 11.12. Twenty-five beams equally spaced in $[1, -1]$

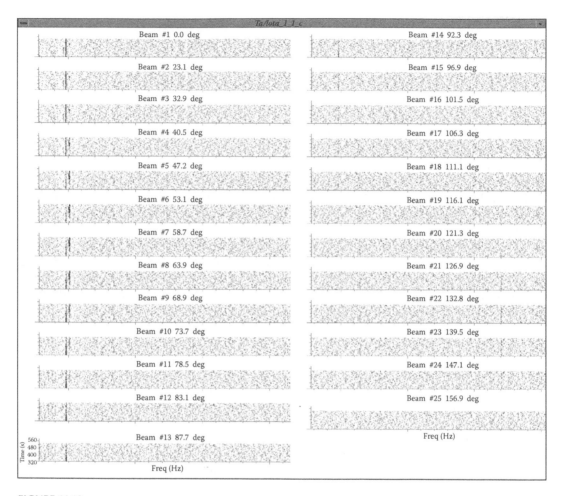

FIGURE 11.10

Conventional beamformer's LOFAR narrowband output. The 25 windows of this display correspond to the 25 steered beams equally spaced in [1, –1] cosine space. The acoustic field included three narrowband signals. Very weak indications of the CW signal of interest are shown in beams #21–24. (Reprinted by permission of IEEE © 1998.)

cosine space were steered for the conventional, the adaptive, and the synthetic aperture beamforming processes. The wavelength λ of the reference CW signal was approximately equal to 1/6 of the aperture size L of the deployed line array. The power level of the CW signal was in the range of 130 dB re 1 μPa, and the distance between the source and receiver was of the order of $O(10^1)$ nm. The water depth in the experimental area was 1000 m, and the deployment depths of the source and the array receiver were approximately 100 m.

Figure 11.10 presents the CBF's LOFAR output. At this particular moment, we had started to lose detection of the reference CW signal tonal. Very weak indications of the presence of this CW signal are shown in beams #21–24 of Figure 11.10. In Figures 11.11 and 11.12, the LOFAR outputs of the synthetic aperture and the partially adaptive subaperture MVDR processing schemes are shown for the set of data and are the same as those of Figure 11.10. In particular, Figure 11.11 shows the synthetic aperture (ETAM algorithm) LOFAR narrowband output, which indicates that the basic difference between the LOFAR-gram results of the CBF in Figure 11.10 and those of the synthetic aperture beamformer is that

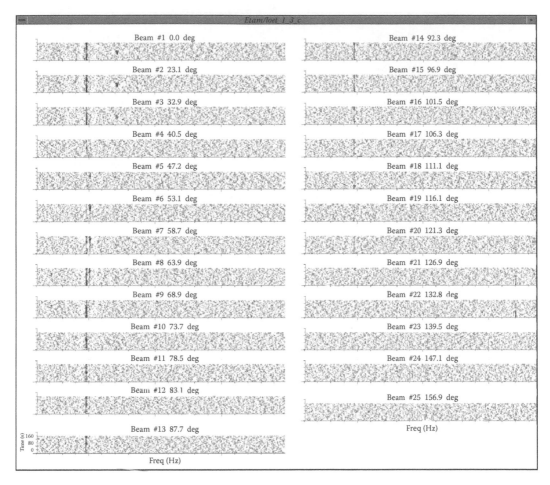

FIGURE 11.11
Synthetic aperture (ETAM algorithm) LOFAR narrowband output. The processed sensor time series are the same as those of Figure 11.10. The basic difference between the LOFAR-gram results of the con beamformer in Figure 11.10 and those of the synthetic aperture beamformer is that the improved directionality (array gain) of the non-CBF localizes the detected narrowband signals in a smaller number of beams than the CBF. For the synthetic aperture beamformer, this is translated into a better tracking and localization performance for detected narrowband signals, as shown in Figures 11.13 and 11.14. (Reprinted by permission of IEEE © 1998.)

the improved directionality (AG) of the non-CBF localizes the detected narrowband signals in a smaller number of beams than the CBF. For the synthetic aperture beamformer, this is translated into a better tracking and localization performance for detected narrowband signals, as shown in Figures 11.13 and 11.14.

Figure 11.12 presents the subaperture MVDR beamformer's LOFAR narrowband output. In this case, the processed sensor time series are the same as those of Figures 11.10 and 11.11. However, the sharpness of the adaptive beamformer's LOFAR output was not as good as the one of the conventional and synthetic aperture beamformer. This indicated loss of temporal coherence in the adaptive beam time series, which was caused by nonoptimum performance and poor convergence of the adaptive algorithm. The end result was poor tracking of detected narrowband signals by the adaptive schemes as shown in Figure 11.14.

FIGURE 11.12
Subaperture MVDR beamformer's LOFAR narrowband output. The processed sensor time series are the same as those of Figures 11.10 and 11.11. Even though the angular resolution performance of the subaperture MVDR scheme in this case was better than that of the conventional beamformer, the sharpness of the adaptive beamformer's LOFAR output was not as good as the one of the conventional and synthetic aperture beamformer. This indicated loss of temporal coherence in the adaptive beam time series, which was caused by nonoptimum performance and poor convergence of the adaptive algorithm. The end result was poor tracking of detected narrowband signals by the adaptive schemes as shown in Figure 11.15. (Reprinted by permission of IEEE © 1998.)

The narrowband LOFAR results from the subaperture GSC and STMV adaptive schemes were almost identical with those of the subaperture MVDR scheme, shown in Figure 11.12. For the adaptive beamformers, the number of iterations for the exponential averaging of the sample covariance matrix was approximately five to ten snapshots ($\mu=0.9$ convergence coefficient of Equation 6.79 in Chapter 6). Thus, for narrowband applications, the shortest convergence period of the subaperture adaptive beamformers was of the order of 60–80 s, while for broadband applications the convergence period was of the order of 3–5 s.

Even though the angular resolution performance of the adaptive schemes (MVDR, GSC, STMV) in element space for the above narrowband signal was better than that of the CBF, the sharpness of the adaptive beamformers' LOFAR output was not as good as that of the conventional and synthetic aperture beamformer. Again, this indicated loss of temporal coherence in the adaptive beam time series, which was caused by nonoptimum performance and poor convergence of the above adaptive schemes when their implementation was in element space.

Loss of coherence is evident in the LOFAR outputs because the generic beamforming structure in Figure 11.4 includes coherent temporal spectral analysis of the continuous beam time series for narrowband analysis. For the adaptive schemes implemented in element space, the number of iterations for the adaptive exponential averaging of the sample covariance matrix was 200 snapshots ($\mu=0.995$ according to Equation 6.79 in Chapter 6). In particular, the MVDR element space method required a very long convergence period of the order of 3000 s. In cases that this convergence period was reduced, then the MVDR

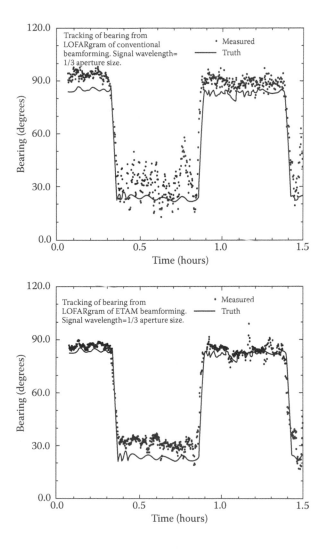

FIGURE 11.13
Signal following of bearing estimates from the conventional beamforming LOFAR narrowband outputs and the synthetic aperture (ETAM algorithm). The solid line shows the true values of source's bearing. The wavelength of the detected CW was equal to one third of the aperture size L of the deployed array. For reference, the tracking of bearing from conventional beamforming LOFAR outputs of another CW with wavelength equal to 1/16 of the towed array's aperture is shown in the lower part. (Reprinted by permission of IEEE © 1998.)

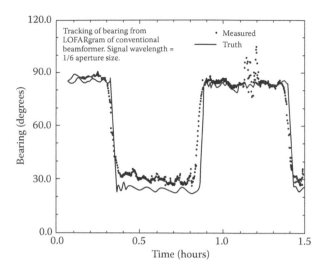

FIGURE 11.13 (Continued)

element space LOFAR output was populated with artifacts.[23] However, the performance of the adaptive schemes of this study (MVDR, GSC, STMV) improved significantly when their implementation was carried out under the subaperture configuration, as discussed in Chapter 6.

Apart from the presence of the CW signal with $\lambda=L/6$ in the conventional LOFAR display, only two more narrowband signals with wavelengths approximately equal to $\lambda=L/3$ were detected. No other signals were expected to be present in the acoustic field, and this is confirmed by the conventional narrowband output of Figure 11.10, which has white noise characteristics. This kind of simplicity in the received data is very essential for this kind of demonstration process in order to identify the presence of artifacts that could be produced by the various beamformers.

The narrowband beam power maps of the LOFAR-grams in Figures 11.10 through 11.12 form the basic unit of acoustic information that is provided at the input of the data manager of our system for further information extraction. As discussed in Section 11.3.3, one basic function of the data management algorithms is to estimate the characteristics of signals that have been detected by the beamforming and spectral analysis processing schemes, which are shown in Figure 11.4. The data management processing includes signal following or tracking[105,106] that provides monitoring of the time evolution of the frequency and the associated bearing of detected narrowband signals.

If the output results from the nonconventional beamformers exhibit improved AG characteristics, this kind of improvement should deliver better system tracking performance over that of the conventional beamformer. To investigate the tracking performance improvements of the synthetic aperture and adaptive beamformers, the deployed source was towed along a straight-line course, while the towing of the line array receiver included a few course alterations over a period of approximately 3 h. Figure 11.14 illustrates this scenario, showing the constant course of the towed source and the course alterations of the vessel towing the line array receiver.

The parameter estimation process for tracking the bearing of detected sources consisted of peak picking in a region of bearing and frequency space sketched by fixed gate sizes in the LOFAR-gram outputs of the conventional and non-CBFs. Details about this estimation process can be found in Reference 107. Briefly, the choice of the gate sizes was based on the observed bearing and frequency fluctuations of a detected

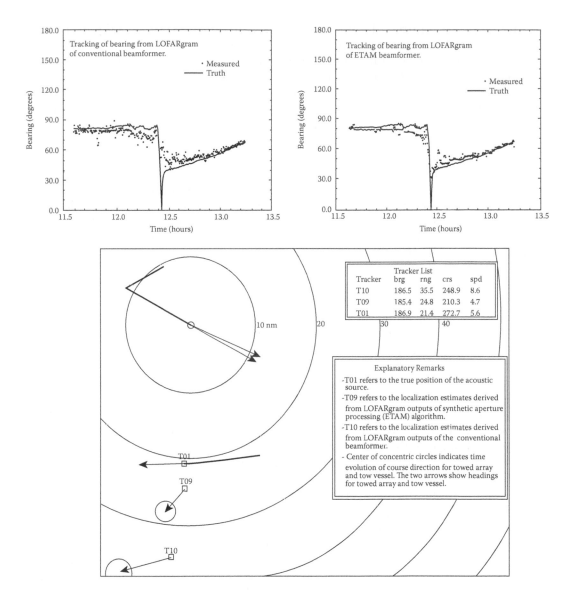

FIGURE 11.14

The upper part shows signal following of bearing estimates from conventional beamforming and synthetic aperture (ETAM algorithm) LOFAR narrowband outputs. The solid line shows the true values of source's bearing. The lower part presents localization estimates that were based on the bearing tracking results shown in the upper part. (Reprinted by permission of IEEE © 1998.)

signal of interest during the experiments. Parabolic interpolation was used to provide refined bearing estimates.[108] For this investigation, the bearings-only tracking process described in Reference 107 was used as a narrowband tracker, providing unsmoothed time evolution of the bearing estimates to the localization process.[105,109] The localization process of this study was based on a recursive extended Kalman filter formulated in Cartesian coordinates. Details about this localization process can be found in References 107 and 109.

Shown by the solid line in Figure 11.13 are the expected bearings of a detected CW signal with respect to the towed array receiver. The dots represent the tracking results of bearing estimates from LOFAR data provided by the synthetic aperture and the CBFs.

The middle part of Figure 11.13 illustrates the tracking results of the synthetic aperture beamformer. In this case, the wavelength λ of the narrowband CW signal was approximately equal to one third of the aperture size of the deployed towed array (directivity index, DI = 7.6 dB). For this very LF CW signal, the tracking performance of the CBF was very poor, as this is shown by the upper part of Figure 11.13. To provide a reference, the tracking performance of the CBF for a CW signal, having a wavelength approximetely equal to 1/16 of the aperture size of the deployed array (DI = 15 dB), is shown in the lower part of Figure 11.13.

Localization estimates for the acoustic source transmitting the CW signal with $\lambda = L/3$ were derived only from the synthetic aperture tracking results, shown in the middle of Figure 11.13. In contrast to these results, the CBF's localization estimates did not converge because the variance of the associated bearing tracking results was very large, as indicated by the results of the upper part of Figure 11.13. As expected, the CBF's localization estimates for the higher frequency CW signal $\lambda = L/16$ converge to the expected solution. This is because the system AG in this case was higher (DI = 15 dB), resulting in better bearing tracking performance with a very small variance in the bearing estimates.

The tracking and localization performance of the synthetic aperture and the conventional beamforming techniques were also assessed from other sets of experimental data. In this case, the towing of the line array receiver included only one course alteration over a period of approximately 30 min. Presented in Figure 11.14 is a summary of the tracking and localization results from this experiment. The upper part of Figure 11.14 shows tracking of the bearing estimates provided by the synthetic aperture and the conventional beamforming LOFAR-gram outputs. The lower part of Figure 11.14 presents the localization estimates derived from the corresponding tracking results.

It is apparent from the results of Figures 11.13 and 11.14 that the synthetic aperture beamformer improves the AG of small size array receivers, and this improvement is translated into a better signal tracking and target localization performance than the CBF.

With respect to the tracking performance of the narrowband adaptive beamformers, our experience is that during course alterations the tracking of bearings from the narrowband adaptive beampower outputs was very poor. As an example, the lower part of Figure 11.15 shows the subaperture MVDR adaptive beamformer's bearing tracking results for the same set of data as those of Figure 11.14. It is clear in this case that the changes of the towed array's heading are highly correlated with the deviations of the adaptive beamformer's bearing tracking results from their expected estimates.

Although the angular resolution performance of the adaptive beamformer was better than that of the synthetic aperture processing, the lack of sharpness, fuzziness, and discontinuity in the adaptive LOFAR-gram outputs prevented the signal following algorithms from tracking the signal of interest.[107] Thus, the subaperture adaptive algorithm should have provided better bearing estimates than those indicated by the output of the bearing tracker, shown in Figure 11.15. In order to address this point, we plotted the subaperture MVDR bearing estimates as a function of time for all the 25 steered beams equally spaced in [1, −1] cosine space for a frequency bin including the signal of interest. Shown in the upper part of Figure 11.15 is a waterfall of these bearing estimates. It is apparent in this case that our bearing tracker failed to follow the narrowband bearing outputs of the adaptive beamformer. Moreover, the results in the upper part of Figure 11.15 suggest signal fading and performance degradation for the narrowband adaptive processing during certain periods of the experiment.

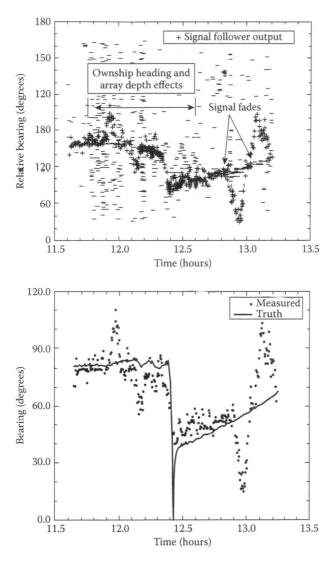

FIGURE 11.15
The upper part provides narrowband bearing estimates as a function of time for the subaperture MVDR and for a frequency bin including the signal of interest. These narrowband bearing estimates are for all the 25 steered beams equally spaced in [1, −1] cosine space. The lower part presents tracking of bearing estimates from subaperture MVDR LOFAR narrowband outputs. The solid line shows the true values of bearing. (Reprinted by permission of IEEE © 1998.)

Our explanation for this performance degradation is twofold. First, the drastic changes in the noise field, due to a course alteration, would require a large number of iterations for the adaptive process to converge. Second, since the associated sensor coordinates of the towed array shape deformation had not been considered in the steering vector $\bar{D}(f_i, \theta)$, this omission induced erroneous estimates in the noise covariance matrix during the iteration process of the adaptive processing. If a towed array shape estimation algorithm had been included in this case, the adaptive process would have provided better bearing tracking results than those shown in Figure 11.15.[77,107] For the broadband adaptive results, however, the situation is completely different, and this is addressed in the following section.

11.3.1.2 Broadband Acoustic Signals

Shown in Figure 11.16 are the conventional and subaperture adaptive broadband bearing estimates as a function of time for a set of data representing an acoustic field consisting of radiated noise from distant shipping in acoustic conditions typical of a sea state 2–4. The experimental area here is different than that including the processed data presented in Figures 11.10 through 11.15. The processed frequency regime for the broadband bearing estimation was the same for both the conventional and the partially adaptive subaperture MVDR, GSC, and STMV processing schemes. Since the beamforming operations in this

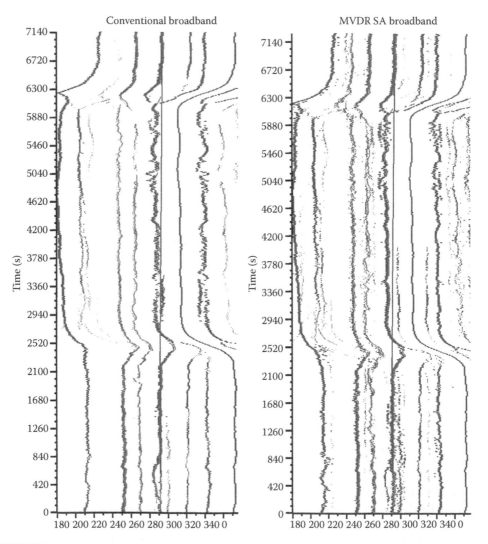

FIGURE 11.16
Broadband bearing estimates for a 2-h-long set of data: left-hand side, output from conventional beamformer; right-hand side, output from subaperture MVDR beamformer. Solid lines show signal tracking results for the broadband bearing estimates provided by the conventional and subaperture MVDR beamformers. These results show a superior signal detection and tracking performance for the broadband adaptive scheme compared with that of the conventional beamformer. This performance difference was consistent for a wide variety of real data sets. (Reprinted by permission of IEEE © 1998.)

study are carried out in the frequency domain, the LF resolution in this case was of the order of $O(10^0)$. This resulted in very short convergence periods for the partially adaptive beamformer of the order of a few seconds.

The left-hand side of Figure 11.16 shows the conventional broadband bearing estimates, and the right-hand side shows the partially adaptive broadband estimates for a 2-h-long set of data. Although the received noise level of a distant vessel was very low, the adaptive beamformer has detected this target in time-space position (240°, 6300 s) in Figure 11.16, something that the CBF has failed to show. In addition, the subaperture adaptive outputs have resolved two closely spaced broadband signal arrivals at space-time position (340°, 3000 s), while the conventional broadband output shows an indication only that two targets may be present at this space-time position.

It is evident by these results that the subaperture adaptive schemes of this study provide better detection (than the CBF) of weak signals in the presence of strong signals. For the previous set of data, shown in Figure 11.16, broadband bearing tracking results (for a few broadband signals at bearing 245°, 265°, and 285°) are shown by the solid lines for both the adaptive and the conventional broadband outputs. As expected, the signal followers of the CBF lost track of the broadband signal with bearing 240° at the time position (240°, 6300 s). On the other hand, the trackers of the subaperture adaptive beamformers did not loose track of this target, as shown by the results at the right-hand side of Figure 11.16. At this point, it is important to note that the broadband outputs of the subaperture MVDR, GSC, and STMV adaptive schemes were almost identical.

It is apparent from these results that the partially adaptive subaperture beamformers have better performance than the CBF in detecting very weak signals. In addition, the sub-aperture adaptive configuration has demonstrated tracking targets equivalent to the CBF's dynamic response during the tow vessel's course alterations. For the above set of data, localization estimates based on the broadband bearing tracking results of Figure 11.16 converged to the expected solution for both the conventional and the adaptive processing beam outputs.

Given the fact that the broadband adaptive beamformer exhibits better detection performance than the conventional method, as shown by the results of Figure 11.16 and other data sets which are not reported here, it is concluded that for broadband signals the subaperture adaptive beamformers of this study provide significant improvements in AG that result in better tracking and localization performance than that of the conventional signal processing scheme.

At this point, questions may be raised about the differences in bearing tracking performance of the adaptive beamformer for narrowband and broadband applications. It appears that the broadband subaperture adaptive beamformers as energy detectors exhibit very robust performance because the incoherent summation of the beam powers for all the frequency bins in a wideband of interest removes the fuzziness of the narrowband adaptive LOFAR-gram outputs, shown in Figure 11.12. However, a signal follower capable of tracking fuzzy narrowband signals[27] in LOFAR-gram outputs should remedy the observed instability in bearing trackings for the adaptive narrowband beam outputs. In addition, towed array shape estimators should also be included because the convergence period of the narrowband adaptive processing is of the same order as the period associated with the course alterations of the towed array operations. None of these remedies are required for broadband adaptive beamformers because of their proven robust performance as energy detectors and the short convergence periods of the adaptation process during course alterations.

11.3.2 Active Towed Array Sonar Applications

It was discussed in Chapter 6 that the configuration of the generic beamforming structure to provide continuous beam time series at the input of a matched filter and a temporal spectral analysis unit forms the basis for integrated passive and active sonar applications. However, before the adaptive and synthetic aperture processing schemes are integrated with a matched filter, it is essential to demonstrate that the beam time series from the output of these non-CBFs have sufficient temporal coherence and correlate with the reference signal. For example, if the received signal by a sonar array consists of FM type of pulses with a repetition rate of a few minutes, then questions may be raised about the efficiency of an adaptive beamformer to achieve near-instantaneous convergence in order to provide beam time series with coherent content for the FM pulses. This is because partially adaptive processing schemes require at least a few iterations to converge to a suboptimum solution.

To address this question, the matched filter and the nonconventional processing schemes, shown in Figure 11.4, were tested with real data sets, including HFM pulses 8-s long with a 100-Hz bandwidth. The repetition rate was 120 s. Although this may be considered as a configuration for bistatic active sonar applications, the findings from this experiment can be applied to monostatic active sonar systems as well.

In Figures 11.17 and 11.18 we present some experimental results from the output of the active unit of the generic signal processing structure. Figure 11.17 shows the output of the replica correlator for the conventional, subaperture MVDR adaptive, and synthetic aperture beam time series. The horizontal axis in Figure 11.17 represents range or time delay ranging from 0 to 120 s, which is the repetition rate of the HFM pulses. While the three beamforming schemes provide artifact-free outputs, it is apparent from the values of the replica correlator output that the conventional beam time series exhibit better temporal coherence properties than the beam time series of the synthetic aperture and the subaperture adaptive beamformer. The significance and a quantitative estimate of this difference can be assessed by comparing the amplitudes of the normalized correlation outputs in Figure 11.17. In this case, the amplitudes of the replica correlator outputs are 0.32, 0.28, and 0.29 for the conventional, adaptive, and synthetic aperture beamformers, respectively.

This difference in performance, however, was expected because for the synthetic aperture processing scheme to achieve optimum performance the reference signal is required to be present in the five discontinuous snapshots that are being used by the overlapped correlator to synthesize the synthetic aperture. So, if a sequence of five HFM pulses had been transmitted with a repetition rate equal to the time interval between the above discontinuous snapshots, then the coherence of the synthetic aperture beam time series would have been equivalent to that of the CBF. Normally, this kind of requirement restricts the detection ranges for incoming echoes. To overcome this limitation, a combination of the pulse length, desired synthetic aperture size, and detection range should be derived that will be based on the aperture size of the deployed array. A simple application scenario, illustrating the concept of this combination, is a side scan sonar system that deals with predefined ranges.

Although for the adaptive beam time series in Figure 11.17 a suboptimum convergence was achieved within two to three iterations, the arrangement of the transmitted HFM pulses in this experiment was not an optimum configuration because the subaperture beamformer had to achieve near-instantaneous convergence with a single snapshot. Our simulations suggest that a suboptimum solution for the subaperture MVDR adaptive beamformer is possible if the active sonar transmission consists of a continuous sequence

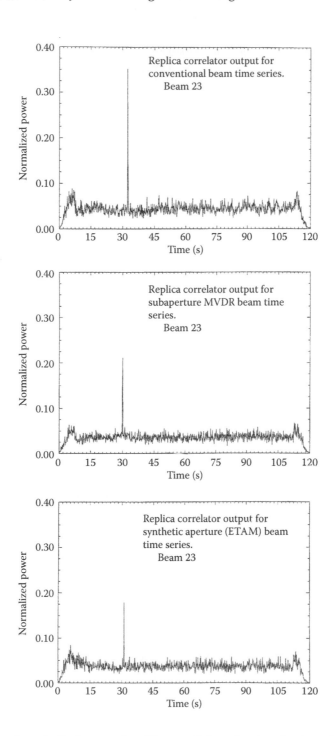

FIGURE 11.17
Output of replica correlator for the beam series of the generic beamforming structure shown in Figures 11.4 and 11.5. The processed hydrophone time series includes received HFM pulses transmitted from the acoustic source, 8-s long with a 100-Hz bandwidth and 120-s repetition rate. The upper part is the replica correlator output for conventional beam time series. The middle part is the replica correlator output for subaperture MVDR beam time series. The lower part is the replica correlator output for synthetic aperture (ETAM algorithm) beam time series. (Reprinted by permission of IEEE © 1998.)

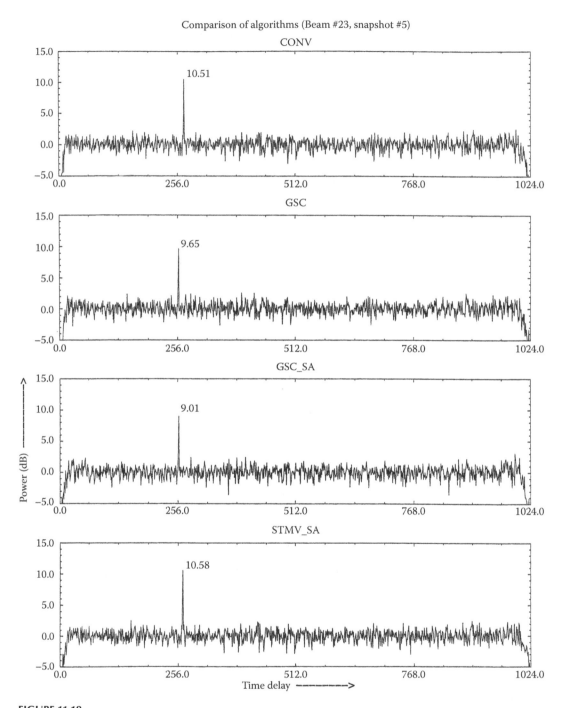

FIGURE 11.18

Output of replica correlator for the beam series of the conventional and the subaperture MVDR, GSC, and STMV adaptive schemes of the generic beamforming structure shown in Figure 11.4. The processed hydrophone time series are the same as those of Figure 11.17. (Reprinted by permission of IEEE © 1998.)

of active pulses. In this case, the number of pulses in a sequence should be a function of the number of subapertures, and the repetition rate of this group of pulses should be a function of the detection ranges of operational interest.

The near-instantaneous convergence characteristics, however, for the other two adaptive beamformers, namely, the GSC and the STMV schemes, are better compared with those of the subaperture MVDR scheme. Shown in Figure 11.18 is the replica correlator output for the same set of data as those in Figure 11.17 and for the beam series of the conventional and the subaperture MVDR, GSC, and STMV adaptive schemes.

Even though the beamforming schemes of this study provide artifact-free outputs, it is apparent from the values of the replica correlator outputs, shown in Figures 11.17 and 11.18, that the conventional beam time series exhibit better temporal coherence properties than the beam time series of the adaptive beamformers, except for the subaperture STMV scheme. The significance and a quantitative estimate of this difference can be assessed by comparing the amplitudes of the correlation outputs in Figure 11.18. In this case, the amplitudes of the replica correlator outputs are 10.51, 9.65, 9.01, and 10.58 for the conventional scheme and for the adaptive schemes: GSC in element space, GSC-SA (subaperture), and STMV-SA (subaperture), respectively. These results show that the beam time series of the STMV subaperture scheme have achieved temporal coherence properties equivalent to those of the CBF, which is the optimum case.

Normalization and clustering of matched filter outputs, such as those of Figures 11.17 and 11.18, and their display in a LOFAR-gram arrangement provide a waterfall display of ranges as a function of beam steering and time, which form the basis of the display arrangement for active systems, shown in Figures 11.8 and 11.9. Figure 11.19 shows these results for the correlation outputs of the conventional and the adaptive beam time series for beam #23. It should be noted that Figure 11.19 includes approximately 2 h of processed data. The detected HFM pulses and their associated ranges are clearly shown in beam #23. A reflection from the sidewalls of an underwater canyon in the area is visible as a second echo closely spaced with the main arrival.

In summary, the basic difference between the LOFAR-gram results of the adaptive schemes and those of the conventional beam time series is that the improved directionality of the non-CBFs localizes the detected HFM pulses in a smaller number of beams than the CBF. Although we do not present here the LOFAR-gram correlation outputs for all 25 beams, a picture displaying the 25 beam outputs would confirm the above statement regarding the directionality improvements of the adaptive schemes with respect to the CBF. Moreover, it is anticipated that the directional properties of the non-CBFs would suppress the anticipated reverberation levels during active sonar operations. Thus, if there are going to be advantages regarding the implementation of the above non-CBFs in active sonar applications, it is expected that these advantages would include minimization of the impact of reverberations by means of improved directionality. More specifically, the improved directionality of the non-CBFs would restrict the reverberation effects of active sonars in a smaller number of beams than that of the CBF. This improved directionality would enhance the performance of an active sonar system (including non-CBFs) to detect echoes located near the beams that are populated with reverberation effects. The real results from an active adaptive beamforming (subaperture STMV algorithm) output of a cylindrical sonar system, shown in Figure 6.31 of Chapter 6, provide qualitative supporting arguments that demonstrate the enhanced performance of the adaptive beamformers to suppress the reverberation effects in active sonar operations.

Real data processing: Beam #23 (Time evolution)

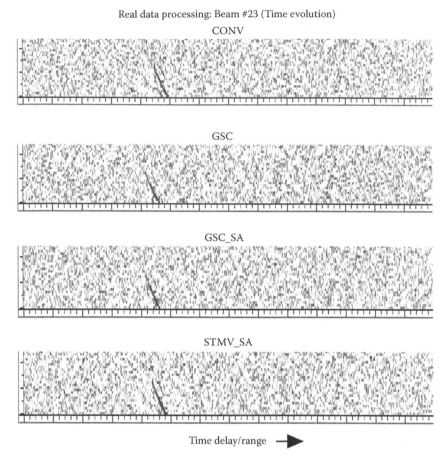

Time delay/range ➔

FIGURE 11.19
Waterfall display of replica correlator outputs as a function of time for the same conventional and adaptive beam time series as those of Figure 11.18. It should be noted that this figure includes approximately 2 h of processed data. The detected HFM pulses and their associated ranges are clearly shown in beam #23. A reflection from the side-walls of an underwater canyon in the area is visible as a second echo closely spaced with the main arrival. (Reprinted by permission of IEEE © 1998.)

11.4 Conclusion

The experimental results of this study were derived from a wide variety of CW, broadband, and HFM types of strong and weak acoustic signals. The fact that adaptive and synthetic aperture beamformers provided improved detection and tracking performance for the above type of signals and under a real-time data flow as the CBF demonstrates the merits of these nonconventional processing schemes for sonar applications. In addition, the generic implementation scheme, discussed in Chapter 6, suggests that the design approach to provide synergism between the CBF and the adaptive and synthetic aperture processing schemes could probably provide some answers to the integrated active and passive sonar problem in the near future.

Although the focus of the implementation effort included only adaptive and synthetic aperture processing schemes, the consideration of other types of nonlinear processing schemes for real-time sonar applications should not be excluded. The objective here was to demonstrate that nonconventional processing schemes can address some of the challenges that the next-generation active-passive sonar systems will have to deal with in the near future. Once a computing architecture and a generic signal processing structure are established, such as those suggested in Chapter 6, the implementation of a wide variety of nonlinear processing schemes in real-time sonar and radar systems can be achieved with minimum efforts. Furthermore, even though the above real results are from an experimental towed array sonar system, the performance of sonar systems including adaptive beamformers and deploying cylindrical or spherical arrays will be equivalent to that of the above experimental towed array sonar. As an example, Figure 6.31 of Chapter 6 reports results from an active sonar system, including adaptive beamformers deploying a cylindrical hydrophone array.

In conclusion, the previous results suggest that the broadband outputs of the subaperture adaptive processing schemes and the narrowband synthetic aperture LOFAR-grams exhibit very robust performance (under the prevailing experimental conditions) and that their AG improvements provide better signal tracking and target localization estimates than the conventional processing schemes. It is worth noting also that the reported improvements in performance of the previous non-CBFs compared with that of the CBF have been consistent for a wide variety of real data sets. However, for the implementation configuration of the adaptive schemes in element space, the narrowband adaptive implementation requires very long convergence periods, which makes the application of the adaptive processing schemes in element impractical. This is because the associated long convergence periods destroy the dynamic response of the beamforming process, which is very essential during course alterations and for cases that include targets with dynamic changes in their bearings.

Finally, the experimental results of this chapter indicate that the subaperture GSC and STMV adaptive schemes address the practical concerns of near-instantaneous convergence associated with the implementation of adaptive beamformers in integrated active-passive sonar systems.

References

1. S. Stergiopoulos. 1998. Implementation of adaptive and synthetic aperture processing in integrated active-passive sonar systems. *Proc. IEEE*, 86(2), 358–396.
2. B. Windrow, et al . 1967. Adaptive antenna systems. *Proc. IEEE*, 55(12), 2143–2159.
3. A.A. Winder. 1975. Sonar system technology. *IEEE Trans. Sonic Ultrasonics*, SU-22(5), 291–332.
4. A.B. Baggeroer. 1978. Sonar signal processing. In *Applications of Digital Signal Processing*, A.V. Oppenheim, Ed. Prentice-Hall, Englewood Cliffs, NJ.
5. American Standard Acoustical Terminology S1.1–1960. American Standards Association, New York, May 25.
6. D. Stansfield. 1990. *Underwater Electroacoustic Transducers*. Bath University Press and Institute of Acoustics .
7. J.M. Powers. 1988. Long range hydrophones. In *Applications of Ferroelectric Polymers*, T.T. Wang, J.M. Herbert, and A.M. Glass, Eds. Chapman and Hall, New York, NY.

8. R.I. Urick. 1983. *Principles of Underwater Acoustics*, 3rd ed. McGraw-Hill, New York, NY.

9. S. Stergiopoulos and A.T. Ashley. 1993. Guest editorial for a special issue on sonar system technology. *IEEE J. Oceanic Eng.*, 18(4), 361–365.

10. R.C. Trider and G.L. Hemphill. 1994. *The Next Generation Signal Processor: An Architecture for the Future, DREA/Ooral 1994/Hemphill/1*. Defense Research Establishment Atlantic, Dartmouth, Nova Scotia, Canada.

11. N.L. Owsley. 1985. *Sonar Array Processing*, S. Haykin, Ed., Signal Processing Series. A.V. Oppenheim, Series Editor. Prentice-Hall, Englewood Cliffs, NJ.

12. B. Van Veen and K. Buckley. 1988. Beamforming: a versatile approach to spatial filtering. *IEEE ASSP Mag.*, 4–24.

13. A.H. Sayed and T. Kailath. 1994. A state-space approach to adaptive RLS filtering. *IEEE SP Mag.*, 18–60.

14. E.J. Sullivan, W.M. Carey, and S. Stergiopoulos. 1992. Editorial special issue on acoustic synthetic aperture processing. *IEEE J. Oceanic Eng.*, 17(1), 1–7.

15. N.C. Yen and W. Carey. 1989. Application of synthetic-aperture processing to towed-array data. *J. Acoust. Soc. Am.*, 86, 754–765.

16. S. Stergiopoulos and E.J. Sullivan. 1989. Extended towed array processing by overlapped correlator. *J. Acoust. Soc. Am.*, 86(1), 158–171.

17. S. Stergiopoulos. 1990. Optimum bearing resolution for a moving towed array and extension of its physical aperture. *J. Acoust. Soc. Am.*, 87(5), 2128–2140.

18. S. Stergiopoulos and H. Urban. 1992. An experimental study in forming a long synthetic aperture at sea. *IEEE J. Oceanic Eng.*, 17(1), 62–72.

19. G.S. Edelson and E.J. Sullivan. 1992. Limitations on the overlap-correlator method imposed by noise and signal characteristics. *IEEE J. Oceanic Eng.*, 17(1), 30–39.

20. G.S. Edelson and D.W. Tufts. 1992. On the ability to estimate narrow-band signal parameters using towed arrays. *IEEE J. Oceanic Eng.*, 17(1), 48–61.

21. C.L. Nikias and J.M. Mendel. 1993. Signal processing with higher-order spectra. *IEEE SP Mag.*, 10–37.

22. S. Stergiopoulos, R.C. Trider, and A.T. Ashley. 1994. Implementation of a synthetic aperture processing scheme in a towed array sonar system. 127th ASA Meeting, Cambridge, MA.

23. J. Riley, S. Stergiopoulos, R.C. Trider, A.T. Ashley, and B. Ferguson. 1994. Implementation of adaptive beamforming processing scheme in a towed array sonar system. 127th ASA Meeting, Cambridge, MA, June.

24. A.B. Baggeroer, W.A. Kuperman, and P.N. Mikhalevsky. 1993. An overview of matched field methods in ocean acoustics. *IEEE J. Oceanic Eng.*, 18(4), 401–424.

25. R.D. Doolitle, A. Tolstoy, and E.J. Sullivan. 1993. Editorial special issue on detection and estimation in matched field processing. *IEEE J. Oceanic Eng.* 18, 153–155.

26. Editorial special issue on neural networks for oceanic engineering systems. *IEEE J. Oceanic Eng.*, 17, 1992.

27. A. Kummert. 1993. Fuzzy technology implemented in sonar systems. *IEEE J. Oceanic Eng.*, 18(4), 483–490.

28. A.D. Whalen. 1971. *Detection of Signals in Noise*. Academic Press, New York, NY.

29. D. Middleton. 1960. *Introduction to Statistical Communication Theory*. McGraw-Hill, New York, NY.

30. H.L. Van Trees. 1968. *Detection, Estimation and Modulation Theory*. Wiley, New York, NY.

31. P.D. Welch. 1967. The use of fast Fourier transform for the estimation of power spectra: a method based on time averaging over short, modified periodigrams. *IEEE Trans. Audio Electroacoust.*, AU-15, 70–79.

32. F.J. Harris. 1978. On the use of windows for harmonic analysis with discrete Fourier transform. *Proc. IEEE*, 66, 51–83.

33. W.M. Carey and E.C. Monahan. 1990. Guest editorial for a special issue on sea surface-generated ambient noise 20–2000 Hz. *IEEE J. Oceanic Eng.*, 15(4), 265–267.

34. R.A. Wagstaff. 1978. Iterative technique for ambient noise horizontal directionality estimation from towed line array data. *J. Acoust. Soc. Am.*, 63(3), 863–869.

35. R.A. Wagstaff. 1993. A computerized system for assessing towed array sonar functionality and detecting faults. *IEEE J. Oceanic Eng.*, 18(4), 529–542.

36. S.M. Flatte, R. Dashen, W.H. Munk, K.M. Watson, and F. Zachariasen. 1985. *Sound Transmission through a Fluctuating Ocean*. Cambridge University Press, New York, NY.

37. D. Middleton. 1989. Acoustic scattering from composite wind-wave surfaces in bubble-free regimes. *IEEE J. Oceanic Eng.*, 14, 17–75.

38. W.A. Kuperman and F. Ingenito. 1977. Attenuation of the coherent component of sound propagating in shallow water with rough boundaries. *J. Acoust. Soc. Am.*, 61, 1178–1187.

39. B.J. Uscinski. 1989. Acoustic scattering by ocean irregularities: aspects of the inverse problem. *J. Acoust. Soc. Am.*, 86, 706–715.

40. W.M. Carey and W.B. Moseley. 1991. Space-time processing, environmental-acoustic effects. *IEEE J. Oceanic Eng.*, 16, 285–301; also In *Progress in Underwater Acoustics*. Plenum Press, New York, NY, 743–758, 1987.

41. S. Stergiopoulos. 1991. Limitations on towed-array gain imposed by a nonisotropic ocean. *J. Acoust. Soc. Am.*, 90(6), 3161–3172.

42. W.A. Struzinski and E.D. Lowe. 1984. A performance comparison of four noise background normalization schemes proposed for signal detection systems. *J. Acoust. Soc. Am.*, 76(6), 1738–1742.

43. S.W. Davies and M.E. Knappe. 1988. Noise background normalization for simultaneous broadband and narrowband detection. Proceedings from IEEE-ICASSP 88, U3.15, 2733–2736.

44. S. Stergiopoulos. 1995. Noise normalization technique for beamformed towed array data. *J. Acoust. Soc. Am.*, 97(4), 2334–2345.

45. A.H. Nuttall. 1982. Performance of three averaging methods, for various distributions. Proceedings of SACLANTCEN Conference on Underwater Ambient Noise, SACLANTCEN CP-32, Vol. II, 16–1. SACLANT Undersea Research Centre, La Spezia, Italy.

46. H. Cox, R.M. Zeskind, and M.M. Owen. 1987. Robust adaptive beamforming. *IEEE Trans. Acoust. Speech Signal Process.*, ASSP-35(10), 1365–1376.

47. H. Cox. 1973. Resolving power and sensitivity to mismatch of optimum array processors. *J. Acoust. Soc. Am.*, 54(3), 771–785.

48. J. Capon. 1969. High resolution frequency wavenumber spectral analysis. *Proc. IEEE*, 57, 1408–1418.

49. T.L. Marzetta. 1983. A new interpretation for Capon's maximum likelihood method of frequency-wavenumber spectra estimation. *IEEE Trans. Acoust. Speech Signal Process.*, ASSP-31(2), 445–449.

50. S. Haykin. 1986. *Adaptive Filter Theory*. Prentice-Hall, Englewood Cliffs, NJ.

51. S. Stergiopoulos and A. Dhanantwari. 2004. High resolution 3D ultrasound imaging system deploying a multi-dimensional array of sensors and method for multi-dimensional beamforming sensor signals. *Assignee: Defence R&D Canada*, US Patent: 6,719,696, 13 April 2004.

52. S. Stergiopoulos. 1996. Influence of underwater environment's coherence properties on sonar signal processing. Proceedings of 3rd European Conference on Underwater Acoustics, FORTH-IACM, Heraklion, Crete, V-I, 453–458.

53. A. Dhanantwari and S. Stergiopoulos. 1998. An adaptive processing structure for integrated active_passive sonars deploying cylindrical arrays. *Proceedings of the 135th ASA Meeting. J. Acoust. Soc. Am.*, 103(5), Pt.2, 2854.

54. R.O. Nielsen. 1991. *Sonar Signal Processing*. Artech House, Norwood, MA.

55. D. Middleton and R. Esposito. 1968. Simultaneous otpimum detection and estimation of signals in noise. *IEEE Trans. Inf. Theory*, IT-14, 434–444.

56. V.H. MacDonald and P.M. Schulteiss. 1969. Optimum passive bearing estimation in a spatially incoherent noise environment. *J. Acoust. Soc. Am.*, 46(1), 37–43.

57. G.C. Carter. 1987. Coherence and time delay estimation. *Proc. IEEE*, 75(2), 236–255.

58. C.H. Knapp and G.C. Carter. 1976. The generalized correlation method for estimation of time delay. *IEEE Trans. Acoust. Speech Signal Process.*, ASSP-24, 320–327.

59. D.C. Rife and R.R. Boorstyn. 1974. Single-tone parameter estimation from discrete-time observations. *IEEE Trans. Inf. Theory*, 20, 591–598.

60. D.C. Rife and R.R. Boorstyn. 1977. Multiple-tone parameter estimation from discrete-time observations. *Bell System Technical J.*, 20, 1389–1410.

61. S. Stergiopoulos and N. Allcott. 1992. Aperture extension for a towed array using an acoustic synthetic aperture or a linear prediction method. *Proc. ICASSP-92.*

62. S. Stergiopoulos and H. Urban. 1992. A new passive synthetic aperture technique for towed arrays. *IEEE J. Oceanic Eng.*, 17(1), 16–25.

63. W.M.X. Zimmer. 1986. High resolution beamforming techniques, performance analysis. SACLANTCEN SR-104. SACLANT Undersea Research Centre, La Spezia, Italy, 1986.

64. A. Mohammed. 1985. Novel methods of digital phase shifting to achieve arbitrary values of time delays. DREA Report 85/106. Defense Research Establishment Atlantic, Dartmouth, Nova Scotia, Canada.

65. A. Antoniou. 1993. *Digital Filters: Analysis, Design, and Applications*, 2nd ed. McGraw-Hill, New York, NY.

66. L.R. Rabiner and B. Gold. 1975. *Theory and Applications of Digital Signal Processing*. Prentice-Hall, Englewood Cliffs, NJ.

67. A. Mohammed. 1983. A high-resolution spectral analysis technique. DREA Memorandum 83/D. Defense Research Establishment Atlantic, Dartmouth, Nova Scotia, Canada.

68. B.G. Ferguson. 1989. Improved time-delay estimates of underwater acoustic signals using beamforming and prefiltering techniques. *IEEE J. Oceanic Eng.*, 14(3), 238–244.

69. S. Stergiopoulos and A.T. Ashley. 1997. An experimental evaluation of split-beam processing as a broadband bearing estimator for line array sonar systems. *J. Acoust. Soc. Am.*, 102(6), 3556–3563, 1997.

70. G.C. Carter and E.R. Robinson. 1993. Ocean effects on time delay estimation requiring adaptation. *IEEE J. Oceanic Eng.*, 18(4), 367–378.

71. P. Wille and R. Thiele. 1971. Transverse horizontal coherence of explosive signals in shallow water. *J. Acoust. Soc. Am.*, 50, 348–353.

72. P.A. Bello. 1963. Characterization of randomly time-variant linear channels. *IEEE Trans. Commun. Syst.*, 10, 360–393.

73. D.A. Gray, B.D.O. Anderson, and R.R. Bitmead. 1993. Towed array shape estimation using Kalman filters — theoretical models. *IEEE J. Oceanic Eng.*, 18(4), 543–556.

74. B.G. Quinn, R.S.F. Barrett, P.J. Kootsookos, and S.J. Searle. 1993 The estimation of the shape of an array using a hidden Markov model. *IEEE J. Oceanic Eng.*, 18(4), 557–564.

75. B.G. Ferguson. 1993. Remedying the effects of array shape distortion on the spatial filtering of acoustic data from a line array of hydrophones. *IEEE J. Oceanic Eng.*, 18(4), 565–571.

76. J.L. Riley and D.A. Gray. 1993. Towed array shape estimation using Kalman Filters — experimental investigation. *IEEE J. Oceanic Eng.*, 18(4), 572–581.

77. B.G. Ferguson. 1990. Sharpness applied to the adaptive beamforming of acoustic data from a towed array of unknown shape. *J. Acoust. Soc. Am.*, 88(6), 2695–2701.

78. F. Lu, E. Milios, S. Stergiopoulos, and A. Dhanantwari. 2003. A new towed array shape estimation scheme for real time sonar systems. *IEEE J. Oceanic Eng.*, 28(3), 552–563.

79. N.L. Owsley. 1987. Systolic array adaptive beamforming. NUWC Report 7981, New London, CT.

80. D.A. Gray. 1982. Formulation of the maximum signal-to-noise ratio array processor in beam space. *J. Acoust. Soc. Am.*, 72(4), 1195–1201.

81. O.L. Frost. 1972. An algorithm for linearly constrained adaptive array processing. *Proc. IEEE*, 60, 926–935.

82. H. Wang and M. Kaveh. 1985. Coherent signal-subspace processing for the detection and estimation of angles of arrival of multiple wideband sources. *IEEE Trans. Acoust. Speech Signal Process.*, ASSP-33, 823–831.

83. J. Krolik and D.N. Swingler. 1989. Bearing estimation of multiple broadband sources using steered covariance matrices. *IEEE Trans. Acoust. Speech Signal Process.*, ASSP-37, 1481–1494.

84. J. Krolik and D.N. Swingler. 1990. Focussed wideband array processing via spatial resampling. *IEEE Trans. Acoust. Speech Signal Process.*, ASSP-38.

85. J.P. Burg. 1967. Maximum entropy spectral analysis. Presented at the 37th Meeting of the Society of Exploration Geophysicists, Oklahoma City, OK.

86. C. Lancos. 1956. *Applied Analysis*. Prentice-Hall, Englewood Cliffs, NJ.

87. V.E. Pisarenko. 1972. On the estimation of spectra by means of nonlinear functions on the covariance matrix. *Geophys. J. Astron. Soc.*, 28, 511–531.

88. A.II. Nuttall 1976. Spectrala analysis of a univariate process with bad data points, via maximum entropy and linear predictive techniques. NUWC TR5303. New London, CT, 1976.

89. R. Kumaresan and W.D. Tufts. 1982. Estimating the angles of arrival of multiple plane waves. *IEEE Trans. Acoust. Speech Signal Process.*, ASSP-30, 833–840.

90. R.A. Wagstaff and J.-L. Berrou. 1982. Underwater ambient noise: directionality and other statistics. SACLANTCEN Report SR-59. SACLANTCEN, SACLANT Undersea Research Centre, La Spezia, Italy.

91. S.M. Kay and S.L. Marple. 1981. Spectrum analysis — a modern perspective. *Proc. IEEE*, 69, 1380–1419.

92. D.H. Johnson and S.R. Degraaf. 1982. Improving the resolution of bearing in passive sonar arrays by eigenvalue analysis. *IEEE Trans. Acoust. Speech Signal Process.*, ASSP-30, 638–647.

93. D.W. Tufts and R. Kumaresan. 1982. Estimation of frequencies of multiple sinusoids: making linear prediction perform like maximum likelihood. *Proc. IEEE*, 70, 975–989.

94. G. Bienvenu and L. Kopp. 1983. Optimality of high resolution array processing using the eigensystem approach. *IEEE Trans. Acoust. Speech Signal Process.*, ASSP-31, 1235–1248.

95. D.N. Swingler and R.S. Walker. 1989. Linear array beamforming using linear prediction for aperture interpolation and extrapolation. *IEEE Trans. Acoust. Speech Signal Process.*, ASSP-37, 16–30.

96. P. Tomarong and A. El-Jaroudi. 1993. Robust high-resolution direction-of-arrival estimation via signal eigenvector domain. *IEEE J. Oceanic Eng.*, 18(4), 491–499.

97. J. Fawcett. 1993. Synthetic aperture processing for a towed array and a moving source. *J. Acoust. Soc. Am.*, 93, 2832–2837.

98. L.J. Griffiths and C.W. Jim. 1982. An alternative approach to linearly constrained adaptive beamforming. *IEEE Trans. Antennas Propagation*, AP-30, 27–34.

99. D.T.M. Slock. 1993. On the convergence behavior of the LMS and the normalized LMS algorithms. *IEEE Trans. Acoust. Speech Signal Process.*, ASSP-31, 2811–1825.

100. A.C. Dhanantwari. 1996. Adaptive beamforming with near-instantaneous convergence for matched filter processing. Master thesis, Department of Electrical Engineering, Technical University of Nova Scotia, Halifax, Nova Scotia, Canada.

101. A. Tawfik and S. Stergiopoulos. 1997. A generic processing structure decomposing the beamforming process of 2-D and 3-D arrays of sensors into subsets of coherent processes. Proceedings of IEEE-CCECE, St. John's, Newfoundland, Canada.

102. W.A. Burdic. 1984. *Underwater Acoustic System Analysis*. Prentice-Hall, Englewood Cliffs, NJ.

103. J-P. Hermand and W.I. Roderick. 1993. Acoustic model-based matched filter processing for fading time-dispersive ocean channels. *IEEE J. Oceanic Eng.*, 18(4), 447–465.

104. Mercury Computer Systems, Inc. 1997. *Mercury News Jan-97*. Mercury Computer Systems, Inc., Chelmsford, MA.

105. Y. Bar-Shalom and T.E. Fortman. 1988. *Tracking and Data Association*. Academic Press, Boston, MA.

106. S.S. Blackman. 1986. *Multiple-Target Tracking with Radar Applications*. Artech House Inc., Norwood, MA.

107. W. Cambell, S. Stergiopoulos, and J. Riley. 1995. Effects of bearing estimation improvements of nonconventional beamformers on bearing-only tracking. Proceedings of Oceans '95 MTS/IEEE, San Diego, CA.

108. W.A. Roger and R.S. Walker. 1986. Accurate estimation of source bearing from line arrays. Proceedings of the Thirteen Biennial Symposium on Communications, Kingston, Ontario, Canada.

109. D. Peters. 1995. Long range towed array target analysis — principles and practice. DREA Memorandum 95/217. Defense Research Establishment Atlantic, Dartmouth, Nova Scotia, Canada.

110. J.G. Proakis. 1989. *Digital Communications*, McGraw-Hill, New York, NY.

111. W.W. Peterson and T.G. Birdsall. 1953. The theory of signal detectability. *Univ. Mich. Eng. Res. Inst. Rep.*, 13.

112. J.T. Kroenert. 1982. Discussion of detection threshold with reverberation limited conditions. *J. Acoust. Soc. Am.*, 71(2), 507–508.

113. L. Moreavek and T.J. Brudner. 1999. USS Asheville leads the way in high frequency sonar. *Undersea Warfare*, 1(3), 22–24.

114. P.M. Baggenstoss. 1994. On detecting linear frequency-modulated waveforms in frequency- and time-dispersive channels: alternatives to segmented replica correlation. *IEEE J. Oceanic Eng.*, 19(4), 591–598.

115. B. Friedlander and A. Zeira. 1996. Detection of broadband signals in frequency and time dispersive channels. *IEEE Trans. Signal Proc.*, 44(7), 1613–1622.

Section III

Medical Diagnostic System Applications

12

Digital 3D/4D Ultrasound Imaging Technology

Stergios Stergiopoulos

Defence R&D Canada Toronto
University of Toronto
University of Western Ontario

CONTENTS

12.1 Background

The fully digital three-dimensional (3D)/(4D: 3D+time) ultrasound system technology of this chapter consists of an advanced beamforming structure that allows the implementation of adaptive and synthetic aperture signal processing techniques in ultrasound systems deploying multidimensional arrays of sensors. The aim with this fully digital ultrasound beamformer is to address the fundamental image resolution problems of current ultrasound systems and to provide suggestions for its implementation into existing 2D and/or 3D ultrasound systems as well as develop a complete stand-alone 3D ultrasound solution. This development has received grant support from the Defence R&D Canada (DRDC) and from the European Commission IST Program (i.e., ADUMS project: EC-IST-2001-34088).

However, to fully exploit the advantages of the present fully digital adaptive ultrasound technology, its implementation in a commercial ultrasound system requires that the system has a fully digital design configuration consisting of A/DC and D/AC peripherals. These peripherals have the capability to fully digitize the ultrasound probe time series, to optimally shape the transmitted ultrasound pulses through a D/A peripheral and to integrate linear and/or planar phase array ultrasound probes.

Thus, the digital ultrasound beamforming technology of this chapter, can replace the conventional (i.e., time delay) beamforming structure of ultrasound systems with an adaptive beamforming processing configuration. The results of this development demonstrate that adaptive beamformers improve significantly (at very low cost) the image resolution capabilities of an ultrasound imaging system by providing a performance improvement equivalent to a deployed ultrasound probe with double aperture size. Furthermore, the portability and the low cost characteristics of the present 3D adaptive ultrasound technology can offer the options to medical practitioners and family physicians to have access of diagnostic imaging systems readily available on a daily basis. As a result, a fully digital ultrasound technology, can revise the signal processing configuration of ultrasound devices to move them away from the traditional hardware and implementation software requirements.

In summary, implementation of an adaptive beamformer is a software installation on a PC-based ultrasound computing architecture with sufficient throughput for 3D and 4D ultrasound image processing. Moreover, the PC-based ultrasound computing architecture, which is introduced in this chapter, can accommodate the processing requirements of the "traditional" linear array 2D scans as well as the advanced matrix-arrays performing volumetric scans.

12.1.1 Limitations of 2D Ultrasound Imaging Technology

It has been well established [2,12,16] that the existing limitations of medical ultrasound imaging systems in poor image resolution, is the result of the very small size of deployed arrays of sensors and the distortion effects by the influence of the human-body's nonlinear propagation characteristics. In particular, some of the limitations (e.g., resolution) of ultrasound imaging are related to the fundamental physical aspects of ultrasound transducers and the interaction of ultrasound with tissues (e.g., aberration effects). In addition to these fundamental limitations are restrictions related to the display of the ultrasound images in an efficient manner allowing the physician to extract relevant information accurately and reproducibly. Specifically, the current state-of-the-art of 3D ultrasound imaging technology attempts to address the following limitations:

- Conventional ultrasound images are 2D, hence, the physician must mentally integrate multiple images to develop a 3D impression of the anatomy/pathology during procedures. This practice is time-consuming, inefficient, and requires a highly skilled operator, all of which can potentially lead to incorrect diagnostic and therapeutic decisions.

- Often the physician requires accurate estimation of tumor and organ volume. The variability in ultrasound imaging and volume measurements using a conventional 2D technique is high, because current ultrasound volume measurement techniques assume an idealized elliptical shape and use only simple measures of the width in two views [5,19].

- It is difficult to localize the thin 2D ultrasound image plane in the organ, and difficult to reproduce a particular image location at a later time, making 2D ultrasound a limited imaging modality for monitoring of disease progression/regression and follow-up patient studies.

12.1.1.1 Limitations of Current Beamforming Structure of Ultrasound Imaging Systems

A state-of-the-art transducer array of a commercial ultrasound system is either linear or curvilinear depending on the application; and for each deployed array (i.e., ultrasound probe), the number of transducers is in the range of 96–256 elements. However, only a small number of transducers of a given array are beamformed coherently to reconstruct a 2D tomography image of interest.

A typical number of transducers that may be included in the beamforming structure of an ultrasound system is in the range of 32–128 elements. Thus, the array gain of the beamformer may be smaller by approximately $10 \times \log_{10}(3) \approx 5$ dBs than that available by the deployed ultrasound probe. Figure 12.1 illustrates the basic processing steps associated with the ultrasound beamforming process. For the sake of simplicity and without any loss of generality, the array in Figure 12.1 is considered to be linear with 96-elements that can be used to transmit and receive the ultrasound energy. As shown in Figure 12.1, for each active transmission, the ultrasound beamformer processes coherently the received signal of 32-elements only, which is a subaperture of the 96-element deployed array. The active transmission takes place approximately every $\tau = 0.3$ ms, depending on the desired penetration depth in the body. The beam steering process is at the broadside.

When an active transmission is completed, the receiving 32-element subaperture is shifted to the left by one element, as shown in Figure 12.1. Thus, to make use of all the 96-elements of the deployed probe, the 32-element beamforming process is repeated 64 times, generating 64 broadside beams. In other words, it takes approximately 64×0.3 ms≈ 20 ms to reconstruct a 2D tomography image of interest. As a result, the resolution characteristics of the reconstructed image are defined by the array gain of the beamformer and the temporal sampling of the beam time series, for analog or digital beamformers, respectively. In the specific case of Figure 12.1, the pixel resolution along the horizontal x-axis of a reconstructed tomography image is defined by the angular resolution along azimuth of the 32-element beamformer. This resolution is usually improved by means of interpolation, which defines the basic difference between beamformers of different ultrasound systems.

The pixel resolution along the vertical y-axis of the reconstructed image is defined by the temporal sampling rate, which is always very high and it is not a major concern in ultrasound system applications. Thus, improvements of image resolution in ultrasound system applications requires mainly higher angular resolution or very narrow beamwidth, which

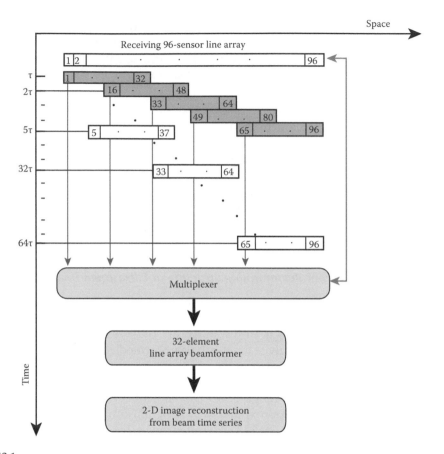

FIGURE 12.1
Typical beamformers for linear arrays of ultrasound systems. The shaded subapertures indicate the selected spatial locations for the formation of synthetic aperture according to ETAM algorithm.

means longer arrays and longer subapertures for the beamforming process with consequent technical and operational implications of hardware complexity and higher system manufacturing cost.

The main advantages of this simplified beamforming structure are the following:

- A broadside beamformer allows the use of frequency regimes that are higher than the corresponding spatial-aliasing frequency of the sensor spacing of the ultrasound probe. This results from the fact that side-lobe artifacts due to spatial aliasing are insignificant for beams with broadside beamsteering. Furthermore, this kind of simplicity in the analog beamforming structure allows for analog high speed hardware design for the beamformers. Then, the A/DC peripherals are used to digitize the beam time series.

- The advantage (i.e., suppression of spatial-aliasing artifacts) provided by the broadside beam-steering process has been used effectively by various types of illumination techniques using higher order-harmonics to achieve deeper penetration with corresponding higher image resolution along the temporal axis.

- The field of view of the probe may be wider than the aperture size required by the broadside beamformer. This approach eliminates the hardware complexity of an A/DC peripheral and it does not require digitization of the probe time series. Instead, it uses a multiplexer combined with an analog beamformer to control the data acquisition process for a probe with a larger number of sensors (e.g., 96-channels in Figure 12.1) than those being used by the broadside focus beamformer (e.g., 32-channels in Figure 12.1).

Until recently, the above beamforming concept (see Figure 12.1) has well served the ultrasound system requirements by providing practical alternatives to technical problems that were due mainly to limitations in the maximum number of channels deployed by A/DC units and the limited capabilities of computing architectures. Presently, these type of technology limitation does not exist. Thus, new advanced technology options have become available to exploit the vast experience from phase array beamformers that have been advanced by the sonar and radar research communities [7,12]. In fact, the introduction of linear phase array probes for cardiac applications is the first successful step toward this direction.

The use, however, of linear arrays introduces another major problem in terms of false targets, a problem that has been identified by both the ultrasound and sonar researchers using towed sonar arrays [7,9,12]. In particular, a linear array provides angular resolution within the tomography plane (B-scan) that the beam steering is formed. The angular resolution, however, of the beam-steering vectors of linear arrays is omnidirectional in the plane perpendicular to the B-scan plane. Thus, reflections from surrounding organs cannot be spatially resolved by the steered beams of a line array; and they appear as false targets in towed array sonars, or false components of a reconstructed image by a linear ultrasound probe.

To address the problem of false components in the reconstructed image, the 1.5D and 1.75D ultrasound array probes have been introduced that consist of linear arrays stacked as partially planar arrays. In particular, the GE 1.75D ultrasound array probe consists of eight linear arrays with 128-sensors each and with 0.2 mm sensor spacing. The linear array spacing is 1.5 mm. Thus, the steered beams are 3D and have the property to resolve the angular components of ultrasound reflected signals along azimuth and elevation. Although the 3D beams of 1.75D arrays may be viewed as the first step for 3D ultrasound imaging, they do not have sufficient angular resolution capabilities along elevation to generate 3D ultrasound volumes. The 3D beamforming structure and the relevant 3D ultrasound experimental system development that is based on a 16×16 planar array probe, which is presented in this chapter, attempts to address the above limitations. At this point, however, it is considered appropriate to briefly review the current state-of-the-art in 3D ultrasound technology.

12.1.1.2 Current Technology Concept of 3D Visualization Methods for Ultrasound Systems

Current 3D ultrasound imaging systems have three components: image acquisition, reconstruction of the 3D image, and display [1,4,5,6,8,10–13,19]. The first component is crucial in ensuring that optimal image quality is achieved. In producing a 3D image, the conventional line array transducer is moved over the anatomy while 2D images are digitized and stored in a microcomputer, as shown in Figure 12.2. To reconstruct the 3D geometry without geometric distortion, the relative position and angulation of the acquired 2D images *must* be known accurately. Over the years there are numerous developments and evaluating techniques for obtaining 3D ultrasound images using the following two approaches: mechanical scanning and freehand scanning [1,4–6,8,10–13,19].

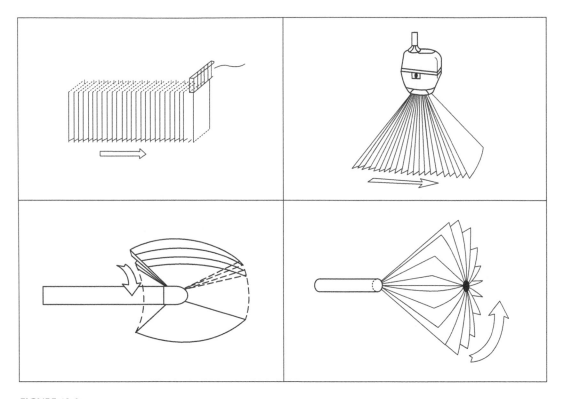

FIGURE 12.2
Mechanical scanning of ultrasound probes for image acquisition of 2D B-scans to obtain 3D ultrasound images through volume rendering.

Mechanical scanning. Based on earlier work, Fraunhofer has developed systems [12] in which the ultrasound transducer is mounted on a special assembly, which can be driven by a motor to move in a linear fashion over the skin or tilted in equal angular steps. The movement can be continuous, cardiac and/or respiratory [11,19]. In addition, the spatial-sampling frequency of the image acquisition can be adjusted based on the elevational resolution of the transducer and the depth of the region-of-interest. For linear scanning, they collect 140 images (336×352 pixels each) at 0.5 mm intervals in a time that depends on the ultrasound machine frame rate and whether cardiac gating is used. All scanning parameters can be adjusted depending on the experiment and type of acquisition. For example for 3D B-mode, they typically use two or three focal zones resulting in about 15 frames/ sec and a total 3D scanning time of 9 sec for 140 images.

Freehand scanning. Although the mechanical scanning approach produces accurate 3D ultrasound images, the mechanical assembly is bulky and not convenient for the operator. In addition, the mechanical constraint does not permit its use in imaging larger structures such as the liver or fetus. An alternative freehand technique has been proposed [12] that maintains the flexibility of the 2D exam yet produces a 3D image. We are investigating a freehand scanning system, in which a magnetic positioning and orientation measurement (POM) device is mounted on the transducer [1,4,6,8,13]. To produce a 3D image, the operator manually moves the hand-held transducer, while the POM device transfers the position and orientation coordinates of the transducer to a microcomputer. At the same time, 2D images are digitized by the same computer and associated with the appropriate coordinates. After the necessary number of 2D images are acquired (typically 60–160), the computer reconstructs the 3D image. Care is taken to scan the patient sufficiently slowly so that the region of interest is scanned with no gaps. Typically, the scan lasts 4–11 sec. while the patient holds their breath. Although this technique does produce useful images, it still suffers from major limitations that precludes its use for general diagnostic procedures. Most importantly, the manual scanning of the 3D space with a linear array does not eliminate the false components of the reconstructed B-scan images that were discussed in the previous section.

12.2 Next Generation 3D/4D Ultrasound Imaging Technology

The experimental 3D/4D ultrasound developments, discussed in this chapter, demonstrate an imaging technology that can lead to a next generation high-resolution diagnostic ultrasound imaging systems. The main components of this technology include:

- Synthetic aperture processing to accommodate digitization requirements for the sensor time series of large size 2D phase array probes for 3D phase array beamforming.
- 3D adaptive beamforming for the full aperture of the deployed probe to effectively maximize the available array gain and improve angular resolution.
- A PC-based computing architecture capable to accommodate the computationally intensive signal processing and data acquisition requirements of fully digital 3D/4D ultrasound imaging systems deploying planar arrays.
- An experimental fully digital ultrasound system with a 16×16 sensor phase planar array with uniform sensor spacing.
- Integration of Fraunhofer's 3D and 4D visualization schemes with the image reconstruction process of the 3D ultrasound beamformer.

The following sections provide technical details for the above components.

12.2.1 Synthetic Aperture Processing for Digitizing Large Size Planar Arrays

Integration of a synthetic aperture processing in the data acquisition structure of a fully digital ultrasound system can accommodate the highly demanding requirements to digitize the channels of a large size planar array by using a single A/DC peripheral, which may have fewer A/D channels than those of the planar array probe. More specifically, let us assume that a planar array probe consists of $N \times N$ sensors and the A/DC peripheral has the capability to digitize 2N channels. Then, a digitization process of the $N \times N$ channels of the planar array probe requires the use of a multiplexer for $N/2$ successive acquisitions (i.e., $N \times N/(2N) = N/2$) of equal size (i.e., 2N-channels) of $N/2$ subapertures of the planar array by the 2N-channel A/DC peripheral. Then each set of 2N digitized time series of each subaperture will be integrated with the remaining digitized time series of the $N/2$ subapertures to form a complete set of digitized $N \times N$ sensor time series representing a snap-shot of the illuminated ultrasound field. However, this successive digitization process of the $N \times N$ channels of the planar array by a 2N channel A/DC peripheral may not be fast enough to ensure that there are no motion artifacts between the successively digitized subapertures. To minimize potential motion artifacts during the digitization process of large size planar arrays, the following synthetic aperture process is recommended.

Shown in Figure 12.1 is the proposed experimental implementation of a synthetic aperture algorithm [14] (i.e., ETAM) for ultrasound applications in terms of the subaperture line array size and sensor positions as a function of time and space. Between two successive positions of the 32-sensor subaperture there are a number of sensor pairs of space samples of the acoustic field that have the same spatial information, their difference being a phase factor [14] related to the time delay these measurements were taken. The optimum overlap size, which is related to the variance of the phase correction estimates, has been shown [14] to be equal to the half size of the deployed subaperture.

For the particular example of Figure 12.1, the spatial overlap size will be 16-sensors. Thus, by cross-correlating the 16-sensor pairs of the sensor time series that overlap, the desired phase correction factor is derived, which compensates for the time delay between these measurements and the phase fluctuations caused by the variability and nonisotropic propagation characteristics of the human body; this is called the overlap correlator. The key parameters in the ETAM algorithm is the time increment τ between two successive sets of measurements. This may be the interval of 0.3 ms between two active ultrasound transmissions. Then, the total number of sets of measurements required by the 32-sensor subaperture to achieve an extended aperture size equal to the deployed array (i.e., 96-sensor array) is five.

Thus, if we consider the subaperture acquisition process in Figure 12.1, the proposed synthetic aperture processing will coherently synthesize the spatial measurements derived from the 32-element subapertures of the ultrasound receiving array into a longer aperture equivalent to the 96-sensor deployed array using only five subaperture measurements instead of 64. In this way, the required hardware modifications of an ultrasound system will be minimized since the A/DC will remain the same. Moreover, the time required to reconstruct a tomography image will be reduced from the current 20 ms time interval to 5×0.3 ms$\approx$1.5 ms. In parallel to this improvement, there will be an increase in the beam-former's array gain by $10 \times \log_{10}(3) \approx 5$ dBs, with improvement also in angular resolution of the ultrasound beamforming structure by a factor of 3 (i.e., $\Delta\theta = \lambda/L$ for the synthetic aperture $\Delta\theta = \lambda/(3L)$, as described in Reference [16]).

However, the experience gained from the development of the experimental fully digital ultrasound design, reported in this chapter, suggests the following: When the allocated period for the digitization process of the subapertures of a large size planar array is very fast and of the order of a few milliseconds, then there is no need for synthetic aperture processing and the best approach is to piece together the subapertures in order to form the fully populated physical aperture for beam-forming and to treat the digitized data of the subapertures as samples that have been acquired instantly.

In particular, during the experimental ultrasound development, reported in this chapter, a triggering mechanism was used to attempt to keep a constant time delay between the subaperture firings and a constant period for the digitization process of subapertures. This also ensured that there was a high correlation between the overlapping segments of successive subapertures. If the triggering mechanism can be designed to be very accurate, meaning the correlation coefficient become exactly one, then there is no need to perform the correlation phase correction process of the ETAM algorithm, because every subaperture is already coherent with respect to the first subaperture. Furthermore since all of the subapertures are coherent, and no correction is needed, there is no need to accumulate overlapping subapertures. This argument is supported also by experimental studies [2] that have shown that when there are no motion effects, the correlation coefficients between subaperture data sets, were consistently computed to be very close to 1.0. This proved that the triggering mechanism is accurate and eliminates the need to reconstruct a synthetic aperture using the ETAM software processing, suggested in this section. Instead with each firing, a section of the aperture is collected and these subapertures are pieced together to create a full synthetic aperture. In the example of Figure 12.1, with the first fire, sensors 1–32 are collected; with the second fire, sensors 33–64 are collected and sensors 64–96 are collected with the third fire. These three segments are pieced together to create the full 96-element aperture. This is the approach used in both the 2D/3D and 3D/4D ultrasound systems, reported in this chapter. In particular, for the linear phase array probe with 64 elements, the A/DC peripheral included 16 channels and therefore, the full 64-sensor aperture was digitized from four successive subaperture acquisitions. In the case of the planar array system with 16×16=256 elements, the A/DC peripheral included 64 channels and the full 256-channel aperture was digitized from four successive subaperture acquisitions.

The validity of this approach was assessed also from the reconstructed images. With a coherent aperture a clear consistent image is created. With any loss of coherence between subapertures, objects in the image are blurred and ghosting appears.

In conclusion, implementation of the synthetic aperture processing by means of the ETAM algorithm, may not be needed when the acquisition period of subapertures is well synchronized, is very fast and of the order of a few milliseconds. In this case, the simplest approach for the acquisition and digitization process of a large size phase array is to piece together the subapertures in order to form the fully populated physical aperture for beamforming. This approach minimizes the number of firings by 50% and it reduces the computational requirements of the overlap correlator.

12.2.2 Ultrasound Beamforming Structure for Line and Planar Arrays

Deployment of planar arrays by ultrasound medical imaging systems has been gaining increasing popularity because of its advantage to provide real 3D images of organs under medical examination. The details of a 3D beamforming structure for planar arrays are provided in Reference [16]. However, commercial ultrasound systems deploying planar arrays are not yet available. Moreover, if we consider that a state-of-the-art line phase array ultrasound system consists of 128 sensors, then a planar array ultrasound system should include at least 128×128=16, 384 sensors in order to achieve the angular resolution performance of a line array system and the additional 3D image reconstruction capability provided by the elevation beam steering of a planar array. Thus, increased angular resolution in azimuth and elevation beam steering for ultrasound systems means larger sensor arrays, with consequent technical and higher cost implications. As it will be shown in this chapter, the alternative is to implement adaptive beamforming in ultrasound systems that deploy a planar array with 32×32=1024 sensors, which consist of 32 line arrays with 32 sensors each. Then, the anticipated array gain improvements by an adaptive beamformer, as defined in the next section, will be equivalent to those provided by a 64-sensor line array for azimuth beam-steering and a 64-sensor vertical line array for elevation beam steering for real 3D ultrasound imaging. In summary, the array gain improvements for an adaptive 1024-sensor planar array will be equivalent to those that could be provided by a conventional 64×64=4096 sensor planar array. This is because for line arrays, a quantitative assessment [15,16] shows that the image resolution improvements of the proposed adaptive beamformers, will be equivalent to a two to three-time longer physical aperture. To achieve an effective implementation of the above adaptive ultrasound beamforming concept and to allow for a flexible system design for line and or planar array ultrasound probes, the discussion in this chapter suggests that the advance beamforming structures, defined in Refs. [15,16], address the above requirements for both line and planar array ultrasound probes.

12.2.3 Adaptive Beamforming Structure for Ultrasound Systems

Details on the implementation of adaptive beamformers [17,18] for ultrasound imaging applications have been presented elsewhere [15,16]. These adaptive beamformers are characterized as dual-use technologies, they have been tested in operational active sonars [15] and are defined in detailed in Reference [16]. Real data results have shown that they provide array gain improvements for signals embedded in anisotropic noise fields, as is the case in the human body.

Despite the geometric differences between the line and planar arrays, the underline beamforming processes for these arrays are time delay beamforming estimators, which are basically spatial filters. However, optimum beamforming requires the beamforming filter coefficients to be chosen based on the covariance matrix of the received data by the N-sensor array in order to optimize the array response [15]. For ultrasound applications, the outputs of the adaptive algorithms are required to provide coherent beam time series to facilitate the postprocessing of the image reconstruction processes. This means that these algorithms should exhibit near-instantaneous convergence and provide continuous beam time series that have sufficient temporal coherence to correlate with the reference replica in matched filter processing [15]. In what follows, the adaptive beamforming structure and its implementation in frequency domain, will be re-defined in Sections 12.2.4 and 12.2.5 to address system design requirements for fully digital ultrasound imaging technologies. In particular, Section 12.2.4 will introduce the design of a fully digital multifocus active beamformer and Section 12.2.5 will analyze the complex structure of a multifocus receiving beamformer for linear and planar phase array probes, respectively.

12.2.4 Multifocus Transmit Beamformer for Linear and Planar Phase Arrays

In ultrasound imaging, the transmitted signals consist of steered beams that illuminate a 2D or a 3D field in the case of a linear array or a planar array, respectively. The beamforming design, discussed in this section, defines beam illuminations that include a set of simultaneously transmitted multifocused beams to cover various depths (ranges) and are characterized by a wide angular width with low side lobes. To illuminate a specific region, several beam transmissions may be required where each transmission illuminates a certain angular sector. The preference in the present design is for a wide angle beamwidth to illuminate a wide region per beam so as to minimize the number of transmissions. Furthermore, the beam's side lobe structure should be downsized in order to minimize the signal power distributed outside the intended illuminated region that may contribute noisy interference in the reflected signals; and its performance should be assessed from the beam power pattern of the illuminated field.

12.2.4.1 Multifocus Transmit Beamformer for Linear Phase Array

Figure 12.3 depicts the configuration of a linear phase array probe in Cartesian coordinates with angle θ showing the angular direction of the active beam steering for the transmitted pulse. The illuminated field is the x–y plane along the positive y-axis with $0 \leq \theta \leq 180°$. The broadband characteristics of the transmitted pulse are defined by a linear

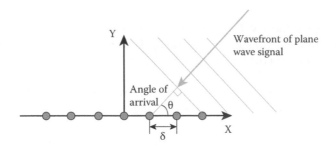

FIGURE 12.3
Geometric configuration and coordinate system for a line array of sensors.

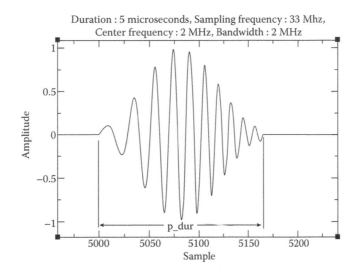

FIGURE 12.4
A linear FM pulse with a temporal window.

frequency modulated (FM) signal, defined by Equation 12.1 below and with a time domain response shown in Figure 12.4.

$$p(t) = \sin(2\pi(f_0 t + kt^2)) \tag{12.1}$$

where:
 f_0 is the lower cut-off frequency of the modulated FM signal,
 k is the linear frequency sweep rate ($k = BW/(2 \times T_0)$),
 BW is the bandwidth of the modulated FM signal and
 T_0 is the pulse duration.
 Then, the temporal characteristics of the transmitted signal by the active sensors of the phase array probe are defined below by the time series of Equation 12.2,

$$\left\{ \begin{array}{l} S_{ref}(\theta_i, R_s, m) = \sin\left(2\pi\left(f_0 + k \cdot \dfrac{m}{f_s}\right)\right) \cdot \dfrac{m}{f_s} \cdot w_p(m - s_{pos}), \text{ when } s_{pos} \leq m \leq (s_{pos} + p_{dur}) \\ \\ S_{ref}(\theta_i, R_s, m) = 0, \text{ otherwise} \end{array} \right. \tag{12.2}$$

where:

f_s is the sampling frequency, f_0 is $f_0(R_s, \theta_i)$ which is the lower cut-off frequency speci-fied for the focal range R_s and steering angle θ_i, S_{ref} refers to the transmitted ultrasound pulse signal, s_{pos} is the starting temporal location in the time series which can be set to zero, as discussed in Section 12.2.4.2, p_{dur} is the duration of the active transmission, defined by the values of focal range R_s and steering angle θ_i, as discussed in Section 12.2.4.2, $w_p(m - s_{pos})$ is the temporal window of length p_{dur} m is the index for the temporal samples.

The temporal window $w_p(m - s_{pos})$ reduces the spectral leak in the transmitted pulse, as discussed in Reference [16], and it can be Hamming or Kaizer window function. The frequency characteristics of the discrete time series, such as the lower cut-off fre-quency, f_0, and the bandwidth are selected according to the focal range R_s and steering angle θ_i of the transmitted pulse.

Then, for the nth transducer of the active aperture of the linear phase array probe, the time series $S_{ref}(\theta_i, R_s, m)$, of Equation 12.1, are modified as follows:

$$S_m(\theta_i, R_s, m) = S_{ref}(\theta_i, R_s, m + \tau_n) \cdot w_n \tag{12.3}$$

where n is the transducer index, τ_n is the time delay for the nth transducer of the active component of the line array probe steered at direction θ_i and focused at range R_s, w_n is the spatial window value for the mth transducer. τ_n is defined by,

$$\tau_n = \left[(R_s^2 + \delta_n^2 - 2R_s \cdot \delta_n \cdot \cos(\theta_i))^{1/2} - R_s \right] \cdot \frac{f_s}{c} \tag{12.4}$$

where, δ_n is the location of the nth transducer with respect to the linear array coordinate system. To allow for multifocus at different focal ranges and along the same steering angle by using a single transmission, the transmitted time series are modified as follows,

$$S_n(\theta_i, m) = \sum_{\text{all} R_s | \theta = \theta_i} S_n(\theta_i, R_s, m) \tag{12.5}$$

To achieve a wide angle illumination coverage with a single transmission, the beamwidth of the radiated pulses need to be as wide as possible and this can be achieved with a spatial window applied as a weighting function along the active transducers of the linear phase array probe. There are two types of windows that have the above desired characteristics and these are the Gaussian and 4-Term Blackman–Harris windows. More specifically, the 4-Term Blackman–Harris window function is defined by the following equation:

$$w_n = a_0 - a_1 \cos\left(2\pi \frac{n}{N-1}\right) + a_2 \cos\left(4\pi \frac{n}{N-1}\right) - a_3 \cos\left(6\pi \frac{n}{N-1}\right) \tag{12.6}$$

where $n = 0, 1, 2, \ldots, N-1$, a's sare the 4-Term Blackman–Harris window coefficients: $a_0 = 0.35875$, $a_1 = 0.48829$, $a_2 = 0.14128$, $a_3 = 0.01168$, and N is the window length, or the number of transducers in the line array probe.

The Gaussian window is expressed by

$$w(n) = \exp\left(-\frac{1}{2}\left(\alpha\frac{n-\frac{N}{2}}{\frac{N}{2}}\right)^2\right) \tag{12.7}$$

where $n = 0, 1, 2, \ldots, N-1$, with N being the number of the active elements of the array, which can be different than the corresponding N in Equation 12.6, α is the alpha parameter, $\alpha \geq 2$. α is the Gaussian window parameter that controls the width of the Gaussian window. The width of the Gaussian window is the inverse of the beamwidth of the radiated pulse and the size of the side lobes with respect to the beam peak magnitude.

Then, the time series defined by Equation 12.5 are converted into voltage levels through a D/AC peripheral to excite the transducers of the linear phase array probe. This kind of excitation generates illumination patterns that their beam power pattern distribution is defined in the next section. Figure 12.5 shows the beam power plot according to Equation 12.5 and the relations 12.1 through 12.4 for a Gaussian spatial window with $\alpha = 3.4$.

As shown in Figure 12.5, there is a considerable three dB beamwidth increase (about 12°) due to the implementation of a spatial window. Moreover, since the Gaussian and the 4-Term Blackman–Harris spatial windows show similar beamwidth improvements, the Gaussian window is preferable due to its flexibility in changing the α parameter to obtain a wider selection of beamwidths.

FIGURE 12.5
2D beam power pattern distribution of an illuminated area with size (10 cm × 10 cm) covering the angular sector $0° \leq \Theta \leq 180°$ for radiated pulses by a linear phase array probe with 12 transducers having 0.4 mm sensor spacing. The beam steering is at 75° and focused at 5 cm. The center frequency and bandwidth of the transmitted signal are 2 MHz and 2 MHz, respectively

12.2.4.2 Beam Power Pattern for Multifocus Transmit Beamformer
for Linear Phase Array

The image of the beam power pattern of the multifocus illumination is obtained by computing the total power of the signals that arrive at each of the pixels in the coordinate system of the linear phase array. The arriving signal at each pixel is the added combinations from all active transducers radiated signals that are time delayed according to Equation 12.5. Equation 12.8 expresses the time delay factor in terms of temporal samples and defines the difference between the distance from the pixel to the center of the array and the distance from the pixel to the transducer,

$$\zeta_n(p_x, p_y, \delta_n) = \left[(p_x^2 + p_y^2)^{1/2} - ((p_x - \delta_n)^2 + p_y^2)^{1/2} \right] \cdot \frac{f_s}{c} \tag{12.8}$$

where p_x and p_y are the x and y locations of the pixel, respectively. Thus, the combined signals from all transducers arriving at a pixel are defined by

$$b(p_x, p_y, \theta_i, m) = \sum_n s_n(\theta_i, m + \zeta_n) \tag{12.9}$$

and the total power level of the combined signals that arrive at a pixel location (x, y) is

$$P(p_x, p_y, \theta_i) = \sum_m \left| b(p_x, p_y, \theta_i, m) \right|^2 \tag{12.10}$$

The beam power pattern image is finally obtained from Equation 12.11

$$P(p_x, p_y) = \sum_{\text{all } \theta_i} P(p_x, p_y, \theta_i) \tag{12.11}$$

which includes the summation of the radiated power from all steering angles. The values of $P(p_x, p_y)$ are further scaled to fit an available grey scale range for image presentation. However, the computations as expressed by the Equations 12.10 and 12.11 require only a small portion of the transducer signals. The temporal starting point of the pulse, s_{pos}, as defined in Equation 12.2, can be arbitrarily set to *zero*. Then, this starting point s_{pos}^{th} of the active pulse is delayed or advanced by a specific time delay, as defined in Equation 12.4, in order to generate the directional and focusing characteristics of the transmitted signals for a specific depth and angular direction. Furthermore, excitation of these signals by the transducers of the active aperture of the probe includes time delays according to Equation 12.8 to get the intensity of the illumination at a specific pixel. The two time delays according to Equations 12.4 and 12.8, determine the total time delay boundaries of how far the pulse on the reference signal in Equation 12.2 might be shifted over time. Beyond these boundaries, the transducer signals have zero values and can be neglected in the signal power computations. These time delay boundaries also determine the temporal length of the segment of the signal that needs to be taken into account for processing. The lower boundary can also be computed as

$$s_{min} = s_{pos} - \tau_{max} \tag{12.12}$$

where,

$$\tau_{max} = 2 \cdot \left(\frac{N}{2} + 0.5 \right) \cdot \delta \cdot \frac{f_s}{c} \tag{12.13}$$

and S_{min} is the lower boundary or the starting temporal sample of the signal in Equation 12.2 that needs to be taken into account for processing, τ_{max} is the maximum time delay, δ is the transducer spacing in the linear array.

The multiplication factor of two in Equation 12.13 corresponds to the two way propagation in Equations 12.4 and 12.8

The beam output of the transmitted pulse by a linear phase array probe is defined by the following equation,

$$B(\theta_i, f) = \sum_n S_n(\theta_i, f) \cdot e^{j2\pi f \tau_n} \tag{12.14}$$

where $S_n(\theta_i, f)$ is the Fourier transform of the nth transducer signal $s_n(\theta_i, m)$. Equation 12.14 is identical to Equation 12.25 in Section 12.2.4.4. The time delay τ_n is a function of the transducer location and the steering angle θ_i and was defined before by Equation 12.4.

12.2.4.3 Multifocus Transmit Beamformer for Planar Phase Array

The left hand side of Figure 12.6 depicts the configuration of a linear phase array probe in Cartesian coordinates. The same coordinate system located at the center of the planar array defines the parameters (i.e., A, B, θ, ϕ) of the 3D beamformer at the right hand side of Figure 12.6. The broadband characteristics of the transmitted pulses are the same as in the case of the linear phase array and are expressed by Equation 12.1, and the time series defining transimitted pulses for the active aperture of the phase array

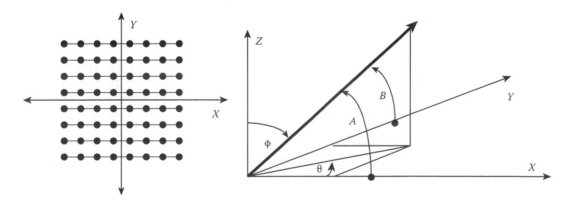

FIGURE 12.6
(Left) Coordinate system and configuration of a planar array transducer symmetrically located on the *x–y* plane. (Right) Coordinate system for 3D beamforming process for the planar array shown at the left hand side diagram. (Reprinted from Dhanantwari A., Stergiopoulos S., Parodi C., Bertora F., Questa A., and Pellegretti P. Adaptive 3D beamforming for ultrasound systems deploying linear and planar array phased array probes. In *IEEE Conference Proceedings, IEEE International Ultrasonics Symposium*, pp. 5–8, Honolulu, HI, 2003. With permission.)

planar array include also time delays to account for their range focusing and angular steering characteristics. These time series are expressed by,

$$S_{l,n}(\theta_i, \phi_q, R_s, m) = S_{ref}(\theta_i, \phi_q, R_s, m + \tau_{l,n}) \cdot w_{l,n} \tag{12.15a}$$

where, $S_{ref}(\theta_i, \phi_q, R_s, m + \tau_{l,n})$, is defined by Equation (12.2) with $\tau_{l,n}(A, B, R_s)$ being the time delay steering defined by the angles (A, B), defined in Figure 12.6, and the focus range Rs and (l, n) are the indexes for a transducer located at the $(l^{th}$ row, n^{th} column) of the planar array, $\tau_{l,n}$ is the shift of the reference signal needed to create the signal for the (l, n) th transducer to achieve focus at R_s and (θ_i, ϕ_q) steering angle in 3D space, defined in Figure 12.6, $w_{l,n}$ is the spatial window value for the $(l, n)^{th}$ transducer, δ_{l,n_x} and δ_{l,n_y} are the x and y locations, respectively, of the $(l, n)^{th}$ transducer. Estimates of $\tau_{l,n}(A, B, R_s)$ are provided from the following expression (Equation 12.15b):

$$\tau_{l,n}(A, B, R_s) = \frac{\sqrt{R_s^2 + \delta_{l,n_x}^2 + \delta_{l,n_y}^2 - 2\delta_{l,n_x}\cos A - 2\delta_{l,n_y}\cos B} - R_s}{c} \tag{12.15b}$$

As in the case of the linear phase array active beamformer, the transmitted signals focused at different ranges, R_s but along the same steering angle can be summed up to be fired by the probe under a single transmission as shown below,

$$S_{l,n}(\theta_i, \phi_q, m) = \sum_{\text{all } R_s | \theta_i, \phi_q} S_{l,n}(\theta_i, \phi_q, R_s, m) \tag{12.16}$$

Implementation of a spatial window $w_{l,n}$ on the planar array requires the application of the decomposition process of the 2D planar array beamformer into two line array beamforming processes that have been introduced in Reference [16]. Thus, implementation of 3D Gaussian and 4-Term Blackman–Harris windows to adjust the beamwidth of the 3D beamstering is reduced into simple linear spatial window as part of the above decomposition process.

An alternative approach to the above decomposition process is to use an approximate implementation of the 2D Gaussian spatial window on the active transducers of the planar array. However, this approximation requires that the planar array is square with equal sizes along the x and y coordinates. This approximation includes the 2D Gaussian window expressed by

$$W(n) = \exp\left(-\frac{1}{2}\left(\alpha\frac{n - \dfrac{N}{2}}{\dfrac{N}{2}}\right)^2\right) \tag{12.17}$$

where $n = 0, 1, 2, \ldots, N$, and N is the window length with $\alpha \geq 2$.

To compute the window values for the transducers in the planar array, a 2D Gaussian window that spans over the diagonal of the planar array is formed. The 2D diagonal Gaussian window is used as a reference to describe the relationship between a transducer's window value versus the distance from the transducer to the center of the planar array. In forming the 2D diagonal Gaussian window, the parameter N needs to be sufficiently large in order to have good window resolution. D in Equation 12.18 describes the diagonal length of the planar array.

$$D = \sqrt{2}(N-1)\cdot d \tag{12.18}$$

where N is the size of the one dimension of a square planar array, $N\times N$, and d is the sensor spacing along both the x and y coordinates of the equally spaced transducers of the planar array. Let us denote $\delta_{l,n}$ as the distance of the $(l,n)^{\text{th}}$ transducer from the center of the planar array,

$$\delta_{l,n} = \sqrt{\delta_{l,n_x}^2 + \delta_{l,n_y}^2} \tag{12.19}$$

Then, the spatial window values, $w_{l,n}$ for the $(l,n)^{\text{th}}$ transducer, can be obtained from the 2D Gaussian window $W(n)$:

$$w_{l,n} = W\left(\text{round}\left(N\cdot\frac{\dfrac{D}{2}-\delta_{l,n}}{D} \right) \right) \tag{12.20}$$

function round(x) rounds the x value to the nearest integer.

Estimates of the spatial window weights for a rectangular planar array of size $N\times M$ can be described from the decomposition of the planar array beamformer into two sets of linear array beamformers as defined in Reference [16].

12.2.4.4 Beam Power Pattern for Multifocus Transmit Beamformer for Planar Phase Array

The image of the beam power pattern of the multifocus illumination is obtained by computing the total power of the signals that arrive at each of the pixels in the coordinate system of the planar phase array. The arriving signal at each pixel is the added combinations from all active transducers radiated signals that are time delayed according to Equation 12.21,

$$\zeta_{l,n}(\delta_{l,n_x}, \delta_{l,n_y}, p_x, p_y, p_z)$$
$$= \left[(p_x^2 + p_y^2 + p_z^2)^{1/2} + ((p_x - \delta_{l,n_x})^2 + (p_y - \delta_{l,n_y})^2 + p_z^2)^{1/2} \right]\cdot\frac{f_s}{c} \tag{12.21}$$

where p_x, p_y, p_z are the x, y, z locations, respectively, of the pixel and $\zeta_{l,n}$ denotes the difference between the distance from the pixel to the center of the array and the distance from the pixel to the $(l, n)^{\text{th}}$ transducer.

As a result of the time delay of Equation 12.21, the combined signal radiated from all the active transducers of the planar array that arrives at a pixel (x, y) can be computed by adding the time shifted transducers signals as follows:

$$b(p_x, p_y, p_z, \theta_i, \phi_q, m) = \sum_{l}^{N} \sum_{n}^{N} S_{l,n}(\theta_i, \phi_q, m + \zeta_{l,n}) \tag{12.22}$$

Then, the intensity of illumination of the pixel located at (x, y) can then be expressed by,

$$P(p_x, p_y, p_z, \theta_i, \phi_q) = \sum_{m} \left| \sum_{l}^{N} \sum_{n}^{N} S_{l,n}(\theta_i, \phi_q, m + \zeta_{l,n}) \right|^2 \tag{12.23}$$

and the 3D beam power pattern image is obtained from

$$P(p_x, p_y, p_z) = \sum_{\text{all}\,\theta_i, \phi_q} P(p_x, p_y, p_z, \theta_i, \phi_q) \tag{12.24}$$

which represents the 3D image of the summation of all the illuminations of the 3D steering angles.

Figure 12.7 shows the 3D beam power pattern distribution from the illumination of a volume of size (10 cm×10 cm×10 cm) by a planar array with 12×12 transducers having 0.4 mm transducer spacing. The beam steering with multiple range focusing is at a 3D angle (75°, 75°) as defined by the parameters depicted in the coordinate system of Figure 12.6. The center frequency and bandwidth of the transmitted signal are 2 MHz and 2 MHz, respectively.

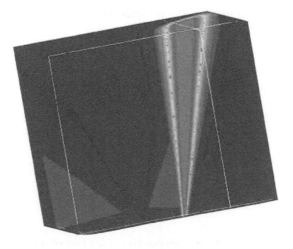

FIGURE 12.7
3D beam power pattern distribution from the illumination of a volume of size (10 cm×10 cm×10 cm) by a planar array with 12 × 12 transducers having 0.4 mm transducer spacing. The beam steering with multiple range focusing is at a 3D angle (75°, 75°) as defined by the parameters depicted in the coordinate system of Figure 12.6. The center frequency and bandwidth of the transmitted signal are 2 MHz and 2 MHz, respectively.

As in the case of the linear array active beamformer, the maximum temporal length of the active beam time series of the planar array can be estimated from:

$$\tau_{max} = 2 \cdot \left(\frac{N}{2} + 0.5 \right) \cdot \sqrt{2} \cdot \delta \cdot \frac{f_s}{c} \qquad (12.25)$$

where the multiplication factor $\sqrt{2}$ takes into account the transducer spacing along the diagonal direction. Then the lower boundary or the starting temporal sample of the synchronized active transducer signals is,

$$s_{min} = s_{pos} - \tau_{max} \qquad (12.26)$$

Finally, the beam output of the transmitted pulse by an active planar array probe is defined by

$$B(\theta_i, \phi_q, R_s, f) = \sum_l \sum_m S_{l,n}(\theta_i, \phi_q, f) \cdot e^{j 2\pi f \tau_{l,n}(\theta_i, \phi_q, R_s)} \qquad (12.27)$$

where $S_{l,n}(\theta_i, \phi_q, f)$ is the Fourier transform of the $(l,n)^{\text{th}}$ transducer signal $S_{l,n}(\theta_i, \phi_q, m)$. The details of the 3D planar array beamformer and its decomposition process into two line array beam-formers are defined in Reference [16].

12.2.5 Multifocus Receiving Beamformer for Linear and Planar Phase Arrays

The main functions of a receiving ultrasound beamformer is to beamform the reflections and backscattering fields that result from the illumination of a field of view (i.e., body organs) with ultrasound beam steering waves, as defined in the previous two sections. Thus, following the steered transmission of directional ultrasound pulses in a specific region, the operation of an ultrasound system reverts to the receiving mode. During the receiving mode, a fully digital ultrasound system digitizes the sensor time series of an ultrasound linear or planar array probe that senses the reflections and backscattering effects of the illuminated field of view. Then, a digital receiving beamformer will beamform the received sensor time series and the output of this process will provide beam time series that will form the basis for the reconstruction of the tomography 2D (B-scan) or 3D volumetric image of the illuminated field of view. In this case, the theoretical analysis on beamforming [2,3,15–18], can be used by system engineers to design a receiving 2D or 3D ultrasound beamformer. However, the receiving ultrasound beamformer is more complex than the corresponding transmit (i.e., active) beamformer and the following two sections will provide details in terms of their multifocus configuration and their time series concatenation to prepare them for the image reconstruction process by taking into consideration the theory in Reference [16].

12.2.5.1 Receiving Ultrasound Beamformer for Linear Phase Array

The coordinate system of a receiving beamformer for linear phase arrays is the same as in the case of the corresponding transmit beamformer. The receiving line array is chosen to be located along the x-axis with the array center at $x=0$, as depicted in Figure 12.8. The

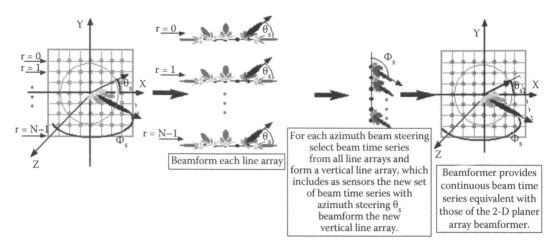

FIGURE 12.8
Coordinate system and geometric representation of the concept of decomposing a planar array beamformer. The sensor planar array beamformer consists of N linear arrays with M being the number of sensors in each linear array.

receiving beams are steered in the 2D plane along the positive y-axis, $0° \leq \theta \leq 180°$. Their beamwidth is defined by Equation 2.29 in Reference [16] and forms the image resolution characteristics of the reconstructed image. Thus, each receiving ultrasound beam $b(\theta_i, r_s, m)$ is characterized by its angular direction, θ_i (i.e., steering angle) and focus range, r_s. In terms of angular resolution, the image characteristics can be improved by either using interpolation techniques or generating adaptive beams of intermediate fine angular directions between two consecutive steerings θ_i and θ_{i+1}. In terms of pixel resolution along the temporal axis (depth), the reconstructed image can be improved by dividing the field of view into focal zones, as shown in Figure 12.9. In other words, for a specific angular direction θ_i (i.e., steering) the receiving beamformer generates a number of different beams focused at different ranges, $b(\theta_i, r_s, m)$, with $s=1, 2, 3, \ldots$ Thus, the focal zones in units of temporal samples can be expressed as,

$$r_s = 2 \times r_{s_{meter}} \times \frac{f_s}{c} \tag{12.28}$$

where,

$$\begin{cases} 0 \leq r_{s_i} \leq r_s\left(i+\frac{1}{2}\right), & i=1 \\ r_s\left(i-\frac{1}{2}\right) \leq r_{s_i} \leq r_s\left(i+\frac{1}{2}\right), & i>1 \end{cases}$$

defines each focal zone by its center, as schematically depicted in Figure 12.9. The factor of two in Equation 12.28 is due to the round trip travel time of the received ultrasound signal.

Although a decrease in the size of a focal zone may improve the image resolution along the axis of range, this choice will increase the number of focal zones (NZONES) and therefore the generation of number of focused receiving beams, a process that will increase the computational load. In other words, for I number of steering angles and S

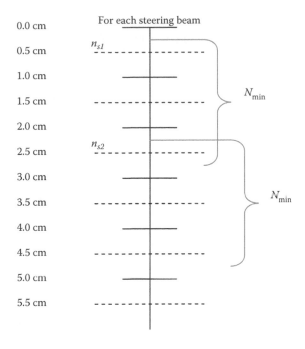

FIGURE 12.9
The blocks show the focal zones, each of size 1.0 cm, along a specific azimuth direction, which is the temporal axis denoting range or depth. N_{min} denotes the smallest amount of temporal samples for the beamforming process.

NZONES, the beamforming process has to be repeated $I \times S$ times. In the experimental system discussed in this chapter, the focal zone size was set to to 1 cm, with angular beamwidth of 0.5°.

It is apparent form the above discussion that the temporal samples of the sensor or beam time series represent range (or depth) and this relationship is defined by:

$$r_m = \frac{m}{f_s} \cdot c \cdot 1/2 \qquad (12.29)$$

where, r_m represents range (or depth) in units of temporal samples, m is the sample index of the input sensor time series.

The factor of 1/2 in Equation 12.29 takes into account the round trip travel time of the received signal as denoted by Equation 12.28.

Since the temporal samples of the sensor and beam time series are related to range, a substantial saving in processing time can be achieved by using the smallest possible set of data samples in the beamforming process. This smallest amount of temporal samples, M_{min} that can be considered in the beamforming process should be sufficient to cover at least one focal zone plus twice the maximum time delay of the beam steering process as defined below,

$$M_{min} \geq r_m + 2 \left| \frac{1}{2}(N-1)\delta \frac{f_s}{c} \right| \qquad (12.30)$$

where M_{min} denotes the lowest number of temporal samples for the beamforming process and N is the number of sensors in the linear array probe.

Let us consider now a beamforming process with M_{min} temporal samples for the sth focal zone. Then, the sensor time series and their data segment of temporal samples associated with the specific focal zone r_s are defined by

$$\begin{cases} x_n(m) = y_n(m + r_{s_i}) & 0 \leq m \leq M_{min} - 1, 0 \leq m + r_{s_i} \leq M \\ x_n(m) = 0 & 0 \leq m \leq M_{min} - 1, (m + r_{s_i} \leq 0, \text{or}, m + r_{s_i} \geq M) \end{cases} \quad (12.31)$$

where $y_n(m)$ is the nth sensor signal of the linear array probe with M being the total number of temporal samples in the digitized sensor time series of the nth sensor. The Fourier transform of $x_n(m)$ with a filtering operation are defined by,

$$X\_\{n\}(f\_\{m\}) = FFT[x\_\{n\}(m)] \quad (12.32)$$

$$S_n(f_m) = X_n(f_m) \cdot H_{r_s}(f_n, r_s) \quad (12.33)$$

where, $m = l, l+1, \ldots, l+L$ and $H_{rs}(f_n, r_s)$ are the band-pass filter coefficients designated for each focal zone. The filtering process of Equation 12.33 is a critical step in the focus receiving beamformer. In particular, high frequency regimes of ultrasound signals provide better image resolution than lower frequency regimes. However, the associated high propagation losses allow only for very short ranges of ultrasound penetration to achieve effective imaging. On the other hand, lower frequency regimes in the ultrasound signals provide deeper penetration at the expense of poor image resolution.

Ultrasound system designers have used the above propagation characteristics of the ultrasound energy in the human body to structure the broadband frequency spectrum of transmitted pulses illuminating a medium of interest. Thus, the high frequency regime of a received ultrasound signal is being used for short ranges (i.e., short range focal zones), while the lower part of the spectrum is being allocated for the deeper focal zones. This kind of design approach requires very wide broadband ultrasound pulses to allow for segmentation of their wide-band frequency spectrum into smaller frequency regimes that can be activated by the receiving focus beamformer through the filtering process of Equation 12.33. Moreover, this filtering process can use a small number of frequency bins $\{X_n(f_m), m = l, l+1, \ldots, l+L\}$ of the sensor signals that represent the frequency regime associated with the focal zone of the receiving beamformer. As a result, the same $X_n(f_m)$ sensor time series can be re-used for as many times in the receiving focus beamforming process as is the NZONES. The design characteristics for the filter coefficients, $H_{rs}(f_n, r_s)$ can be Hamming or Kaizer FIR filters. An additional advantage of the filtering process defined in Equation 12.33 is that it minimizes the computational load by minimizing the summation processes of the receiving beamformer (i.e., see Equation 12.34) to be equal with the number, L, of the frequency bins that are considered in the filtering process in Equation 12.33.

The beamforming for each focal zone and steering angle is then performed by phase shifting the sensor signals $S_n(f_m)$ in the frequency domain through multiplications with the steering vector $D^-(f, r, \theta)$ with its nth element being $d_n(f_m, r_s, \theta_l)$ and summations as defined by Equation 12.25 in Reference [16], which are rewritten below,

$$B(f_m, r_s, \theta_i) = \sum_n S_n(f_m) \cdot d_n(f_m, r_s, \theta_i) w_n \tag{12.34}$$

The nth steering element $d_n(f_m, r_s, \theta_i)$ of the steering vectors is defined by

$$d_n(f_m, r_s, \theta_i) = \exp\left[j2\pi \frac{(m-1)f_s}{M} \frac{\tau_n(r_s)}{c} \right] \tag{12.35}$$

and

$$\tau_n(r_s) = \sqrt{r_s^2 + \delta_n^2 - 2r_s\delta_n \cos\theta_i} \tag{12.36}$$

where: δ_n is the location of the nth transducer.

For the image reconstruction process, however, the beams in Equation 12.34, which are in the frequency domain, need to be converted into beam time series as follows.

$$b(\theta_i, r_s, m) = IFFT\{B(f_m, r_s, \theta_i)\} \tag{12.37}$$

Furthermore, the beam time series $b(\theta_i, r_s, m)$ of different focal zones, $(r_s, s = 1, 2, 3, ...)$ but along the same steering angle θ_i need to be concatenated to reconstruct a new single beam time series for the steering angle θ_i and with a total number of temporal samples covering the full range (i.e., depth) of the image reconstruction process. For a given focus range, the minimum number of temporal samples from the output of the focus receiving beamformer have already been defined by Equation 12.30. Thus, the concatenation process is a simple operation of cut and paste the time samples of $b(\theta_i, r_s, m)$ from each sth focal zone. This process will generate a new beam time series $b(\theta_i, m)$ with typical maximum number of temporal samples M, equal to the number of samples in the sensor time series $x_n(m)$ of the line array probe. To comply also with display requirements, the beam time series $b(\theta_i, m)$ may need to be compressed up to a certain length. The compression is done through an or-ing operation as expressed by Equation 12.38.

$$b_l(\theta_s, m) = (b_{l-1}^6(\theta_i, 2m) + b_{l-1}^6(\theta_i, 2m+1))^{1/6} \tag{12.38}$$

12.2.5.2 Receiving Ultrasound Beamformer for Planar Phase Array

The focus in the design concept of the energy transmission or reception beamforming module for planar arrays is to illuminate or receive the entire volume of interest with a few firings. This is shown in Figure 12.10. Here the volume is illuminated in 3×3 sectors, which includes a total of nine firings. The transmitted signals are all broadband FM (chirp) signals as discussed in the previous sections. They are fired with inter-element delays to allow the transmitted energy to be focused at specific regions in space, which is the space highlighted by the square shaded areas of Figure 12.10. The beamforming energy transmission is done through the 12×12 elements at the center

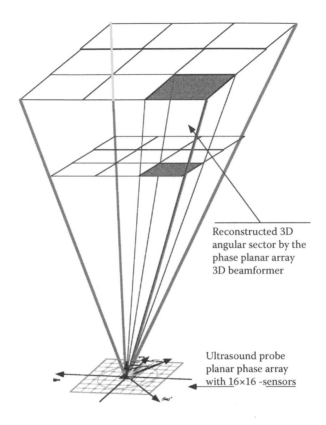

Reconstructed 3D
angular sector by the
phase planar array
3D beamformer

Ultrasound probe
planar phase array
with 16×16 -sensors

FIGURE 12.10
Volume 3D/4D digital scanning for ultrasound applications using a planar array phase array probe. (Reprinted from Dhanantwari A., Stergiopoulos S., Bertora F. Parodi C., Pellegretti P., and Questa A. An efficient 3D beam-former implementation for real-time 4D ultrasound systems deploying planar array probes. In *Proceeding of the IEEE UFFC'04 Symposium*. Montreal, Canada, 2004. With permission.)

of the array, while the receiving 3D beamformer through the 32×32 elements of the full phase array. The receiving data acquisition unit digitizes the sensed ultrasound reflections via the A/DC peripheral of the unit, under a similar arrangement as was defined in the previous sections. The angular subsectors depicted in Figure 12.10, are arranged in column-row configuration. Each angular subsector occupies the region bounded by the A and B angles, as defined in Figure 12.6. In Figure 12.10, the angular sector $70° \le A \le 110°$, $70° \le B \le 110°$ is shown to be divided into nine angular subsectors consisting of three rows and three columns. Each subsector occupies 10° of A angle and 10° of B angle, with their coordinate system defined in Figure 12.6. As discussed in the previous sections, each subsector is illuminated by separate transmissions. There are, therefore, the same number of sets of received signals as the number of subsectors. The image of each angular subsector is also reconstructed separately using the corresponding set of received data. Volumetric reconstruction of each subsector is derived from the 3D beamforming process applied on all the received sensor time series of the planar array. The number of beams to be formed by the receiving beamformer is specified by the size of the angular subsector. For example, the receiving beamformer is specified to form ten beams in the angular direction A and ten beams in the angular direction B. This means that 100 (10×10) beams will be used to fill and reconstruct the image of each angular subsector.

As in the case of the receiving linear array beamformer, the temporal samples of the sensor and beam time series are related to range (i.e., depth of ultrasound penetration) and therefore a substantial saving in processing time can be achieved by using the small-est possible set of temporal samples in the beamforming process. This smallest amount of temporal samples, M_{min} that can be considered in the beamforming process should be sufficient to cover at least one focal zone plus twice the maximum time delay of the beam steering process as defined before by Equation 12.30.

Thus, Equation 12.31 for the linear array is modified for the planar array as follows:

$$\begin{cases} x_{l,n}(m) = y_{l,n}^{v,w}(m+r_{s_i}) & 0 \le m \le M_{min}-1, 0 \le m+r_{s_i} \le M \\ x_{l,n}(m) = 0 & 0 \le m \le M_{min}-1, (m+r_{s_i} \le 0 \quad \text{or} \quad m+r_{s_i} \ge M) \end{cases} \tag{12.39}$$

where l,n are the transducers indexes (l^{th} row and n^{th} column) of the array, v,w are the angular subsector index (v^{th} row and wth column), $y_{l,n}^{v,w}(m)$, is the input signal of the $(l,n)^{\text{th}}$ transducer of the $(v,w)^{\text{th}}$ angular subsector, r_{si} is the starting sample for the Si^{th} focal zone, as defined below by Equations 12.40 and 12.28, M is the number of temporal samples of the input sensor time series, $y_{l,n}^{v,w}(m)$, which is the length of one snapshot.

$$M_{min} \ge r_{s_i} + 2\sqrt{2} \left| \frac{1}{2}(N-1)\delta \frac{f_s}{c} \right|. \tag{12.40}$$

Then, the 3D beamforming process implemented in the frequency domain for the digitized sensor time series $x_{l,n}(t_m)$, $l=1, 2, \dots, N$, $n=1, 2, \dots, N$, $m=1, 2, \dots, M$ of the $N \times N$ detectors of the phase planar array with M time samples for each sensor collected, is given below by Equation 12.41, which is identical with Equation 12.27 for the transmit beamformer, with the coordinates defined in Figures 12.5 and 12.8.

$$B(f_m, A, B, r_s) = \sum_{l=0}^{N-1} \sum_{n=0}^{N-1} X_{l,n}(f_m) \cdot H_s(f_m, r_s) \cdot S_{l,n}(f_m, A, B, r_s) \tag{12.41}$$

where $X_{l,n}(f_m) = FFT[x_{l,n}(t_m)]$ and $H_s(f_m, r_s)$ are the FIR filter coefficients as in Equation 12.33 with the steering vector expressed by $S_{l,n}(f_m, A, B, r_s) = \exp(j2\pi f_m \tau_{l,n}(A, B, r_s))$ and the inter-element time delays defined below by Equation 12.42.

$$\tau_{l,n}(A, B, r_s) = \frac{\sqrt{r_s^2 + \delta_{l,n_x}^2 + \delta_{l,n_y}^2 - 2\delta_{l,n_x} \cos A - 2\delta_{l,n_y} \cos B} - r_s}{c} \tag{12.42}$$

where the sensor element (l,n) is located at position $(\delta_{l,n_x}, \delta_{l,n_y})$.

Equation 12.42 indicates that for each beam (A, B) and focal depth r_s there needs to be $N \times N$ complex steering vectors computed for each frequency bin of interest. Furthermore, each set of steering vectors $S_{l,n}(f_m, A, B, r_s)$ is unique, with each of the four function variable independent and not separable. Because of this independence, it is not possible to decompose this beamformer in an efficient manner.

Reference [16] introduces a decomposition process for the 3D planar array beamformer that presents an alternative to the beamformer in Equation 12.41 [17,18]. This decomposition process allows the beamforming equation in Equation 12.41 to be divided, and hence decomposed, which in turn allows for it to be easily implemented on a parallel architecture, discussed in Section 12.3 of this chapter. However, it is important to note that this decomposition process, is a very close approximation of Equation 12.41, resulting in a simplified two-stage linear beamforming implementation. For plane wave arrivals, (i.e., $r_s \to \infty$), the approximation below in Equation 12.43 can be directly derived from Equation 12.41 and the derivation is exact. The approximation in the decomposition process for the time delay parameter in Equation 12.42 is defined by,

$$\tau_{l,n}(A,B,r_s) = \frac{\sqrt{r_s^2 + \delta_{l,n_x}^2 - 2\delta_{l,n_x} r_s \cos A} - r_s}{c} + \frac{\sqrt{r_s^2 + \delta_{l,n_y}^2 - 2\delta_{l,n_y} r_s \cos B} - r_s}{c} \tag{12.43}$$

This approximation leads to the decomposition of the 3D beamforming into two linear steps, expressed by:

$$B(f_m,A,B,r_s) = \sum_{l=0}^{N-1} S_l(f_m,B,r_s) \cdot \left[\sum_{n=0}^{N-1} X_{l,n}(f_m) \cdot H_s(f_m,r_s) \cdot S_n(f_m,A,r_s) \right] \tag{12.44}$$

with the two separated steering vectors expressed as:

$$S_l(f_m,A,r_s) = \exp\left\{ j2\pi f_m \left(\frac{\sqrt{r_s^2 + \delta_{l,n_x}^2 - 2\delta_{l,n_x} r_s \cos A} - r_s}{c} \right) \right\}$$

$$S_n(f_m,B,r_s) = \exp\left\{ j2\pi f_m \left(\frac{\sqrt{r_s^2 + \delta_{l,n_y}^2 - 2\delta_{l,n_y} r_s \cos B} - r_s}{c} \right) \right\}$$

In Equation 12.44, the summation in square brackets is equal to a line array beamformer along the x-axis of the coordinate system in Figure 12.6. This term is a vector which can be denoted as $B_n(f_m, A, r_s)$. Then, Equation 12.43 can then be rewritten as follows:

$$B(f_m,A,B,r_s) = \sum_{n=0}^{N-1} B_n(f_m,A,r_s) \cdot S_n(f_m,B,r_s) \tag{12.45}$$

which defines a linear beamforming along the y-axis, with the beams $B_n(f_m,A,r_s)$ treated as the input time series. This kind of two stage implementation is easily parallelized and implemented on a multinode system, as discussed in Section 12.3 and schematically illustrated by Figure 12.8 or Figure 3.6 in Reference [16]. In this approximate implementation {e.g., Equation 12.41 compared with Equations 12.44 and 12.45}, the error introduced at angles A and B close to broadside is negligible. Side by side comparisons show that there is no degradation in image quality over the exact implementation for planar array ultrasound application.

The rest of the beamforming and image reconstruction processes for the planar array ultrasound system are identical with those defined for the linear array receiver beamformer in the previous section and expressed by Equations 12.32, 12.33 and

Equations 12.37, 12.38. Thus, the formation of beam time series, according to Equation 12.37, $b(A_i,B_i,r_s,m)$ of different focal zones, $(r_s, s=1, 2, 3, ...)$, but along the same steering angle (A_i,B_i) need to be concatenated (e.g., cut and paste operations) to reconstruct a new single multifocus beam time series for the steering angle (A_i,B_i) and with a maximum total number of temporal samples to be equal with the temporal length of the input sensor time series, covering the full range (i.e., depth) of the image reconstruction process.

To comply also with display requirements, the beam time series $b(A_i,B_i,r_s,m)$ may need to be compressed up to a certain length. The compression is done through an or-ing operation defined by Equation 12.38; and when the formation of the multifocus beam time series is complete, the 3D volumetric ultrasound image, as depicted in Figure 12.10, can be reconstructed. The volume is divided into pixels. A pixel value is determined through an interpolation process of the beam time samples, as follows. First, the radial distance r_p of a pixel, (x_p, y_p, z_p) from the center of the array, is determined. The angular location (A_p, B_p) of the pixel is also computed from,

$$r_p = \sqrt{x_p^2 + y_p^2 + z_p^2} \tag{12.46a}$$

$$A_p = \tan^{-1} z_p/x_p \tag{12.46b}$$

$$B_p = \tan^{-1} z_p/y_p \tag{12.46c}$$

Then, the following eight beam-time-samples

$$b(r_{p-},A_{p-},B_{p-}) \quad b(r_{p+},A_{p-},B_{p-})$$

$$b(r_{p-},A_{p-},B_{p+}) \quad b(r_{p+},A_{p-},B_{p+})$$

$$b(r_{p-},A_{p+},B_{p-}) \quad b(r_{p+},A_{p+},B_{p-})$$

$$b(r_{p-},A_{p+},B_{p+}) \quad b(r_{p+},A_{p+},B_{p+})$$

are used to interpolate each of the pixel values, where

$$r_{p-} \leq r_p \leq r_{p+} \tag{12.47a}$$

$$A_{p-} \leq A_p \leq A_{p+} \tag{12.47b}$$

$$B_{p-} \leq B_p \leq B_{p+} \tag{12.47c}$$

with r_{p-} and r_{p+} representing the two closest beam time samples to the pixel location r_p. Similarly, (A_{p-},A_{p+}) are the two closest angles to the angular location, A_p and (B_{p-},B_{p+}) are the two closest angles to the angular location, B_p. If the pixel is located outside the angular sector being illuminated, its value is set to zero.

12.3 Computing Architecture and Implementation Issues

Implementation of a fully digital 3D adaptive beamforming structure in ultrasound systems is a not a trivial issue. In addition to the selection of the appropriate algorithms, success is heavily dependent on the availability of suitable computing architectures.

Past attempts to implement matrix based signal processing methods, such as adaptive beamformers, were based on the development of systolic array hardware because systolic arrays allow large amounts of parallel computation to be performed efficiently since communications occur locally. None of these ideas are new. Unfortunately systolic arrays have been much less successful in practice than in theory. The fixed size problem for which it makes sense to build a specific array is rare. Systolic arrays big enough for real problems cannot fit on one board, much less one chip, and interconnects have problems. A 2D systolic array implementation will be even more difficult. So, any new computing architecture development should provide high throughput for vector as well as matrix based processing schemes.

A fundamental question, however, that must be addressed at this point is whether it is worthwhile to attempt to develop a dedicated architecture that can compete with a multiprocessor using stock microprocessors. However, the experience gained from sonar computing architecture developments [15] suggests that a cost effective approach in that direction is to develop a PC-based computing architecture that will be based on the rapidly evolving microprocessor technology of the CPUs of PCs. Moreover, the signal processing flow of advanced processing schemes that include both scalar and vector operations should be very well defined in order to address practical implementation issues. When the signal processing flow is well established, such as in Figures 13, 19, 20 and 21 in Reference [16], then distribution of this flow in a number of parallel CPU's will be straightforward. In the following sections, we address the practical implementation issues by describing the current effort of developing an experimental fully digital 3D/4D ultrasound system deploying a planar array to address the requirements of the Canadian Forces for noninvasive portable diagnostic devices deployable in fields of operations.

12.3.1 Technological Challenges for Fully Digital Ultrasound System Architecture

The current state-of-the-art in high-resolution, digital, 3D ultrasound medical imaging faces two main challenges.

First, the ultrasound signal processing structures are computationally demanding. Traditionally, specialized computing architectures and hardware have been used to provide the levels of performance *and* I/O throughput required, resulting in high system design and ownership costs. With the emergence of high-end workstations and low-latency, high bandwidth interconnects [20], it now becomes interesting and timely to investigate if such technologies can be used in building low-cost, high-resolution, 3D ultrasound medical imaging systems.

Second, although beamforming algorithms in digital configuration have been studied in the context of other applications [15], little is known about their computational characteristics with respect to ultrasound-related processing, and medical applications in general. It is not clear which parts of these algorithms are the most demanding in terms of processing or communication and how exactly they can be mapped on modern parallel PC-based architectures. In particular, although the algorithmic complexity of different sections can be calculated, little has been done in terms of actual performance analysis on real systems. The lack of such knowledge inhibits further progress in this area, since it is not clear how these algorithms should evolve to lead to applicable solutions in the area of ultrasound medical imaging.

Reference [20] addresses both these two issues by introducing a design of a parallel implementation of advanced 3D beamforming algorithms [16] and studying its behavior and requirements on a generic computing architecture that consists of commodity components. This design concept provides an efficient, all-software, sequential implementation that shows considerable advantages over hardware-based implementations of the past. It provides also an efficient parallel implementation of advanced 3D beamforming [16] for a cluster of high-end PCs connected with a low-latency, high-bandwidth interconnection network that allows also for an analysis of its behavior. The emphasis in this design has been placed also on the identification of parameters that critically affect both the performance and cost of ultrasound system.

The end result [20] reveals a number of interesting characteristics leading to conclusions about the prospect of using commodity architectures for performing all related processing in ultrasound imaging medical applications. A brief summary of these findings suggests the following:

- A PC-based multiprocessor system today can achieve close-to-real-time performance for high-end ultrasound image quality and is certainly expected to do so in the near future.
- The major components of a digital 3D ultrasound beamforming signal processing structure [16,20] consists of:
  - 85–98% of the time is spent in FFT and beam steering functions.
  - The communication requirements in the particular implementation are fairly small, localized, and certainly within the capabilities of modern low-latency, high-bandwidth interconnects.
  - The results in Reference [20] provide an indication of the amount of processing required for a given level of ultrasound image quality and number of channels in a probe that can be used as a reference in designing computing architectures for ultrasound systems.

12.4 An Experimental Planar Array Ultrasound Imaging System

12.4.1 System Overview

The experimental configuration of the fully digital ultrasound imaging concept that was used to assess the fully digital 3D ultrasound beamforming structure of this chapter, included two versions. The first was configured to be integrated with a linear phase array ultrasound probe and it is depicted in Figure 12.11. This is a fully digital ultrasound system which includes the linear array probe (64 elements) and the two-node data acquisition unit including two mini-PCs that control the transmitting and receiving functions. This linear array ultrasound system was configured also to provide 3D images through

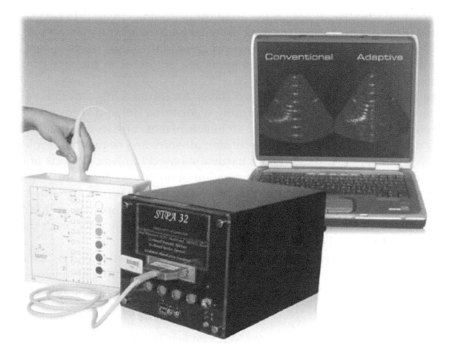

FIGURE 12.11
The experimental linear phase array (64-element) ultrasound imaging system integrated with a laptop computer to provide visualization functionalities.

volume rendering of the B-scan outputs. Figure 12.12 shows this 3D configuration which is defined by the integration of the experimental linear phase array ultrasound system with a portable PC and a tracking device. A USB communication protocol allowed the transfer of the B-scans (2D digital images) as inputs to the visualization software for 3D volume rendering (see Chapter 4) installed in the portable PC. More specifically, the experimental linear array ultrasound system can provide 3D volumetric images from a series of B-scan (2D) outputs. The magnetic tracker system, shown schematically in Figure 12.12, provides the co-ordinates of the probe for each of the acquired image frames (B-scans). This tracker provides translational (x, y, and z) as well as rotational co-ordinates with respect to the x, y, and z, axes.

The second configuration is a fully digital planar phase array volumetric ultrasound imaging system, depicted in Figure 12.13. The multinode computing cluster that allows for an effective implementation of the 3D beamforming structure is shown at the lower part of this figure. The top left image in Figure 12.13 shows the planar phase array probe and the top right image presents the data acquisition unit with the A/D and D/A peripherals controlled by the multinode cluster.

Implementation of the 3D beamforming structure and communication requirements relevant with the system configuration of Figure 12.13, have been discussed already in Reference [20].

Figure 12.14 shows a schematic representation of the main components of the fully digital real time planar array ultrasound imaging system that summarizes the developments that have been presented in the previous sections. It depicts a 256 (16×16) element phased array probe, a A/DC with 64-channel data acquisition unit that through multiplexing

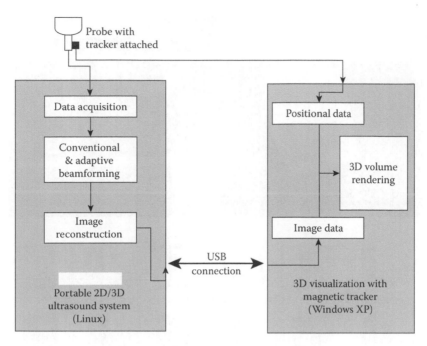

FIGURE 12.12
Integration of the experimental linear phase array ultrasound system with a portable PC, a tracking device and a USB communication protocol for 2D digital image transfers provided as inputs to Fraunhofer's visualization software for 3D volume rendering.

acquires time series signals for the 256 channels, and a computing architecture to process the acquired time series into ultrasound volumes, shown also in Figure 12.13. In addition the system uses a 36-channel digital to analog converter (D/AC) to excite the center (6×6) transducers of the planar array during the illumination process. The transmit functionality that address the pulse design to illuminate at various depths simultaneously is addressed in a subsequent section. The inter-element spacing of the probe is 0.4 mm in both directions. This combination forms the front end of the 3D ultrasound system that will support the transmit functions and the receiving functions required for the 3D beamforming. The probe is attached to the data acquisition unit via an interface card. This card provides the means of data flow into and out of the probe through the data acquisition system.

The computing cluster that implements the 3D beamformer software has already been introduced in References [16,20]. This is the multinode cluster that was designed to allow for easy implementation of the 3D beamformer algorithms—both conventional and adaptive. The integrated hardware platform in Figures 12.13 and 12.14 brings together the planar array probe, the data acquisition unit for the planar array probe and the multinode PC cluster.

The A/DC is well grounded and capable to sample the 64 channels with an equivalent 14-bit resolution and 33 MHz sampling frequency per channel. Moreover, the unit has dedicated memory and separate bus-lines. The D/AC is capable to drive 36 channels with 12-bit resolution and 33 MHz sampling frequency. The period between two consecutive active transmissions is in the range of 0.2 ms. Moreover, the local memory of the D/AC unit has the capability to store the active beam time series with total memory size of 1.35 Mb, being generated by the main computing architecture for each focus depth and transferred to the local D/AC memory when the transmission-acquisition process begins.

FIGURE 12.13
The multinode computing cluster that allows the implementation of the parallel beamforming structure of the experimental planar phase array ultrasound imaging system is shown at the lower part of this figure. The top left image shows the planar phase array probe and the top right image depicts the data acquisition unit with the A/D and D/A peripherals (with probe attached) controlled by the multinode cluster.

The digitization process of the 16×4 subapertures by the 14-bit 64-channel A/D unit, provide the signals to a system of pin connectors-cables with suppressed cross-talk characteristics (minimum 35 dB). The sampling frequency is 33 MHz for each of the channels associated with a receiving single sensor. The multiplexer associated with the A/DC allows the sampling of the 16×4-sensors of the planar array in four consecutive active transmissions to be able to digitize the 16×16 planar array channels.

The computing architecture, discussed in Reference [20], includes sufficient data storage capabilities for the sensor time series. The A/DC and signal conditioning modules of the data acquisition process and the communication interface are controlled through S/W drivers that form an integral part of the computing architecture.

It has been assessed that the ultrasound adaptive 3D beamforming structure, defined in Reference [16], provides an effective beam-width size, which is equivalent to that of a two to three times longer aperture along azimuth and elevation of the deployed planar array. Thus, for the deployed receiving 16×16 planar array, the adaptive beamformer's beamwidth characteristics will be equivalent with those of a (16×2)×(16×2) size planar array. For example, the beamwidth of a receiving 16×16 planar array with element spacing

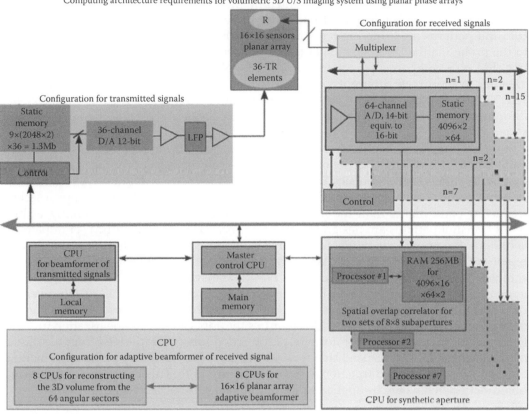

Computing architecture requirements for volumetric 3D U/S imaging system using planar phase arrays

FIGURE 12.14
Structure of the PC-based computing schematic representation of the main components including the data acquisition units of a fully digital real time planar array ultrasound imaging system.

of 0.5 mm for a 3-MHz center frequency, is approximately 7.4°, with effective angular resolution by the adaptive beamformer in terms of beamwidth size, to be less than $3.7° \times 3.7°$. As a result, the receiving adaptive beams along azimuth will have the following image resolution capabilities:

- For C-scan and for depth of 10 cm, the $3.7° \times 3.7°$ angular resolution sector corresponds to a $(0.64 \text{ cm}) \times (0.64 \text{ cm}) = 0.41 \text{ cm}^2$ size of tissue resolution, or for depth of 5 cm to a $(0.32 \text{ cm}) \times (0.32 \text{ cm}) = 0.1024 \text{ cm}^2$ size of tissue resolution.

- For B-scan the line resolution will be equivalent to the wavelength of the transmitted center frequency , which is 0.5 mm.

- Thus, the volume resolution of the 3D adaptive beamforming structure will be equivalent to $(10.24 \text{ mm}^2) \times (0.5 \text{ mm}) = 5.12 \text{ mm}^3$, in 3D tissue size at a depth of 5 cm.

The concept of the energy transmission module to illuminate the entire volume of interest with a few firings, has already been depicted in Figure 12.10. Here the volume is illuminated in subsectors. The transmitted signals are all broadband FM (chirp) signals. They are fired with inter-element delays to allow the transmitted energy to be focused at specific regions in space, (e.g., the space highlighted by the square shaded areas of Figure 12.10). The energy transmission is done through the 6×6 elements at the center of the array. The transmit patterns are loaded into the memory of the data acquisition unit and delivered to the probe via the D/AC portion of the unit when a trigger signal is received. In addition, FM pulses that occupy different nonoverlapping frequency regimes may be coded together to illuminate different focal depths with a single firing, as defined in Section 12.2.5.1. This means that it can be arranged so that one frequency regime can focus and illuminate the lower shaded square, and a second frequency regime the upper shaded square in Figure 12.10.

Suppose three focal depths $d_1<d_2<d_3$ are desired. Then, three separate sets of transmit patterns are created. The first set of transmit patterns with the appropriate delay profiles that occupy the lowest frequency band are designed to focus at depth d_3. A second set of transmit patterns with delay profiles to focus at depth d_2 are designed to occupy a second frequency band higher than the first and with no frequency overlap. Similarly, a third set of transmit patterns are designed to focus at focal depth d_1. This third set of patterns is designed to occupy the highest frequency band since they will illuminate the shallow regions of the medium of interest. When the design of the three sets of transmit patterns is complete they are superimposed to create a single transmit pattern. This composite transmit pattern is used to provide the illumination as described in Figure 12.10. However, the requirement to use the smallest possible number of illumination beams leads to a nonuniform energy distribution in space. This requires the application of a linearization function to correct for this type of nonuniformity. A correction function is derived from the illuminating beam shapes and is used later for linearization of the results of the beamformer. An example of a linearization function is shown in Figure 12.15. This figure shows the correction function that would be applied to the output of a 4×4 subsector illuminations.

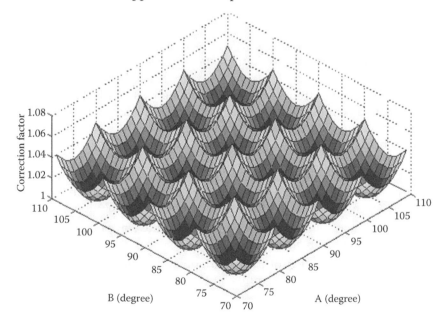

FIGURE 12.15
Correction function for 4×4 sectors illumination pattern, as depicted in Figure 12.10. (Reprinted from Dhanantwari A., Stergiopoulos S., Bertora F. Parodi C., Pellegretti P., and Questa A. An efficient 3D beamformer implementation for real-time 4D ultrasound systems deploying planar array probes. In *Proceeding of the IEEE UFFC'04 Symposium*. Montreal, Canada, 2004. With permission.)

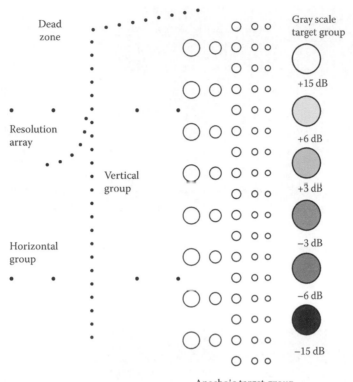

FIGURE 12.16
Cross sectional view of the experimental phantom.

12.4.2 Performance Results

Presented in this section are the image output results from the linear phase array (e.g., 64 elements, 2D/3D) and the planar phase array probes (e.g., 16×16, 3D/4D). While numerous experiments were carried out, only a few typical image/volume outputs for both the adaptive and conventional image results are presented here. All these image results are from the standard ultrasound test target called "phantom", the cross section of which is shown in Figure 12.16.

12.4.2.1 *Portable 2D/3D Experimental System with Linear Phase Array Probe*

B-scan results
Figure 12.17 shows two typical images from the portable 2D/3D system operating in B-scan mode. The left hand image shows the image obtained using the conventional beamforming technique and the right hand image shows the image output from the adaptive beamformer. Both images are obtained by placing the probe on the top of the phantom just above the "dead zone" label. The phantom's vertical row of reflectors and the curved arrangement are depicted in both images that show no "ghosting" or blurring of there strong reflectors. This assessment is consistent with the triggering mechanism characteristics being sufficient to create the synthetic aperture processing without having the need to implement the overlap correlation software synthetic aperture technique, as this was substantiated in Section 12.2.2.

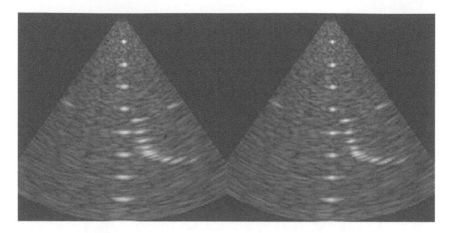

FIGURE 12.17
B-scan image results for both the conventional (left hand side image) and adaptive (right hand side image) beamformers of the experimental linear phase array (64-element) ultrasound imaging system, depicted in Figure 12.11.

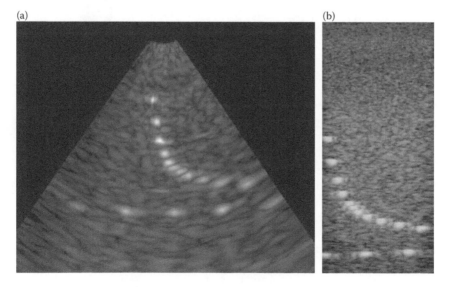

FIGURE 12.18
The left hand side image (a) shows the B-scan output of the adaptive beamformer for the same phantom as in Figure 12.17, illuminated at 2 MHz. The right hand side image (b) shows the output of a commercial system using the same probe at 4 MHz.

Figure 12.18 provides a comparison of the B-scan outputs from a commercial ultrasound imaging system and the experimental prototype of this investigation depicted in Figure 12.11. The images for both systems are obtained by placing the probes on the left side of the phantom just beside the "resolution array" label. Both systems use the same probe (64 channels). However, the commercial system operates at 4 MHz while the experimental system in Figure 12.11 operates at 2 MHz. The left hand side image (Figure 12.18a) shows the output of the adaptive beamformer while Figure 12.18b shows the image output from the commercial 4 MHz ultrasound system.

It can be seen from the image results of Figure 12.18 that the adaptive beamforming scheme of this investigation performs quite as good, if not even better, with respect to the ultrasound commercial system, in terms of both resolution of the strong scatterers, and noise in the bulk image. It should furthermore be considered that the images obtained with the prototype of this investigation have been acquired using a 2 MHz probe, whereas a 4 MHz probe has been employed for the image output (Figure 12.18b) of the commercial system.

The transmission/acquisition schemes adopted by the experimental 2D/3D prototype system allow to cover a wide scanning angle (90–120°), a characteristic which can be very useful in cardiac imaging applications. Furthermore, the deployment of low frequencies in the range of 2 MHz can achieve the deep penetration depths which are required in cardiologic ultrasound applications.

Volumetric 2D/3D imaging

The volumetric images created from the 2D/3D system are shown in Figure 12.19. This figure shows cross sections of the 3D output from volume rendering of B-scan images using the techniques discussed in Reference [12]. In particular, Figure 12.19 shows the volume obtained using the standard ultrasound phantom of Figure 12.16. In this experiment the volume is taken along the top surface of the phantom. The top right panel in Figure 12.20 shows the full reconstructed volume. The remaining three panels show three orthogonal views of the volume scanned. Although the experimental 2D/3D system in Figure 12.11 includes a fully digital ultrasound technology with an advanced beamforming structure

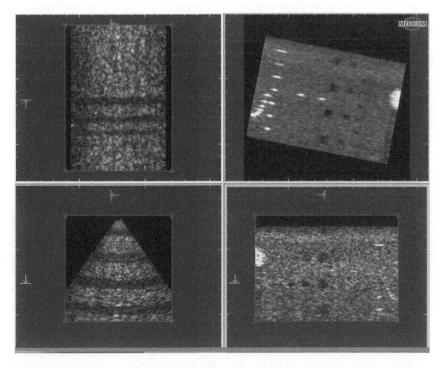

FIGURE 12.19
The 3D volume is taken along the top surface of the phantom shown in Figure 12.16. The top right panel shows the full reconstructed volume. The remaining three panels show three orthogonal views of the volume scanned.

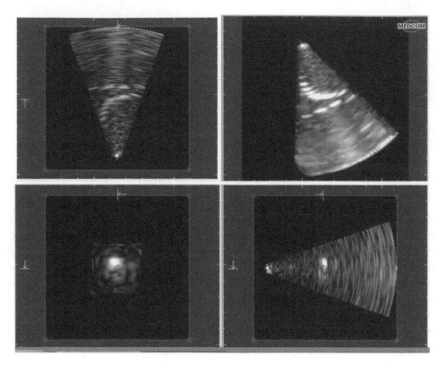

FIGURE 12.20
The top right panel shows the full reconstructed volume from the 3D conventional beamformer implemented
on the 16×16 planar array probe. The remaining three panels show three orthogonal views of the volume
scanned, which represent reconstructed B-scans of cross sections of the 3D output of the volumetric data of the
planar array beamformer, discussed in Section 12.2. The reconstructed volume was for the standard ultrasound
phantom of Figure 12.16. In this experiment the volume is taken along the top surface of the phantom.

implemented in frequency domain, this system has been reduced to a portable size, compared to nowadays commercial ultrasound units.

12.4.2.2 3D/4D Experimental System with Planar Phase Array Probe

The volumes created from the 3D/4D system, deploying the planar (16×16) phase array
probe, are shown in Figures 12.20 and 12.21. Figure 12.20 shows the volume derived from
the 3D conventional beamformer and Figure 12.21 presents the output of the 3D adaptive
beamformer. The volumes in both figures are obtained by placing the probe on the left
side of the phantom just beside the "resolution array" label. The top right panel in Figures
12.20 and 12.21 show the full reconstructed volume in each case. The remaining three
panels show three orthogonal views of the volume scanned. Like the portable ultrasound
2D/3D unit, the experimental 3D/4D system deploying a planar phase array probe has the
capability of performing both conventional as well as adaptive beamforming, and both
the modules are integrated into the parallel processing scheme discussed in this chapter.
Furthermore, the 3D/4D system is also capable of frequency coding for multizone focusing, and of applying deblurring algorithms for enhancing the image quality. Both experimental systems (e.g., 2D/3D, 3D/4D) show good performances in terms of scanning angle
apertures, penetration depth and image quality. As expected, the adaptive beamforming
scheme implemented into the 3D parallel architecture seems to allow for a higher image

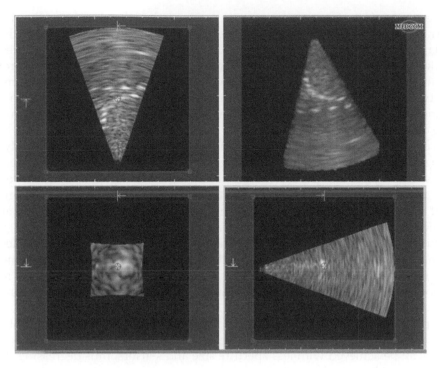

FIGURE 12.21
The results are for the 3D adaptive beamformer implemented on the same planar array time series being used also in Figure 12.20. The top right panel shows the full reconstructed volume from the 3D adaptive beamformer implemented on the 16×16 planar array probe. The remaining three panels show three orthogonal views of the volume scanned, which represent reconstructed B-scans of cross sections of the 3D output of the volumetric data of the planar array beamformer, discussed in Section 12.2. The reconstructed volume was for the standard ultrasound phantom of Figure 12.16. In this experiment the volume is taken along the top surface of the phantom.

quality with respect to conventional beamforming for what concerns the axial and contrast resolutions.

12.5 Conclusion

The fully digital ultrasound system technology discussed in this chapter consists of a set of unique adaptive ultrasound beamformers [17,18], a PC-based computing architecture and a set of visualization tools [12], addressing the fundamental image resolution problems of current 3D ultrasound systems. The results of this development can be integrated into existing 2D and/or 3D ultrasound systems or they can be used to develop a complete stand-alone 3D ultrasound system solution.

It has been well established [2,3,12,15,16] that the existing limitations of medical ultrasound imaging systems in poor image resolution, is the result of the very small size of deployed arrays of sensors and the distortion effects by the influence of the human-body's nonlinear propagation characteristics. The ultrasound technology, discussed in this chapter, replaces the conventional (time delay) beamforming structure of ultrasound systems with an adaptive beamforming processing configuration that has been developed for sonar array systems. The results of this development [2,3] have demonstrated that these novel adaptive beamformers improve significantly (at very low cost) the image resolution capabilities of an ultrasound imaging system by providing a performance improvement equivalent to a deployed ultrasound probe with double aperture size. Furthermore, the portability and the low cost for the 3D ultrasound systems offer the options to medical practitioners and family physicians to have access of diagnostic imaging systems readily available on a daily basis.

At this point, however, it is important to note that in order to fully exploit the advantages of this digital adaptive ultrasound technology, its implementation in a commercial ultrasound system requires that the system has a fully digital design configuration consisting of A/DC and D/AC peripherals that would fully digitize the ultrasound probe time series, they will optimally shape the transmitted ultrasound pulses through a D/A peripheral and they will use phase array linear or matrix ultrasound probes.

This kind of fully digital ultrasound configuration, revises the system architecture of ultrasound devices and moves it away from the traditional hardware and implementation software requirements. Thus, implementation of the adaptive beamformer is a software installation on a PC-based ultrasound computing architecture with sufficient throughput for 3D and 4D ultrasound image processing.

In addition, the use of adaptive ultrasound beamformers, provides significantly better image resolution than the traditional time delay based beamformers. Thus, a good image resolution can be achieved with less aperture size and sensors, thus decreasing the hardware costs of an ultrasound system.

In summary, the PC-based ultrasound computing architecture of this chapter, its adaptive 2D and 3D ultrasound beamforming structure and the set of visualization tools allow for a flexible cost-to-image quality adjustment. The resulting product can be upgraded on a continuous base at very low cost by means of software improvements and by means of hardware by taking advantage of the continuous upgrades and CPU performance improvements of the PC-based computing architectures. Thus, for a specific image resolution performance, a complete re-design or product upgrade can be achieved by means of software improvement, since the digital hardware configuration would remain the same.

References

1. P.R. Detmer, G. Bashein, T. Hodges, K.W. Beach, E.P. Filer, D.H. Burns, and D.E. Strandness. 1994. 3D ultrasonic image feature localization based on magnetic scanhead tracking: vitro calibration and validation. *Ultrasound Med. Biol*, 20:923–936.
2. A. Dhanantwari, S. Stergiopoulos, F. Bertora, C. Parodi, P. Pellegretti, and A. Questa. 2004. An efficient 3D beamformer implementation for real-time 4D ultrasound systems deploying planar array probes. In *Proceeding of the IEEE UFFC'04 Symposium*, Montreal, Canada.
3. A. Dhanantwari, S. Stergiopoulos, C. Parodi, F. Bertora, A. Questa, and P. Pellegretti. 2003. Adaptive 3D beamforming for ultrasound systems deploying linear and planar array phased array probes. In *IEEE Conference Proceedings, IEEE International Ultrasonics Symposium*, pp. 5–8, Honolulu, HI.
4. D. Downey and A. Fenster. Three-dimensional ultrasound: a maturing technology. 1998. *Ultrasound Quart.*, 14(1):25–39.

5. T.L. Elliot, D.B. Downey, S. Tong, C.A. Mclean, and A. Fenster. 1996. Accuracy of prostate volume measurements in vitro using three-dimensional ultrasound. *Acad. Radiol.*, 3:401–406.

6. S.W. Hughes, T.J.D. Arcy, D.J. Maxwell, W. Chiu, A. Milner, R.J. Saunders, and J.E. Shepperd. 1996. Volume estimation from multiplanar 2D ultrasound images using a remote electromagnetic position and orientation. *Ultrasound Med. Biol.*, 22:561–572.

7. J.A. Jensen. 1996. Field: a program for simulating ultrasound systems. *Med. Biol. Eng. Comput.*, 34:351–353.

8. D.F. Leotta, P.R. Detmer, and R.W. Martin. 1997. Performance of a miniature magnetic position sensor for three-dimensional ultrasound imaging. *Ultrasound Med. Biol.*, 23:597–609.

9. Feng Lu, E. Milios, S. Stergiopoulos, and A. Dhanantwari. A new towed array shape estimation scheme for real-time sonar systems. *IEEE J. Oceanic Eng.*, 28(3):552–563, 2003.

10. T.R. Nelson, D. Downey, D.H. Pretorius, and A. Fenster. 1999. *Three-Dimensional Ultrasound*. Lippincott, Williams and Wilkins, Philadelphia, PA.

11. P.A. Picot, D.W. Rickey, R. Mitchell, R.N. Rankin, and A. Fenster. 1993. Three-dimensional colour Doppler imaging. *Ultrasound Med. Biol.*, 19:95–104.

12. G. Sakas, G. Karangelis, and A. Pommert. 2000. Advanced applications of volume visualisation methods in medicine. In S. Stergiopoulos, editor. *Handbook on Advanced Signal Processing for Sonar, Radar and Medical Imaging Systems*. CRC Press LLC, Boca Raton, FL.

13. S. Sherebrin, A. Fenster, R. Rankin, and D. Spence. 1996. Freehand three-dimensional ultrasound: implementation and applications. *SPIE: Phys. Med. Imaging*, 2708:296–303.

14. S. Stergiopoulos. 1990. Optimum bearing resolution for a moving towed array and extension of its physical aperture. *J. Acoust. Soc. Am.*, 87(5):2128–2140.

15. S. Stergiopoulos. 1998. Implementation of adaptive and synthetic aperture beamformers in sonar systems. *Proc. IEEE*, 86:358–396.

16. S. Stergiopoulos. 2000. Advanced beamformers. In S. Stergiopoulos, editor. *Handbook on Advanced Signal Processing for Sonar, Radar and Medical Imaging Systems*. CRC Press LLC, Boca Raton, FL.

17. S. Stergiopoulos and A. Dhanantwari. 2002. High Resolution 3D Ultrasound Imaging System Deploying a MultiDimensional Array of Sensors and Method for MultiDimensional Beamforming Sensor Signals. Assignee: Defense R&D Canada, US Patent: 6,482,160.

18. S. Stergiopoulos and A. Dhanantwari. 2004. High Resolution 3D Ultrasound Imaging System Deploying a MultiDimensional Array of Sensors and Method for Multi-Dimensional Beamforming Sensor Signals. Assignee: Defense R&D Canada, US Patent: 6,719,696.

19. S. Tong, D.B. Downey, H.N. Cardinal, and A. Fenster. 1996. A three-dimensional ultrasound prostate imaging system. *Ultrasound Med. Biol.*, 22:735–746.

20. F. Zhang, A. Bilas, A. Dhanantwari, K.N. Plataniotis, R. Abiprojo, and S. Stergiopoulos. 2002. Parallelization and performance of 3D ultrasound imaging beamforming algorithms on modern clusters. In *Proceedings of the 16th International Conference on Supercomputing (ICS'02)*, New York, NY.

13

Magnetic Resonance Tomography— Imaging with a Nonlinear System

Arnulf Oppelt*

Siemens AG

CONTENTS

13.1 Introduction

Since its introduction to clinical routine in the early 1980s magnetic resonance imaging (MRI) or tomography (MRT) has developed to a preferred imaging modality in many diagnostic situations due to its unparalleled soft tissue contrast, combined with high spatial resolution, and its capability to generate images of slices in arbitrary orientation or even of entire volumes. Furthermore, the possibility to display blood vessels, to map brain functions, and to analyze metabolism is widely valued.

Magnetic resonance (MR) is the phenomenon according to which particles with an angular and a magnetic moment precess in a magnetic field, thereby absorbing or emitting electromagnetic energy. This effect is called electron spin resonance (ESR) or electron paramagnetic resonance (EPR) for unpaired electrons in atoms, molecules, and crystals and nuclear magnetic resonance (NMR) for nuclei. ESR was discovered in 1944

---

* Retired.

by the Russian scientist Zavoisky,[1] but until now has not yet gained any real significance for medical applications. NMR was observed independently in 1945 by Bloch et al.[2] at Stanford University in California and by Purcell et al.[3] in Cambridge, MA. The Nobel Prize for physics was awarded in 1952 to these two groups.

In 1973, Lauterbur[4] described how magnetic field gradients could be employed to obtain images similar to those recently generated with X-ray computed tomography. The limits placed on spatial resolution by the wavelength in the imaging process with waves are circumvented in MRI by superposing two fields. With the aid of a radio frequency (rf)-field in the megahertz (MHz) range and a locally variable static magnetic field, the sharp resonance absorption of hydrogen nuclei in biological tissue is used to obtain the spatial distribution of the nuclear magnetization. Contrary to other imaging modalities in medicine such as X-rays or ultrasound, imaging with NMR employs a nonlinear system. The signal used to construct an image does not depend linearly on the rf-energy applied to generate it and can be influenced in a very wide range by the timing of the imaging procedure. In the following, we will summarize the basic principles of MR and the concepts of imaging.

13.2 Basic NMR Phenomena

All atomic nuclei with an odd number of protons or neutrons, i.e., roughly two thirds of all stable atomic nuclei, possess an intrinsic angular momentum or spin. This is always coupled with a magnetic dipole moment, which is proportional to the angular momentum. As a consequence, these particles align in an external magnetic field. As in matter, many atomic nuclei exist, e.g., 1 mm^3 water contains $6.7 \cdot 10^{19}$ hydrogen nuclei, a small but measurable angular momentum per unit volume and an associated macroscopic magnetization with results proportional to the external magnetic flux density and inversely proportional to temperature.

At thermal equilibrium, the nuclear magnetization of a sample with nuclear spins is aligned parallel to an applied magnetic field. However, if this parallel alignment is disturbed, e.g., by suddenly changing the direction of the field, a torque acts on the magnetic moment of the sample. According to the law of conservation of angular momentum, this torque causes a temporal change of the angular momentum of the sample, resulting in a precession of the magnetization with the (circular) Larmor frequency

$$\omega = \gamma B_Z. \tag{13.1}$$

This precession (NMR) can be detected by measuring the alternating voltage induced in a coil wound around the sample. For hydrogen nuclei or protons, which represent the most frequently occurring nuclei in nature, a sharp resonance frequency of 42.577 MHz is observed at a magnetic flux density of 1 T.

In an NMR experiment, the precession of the nuclear magnetization is often stimulated by disturbing the alignment of the nuclear magnetization parallel to the static magnetic field by an rf-field having a frequency similar to the Larmor frequency. It is provided by a coil wound around the sample with its field axis orthogonal to the static magnetic field. This coil can also be used for signal detection. The linearly polarized rf-field in this coil can be thought of as the superposition of two circularly polarized fields rotating in opposite directions. Thus, there is always the same direction of rotation as the Larmor precession, and in this reference frame (the rotating frame), there is a constant magnetic flux density B_1. The resulting torque causes the nuclear magnetization to precess around the axis of the B_1 field in the rotating frame in the same way as around the static magnetic field B_Z in the laboratory frame. The combined precession movement around the static and the rf-field in the laboratory system causes the tip of the nuclear magnetization vector to execute a spiral path on the surface of a sphere (Figure 13.1). The contra-rotating rf-field, having twice the Larmor frequency in the rotating frame, acts as a perturbation and is effectively averaged out.

Under the influence of the rotating rf-magnetic flux density B_1, an angle α between the static magnetic field and the nuclear magnetization emerges, which is proportional to the duration t of the rf-field:

$$\alpha = \gamma B_1 t. \tag{13.2}$$

For reasons of simplicity, we shall consider in the following the transverse components in the rotating frame, i.e., the coordinate system rotating with ω_L in the laboratory system. With a B_1 field that confines an angle φ with the x-axis, the nuclear magnetization attains the components (Figure 13.2).

$$M_X = -M_0 \sin\alpha \sin\varphi$$

$$M_Y = M_0 \sin\alpha \cos\varphi \tag{13.3a}$$

$$M_Z = M_0 \cos\alpha.$$

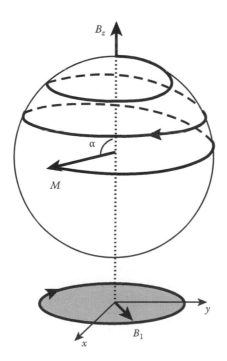

FIGURE 13.1
Motion of the nuclear magnetization vector M_0 under the influence of a static magnetic field B_z and a circularly polarized rf-field B_1 with Larmor frequency γB_0. The initial position of M_0 is parallel to B_z; after time t, M_0 is oriented with an angle $\alpha = \gamma B_1 t$ along the direction of B_z. (The direction of precession depends on whether the angular and the magnetic moment of the nuclei are parallel or antiparallel. For protons, a clockwise rotation follows.)

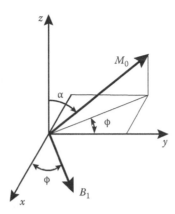

FIGURE 13.2
Tilting of the magnetization M_0 by B_1 field in the rotating frame.

The x and y components of the transverse magnetization can be combined into the complex quantity

$$M_\perp = M_X + iM_Y \quad i = \sqrt{-1},$$ (13.4)

giving

$$M_\perp = iM_0 \sin \alpha e^{i\varphi}$$ (13.3b)

Switching off the rf-pulse after the nuclear magnetization is aligned orthogonally to the static magnetic field ($\alpha = 90°$, hence a 90° pulse) induces the maximum signal in the sample coil.

13.3 Relaxation

Relaxation describes the effect that the precession of the nuclear magnetization decays with time; the original state of equilibrium is reestablished, with the magnetization aligned parallel to the static magnetic field. This phenomenon is described by two separate relaxation time constants, T_1 and T_2, in which the equilibrium states, M_0 and 0, respectively, of the nuclear magnetization M_Z parallel and $M_\perp$ perpendicular to the static magnetic field are obtained again; T_1 is always $\geq T_2$. Longitudinal relaxation is associated with the emission of energy to the surroundings, i.e., the lattice in which the nuclei are embedded, and is therefore also referred to as spin-lattice relaxation. Transverse relaxation is caused by collisions of the nuclear spins and, thus, is often referred to as spin–spin relaxation. In the latter case, since the longitudinal component of the magnetization remains unchanged, the energy of the nuclear ensemble does not change; only the relationship of the phases between the individual spins is lost. T_1 results from an energy effect, and T_2 results from an entropy effect.

The behavior of the nuclear magnetization in an external magnetic field B_Z undergoing relaxation was described by Bloch et al.[5] by adding empirical terms to the classical law of motion conservation:

$$\frac{dM_Z}{dt} = \gamma(\vec{M} \times \vec{B})_Z + \frac{(M_0 - M_Z)}{T_1}$$

$$\frac{dM_\perp}{dt} = \gamma(\vec{M} \times \vec{B})_\perp - \frac{M_\perp}{T_2}.$$

(13.5)

It is the wide range of relaxation times in biological tissue that makes NMR so interesting in medical diagnostics. T_1 is of the order of magnitude of several 100 ms, while T_2 is in the range 30–100 ms. T_1 of biological tissue decreases, when temperature increases. This effect is investigated with MRI to probe temperature changes in the human body.

13.4 NMR Signal

From Bloch's equations (Equation 13.5), one obtains for the precessing nuclear magnetization after a 90° pulse around the x-axis in the rotating frame

$$M_\perp(t) = iM_0 e^{\frac{-t}{T_2}}$$

(13.6)

for the transverse component and

$$M_Z(t) = M_0\left(1 - e^{\frac{-t}{T_1}}\right)$$

(13.7)

for the longitudinal component.

The precessing two components of tranverse magnetization can be measured independently in the laboratory system with two induction coils oriented perpendicular to each other; this is named a free induction decay (FID). An oscillating signal with Larmor frequency ω_L is observed that follows from Equation 13.6 by multiplication with $e^{-1\omega_L t}$.

The frequency dependence of the transverse magnetization is given by its Fourier transformation (Figure 13.3). The imaginary part of the Fourier transformation describes the so-called absorption line

$$M_y(\omega) = M_0 \frac{T_2}{1 + \omega^2 T_2^2},$$

(13.8)

and the real part the dispersion line

$$M_x(\omega) = M_0 \frac{-\omega T_2^2}{1 + \omega^2 T_2^2}.$$

(13.9)

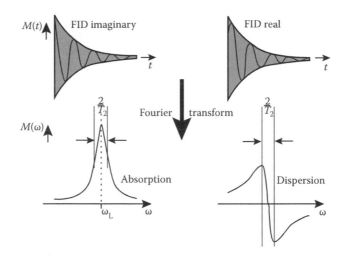

FIGURE 13.3
FID after a 90° pulse with its Fourier transform, representing the NMR absorption and dispersion line. In the laboratory system, resonance is oberserved at $\omega=\omega_L$, compared to $w=0$ in the rotating frame.

Instead of taking the Fourier transformation of the FID, it is also possible to measure absorption and dispersion directly by recording the change of resistance and inductance of the signal coil surrounding the sample as a function of frequency (or as a function of the flux density of the static magnetic field). Such continuous wave (cw) methods, however, are much slower than pulse methods and are therefore hardly used anymore.

The full width at half maximum of the absorption and the distance between the extreme points of the dispersion line are given by the transverse relaxation time

$$\Delta\omega_{1/2} = \frac{2}{T_2}. \tag{13.10}$$

Protons in distilled water exhibit a transverse relaxation time $T_2\approx1\,\mathrm{s}$. The measurement of T_2 through the FID, however, is only possible in very homogenous magnetic fields. In practice, the static magnetic field varies in space, resulting in different precession frequencies of the nuclear magnetization. Because of destructive interference, a shortened FID is observed, resulting in an inhomogeneously broadened resonance line. The line shape depends on the spatial distribution of the static magnetic field deviations ΔB_Z over the entire sample. Reference is often made to an effective transverse relaxation time

$$\frac{1}{T_2^*} = \frac{1}{T_2} + \frac{\gamma\Delta B}{2} \tag{13.11}$$

which, however, can only coarsely describe the effect of magnetic field inhomogeneities since the signal clearly no longer decays exponentially.

The signal loss in an inhomogeneous static magnetic field can be recovered by means of a refocusing or a 180° rf-pulse.[6] The diverging transverse magnetization after a 90° pulse due to field inhomogeneities converges again, since the 180° pulse reverses the order of the spins (Figure 13.4). Thus, slowly precessing spins which have lagged behind now move ahead and realign themselves with the faster precessing spins after the interval between

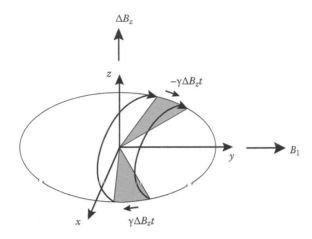

FIGURE 13.4
Dephasing by the angle $\gamma\Delta B_z t$ to transverse nuclear magnetization in the rotating frame due to field inhomogeneities is reversed with a 180° rf-pulse. A spin echo is created.

the two rf-pulses. A spin echo is observed, the amplitude of which is determined by transverse relaxation.

In this context, we will mention another type of echo, the so called stimulated echo.[6] When applying two 90° pulses instead of a 90°/180° pair, an echo also occurs, but only with half the amplitude resulting from using a 180° pulse. The missing magnetization is stored along the z-axis, thereby undergoing longitudinal relaxation. It can be tilted again into the transverse plane with a third 90° pulse and can manifest itself as a stimulated echo with a distance from the third 90° pulse corresponding to that of the first two 90° pulses. The amplitude of the stimulated echo is determined by the longitudinal relaxation.

Relaxation is not the only mechanism that affects the amplitudes of the echoes. Especially the molecules of liquids move stochastically (diffuse) during the time between excitation and observation of the echo from one position in the inhomogeneous static magnetic field to another; in accordance with the field difference, the nuclei precess there with a different Larmor frequency and thus no longer contribute fully to the echo amplitude.[7] The influence of spin diffusion can be enhanced by applying a strong, magnetic field gradient pulse symmetrically before and after the refocusing rf-pulse. In MRI, such types of experiments are of interest because diffusion is anisotropic due to tissue microstructure. This offers the possibility to get information, e.g., about the fiber structure of tissue. Diffusion depends on temperature and therefore provides an alternative to the precise measurement of T_1 for noninvasive of the monitoring temperature *in vivo*.

The gyromagnetic ratio determining the Larmor frequency of nuclei in the static magnetic field is a fixed constant for each nuclear species. In NMR experiments with nuclei embedded in different molecules, however, slightly different resonance frequencies are observed. This effect is caused by the molecular electrons responsible for chemical bonding. These electrons screen the static magnetic field, with the result that the atomic nucleus "sees" different magnetic fields (chemical shift) depending on the nature of the chemical bond. In a molecular complex, often several resonance lines attributable to individual groups of molecules are observed. Quantitatively, the chemical shift is usually given in parts per million (ppm) relative to a reference line.

Besides the chemical shift, a fine splitting of the MR lines is also frequently observed. This is caused by the magnetic interaction (spin-spin coupling) between the nuclei, which again acts indirectly via the valence electrons. Therefore, in chemistry, molecular structure is often investigated amenable to study with NMR spectroscopy.

13.5 Signal-to-Noise Ratio

To obtain biological or medical information from a living being with MR, it is necessary to attribute the recorded nuclear magnetization to the site of its origin. Before discussing methods of spatial resolution, however, we will turn first to the fundamental restriction for such measurements. The signal induced by the precessing nuclear magnetization in the pick-up coil around the sample must compete with the noise generated by the thermal motion of the electrons in the coil and the Brownian molecular motion in the object under investigation, i.e., the human body.

Noise being generated thermally in the coil provided to pick up the NMR signal according to Nyquist is given by

$$U_{\text{Noise}} = \sqrt{\frac{2}{\pi} K_B (R_{\text{Coil}} T_{\text{Coil}} + R_{\text{Sample}} T_{\text{Sample}}) \Delta \omega}, \tag{13.12}$$

$\Delta \omega$ = detection bandwidth for the signal; R_{Coil} = resistance of the signal coil without sample; R_{Sample} = contribution of the sample to the resistance of the signal coil; T_{Coil} = temperature of signal the coil; T_{Sample} = temperature of the sample.

While with small samples of a few cubic millimeters, as commonly investigated in the laboratory, the noise contribution $R_{\text{Coil}} T_{\text{Coil}}$ dominates, this is not true for samples as large as the human body. With a conductive sample, the resistance of the signal coil can be derived from the power distributed in the sample by a mean rf-field B_1 that would be generated by a current i in that coil:[8]

$$P_{\text{Sample}} = R_{\text{Sample}} i^2$$

$$= \frac{1}{4} \sigma B_1^2 \omega^2 \int r_\perp^2 dv, \tag{13.13}$$

$r_\perp$ = radius coordinate orthogonal to the field axis of the signal coil; σ = electrical conductivity of the sample; ω = signal frequency, i.e., Larmor frequency.

When the rf-field B_1 is replaced by the "field per unit current B_i''", which describes the dependence of the magnetic field on the geometry of the coil, Equation 13.13 gives the resistance due to the coupling to the sample.

The voltage that is induced in the signal coil by the precessing magnetization in a volume element (voxel) Δv of the sample after a $90°$ pulse is given by

$$U_{\text{Signal}} = B_i \omega_L M_0 \Delta v, \tag{13.14}$$

where $M_0 \propto (B_Z / T)$ is the transverse nuclear magnetization.

The signal-to-noise (S/N) ratio follows after a succession of steps omitted here:

$$\frac{U_{\text{Signal}}}{U_{\text{Noise}}} = \sqrt{\frac{\pi}{24K_B^3} \gamma N_v \frac{\omega_L}{T_{\text{Sample}}\sqrt{\frac{1}{2\mu_o}\frac{T_{\text{Coil}}V_{\text{Sample}}}{\eta Q \omega_L} + \frac{\sigma}{4}T_{\text{Sample}}\int r_\perp^2 dv}}\frac{\Delta v}{\Delta \omega}}, \qquad (13.15)$$

where N_V is the density of H^1 nuclei and μ_o is the permeability in vacuum.

Thus, the "filling factor" η, which is a measure of the ratio of the sample volume to the signal coil volume, and the "coil quality factor" Q, which gives the ratio of the energy stored in the coil to the energy loss per oscillation cycle, have been introduced

When smaller volume elements in a large sample are to be resolved, a worse S/N ratio is obtained. Since the nuclear magnetization increases with the Larmor frequency, the S/N ratio improves with increasing strength of the static magnetic field. For small values of the "moment of inertia" $\int r_\perp^2 dv$ (i.e., small samples) and for small filling and coil quality factors, the S/N ratio is proportional to $\omega^{3/2}$; for a high coil quality factor or a low coil temperature, it is directly proportional to the Larmor frequency, i.e., the flux density of the static magnetic field.

Since the temperature cannot be lowered with living samples, once the NMR apparatus is set up (i.e., static magnetic field and antennas are chosen), the S/N ratio can only be influenced by voxel size Δv and bandwidth S/N. Repeating the NMR experiment n times and adding the single signals together results in an S/N ratio improvement by a factor of $\sqrt{n}$. However, since a reduction in spatial resolution is as undesirable as a longer time of measurement, these two choices are normally avoided.

13.6 Image Generation and Reconstruction

To derive an MR signal from a localized small volume within as larger object, at least one of the two fields (i.e., the static magnetic field and the rf-field) required for the NMR measurement has to vary over space. It has been proposed that a sharp maximum or minimum be generated in space for these fields so that the MR signal observed would originate mainly from that region.[9] However, along with the technical difficulties of generating sufficiently sharp field extrema, to yield information from other regions would require the movement of the area sensitive to MR through the object under investigation, leading to a long time of measurement, when each voxel has to be measured several times in order to obtain a sufficient S/N ratio.

The utilization of the signal from the entire object rather than from only a single voxel is achieved with the use of magnetic field gradients G, for which the Larmor frequency varies linearly along one direction in space, giving in the laboratory system

$$\omega L = \gamma(B_Z + Gr) \qquad (13.16a)$$

and in the rotating frame

$$\omega = \gamma Gr \qquad (13.16b)$$

The amplitude of the NMR signal as a function of frequency then corresponds to the sum of all spins in the planes orthogonal to the direction of the magnetic field gradient,[14] i.e., the projection of the nuclear magnetization (Figure 13.5)

$$P(r, j, \vartheta) = \int M_\perp(x, y, z) dl dq. \tag{13.17}$$

Normally, the NMR signal is measured as a function of time rather than a function of frequency, and the projection can then be obtained from the Fourier transformation of the time-dependent NMR signal

$$P(\omega, \varphi, \vartheta) = \frac{1}{2\pi} \int M_\perp(t) e^{i\omega t} dt. \tag{13.18}$$

Lauterbur's original suggestion was to collect a set of projections onto different gradient directions φ, ϑ and reconstruct an image of the nuclear magnetization with the same methods as in X-ray computed tomography.[10,11] It emerges, however, that a modification of his proposal by Kumar et al.[12] offers greater flexibility and simpler image reconstruction. Their method is now used routinely in MRI.

For simplicity, we will restrict ourselves for the moment to a two-dimensional (2d) object. The sample is excited with an rf-pulse so that transverse nuclear magnetization, $M_\perp(x, y)$, is

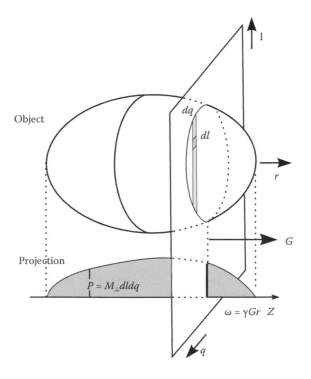

FIGURE 13.5
The NMR signal amplitude as a function of frequency for an object in a linear magnetic field gradient, representing the plane integral of the transverse magnetization (number of spins) in planes orthogonal to the field gradient (projection).

generated in the rotating frame. The phase of the magnetization is then made to vary along the y-direction with a gradient G_y switched on for a time T_y:

$$M_\perp(x,y,G_yT_y) = M_\perp(x,y)e^{-i\gamma G_y yT_y}. \tag{13.19}$$

Then a gradient in the x-direction, G_X, is applied, and the NMR signal, which is the integral of the magnetization precessing differently over the object, is recorded as a function of the time t:

$$M_\perp(G_y,t) = \iint dxdy M_\perp(x,y)e^{-i\gamma G_y y + B_0 T_y}e^{-IG_x xt}. \tag{13.20}$$

When longitudinal magnetization has been reestablished due to longitudinal relaxation, again transverse magnetization is generated with an rf-pulse, and the encoding procedure is repeated. Thus, with a variety of gradients, G_y, a 2d set of signals as a function of G_y and the recording time t is obtained. One can consider this signal set as an interferogram. A 2d Fourier transformation yields the distribution of the local transverse nuclear magnetization:

$$M_\perp(x,y) = \frac{1}{(2\pi)^2} \int_{-T}^{T_X} \int_{x-G_y^{max}}^{G_y^{max}} M_\perp(G_y,t)e^{|y(yT_yG_y+XG_Xt)}dG_ydt. \tag{13.21}$$

Since the spatial distribution of the transverse nuclear magnetization is given by the Fourier transformation of a 2d data set, it is usual to view these data as being acquired in Fourier or k-space (Figure 13.6b) with spatial frequency coordinates

$$K_X = \gamma G_x t \quad \text{and} \quad K_y = \gamma T_y G_y \tag{13.22a}$$

or in the more general case of time-dependent gradients

$$K_x(t) = \gamma \int_0^t G_x(t)dt \quad \text{and} \quad K_y(t) = \gamma \int_0^t G_y(t)dt \tag{13.22b}$$

The MR image signal is sampled along parallel lines in Fourier space that are addressed by a combination of rf- and gradient pulses (Figure 13.6a). When the scanning trajectory crosses the axis $k_y = 0$, a maximum signal named gradient echo is obtained. The Fourier transformation of the amplitudes (real and imaginary part) along a line through the center of the Fourier space results in the projection (Equation 13.17) of the investigated object onto the direction of this line. This is known as the central slice or projection slice theorem.

In digital imaging, an object is sampled with image elements (pixels) of size Δx, Δy, corresponding to the spatial sampling frequencies

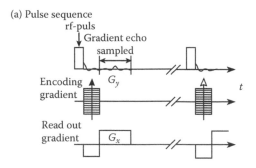

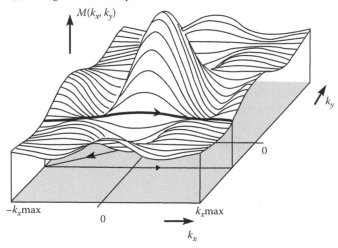

FIGURE 13.6

Scanning 2d object in k-space: the negative lobe of G_y and the encoding gradient G_y address the Fourier line to be recorded (b). The NMR signal is sampled during the positive lobe of gradient G_x (a); the preceding negative lobe ensures sampling to start always at $-k_y^{\max}$.

$$K_X^s = \frac{2\pi}{\Delta X} \quad \text{and} \quad K_y^s = \frac{2\pi}{\Delta y}. \tag{13.23}$$

According to the sampling theorem, the object can then be completely reconstructed, when it does not contain spatial frequencies higher than the (spatial) Nyquist frequency, which is half the sampling frequency

$$K_X^{\max} \le \frac{k_X^s}{2} \quad \text{and} \quad k_y^{\max} \le \frac{k_y^s}{2}. \tag{13.24a}$$

If the object contains information at spatial frequencies larger than the Nyquist frequency, truncation or aliasing leads to typical artifacts; e.g., sharp intensity borders in an object are displayed with parallel lines (Gibbs ringing). So the pixel size has to be chosen according to the spatial resolution.

The spatial frequency interval necessary to image an object with size $\pm x^{\max}$ and $\pm y^{\max}$ is given accordingly by

$$\Delta k_X = \frac{\pi}{X^{\max}} \quad \text{and} \quad \Delta k_y = \frac{\pi}{y^{\max}}. \tag{13.24b}$$

Hence, in order to image an object with diameter $2y^{\max}$ in y-direction (and $2X^{\max}$ in x-direction) with the required spatial resolution Δy, $N_y = (2y^{\max}/\Delta y)$ phase encoding steps are necessary, during which the gradient $-G_y^{\max} \leq G_y \leq G_y^{\max}$ is stepped through, whereby each time $N_X = (2x^{\max}/\Delta x)$ samples have to be taken of the NMR signal. Usually, N_Y and N_X are chosen to be a power of two in order to employ the fast Fourier transform (FFT) algorithm for image reconstruction.[13] As during the image procedure, each voxel in the object under investigation is measured $N_x N_y$ times compared to a (hypothetical) sequential scan, an improvement in the S/N of $\sqrt{N_x N_y}$ results related to an identical measurement time.

Spatial resolution depends on the magnetic field gradient strength. To derive a condition for the gradient G_X in the readout direction, we assume two spins separated by the distance d_X. In order to distinguish the two points, the frequency difference due to the gradient must be greater than that due to the natural line width and static magnetic field inhomogeneities:

$$\gamma G_X d_X \geq \frac{2}{T_2} + \gamma \Delta B_Z. \tag{13.25a}$$

This condition is equivalent to

$$T_X < T_2^*, \tag{13.25b}$$

which ensures that the loss of signal intensity due to the signal decay in a voxel caused by transverse relaxation and field inhomogeneities remains acceptable. Pixel size then should be chosen to be about

$$\Delta X \approx \frac{1}{2} d_X \quad \text{and} \quad \Delta y \approx \frac{1}{2} d_y \tag{13.26}$$

in order to avoid truncation.

Since signal loss due to field inhomogeneities during the encoding interval can be recovered with a 180° pulse, the duration of the encoding gradient has only to be smaller than T_2 rather than T_2^*:

$$T_y < T_2, \tag{13.27a}$$

hence

$$\gamma G_Y d_Y \geq \frac{2}{T_2}. \tag{13.27b}$$

An additional restriction for spatial resolution in MRI is given by self-diffusion. Phase variation over a pixel due to diffusion of the water molecules has to be smaller than that caused by the gradient leading to the constraint

$$d_{x,y} > \sqrt[3]{\frac{D}{\gamma G_{x,y}}}, \tag{13.28}$$

where D is the diffusion coefficient (e.g., in tissue $D = 10^{-5}$–$10^{-6}\,\mathrm{cm^2/s}$). Though this restriction can be neglected in normal imaging experiments, it is of importance for MR microscopy.

When reconstructing the image from the MR signal set with the 2d FFT (Equation 13.21), it is usual to display the magnitude

$$|M_\perp(x,y)| = \sqrt{M_x^2(x,y) + M_y^2(x,y)} \tag{13.29}$$

in order to get rid of phase factors mixing the real and imaginary parts of the magnetization that might arise from sampling delays and the phase of the exciting rf-pulses. Because transverse relaxation poses a multiplicative term

$$e^{-\frac{t}{T_2(X,y)}}$$

on the acquired signal, the local image signal according to Equation 13.29 has to considered as being convoluted with the magnitude of the absorption and dispersion NMR line described by Equations 13.8 and 13.9. Blurring due to this effect is avoided when the condition in Equation 13.25 is observed.

The imaging principle described can be easily extended to three dimensions by adding and stepping through a gradient G_z during the encoding phase. However, this requires a longer time of measurement. Since information for the complete three-dimensional (3d) object is not always required, a 2d object is often generated from the 3d object by selective excitation.

13.7 Selective Excitation

To obtain an image from a slice through a 3d object, a gradient perpendicular to that slice is applied during excitation with the rf-pulse. In this way, the spins are tilted only in a plane, where the precession frequency is identical with the pulse frequency (selective excitation), whereby the bandwidth of the rf-pulse determines the thickness of the excited slice. Assuming a spin system with nuclear magnetization M_0 being exposed to a magnetic field gradient G_z and to a "long" amplitude-modulated rf-pulse $B_1(t)$ of duration $2T_z$, transverse magnetization is generated according to Equation 13.3b that precesses during the time $2T_z$ in the gradient G_z. The distribution of the transverse magnetization along the z-direction can be approximated in the rotating frame as

$$M_\perp(z, 2T_z) = iM_0 \sin|\alpha(\omega)| e^{i\varphi(\omega)} e^{-i\omega 2T_z} \tag{13.30}$$

where $\omega = \gamma G_z z$.

The flip angle $|\alpha(\omega)|$ is determined by the spectral amplitude of the rf-pulse $B_1(t)$ given by its Fourier transformation

$$\alpha(\omega) = \gamma \int B_1(t)e^{i\omega t}dt, \tag{13.31a}$$

and the azimuth $\varphi(\omega)$ of the axis around which the magnetization is tilted (measured against the x-axis in the rotating frame) follows from

$$\varphi(\omega) = \arctan\left(\frac{\mathrm{Im}(\alpha(\omega))}{\mathrm{Re}(\alpha(\omega))}\right). \tag{13.31b}$$

In order to obtain a rectangular distribution of the transverse magnetization along the slice thickness d at position z_0, the shape of the rf-pulse must be selected so as to give a rectangular frequency distribution with spectral width $\Delta\omega = \gamma G_z d$ and a center frequency $\omega_0 = \gamma G_z z_0$ (recalling that we are looking at the spins from the frame rotating with Larmor frequency ω_L, so that in the laboratory system a pulse with frequency $\omega_0 + \omega_L$ must be applied to the coil surrounding the sample). Since an rf-pulse with the shape of a sinc (i.e., $(\sin\pi x)/(\pi x)$) function has such a rectangular spectrum, $B_1(t)$ is chosen to be a sinc function modulated with the center frequency ω_0:

$$B_1(t) = B_1(T_Z)\left(\sin c\,\frac{\Delta\omega}{2\pi}(t - T_Z)\right)e^{-i\omega_0(t-T_Z)} \tag{13.32}$$

where $\Delta\omega = \gamma G_z d$; $\omega_0 = \gamma G_z z_0$; $0 \leq t \geq 2T_Z$; $2Tz > 2\pi/(\gamma G_z d)=$pulse duration

The sinc pulse is restricted here to a duration of $2T_Z$ in order to obtain a selective rf-pulse of finite length and shifted by the interval T_Z in order to expose the spins to the signal part left of the pulse maximum as well; it should be adjusted to extend over several sinc oscillations in order to approximate a rectangular slice profile. With the phase of the rf-pulse chosen to be aligned along the x-axis in the rotating frame, for the transverse magnetization, it follows that

$$M_\perp(z, 2T_z) = iM_0 \sin(\gamma|\int B_1(t)e^{i\omega t}dt|)e^{-iT_z\omega}$$

$$= iM_0 \sin\alpha_0 rect\left(\frac{z - z_0}{d}\right)e^{-i\gamma G_z z T_z} \tag{13.33}$$

$$rect(x) = 1 \quad \text{for} - 1/2 < X < 1/2, \quad rect(X) = 0 \quad \text{for } |X| > 1/2$$

$$\alpha_0 \approx 2\pi\frac{B_1(T_z)}{G_z d}.$$

An oscillating function of the transverse nuclear magnetization along the slice thickness then results (Figure 13.7). The selective rf-pulse has flipped each spin in the slice addressed into the transverse plane, but in its own rotating frame. Since the resonance frequency changes over the slice thickness due to the applied gradient, the transverse nuclear magnetization is twisted. Almost no signal can be observed in a following FID, since the effective value of the

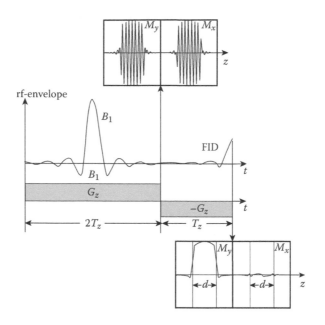

FIGURE 13.7

x and y component of the transverse nuclear-magnetization in the rotating frame after excitation with an rf-pulse of duration $2T_z$ in a magnetic field gradient G_z that is applied along the y-axis. The twisted transverse magnetization resulting immediately after the selective pulse realigns in a refocusing interval of duration T_z with a reversed gradient. The remaining oscillations at the edges of the refocused magnetization originate from the truncation of the sinc pulse. To minimize them, the sinc pulse has been multiplied with a Hanning function.

spiral-shaped nuclear magnetization cancels out. Reversing the polarity of the field gradient during a period T_z following the rf-pulse[14] refocuses all those spins, resulting in an FID corresponding to the full transverse magnetization in the excited slice (Figure 13.7).

It should be mentioned that Equation 13.33 is only an approximation, since Equation 13.3 and hence Equation 13.30 is only valid for those nuclei for which the frequency of the rf-pulse equals the Larmor frequency. Taking exactly into account the Bloch equations requires numerical methods. It can be seen that a residual nuclear magnetization $M_y(z)$ remains even after refocusing, which can be minimized, however, by tuning the refocusing interval and by modified rf-pulse shapes.

For the necessary strength of the slice selection gradient, a similar argumentation holds as for the encoding gradient, i.e., the duration can be chosen according to T_2 rather than T_2^*, when a 180° pulse is used instead of gradient reversal for refocusing; however, the slice profile is influenced by the static magnetic field inhomogeneities (i.e., becomes curved) when the field variations over the slice thickness due to the presence of the gradient are comparable with or less than those due to the field inhomogeneities.

13.8 Pulse Sequences

For imaging with MR, the object to be investigated must be exposed to a sequence of rf- and gradient pulses. Many modifications of these pulse sequences have been designed to optimize the experiment with respect to special problems such as tissue contrast, display of flow, diffusion, susceptibility, or data acquisition time.

For an image, transverse magnetization has to be generated and phase encoded. If the repetition time of the rf-pulses necessary is made short with respect to the longitudinal relaxation time T_1 in order to speed up imaging, after some rf-pulses a dynamic equilibrium or steady state is established at which the nuclear magnetization is the same after each rf-pulse, i.e., in a distance t from the rf-pulse one observes

$$\overrightarrow{M}(n\,1_R + t) = \overrightarrow{M}((n+1)T_R \mid t) = \overrightarrow{M}(t) - \begin{pmatrix} M_X(t) \\ M_y(t) \\ M_z(t) \end{pmatrix}, \qquad (13.34)$$

where T_R is the repetition time and n is the number of rf-pulses.

The magnetization immediately after the rf-pulse $\overrightarrow{M}(0)$ follows from that directly before $\overrightarrow{M}(T_R)$ from the rotation by tip angle α caused by the rf-field as

$$\overrightarrow{M}(0) = Q(\alpha)\overrightarrow{M}(T_R), \qquad (13.35)$$

with the rotation matrix

$$Q\alpha = \begin{pmatrix} 1 & 0 & 0 \\ 0 & \cos\alpha & \sin\alpha \\ 0 & -\sin\alpha & \cos\alpha \end{pmatrix}. \qquad (13.36)$$

On the other hand, because of precession and relaxation between the rf-pulses, $\overrightarrow{M}(T_R)$ and $\overrightarrow{M}(0)$ are related by

$$\overrightarrow{M}(T_R) = R(T_R)\overrightarrow{M}(0) + M_o \begin{pmatrix} 0 \\ 0 \\ 1 - e^{-\frac{T_R}{T_1}} \end{pmatrix}, \qquad (13.37)$$

whereby the matrix $R(T_R)$ describes precession and relaxation in the rotating frame

$$R(T_R) = \begin{pmatrix} \cos\omega t e^{-\frac{T_R}{T_2}} & \sin\omega t e^{-\frac{T_R}{T_2}} & 0 \\ \sin\omega t e^{-\frac{T_R}{T_2}} & \cos\omega t e^{-\frac{T_R}{T_2}} & 0 \\ 0 & 0 & e^{-\frac{T_R}{T_1}} \end{pmatrix}, \qquad (13.38)$$

where ω is the precession frequency in the rotating frame.

In the following, we will analyze the steady-state magnetization for two cases relevant in MR imaging. First, we shall assume that directly before the rf-pulse no transverse magnetization has remained:

$$M_X(T_R) = M_y(T_R) = 0$$

$$M_X(0) = M_z(T_R)\cos\alpha \tag{13.39}$$

$$M_z(T_R) = M_0 - (M_0 - M_z(0))e^{\frac{T_R}{T_1}}.$$

The transverse magnetization immediately after the rf-pulse then follows to

$$M_y(0) = M_z(T_R)\sin\alpha = M\sin\alpha \frac{\left(1 - e^{\frac{T_R}{R_1}}\right)}{1 - \cos\alpha e^{\frac{T_R}{T_1}}} \tag{13.40}$$

and reaches a maximum for the so-called Ernst angle

$$\cos\alpha_{opt} = e^{\frac{T_R}{T_1}}. \tag{13.41}$$

For a given repetition time and flip angle, the NMR signal intensity is determined by the local longitudinal relaxation time T_1. Since at short repetition times with small flip angles a large signal can still be obtained, this pulse sequence is often referred to as FLASH (fast low angle shot).[15] With FLASH imaging, it is assumed that the phase memory of the transverse nuclear magnetization has been lost at the end of the repetition interval; since this is not true when the repetition interval is shorter than the transverse relaxation time, spoiling gradient pulses are applied at the end of each interval in order to prevent the emergence of coherent image artifacts or a stochastically varying jitter is added to the repetition time.[16]

Next, we shall assume that between $M(0)$ and $M(T_R)$ a relation

$$M_y(T_R) = -M_y(0)e^{\frac{T_R}{T_2}} \tag{13.42}$$

exists, which can either be assured by adjusting the frequency of the rf-pulses in the rotating frame to

$$\omega = \pm\frac{\pi}{T_R} \tag{13.43}$$

or alternating their phase between adjacent pulses by 180° (and keeping $\omega=0$).

Then for the other magnetization components,

$$M_y(0) = M_y(T_R)\cos\alpha + M_z(T_R)\sin\alpha$$

$$M_z(0) = -M_y(T_R)\sin\alpha + M_z(T_R)\cos\alpha \tag{13.44}$$

$$M_z(T_R) = M_0 - (M_0 - M_z(0))e^{\frac{T_R}{T_1}}$$

and the transverse magnetization immediately after the rf-pulse yields

$$M_y(0) = \frac{M_0\left(1 - e^{\frac{T_R}{T_1}}\right)\sin\alpha}{\left(1 - e^{-T_R\left(\frac{1}{T_1}+\frac{1}{T_2}\right)}\right) + \left(e^{\frac{T_R}{T_2}} - e^{-\frac{T_R}{T_2}}\right)\cos\alpha}. \tag{13.45}$$

For $(T_R/T_{1,2}) \ll 1$,

$$M_y(0) = \frac{M_0\sin\alpha}{\left(1 + \frac{T_1}{T_2}\right) + \left(1 - \frac{T_1}{T_2}\right)\cos\alpha} \tag{13.46}$$

is obtained, which reduces for $\alpha = 90°$ to

$$M_y(0) = \frac{M_0}{1 + \frac{T_1}{T_2}}. \tag{13.47}$$

Such a sequence is referred to as true FISP (fast imaging with steady precession)[17] or TRUFI (true fast imaging) and has to be constructed with completely balanced gradients (Figure 13.8). Signal intensity is determined by the $T_1:T_2$ ratio. For fluids such as water, where $T_1 \approx T_2$, a signal equivalent to half of the maximum nuclear magnetization can be obtained using 90° pulses with rapid pulse repetition. However, the prerequisite for such a strong signal is that no dephasing of transverse magnetization due to field inhomogeneities occurs in an image voxel, i.e.,

$$G_{x,y,z}\Delta x, y, z \gg \Delta B, \tag{13.48}$$

where $\Delta x, y, z$ are the lengths of the edges.

If this condition cannot be maintained, image artifacts will arise. However, in order to obtain a steady state, it is not necessary that the net precession angle or phase of the transverse magnetization by 180° between the rf-pulses as assumed in Equation 13.43, a constant phase between the pulse is sufficient. Such a pulse sequence, dubbed FISP, reverses only the encoding gradient G_Y at the end of the repetition interval and is much more insensitive to field inhomogeneities. Signal behavior is somewhere between FLASH and TRUFI.

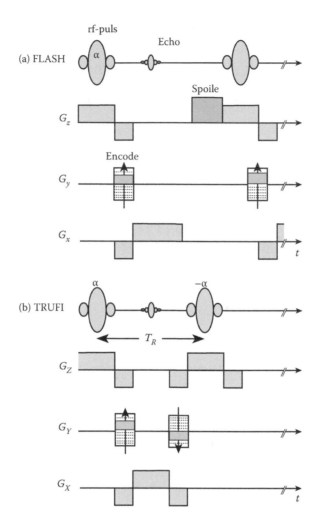

FIGURE 13.8

Examples of steady-state sequences: at FLASH, the transversal magnetization at the end of the repetition interval has to be destroyed, e.g., with a spoiler gradient (a); at true FISP (TRUFI), it is rewound to the state immediately at the end of the rf-pulse (b).

Without preparation gradients (i.e., the gradient pulses applied before the data are sampled) in *x*- and *z*-directions one would observe in a steady-state sequence a focused magnetization before and after the rf-pulse.[16] Graphically, one can consider the signal after the rf-pulse as an FID and before as one half of an echo with an amplitude reduced by the factor $e^{-(T_R/T_2)}$ compared to the FID. In FISP, one is using the steady-state FID to obtain a projection, but one can also utilize the steady-state half echo signal to get an FISP-like image with additional T_2 weighting (though this is not a strong effect at the short repetition times applied in steady-state sequences). In this case, the time course of the imaging sequence has to be reversed, therefore being dubbed PSIF (Figure 13.9). It is even possible to combine FISP and PSIF in a single sequence giving two images differing in T_2 contrast (DESS, double echo in the steady state). In very homogeneous fields or with very strong gradients using a short repetition time, the TRUFI sequence can be set up, where the FISP and the PSIF signals are superimposed.

With steady-state sequences—referred to as gradient echo in contrast to spin echo sequences, because the signal (or echo) is formed without refocusing rf-pulses—fast image

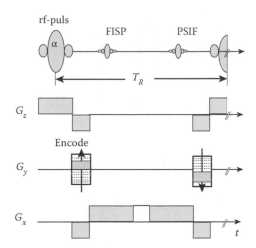

FIGURE 13.9
DESS sequence employing a combination of FISP and PSIF.

acquisition within less than 1 s is possible. To enhance signal contrast between different tissues, the nuclear equilibrium magnetization can be inverted with a 180° rf-pulse before the imaging sequence is started.[18] Thus, during the fast imaging experiment the longitudinal magnetization undergoes relaxation back to its equilibrium state, producing image contrast with respect to tissues having different longitudinal relaxation times T_1.

Because gradient echo sequences are so fast, they are very well suited for 3d data acquisition. Either the rf-pulses are applied nonselectively, i.e., without a slice selection gradient, or a very thick slice is excited. Spatial resolution is then achieved by successively encoding the nuclear magnetization with the gradients G_z and G_y in the y- and z-directions and reading out the signal in the projection gradient G_x. Since each volume element is repeatedly measured according to the number of phase encoding steps, the S/N ratio improves considerably. Image reconstruction is performed with a fast 3d Fourier transformation, resulting in a block of images that can be displayed as slices through the three main coordinate axes. Image postprocessing also allows the display of images in arbitrary projections (multiplanar reformatting = MPR).

With so-called spin echo sequences utilizing an additional 180° pulse, there is greater flexibility with respect to the manipulation of image contrast than in gradient echo sequences, though generally at the expense of acquisition time. Spin echo sequences are also much more stable against static field inhomogeneities of the static magnetic field, since dephasing of transverse magnetization during the encoding interval is reversed; compared to gradient echo sequences, signal loss is less. In this context, we want to mention that in the direction of the phase encoding gradient field, inhomogeneities cause no image distortions.

In the standard spin echo imaging sequence, slice selective 90° and 180° rf-pulses are used with encoding gradients between them (Figure 13.10); the echo signal is read out in the projection gradient. Two parameters are available for signal manipulation, the sequence repetition time T and the echo time T_E. The use of a long echo time allows transverse relaxation of the spin system before signal acquisition, whereas rapidly repeating the pulse sequence prevents longitudinal magnetization from reestablishing. This effect is called saturation. The signal intensity in a picture element is given by

$$M_\perp(x,y) = M_0(x,y)\left(1 - e^{\frac{-T_R}{T_1(x,y)}}\right)e^{\frac{-T_R}{T_2(x,y)}} \tag{13.49}$$

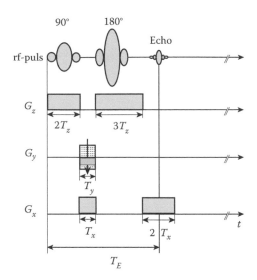

FIGURE 13.10

Standard spin echo imaging sequence. The compensation of the twisted transversal magnetization caused by the selective 90° pulse is achieved here with the prolonged, slice selecting gradient pulse after the selective 180° pulse.

The repetition time and the echo time can be adjusted so that the image contrast due to different types of tissue is determined by either M_o, T_1, or T_2. Short values of T_E and T_R give T_1-weighted images, while a long T_E and a short T_R give spin density or M_o-weighted images, and long values of both T_E and T_R give T_2-weighted images. Thus, in MR, the contrast-to-noise ratio is determined and can be changed in a wide range by the pulse sequence. This is a unique feature for this image modality and cannot not be obtained by retrospective filtering or postprocessing.

Contrast between adjacent anatomic structures can be further enhanced by means of contrast agents. Since the addition of other magnetic moments increases magnetic interactions during the collisions between fluid molecules, a paramagnetic agent dispersed in the tissue accelerates the relaxation of excited spins, longitudinally as well as transversely. A common contrast agent is Gd-DTPA (gadolinium diethylenetriaminepentaacetic acid).[19] Administered to the blood stream, the contrast agent will accumulate at various levels in tissue due to the different microvascular structures.

The standard spin echo imaging sequence can be modified in several ways. At long repetition times, images of several different slices can be acquired in a shorter acquisition time than for a single slice, when the different slices are addressed during the waiting interval. To obtain information on transverse relaxation, the NMR signal can be recovered several times by repeating the 180° pulses. Thus, several images are reconstructed with varying T_2 weighting for a single slice; the transverse relaxation time can be calculated in each pixel and even displayed as an image.

When each echo of a multiecho sequence is encoded in the y-direction with a gradient pulse, several lines in Fourier space are recorded during one pulse sequence, significantly reducing the time of measurement (TSE, turbo spin echo); it is even sufficient to scan only half of the Fourier space (HASTE, half Fourier acquired single shot turbo spin echo). When the 180° pulses are omitted and the polarity of the projection gradient is alternatively reversed[20] to generate gradient echoes, a complete image can be acquired in less than 100 ms (EPI, echo planar imaging).

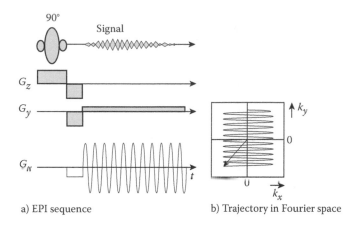

a) EPI sequence b) Trajectory in Fourier space

FIGURE 13.11
EPI employing a sine wave readout gradient (a) and the according trajectory in Fourier space (b).

Since rapid switching of strong magnetic field gradients is not a simple technical task, in EPI the encoding gradients can be kept switched on during the total time of data acquisition and a sine wave oscillating readout gradient can be employed (Figure 13.11a). The resulting trajectory in Fourier space is shown in Figure 13.11b.

13.9 Influence of Motion

If the object to be imaged or parts of it are moving during data acquisition, artifacts occur, since a moving volume element acquires another phase in the applied gradient fields than if it were resting. Image reconstruction then attributes the position of that voxel to other origins that give rise to typical image distortions such as blurring or mirror images. However, depending on the type of movement, data acquisition strategies can be developed to avoid those artifacts or even to get information on parameters of the motion as, e.g., in the case of flow on the velocity distribution.

In principle, a moving object can be described as a four-dimensional (4d) distribution of transverse magnetization with three coordinates in space and one in time. Image artifacts can then be explained as an incomplete data sampling process in the 4d space with time and space or their Fourier conjugates frequency and spatial frequency as coordinates. For illustrative reasons, we will restrict our discussion in the following to a 2d distribution $M(x, y, t)$ of transverse magnetization moving in time as it might be generated by selective excitation. In Fourier or k-space, such an object is described by

$$M(k_x k_y, t) = \iint M(x, y, t) e^{-i(k_x x + k_y y)} \, dx \, dy, \tag{13.50}$$

putting up a 3d space consisting of the familiar two coordinates k_x, k_y of spatial frequency and the time coordinate t. When a moving object is successively sampled line by line in Fourier space, the k, t-space is crossed along a tilted plane (Figure 13.12)

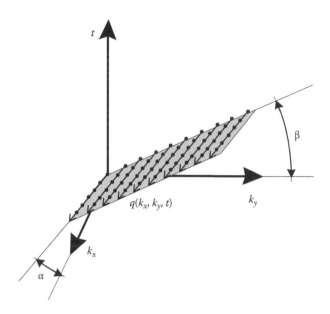

FIGURE 13.12
A moving object is scanned along a tilted plane in Fourier space.

$$q(k_x, k_y, t) = ak_x(t) + bk_y(t) + t$$

$$= \frac{k_x}{\gamma G_x}\cos\alpha + \frac{k_y}{\gamma G_y}\sin\beta + t = 0 \tag{13.51}$$

with angles α and β with respect to the k_x- and k_y-axes (scaled with the factors $\gamma G_{x,y}$ to give them the same dimension as the t-axis) that pass the k_x, k_y, $t = 0$ center of origin.

Reconstructing the image from the samples on this plane must lead to artifacts, since only sampling along the $q(k_x, k_y, t = 0)$ plane would result in a reconstruction of the object $M(x, y)$ at $t = 0$ and only a complete scan of k, t-space would reconstruct the complete time course $M(x, y, t)$ of the object.

The nature of these artifacts is revealed when one considers the moving object not in x, y, t-space, but in x, y, ω-space:

$$M(x, y, \omega) = \frac{1}{2\pi}\int M(x, y, t)e^{-i\omega y}dt, \tag{13.52}$$

in which the spatial dependence of the harmonics of the moving object is displayed. Since x, y, ω are Fourier conjugates to k_x, k_y, t, one can apply the projection slice theorem, which states that the data $M(t)$ sampled along the plane $q(k_x, k_y, t)$ through the center of Fourier (i.e., k_x, k_y, t) space characterized by the angles α and β correspond to the projection of the data in the original (i.e., x, y, ω) space on a plane

$$p(x, y, \omega) = \gamma G_x X\cos\alpha + \gamma G_y Y\cos\beta + \omega = 0 \tag{13.53}$$

with the same direction α and β. For example, in the case of an object moving periodically, the occurrence of replication or ghosting is explained by the projection of the spectral island occurring in the x, y, ω plane onto $p(x, y, \omega)$[21] (Figure 13.13).

Of course, no motion artifacts will occur if the tilting angles of planes $q(k_x, k_y, t)$ and $p(x, y, t)$ are $\alpha\beta=0$, i.e., no movement would occur during scanning. This implies the application of very rapid pulse sequences using gradient echoes as steady-state sequences or EPI. Unfortunately, rapid imaging often results in a low S/N ratio and/or low contrast. Imaging moving parts and organs of the human body with high contrast, e.g., with a spin echo sequence, requires long acquisition times due to the time interval between the phase encoding steps. Ghosting can be avoided in this case when data acquisition is triggered or gated by the movement so that the single phase encoding steps always occur at the same position of the object. Triggering is a prospective method, which puts restrictions on the pulse repetition time, while gating works retrospectively, and with proper reordering of the acquired phase encoding steps, it allows in principle a complete scan of the k_x, k_y, t volume in case of a periodically moving object. For imaging of the heart, trigger and gating pulses can be provided by an electrocardiogram (ECG) run simultaneously during the MRI investigation, while respiratory gating asks for a pressure transducer. An alternative is the use of navigator echoes, which are created during the repetition intervals. A scanning line is laid through the organ along the main movement direction by selective 90° and

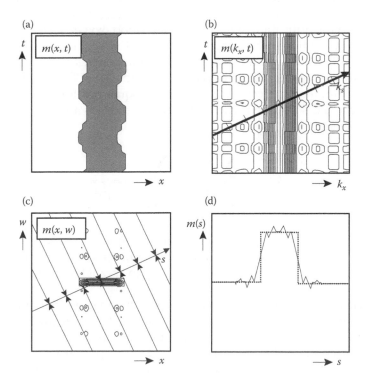

FIGURE 13.13
Illustrating the influence of movement at the example of an oscillating one-dimensional (1d) object (a). In the spatial frequency domain, the moving edges of the object lead to blurring and harmonics (c). With MRI, the object is sampled along an inclined line in the time-spatial frequency domain (b). The imaged transversal magnetization (d) is yielded after Fourier-transforming these samples and represents the projection in the space frequency domain on the direction of this line (d).

$180°$ pulses each with a different gradient direction, and the spin echo is read out in the perpendicular gradient. Fourier transformation of the echo monitors the position of the organ on the projection line, from which a gating signal can be derived.

13.10 Correction of Motion During Image Series

The sensitivity of gradient echo sequences as steady-state sequences or EPI to magnetic field inhomogeneities can be utilized to image effects in the human body which respond sensitively to changes in magnetic susceptibility. Thus, e.g., the perfusion of the cortex varies with the performing of certain actuating or perceptive tasks. If a certain area of the brain becomes activated, e.g., the visual cortex, when the person under study sees light flashes, the local oxygen requirement increases. The circulatory system reacts by increasing the local blood supply even more than necessary. Consequently, the activation process results in an increased oxygen content of the venous blood flow. Oxygen is transported by hemoglobin in the red blood cells. Oxygenated hemoglobin (HbO_2) is diamagnetic, while deoxygenated hemoglobin (Hb) is paramagnetic due to the presence of four unpaired electrons. These different magnetic susceptibilities lead to different signal intensities in a properly weighted sequence. So Hb acts as a natural contrast agent, its effect varying with the oxygen supply and the utilization of the brain cells. Blood oxygen level dependent (BOLD) contrast can thus be studied with a susceptibility-sensitive sequence (functional MRI[21]).

In functional MRI (fMRI), an analysis of voxel time courses is performed in order to detect changes in the regional blood oxygenation of the brain.[22] Enhanced blood supply happening upon stimulation of certain areas of the eloquent cortex is detected from signal changes between images acquired with and without stimulus applied. By calculating the correlation coefficient

$$CC = \frac{\sum_t (X_t - \bar{X}) \sum_i (Y_t - \bar{Y})}{\sqrt{\sum_t (X_t - \bar{X})^2 \sum_i (Y_t - \bar{Y})^2}} \tag{13.54}$$

where X_t is the time course of a given voxel during no stimulation and Y_t is the time course of the same voxel during stimulation, brain stimulation is assumed, when CC exceeds a threshold defined from the probability one is willing to accept for an accidental correlation (t-test).

Though in fMRI such fast image sequences, e.g., EPI, are used that one can assume no motion during data acquisition, the head might move between acquiring one image and another. This movement can occur even when the head is fixed and might result from an involuntary reaction on the applied stimuli.[23] Due to partial volume effects, motion shifts even less than a voxel size can lead to differences in signal intensity misleading the BOLD effect. Efforts to record head motion with MR or optical monitored fiducials fixed to the patient's head or with orbital navigator echoes were only partially successful, because there is not a tight enough correlation between the motion of the brain and the scalp. An algorithm that detects if motion has occurred between two images and corrects for it with postprocessing, therefore, seems to be a better solution.

The region of interest in a reference image is used to define a set of voxels to form a vector $\vec{X} = X_1, ..., X_n)$ and a vector $\vec{Y}$ from the corresponding region in the actual image that might have moved with respect to the reference image and therefore, is to be corrected. A linear transformation

$$\vec{Y} = \vec{X} + \overrightarrow{Ap} \qquad (13.55)$$

is assumed between both images, where the parameters p_j describe the motion parameters (three rotational, three translational) and the transformation matrix contains the derivatives $(\partial X_i/\partial p_j)$ of voxel intensity X_i with respect to parameter p_j. The coefficients of matrix **A** are known, as they can be determined for the object under investigation by exposing the reference image to virtual motions.

The parameter vector p describing motion between the reference and the actual image can be estimated in first order from the Moore–Penrose or pseudo-inverse

$$A^+ = (A^T A)^{-1} A^T \qquad (13.56)$$

of the transformation matrix to be

$$\vec{p} = A^+(\vec{Y} - \vec{X}). \qquad (13.57)$$

Then the values of p_i obtained are used for a coordinate transformation (describing shift and rotation) of the actual image to yield a corrected image $\vec{Y}_1$. If a transformed voxel falls between the points of the sampling grid of the reference image, its intensity can be distributed into the neighboring grid points, e.g., with a linear or with a sinc interpolation. This procedure is often referred to as regridding.

With $\vec{Y}_1$ it can again be checked if a shift or rotation still exists with respect to the reference image. Then the described algorithm can be repeated until no further changes in $\vec{p}$ are observed. Procedures of this kind be can assumed as sensitive to motion down to some 10 μm.[24]

13.11 Imaging of Flow

Flow-related phenomena are used in magnetic resonance angiography (MRA). Two effects have to be considered, namely, time of flight and phase changes. In standard spin echo sequences, it is often observed that the NMR signal of flowing nuclei is enhanced at low and reduced at high flow velocities, compared with stationary tissue.[25] The increase in intensity (sometimes referred to as a paradoxical phenomenon) is explained by introducing nuclei to the imaged slice which are not magnetized from previous excitations, while the signal void at high velocities occurs because part of the nuclei leave the slice between excitation and the echo measurement.

Time of flight effects can be utilized to create images similar in their appearance to those produced in X-ray angiography. With nonselective excitation pulses or pulses exciting a thick slice, a rapidly repeating 3d gradient echo sequence is set up, which gives a weak signal. Flow introduces fully relaxed spins to the imaged volume, giving rise to a stronger signal. Signal intensity increases with flow velocity. Contrary to a spin echo sequence, no signal void is observed at gradient echo sequences because the refocusing gradient effects all spins, those staying in the excited volume and those moving out. In order to visualize the vascular structure, the method of maximum intensity projection (MIP) is often used.[26] In image postprocessing, the acquired 3d image volume can be regarded as illuminated with parallel rays. Along each ray, the image element with the highest signal intensity is searched and displayed in the projection plane.

Phase-sensitive MRA makes use of the fact that moving spins acquire different transverse phases, according to their velocity in magnetic field gradients, than stationary spins. When the volume element is sufficiently small, so that it contains only spins with similar flow velocities, the flow can be quantitatively measured by analyzing the signal phase (phase contrast angiography[27]). In comparison with time of flight, phase contrast angiography is well suited to slow flow velocities.

When, on the other hand, a volume element large enough to contain spins with a variety of velocities is chosen, this will give only a weak signal because the individual contributions of the different spins cancel.

The transverse phase angle of an ensemble of stationary and moving spins follows (in the rotating frame) from the integration of Equation 13.16b:

$$\varphi(t) = \gamma \int_0^t G(t) r(t) dt. \tag{13.58}$$

Assuming uniform flow in a constant gradient,

$$r(t) = r_o + v_{t'} \tag{13.59}$$

hence

$$\varphi(t) = \varphi_o + \gamma G r_o t + \frac{1}{2} \gamma G v t^2. \tag{13.60}$$

As the phases of the stationary spins are recovering following the application of a bipolar gradient, the same happens at a uniform flow velocity with two bipolar gradients back to back (Figure 13.14). The signal loss due to dephasing spins in the readout gradient, e.g., in a gradient echo sequence, can therefore be avoided by applying a bipolar gradient pulse before the data readout phase. Such an imaging sequence is often referred to as a motion-compensated or flow-rephased pulse sequence. Subtraction between a flow-dephased and a flow-rephased image cancels out the signal from the stationary tissue, leaving only the vascular structure.[28] For this reason, the method can be viewed as magnitude contrast angiography.

Though flow-sensitive MRI enables MRA without contrast agents, investigation time can be rather long and spatial resolution limited. Recently, contrast enhanced (ce) MRA has been introduced where relaxation time shortening contrast agents are administered intravenously. The high contrast-to-noise ratio yielded, e.g., allows one to follow the contrast bolus from the arterial to the venous phase through the whole body, finding arterial and venous occlusions.

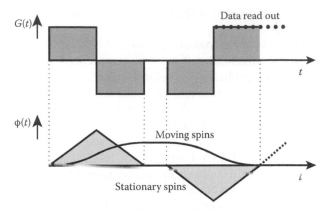

FIGURE 13.14
Gradient pulses for phase and motion compensation and time dependence of the phase stationary and moving spins. Applying a bipolar gradient pulse before the data readout phase minimizes the influence of uniform flow.

13.12 MR Spectroscopy

The signal displayed in MR images derives mostly from the hydrogen nuclei in water and fat. The different chemical shift of fat with respect to water can cause typical image artifacts, e.g., because slightly different slices are excited with a selective rf-pulse. Also, fat and water appear shifted with respect to each other in the direction of the readout gradient. Although these effects can be masked using stronger slice selection and readout gradients, this requires a greater rf-power and a larger bandwidth, in turn leading to an increase in image noise. Therefore, various pulse sequences have been developed to obtain pure fat or pure water images. Such sequences are employed when the strong signal of one compound obscures the weak signal of the other.

One possibility is to suppress the fat or the water signal pulse before beginning the imaging sequence, using an initial 90° pulse having exactly the frequency of the undesired compound and dephasing (spoil) the transverse magnetization of this compound with a strong gradient pulse. The subsequent imaging sequence then acts only on the desired compound.[29] When applied to suppress the signal of fat, this method is often referred to as fat saturation because fat does no longer gives a signal.

In some cases, pure fat and water images are generated using two pulse sequences with different readout delays. The difference is chosen so that the magnetization of fat and water is parallel in one case and antiparallel in the other case.[30] Adding and subtracting the signals from the two sequences yield either a pure fat or a pure water image. The technique of identifying metabolites by employing echo times that generate defined in-phase or opposite-phase alignment of the transverse nuclear magnetization is called spectral editing. It can be used, e.g., to separate lactate from lipid resonances in proton spectroscopy.

Of special interest, however, is the detection of metabolites either by the NMR of hydrogen ^{1}H or of other nuclei, such as phosphorus ^{31}P. Due to the very low concentration of metabolites, their NMR signal is much weaker than that of water and fat. To obtain a sufficient S/N ratio, it is therefore necessary to work with lower spatial resolution and longer acquisition times than with normal imaging.

Measuring chemical shifts requires a static magnetic field with a high field flux density ($>1\,T$) and very good homogeneity. Either the spectrum of a single volume element can be measured (single voxel spectroscopy) or the spatial distribution of spectra in the object can be acquired (chemical shift imaging).

Since the intensity of the water signal can be several orders of magnitude greater than that of the signals from the metabolites, it must be suppressed, e.g., with a narrow bandwidth 90° pulse and a spoiler gradient. The volume selection sequence, e.g., a selective 90° pulse with a gradient in the *x*-direction and two selective 90° or 180° pulses with a gradient in the *y*- and *z*-directions, then acts only on the metabolites, the spectral lines of which result from a Fourier transformation of the FID.[31]

Although NMR spectroscopy in living beings at first seems to be very attractive since it should permit immediate insight into cell metabolism, its clinical importance has remained limited up to now. Compared with nuclear medicine, where radioactive tracers with very low concentration can be detected, NMR is a very insensitive method. This restricts its application concerning components other than fat and water to the detection of volume elements with a size of several cubic centimeters and to metabolites of limited biological or medical usefulness.

13.13 System Design Considerations and Conclusions

For MRI, processor-based Fourier NMR spectrometers used routinely in analytical chemistry have been adapted to the size of a human patient, and components and software have been structured to meet imaging requirements. Figure 13.15 shows a block diagram of an MR imager. Of special importance is the system, or host computer which controls all components of the system often applying digital processors themselves, and acts as

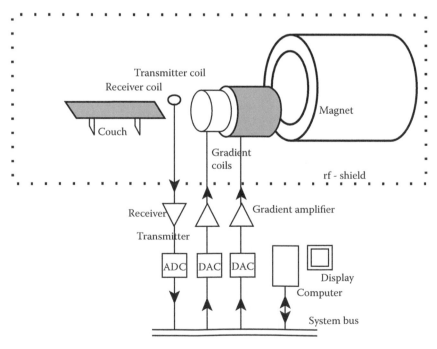

FIGURE 13.15
Block diagram of an MRI apparatus (ADC = analog-to-digital converter, DAC = digital-to-analog converter).

the interface to the user. The mighty software has to control the system, run the imaging sequences, perform image reconstruction, interact with the user, perform archiving, and perform increasingly morepostprocessing tasks.

The magnet is by far the most important (and expensive) component. The optimum field strength for MRI is still a matter of controversy. Flux densities above 0.5 *T* in a volume suitable for patient investigations can only be obtained with superconducting magnets, while permanent and resistive magnets with iron yoke flux return paths are applied at field strengths <0.5 *T*. Though clinical MR systems are restricted to flux densities not larger than 1.5 *T*, experimental instruments apply magnets up to 8 *T*. High magnetic fields are often employed in order to obtain a better S/N ratio, but as the rf-power required for spin excitation increases with B_z^2 (Equation 13.13), limitations are placed on image sequences in order not to generate excessive heat in the patient. The different contrast behavior of tissue and possibilities to better distinguish metabolites by spectroscopy are reasons to explore very high magnetic fields. Here, the influence of decreasing penetration of the rf-field into the sample (skin effect) and possible dielectric resonances occurring at tissue borders are new experimental challenges.

Other than by the static magnetic field, the S/N ratio can be influenced with the rf-antennas. Using arrays of small coils rather than a single large coil leads to a significant improvement[32] because a small coil picks up much less noise from the body.

The gradient strength is another important factor affecting the performance of an MRI system. In order to enable short image acquisition times with steady state sequences or by acquiring several lines in Fourier space, as in TSE or EPI, strong gradients are also necessary for sequences sensitive to flow, perfusion, or diffusion. The switching time for the gradients must always be as short as possible, since during switching intervals undesired signal decay occurs.

The gradient strength and switching time are ultimately limited by the electrophysiology of the patient, who will eventually experience stimulation of the peripheral nerves; even though painful, they are harmless since they already occur at much lower values than would have any effect on the cardiac system.[33] Gradient strengths up to 40 (*mT/m*) with a rise time of 5 (µs/(*mT/m*)) are now in clinical use.

13.14 Conclusion

Since the first availability of commercial instruments at the beginning of the 1980s, clinical MR has expanded rapidly in terms of both medical applications and the number of units installed. First considered to be an expensive method to create images of inferior quality, it has since established itself as a clinical tool for diagnosis in previously inconceivable applications, and the potential of the method is still not exhausted. MRI has led to the first large-scale industrial application of superconductivity and has brought about a far greater public awareness of a physical effect previously known only to a handful of scientists.

Up to now, the growth and spectrum of applications of MR have exceeded all predictions. The most recent development is that of rendering brain functions visible. Cardiac MR can display coronaries and analyze perfusion of the myocardium and hemodynamics of the heart. Thus, MRI is entering the domain of nuclear medicine.

An interesting new application of MRI is its use as an imaging modality during minimal invasive procedures such as rf-ablation, interstitial laser therapy, or high intensity focused ultrasound. With temperature-sensitive sequences, the development of temperature and tissue damage can be checked during heating and destroying of diseased tissue. The sensitivity of MRI to flow helps the physician to stay away from vessels during an intervention. MRI is also used for image-guided surgery, e.g., resection of tumors in the brain. Special open systems have been designed for such purposes, and dedicated nonmagnetic surgery tools have already been developed.

References

1. Zavoisky, E. 1945. *J. Phys. USSR*, 9, 211.
2. Bloch, F.W., W.W. Hansen, and M. Packard. 1946. Nuclear induction. *Phys. Rev. (L)*, 69, 127.
3. Purcell, E.M., H.C. Torrey, and R.V. Pound. 1946. Resonance absorption by nuclear magnetic moments in a solid. *Phys. Rev. (L)*, 69, 37.
4. Lauterbur, P.C. 1955. Image formation by induced local interactions: Examples employing nuclear magnetic resonance. *Nature*, 242, 469.
5. Bloch, F., W.W. Hansen, and M. Packard. 1946. The nuclear induction experiment. *Phys. Rev.*, 70, 474.
6. Hahn, E. 1950. Spin echos. *Phys. Rev.*, 80, 580.
7. Carr, H.Y. and E.M. Purcell. 1954. Effects of diffusion on free precession in nuclear magnetic resonance experiments. *Phys. Rev.*, 94, 630.
8. Hoult, D.I. and P.C. Lauterbur. 1979. The sensitivity of the Zeugmatographic experiment involving human samples. *J. Magn. Reson.*, 343, 425.
9. Abe, Z., K. Tanaka, K. Hotta, and M. Imai. 1974. Noninvasive measurements of biological information with application of nuclear magnetic resonance. In *Biological and Clinical Effects of Low Magnetic and Electric Fields*. Charles C. Thomas, Springfield, IL, 295–313.
10. Cormack, A.M. 1963/1964. Representation of a function by its line integrals, with some radiological applications. I. *J. Appl. Phys.*, 34(1), 2722 and 35(2), 2908.
11. Hounsfield, G.N., J. Ambrose, J. Perry, et al. 1973. Computerized transverse axial scanning (tomography). *Br. J. Radiol.*, 46(1,2), 1016.
12. Kumar, A., D. Welti, and R. Ernst. 1975. NMR Fourier Zeugmatography. *J. Magn. Reson.*, 18, 69.
13. Cooley, J.W. and J.W. Tukey. 1965. An algorithm for machine calculation of complex Fourier series. *Math. Computation*, 19, 297.
14. Hoult, D.I. 1977. Zeugmatography: A criticism of the concept of a selective pulse on the presence of a field gradient. *J. Magn. Reson.*, 26, 165.
15. Haase, A., J. Frahm, D. Matthaei, W. Hänicke, and K. Merboldt. 1986. FLASH imaging: Rapid NMR imaging using low flip angle pulses. *J. Magn. Reson.*, 67, 217.
16. Freeman, R. and H.D.W. Hill. 1971. Phase and intensity anomalies in Fourier transform NMR. *J. Magn. Reson.*, 4, 366.
17. Oppelt, A., R. Graumann, H. Barfuß, H. Fischer, W. Hartl, and W. Schajor. 1986. FISP—eine neue schnelle Pulssequenz für die Kernspintomographie. *Electromedica*, 54, 15.
18. Haase, A., D. Matthaei, R. Bartkowski, E. Dühmke, and D. Leibfritz. 1989. Inversion recovery snapshot FLASH MRI: Fast dynamic T1 contrast. *J. Comput. Assist. Tomogr.*, 13, 1036.
19. Weinmann, H.J., R.C. Brasch, W.R. Press, and G.E. Wesbey. 1984. Characteristics of gadolinium-DTPA complex: A potential NMR contrast agent. *Am. J. Radiol.*, 143, 619.
20. Mansfield, P. 1977. Multi planar image formation using NMR spin echos, *J. Phys. C*, 10, L55.
21. Lauzon, M.L. and B.K. Rutt. 1993. Generalized k-space analysis and correction of motion effects in MR imaging. *Magn. Reson. Med.*, 30, 438.
22. Ogawa, S., D.W. Tank, R. Menon, J.M. Ellermann, S.G. Kim, H. Merkle, and K. Ugurbil. 1992. Intrinsic signal changes accompanying sensory stimulation: Functional brain mapping with magnetic resonance imaging. *Proc. Natl. Acad. Sci. USA*, 89, 5951.

23. Hajnal, J.V., R. Myers, A. Oatridge, J.E. Schwieso, I.R. Young, and G.M. Bydders. 1994. Artifacts due to stimulus correlated motion in functional imaging of the brain. *Magn. Reson. Med.*, 31, 283.

24. Friston, K.J., S.R. Williams, R. Howard, R.S.J. Frackowiak, and R. Turner. 1996. Movement-related effects in fMRI time-series. *Magn. Reson. Med.*, 35, 346.

25. Crooks, L., P. Sheldon, L. Kaufmann, W. Rowan, and T. Millert. 1982. Quantification of obstruction in vesssels by nuclear magnetic resonance (NMR). *IEEE Trans. Nucl. Sci.*, NS-29, 1181.

26. Koenig, H.A. and G.A. Laub. 1988. The processing and display of three-dimensional data in magnetic resonance imaging. *Electromedica*, 56, 42.

27. Dumoulin, C.L., S.P. Souza, M.F. Walker, and W. Wagle. 1989. Three-dimensional phase contrast angiography. *Magn. Reson. Med.*, 9, 139

28. Laub, G.A. and W.A. Kaiser. 1988. MR angiography with gradient motion refocussing. *J. Comput. Assist. Tomogr.*, 12, 377.

29. Haase, A., J. Frahm, W. Hänicke, and D. Matthaei. 1985. 1H NMR chemical shift selective (CHESS) imaging. *Phys. Med. Biol.*, 30, 341.

30. Dixon, W.T. 1986. Simple proton spectroscopic imaging. *Radiology*, 153, 189.

31. Frahm, J., H. Bruhn, M.L. Gyngell, K.D. Merboldt, W. Hänicke, and R. Sauter. 1989. Localized high-resolution protron NMR spectroscopy using stimulated echos: initial applications to human brain in vivo. *Magn. Reson. Med.*, 9, 79.

32. Roemer, P.B., W.A. Edelstein, C.E. Hayes, S.P. Souza, and O.M. Mueller. 1990. The NMR phased array. *Magn. Reson. Med.*, 16, 192.

33. Budinger, T.F., H. Fischer, D. Hentschel, H.E. Reinfelder, and F. Schmitt. 1991. Physiological effects of fast oscillating magnetic field gradients. *J. Comput. Assist. Tomogr.*, 15, 909.

14

Organ Motion Effects in Medical CT Imaging Applications

Ian Cunningham

University of Western Ontario

Stergios Stergiopoulos and Amar Dhanantwari

Defence R&D Canada Toronto

CONTENTS

14.1 Introduction

X-ray computed tomography (CT) was developed in the 1960s and early 1970s as a method of producing transverse tomographic (cross-sectional) images of the human body.* It has since become widely accepted as an essential diagnostic tool in medical centers around the world.

A summary of CT imaging and reconstruction concepts has been described by Martz and Schneberk in Chapter 15. Recall that a CT image is essentially a tomographic "map" of the calculated X-ray linear attenuation coefficient, $\mu(x, y)$, expressed as a function of position (x, y) in the patient. The attenuation coefficient is a function of X-ray energy, and CT images are produced using a spectrum of X-ray energies between approximately 30 and 140 keV, although this varies slightly with manufacturer and sometimes with the type of examination being performed. In order to provide a consistent scale of image brightness for medical uses between vendors and scan parameters, CT images are calibrated and expressed in terms of "CT number" (CT#) in "Hounsfield" units (HU) defined as

$$CT\# = \frac{\mu_t - \mu_w}{\mu_w} \times 1000 \text{ HU} \tag{14.1}$$

where μ_t and μ_w are the attenuation coefficients of tissue in a specified image pixel and in water, respectively. In this way, air corresponds to a CT# of –1000 HU, water to 0 HU, and bone to approximately 1000 HU. Most soft tissues of medical interest are in the range of approximately –20 (fat) to 60 HU (muscle) for all CT systems. A change of 1 HU corresponds to a 0.1% change in the attenuation coefficient. The standard deviation image noise in most CT scanners is approximately 3–8 HU, depending on the scan protocol. A CT image is thus a map of the linear attenuation coefficient expressed in HU.

14.1.1 CT Systems

Images are calculated from a large number of measurements made of X-ray transmission through the body within a specified plane. Figure 14.1 is an illustration of a typical medical CT installation. The important system components include (1) a gantry containing an X-ray tube and an array of detector elements that rotate about the patient, (2) a patient table to position the patient within the gantry, and (3) a control console with associated computer hardware to control the data acquisition process and perform image reconstruction.

As CT scanners evolved, different configurations for measuring X-ray transmission through the patient were adopted.[1] The first scanners used what is known as first-generation geometry consisting of an X-ray tube and a single X-ray detector. The tube and detector were translated across the patient, taking parallel projections. The tube and detector were then rotated by a small angle (typically $\leq 1°$), and another translation was performed. In such a manner, projections through the subject were measured for angles spanning 180°.

---

* Read Webb[58] for an interesting historical account of the development of CT.

Today, several configurations are in clinical use. Third-generation CT scanners use a diverging "fan beam" of X-rays produced at the focal spot of an X-ray tube that encompasses the patient, as illustrated in Figure 14.2. X-ray transmission is measured using an array of detector elements, which eliminates the need for a translation motion, and the X-ray source and detector array simply rotate about the patient. Modern, third-generation CT scanners complete one rotation in less than 1 s.

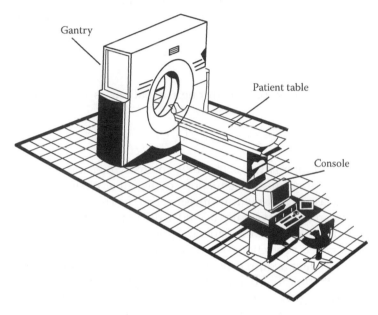

FIGURE 14.1
Schematic illustration showing a typical medical CT installation consisting of the gantry, patient table, and control console. (Courtesy of Picker International Inc.)

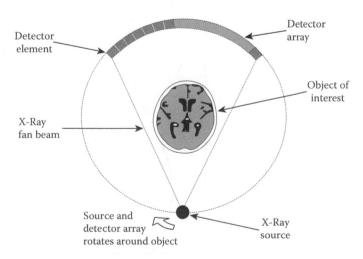

FIGURE 14.2
Typical arrangement of a third-generation X-ray CT.

Fourth-generation CT scanners make use of a stationary circular ring of detectors that surround the patient within the gantry and an X-ray source that moves on a circular track inside the ring. At each X-ray source position, projection data are measured by the opposing arc of detectors. Scan times are similar to those of third-generation geometry, and both systems are widely used at present.

Slip-ring scanners transfer signals and power to the rotating part of the gantry through brush couplings that slide along stationary conductive rings. This allows for multiple continuous rotations of the gantry without the need to rewind cables and leads to the development of "spiral" or "helical" scanning in which the patient moves through the gantry with a continuous motion while the X-ray tube rotates about the patient.

14.1.2 The Sinogram

Transmission of X-rays along path L through a patient is described by the integral expression

$$I = I_o e^{-\int_L \mu(l)dl} \tag{14.2}$$

where I is the measured intensity, I_o is the X-ray intensity measured in the absence of the patient, and $\mu(l)$ is the X-ray linear attenuation coefficient at position l. The detector array is used to obtain a measure of X-ray transmission $T = I/I_o$. The data acquisition process for a single tomographic image is typically 0.5–1.0 s, during which the source and detectors generally rotate a full circle about the patient. Approximately 500,000–1,000,000 transmission measurements, called projections, are used to reconstruct a single image.

The set of projection data acquired during a CT scan can be presented as a grey-scale image of the relative attenuation coefficient $\ln(1/T)$ as a function of θ_i, the angular position of the X-ray source during the ith projection, and σ_n, the angle of the nth detector element within the fan beam. This representation is called a sinogram, as illustrated in Figure 14.3.

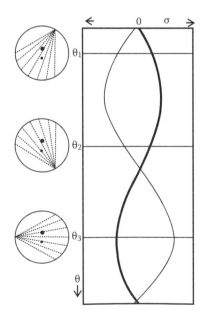

FIGURE 14.3
Sinogram data representation: each line consists of projections measured at one source angular position θ and many angles σ.

Each horizontal line displays one fan beam of projection data acquired at a particular source angle. The projections of each point in the image plane trace out a quasi-sinusoidal curve when using fan-beam geometry. In parallel-beam geometry, the curves are true sinusoids, hence the name sinogram, as shown in Figure 14.3.

14.1.3 Image Reconstruction

Several methods of reconstructing CT images have been proposed over the years and described by many authors,[1-4] including iterative algebraic techniques, the direct Fourier transform technique, and convolution-backprojection. The direct Fourier method is perhaps the simplest method conceptually. It is based on the central section theorem, which states that the one-dimensional (1-D) Fourier transform of a projection at angle θ is equal to a line through the two-dimensional (2-D) Fourier transform of the image at the same angle. This relationship is illustrated in Figure 14.4. When a sufficient number of projections are acquired, the complete 2-D Fourier transform is interpolated from the samples, and the image is obtained as the inverse Fourier transform. Although the technique is conceptually simple, it is computationally complex.

The reconstruction technique of greatest practical importance is known as convolution backprojection (or filtered backprojection). Backprojection refers to the distribution of projections back across the image plane in the direction from which they were measured.

The data acquisition process for first-generation, fan-beam CT systems is depicted in Figure 14.5. The projection measurements $\{p_n(r_n, \theta_i); n=1, \ldots, N\}$ are defined as line integrals of the attenuation coefficient through the object $f(x, y)$. For a given detector n and projection angle θ_i, the projections are given by

$$p_n(r_n, \theta_i) = \iint f(x, y)\delta(x\cos\theta_i + y\sin\theta_i - r_n)dxdy \tag{14.3}$$

The angular increment between two projections of the X-ray scanner is $\Delta\theta=2\pi/M$, where M is the number of projections taken during the period T required for one full rotation of

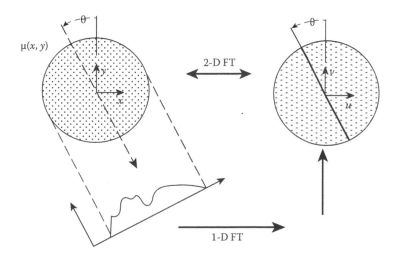

FIGURE 14.4
Central-slice theorem: Fourier transform of a set of projections of μ taken at angle θ equals a line in the 2-D Fourier transform of the image oriented at angle θ.

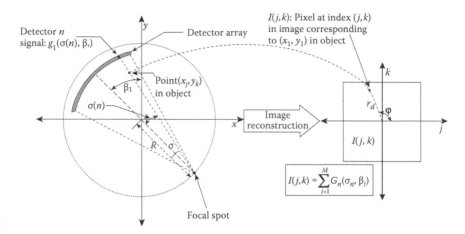

FIGURE 14.5
Schematic diagram of projection function for CT X-ray imaging systems. (Reprinted from IEEE © 2000. With permission.)

the source around the object. The transformations $r_n = R \sin\sigma_n$ and $\theta_i = \sigma_n + \beta_i$ are required to account for the geometry of the fan beam as shown in Figure 14.5. The projection function is then defined by

$$g_n(\sigma_n, \beta_i) = p_n \{[r_n = R \sin \sigma_n], [\theta_i = \sigma_n + \beta_i]\} \tag{14.4}$$

where g_n is the signal from the nth element of the detector array.

In the image reconstruction process, the pixel $I(j, k)$ in the actual image, shown at the right of Figure 14.5, corresponds to the Cartesian point (x_j, y_k) in the CT scan plane. The pixel value $I(j, k)$ is given by

$$I(j, k) = \sum_{i=1}^{M} G_n(\sigma_n, \beta_i), \tag{14.5}$$

where $G_n(\sigma_n, \beta_i)$ is the filtered version of the projection $g_n(\sigma_n, \beta_i)$ that has been adjusted to account for geometric effects. The angle σ_n defines the detector that samples the projection through a point (x_j, y_k) for a given projection angle β_i and is provided by Equation 14.6, where (r_d, φ) is the polar representation of (x_j, y_k):

$$\sigma_n = Tan^{-1}\left[\frac{r_d \sin(\varphi - \beta_i)}{R + r_d \cos(\varphi - \beta_i)}\right] \tag{14.6}$$

14.2 Motion Artifacts in CT

For an image to be reconstructed successfully from a data set using convolution back-projection, the data set must be complete and consistent. For a data set to be complete, projection data must be acquired over a sufficient range and with uniform angular spacing. No angular views should be missing. In parallel-beam geometry, projection data must be acquired over a range of 180°. In fan-beam geometry, data must be acquired over a range of 180° plus the fan angle.

Obtaining a consistent projection data set also requires that the patient not move during the entire data acquisition process, which may last 15–50 s for a multislice study. If the examination includes the chest or upper abdomen, the patient must hold their breath, and it may be necessary for multislice studies to be acquired in several single-breath-hold sections.

Motion during a scan can be in three dimensions and generally results in artifacts that appear as streaks or distorted semi-transparent structures in the general vicinity of the motion. It may be caused by a variety of reasons. Young or infirm patients are often restless and may not be cooperative or able to hold their breath. On occasion, general sedation may be required, and in extreme cases, muscle paralysis has been used.[5] Injection of a vascular contrast agent may result in involuntary patient motion[6–8] or blood flow artifacts.

14.2.1 Clinical Implications

These artifacts may obscure diagnostically important details or, in some circumstances, give a false indication of an unrelated condition. Respiratory artifacts may cause a ghosting of pulmonary nodules[9] or interfere with the assessment of interstitial lung disease.[10] Cardiac motion may produce an artifact that can be misdiagnosed as an aortic dissection.[11–15] Although such cases can be controversial, an additional angiographic examination may be required to confirm or exclude this diagnosis. Transesophageal ultrasound (echocardiography) can also be used to help determine whether a dissection is present. Cardiac and respiratory motion artifacts may hinder the visualization and clinical scoring of coronary calcifications, an important component of disease assessment and risk management.[16,17] Figure 14.6 illustrates an example of a cardiac motion artifact in a trauma victim that could be misinterpreted as a dissection (separation of the arterial wall) of the ascending aorta. In particular, the curvilinear shape of the artifact, in this example running approximately parallel to the aortic wall, and the fact that the artifact is restricted to the interior of the vessel, are suggestive of dissection and complicate the diagnosis. The ability to properly diagnose an aortic rupture or dissection is critical in the examination of many clinical settings.

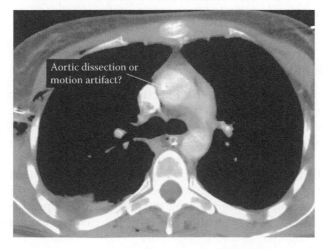

FIGURE 14.6
A CT image with a cardiac motion artifact in the ascending aorta that mimics an aortic dissection.

14.3 Reducing Motion Artifacts

A seemingly straightforward approach to reducing motion artifacts is to minimize the data acquistion time. However, the ability to rotate a conventional X-ray tube rapidly about the patient is limited by several factors including large forces that would be exerted on bearings supporting the rotating anode and the maximum output exposure rate. It is unlikely that acquistion times significantly less than 0.5 s will be achieved without major design changes to X-ray tubes. For this reason, many alternative methods have been developed for the purpose of reducing motion artifacts, and each has been successful under particularly circumstances. In this chapter, a brief summary is presented of (1) established methods, particular underscanning; (2) fast-acquisition methods using high-speed scanners; (3) respiratory gating; and (4) electrocardiographic (ECG) gating. Somewhat more detail is given in a description of new image processing methods for ECG gating.

14.3.1 Established Methods

It is often true that motion is relatively continuous, in which case the maximum discrepancy caused by motion occurs between the beginning and the end of a scan, which is the longest time span between two projection views.[18] These views are considered to be less reliable, and Pelc and Glover[19] developed a method of minimizing motion artifacts by applying a weighting factor to minimize the contributions of the most inconsistent projections. They showed that the weighting factor was a function of view angle and fan angle. Motion artifact reduction using this "underscanning" method has gained wide acceptance for routine clinical use.

In general, the extent of the artifact is dependent on the direction of motion. When the motion is in a direction parallel to the X-ray beam, the motion does not cause a change in the measured X-ray transmission and has no effect on the projection values. However, when the motion is in a perpendicular direction, it is likely that the motion will affect projection measurements. By choosing the first and last views to be obtained when motion is parallel to the X-ray beam, motion artifacts can be reduced. Respiratory motion tends to be in the vertical direction, and hence artifacts can sometimes be reduced by starting a scan with the X-ray tube either directly above or below the patient.

14.3.2 Fast Acquisition

Early work by Alfidi et al.[41] suggested that effective scan times of approximately 50 ms are required to reduce artifacts to an acceptable level. There are two approaches to meeting this 50-ms criterion. The first involves using high-speed CT scanners with very short scanning times with respect to the period of the cardiac cycle. The second requires ECG gating to synchronize the data acquisition process with the beating heart.

The earliest high-speed scanner was the dynamic spatial reconstructor (DSR) at the Mayo Clinic.[20,21] It employed 14 X-ray tubes, fired in rapid succession, to reduce scan times to 10 ms. While successful at demonstrating the need for high-speed volume CT systems for cardiac imaging and other applications, it was primarily a research tool and was never commercialized.

An alternative approach to using conventional X-ray tubes was the development of the electron-beam CT (EBCT).[17,22–24] It uses an electron beam deflected within the gantry to produce X-rays from a focal spot that rotates about the patient. Projection data are acquired in approximately 50 ms, which is fast enough to avoid both respiration and cardiac motion artifacts. It has been used successfully for measuring ventricular mass and border definition (essential for quantification of ventricular anatomy and function),[17] for detection of thrombi,[25] and for the management of stroke patients.[26] It provides an accurate method for scoring calcification of the coronary arteries.[16,17] However, EBCT is relatively complex and has not yet demonstrated images that can compete with the quality of conventional CT systems for noncardiac applications.

14.3.3 Respiratory Gating

Respiratory gating has been used with an algorithm to predict when a motionless period is about to occur, which is then used to trigger an acquisition.[27–30] Adaptive prediction schemes were developed to accommodate variable respiration patterns.[31] These methods were successful at reducing motion artifacts, but they increase the data acquisition times for multislice studies, increasing the probability of misregistration between slices.

Active breathing control methods have also been developed that can be used to control a patient's respiratory motion in an attempt to ensure reproducibility of respiratory motion,[32] but they are generally inappropriate for use in diagnostic procedures.

14.3.4 ECG Gating

The earliest attempts at ECG gating used nonslip-ring scanners.[33] Projection data was acquired only during diastole when the heart is moving the least. In order to keep scan times to within 20 ms, a small number of angles were used. Morehouse et al.[34] introduced retrospective gating in 1980 with a technique that involved measuring projection data continuously during four rotations of the X-ray source and selecting the projection data acquired during a specified window of the cardiac cycle. The technique resulted in missing views, causing artifacts of a different nature. These artifacts were reduced by the techniques of reflection, augmentation, and interpolation.[35,36] Joseph and Whitley[37] proposed that a small number of views may be adequate, provided that the heart can be isolated and centered in the field of view. Johnson et al.[38] performed eight scans per level on 32 contiguous levels of the heart, and reformatted the images into an early three-dimensional (3-D) data set. Although the images were improved over previous ECG-gating attempts, imaging time approached 1 h. In 1983, Moore et al.[27] used prospective ECG gating to determine the optimal start time for each rotation, later including coincident-ray considerations.[39] They were able to bring the time resolution down to 100 ms.

The myocardium of the heart behaves like a spring that is restrained from contracting by an electric potential, controlled mainly by concentrations of sodium and potassium ions. When discharged, or depolarized, these concentrations change, and the myocardium contracts vigorously. Repolarization causes the fibers to lengthen again, allowing the heart to fill with blood. It is the motion of the ions during depolarization and repolarization that provides the electrical signal detected by the ECG (Figure 14.7). In an ECG waveform, the *P* wave corresponds to atrial depolarization, the QRS complex corresponds to ventricular depolarization, and the *T* wave corresponds to ventricular repolarization. The period of inactivity between the *P* wave and the QRS complex is caused by the delay in conduction through the AV node. In the term "cardiac phase", τ is used loosely here to represent the time following the *R* wave (ms) within one cycle.

14.3.5 Single-Breath-Hold ECG Gating

The development of slip-ring CT scanners made possible single-breath-hold ECG gating. Nolan[40] used a continuous rotation of the X-ray tube and acquisition of an ECG waveform for 12–16 s during a single breath hold. They used a "data space" diagram as shown in Figure 14.8 to represent the relationship between the cardiac phase τ and the angular position of the X-ray focal spot for each rotation. After multiple rotations, data space is occupied by a series of diagonal lines (Figure 14.9). Once the data space is filled for all source angles covering 180° plus fan angle and any specified cardiac phase, a complete set of projection data exists, and an image can be reconstructed. In practice, it is generally necessary to reconstruct an image without a complete data set.

An improvement on this technique is to select data from each source position spanning 360° that is closest in phase to the desired cardiac phase. Coincident-ray replacement[39] was used to compress the 360° sinogram into one spanning only 180° plus the fan angle.

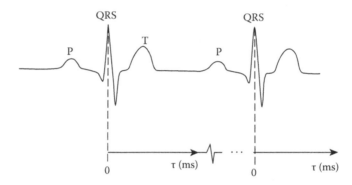

FIGURE 14.7
Components of an ECG waveform.

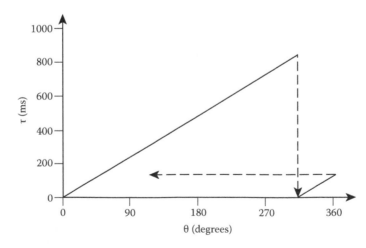

FIGURE 14.8
Filling of data space showing the relationship between the cardiac phase and the source position for one rotation.

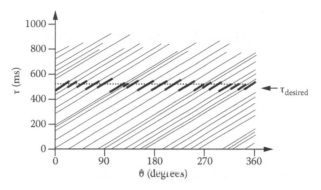

FIGURE 14.9
An image can be reconstructed when projection data have been acquired for source positions covering 180° plus fan angle for any specified cardiac phase. This generally results in a residual temporal inconsistency.

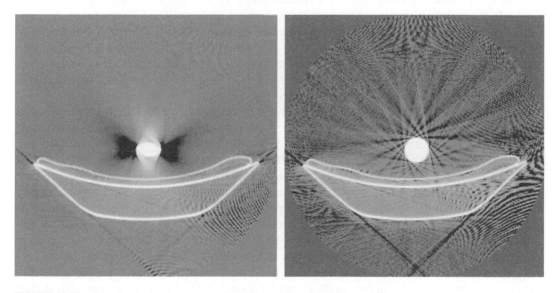

FIGURE 14.10
Comparison of a conventional CT scan of a cardiac phantom simulating ventricle expansion with an ECG-gated scan acquired in 16 s.

An image of a cardiac phantom simulating ventricle expansion is shown in Figure 14.10. The motion artifact, characterized by distortion of the circular phantom and background streaks, is significantly reduced in the ECG-gated image. Images such as this can be reconstructed for each of approximately 16 different cardiac phases. Using an ECG study of normal volunteers, Nolan suggested that a residual temporal inconsistency of 50 ms could be achieved with 12–16 rotations of the X-ray tube for approximately 95% of a normal population if the operator could choose prospectively between two rotation speeds.[40]

The ECG-gating methods described here require some degree of reproducibility of cardiac motion. While this is potentially a limitation in some circumstances, most of the variability in the cardiac cycle is due to variations in the P-QRS interval (Figure 14.7). By synchronizing data acquisition and reconstruction relative to τ, the time following the R peak, very little variability is observed from one cardiac cycle to the next.

14.4 Reducing Motion Artifacts by Signal Processing—A Synthetic Aperture Approach

Early work by Alfidi et al.[41] suggested that effective scan times of approximately 50 ms are required to reduce artifacts to an acceptable level. This is consistent with the scan times of EBCT and with the residual temporal inconsistency of the ECG-gated technique of Nolan.[40] However, Ritchie et al.[42] have suggested more recently that 50 ms is not fast enough for many applications of clinical importance, and additional methods are required for motion correction. It is unlikely that scan times can be reduced significantly, but there may be additional benefits from sophisticated new image processing techniques.

Several mathematical techniques have been proposed as solutions to this problem. In some specific cases, 3-D reconstructions have been used to assist in distinguishing motion artifacts from physical dissections in the descending aorta.[43] Most methods require a simple model of organ motion, such as a translational, rotational, or linear expansion.[44] In most situations of practical importance, such simplifications are not very useful. A more general technique attempts to iteratively suppress motion effects from projection data[45] by making assumptions regarding the spectral characteristics of the motion. However, this depends on knowing some properties of the motion *a priori* and requires a number of iterations to converge. This is generally undesirable for CT imaging, as it results in additional radiation doses to the patient from the X-ray exposure.

Motion artifacts have been reduced in magnetic resonance imaging (MRI) for chest scans by first defining the motion with a parametric model and then adapting the reconstruction algorithm to correct for the modeled motion.[46] Ritchie[47] attempted to address cardiac and respiratory motion in CT using a pixel-specific backprojection algorithm that was conceptually influenced by this MRI approach. Motion of a frame of reference was specified by making an estimate of where each pixel location was when each projection was acquired. These maps then formed the basis of a backprojection algorithm that reconstructed each pixel in a frame of reference that moved according to the information provided by the maps. The method requires manual efforts to describe the motion of each pixel and is therefore not practical for routine clinical use at present.

The problem of motion artifacts has also been addressed in other types of real-time imaging systems such as radar satellites and sonars.[48,49] In this case, it was found that application of synthetic aperture processing increases the resolution of a phased array imaging system as well as corrects for the effects of motion. Reported results showed that the problem of correcting motion artifacts in sonar synthetic aperture applications is centered on the estimation of a phase correction factor. This factor is used to compensate for the phase differences between sequential sensor array measurements in order to coherently synthesize the spatial information into a synthetic aperture.

Dhanantwari et al.[50,51] described a synthetic aperture approach to correct for motion artifacts in CT, which is described here in more detail. Their approach consists of three components:

1. Detection of changes in the CT projection data caused by organ motion using a spatial overlap correlator approach, resulting in a "motion" sinogram that reflects changes in the projection data.

2. Use of an adaptive interference canceller (AIC) approach to isolate the effects of organ motion using the motion sinogram and the conventional sinogram corrupted by motion to make an estimate of a "stationary" sonogram.

3. Use of a "coherent sonogram synthesis" technique that identifies through a replica correlation process the segments of the continuous sinograms that have identical phases of the motion effects.

These are described in more detail in the following sections.

14.4.1 Spatial Overlap Correlator to Identify Motion Effects

The spatial overlap correlator[48] makes use of two X-ray sources that rotate about the patient separated by a very small time delay δ, where $\delta = T/M$, with T as the total acquisition time for one slice (typically 1 s) and M as the number of angular projections. Source #1 trails source #2, so that if t_0 is the starting time of one source rotation, a view acquired by source #2 at time $t = t_0 + n\delta$ will be sampled again by source #1 at time $t = t_0 + (n+1)\delta$, as illustrated in Figure 14.11. Comparison of any two spatially overlapping measurements will provide a measurement that is associated with any organ motion that may have occurred during the elapsed time δ. For X-ray CT systems, differences between the two sets of samples are caused by organ motion or system noise.

Figure 14.12 shows a graphical representation of the above 2-D space-time sampling process. The vertical axis shows the times associated with the angular positions of the source-array receiver shown on the horizontal axis. Line segments along the diagonal represent the measurements of an X-ray CT scanner. Darker line segments show positions of the first source-detector pair and lighter lines show the second pair. Image reconstruction algorithms work best when the object is stationary, corresponding to horizontal lines of Figure 14.12.

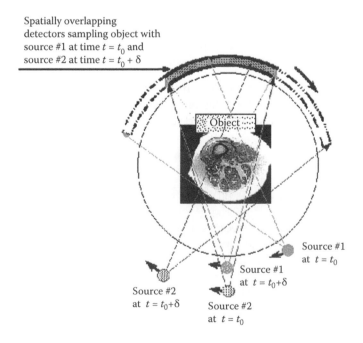

Spatially overlapping detectors sampling object with source #1 at time $t = t_0$ and source #2 at time $t = t_0 + \delta$

Source #1 at $t = t_0$

Source #1 at $t = t_0 + \delta$

Source #2 at $t = t_0 + \delta$

Source #2 at $t = t_0$

FIGURE 14.11
Two-source concept of spatial overlap correlator for CT X-ray imaging systems. (Reprinted from IEEE © 2000. With permission.)

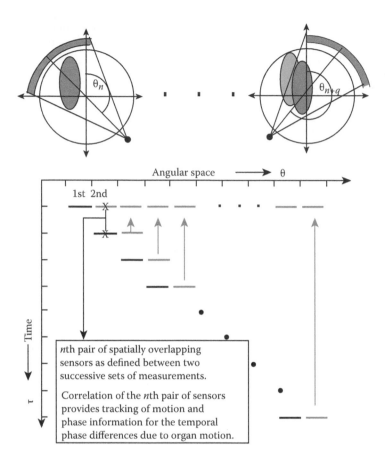

FIGURE 14.12
Graphical representation of the space-time sampling process of the spatial overlap correlator for CT X-ray imaging systems. (Reprinted from IEEE © 2000. With permission.)

In the following, let the projection measurement be given by $g_n(\sigma(n),\ \beta(t),\ t)$, given as a function of the fan angle $\sigma(n)$ on the detector arc, projection angle $\beta(t)$, and time t. The projection measurement for the first and second detector-source pair is given by

$$\{gn_{s1}\ (\sigma(n_{s1}),\ \beta(t),\ t)\ (n_{s1}=q,\ q+1,\ \dots,\ N)\} \tag{14.7}$$

$$\{gn_{s2}\ (\sigma(n_{s2}),\ \beta(t),\ t+\partial),\ (n_{s2}=1,\ 2,\ \dots,\ N-q)\} \tag{14.8}$$

where there are $N-q$ overlapping detectors. The source locations are identical for both acquisitions, and hence the projection angles for each are given by $\beta(t)=\beta_{s2}(t)=\beta_{s1}(t+\delta)$. In Figure 14.12, these spatially overlapping measurements are depicted by the pair of lines overlapping in angular space, but in two successive time moments. The difference between the two data acquisitions for a given spatial location defined by Equations 14.7 and Equation 14.8 is

$$\Delta g_n(\sigma(n),\ \beta(t),\ t)=g_{n_{s2}}(\sigma(n_{s2}),\beta(t),\ t+\delta)$$
$$-g_{n_{s1}}(\sigma(n_{s1}),\ \beta(t),\ t)\ \text{for}\ (n=1,\ 2,\dots,\ N-q) \tag{14.9}$$

where it is necessary that $\sigma(n)=\sigma(n_{s1})=\sigma(n_{s2})$ to ensure spatial overlap.

With reference to Equations 14.3 and 14.4, the time dependent projection measurement for a fan-beam X-ray CT scanner is given by

$$g_n(\sigma(n), \beta(t), t) = \iint f(x, y, t)\delta\{x\cos[\sigma(n)+\beta(t)] - R\sin\sigma(n)\}dxdy \tag{14.10}$$

Therefore, the time dependent projection measurement may be rewritten as

$$\Delta g_n(\sigma(n), \beta(t), t) - \iint [f(x, y, t) - f(x, y, t+\delta)]$$

$$\{\delta\{x\cos[\sigma(n)+\beta(t)]\} + y\sin[\sigma(n)+\beta(t)] - R\sin\sigma(n)\}dxdy \tag{14.11}$$

where $f(x, y, t)$–$f(x, y, t+\delta)$ indicates differences within the image plane caused by motion. It is clear that a stationary object will result in a zero output from the spatial overlap correlator.

If the projection measurement $g_n(\sigma(n), \beta(t), t)$ consists of both stationary and moving components, $g_{ns}(\sigma(n), \beta(t), t)$ and $g_{nm}(\sigma(n), \beta(t), t)$, respectively, then

$$g_n(\sigma(n), \beta(t), t) = g_{ns}(\sigma(n), \beta(t), t) + g_{nm}(\sigma(n), \beta(t), t) \tag{14.12}$$

and

$$\Delta g_n(\sigma(n), \beta(t), t) = g_{nm_{s2}}(\sigma(n_{s2}), \beta(t), t+\delta)$$

$$-g_{nm_{s1}}(\sigma(n_{s1}), \beta(t), t) \text{ for } (n = 1, 1, \ldots, N-q) \tag{14.13}$$

If the motion is oscillatory, then the motion may be represented as

$$g_{nm_{s2}}(\sigma(n), \beta(t), t) = \sin(2\pi f_0 t) \tag{14.14}$$

$$g_{nm_{s1}}(\sigma(n), \beta(t), t) = \sin(2\pi f_0(t+\delta)) \tag{14.15}$$

$$\sin(2\pi f_0 t) - \sin(2\pi f_0(t+\delta)) = -2\left[\sin(\pi f_0\delta)\cos\left(2\pi f_0\left(t+\frac{\delta}{2}\right)\right)\right] \tag{14.16}$$

$$\Delta g_n(\sigma(n), \beta(t), t) = -2\left[\sin(\pi f_0\delta)\cos\left(2\pi f_0\left(t+\frac{\delta}{2}\right)\right)\right] \tag{14.17}$$

Nonperiodic motion may be considered as being piecewise periodic, where the motion is broken into small periodic segments. In such a case, the scale factor $\sin(\pi f_0\delta)$ will vary as f_0 varies, while the piecewise periodic signal

$$\cos\left(2\pi f_0\left(t+\frac{\delta}{2}\right)\right)$$

will continue to track the motion. The resulting time series from the spatial overlap correlator will yield a signal that tracks the organ motion. This signal will be scaled depending on the frequency of the organ motion.

Although motion is assumed to be periodic for mathematical evaluations, the spatial overlap correlator is capable of tracking any form of organ motion, including transients if they are present, limited by sampling considerations as the sources rotate about the patient.

14.4.1.1 Simulations

A simulation study illustrates operation of the spatial overlap correlator using a Shepp–Logan phantom with the dark lower ellipse on the left-hand side of the phantom undergoing deformation. The sinogram from the standard data acquisition process is shown on the left of Figure 14.13. The right-hand side of Figure 14.13 shows motion tracking by the spatial overlap correlator. It is expected that the spatial overlap correlator will track the boundaries of the deforming ellipse, since that is where the motion occurs.

Reconstruction using these sinograms is shown in Figure 14.14. The image on the left of Figure 14.14 corresponds to the sinogram on the left of Figure 14.13 and shows the reconstructed phantom with the motion artifacts caused by the dark lobe on the left-hand side moving. The image on the right of Figure 14.14 corresponds to the sinogram on the right of Figure 14.13. The image shows only the moving object and no indication of any of the

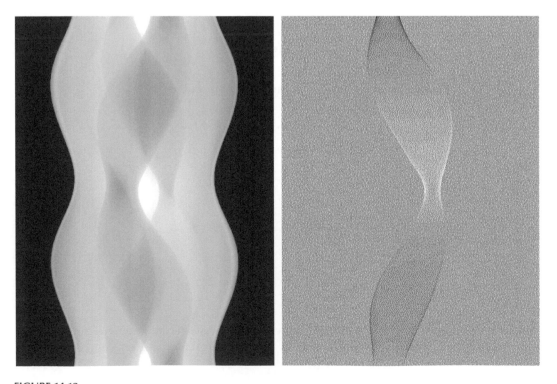

FIGURE 14.13
Simulated sinograms of the Shepp–Logan phantom (left conventional, right hardware spatial overlap correlator). A reconstructed image is in Figure 16.14.

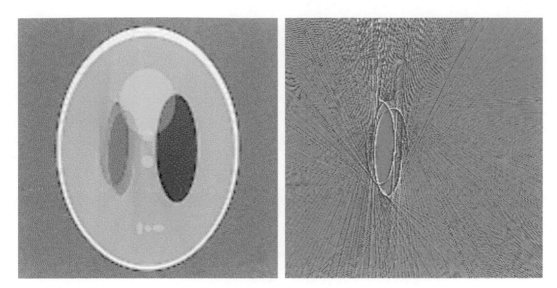

FIGURE 14.14
Reconstructed images from standard X-ray CT data acquisition and the hardware spatial overlap correlator.

stationary objects. Since the motion of the object is tracked in an incremental fashion, the image shows the changes that occur along the boundary of the ellipse.

14.4.1.2 Hardware Implementation

Some currently available X-ray CT scanners (Siemens, GE, and Elscint) use a dual focal spot technique that can be modified for implementation of the spatial overlap correlator. Conventional projections are identified as the spatial locations $\beta(t)$, $\beta(t+\delta)$, $\beta(t+2\delta)$, and so on. The second focal spot position is adjusted so that it will coincide with location $\beta(t)$ at time $t+\delta/2$. In this fashion, the two-source concept of the spatial overlap correlator may be achieved.

However, with third-generation scanners, there is also a shift of the detector array over the time interval between projections from source #1 and #2, $\delta/2$, that does not correspond to an integer number of detector elements. This prevents proper alignment without an expensive hardware modification.

14.4.1.3 Software Implementation

A second approach to implementing the spatial overlap correlator makes use of the rotation of a single X-ray source.[50] Rather than using two projections from different sources that differ in time by the small value δ, two projections from the same source over two rotations that differ in time by $T = M\delta$ are used. With this approach, the subtraction process is sensitive to motions on a time scale of approximately 1 s and requires a minimum of two full gantry rotations. As such, this approach is not sensitive to transient motions. In addition, the assumption that a signal be viewed as being piecewise periodic is often acceptable for motions sampled on an interval of approximately 1 ms, but less likely to be acceptable when the same motion is sampled on an interval of approximately 1 s. However, cardiac motions are remarkably periodic, and good results are still obtained.

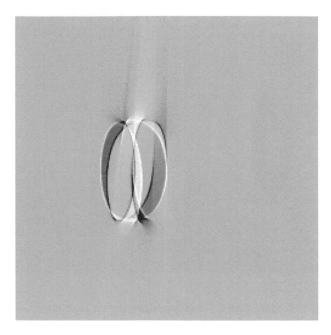

FIGURE 14.15
Reconstructed images of motion using the software spatial overlap correlator.

Figure 14.15 shows an image obtained from the software spatial overlap correlator (SSOC) for the example shown in Figure 14.14. While sampling at intervals of approximately 1 s does not do as well as sampling at intervals of approximately 1 ms, the method still identifies the important motion components.

14.4.2 Adaptive Processing to Remove Motion Artifacts

The AIC approach is used to remove the motion artifacts identified by the spatial overlap correlator. It has been used extensively for isolation of signals that were originally measured in the presence of noise.[49,52] The data sequence from each detector is treated as a time series. The sequence from the conventional CT acquisition is treated as the signal including unwanted interference due to motion effects, and the sequence from the spatial overlap correlator is treated as the interference in the AIC processing scheme. Both the AIC and the coherent sinogram synthesis (CSS) methods, which are discussed in Section 14.4.3, require that data be acquired over a number of revolutions of the X-ray source. In the case of the AIC algorithm, the number of rotations is a function of the convergence rate of the adaptive algorithm and requires at least two rotations. For the CSS method, this number is simply a function of desired image quality. There is a direct relationship between the length of the data sequence acquired and the exactness of the synthesized sinogram to the desired sinogram of a stationary object and the speed of the object's motion. Both AIC and CSS algorithms may be implemented with the hardware spatial overlap correlator (HSOC), but only the CSS algorithm can be used with the SSOC.

At this point, the physical significance of the incremental tracking of organ motion by the HSOC is analyzed. Also of importance is the relationship between this incremental motion tracked and the standard X-ray CT measurements. First, the ideal case where the motion artifacts are readily available for removal is described, then the relationship between such a system and the HSOC data acquisition scheme is developed.

Let the projections for a single sinogram of M projections acquired without motion be represented as

$$g_{n_{ct}}(\sigma(n_{ct}), \beta(t_j), t_j), (n_{ct} = 1, \ldots, M), \quad (j = 1, \ldots, M) \tag{14.18}$$

This is a sinogram for a stationary object. Let the projections for a single sinogram for the same object, but with motion effects present, be given by

$$g_{n_{mov}}(\sigma(n_{mov}), \beta(t_j), t_j), (n_{mov} = 1, \ldots, N), \quad (j = 1, \ldots, M) \tag{14.19}$$

The difference between these two sinograms provides information about the accumulated motion over the data acquisition period, given by

$$\Delta f_n(\sigma(n), \beta(t_j), t_j) = g_{n_{mov}}(\sigma(n_{mov}), \beta(t_j), t_j) - g_{n_{ct}}(\sigma(n_{ct}), \beta(t_j), t_j);$$
$$(n = n_{ct} = n_{mov} = 1, \ldots, N), \quad (j = 1, \ldots, M) \tag{14.20}$$

Alternatively stated, the optimum sinogram is the difference between the sinogram with organ motion effects included and that of only the organ motion effects, given by

$$g_{n_{ct}}(\sigma(n_{ct}), \beta(t_j), t_j) = g_{n_{mov}}(\sigma(n_{mov}), \beta(t_j), t_j) - \Delta f_n(\sigma(n), \beta(t_j), t_j);$$
$$(n = n_{ct} = n_{mov} = 1, \ldots, N), \quad (j = 1, \ldots, M) \tag{14.21}$$

Thus, an artifact-free image can be reconstructed if an estimate of $\Delta f_n(\sigma(n), \beta(t_j), t_j)$ can be obtained and subtracted from the sinogram measured in the presence of motion.

From Equation 14.9, the measurements obtained from the HSOC are written in discrete form as

$$\Delta g_n(\sigma(n), \beta(t_j), t_j) = g_{n_{s2}}(\sigma(n_{s2}), \beta(t_j), t_j + \delta - g_{n_{s1}}(\sigma(n_{s1}), \beta(t_j), t_j)$$
$$\text{for } (n = 1, 2, \ldots N), \quad (j = 1, \ldots, M) \tag{14.22}$$

It is evident that the measurement described by Equation 14.9 is that of any motion that occurred over the time interval from t_j to $t_j + \delta$ from view angle $\beta(t_j)$. This means that over the complete data acquisition period of $M\delta$ seconds, the spatial overlap correlator will sample the motion M times at the angles $\beta(t_j)$ ($j = 1, \ldots, M$). Therefore, Equation 14.22 can be rewritten as

$$\Delta g_n(\sigma(n), \beta(t_j), t_j) = \left(\frac{g_{n_{s2}}(\sigma(n_{s2}), \beta(t_j), t_j + \delta) - g_{n_{s1}}(\sigma(n_{s1}), \beta(t_j), t_j)}{\delta} \right) \delta \tag{14.23}$$

Recall that $\Delta g_n(\sigma(n), \beta(t_j), t_j)$ is simply a measure of the motion present, since the constant terms due to the stationary components disappear:

$$\Delta g_n(\sigma(n), \beta(t_j), (t_j) = g_{nm_{s2}}(\sigma(n_{s2}), \beta(t_j), t_j + \delta) - g_{nm_{s1}}(\sigma(n_{s1}), \beta(t_j), t_j)$$
$$\text{for } (n = 1, 2, \ldots N), \quad (j = 1, \ldots, M) \tag{14.24}$$

It follows directly that integration of the time series derived from the HSOC gives the compound motion at the mth projection:

$$\int_{t_0}^{t_{m-1}} \Delta g_n(\sigma(n), \beta(t_j), m\delta)dt$$

$$= \sum_{j=1}^{m} \left\{ g_{nm_{s2}}^{t_{j+1}}(\sigma(n), \beta(t_j), t_j + \delta) - g_{nm_{s1}}^{t_j}(\sigma(n), \beta(t_j), (t_j)) \right\}$$

$$\text{for } (n = 1, \ldots, N), \quad (m = 1, \ldots, M) \tag{14.25}$$

In Equation 14.25, the projection index j specifies the time t_j, and the detector index n specifies the detector fan angle $\sigma(n)$. For simplicity, $g_n^j(\sigma(n)), \beta(t_j), t_j)$ is expressed as g_n^j.

The HSOC makes measurements continuously over the time period $M\delta$, but at each moment the measurement corresponds to different projection numbers, or view angles β. Rewriting Equation 14.25 leads to Equation 14.26, where the factors ρ_m, $(m=1, \ldots, M)$ compensate for the views being from different angles:

$$\int_{t_0}^{t_{m-1}} \Delta g_n(\sigma(n), \beta(t_j), t_j)dt \tag{14.26}$$

$$= (g_{nm_{s2}}^{j=1} - g_{nm_{s1}}^{j=0})\rho_1 + (g_{nm_{s2}}^{j=2} - g_{nm_{s1}}^{j=1})\rho_2 + \ldots + (g_{nm_{s2}}^{j=m} - g_{nm_{s1}}^{j-m-1})\rho_m$$

The factors $\left(g_{nm_{s2}}^{j-m} - g_{nm_{s1}}^{j-m-1}\right)$ represent measurements from the HSOC as described in Equation 14.24. Alternatively, this may be represented in an incremental form as in Equation 14.27, with the references to source positions $s1$ and $s2$ omitted but assumed. This relationship suggests that there is no interdependency between the individual ρ_m, $(m=1, \ldots, M)$.

$$\int_{t_0}^{t_{m-1}} \Delta g_n(\sigma(n), \beta(t_j), t_j)dt = \int_{t_0}^{t_{m-2}} \Delta g_n(\sigma(n), \beta(t_j), t_j)dt + (g_{nm}^{m-1} - g_{nm}^{m-2})\rho_{m-1} \tag{14.27}$$

Defining the function $\Delta h_n^m(\sigma(n), \beta(t_j), t_j)$ to be an estimate of the function $\Delta f_n(\sigma(n), \beta(t_j), t_j)$, of Equation 14.20 and using the definition in Equation 14.28, the relationship between HSOC measurements and the desired organ motion effect measurements may be expressed as

$$\Delta h_n^m(\sigma(n), \beta(t_j), t_j) = \int_{t_0}^{t_{m-1}} \Delta g_n(\sigma(n), \beta(t_j), t_j)dt \tag{14.28}$$

$$\Delta h_n^m(\sigma(n), \beta(t_j), t_j) = \Delta h_n^{m-1}(\sigma(n), \beta(t_j), t_j) + (g_n^{m-1} - g_n^{m-2})\rho_{m-1} \tag{14.29}$$

The initial condition is $\rho_1 = 1$. In cases where the motion is independent of view angle, as would be the case for a deforming object at the center of the field of view, $\rho_m = 1$ $(m=1, \ldots, M)$. In such a case, Equation 14.26 (or Equation 14.27) reduces to Equation 14.30:

$$\int_{t_o}^{t_{m-1}} \Delta g_n(\sigma(n), \beta(t_j), t_j)dt = g_n^{j-m} - g_n^{j-0} \tag{14.30}$$

Because of Equation 14.31, Equation 14.30 may be represented as two separate sinograms.

$$g_n(\sigma(n), \beta(t), t) = \iint f(x, y, t)\delta \{x \cos[\sigma(n) + \beta(t)] + y \sin[\sigma(n) + \beta(t)] - R \sin \sigma(n)\} \, dxdy \tag{14.31}$$

The first sinogram is the standard X-ray CT sinogram with motion artifacts present, defined by Equation 14.32. The second corresponds to no motion and defined in Equation 14.33.

$$g_n^{j-m}(\sigma(n), \beta(t), t) = \iint f(x, y, t)\delta\{(x, y), (\sigma(n) + \beta(t), R(t))\}dxdy \tag{14.32}$$

$$g_n^{j-0}(\sigma(n), \beta(t_0), t_0) = \iint f(x, y, t_0)\delta\{(x, y), (\sigma(n), \beta(t_0), R(t_0))\}dxdy \tag{14.33}$$

The first sinogram is derived over the complete data acquisition period, whereas the second sinogram is derived from the time that the first projection is taken; in effect, all motion has been frozen. Comparing Equations 14.30, 14.32, and 14.33 with Equation 14.20, it is evident that the sinograms are identical in the two representations. Specifically, these relationships are defined as

$$g_n^{j=0}(\sigma(n), \beta(t_0), t_0) = g_{n_{ct}}(\sigma(n), \beta(t), t) \tag{14.34}$$

$$g_n^{j=M}(\sigma(n), \beta(t_M), t_M) = g_{n_{mov}}(\sigma(n), \beta(t), t) \tag{14.35}$$

As a result, for the case where the factors $\rho_m = 1$, $(m=1, \ldots, M)$ the estimate of $\Delta f_n(\sigma(n), \beta(t), t_j)$ is given by

$$\Delta f_n(\sigma(n), \beta(t_j), (t_j)) = \int_{t_0}^{t_j} \Delta g_n(\sigma(n), \beta(t), dt \tag{14.36}$$

For X-ray CT applications, this integral expression is not a simple problem because the scalar factors ρ_m, $(m=1, \ldots, M)$, introduce dc offsets. The impact of these factors on the integration process of Equation 14.27 may be removed by means of a normalization process[53] or nonlinear adaptive processing.[52,54,55] Moreover, from Equation 14.36, the derivative of the ideal set of measurements, $\Delta f_n(\sigma(t_j), \beta(t_j), t_j)$, which define the difference between sinograms corresponding to projections with motion effects present and those acquired without motion effects, should predict the measurements of the spatial overlap correlator:

$$\Delta g_n(\sigma(t_j), (t_j), t_j) \approx \frac{d(\Delta f_n(\sigma(n), \beta(t_j), t_j))}{dt} \Delta t \tag{14.37}$$

This relationship suggests that the HSOC provides the derivative of the motion effects, and it follows directly that measurements of the spatial overlap correlator form the basis of a new processing scheme to remove motion artifacts associated with the CT data acquisition process. In particular, motion effects are defined by temporal integration of the HSOC measurements, as given by

$$\Delta f(\sigma(n), \beta(t_j), t_j \approx \sum_{j=1}^{J} \Delta g_n(\sigma(n), \beta(t_j), t_j)\rho_j, \quad \text{for } (n = 1, 2, \ldots, N)$$

(14.38)

In practical terms, this suggests the possibility of generating two types of sinograms. The first is the standard X-ray CT measurements $G_{n_{mov}}(\sigma(n), \beta(t_j), t_j)$, with organ motion effects present. The second is the HSOC measurements $\Delta g_n(\sigma(n), \beta(t_j), t_j)$, which provide estimates of $\Delta f_n(\sigma(n), \beta(t_j), t_j)$ as described in Equation 14.38.

In the case of nonlinear effects, which require a normalization process,[53] or an estimation scheme for the scalar factors ρ_m, alternative optimum estimates of $\Delta f_n(\sigma(n), \beta(t_j), t_j)$ can be provided by an AIC process with inputs as the sinograms $g_{n_{mov}}(\sigma(n), \beta(t_j), t_j)$ and $\Delta g_n(\sigma(n), \beta(t_j), t_j)$, as discussed in the next section.

14.4.2.1 Adaptive Interference Cancellation

The concept of adaptive interference cancellation is particularly useful for isolating signals in the presence of additive interferences.[52,54,55] The AIC scheme is shown in Figure 14.16. The detector signal with interference due to motion is represented as $y(j\delta) = s(j\delta) + n(j\delta)$, where $s(j\delta)$ and $n(j\delta)$ are the signal and interference components, respectively. In an AIC system with performance feedback, it is essential that the interference component $x(j\delta)$ is either available or measured simultaneously with the received noisy signal $y(j\delta)$.[52]

We assume that an adaptation process with performance feedback provides the weight vector $w(i)$ and through linear combination generates estimates of the interference $n(j\delta)$. In general, the adaptation process includes a minimization of the mean square value of the error signal defined by the performance feedback. Optimization by this criterion is common in many adaptive and nonadaptive applications. In the case of the X-ray CT system, measurements provided by the spatial overlap correlator $\Delta g_n(\sigma(n), \beta(t_j), t_j)$ form the basis of the interference estimates for the adaptation process, while the standard X-ray CT projection measurements $\Delta g_{mov}(\sigma(n), \beta(t_j), t_j)$ represent the noisy signal.

We also assume that interference measurements at the input of the adaptation process in Figure 14.16 are provided by the input $\overline{X}$ with terms $[x(\delta), x(2\delta), \ldots, x(L\delta)]^T$. Furthermore, the

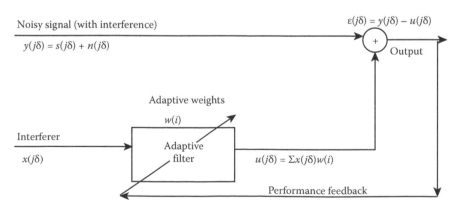

FIGURE 14.16
Concept of an AIC. (Reprinted from of IEEE © 2000. With permission.)

output of the adaptation process, $u(j\delta)$, is a linear combination of the input measurements $\bar{X}$ and the weight adaptive coefficients $[w(1), w(2), ..., w(L)]^T$ of the vector $\bar{W}$. In general, the weights $\bar{W}$ are adjusted so that the system descends toward the minimum of the surface of the performance feedback.[52,56] The output of the AIC is given by Equation 14.39, where $\bar{Y}$ is the input vector of the interference measurements, $y(j\delta)$ for $j = 1, 2, ..., M$ and M is the maximum number of samples to be processed. Since M is generally larger than L, segments of the inputs of length L are selected in a sliding window fashion and processed.

$$\bar{\varepsilon} = \bar{Y} - \bar{X}^T \bar{W} \qquad (14.39)$$

The AIC concept is based on the minimization of Equation 14.39 in the least mean square (LMS) sense, giving

$$E_{\min}[(\bar{\varepsilon})^2] = E_{\min}[(\bar{S} + \bar{N} - [\bar{X}^T \bar{W}])^2] \qquad (14.40)$$

When $E[(\bar{\varepsilon})^2]$ is minimized, the signal power $E[\bar{S}]$ is unaffected, and the term $E[\bar{N} - \bar{X}^T \bar{W}]$ is minimized. Thus, the output $u(j\delta)$ of an adaptive filter with L adaptive weights ($w_1, w_2, ..., w_L$) and for an interference input vector $x(j\delta)$, of arbitrary length, at time $j\delta$ are given by Equation 14.41, where the output of the adaptive filter depends on some history of the interference. The number of past samples required is determined by the length of the adaptive filter.

$$u(j\delta) = \sum_{i=1}^{L} w_i^{j\delta} \times x((j + i - L)\delta) \qquad (14.41)$$

The weights of the adaptive filter are adjusted at each time interval, and the update method depends on the adaptive algorithm. If μ is the adaptive step size, the update equation for the LMS adaptive filter is given as[52,55,57]

$$x_i^{(j+1)} = x_i^{j\delta} + (\mu \times x((j + 1 - L)\delta)) \times u(j\delta), \quad (i = 1, 2, ..., L) \qquad (14.42)$$

Similarly, for the normalized least mean square (NLMS) algorithm, the update equation is given by Equation 14.43. In this update equation, λ is the adaptive step size parameter, α is a parameter included for stability, and $|n|$ is the Euclidean norm of the vector input interference vector $[x((j + 1 - L)\delta), x((j + 2 - L)\delta), ..., x((j)\delta)]$.

$$W_i^{(j+1)\delta} = W_i^{j\delta} + \left(\frac{\lambda}{\alpha + |n|} \times x((j + i - L)\delta) \times u(j\delta) \right), \quad (i = 1, 2, ..., L) \qquad (14.43)$$

Figure 14.17 shows the AIC processing structure that has been modified to meet the requirements of the X-ray CT motion artifact removal problem. The CT sinogram can then be expressed as

$$\bar{g}_{n_{CT}} = \bar{g}_{n_{mov}} - P_n \overline{\Delta g_n} \qquad (14.44)$$

where vectors $\bar{g}_{n_{CT}} = [gnCT = [g_{n_{CT}}(t_1), g_{n_{CT}}(t_2), ..., g_{n_{CT}}(t_M)]^T$ and $\bar{g}_{n_{mov}} = [g_{n_{mov}}(t_1), g_{n_{mov}}(t_2), ..., g_{n_{mov}}(t_M)]^T$ are defined for each one of the detector elements $n = 1, 2, ..., N$, of the CT detector array. The vector $\overline{\Delta g_n} = [\Delta g_n(t_1), \Delta g_n(t_2), ..., \Delta g_n(t_M)]^T$ represents the HSOC measurements,

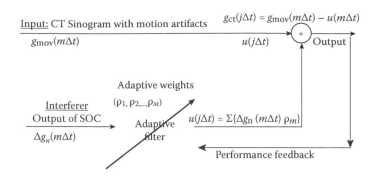

FIGURE 14.17
Concept of an AIC and spatial overlap correlator for removing motion artifacts in CT medical imaging applications. (Reprinted from IEEE © 2000. With permission.)

which represent the interference measurements. The matrix $\underline{P_n}$ includes the adaptive weights defined by

$$\underline{P_n} = \begin{bmatrix} \rho_1 & 0 & 0 & 0 \\ \rho_1 & \rho_2 & 0 & 0 \\ . & . & . & . \\ \rho_1 & \rho_2 & . & \rho_M \end{bmatrix} \tag{14.45}$$

The adaptive filter weights $w(j)$ are replaced with ρ_m, $(m=1, \ldots, M)$. Recall also that $\rho_1=1$. The optimization process by the AIC processor now includes optimizing ρ_m, to reduce the effects of the motion artifacts. Thus, the adaptive algorithms form the basis of an iterative estimation and accumulation process for the terms $\left(g_n^{m-1} - g_n^{m-2}\right)\rho m-1$ that defines the nonlinear temporal integration of the HSOC output, as defined in Equation 14.27. The output of the adaptive filter $u(j\delta)$ is a predictive estimate of the term $\Delta f_n(\sigma(n), \beta(t_j), t_j)$ of Equation 14.20. The output of this AIC process provides predictive estimates for sinograms that have been corrected for motion artifacts according to the information provided by the measurements of the spatial overlap correlator.

14.4.3 CSS from SSOC

An alternative approach is to assemble a sinogram using the CSS method, which uses the time series produced by either version of the spatial overlap correlator. When the object, or heart in the case of cardiac X-ray CT imaging, is at a specified phase in its motion cycle, the CSS technique defines this as the phase of interest. It then isolates every subsequent time moment during the data acquisition period when the object is again at that phase of its motion cycle. A number of projections are selected at each of these time moments and assembled into a sinogram. Figure 14.18 depicts the complete process using the curves obtained from using the software spatial overlap correlator (SSOC) scheme for the purpose of illustration. The process is identical with the time series from the HSOC.

14.4.3.1 Phase Selection by Correlation

A sliding window correlation process is used to isolate the moments during the data acquisition that include the same information and phase as the phase of interest. In general, the

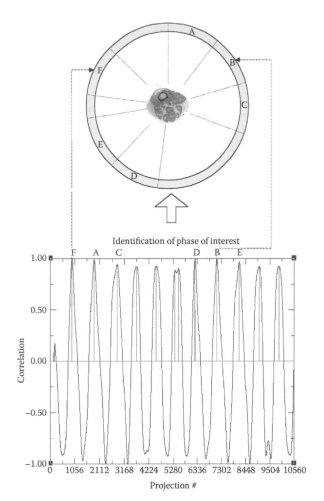

FIGURE 14.18
Operation of the CSS method. (Reprinted from IEEE © 2000. With permission.)

cross-correlation coefficient, CC_{sr}, between two time series, s_i and r_i of length L, is given by Equation 14.46.[52]

$$CC_{sr} = \frac{\sum_{i=1}^{L} s_i r_i}{\sqrt{\sum_{i=1}^{L} s_i^2 \sum_{i=1}^{L} r_i^2}} \tag{14.46}$$

where s_i is the signal from the spatial overlap correlator and r_i is the replica kernel, which is a short segment of the motion cycle extracted from s_i. The sliding window correlation technique uses a subset of the signal near the phase of the interest as the replica kernel and then correlates this replica with segments of the continuous signal to compute a time varying correlation function. The segments of the signal used in the cross-correlation function are selected in a sliding window fashion. The time varying cross-correlation function CC_i

is given by Equation 14.47, where L is the length of the segment used in the cross-correlation function and N is the length of the complete time series from the spatial overlap correlator.

$$CC_i = \frac{\sum_{j=-L/2}^{L/2} s_{i+j} r_{j+L/2}}{\sqrt{\sum_{j=-L/2}^{L/2} s_{i+j}^2 \sum_{j=-L/2}^{L/2} r_{j+L/2}^2}}, \, i = \frac{L}{2}, \frac{L}{2}+1, ..., N - \frac{L}{2} \tag{14.47}$$

The time moments at which the maxima of the correlation function are sufficiently close to 1 are considered as the time moments at which the phase of interest reoccurs. The time moments directly define a projection number, since the projections are acquired sequentially at a known sampling rate. With all of the projections that define a phase of interest known, the sinogram may now be assembled.

14.4.3.2 Assembling the Coherent Sinogram

In the final stage, segments of the continuous sinogram are selected and used to assemble a sinogram for a single image. The selection criterion is to use projections that occur when the organ is at the desired point in its motion cycle. In other words, the time moments at which the level or correlation approaches one in the curve of Figure 14.18 are the time moments that are used as a selection criterion to identify the projections for coherent synthesis of the sinogram. The curve in the lower segment of Figure 14.18 shows the time moments selected because of the sufficiently high level of correlation at these times. These time moments correspond to angular locations of projections, as shown in the upper panel of Figure 14.18. The angular locations they map to are determined by the physical locations of the X-ray source and detector array at the time the organ is at the desired point in its phase.

Since the X-ray source rotates about the patient, spatial locations distributed on a circle are sampled repeatedly. A projection number P maps to a physical projection number p, where P is referenced to the entire data acquisition period and p is referenced to the physical location on the circle along which data are acquired. The relationship between P and p is given by

$$P = nN + p; \quad 0 \le P \le \infty; \quad 0 \le p \le N; \quad n = 1, 2, ..., \infty \tag{14.48}$$

In effect, the segments chosen are synchronized to the organ motion cycle, and valid projections are only extracted when the organ is exactly at the desired point in its motion cycle. The difference is that data are acquired continuously, and not all projections are used in generating a single image. However, there is the need for additional data for the interpolation process, and there may be a need for images from a number of phases. Therefore, although not all of the data are used for a single image, the complete processing scheme requires all of the data acquired to produce a complete image set.

Under ideal conditions, one view would be taken each time the organ reaches the desired point in its motion cycle, and after N cycles of motion, a complete sinogram would be available. Since there is no physical synchronization between the data acquisition process and the organ motion, there may be a repetition of some views, while other views may be missing.

14.4.3.3 Interpolation to Complete the Sinogram

Interpolation is used to account for any missing angular segments data acquired by the X-ray CT system. The first option is to take data from one complete revolution and use those data as a basis for the final image. The idea is to try to improve the image that would be produced by this standard X-ray CT sinogram. Using this sinogram as a starting point, projection windows are selected in the same manner as described in the previous section. Whenever a suitable projection window is found, it is used to overwrite the original projections of the sinogram.

The second option synthesizes a new sinogram. This method does not limit the number of projections in a projection window. Rather, windows are allowed to be as large as necessary to fill in the entire sinogram. The center of the initial window is defined as the desired point, and missing projections are filled in using appropriate projections from the data acquired by choosing projections as close as possible to the desired point. This method is preferred when the original image quality is poor, since it attempts to create a completely new image.

14.4.4 Signal Processing Structure for an AIC

A block diagram representation of the signal processing structure that implements the AIC and CSS schemes are shown in Figures 14.19 and 14.20, respectively. Both structures consists of three major blocks: the data acquisition system, the signal processor, and the display functionality.

The data acquisition system requires specialized CT hardware to support the HSOC. This means that the data acquisition system will effectively provide two data streams, with both streams consisting of samples from the same spatial locations, but at different times on the order of δ.

14.4.5 Phantom Experiment

The phantom shown in Figure 14.21, consisting of a hollow Plexiglas™ cylinder and an inner solid Teflon™ cone, was constructed to demonstrate the motion-artifact reduction potential of the synthetic aperture approach. The cone moves back and forth through the image plane in the gantry, simulating an expanding and contracting ventricle. Seven metal wires were placed on the outside of the cone to simulate arterial calcifications. A conventional CT image of this phantom when stationary is shown in Figure 14.22. This image has no motion artifacts and represents the target image for any motion-artifact reduction method. A conventional CT image of the phantom operating with a period of 0.6 s, simulating a heart rate of 100 beats per minute, is shown Figure 14.23. Severe motion artifacts are evident.

A sequence of six CSS/SSOC images were generated at equally spaced points covering one cycle of the phantom's motion. Figure 14.24 shows images obtained using projection data acquired over 9 s. Motion artifacts are reduced in all images relative to the conventional image, showing expansion and contraction of the simulated ventricle. Artifacts exist due to motion of the simulated calcifications, but they are relatively minor in the top-right image corresponding to end diastole where motion is the least. Figure 14.25 shows a sequence of images obtained over a period of 3 s. Residual motion artifacts are more pronounced, but all images are superior to the conventional image.

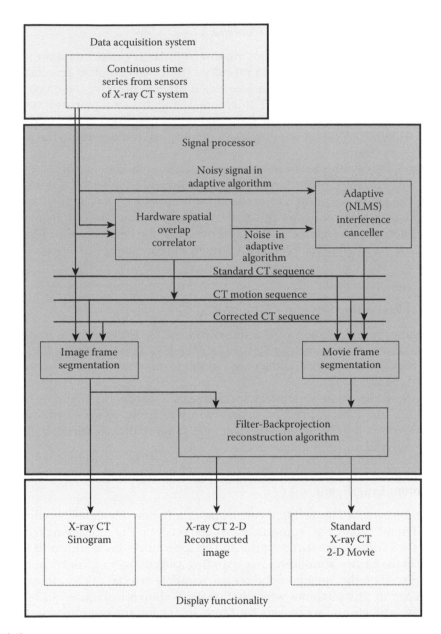

FIGURE 14.19
Signal processing structure for AIC implementation.

14.4.6 Human Patient Results

The CSS/SSOC method was evaluated in a cardiac study of a middle-aged female. The patient's heart rate was approximately 72 beats per minute. CT projection data were acquired during multiple rotations of the X-ray tube. No restriction was placed on the patients' breathing. The conventional CT image is shown in Figure 14.26. There is no indication of any calcification in the arteries of the heart in this image, and the effect of respiratory motion is evident. The chest walls and sternum are not clearly defined, appearing as dual images.

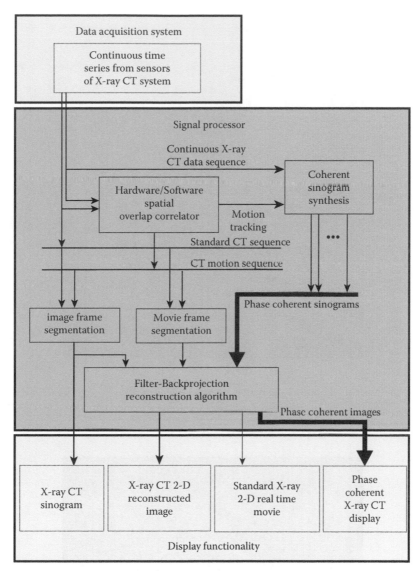

FIGURE 14.20
Signal processing structure for CSS implementation.

Figure 14.27 corresponds to the synthesized sinogram output of the SSOC and CSS processes. In this case, motion artifacts due to breathing effects have been removed, as indicated by the clarity of the image near the area of the sternum. This was achieved with the CSS process by selecting segments of the sinogram corresponding to the same phases of the heart and breathing motion cycles. Since the period of the breathing motion is long (2–3 s) compared to that of the heart's periodic motion (0.5–1 s), another method to remove the breathing motion effects from the SSOC time series is by applying a band pass filter on the SSOC time series.

Another improvement of diagnostic importance, in this case, is the better and brighter definition compared to the conventional CT image (Figure 14.26) of the bright region

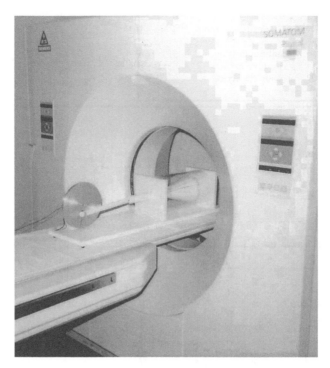

FIGURE 14.21
Experimental phantom in a CT scanner simulating an expanding ventricle.

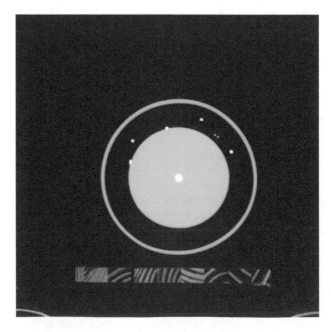

FIGURE 14.22
Conventional CT image of a stationary phantom. (Reprinted from IEEE © 2000. With permission.)

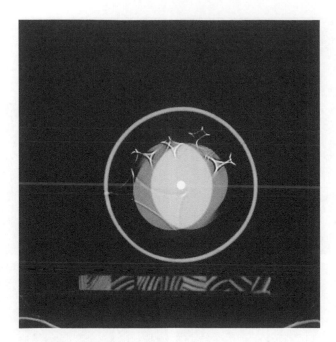

FIGURE 14.23
Conventional CT image of a moving phantom showing severe motion artifacts. (Reprinted from IEEE © 2000. With permission.)

corresponding to a coronary calcification in the top-right area of the heart. Overall quality of the image in Figure 14.27 is superior to the conventional CT image in Figure 14.26.

14.5 Conclusions

The problem of motion artifacts is well known in X-ray CT systems. Some types of motion are generally considered to be controllable for most patients, such as respiratory motion and patient restlessness, although there are many exceptions. Other forms of motion are not controllable, such as cardiac motion and blood flow. Patient motion during the data acquistion process results in an inconsistent set of projection data and a degradation of image quality due to motion artifacts. These artifacts appear in the form of streaking, distortions, and blurring. Anatomical regions that are moving appear distorted, while nearby stationary structures may be corrupted by overlying artifacts. These artifacts may result in inaccurate or misleading diagnoses.

Numerous techniques have been developed as partial solutions to this problem. They include (1) simple methods that restrict controllable forms of motion, (2) techniques that modify the data acquisition and reconstruction processes to minimize motion effects, and (3) more generalized techniques that perform data acquisition in a conventional fashion and reduce the effects of motion using retrospective signal processing.

The synthetic aperture signal processing approach has been treated in detail. It provides a measure of organ motion from an analysis of the projection data using a spatial overlap correlator. This information is then used to estimate the effects of this motion and to provide tomographic images with suppressed motion artifacts.

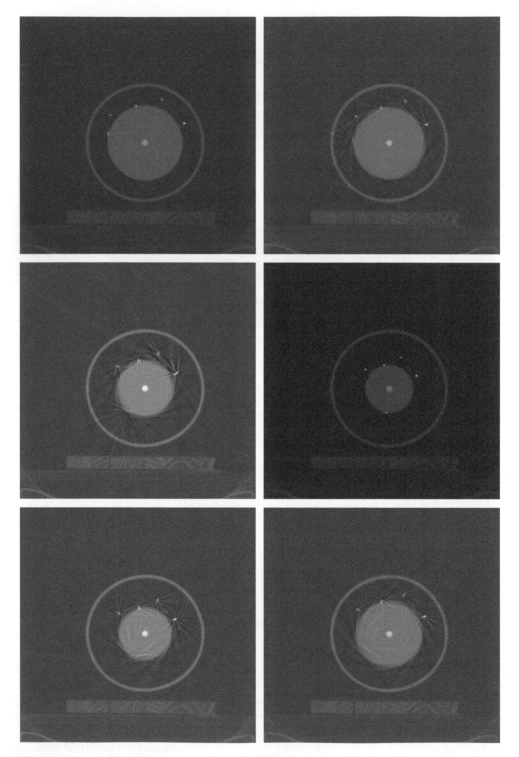

FIGURE 14.24
Resulting images of the CSS/SSOC method with a phantom motion cycle period of 0.6 and 9 s of total data acquisition. (Reprinted from IEEE © 2000. With permission.)

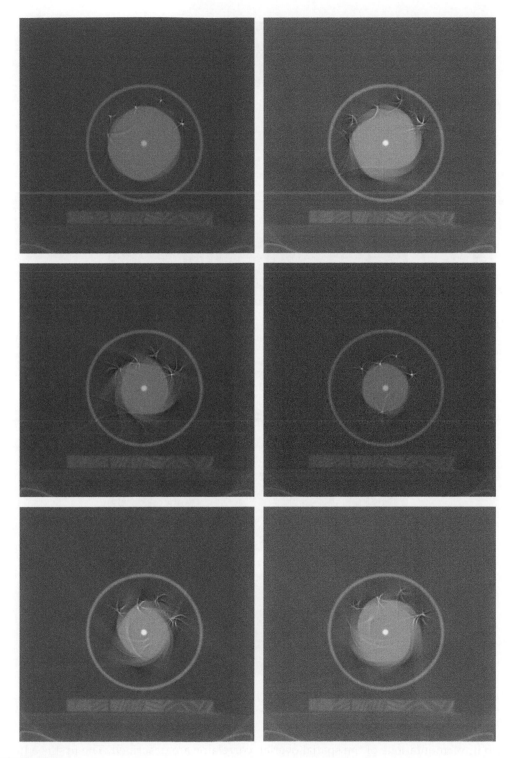

FIGURE 14.25
Resulting images from the CSS/SSOC method with a phantom motion cycle period of 0.6 and 3 s of total data acquisition.

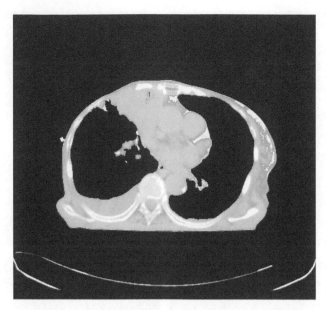

FIGURE 14.26
Conventional CT image including cardiac and respiratory motions. (Reprinted from IEEE © 2000. With permission.)

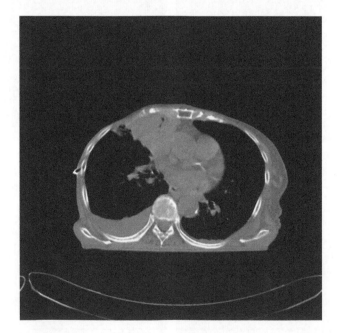

FIGURE 14.27
Corrected image from the SSOC/CSS motion correction scheme. (Reprinted from IEEE © 2000. With permission.)

Two implementations of the spatial overlap correlator are described. The first is a hardware implementation. While it provides the best estimate of organ motion, it generally requires a hardware modification to conventional CT systems. The second is a software implementation that provides an approximate measure of organ motion. It works very well with most physiologic motions and can be implemented on existing slip-ring CT systems.

Two motion-correction schemes are described. The first is an AIC that uses organ motion information from the HSOC as an "interference" signal. Motion artifacts are suppressed by removing this interference from the conventional CT sinogram. The second scheme is called a CSS method. It uses organ motion as determined by either the HSOC or SSOC to generate "phase-coherent" sinograms. These are estimated motion-free sinograms from which motion-free images are reconstructed for any specified moment in the motion cycle. Multiple images can be reconstructed to generate "cine-loop" movies of the patient, showing both cardiac and respiratory motions.

The effectiveness of the CSS method along with the spatial overlap correlator was demonstrated with a simulation study, experiments using a moving phantom, and a clinical patient study. The clinical images showed significantly reduced motion artifacts in the presence of both cardiac and respiratory motions.

References

1. Brooks, R.A. and Di-Chiro, G. 1976. Principles of computer assisted tomography (CAT) in radiographic and radioisotopic imaging. *Physics in Medicine & Biology*, 21, 689–732.
2. Edelheit, L.S., Herman, G.T., and Lakshminarayanan, A.V. 1977. Reconstruction of objects from diverging x rays, *Medical Physics*, 4, 226–31.
3. Herman, G.T. 1981. Advanced principles of reconstruction algorithms. In Newton, T.H. and Potts, D.G. (Eds.), *Radiology of the Skull and Brain*. C.V. Mosby, St. Louis, MO.
4. Macovski, A. 1981. Basic concepts of reconstruction algorithms. In Newton, T.H. and Potts, D.G. (Eds.), *Radiology of the Skull and Brain*. C.V. Mosby, St. Louis, MO.
5. Hutchins, W.W., Vogelzang, R.L., Fuld, I.L., and Foley, M.J. 1984. Utilization of temporary muscle paralysis to eliminate CT motion artifact in the critically ill patient. *Journal of Computer Assisted Tomography*, 8(1), 181–183.
6. Stockberger, S.M., Jr., Hicklin, J.A., Liang, Y., Wass, J.L., and Ambrosius, W.T. 1998. Spiral CT with ionic and nonionic contrast material: evaluation of patient motion and scan quality. *Radiology*, 206(3), 631–636.
7. Stockberger, S.M., Jr., Liang, Y., Hicklin, J.A., Wass, J.L., Ambrosius, W.T., and Kopecky, K.K. 1998. Objective measurement of motion in patients undergoing spiral CT examinations. *Radiology*, 206(3), 625–629.
8. Foley, W.D. 1998. Contrast-enhanced hepatic CT and involuntary motion: an objective assessment [editorial]. *Radiology*, 206(3), 589–591.
9. Luker, G.D., Bae, K.T., Siegel, M.J., Don, S., Brink, J.A., Wang, G., and Herman, T.E. 1996. Ghosting of pulmonary nodules with respiratory motion: comparison of helical and conventional CT using an in vitro pediatric model. *American Journal of Roentgenology*, 167(5), 1189–1193.
10. Mayo, J.R., Muller, N.L., and Henkelman, R.M. 1987. The double-fissure sign: a motion artifact on thinsection CT scans. *Radiology*, 165(2), 580–581.
11. Qanadli, S.D., El Hajjam, M., Mesurolle, B., Lavisse, L., Jourdan, O., Randoux, B., Chagnon, S., and Lacombe, P. 1999. Motion artifacts of the aorta simulating aortic dissection on spiral CT. *Journal of Computer Assisted Tomography*, 23(1), 1–6.
12. Loubeyre, P., Grozel, F., Carrillon, Y., Gaillard, C., Guyard, F., Pellet, O., and Minh, V.A. 1997. Prevalence of motion artifact simulating aortic dissection on spiral CT using a 180 degree linear interpolation algorithm for reconstruction of the images. *European Radiology*, 7(3), 320–322.
13. Duvernoy, O., Coulden, R., and Ytterberg, C. 1995. Aortic motion: a potential pitfall in CT imaging of dissection in the ascending aorta. *Journal of Computer Assisted Tomography*, 19(4), 569–572.

14. Mukherji, S.K., Varma, P., and Stark, P. 1992. Motion artifact simulating aortic dissection on CT [letter; comment]. *American Journal of Roentgenology*, 159(3), 674 .

15. Burns, M.A., Molina, P.L., Gutierrez, F.R., and Sagel, S.S. 1991. Motion artifact simulating aortic dissection on CT [see comments]. *American Journal of Roentgenology*, 157(3), 465–467.

16. Kaufman, R.B., Sheedy, P.F., Breen, J.F., Kelzenberg, J.R., Kruger, B.L., Schwartz, R.S., and Moll, P.P. 1994. Detection of heart calcification with electron beam CT: interobserver and intraobserver reliability for scoring quantification. *Radiology*, 190, 347–352.

17. Lipton, M.J. and Holt, W.W. 1989. Value of ultrafast CT scanning in cardiology. *British Medical Bulletin*, 45, 991–1010.

18. Hsieh, J. 1995. Image artifacts, causes, and correction. In Goldman, L.W. and Fowlkes, J.B. (Eds.), *Medical CT and Ultrasound: Current Technology and Applications*. Advanced Medical Publishing for the American Association of Physicists in Medicine, Madison, WI, 487–518.

19. Pelc, N.J. and Glover, G.H. 1986. Method for reducing image artifacts due to projection measurement inconsistencies. U.S. patent #4,580,219.

20. Ritman, E.L. 1980. Physical and technical considerations in the design of the DSR, and high temporal resolution volume scanner. *American Journal of Roentgenology*, 134, 369–374.

21. Ritman, E.L. 1990. Fast computed tomography for quantitative cardiac analysis—state of the art and future perspectives. *Mayo Clin Proceedings*, 65, 1336–1349.

22. Boyd, D.P. 1979. A proposed dynamic cardiac 3D densitometer for early detection and evaluation of heart disease. *IEEE Transactions in Nuclear Science*, 2724–2727.

23. Boyd, D.P. and Lipton, M.J. 1983. Cardiac computed tomography. *Proceedings of the IEEE*, 198–307.

24. Lipton, M.J., Brundage, B.H., Higgins, C.B., and Boyd, D.P. 1986. Clinical applications of dynamic computed tomography. *Progress Cardiovascular Disease*, 28(5), 349–366.

25. Nakanishi, T., Hamada, S., Takamiya, M., Naito, H., Imakita, S., Yamada, N., and Kimura, K. 1993. A pitfall in ultrafast CT scanning for the detection of left atrial thrombi. *Journal of Computer Assisted Tomography*, 17, 42–45.

26. Helgason, C.M., Chomka, E., Louie, E., Rich, S., Zajac, E., Roig, E., Wilbur, A., and Brundage, B.H. 1989. The potential role for ultrafast cardiac computed tomography in patients with stroke. *Stroke*, 20, 465–472.

27. Moore, S.C., Judy, P.F., Garnic, J.D., Kambic, G.X., Bonk, F., Cochran, G., Margosian, P., McCroskey, W., and Foote, F., Prospectively gated cardiac computed tomography, *Medical Physics*, 10, 846–855, 1983.

28. Kalender, W., Fichie, H., Bautz, W., and Skalej, M. 1991. Semiautomatic evaluation procedures for quantitative CT of the lung. *Journal of Computer Assisted Tomography*, 15, 248–255.

29. Crawford, C.R., Goodwin, J.D., and Pelc, N.J. 1989. Reduction of motion artifacts in computed tomography. *Proceedings of the IEEE Engineering in Medicine and Biological Society*, 11, 485–486.

30. Ritchie, C.J., Hsieh, J., Gard, M.F., Godwin, J.D., Kim, Y., and Crawford, C.R. 1994. Predictive respiratory gating: a new method to reduce motion artifacts on CT scans. *Radiology*, 190(3), 847–852.

31. Hsieh, J. 1994. Generalized adaptive median filters and their application in computed tomography. *Applications of Digital Image Processing XVII. Proceedings of the SPIE*, 662–669.

32. Wong, J.W., Sharpe, M.B., Jaffray, D.A., Kini, V.R., Robertson, J.M., Stromberg, J.S., and Martinez, A.A. 1999. The use of active breathing control (ABC) to reduce margin for breathing motion. *International Journal of Radiation Oncology, Biology, Physics*, 44(4), 911–919.

33. Sagel, S.S., Weiss, E.S., Gillard, R.G., Hounsfield, G.N., Jost, R.G.T., Stanley, R.J., and Ter-Pogossian, M.M. 1977. Gated computed tomography of the human heart. *Investigative Radiology*, 12, 563–566.

34. Morehouse, C.C., Brody, W.R., Guthaner, D.F., Breiman, R.S., and Harell, G.S. 1980. Gated cardiac computed tomography with a motion phantom. *Radiology*, 134(1), 213–217.

35. Nassi, M., Brody, W.R., Cipriano, P.R., and Macovski, A. 1981. A method for stop-action imagingof the heart using gated computed tomography. *IEEE Transactions in Biomedical Engineering*, 28, 116–122.

36. Cipriano, P.R., Nassi, M., and Brody, W.R. 1983. Clinically applicable gated cardiac computed tomography. *American Journal of Roentgenology*, 140, 604–606.

37. Joseph, P.M. and Whitley, J. 1983. Experimental simulation evaluation of ECG-gated heart scans with a small number of views. *Medical Physics*, 10, 444–449.

38. Johnson, G.A., Godwin, J.D., and Fram, E.K. 1982. Gated multiplanar cardiac computed tomography. *Radiology*, 145, 195–197.

39. Moore, S.C. and Judy, P.F. 1987. Cardiac computed tomography using redundant-ray prospective gating. *Medical Physics*, 14, 193–196.

40. Nolan, J.M. 1998. Feasibility of ECG gated cardiac computed tomography. MSc thesis, University of Western Ontario.

41. Alfidi, R.J., MacIntyre, W.J., and Haaga, J.R. 1976. The effects of biological motion on CT resolution. *American Journal of Roentgenology*, 127, 11–15.

42. Ritchie, C.J., Godwin, J.D., Crawford, C.R., Stanford, W., Anno, H., and Kim, Y. 1992. Minimum scan speeds for suppression of motion artifacts in CT. *Radiology*, 185(1), 37–42.

43. Posniak, H.V., Olson, M.C., and Demos, T.C. 1993. Aortic motion artifact simulating dissection on CT scans: elimination with reconstructive segmented images. *American Journal of Roentgenology*, 161(3), 557–558.

44. Srinivas, C. and Costa, M.H.M. 1994. Motion-compensated CT image reconstruction. *Proceedings of the IEEE Ultrasonics Symposium*, 1, 849–853.

45. Chiu, Y.H. and Yau, S.F. 1994. Tomographic reconstruction of time varying object from linear time-sequential sampled projections. *Proceedings of the IEEE Conference on Acoustic, Speech and Signal Processing*, 1, V307–V312.

46. Hedley, M., Yan, H., and Rosenfeld, D. 1991. Motion artifacts correction in MRI using generalized projections. *IEEE Transactions in Medical Imaging*, 10(1), 40–46.

47. Ritchie, C.J. 1996. Correction of computed tomography motion artifracts using pixel-specific backprojection. *IEEE Transactions in Medical Imaging*, 15(3), 333–342.

48. Stergiopoulos, S. 1990. Optimum bearing resolution for a moving towed array and extension of its physical aperture. *Journal of the Accoustical Society of America*, 87(5), 2128–2140.

49. Stergiopoulos, S. 1998. Implementation of adaptive and synthetic aperture processing in real-time sonar systems. *Proceedings of the IEEE*, 86(2), 358–396.

50. Dhanantwari, A.C., Stergiopoulos, S., and Iakovides, I. 2001. Correcting organ motion artifacts in X-ray CT medical imaging by adaptive processing (Part-I: Theory). *Medical Physics*, 28(8), 1562–1576.

51. Dhanantwari, A.C. 2000. Synthetic aperture and adaptive processing to track and correct for motion artifacts in X-ray CT imaging systems. Ph.D. thesis, University of Western Ontario.

52. Widrow, B. and Steams, S.D. 1985. *Adaptive Signal Processing*. Prentice-Hall, Engelwood Cliffs, NJ.

53. Stergiopoulos, S. 1995. Noise normalization technique for broadband towed array data. *Journal of the Accoustical Society of America*, 97(4), 2334–2345.

54. Widrow, B., Glover, J.R., McCool, J.M., Kaunitz, J., Williams, C.S., Hearn, R.H., Zeidler, J.R., Dong, E., Jr., and Goodlin, R.C. 1975. Adaptive noise cancelling: principles and applications. *Proceedings of the IEEE*, 63(12), 1692–1716.

55. Haykin, S. 1986. *Adaptive Filter Theory*. Prentice-Hall, Engelwood Cliffs, NJ.

56. Chong, E.K.P. and Zak, S.H. 1996. *An Introduction to Optimization*. John Wiley & Sons, New York, NY.

57. Slock, D.T.M. 1993. On the convergence behavior of the LMS and the normalized LMS algorithms. *IEEE Transactions on Signal Processing*, 41(9), 2811–2825.

58. Webb, S. 2000. *From the Watching of Shadows: The Origins of Radiological Tomography*. Adam Hilger, New York, NY.

15

Cardiac Motion Effects in MultiSlice Medical CT Imaging Applications

Stergios Stergiopoulos
Defence R&D Canada Toronto
University of Toronto
University of Western Ontario

Waheed A. Younis
Defence R&D Canada Toronto

CONTENTS

15.1 Introduction

Chapter 14 has introduced already single-slice x-ray CT systems as medical imaging diagnostic tools that provide cross-sectional images or "slices" of the human body. The patient is placed in between an x-ray source and an array of detectors that measure the amount of radiation that passes through the body. During the data acquisition process, the source and detectors rotate around the patient acquiring a large number of x-ray projections, which are used by the image reconstruction algorithms to produce cross section images of the patient's body. However, during the CT data acquisition process, the assumption of stationarity of the object being scanned (i.e., patient) is violated due to organ motion

[1–3] (e.g., cardiac motion, blood flow, respiratory motion or patient restlessness). Thus the image quality is adversely affected by these motion artifacts, which appear as blurring, doubling and/or streaking, which may lead to an erroneous diagnosis.

The most straightforward solution to avoid this kind of image distortion would be to speed up the data acquisition process so that motion effects become negligible. This requirement was partly fulfilled by the electron beam CT (EBCT) scanner, which is considered as a fifth generation CT system. There are no moving parts in this scanner and is capable of scanning one slice in 50 milliseconds. Though this speed is fast enough, these scanners are not cost effective and the major medical manufacturers have introduced the concept of multislice CT scanners having periods of rotation speeds lower than 250 milliseconds and with a number of slices ranging from 64 to 256. Thus, the multislice CT concept combined with helical data acquisition and fast rotation speeds attempts to provide volumetric 4D (3D dimension + 1D-time) images as a function of time for a patient's chest area around the heart.

In-spite of the above technological advancements, radiologists have noted [4] that 64-slice CTs can suffer from motion artifacts when patients reach 80–90 ppm (pulses per minute) heart rate pulses. These heart rates currently require premedication with beta blocker therapy to reduce heart rates to 60–70 beats per minute range. Not all patients can be safely placed under the beta blocking drug therapies due to asthma or congestive heart failure and therefore, may not be able to benefit from multislice cardiac CT studies. Thus, the purpose of the algorithmic development, detailed in this chapter, was to assess software solutions that can aid in obtaining diagnostic information from such patients where a multislice CT scan can be performed without beta blockade. Moreover, these software solutions have the potential to address the following problems:

- Detecting coronary calcification with motion artifact correction.
- Improving calcium scoring accuracy by correcting for the effects of motion prior to calculations of scoring.
- Providing 3D visualization functions with tools for planning XRT operations through the acquired volume of CT scans.
- Identifying additional diagnostic capabilities with the use of cardiac motion correction CT software in suppressing motion artifacts in various single and multislice CT scanners, such as four-slice, eight-slice 16-slice, 40-slice and 64-slice CT scanners.

Radiologists [4] have also suggested that it maybe possible to achieve motion compensation for cardiac motion induced artifacts without deploying ECG gating across a broad array of diagnostic CT exams, which currently do not employ cardiac gating. These include standard pulmonary embolism (PE) CT examinations of the chest and aortic dissection scans of the chest. It is also conceivable that these software solutions could aid in the detection of pulmonary nodules in proximity to the heart where cardiac motion might otherwise blur the margins of a small pulmonary nodule. In addition, the use of software motion correction algorithms has the potential to aid in accurate measurement of cardiac ejection fractions. The current standard of ten phase, multiphase reconstructions of gated CT acquisition data for coronary CTA introduces a 5–8% overestimation of ejection fraction. This is due to the limitations in temporal resolution introduced by the current gantry rotations and overall 20 millisecond barrier in temporal resolution. A software solution may be able to overcome this limitation by allowing for the most accurate reconstruction of the end-systolic and end-diastolic periods based on the cardiac motion compensation schemes whereby reconstruction endpoints will be derived at motion maxima and minima corresponding to the respective end-systolic and end-diastolic periods of the cardiac cycle.

Several techniques have been proposed as solutions to the single-slice CT motion correction problem. Some of these techniques assume a linear model for the motion [5] while some others model the motion as periodic and take projection at a particular point in the motion cycle to produce the effect of stroboscope [6]. However, cardiac motion is much more complex and these techniques prove to have limited success. A more general technique attempts to iteratively suppress the motion effects from the projection data [7]. Besides some other shortcomings, this algorithm requires a proper initialization without which the convergence period could be long. Another technique based on retrospective gating [8], employed ECG signals to identify diastole phases of cardiac cycles and use the projection data collected during that phase to reconstruct the tomographic images.

The basic concept for the two new signal processing techniques, presented in this chapter, is to provide tracking of the phases of cardiac motion without ECG gating. This phase information is later used to remove the cardiac motion artifacts and reconstruct images with better quality and diagnostic value. These techniques can be implemented on data sets acquired from single and multislice helical CT scanners. In the case of helical scans, some additional data processing (i.e., interpolation) will be required to produce planner measurements between the helical projections. Two algorithms with corresponding advantages and limitations will be described for this interpolation. The major advantage of multislice helical CT scans is the acquisition of continuum of data sets, which can be used to reconstruct 3D volumes without additional x-ray exposure [9,10]. Using the cardiac phase tracking techniques mentioned earlier, the 3D volume can be reconstructed without motion artifacts from the multislice helical CT scan data.

It is reasonable to assume that during a single cardiac cycle, the speed of heart motion is not uniform [11,12]. In other words, during a fraction of the cardiac cycle (i.e., diastole phase), the heart is less active as compared to the rest of the cycle. This less active phase lies between "T" and "P" waves of an ECG signal [11]. The signal-processing techniques presented in this chapter can be used to identify the projection data that corresponds to the "less active" phase of the cardiac cycle (i.e., diastole phase). Once the projection data sets are correctly identified, they can be fed into the tomographic image reconstruction algorithms to produce images that will be free of cardiac motion artifacts.

The two proposed signal-processing techniques to identify the less active phase of the cardiac cycle are: (1) the *spatial overlap correlator* (SOC) *with unwrapping filter* and (2) a technique based on *constant attenuation property of the Radon transform*. The technique of the SOC has been derived from a synthetic aperture concept for sonar system applications [13], [14] and its implementation in single slice CT imaging applications as a two x-ray source concept, has been discussed in detail in Chapter 14 and by Dhanantwari et al. [2,3]. The main idea of this technique is to compare/subtract the projection data recorded at two different time moments but from the same source-detector physical location. In case of no motion, the results of this subtraction process would be zero whereas a nonzero value indicates the presence of motion. This is the theoretical foundation in identifying projections associated with the cardiac diastole phase. During the process of computing the SOC signals, the information about time reference is lost; which is recovered by using a processing step, called *unwrapping filter* [15]. The output of this unwrapping filter is then used to define the relaxed and more Radon transform, which states that for a given object the integral of projection data is constant for all projection angles. If the object starts deforming during a CT data acquisition process (e.g., due to organ motion), this integral will not remain constant over time. Using this property, the phases of periodic or aperiodic (e.g., arrhythmia) cardiac motion can be identified.

15.2 Theoretical Remarks

This section presents three introductory topics, which are related to the correction of motion artifacts in CT images. The first section provides a brief description of the CT image reconstruction process from projection data sets. Next, we detail the techniques that have been developed under this investigation for the tracking of cardiac motion phases; and finally we present a review of the interpolation algorithms for helical CT data.

15.2.1 CT Image Reconstruction from Projection Data

As detailed in Chapter 14, when x-ray beams pass through an object, which is modeled as two dimensional distribution of the x-ray attenuation constant $f(x,y)$, the total attenuation suffered by the x-ray beams is a measure of the line integral of attenuation constant along the straight line (see Figure 15.1) which is given by $x \cos \theta + y \sin \theta = t$, whereas the attenuation is defined by

$$P(\theta,t) = \int\limits_{-\infty}^{\infty} \int\limits_{-\infty}^{\infty} f(x,y)\delta(x\cos\theta + y\sin\theta - t)dxdy \qquad (15.1)$$

Here $\delta(\cdot)$ is the Dirac delta function and $P(\theta,t)$ is known as the *Radon transform* of the function $f(x,y)$ [16]. A projection is formed by combing a set of these line integrals. The simplest projection is a collection of parallel ray integrals as is given by $P(\theta,t)$ for a constant θ and varying t and is called parallel projection. In another type of projection known as fan beam projection, the line integrals are measured along fan lines as shown in Figure 15.2, $x \cos(\gamma+\sigma) + y \sin(\gamma+\sigma) = D \sin\sigma$. For third generation CT scanners utilizing fan beam projection, the data acquisition process takes place through discrete detectors, as depicted in Figure 15.2. The projection measurements, $g_m(\sigma_m,\beta_i)$, $(m=1, ..., M)$, $(i=1, ..., N)$, shown schematically in this figure, are defined as the line integrals along lines passing through the object $f(x,y)$. For a given detector m, projection i and projection angle β_i, $g_m(\sigma_m,\beta_i)$ is defined by Equation 15.2. In Equation 15.2, σ_m represent the angle that detector m makes with the center of the M-detector array. The line traced by the ray from the source to the detector is given in Equation 15.3, and the integration along this path is reflected in Equation 15.2.

$$g_m(\sigma_m,\beta_i) = \iint f(x,y)\delta[x\cos(\beta_i + \sigma_m) + y\sin(\beta_i + \sigma_m) - R\sin(\sigma_m)]dxdy \qquad (15.2)$$

$$x \cos(\beta_i + \sigma_m) + y \sin(\beta_i + \sigma_m) = R \sin(\sigma_m) \qquad (15.3)$$

$f(x,y)$ is also the function we wish to approximate through the image reconstruction process. The angular step increment between two successive projections of the x-ray scanner

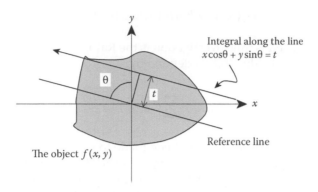

FIGURE 15.1
Line integral in parallel beam projection.

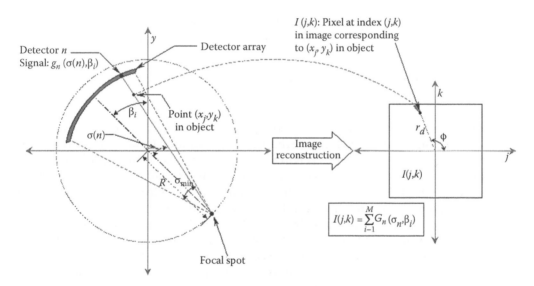

FIGURE 15.2
Schematic diagram of projection function for CT X-ray imaging systems. The projection measurements are denoted by $g_n(\sigma_n, \beta_i)$, $(n=1, \ldots, N)$, $(i=1, \ldots, M)$, with the angular step increment between two successive projections of the x-ray scanner defined by: $\Delta\beta=2\pi/M$, and Δt is the time elapsed between these successive projections, where M is the number of projections taken during the period $T=M\Delta t$ that is required for one full rotation of the source and receiving N-detector array around the object $f(x,y)$ that is being imaged. (Reprinted by permission of IEEE © 2000.)

is defined by: $\Delta\beta=2\pi/N$, and Δt is the time elapsed between these successive projections, where N is the number of projections taken during the period $T=N\Delta t$ that is required for one full rotation of the source and receiving array around the object $f(x,y)$ that is being imaged.

In the image reconstruction process the pixel $I(j,k)$ in the actual image, shown at the right of Figure 15.2, corresponds to the Cartesian point (x_j, y_k) in the CT scan plane. Given a filter with frequency response $F(m)$, $(m=1, \ldots, M)$, the filtered projection data are defined by Equation 15.4.

$$G_m(\sigma_m, \beta_i) = Ifft[fft(g_m(\sigma_m, \beta_i) \cdot F(m))] \tag{15.4}$$

The filtering function used is this investigation is the Ram–Lak filter [16] cascaded with the Parzen window. The pixel value $I(j,k)$ is defined by:

$$I(j,k) = \sum_{i=1}^{N} G_m(\sigma_m, \beta_i), \tag{15.5}$$

where the appropriate modifications have been made to $G_m(\sigma_m, \beta_i)$ to account for geometric effects. For each β_i in Equation 15.4, a detector, m, is selected. The angle σ_m, which defines the detector, m, that samples the projection through a point (x_j, y_k), for a given projection angle β_i, is provided by Equation 15.6, where (r_d, φ) is the polar representation of (x_j, y_k).

$$\sigma_m = \tan^{-1}\left[\frac{r_d \sin(\beta_i - \phi)}{R + r_d \cos(\beta_i - \phi)}\right], \quad \frac{-\pi}{2} \le (\beta_i - \phi) < \frac{\pi}{2}$$

$$\sigma_m = \tan^{-1}\left[\frac{r_d \sin(\phi - \beta_i)}{R + r_d \cos(\phi - \beta_i)}\right], \quad \frac{\pi}{2} \le (\beta_i - \phi) < \frac{3\pi}{2}, \tag{15.6}$$

15.2.2 Tracking the Phases of Cardiac Motion

15.2.2.1 SOC and Unwrapping Filter

It has been shown elsewhere [13,14] that the concept of the SOC can be used in synthetic aperture SONAR applications to increase the angular resolution (array gain) and to reduce the artifacts that are caused by the receiving array's motion effects. Implementation of the concept of the SOC to track organ motion in x-ray CT scanners has been detailed elsewhere [2,3] and in Chapter 14. The present investigation has expanded this concept and included an unwrapping filter to accurately detect the less-active phases of cardiac motion.

The SOC process can be understood by a simple example. Let us take two photographs at time t_0 and $t_0 + \Delta t$ of a field of view that includes a moving train and other stationary objects such as trees and houses. If the images from these two photographs are subtracted at the pixel level, the resulting image will include information, only about the motion of the train during the time interval, Δt. In a similar fashion, comparison of CT measurements taken at two different times (t_0 and $t_0 + \Delta t$), but for the same physical location of the object being scanned, can detect the presence of motion during Δt [2].

Let us consider a fixed detector m_0 in the detector array of a CT scanner; and assume that this detector makes p rotations around a given stationary object and collects projection data. Let us call these project data, $x[n]$, where $t_n = n\Delta t$, and n denotes time index. As expected, $x[n]$, will be a periodic signal with period N which is the total number of projections during a full rotation around the object; and $x[n]$ will be zero beyond the interval $0 \le n < pN$. Let us define now a *circularly shifted* version, $x_s[n]$ of $x[n]$ as follows:

$$x_s[n] = \begin{cases} x[n+N] & \text{for} & 0 \le n < (p-1)N \\ x[n-(p-1)N] & \text{for} & (p-1)N \le n < pN \end{cases} \tag{15.7}$$

In other words, we shifted $x[n]$ to the left by one period and wrapped around the first period at the end on right. This is depicted in Figure 15.3. But since $x[n]$ is periodic with

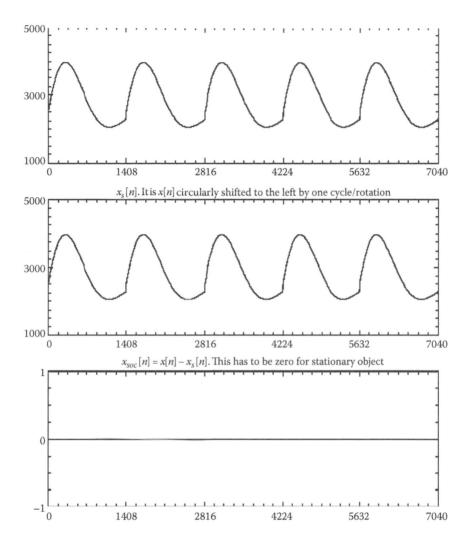

FIGURE 15.3
Simulated projections for a stationary object. Number of rotations $p=5$, number of projections per rotation $M=1408$.

period N, the difference between $x[n]$ and $x_s[n]$ can not be detected. It should be noted also that $x_s[n]$ will be zero beyond the interval $0 \le n < pN$. We define SOC of $x[n]$ as

$$x_{soc}[n] = x[n] - x_s[n] \tag{15.8}$$

For a given stationary object, the signal in Equation 15.8 will be zero for all n; and this can be interpreted as follow: for a given projection $n=n_0$, the two signal values $x[n_0]$ and $x_s[n_0]$ represent the line integral of the given object through the same location of line but at different times. This is because the x-ray source and detector m_0 occupy the same physical positions at both of these time instances. If the object is stationary, both measurements have to be identical.

Now consider the case when a weaker signal $y[n]$ is superimposed on $x[n]$, as depicted in Figure 15.4. The signal $y[n]$ represents an arbitrary motion, such as cardiac motion,

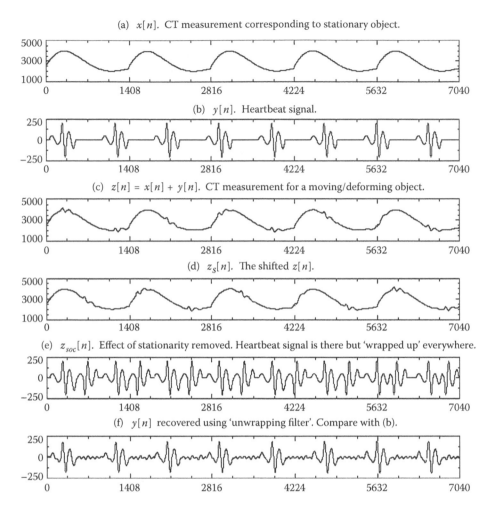

FIGURE 15.4
Demonstration of the SOC and unwrapping filter technique to detect phases of cardiac motion. (a) CT signal observed by a detector, scanning a patient with no organ motion. It is periodic with a period equivalent to a full rotation (1408 projections). (b) Heartbeat signal; 95 beats per minute (bpm). (c) Signal observed by the detector, scanning the patient with heartbeat. (d) and (e) Signal processing steps to remove the detector measurement associated with stationary organs. (f) Heartbeat signal recovered after applying unwrapping filter. Comparison with (b) reveals that heartbeat motion phases have been correctly identified.

with multiple spectra periods being different than the period N of $x[n]$. Let us call this new signal,

$$z[n] = x[n] + y[n] \tag{15.9}$$

Our immediate goal is to recover $y[n]$ from $z[n]$. If we apply the concept of the SOC on $z[n]$, as defined by Equation 15.8, we will get

$$z_{soc}[n] = z[n] - z_s[n] = x[n] + y[n] - x_s[n] - y_s[n]$$

$$= \underbrace{x[n] - x_s[n]}_{= 0} + y[n] - y_s[n] = y[n] - y_s[n]. \tag{15.10}$$

The term $(y[n]-y_s[n])$ is not zero since the period of the superimposed signal $y[n]$ is not N. It is obvious that by taking the SOC of $z[n]$, we can get rid of the periodic part of the $z[n]$ (i.e., $x[n]$) and we are left with $y[n]-y_s[n]$, as depicted in Figure 15.4e.

Next we would like to recover $y[n]$ from $y[n]-y_s[n]$. Let us consider the pN point Fast Fourier Transforms (FFTs) of $y[n]$, $y_s[n]$ and $y[n]-y_s[n]$, as defined below,

$$y[n]\xleftrightarrow{FFT} Y[k]=\sum_{n=0}^{pN-1} x[n]^{-j(\frac{2\pi}{pN})kn} \quad \text{for} \quad k=0,1,\ldots,pN-1$$

$$y_s[n]\xleftrightarrow{\ FFT\ } Y[k]e^{j(\frac{2\pi}{pN})kN}$$

$$z_{soc}[n]=y[n]-y_s[n]\xleftrightarrow{FFT} Z_{soc}[k]=Y[k]\left(1-e^{-j(\frac{2\pi}{p})k}\right).$$

Thus, by using the above expressions, $y[n]$ can be recovered from $z_{soc}[n]=y[n]-y_s[n]$, as follows,

$$y[n]\xleftrightarrow{FFT} Y[k]=\frac{Z_{soc}[k]}{\left(1-e^{-j(\frac{2\pi}{p})k}\right)}=\frac{Z_{soc}[k]}{\left(1-\cos(\frac{2\pi k}{p})\right)+j\sin(\frac{2\pi k}{p})} \tag{15.11}$$

Figure 15.4f shows schematically the recovery of $y[n]$ from $z[n]=x[n]+y[n]$. It should be noted also that some caution needs to be observed in implementing Equation 15.11 as the denominator becomes zero whenever k/p is an integer (which will happen N times). In order to avoid the "divide-by-zero" cases in Equation 15.11, we set the denominator to be $\varepsilon+j\sin(\frac{2\pi k}{p})$ whenever $1-\cos(\frac{2\pi k}{p})\le\varepsilon$.

It is of interest to note that the denominator of Equation 15.11, (i.e., unwrapping filter) traverses a circle of unit radius, as shown in Figure 15.5, at discrete points, exactly N times. Moreover, if the amount of shift to produce $z_s[n]$ is one half the total length of $z[n]$, in other words if $p=2$ in current setting, the denominator of Equation 15.11 will reduce to $1-(-1)^k$, (i.e., 0, 2, 0, 2, 0, … for k=0, 1, 2, 3, 4, … respectively). This corresponds to the two points shown by "×" in Figure 15.5 and simply scales down the FFT of $z_{soc}[n]$. As a result, $y[n]$ cannot be recovered from $y[n]-y_s[n]$.

It should be noted also that though $z_s[n]$ was created by shifting $z[n]$, by one rotation (i.e., N projections), it could have been shifted by half rotation (i.e., $N/2$) for the case of the middle detector. This is due to the fact that the middle detector measures the line integral along the same line after every half rotation. In other words, the signal received by the middle detector is periodic with period $N/2$. For all the other detectors, we can exploit the symmetry property of transform [16,17] and then we can estimate the SOC according to Equations 15.10 and 15.11 for half rotation of the SOC. This assumes that we use parallel beam sinogram data. Therefore, implementation of this SOC concept fan beam data, requires their conversion to an equivalent parallel beam data set.

We call the processing of Equation 15.11 "unwrapping filter", since its effect is to restore the temporal contents of $y[n]$, which were lost (or wrapped up) in $y[n]-y_s[n]$. Although the temporal contents are restored, the constant amplitude component of $y[n]$ is lost because $y[n]-y_s[n]$ was a zero mean signal. Note also that we need to take exactly a pN point FFT in order to account for the shifting and wrapping effects as defined by $y_s[n]$ in Equation 15.7.

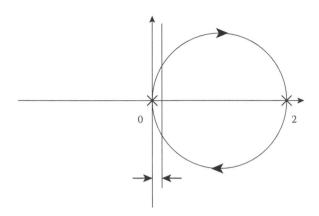

FIGURE 15.5
Denominator of Equation 15.11, (unwrapping filter) traverses a circle of unit radius.

Let us assume now that the weak signal $y[n]$ in Equation 15.9 is a component of the projection signal $z[n]$ caused by the cardiac motion, superimposed on the regular projection signal due to stationary organ $x[n]$. Then the SOC processing Equation 15.11 can be used in this case to remove the signal component caused by cardiac motion imposed on the stationary component $x[n]$. Figure 15.4 depicts all these processing steps in recovering $y[n]$ (i.e., Figure 15.4f).

In the implementation concept of SOC as defined by Dhanantwari et al. [2], the information about time reference is lost, since $z_{soc}[n]$ in Equation 15.10 contains information about the signal component associated with the organ motion only (i.e., see Figure 15.4e); and we have to use the unwrapping filter to recover it. Thus, the output of the unwrapping filter and the recovery of the reference time are the critical processing steps to identify the phases of the cardiac motion cycles.

In summary, the unwrapping filter should be considered as a phase shift $j[e^{-j\pi f_k \Delta t}]$ and amplification $1/[2 \sin (\pi f_k \Delta t)]$ to the SOC measurements in the frequency domain in order provide tracking of the phases of the organ motion. Since the SOC is a sampling process [2], the upper limit of the organ motion frequency is dependent of the time interval Δt between two successive projections in order to satisfy the Nyquist sampling rate. The upper limit of the product $(f_k, \times \Delta t)$ should be 0.5 to ensure that there is no aliasing in the SOC measurements. When the product $(f_k, \times \Delta t)$ reaches 1.0, or is any integer, it corresponds to the case where the frequency of the motion corresponds exactly to the periodicity of the sampling of the motion. In this case, the SOC does not track the motion, and the system appears stationary, which is an indication that the Nyquist sampling criterion has been violated. In cardiac x-ray CT imaging applications, this occurs when the period of the subject's cardiac cycle coincides exactly with the rotational period of the x-ray CT scanner.

At this point, it is important to note that, since the frequency spectrum of the heart motion has broadband characteristics, then the selected frequency bins of the above phase correction process of the unwrapping filter will select a lower frequency spectrum of the heart motion to satisfy the Nyquist criterion. Moreover, this lower frequency spectrum will be sufficient to represent the motion characteristics of the original heart motion spectrum. This argument has been confirmed by the controlled experiments with moving phantoms that are discussed by Dhanantwari et al. [3].

15.2.2.2 Constant Attenuation Property of the Radon Transform

The various phases of a cardiac cycle can also be identified by using a specific property of transform [16,17].

Theorem:

Let $P(\theta,t)$ be transform of an object $f(x,y)$, which can be assumed without any loss of generality to be contained within a unit circle. The integral $\int_{-1}^{1} P(\theta,t)t^k dt$ is a homogeneous polynomial of degree k in $\cos\theta$ and $\sin\theta$, where $k=0, 1, \dots$. That is

$$\int_{-1}^{1} P(\theta,t)t^k dt = \sum_{j=0}^{k} \alpha_j \cos^{k-j}\theta \sin^j\theta \qquad \forall k \geq 0 \ (k \text{ integer}). \tag{15.12}$$

If $k=1$, then,

$$\underbrace{\int_{-1}^{1} P(\theta,t)t\,dt}_{\text{Center of mass}} = \alpha_0 \cos\theta + \alpha_1 \sin\theta = \sqrt{\alpha_0^2 + \alpha_1^2}\,\sin\left(\theta + \tan^{-1}(\alpha_0/\alpha_1)\right). \tag{15.13}$$

Equation 15.13 shows that the center of mass varies sinusoidely with θ. Also if $f(x,y)$ consists of a single point, it will be mapped as a sinusoid in $P(\theta,t)$; and this is why the function $P(\theta,t)$ is called a "sinogram" in CT medical imaging applications.

In case $k=0$, we have,

$$\int_{-1}^{1} P(\theta,t)dt = \alpha_0 \quad \text{(i.e. constant)} \qquad \forall\theta \tag{15.14}$$

The constant on the right side of Equation 15.14 is also called the "mass of the object" and it is a measure of the total x-ray attenuation constant of the object $f(x,y)$. Intuitively, the term mass of a given object (or the total x-ray attenuation constant) should be constant no matter which projection angle it is looked upon.

The next step is to exploit the above property in Equation 15.14 to track the cardiac phases. In particular, Equation 15.14 states that for a stationary object the integral of the Radon transform with respect to t is constant for all θ. Then, the associated data acquisition process for parallel beam projection results into a 2-D array of projection data, called sinogram matrix. The dimensions of this matrix are pN (total numbers of projections) by M (number of detectors in the array). According to Equation 15.14, the sum of this matrix along M will result into a constant vector of length pN, provided that the object was constant/stationary. If the object starts deforming at some projections, then the total x-ray attenuation constant will have changes and as a result this vector will not be a contact at the above projections. Thus, if this vector starts deviating from its constant value at some projection n_0, it will be an indication of the presence of motion during the CT data acquisition process, as depicted in Figure 15.6.

At this point, it is important to note that this theorem (Equation 15.14) is valid for parallel beam projections only and not for fan beam projection data. In order to apply this technique to track the phases of a cardiac motion on fan beam data sets, it is required that the

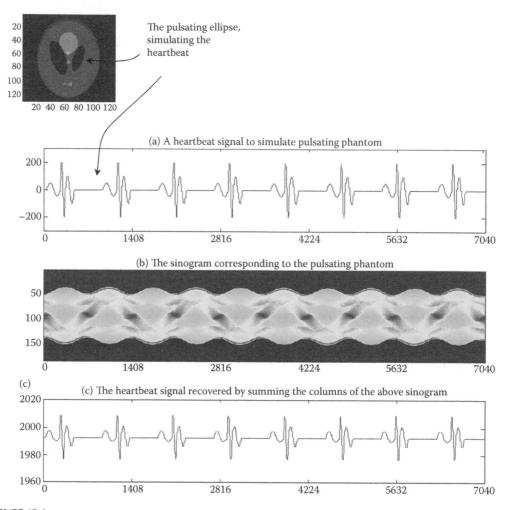

FIGURE 15.6

Demonstration of the cardiac phase tracking using the constant attenuation property of the Radon transform. (a) Variations in the size of pulsating ellipse to simulate heartbeat. (b) Parallel projection sonogram of pulsating phantom. (c) Heartbeat signal recovered by using Equation 15.14.

fan beam data should be converted into equivalent parallel beam data sets. This process is known as "rebinning," as depicted schematically in Figure 15.7 and it takes place according to the following relationship [16]:

$$t = D \sin\sigma \quad \text{and} \quad \theta = \gamma + \sigma, \tag{15.15}$$

where the parameters in these relationships are defined in Section 15.2.2 and in Figures 15.1 and 15.2. It is apparent also from the simulations in Figure 15.6 that as a result of the rebinning process, the data corresponding to a single fan beam projection, spread out over some range (equivalent to fan angle) in the corresponding parallel projection [16]. This kind of spread may cause a problem in a sense that the effect of a motion event starting at some instant t_0, when converted from fan to parallel projection data set is spread out over some range and thus the starting instant t_0 may not be detected precisely.

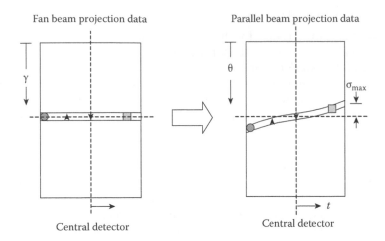

FIGURE 15.7
"Rebinning" conversion of fan-beam-data into parallel-beam-data results in re-arrangement at different location.

Before closing this section, it is worth comparing the two techniques mentioned above. First of all, it is clear from the above analysis that the technique using the constant attenuation property of the Radon transform seems simpler and straight forward as compared to the SOC process with unwrapping filter. It simply involves the rebinning from fan beam data to parallel beam data sets and then summing them along one dimension. Note that the unwrapping filter is very sensitive to noise levels due to its *nearly divide-by-zero* property defined by Equation 15.11. Moreover, simulations have shown that a small amount of noise in $z_{soc}[n]$, may completely conceal the desired signal. Furthermore, if the heartbeat rate coincides with the frequency of rotation, it cannot be detected using the SOC process, as discussed in the previous section. On the other hand, if the total x-ray attenuation of the scanned object does not change significantly due to the cardiac motion, it will be difficult to detect the motion phase using the constant attenuation property of the Radon transform.

15.2.2.3 Interpolation Algorithms

This section provides a brief review for two interpolation algorithms that can be implemented in multislice helical CT scanners. Both algorithms have their own advantages as well as limitations. To simplify the discussion on the interpolation algorithms, only single slice helical CT scanner measurements will be considered. However, both these two algorithms have been applied successfully in synthetic multislice helical CT data sets that are presented in Section 15.5.

In helical CT scanners there is a translation motion of the patient through the gantry while the x-ray source rotates around the patient such that continuous helical data are acquired throughout the volume of interest. The x-ray source and detectors trace a helical path around the patient and detect raw projection data from which planar images must be generated. Each rotation of the x-ray tube can be considered as of generating data specific to an angled plane of a body section [18]. To achieve a true trans-axial image, data points above and below the desired plane of a given cross-section must be interpolated to estimate the data value in the trans-axial plane. As a benefit, the interval between reconstructed trans-axial images can be chosen arbitrarily and retrospectively [19].

15.2.2.3.1 360° Interpolation Algorithm

Figure 15.8 shows a scan diagram for the 360° interpolation algorithm. The slanted lines represent the helical path taken by the x-ray source around the patient's body as it slides through the gantry. It is shown that for any given reconstructed plane, which is normal to the direction of patient's motion (i.e., Δz), there is only one direct measurement available; the rest of the data is generated by performing interpolation between the measurements that are 360° (or Δz, the pitch size of the CT scanner) apart. These interpolated measurements are recorded from the same projection angle but different axial positions, around the given reconstructed plane.

In order to generate the planner data set for one 360° rotation, the data measurements from two full helical rotations (or 720°) of the detector array around the patient's body are required. In the case of parallel beam x-ray projection measurements, the planner data for only half rotation (180°) of the detector array around the patient's body is sufficient for image reconstruction. However, in order to generate the half rotation data, helical data collected from two rotations (or 720°) of the detector array are required, though not all of this data will be utilized. Whereas in the case of fan beam projections, planner data measurements obtained from 180°+*fan* -angle are needed for image reconstruction. Again, measurements obtained from two 360° rotations (or 720°) of the detector array are needed to generate this amount of planner data.

15.2.2.3.2 180° Interpolation Algorithm

In the 360° interpolation algorithm, the interpolated measurements span a range of 720°. The interpolation errors are not severe if the pitch size of the helical path is relatively small

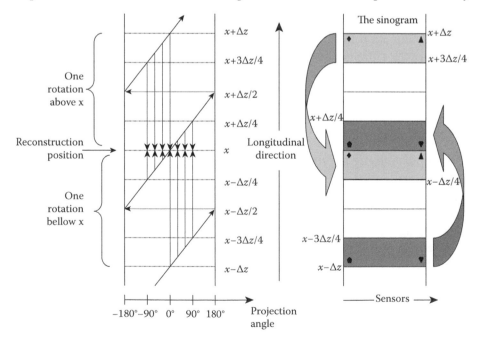

FIGURE 15.8
Scan diagram from helical measurements using 360° interpolation. Parallel beam projection data sets are assumed. It is obvious that two rotation data (one rotation on each side of the reconstructed plane) is needed. Only half rotation data is produced. The figure on the right hand side shows the pairs of sinograms that are required to be used for interpolation.

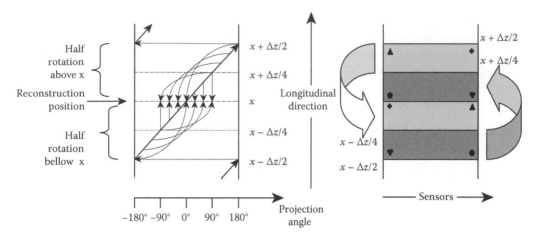

FIGURE 15.9
Scan diagram from helical measurement for 180° interpolation. A parallel beam projection data is assumed. It is obvious that one rotation data (e.g., half rotation on each side of the reconstructed plane) is needed. Only half rotation data is produced. The figure on the right hand side shows the pairs of sinograms that are required for interpolation. One set of interpolated measurements have to be "flipped" (e.g., symmetry property of the Radon transform) before interpolation.

(i.e., the spiral is "tightly" wound). However, for larger pitch size, the interpolation errors become more prominent in the reconstructed images. These errors can be reduced when the 180° interpolation algorithm is used (i.e., Figure 15.9). In this technique, the interpolated measurements are only 180° apart, which is applicable only on parallel beam projection measurements whereas for the fan beam projection, this is slightly different except for the central sensor. According to the symmetry property of the Radon transform, projection measurements that are 180° apart give the line integral along the same physical line through the object and hence they should be the same. In the case of helical scanning, measurements that are 180° apart have less longitudinal separation as compared to the measurements that are 360° apart and hence introduce less interpolation artifacts. It is of interest to note that in order to generate a data set either from 360° rotation data sets, or 180° (e.g., for image reconstruction from parallel projections), or from a 180°+*fan* beam rotation (e.g., for image reconstruction from fan beam projections), only one full (360°) rotation of helical interpolated data is needed.

15.3 System Implementation

15.3.1 Single-slice CT Scanner System Applications

An assessment of the effectiveness of the two proposed organ motion tracking algorithms on single-slice CT scanners was carried out with simulations using the Shepp–Logan phantom [20] and real projection data sets from a middle-aged female using CT medical imaging system.

The 2D Shepp–Logan phantom, shown in the upper left corner of Figure 15.6, consists of ten ellipses and one of them was made to pulsate to simulate cardiac motion. Projection data for five rotations of a CT scanner were simulated with 1056 projections per rotation. The pulsating ellipse was simulated to have period equivalent to 924 projections. The duty cycle of this pulsating movement was set to 43%, or the simulated organ motion included a 57% segment of the total pulsating period to be stationary, emulating the diastole phase of a heart.

15.3.1.1 Simulations with 2D Shepp–Logan Phantom

Implementation of the SOC process and the unwrapping filter on the Shepp–Logan synthetic projection data is depicted in Figure 15.10. More specifically, Figure 15.10a, shows the sum of all the detector projections, Figure 15.10b the output of the SOC process and Figure 15.10c the output of the unwrapping filter. Figure 15.10d shows the actual phases of the simulated organ motion, which suggests that the above processing in Figure 15.10c has successfully recovered the motion phases.

Figure 15.11 shows the image reconstruction of the Shepp–Logan phantom from selected sections of projection data that correspond to phasés with the lowest amplitude of organ motion, as defined in Figure 15.10c and d. Figure 15.11a, presents the reconstructed image from a segment of projection data that correspond to the simulated diastole phase of the cardiac cycle. For comparison, Figure 15.11b shows the reconstructed image from the simulated projection data that include organ motion.

An assessment of the effectiveness of the constant attenuation property of the Radon transform to recover the organ motion phases was carried out with the same simulated data sets discussed in the Figure 15.11. The first step was to convert the fan beam projection data into parallel beam projections and then sum up along the detectors. This processing step is shown in Figure 15.12. From Figure 15.12b and c, it is clear that the phases of the organ motion have been recovered successfully. Image reconstruction from the above projections revealed identical image results as those shown in Figure 15.11. The signal distortion at the two ends of Figure 15.12b is due to the rebinning process, which creates unreliable data points in the parallel projection matrix (see Figure 15.7). The length of this unreliable signal segment is half the fan angle on each end.

15.3.1.2 Results with Human Patients

The projection data from two human subjects were obtained with a Siemens Somatom Plus 4 x-ray CT scanner. These are the same data sets that have been used for the assessment of the motion correction algorithms discussed in Chapter 14 and by Dhanantwari et al. [2,3]. During the data acquisition process, the CT scanner was positioned to obtain a patient's heart ventricles. The first patient was asked to follow breath-holding procedures. The rotation period of the CT was set to 1.0 second (1408 projections). Projection data sets were recorded for 20 rotations. For this test, the patient's ECG signal was also recorded synchronously with the CT projection data. The heartbeat was measured to be 75 beats per minute (bpm). Figure 15.13 shows all signal processing steps to identify the phases of the organ motion using the two techniques of this investigation.

Figure 15.13a presents the sum of rows of parallel beam sinograms, which have been converted from regular fan beam projection data sets by using the rebinning process. Since the parallel beam projections are available now, implementation of the SOC process can be carried out with a delay of half rotation only, instead of f ull rotation as mentioned in Section 15.2.2.1. This is of importance, as lesser amount of projection data is needed resulting in lesser radiation levels for the patient. The next, Figure 15.13c shows the half

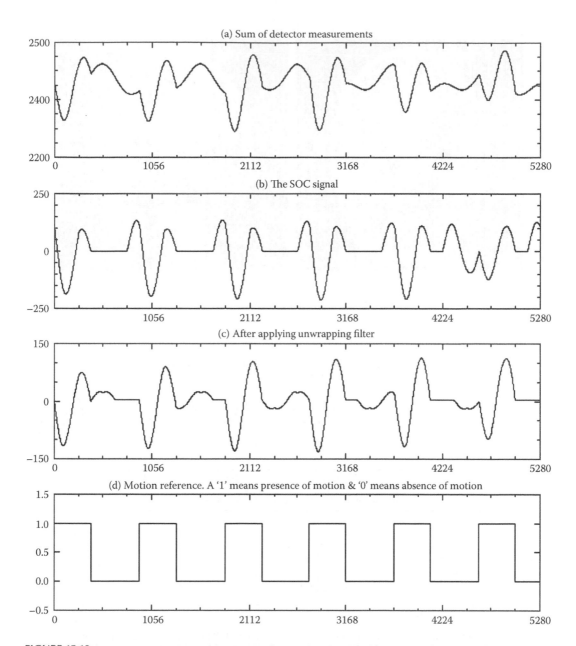

FIGURE 15.10
Implementation of the SOC and unwrapping filter techniques on the sum of all detector measurements demonstrating their effectiveness in identifying the cardiac motion phases on the sum of all detector measurements. Note the high amplitude in (c), which is in agreement with the motion amplitude indicated in (d).

rotation SOC process for the signal in Figure 15.13a. The results of the unwrapping filter are depicted in Figure 15.14d. For the real CT data sets we have introduced one more processing step, shown in Figure 15.13b, which plots the standard deviation of a sliding window of a length equivalent to $180° + $ fan projections (i.e., 908 projections). In other words, the value shown at an arbitrary projection n_0 in Figure 15.13b is the standard deviation of a window from n_0 to $n_0 + 908$ of Figure 15.13a. The same processing step was applied also

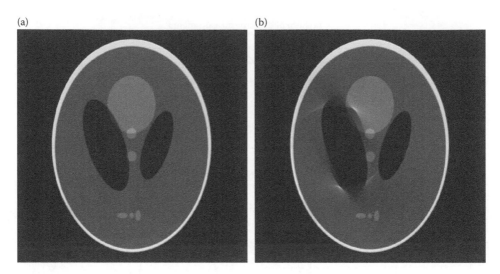

FIGURE 15.11
Image reconstruction based on motion phase information provided from Figures 15.10 and 15.12 (a) Reconstructed image from projections corresponding to the lowest amplitude of organ motion. (b) Reconstructed image that includes projections with organ motion.

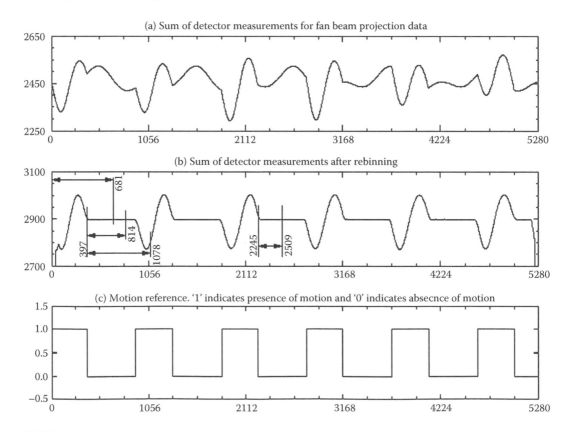

FIGURE 15.12
Implementation of the constant attenuation property of the Radon transform in an effort to identify the motion phase. Note the agreement between (b) and (c).

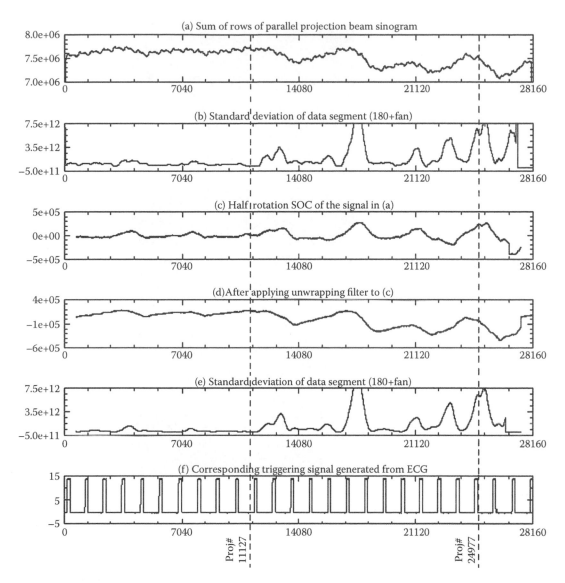

FIGURE 15.13

Signal processing steps to identify cardiac motion phase for the data of first patient. (a) Results from the technique based on constant attenuation property of the Radon transform. (b) A measure of the "amount of motion" in projection data sets for 180°+fan-angle. (c) Half rotation SOC of (a). (d) Output of unwrapping filter. (e) Measure of "amount of motion" in (d). (f) Square pulses derived from ECG signal used for RG-ECG. Vertical dotted lines indicate the starting points of projection data segments used to reconstruct images shown in Figure 15.14. It is expected that the images reconstructed from the data segment starting from projection #11127 will not be affected by motion artifacts and the one reconstructed from the data segment starting from projection #24977 will be affected by motion artifacts. This is confirmed in Figure 15.14.

on the results of Figure 15.13d and the output is shown in Figure 15.13e. Both these two curves (i.e., Figure 15.13b and e) convey very valuable information about organ motion and they are very close in identifying the same information as in Dhanantwari et al. [3, Figures 16 and 20]. More specifically, large amplitudes at some projection point in Figure 15.13b and e will indicate large fluctuations in a data window, which in turn will predict

(a)

Motion artifacts
removed.

(b)

Motion artifacts.

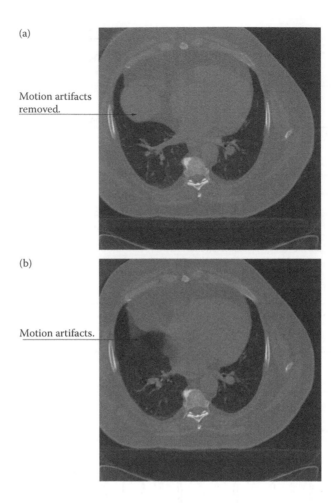

FIGURE 15.14
Image reconstruction based on motion phase information provided by Figure 15.13. (a) Image reconstructed from a data segment starting from projection 11127. (b) Image reconstructed from a data segment starting from projection 24977.

the presence of motion artifacts in the corresponding reconstructed image. In the same way, smaller amplitudes will indicate minor motion effects in the data window and this may act as a predictor whether the reconstructed images will suffer from motion artifacts. It is of interest to note also that the amplitude of the overall motion effects increases significantly during the latter part of the CT data acquisition process in Figure 15.13. This gradual increase over time in overall movement is due to the patient's restlessness because of breath-holding. Finally, the last Figure 15.13f shows the patient's ECG presented as square wave pulses, which coincide with the QRS-peaks of the patient's ECG pulses indicating the start of diastole phases [11]. This signal can be used to implement the ECG retrospective gating (RG-ECG) technique [8] to identify phases of the cardiac motion.

Figure 15.13b and e shows a low value at projection #11127 (e.g., see the vertical dotted line), indicating lesser motion near the apical portion of the ventricles in a data segment starting at this projection. Image reconstructed from this data segment, shown in

Figure 15.14a, presents the heart's pericardium, while its thickness can be identified clearly at the upper part of the image. In addition, the circular lobe of the liver is visible at the left part of the image.

At this point, it will be of interest to compare the performance of our technique with that of RG-ECG in identifying the motion phase. According to Figure 15.13f, the data segment starting at projection #11127 includes the QRS phase of the cardiac phases, suggesting the strongest heart motion. This is contrary to our processing results shown by Figure 15.13d and e, which demonstrate that the reconstructed image from this data segment is not affected by motion artifacts. Another interesting observation can be made at projection #24977. Here our techniques indicate large motion amplitude (i.e., Figure 15.13b and e) whereas the ECG signal suggests "cardiac diastole phase" (e.g., Figure 15.14f) in a data segment starting from this projection. The reconstructed image (i.e., Figure 15.14b) from this data set at projection #24977 indicates motion artifacts, as image clarity is inferior to that in Figure 15.14a.

For the second patient, the CT period of rotation was again set to 1.0 second (1408 projections) and the projection data sets were recorded for 28 rotations. During the data acquisition process, the patient was asked to breathe freely. Patient's heartbeat was recorded to be 72 bpm. The signal processing steps for the CT projection data sets from this patient are similar to that of the previous patient, except that the ECG signal was not recorded, and are shown in Figure 15.15.

The sum of rows of parallel projections is shown in Figure 15.15a. This gives the motion phase information directly. The output of the SOC and unwrapping process is shown in Figure 15.15c and d, respectively. Figure 15.15d also gives the motion phase information, which is nearly identical with Figure 15.15a. The standard deviation of a sliding window of length equivalent to $180° +$ fan projections was computed for the signals in Figures 15.15a and d and are plotted in Figures 15.16b and 15.15e, respectively.

The phase information related to cardiac motion provided by Figure 15.15, formed the basis to select segments of projection data sets for the image reconstruction process. For the first segment starting from projection number #17429, Figure 15.15b and e indicate "large amplitude" cardiac motion. The reconstructed image from this data set at projection point #17429 is shown in Figure 15.16a and it confirms our prediction showing large motion artifacts. Similarly the data segment starting from projection number #16763 indicates low amplitude cardiac motion (i.e., Figure 15.15b and e) . The reconstructed image from this data segment is shown in Figure 15.16b and it confirms our expectations by showing lesser motion artifacts that can allow detection of coronary calcification, which is clearly visible in this figure.

15.3.2 Multislice Helical CT System Applications

In this section, the performance of the developments of this investigation will be tested with synthetic projection data sets derived from a 3D Shepp–Logan phantom for multislice helical CT scanners. In the simulations, we consider the case of a multislice helical CT scanner having four sets of detector arrays, as shown in Figure 15.17. We also assume that the distance between the two consecutive sets is Δz and the linear speed of the patient's table through the gantry is such that it covers a distance of Δz during one rotation of the gantry. This speed causes an overlap of measurements through each cross-section. In fact, each helical path is traced four times by the detectors. Alternatively, each cross-section goes through the x-ray CT measurements for four times.

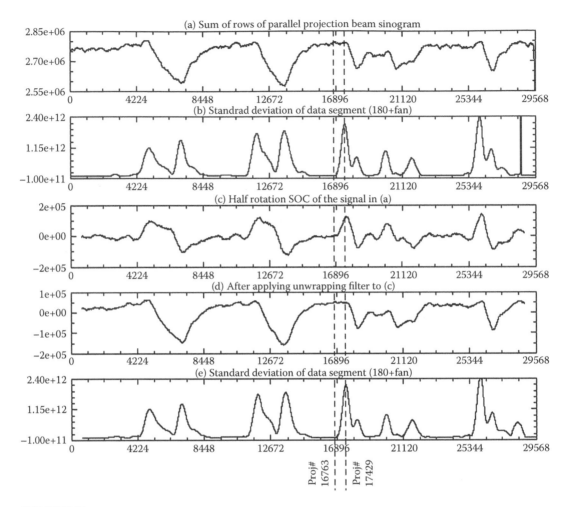

FIGURE 15.15

Signal processing steps to identify cardiac motion phase for patient 2. (a) Results from the technique based on constant attenuation property of the Radon transform. (b) A measure of the "amount of motion" in projection data sets for 180°+fan-angle. (c) Half rotation SOC of (a). (d) Results of unwrapping filter. (e) Measure of "amount of motion" in (d). Vertical dotted lines indicate the starting points of projection data segments used to reconstruct images shown in Figure 15.16. It is expected that the image reconstructed from the data segment starting from projection #16763 will not be affected by motion artifacts and the one constructed from the data segment starting from projection #17429 will be affected by motion artifacts. This is confirmed in Figure 15.16.

Method 1: Figure 15.18 illustrates schematically our first approach to interpolate the helical CT projection data sets before the application of the motion phase identification algorithms for the recovery of the diastole phase of the cardiac cycle. The first step is to identify the date sets for the interpolation process.

More specifically, for each selected cross-section, projection data sets are identified from helical measurements that have one rotation on each side of the selected cross-section in order to be used for interpolation. Because of the above predetermined speed of the patient's table, the three physical rotations of the gantry and the presence of four sets of detector arrays, provide projection data sets equivalent to data collected during twelve rotations by a single-slice CT scanner. Out of these twelve rotations of data sets, only six rotations of data correspond to the selected cross-sections, shown by thick lines in Figure 15.18. Out of these

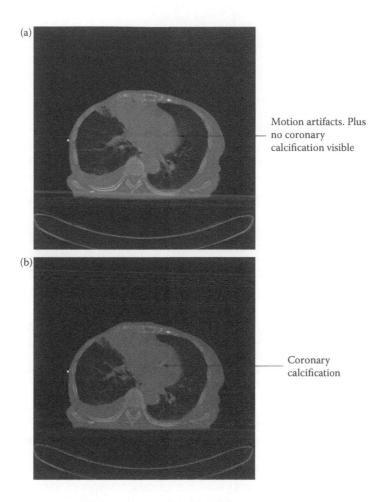

FIGURE 15.16
Image reconstruction based on motion phase information provided by Figure 15.15. (a) Image reconstructed from a data segment starting from projection 17429. (b) Image reconstructed from a data segment starting from projection 16763.

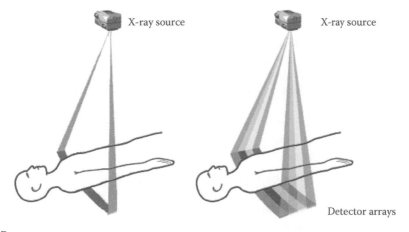

FIGURE 15.17
Right hand side shows a multislice CT scanner with four sets of detector arrays as opposed to a single slice scanner (left hand side).

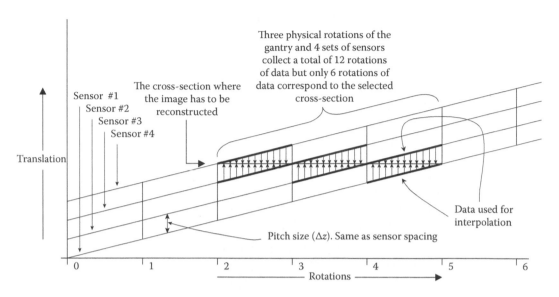

FIGURE 15.18
Schematic diagram showing implementation of *interpolation of helical projections followed by identification of motion phase* (Method 1). (a) A total of six rotations of data (shown by thick lines) are used for a given cross-section. The data were spread from $-\Delta z$ to Δz around the cross-section. (b) The 360° interpolation method produced three rotation data for the selected cross-section. (c) The interpolated data are used by *motion phase identification algorithms* to detect the diastole phase of the cardiac cycle. (d) Finally, a "clean" (180° + *fan*) data set can be used for image reconstruction.

six rotation data sets, three rotation data sets, from detector arrays # 1, 2, and 3, are on one side of the selected cross-section; and the rest of the three rotations of data sets, from detector arrays # 2, 3, and 4, are on the other side of it.

The data sets, collected from the previous identification step, are used by the 360° interpolation algorithm to generate the projection data for the selected cross-sections. At any given instant, the interpolating measurements belong to different (but consecutive) sets of detectors on either side of the selected cross-section, as shown by Figure 15.18. Thus, the helical interpolated data sets collected from the above six rotations will generate interpolated data, which are equivalent to data sets generated by a single-slice CT scanner, scanning for three rotations in one plane. In other words, after interpolation, the data generated will be equivalent to a sinogram of three rotations.

Then, the interpolated data sets from the above step are fed into the input of the *motion phase identification* algorithms discussed in Section 15.2.2. These algorithms identify the segments of the projection data sets that correspond to the diastole phase of the cardiac motion.

Following the identification of the segments of the projection date sets corresponding to cardiac diastole phase, a contiguous data set of size 180° + *fan* is used to reconstruct the images of the selected cross-sections by using the *filtered back projection* algorithm [16].

Method 2: The second approach that has been developed under this investigation includes the following steps. The phases of the organ motion associated with the acquired multislice CT projections are first being identified and then we follow with the implementation of the interpolation algorithms. Finally, both approaches provide 3D images from the reconstructed images that are generated for closely spaced but varying z-location.

The detailed procedures for the second method are as follows.

The identification process for the projection data sets used in this method are different from those of the previous method 1. Here, for a given cross-section, data from helical measurements that are half rotation on either side of a selected cross-section are identified. It should be noted that due to four physical rotations of the gantry and the presence of four sets of detector arrays, the collected data are equivalent to data collected by 16 rotations of a single-slice CT scanner. Out of these 16 rotations of data, only four rotations of data correspond to the selected cross-sections, shown by thick lines in Figure 15.19. Half of this data set are on one side of the selected cross-section and the rest are on the other side, as depicted in Figure 15.19. In other words, this set of helical data, collected by a four detector array during the four physical rotations of the gantry are equivalent to the data generated by a single-slice CT scanner, scanning for four rotations in one plane. In other words, this data are equivalent to a sinogram of four rotations.

Next, the projection data sets identified by the previous processing step are provided at the input of the motion phase identification algorithms of this investigation, which generate sections of projection data sets corresponding to the diastole phase of cardiac motion. These segments are provided at the input of the 180° interpolation algorithm, which creates 180°+*fan* interpolated projection data sets for the image reconstruction process (i.e., *filter back projection algorithm*) to generate the image of the selected cross-sections.

As discussed in Section 15.2.2.3, both the above two implementation methods have positive aspects. In particular, Method 1 performs better when dealing with organ motion artifacts whereas Method 2 is better for dealing with interpolation artifacts. Their assessment was carried out with simulation experiments using the 3D Shepp–Logan phantom containing several ellipsoids, which included the following:

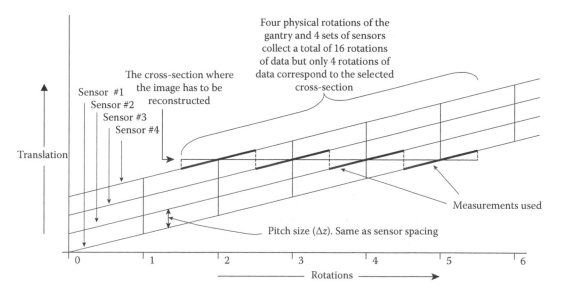

FIGURE 15.19
Schematic diagram showing implementation of *identification of motion phase followed by interpolation* (Method 2). (a) A total of four rotations of data (shown by thick line) are used for a given cross-section. The data were spread from −(Δz/2) to (Δz/2) around the selected cross-section. (b) The motion phase identification algorithm implemented on this data set to detect the diastole phase of the cardiac cycle. (c) A 360° of projection data sets corresponding to the identified diastole phase are selected and the 180° interpolation is performed to produce data of length 180°+fan-angle. (d) Finally, this "clean" and interpolated data set is used for image reconstruction.

First, the simulations generated 3D projection data sets for planner cross sections (i.e., not-helical projections) of a stationary 3D Shepp–Logan phantom. The aim here was to eliminate the complications of the helical projections and their interpolation when we are assessing the motion tracking algorithms.

Next, the simulations were extended to include the 3D Shepp–Logan phantom having all the ellipsoids stationary except for one ellipsoid which was considered pulsating with a period different than the rotational period of the simulated multislice CT scanner, which did not include helical acquisition.

Finally, the simulations included the case of a pulsating 3D phantom for a complete four-slice helical CT scanner and with experimental parameters equivalent with those of an operational helical, multislice CT imaging system.

The parameters of the 3D Shepp–Logan phantom are summarized in Table 15.1 and include several 3D ellipsoids, with one of them assumed to pulsate in order to simulate the cardiac motion. The period of the pulsating ellipsoid was 1.2 Hz and with a duty cycle of 15%. Thus, 85% of the time it was stationary, which simulates the diastole phase.

The projection measurements of the simulated multislice helical CT scanner around the 3D phantom were synthesized for a section of length 5 cm (from $z=-5$ cm to $z=0$ cm). The rotational speed was 0.75 seconds per rotation with the linear speed of the phantom through the gantry set at 0.15 cm per rotation and the spacing between consecutive detector arrays set also to 0.15 cm. This set of parameters allowed the projections measurements to overlap along the z-axis. Thus, in order for the gantry to scan 5 cm of the phantom along the z-axis, it requires to execute 34 rotations during 25.5 seconds. If the simulated multislice CT scanner will be considered to include higher number (i.e., 64-slices) of detector arrays than the current case of four detector arrays, then the number of rotations to scan the 5-cm distance along the z-axis will be reduced significantly. In fact, this is the main supporting argument that drives the medical manufactures to introduce multislice CT scanners that currently have reached 128-detector arrays with a trend to be increased to 256-slice CTs, which is at the expense of higher levels of radiations for the patients.

Figures 15.20 through 15.23 summarize the results from the above simulations. For the case of the stationary phantom, Figure 15.20a shows, as expected, the tracking of organ

TABLE 15.1

Parameters of the Ellipsoids in the 3D Shepp–Logan Phantom

Coordinates of the Center of the Ellipsoid			Size of the Ellipsoids			Inclination of the Ellipsoid in xy-plane	Density of the Ellipsoid
x (cm)	y (cm)	z (cm)	x-axis (cm)	y-axis (cm)	z-axis (cm)		
0.0	0.00	0.000	6.900	9.20	9.00	0°	0.05
0.0	0.00	0.000	6.624	8.74	8.80	0°	−0.03
−2.5	−0.50	−2.500	3.500	1.50	1.50	108°	−0.02
2.5	−0.50	−2.500	3.500	1.50	3.50	72°	−0.02
0.0	5.00	−2.500	3.000	2.00	3.00	0°	0.01
0.0	−5.50	−2.500	0.460	0.46	0.46	0°	0.01
−0.8	−6.50	−2.500	0.460	0.23	0.23	0°	0.01
0.8	−6.50	−2.500	0.460	0.23	0.23	0°	0.01
0.6	−1.05	−0.625	0.560	0.40	1.00	90°	0.01
0.0	1.00	−0.625	0.560	0.56	1.00	0°	−0.01

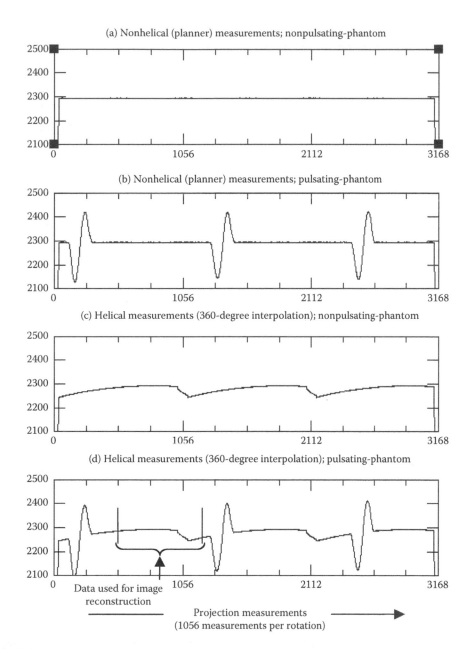

FIGURE 15.20
Results from the implementation of the constant attenuation property of the Radon transform (CART) for motion phase identification on simulated helical-measurements for pulsating-phantom using *interpolation followed by identification of motion phase* (Method 1). Units along *y*-axis are arbitrary; simply show relative amplitude of organ motion. (a) Planner measurements from stationary phantom. No interpolation was needed. CART shows no motion. (b) Planner measurements for pulsating phantom. No interpolation was needed. CART identifies pulsating and nonpulsating phases of phantom correctly. (c) Helical measurements from stationary phantom. Implementation of CART technique on Method 1 followed by 360° interpolation. No motion was present but still some artifacts due to the 360° interpolation are present. (d) Helical measurements from pulsating phantom. Implementation of CART technique followed by 360° interpolation. The pulsating and nonpulsating phases have been identified but have been affected by the 360° interpolation artifacts shown in (c).

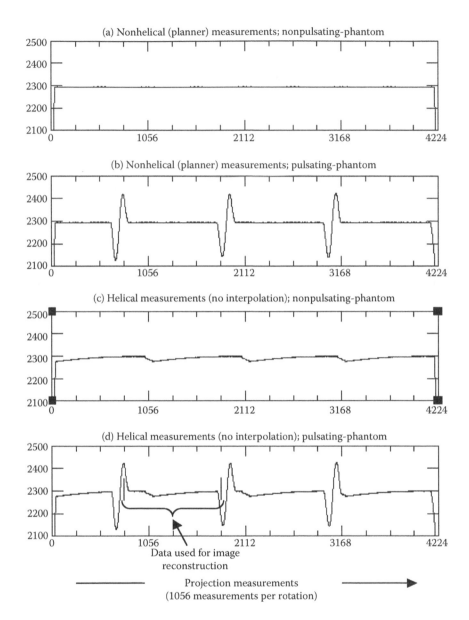

FIGURE 15.21
Results from the constant attenuation property of the Radon transform (CART) for motion phase identification on simulated helical-measurements for pulsating-phantom using *identification of motion phase followed by interpolation* (Method 2). Units along *y*-axis are arbitrary; simply show relative amplitude of organ motion. (a) Planner measurements from stationary phantom. No interpolation was needed. CART shows no motion. (b) Planner measurements from pulsating phantom. No interpolation was needed. CART identifies pulsating and nonpulsating phases of the phantom correctly. (c) Helical measurements from stationary phantom. Implementation of CART technique on projection data (Method 2). No organ motion was present but still artifacts are present. (d) Helical measurements from pulsating phantom. Implementation of CART technique on Method 2, without any interpolation. It shows detection of the pulsating and nonpulsating phases, which are not affected by artifacts due to interpolation. Next, a selection of 180°+fan-angle data for a complete 360° of projections corresponding to the diastole phase are used for 180° interpolation.

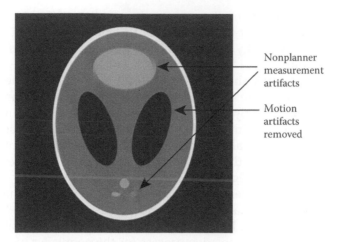

FIGURE 15.22
Reconstructed image from helical data using *interpolation followed by motion phase identification* (Method 1).

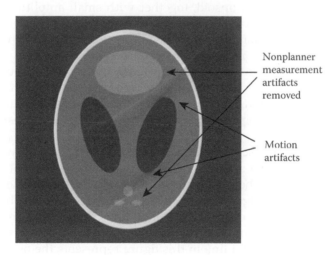

FIGURE 15.23
Reconstructed image from helical data using *identification of motion phase followed by interpolation* (Method 2).

motion to be a straight horizontal line, which indicates stationarity. Next, Figure 15.20b shows that the motion of the pulsating ellipsoid has been identified correctly. The results from the above two figures do not include interpolation, since the projection measurements were simulated for nonhelical motion.

Figure 15.20c and d show the effects of 360° interpolation of Method 1 on the algorithm that exploits the constant attenuation property of the random transform (see Section 15.2.2.2) in tracking organ motion, for the cases of stationary and pulsating phantoms, respectively. It is clear from Figure 15.20c that although there is no organ motion, the algorithm senses motion with periodicity equal to the gantry's rotation period. This is clearly an artifact due to the 360° interpolation algorithm, which is present also in Figure 15.20d for the case of the pulsating phantom. The results of Figure 15.20d indicate also that the tracking of the motion of the pulsating ellipsoid has been recovered successfully, but it has been affected by the artifacts of the 360° interpolation algorithm.

Image reconstruction results, shown in Figure 15.22, for the case of the pulsating phantom from the data sets of Figure 15.20d, demonstrate that the algorithms of this investigation have been able to correctly identify the diastole phase of the simulated pulsating ellipsoid and the reconstructed images do not suffer from motion artifacts. However, the artifacts of the 360° interpolation algorithm (i.e., nonplanner measurement artifacts) appear as motion artifacts on ellipsoids that are stationary and should not have been affected by motion artifacts, as depicted in Figure 15.22.

Figure 15.21 has the same presentation arrangements and for the same projection measurements, as in Figure 15.20, that have been processed with Method 2, which includes the 180° interpolation algorithm. In this case it is important to note that in Figure 15.21c the artifacts, due to the 180° interpolation process for the nonplaner measurements, have smaller amplitude and are not so pronounced as in the case for the 360° interpolation process in Figure 15.20c. Figure 15.21d shows successful tracking results of the simulated organ motion for the pulsated ellipsoid, together with small amplitude artifacts due to the 180° interpolation process, discussed also in Figure 15.21c. However, due to the fact that for the image reconstruction process a longer (i.e., 360° projections with no motion—"clean") data set was needed for the 180° interpolation algorithm, this set of contiguous "clean" data set was not possible to be obtain due to the insufficient duration of the diastole phase, as shown in Figure 15.21d. As a result, the reconstructed image from this data set, presented in Figure 15.23, contains motion artifacts for the pulsating ellipsoid, though the artifacts due to the 180° interpolation process are insignificant, as compared with those of the 360° process in Figure 15.22.

Finally, the organ motion tracking algorithms of this investigation were applied on a complete set of projection data sets from 128 cross-sectional slices that cover the helical scanning of the complete 3D Shepp–Logan phantom. Figure 15.24 presents the output of the *motion phase identification* algorithm for all the 128 cross-sections. The diagonal white stripe in this figure is made up of small horizontal lines. These lines are stacked on each other such that each line is shifted to the right and on top of the previous one, along the diagonal axis. Each horizontal line in this figure represents the output of the motion phase identification algorithm for one cross-section, which are data sets similar with those in Figure 15.20d. Thus, the data in Figure 15.24 present the amplitude of the pulsating and diastole phases as variations in gray scale. It should also be noted that as the measurements are collected for different cross-sections along increasing values of the z-axis, the time reference also increases for each cross-section. As a result, each horizontal line in Figure 15.24 is shifted to the right along the time axis. This shift gives Figure 15.24 the appearance of a diagonal white stripe, while the black triangles above and below this stripe do not represent any data.

It should be noted also that since the phases of the identified cardiac cycles in consecutive cross-sections occur at the same time, these horizontal lines, in Figure 15.24, coincide in identifying the phases of cardiac cycles, which demonstrates the correct alignment of the projections from the 128 cross-sectional slices.

Finally, consecutive data sets of 180°+*fan* size of projections, representing the identified diastole phases from the aligned 128 cross-sectional slides, were provided as inputs to the image reconstruction algorithm, which produced cross-sectional images of the 3D

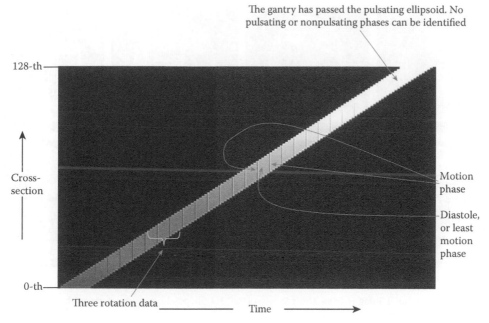

The gantry has passed the pulsating ellipsoid. No pulsating or nonpulsating phases can be identified

128-th

Cross-section

Motion phase

Diastole, or least motion phase

0-th

Three rotation data — Time ⟶

FIGURE 15.24
Identification of cardiac motion phase for 128 cross-sectional provided from a simulated four-slice CT scanner implemented on a 3D Shepp–Logan phantom.

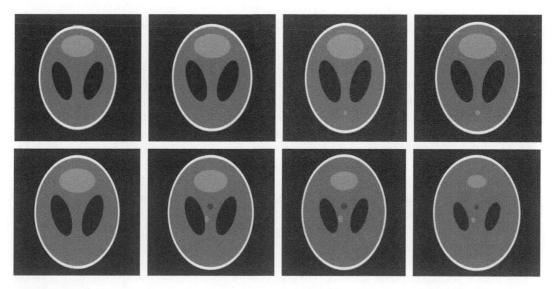

FIGURE 15.25
Selected images from the reconstructed 3D Shepp–Logan phantom after removing motion artifacts.

Shepp–Logan phantom of size $512 \times 512 \times 128$ without any "motion artifacts." A selection of these 128 "clean" cross-sectional images is shown in Figure 15.25 for slices without any motion artifacts. As a comparison, Figure 15.26 shows the same slices with motion artifacts present.

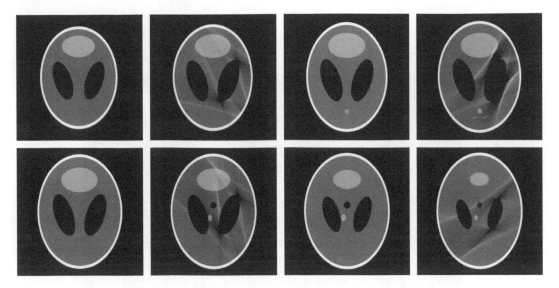

FIGURE 15.26
Selected cross-sectional images from the reconstructed 3D Shepp–Logan phantom without correcting for motion artifacts.

15.4 Conclusion

The algorithmic developments of this chapter suggest a novel approach to track the phases of cardiac motion using two signal-processing techniques, without ECG gating. Selected sets of projections, that correspond to the identified phases of the cardiac motion, form the basis for the 3D image reconstruction process that is not affected by cardiac motion artifacts. The first technique for tracking phases of cardiac motion consists of a modified version of the SOC [2] integrated with an unwrapping filter, as defined in Section 15.2.2.1. The second technique is much simpler and efficient and it is based on the constant attenuation property of the Radon transform, as discussed in Section 15.2.2.2. Both techniques were tested with simulated as well as real patient data and their performance was nearly identical. Once cardiac motion phases are identified by the above two techniques, then image reconstruction algorithms can be applied on CT projection data sets that correspond to selected phases of the cardiac motion to reconstruct images without motion artifacts.

The above techniques for tracking cardiac motion phases were implemented also on multislice helical CT imaging applications to reconstruct 3D cardiac volumetric images without motion artifacts. Their effectiveness was demonstrated with simulated data sets for a four-slice helical CT scanner. Additional simulation results from 16-slice helical CT scanners, that are not presented in this chapter, have shown that the performance of the above two techniques improves significantly, as expected, when implemented in CT systems with large number of detector arrays.

References

1. Bushberg, J. T., Seibert, J. A., Leidholdt, E. M. and Boone, J. M. 1994. *The Essential Physics of Medical Imaging*. Baltimore, MD: Williams & Wilkins.
2. Dhanantwari, A. C., Stergiopoulos, S., Iakovides, I. 2001. Correcting organ motion artifacts in x-ray CT medical imaging by adaptive processing (Part I: theory). *Med. Phys.*, 28(8), 1562–1576.
3. Dhanantwari A.C., Stergiopoulos S., Zamboglou, N., Baltas, D., Vogt, H. and Karangelis, G. 2001. Correcting organ motion artifacts in x-ray CT systems based on tracking of motion phase by the spatial overlap correlator (Part II: experimental study). *Med. Phys.*, 28(8), 1577–1596.
4. Dr. Samaraweera (email communication), Sparrow Health Care Radiology Department, Lansing, MI.
5. Crawford, C. R., King, K. F., Ritchie, C. J. and Godwin, J. D. 1996. Respiratory compensation in projection imaging using a magnification and displacement model. *IEEE TMI*, 15(3), 327–332.
6. Ritchie, C. J., Hsieh, J., Gard, M. F., Godwin, J. D. Kim, Y. and Crawford, C. R. 1994. Predictive respiratory gating: a new method to reduce motion artifacts in CT scans. *Radiology*, 190, 847–852.
7. Chiu, Y. H. and Yau, S. F. 1994. Tomographic reconstruction of time-varying object from linear time-sequential sampled projections. *Proc. IEEE Conf. ASSP*, 1, V309–V312.
8. Morehouse, C. C., Brody, W. R., Guthaner, D. F., Breiman, R. S. and Harell, G. S. 1980. Gated cardiac computed tomography with a motion phantom. *Radiology*, 134(1), 213–217.
9. Kalender, W. A., Seissler, W., Klotz, E. and Vock, P. 1990. Spiral volumetric CT with single-breath-hold technique, continuous transport, and continuous scanner rotation. *Radiology*, 176, 967–982.
10. Urban, B. A., Fishman, E. K., Kuhlman, J. E., Kawashima, A., Hennessey, J. G. and Siegelman, S. S. 1993. Detection of focal hepatic lesions with spiral CT: comparison of 4- and 8-mm inter-scan spacing. *Am. J. Radiol.*, 160, 783–785.
11. Katz, A. M. 1977. *Physiology of the Heart*. New York, NY: Raven Press.
12. Opie, L. H. 1991. *The Heart: Physiology and Metabolism*, 2nd edition. New York, NY: Raven Press.
13. Stergiopoulos, S. 1990. Optimum bearing resolution for a moving towed array and extension of its physical aperture. *J. Acoust. Soc. Am.*, 87(5), 2128–2140.
14. Stergiopoulos, S. 1998. Implementation of adaptive and synthetic aperture processing in real-time sonar systems. *Proc. IEEE*, 86(2), 358–396.
15. Waheed, Y. and Stergiopoulos, S. 2006. Method for tracking motion phase of an object for correcting organ motion artifacts in x-ray CT systems. US Patent: 7,085,42.
16. Kak, A. C. and Slaney M. 1999. *Principles of Computerized Tomographic Imaging*, New York, NY: IEEE Press.
17. Helgason, S. 1999. *Transform*. Birkhauser Verlag, Berlin.
18. Bresler, Y. and Skraba, C. Z. 1989. Optimal interpolation in helical scan computed tomography. *Proc Int Conference Acoustics Speech Signal Processing*, 3, 1472–1475.
19. Brink, J. A., Heiken, J. P., Wang, G., McEnery, K. W., Schlueter, F. J and Vannter, M. W. 1994. Helical CT: principles and technical considerations. *Radio Graph.*, 14(4), 887–893.
20. Shepp, L. A. and Logan, B. F. 1974. The Fourier reconstruction of a head section. *IEEE Trans. Nucl. Sci.*, NS-21, 21–43.

16

NonInvasive Monitoring of Vital Signs and Traumatic Brain Injuries

Stergios Stergiopoulos

Defence R&D Canada Toronto
University of Toronto
University of Westorn Ontario

Andreas Freibert and Jason Zhang

Defence R&D Canada Toronto

Dimitrios Hatzinakos

University of Toronto

CONTENTS

16.1 Background

The emerging trends in the field of medical diagnostic technologies aim to address the challenging requirements for noninvasive monitoring of vital signs in hospital emergency departments, ambulances, hospital intensive and after care units, home care, medical clinics, and nursing homes. It has been assessed that these requirements can be fulfilled with the development of compact, portable, and field-deployable medical monitoring and imaging systems. Recent announcements of major R&D efforts supported by the US Armed Forces [1] and the European Commission [2] for novel noninvasive system concepts provide the basis to identify the technological trends in this field that include:

- The smart shirt for continuous remote monitoring of vital signs.
- The wearable computer to address the technology and computational requirements for integrating various sensor technologies in the functionality of the smart shirt.

The technological trends in this field are divided into two distinct areas:

I. Advanced next generation medical 4D (3-dimensional + temporal) imaging modalities (e.g., 3D/4D ultrasound, x-ray CT and MRI 3D tomography imaging systems)

II. Noninvasive monitoring of vital signs such as:
  - Blood pressure and heart rate.
  - Electrocardiogram (ECG) for monitoring cardiac abnormalities.
  - Pulse oxymeter for monitoring the oxygen content in live tissue.
  - Electroencephalography (EEG) for monitoring brain injuries and abnormalities.
  - Core temperature.
  - Intracranial pressure (ICP) and brain density variations that may be caused by traumatic brain injuries (TBI), brain hemorrhage, stroke, variations of blood flow in the brain due to drug effects and pain.

The essential R&D efforts to address the system development requirements relevant with the above two distinct areas include:

- Advanced signal processing that would enhance the information of interest by suppressing noisy interferences.
- A common based computing architecture with advanced digital signal processors (DSPs) to address highly demanding signal processing requirements.
- Novel sensor technologies that would improve the diagnostic capabilities of the existing systems and would generate new procedures for monitoring vital signs that cannot be accomplished with the existing system technologies.
- Tele-medicine with minimum electromagnetic EMC interference for remote monitoring the sensor functionalities and transfer of information to monitoring stations via wire and/or wireless communication channels.
- User friendly graphic interfaces.

The previous chapters have already introduced the signal processing structures of advanced 3D medical imaging modalities such as 3D ultrasound, cardiac CT and MRI. In this chapter we will focus on the challenging signal processing problems in monitoring blood pressure, nonvisible brain traumatic injuries and other brain abnormalities. The system technologies for monitoring the remaining vitals signs discussed above (i.e., EEG, ECG, core temperature, pulse oxymeter) have reached already a mature state of development and their signal processing structures are simple and therefore, are beyond the scope of this chapter.

16.2 A Dispersive Ultrasound Technology for Monitoring TBI

ICP monitoring is a critically important diagnostic tool for trauma patients and patients undergoing neurosurgery. Elevated ICP is a pathological state and is an indicator of serious neurological damage and illness. Several pathological conditions and nonvisible head injuries cause the volume within the skull to increase, but the inability of the skull to expand significantly causes the ICP to increase exponentially [3]. The primary concern caused by increased ICP is that the brain will become herniated. Brain herniation is usually the result of cerebral edema, which is the medical term for the condition when the brain is swollen and edema is usually caused by head injury. This will result in progressive damage to the brain and can ultimately be fatal. Other metabolic, traumatic, infectious conditions such as hypoxia, ischemia, brain hemorrhage, tumor and meningitis may all cause elevated ICP. Hypoxia occurs when there is a lack of oxygen supplied to the brain, usually due to cardiac arrest. Hypoxia often leads to brain edema. Ischemia is the state in which there is a deprivation of blood flow to the brain. Ischemia often leads to stroke because blood flow is interrupted or blocked, thus glucose and oxygen cannot nourish the brain. Ischemia is usually caused by formation of a blood clot (thrombus) and can lead to the death of all or parts of the brain (cerebral infarction). Intracerebral hemorrhage is an increase in blood volume within the cranium. Increased blood flow is positively correlated with an increase in ICP. The main cause of intracerebral hemorrhage is a ruptured blood vessel in the brain. Ruptured blood vessels can be the result of a blow to the head, but is usually due to hypertension. Ischemic and hemorrhagic strokes can lead to variations in the ICP [4]. The added volume of a brain tumor can also cause an increase in ICP. In fact, an increase in ICP that occurs without any head injury is often a sign of the presence of a brain tumor. Meningitis is a bacterial or viral infection of the meninges, a three-layer membrane that surrounds the brain and spinal cord. Meningitis causes the meninges to swell, press against the skull and push down on the brain. As a result, intense pressure buildup occurs, and a rise in ICP is notable. If not treated rapidly, meningitis can lead to herniation.

Therefore, monitoring of ICP as a vital sign is an essential process for the early detection of brain injuries, some of which have been discussed above. Early detection of ICP fluctuations is recognized as an important tool in improving the condition of a patient. Currently, both invasive and noninvasive modalities are used to provide information about ICP. However, at the present time, accurate recordings of ICP can only be obtained with the use of invasive modalities. Invasive methods involve exposing brain tissue, which increases the risk of infection, hemorrhage, leakage of cerebral spinal fluid and therefore also increases the possibility of further aggravating the condition being treated.

The current noninvasive modalities for ICP monitoring frequently sacrifice accuracy or in other cases are too expensive and cumbersome. Noninvasive techniques include computed tomography (CT), magnetic resonance imaging (MRI), positron emission tomography (PET), single photon emission computed tomography (SPECT) and other techniques such as transcranial doppler ultrasonography (TCD), transcranial near-infrared spectroscopy (TNIRS), ophthalmodynamomery (OMD) and impedance audiometry have been attempted or are still in development.

Recent investigations [5,6] have been shown that monitoring of the intracranial density of the brain can provide estimates (e.g., monitoring) of ICP. This can be achieved by probing the brain with ultrasonic pulses. The propagation speed of the pulses varies with the brain density, and density changes are assessed through the continuous monitoring of the time delay between transmitted and reflected pulses [5,6].

However, accurate estimates of ICP variations are not sufficient to differentiate the numerous pathological cases of brain injuries and abnormalities that have been discussed earlier and therefore, monitoring of ICP is not sufficient to provide diagnosis for the various brain injuries. Furthermore, the above ultrasound techniques [5,6] provide time delay estimation (or monitoring of density fluctuations) by probing the brain with ultrasound pulses at a specific frequency region. In doing so, they ignore the dependency of the density fluctuations on a wide range of frequencies, and so disregard information essential to accurately account for the dispersive properties of the human brain.

An alternative approach [7], which is unique with respect to on going R&D investigations in the same field, attempts to exploit the dispersive properties of the brain-medium and as such it offers a wide range of degrees of freedom to differentiate the system response for various diagnostic cases.

More specifically, for biological structures, such as the brain, that are heterogeneous the important parameters affecting the velocity of sound are the elastic constant, the bulk modulus (i.e., $B(f)$), which is frequency dependent and slightly temperature dependent. As a result, the propagation speed of ultrasonic waves varies with brain density, and therefore monitoring of density variations can be achieved by monitoring changes in the velocity of ultrasound for a sequence of ultrasonic pulses radiated in the brain through a small area of the skull near the human ear (the "temporal bone") that permits ultrasound penetration into the skull. At this point, it is important to note that for accurate diagnosis, monitoring of the density fluctuations and their correlation with various conditions of the brain requires estimates of density fluctuations for a wide range of frequencies, to account for dispersive properties, as expressed by $c(f)=(\gamma B(f)/\rho)^{1/2}$. This is in fact what differentiates this new approach [7] from other noninvasive techniques that have been announced recently [5,6].

A system concept for monitoring intracranial density fluctuations that is based on a time delay estimation scheme, which is the traveling time $\Delta\tau=D/c(f)$, of an ultrasound wave propagating in a confined medium, such as the cranium with fixed dimensions, D, requires very accurate time delay estimation of the order of micro-seconds in order to account for small variations of the intracranial density. The new approach [7] accounts for the change in D (i.e., the dimension of the skull) by using a set of two monochromatic frequencies to estimate the potential changes in D (e.g., changes in the dimensions of the cranium due to thermal or other effects) while time estimation processing would account only for variation in c. The observed fluctuations in traveling time τ between the transmitted and reflected acoustic pulse are therefore directly related to fluctuations in the density of the brain because sound velocity c is directly related to density ρ. Moreover, this new methodology [7], monitoring of the brain density fluctuations is obtained through a phase estimation approach between a transmitted and reflected monochromatic ultrasound waves, instead of a time delay estimation process. The details about the proposed phase estimation scheme are discussed in this chapter and the accuracy of this phase estimation approach has been demonstrated with the experimental system of Figure 16.1 and the preliminary results shown in the next section.

16.2.1 Preliminary Test Results with Phantoms

Testing of the experimental prototype, shown in Figure 16.1, has been carried out with water phantoms. The current prototype employees a pair of transducers to transmit and receive monochromatic (i.e., narrowband) ultrasound pulses in the frequency range of 0.5–4.0 MHz.

The experimental set-up in the upper part of Figure 16.2 includes water inside a cylindrical glass container that was approximately 10 cm in width. The pair of transducers was placed on both sides of the container and the coupling between the container and transducers was maximized using standard ultrasonic probe gel. To demonstrate the system's sensitivity in phase estimates $\Delta\varphi$ that correspond to changes in density, the temperature of the water was slowly changed. In this case, hot water was gradually allowed to cool, increasing the density of the water. The lower part of Figure 16.2 shows estimates of $\Delta\varphi$ that correspond to the linear decrease in water temperature from 41.2°C to 36.9°C. For constant water temperature, estimates of $\Delta\varphi$ were nearly constant as a function of time, as expected.

The results of Figure 16.3 shows the classification capabilities of this new dispersive ultrasound approach [7] in probing various fluids for industrial quality control applications and nondestructive identification of contained substances. In this case, samples from two types of fluids with approximately equal densities were used to train the system by using three different ultrasound frequencies. Figure 16.3 shows the time delay estimates, denoted by cluster islands shown by different symbols, for the two different fluids. Identification of an unknown sample of fluid was obtained through time delay estimates denoted by the symbol "x" to be identical with the known substance, as defined in Figure 16.3. This identification process was later confirmed to be correct. This is an important result; and it demonstrates the consistency of this dispersive ultrasound technology to

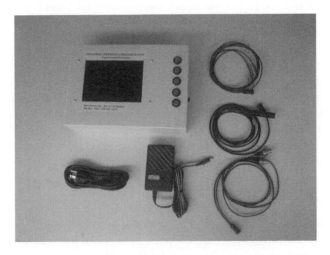

FIGURE 16.1
Laboratory prototype of the dispersive ultrasound technology.

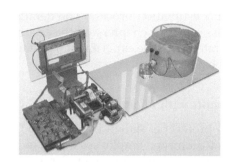

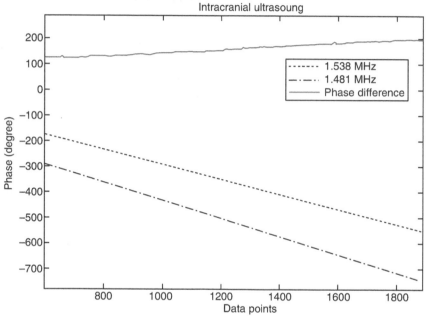

FIGURE 16.2
Upper part shows the experimental arrangement with water inside a cylindrical plexiglass container that was approximately 20 cm in width. The pair of transducers was placed on both sides of the container and the coupling between the container and transducers was maximized using standard ultrasonic probe gel. Lower part shows estimates of $\Delta\varphi$ that correspond to the linear decrease in water temperature from 41.2°C to 36.9°C.

track and classify the complex properties of fluids. This kind of classification capability will be translated also through animal experimentation and clinical trials into a diagnostic capability to track and differentiate the various pathological cases of brain injuries and pathology. However, for complex cases such as the brain, it is essential that a large number of frequencies is been used to be able to have enough degrees of freedom to describe more accurately the complexity of the brain pathology.

16.2.2 Results from Animal Experimentation

A preliminary assessment of the capabilities of the experimental system in Figure 16.1, to provide diagnostic information relevant with brain injuries, was provided from a limited set of animal experimentation. The results from these experiments are presented in Figure 16.4. All experiments were approved by a University Ethics Committee and adhered

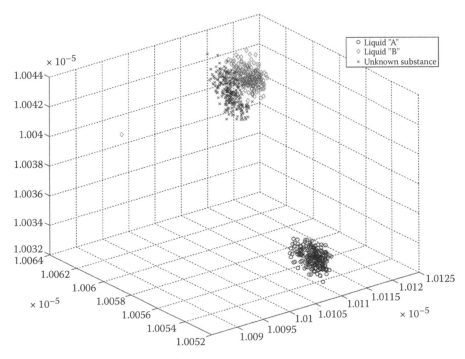

FIGURE 16.3

Samples from two fluid substances with approximately equal densities were used to train the system in Figure 16.1 by using three different ultrasound frequencies. Estimates of the time delays for the three different frequencies and for the two different fluids are denoted by clusters of islands shown by diamonds and circles. Identification of an unknown sample of fluid was confirmed through time delay estimate denoted by symbol "x" to be identical with the known substance denoted by diamonds.

to International Association for the Study of Pain guidelines (Zimmermann reported by Freibert et al. [8]).

The animal experimentation included testing of the diagnostic information of the dispersive ultrasound device in Figure 16.1 with male rats with a weight of 430–500 g. The entire narcosis and circulation monitoring procedures were conducted and/or supervised by Dr Matthias Pawlak in accordance with international standards [8]. The results in Figure 16.4 show that the system response has a maximum frequency response at different frequencies and for various brain injuries, which demonstrate the capabilities of this technology to track dispersive properties of brain tissues.

Figure 16.4, shows the system response at various frequencies for the case of stroke induced on the animal. These results and those from similar experiments reported by Freibert et al. [8] are indicative of the capabilities of the prototype dispersive ultrasound system to differentiate the case of stroke from other type of TBI by having a maximum frequency system response at 4.0 MHz. This kind of differentiation needs to be mapped under a similar graphic configuration as in the case of fluid identification (i.e., Figure 16.3). Then, a mapping process for known type of brain injuries will serve as a diagnostic tool to provide diagnosis for the type of TBI that the system has been trained to detect.

In summary, the results from animal experimentations [8] provide a preliminary assessment of the capabilities of a dispersive ultrasound system concept in detecting various cases of brain trauma (i.e., hemorrhage, stroke) and brain metabolic effects (i.e., pain). However,

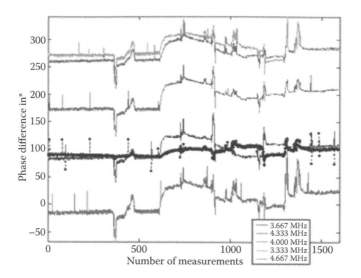

FIGURE 16.4
After an intra-arterial injection of adrenalin, the volume of blood in the brain significantly increased due to the rising blood pressure (case of stroke). The maximum system response is at 4.0 MHz. Changes of the blood tissue- ratio caused an alteration in the brain density.

these capabilities need to be demonstrated with clinical trials, which may define the most important phase of a development effort in designing a complete dispersive ultrasound system for TBI diagnostic applications. In what follows, the signal processing challenges relevant with the dispersive ultrasound system concept of Figure 16.1 are discussed as part of a system development effort.

16.2.3 Signal Processing Concept for Intracranial Dispersive Ultrasound

The concept of utilizing ultrasonic waves to measure density variation in the brain is not new [4,5,6]. Density variations are recorded by monitoring changes in the velocity of sound. The velocity of sound is dependent on a number of parameters, including the state of the propagation properties of the medium. Biological structures consist of substances in all states of matter; bones and organs are solid, blood and other fluids are liquid, and gas exists in the respiratory and digestive system. The brain is composed primarily of liquid matter with many dissolved organic substances.

The equation of the velocity of sound in a homogenous liquid is given below.

$$c(f) = \sqrt{\frac{K(f)}{\rho_o}} \tag{16.1}$$

where f=frequency; $K(f)=\gamma B_T(f)$ (constant for the same medium and frequency); $C(f)$=velocity of sound; γ=ratio of specific heat at constant pressure to that at constant volume; ρ_o=density; and $B_T(f)$=isothermal bulk modulus.

The biological structures that are heterogeneous greatly complicate the equation, but the important parameters affecting the velocity can still be distinguished. The elastic constant, the bulk modulus (i.e., $B_T(f)$), is frequency dependent and slightly temperature dependent, but the main parameter that changes the velocity of sound is the density, which is also frequency dependent [8]. At this point, it is important to note that for accurate diagnosis, monitoring of the density fluctuations and their correlation with various conditions of the brain requires estimates of density fluctuations for a wide range of frequencies, to account for dispersive properties, as expressed by Equation 16.1.

For a device monitoring the density variations of a fluid substance enclosed in a container with fixed dimension, D, the traveling time $\Delta\tau$, of a propagating ultrasound wave, is,

$$\Delta\tau = \frac{D}{c}. \tag{16.2}$$

For uniform and with no dynamic variations of brain density distributions of a healthy person, estimates of the relative variations of $\Delta\tau$ are related mainly to changes in core temperature. Similarly, for constant core temperature, estimates of $\Delta\tau$ between successive ultrasound pulses remain constant or exhibit very little variation over time. However, in cases of visible or nonvisible brain injuries, the distribution characteristics or variation pattern of $\Delta\tau$ as a function of a set of different frequencies of radiated ultrasound pulses will generate patterns of frequency dependent time delay estimates that will correlate with observed brain abnormalities that can be established through clinical trials. This kind of pattern recognition approach can lead to noninvasive assessment of patient brain trauma or other pathological cases discussed above.

However, accurate estimates of $\Delta\tau$ is a challenge and the approach that is suggested by Stergiopoulos and Wrobel [7] is to use a trigger threshold method as depicted in Figure 16.5. Furthermore, time delay estimation techniques developed for sonar and radar applications may be applied to this technology to make the $\Delta\tau$ estimation process more robust [9,10]. The precision of these methods cannot exceed the sampling period. For example, a sampling rate of 10 MHz will have a sampling period of 0.1 µs. For long distance and therefore large time scale applications such as radar or sonar, this is not significant. For intracranial applications, however, the path traveled by an ultrasonic pulse signal is relatively small (twice the diameter of the brain) and the error may be significant unless an extremely large sampling frequency is used. This would make system implementation very costly due to the hardware complexities that would be required to achieve a very high sampling frequency

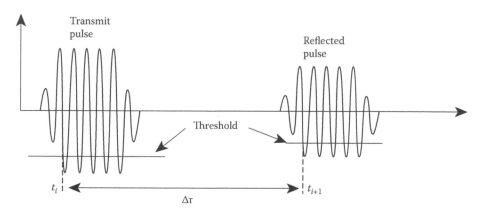

FIGURE 16.5
Trigger method for ultrasound pulses that can provide estimates of $\Delta\tau$.

for ultrasonic pulses. Equation 16.3 is a simplistic expression deriving $\Delta\tau$ for two successive (the ith and $(i+1)^{th}$) ultrasonic pulses.

$$\Delta\tau = t_{i+1} - t_i \tag{16.3}$$

Error $= \pm t_{sp}$ (sampling period)

However, instead of attempting to measure directly the parameter $\Delta\tau$, we select to compute the relative phase difference $\Delta\varphi$, as defined below by Equation 16.4,

$$\Delta\varphi = \varphi(r) - \varphi(t), \tag{16.4}$$

between the phases of transmitted $\varphi(t)$ and reflected $\varphi(r)$ pulses, which can provide a simpler and more accurate estimation approach for the parameter $\Delta\tau$. More specifically, this kind of phase estimation approach does not require a high sampling frequency, and it provides a better accuracy than the directly measured time delay estimation approach [5,6]. A slight change in density ρ will therefore appear as a change in $\Delta\varphi$ and a theoretical output of the system response is depicted in Figure 16.6.

16.2.3.1 Estimation of Phase Variations

The phase fluctuations, $\Delta\varphi_{change}$ shown in Figure 16.6, can be directly related to a density fluctuation $\Delta\rho$ in the brain tissue. First, $\Delta\varphi_{change}$ can be related to the ultrasound pulse traveling time by the following equation,

$$\Delta\tau_{change} = \frac{\Delta\phi_{change}}{2\pi f}. \tag{16.5}$$

If we assume the dimensions, D, of the skull that remains constant, the traveling time fluctuation can only be caused by an intracranial change in the ultrasound velocity.

$$\Delta\tau_{change} = \left(\frac{2D}{c_{i+1}}\right)_{t_{i+1}} - \left(\frac{2D}{c_i}\right)_{t_i} \tag{16.6}$$

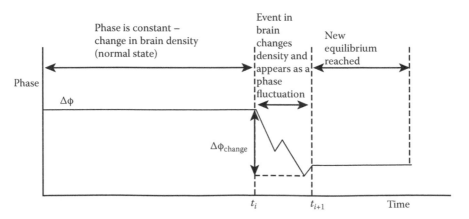

FIGURE 16.6
System response as defined by an estimation process of phase ($\Delta\varphi$) variations relative with the monitoring relative density fluctuations.

Thus, a change in density (i.e., $\Delta\rho$) is responsible for the change in velocity, c, since $K(f)$ in Equation 16.1 is constant for a specific frequency and medium. Equation 16.6 can be modified to include the parameters $\Delta\rho$ and $\Delta\varphi_{change}$,

$$\Delta\tau_{change} = \frac{2D}{\sqrt{K/\rho+\Delta\rho}} - \frac{2D}{\sqrt{K/\rho}} \tag{16.7}$$

and because of Equation 16.5, we have

$$\Delta\phi_{change} = 2\pi f \frac{2D}{\sqrt{K}}\left[\sqrt{\rho+\Delta\rho} - \sqrt{\rho}\right] \tag{16.8}$$

Although estimates of changes in phase (i.e., $\Delta\varphi_{change}$ in Equation 16.8) provide greater sensitivity than the corresponding changes in $\Delta\tau$, they cannot always determine $\Delta\tau_{change}$ accurately because of phase wrapping. In particular, if the measured phase change, $\Delta\varphi_{change}$ is greater than plus or minus half a wavelength, Equation 16.8 will not be sufficient in monitoring density variation. $\Delta\varphi_{change}$ is only reliable for a small range of $\Delta\tau_{change}$ in which the phase does not wrap.

As an example, consider a narrowband ultrasonic pulse with a frequency of 1.5 MHz, traveling through a medium with a phase velocity of 1500 m/s and with period of 0.667 µs. By considering the dimensions of an average size skull to be, $D = 10$ cm and assuming $\Delta\varphi$ is initially half a wavelength (π or 180°), the percentage density change that can be detected accurately before phase wrapping can be calculated from Equation 16.8. As a result, the range of 0.667 µs relative time delay fluctuations correspond to a ±0.5% change in brain density. This range cannot sufficiently cover larger changes in brain density; in fact, the changes in $\Delta\varphi$ (assumed to be π in this example) are essentially random so this range is the best-case scenario. If the variations in $\Delta\varphi$ are 350° for example, only an increase of 10° can be detected accurately. An example of the phase wrapping effect and the limitations and a system's sensitivity in providing estimates of $\Delta\varphi$ is shown in Figure 16.7.

Hence, in order to determine accurately $\Delta\tau_{change}$ from $\Delta\varphi_{change}$, the number of wavelength shifts (i.e., phase wrapping effect expressed by N_{change}) needs to be accounted for. Thus, Equations 16.5 and 16.8 can be modified to form Equations 16.9 and 16.10 below to account for the number of wavelength shifts.

$$\Delta\tau_{change} = \frac{2\pi N_{change} + \Delta\phi_{change}}{2\pi f}, \tag{16.9}$$

$$\Delta\phi_{change} = 2\pi f \frac{2D}{\sqrt{K}}\left[\sqrt{\rho+\Delta\rho} - \sqrt{\rho}\right] - 2\pi N_{change}. \tag{16.10}$$

Estimates of N_{change} can be derived from two transmitted ultrasound pulses with frequencies that meet the following criteria.

$$f_2 = \frac{n}{m} f_1, \quad m,n \in N, \tag{16.11}$$

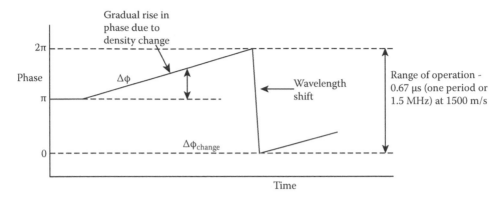

FIGURE 16.7
Range of phase variations with phase wrapping effect for a monochromatic frequency at 1.5 MHz and velocity 1500 m/s.

and

$$\Delta\phi_{change}(f_1) = 2\pi f_1 \frac{2D}{\sqrt{K(f_1)}}\left[\sqrt{\rho + \Delta\rho} - \sqrt{\rho}\right] - 2\pi N_{change} \qquad (16.12)$$

$$\Delta\phi_{change}(f_2) = 2\pi f_2 \frac{2D}{\sqrt{K(f_2)}}\left[\sqrt{\rho + \Delta\rho} - \sqrt{\rho}\right] - 2\pi N_{change} \qquad (16.13)$$

The main concept here is that the beating effect between the two transmitted frequencies forms the basis to extend the sensitivity and operating range of the system through the beat period, T_{beat} in order to derive accurate estimates of N_{change}. The beat period between the two transmitted frequencies, in Equation 16.11 is defined by,

$$T_{beat} = \frac{1}{|f_2 - f_1|}. \qquad (16.14)$$

Equations 16.11 and 16.14 can be combined to form the following relationship:

$$T_{beat} = \frac{m}{n-m}T_1, \quad \text{assuming that } n > m, \qquad (16.15)$$

The following example demonstrates the process of using the phase of the beat period of two frequencies to calculate N_{change}. In this example and according to Equation 16.11, the two transmitted frequencies have been chosen to be, $f_2 = (5/4)f_1$. Figure 16.8 provides a schematic representation of the phases for f_1, f_2 and f_{beat}. In this example the phase of the beat frequency can be derived also from the following relationship,

$$\Delta\phi_{beat} = \Delta\phi_2 - \Delta\phi_1 \bmod 2\pi. \qquad (16.16)$$

Figure 16.8 illustrates schematically the transition range of the wavelength shifts for the two beating frequencies f_1 and f_2 as the beat frequency goes from 0 to 2π, during one beat period, T_{beat}. In particular, the transmit frequency f_1 shifts from 2π to 0 four times, and the other transmit frequency f_2 shifts five times. In other words, for f_1 there are four wavelength

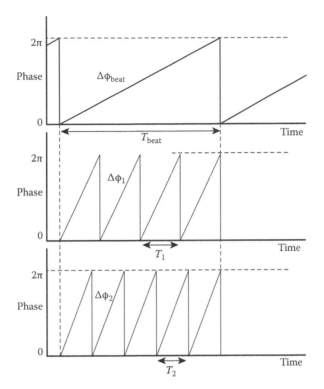

FIGURE 16.8
Schematic representation of phase configurations for two frequencies and their beating effect when the two frequencies are related by $f_2 = (5/4)f_1$.

shifts ($N_{change} = 4$) and for f_2 there are five wavelengths shifts ($N_{change} = 5$). These changes occur at unique points in the beat period T_{beat} so that the corresponding wavelength shifts (N_{change}) can be counted by using a look-up table. Moreover, the system's sensitivity to extend the range of operations in $\Delta\tau_{change}$ is extended to T_{beat} since the wavelength shifts in that range will be known. The values of the phase of the beat frequency at which wavelength shifts occur for either one of the two transmitted frequencies are shown in Figure 16.9.

Two important issues must be considered when choosing T_{beat}, according to Equations 16.11 and 16.14. Although a long beat period has the advantage of creating a wider range of operations, more wavelength shifts of f_1 and f_2 must be resolved over the same range of 2π. If the spacing between wavelength shifts becomes smaller than the accuracy of the phase measurements, the measurement will no longer be reliable.

Moreover, to fully exploit the entire T_{beat} range, the initial $\Delta\varphi_{beat}$ should be close to π or 180°. In Figure 16.9, it is evident that the f_1 and f_2 wavelength shifts have greater angular difference when near the center of the π range. The initial beat frequency phase is dependent on the initial phases of the individual frequencies, which are in turn dependent on the frequencies of the ultrasound signals used, the density of the medium interrogated and its dimensions. As a result, a practical approach is to transmit variable frequencies, and then adjust the frequencies to make the initial beat frequency phase π, while still meeting the criteria described by Equation 16.11.

A practical approach to create fixed or variable transmit frequencies is through the use of a direct digital synthesizer (DDS). With the use of a DDS, the system is not limited to using only two transmit frequencies, but can utilize a number of transmission frequencies

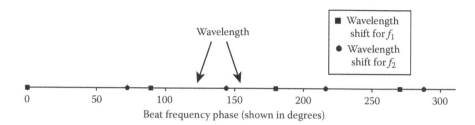

FIGURE 16.9

Beat frequency values for wavelength shifts. Square points represent wavelength shifts for transmit frequency f_1 and circular points are for transmit frequency f_2, when the two frequencies are related by $f_2=(5/4)f_1$.

$(f_n, n=1, 2, ..., N)$. Thus, a DDS based system concept can have more than one T_{beat} for as long as all the transmitted frequencies meet the criteria defined in Equation 16.17, which is an extension of Equation 16.11.

$$f_1 = \frac{n_2}{m_2} f_2 = \frac{n_3}{m_3} f_3 = ..., \quad \text{where} \quad n_x, m_x \in N \tag{16.17}$$

The advantages of having a large number of different T_{beat} periods permits different operating ranges and allows for multiple estimates of N_{change} and $\Delta\varphi_{\text{change}}$, thereby improving the precision of the system and allowing for accurate representation of the dispersive properties of the brain (i.e., $K(f)=\gamma B_T(f)$), expressed by Equation 16.1. In fact this kind of system implementation concept in using a DDS device to accommodate the requirements illustrated by Figure 16.9 and Equation 16.17 differentiates the present approach [7] from other innovations [5,6]. Along the same lines and since the phase of the beat frequency is determined by the phase of the individual frequencies, the ultrasonic pulse signals can be transmitted separately, as shown in Figure 16.10.

16.2.3.2 Selection Criteria for Transmitted Frequencies and their Sampling Rate

As discussed in the previous section, the beat period is an important consideration in frequency selection, but other parameters limit the choice of the transmitted frequencies. In particular, the very high attenuation rate of ultrasound waves, travelling through the human skull to probe the brain, imposes a strict limit on the frequency range of the ultrasound pulses that can penetrate the skull. This limited frequency range is between 0.5 MHz and 2.5 MHz, which is sufficiently wide to allow for the proper selection of a large number of transmission frequencies for an effective excitation of the dispersive nature of the interrogated medium with ultrasound pulses.

Another selection criterion for the transmitted frequencies is determined by the sampling rate. More specifically, an analog-to-digital converter (ADC) will convert the ultrasonic pulses into digital form to allow a microprocessor to calculate the relevant phase information from the transmitted and received pulses, as discussed in the previous section; and the sampling frequency, f_{samp}, of the ADC needs to satisfy the Nyquist criteria. It has been suggested, however, that for the particular application discussed in this chapter, there is no need to satisfy the Nyquist criteria because only the phase information is needed, making under-sampling a possibility [11]. The condition for the

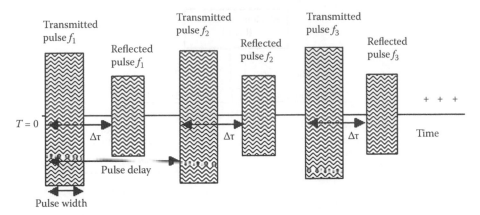

FIGURE 16.10
Schematic representation of multiple set of transmit and receive pulses.

sampling frequency is that complete periods of the respective signals are sampled. This condition, is given in Equation 16.18,

$$f_{\text{samp}} = \frac{n_1}{m_1} f_1 = \frac{n_2}{m_2} f_2 = \frac{n_3}{m_3} f_3 = ..., \quad \text{where } n_x, m_x \in N \tag{16.18}$$

which is the same with that for the transmitted frequencies defined by Equation 16.17, but their difference is the addition of f_{samp}. Since the transmit frequencies, according to Equation 16.18 are variables, with f_{samp} being a fixed parameter, this restriction limits the number of transmit frequencies that satisfy Equation 16.18. Thus, an ADC peripheral that can provide variable sampling frequencies will allow unlimited frequency choices, but will add to the hardware complexity of the system.

There is also an obvious limitation on the pulse width of the transmitted signal, as defined by $t_i < D/c$, in order to avoid overlap between the transmitted and reflected pulse. The pulse repetition rate must also be chosen to avoid overlap between the reflected and transmitted pulses of different frequencies.

Thus, a system implementation approach needs to consider a wide range of parameters to account for requirements relevant with the number of pulses, as well as pulse frequency, width, delay and repetition rate, f_{rep}. The repetition process of the transmitted pulses is essential to continuously monitor the phase relationship ($\Delta \varphi$) with respect to the first observation and therefore track relative brain density fluctuations as functions of frequencies in order to identify the dispersive properties of the interrogated medium. Figure 16.11 illustrates schematically a simplified overview of the basic peripherals that are required to accommodate the processing for the dispersive ultrasound concept discussed in this chapter.

Another practical issue is the coupling between the transmit-receive transducer and the medium to be interrogated with the ultrasound pulses. When a single transducer is deployed, a requirement for a high signal to noise ratio for the reflected signal can be satisfied when both the opposing walls of the confined medium are parallel. This kind of restriction creates a limited selection for the locations of the transducer's position on the human skull (i.e., over the temporal bone) in order to achieve a good transducer-medium coupling interface that allows optimal reception of the reflected signals.

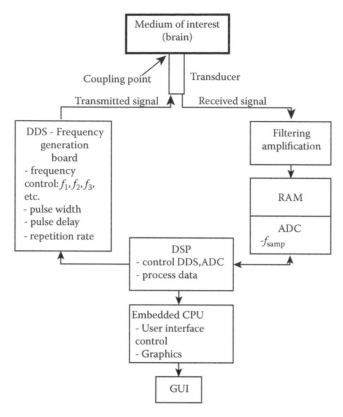

FIGURE 16.11
A simplified system concept of the dispersive ultrasound technology.

To minimize the impact of these restrictions, an alternative option may include the deployment of a circular transducer array that includes multiple detectors. This kind of array-sensor configuration allows for the formation of digital 3-dimensional (3D) steering beams using beamformers that are defined in Stergiopoulos [9] and in Chapter 3. As a result, the digital 3D beam steering would allow detection of the reflected signals from different directions, as depicted schematically in Figure 16.12.

16.2.3.3 Estimation Process for Phase Variations for Dispersive Ultrasound

The processing steps, discussed in this chapter, for monitoring phase variations require accurate estimation of the phase properties of transmitted and received pulses. Thus, the resolution of an ADC peripheral will directly influence the accuracy of a phase estimation process and as an example, a 12-bit ADC peripheral will provide sufficient resolution to keep errors in a phase estimation process smaller than 1°. Moreover, the width of the ultrasonic pulses and the sampling frequency will determine the number of data points that can be used to calculate the phase of the pulse, with the pulse-width being limited by $t_i < D/c$, which suggests that for average human skull dimensions of 10 cm and a phase velocity of 1540 m/s for the brain medium, the maximum pulse width of an ultrasound pulse should be approximately 65 μs.

In what follows, we present two estimation processes for the phase of the transmitted and reflected pulse. The first step in both procedures is to find the position of the transmitted and reflected pulses, in order to calculate the phase.

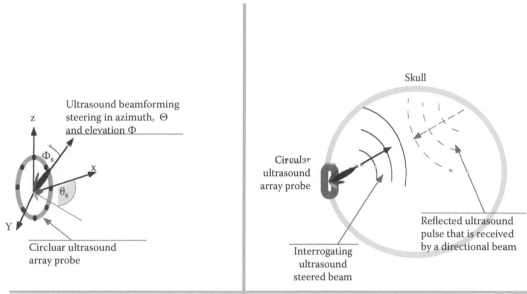

FIGURE 16.12

Sensor configuration for ultrasound circular array for 3D active beam steering and reception of reflected ultrasound pulses propagating in the cranium.

For the first method, the positions of the transmitted pulses are predetermined because pulse width and pulse delay are controlled by the DDS frequency generation peripheral. The location of the reflected pulses will vary depending on the physical properties (i.e., the speed of sound and the density) and dimensions of the medium of interest. Then we use the Hilbert transform to calculate the envelope of the signal, which will assist to identify the edges of the reflected pulses and therefore to locate the position of all the reflected pulses. The next step requires an interpolation method that includes FFT processing to select sample points from the center of each pulse. For example and according to Equation 16.18, if $f_{samp=}(7/3)\,f_1$, then seven sampling points would represent three complete periods of f_1. Interpolation for the seven sample points is achieved with zero padding and a hamming window applied on the data. Next, by taking the FFT of the segmented data, we choose the frequency bin closest to the transmitted frequency, f_1, f_2, f_3, etc., which is used to determine the phase of each pulse. A summary of the processing steps of this method is shown in Figure 16.13.

The second alternative method uses an analytic Fourier transform described in Szustakowski et al. [11]. The algorithm determines the DFT coefficients of the transmitted frequencies, which are used to calculate the phase of the wave. It also requires samples of complete periods to determine the phase. This method reduces the computational complexity without sacrificing accuracy, when compared to the FFT method, because only one coefficient is calculated. In addition, a specific frequency needs to be chosen instead of finding the closet bin, as in the FFT method, shown in Figure 16.13.

There are, however, inherent problems with both methods in that a phase estimation approach for CW signals requires a very high signal to noise ratio for both the transmitted and reflected pulses. The alternative is to use broadband pulses, such as frequency modulated (FM), instead of CW signals. Then both phase estimation approaches for dispersive ultrasound systems can be robust by having phase calculations in the frequency domain

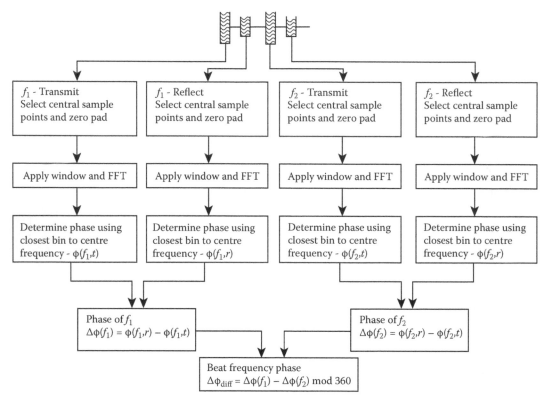

FIGURE 16.13
Processing steps of an FFT based phase estimation approach for dispersive ultrasound system concepts.

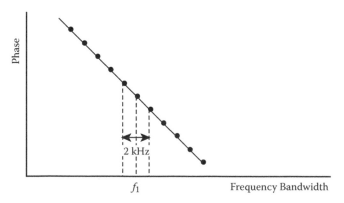

FIGURE 16.14
Schematic representation of the linear relationship between phase and frequency.

for many frequency bins around the center frequency of the FM pulses. For example, the FFT based algorithm in Figure 16.13, can use eleven phase estimations for eleven frequency bins with 2 kHz spacing in bandwidth. Since the relationship between the eleven phase estimates and the relevant frequencies is linear, as shown in Figure 16.14, this linearity can be used to test whether any noise or interference has influenced the system. For example, if one or more of the data points deviate from the expected linear relationship,

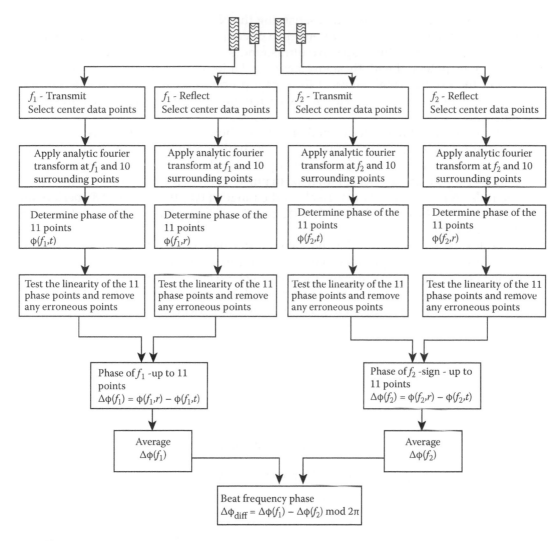

FIGURE 16.15
Processing steps for the analytic Fourier method for phase estimation.

the erroneous point is removed, increasing the robustness of the system. The remaining data points are averaged to determine the phase. In addition, the accuracy of the measurement can be estimated by correlating the data points with the expected linear relationship. A summary of the processing steps of the analytic Fourier transform method is shown in Figure 16.15.

16.2.3.4 Experimental Testing with Water Phantoms

Testing of the experimental prototype using water phantoms has already been discussed in Section 16.2.1. The experimental prototype in Figures 16.1, that has been used to evaluate the dispersive ultrasound processing concepts discussed in this chapter, has limited capabilities; it does not use a wide range of frequencies, nor does it employ a circular transducer array for 3D digital beam steering. It uses only two fixed transmit frequencies with a fixed sampling rate satisfying the Nyquist criteria and a single trans-ducer. However, the test results, presented in Sections 16.2.1 and 16.2.2, have demon-strated the capabilities of the dispersive ultrasound system concept, while providing an indication of its potential with the abovementioned additional features. In what follows, we present the experimental values of the signal processing parameters, defined in the above sections that are relevant with the experimental water phantom results presented in Figure 16.2.

The characteristics of the experimental prototype, shown in Figure 16.1 consist of a par-tially omni-directional ultrasonic transducer with both transmitting and receiving capa-bilities, a multifrequency signal generator peripheral that was set to transmitted two CW pulses at $f_1 = 1.538$ MHz and $f_2 = 1.481$ MHz, with pulse width in the range of 0.1 μs to 13.1072 ms. The pulse repetition rate was variable in the range of 0.1–10 Hz. The ADC peripheral (i.e., Gage CS12100 ADC card installed in PC-based computer) had 12-bit resolution with 10 MHz sampling rate.

This prototype was tested under the scenario discussed in Section 16.2.1 by using a water phantom cylindrical plexiglass container with 10 cm diameter. The transducer was placed on one side of the container and the coupling between the container and transducer was maximized using standard ultrasonic probe gel. The signal response from this experi-mental setup is shown in Figure 16.16. The expected time delay between the transmitted and received signals was,

$$\Delta \tau \approx \frac{2(0.1\,\text{m})}{1481\,m/s\,(\approx \text{speed of sound in water})} = 135.04\,\text{μs}\,(\approx 1350\,\text{points}),$$

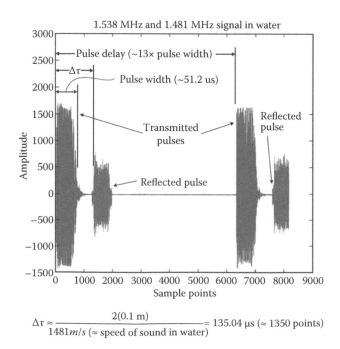

$$\Delta \tau \approx \frac{2(0.1\,\text{m})}{1481\,m/s\,(\approx \text{speed of sound in water})} = 135.04\,\text{μs}\,(\approx 1350\,\text{points})$$

FIGURE 16.16
Dispersive ultrasound signal response for water phantom.

and the results in Figure 16.2 provide estimates of $\Delta\varphi$, that have been calculated according to the FFT based processing methods shown in Figures 16.13 and 16.15. These $\Delta\varphi$ variations correspond to the linear decrease in water temperature from 41.2°C to 36.9°C.

16.3 NonInvasive Blood Pressure Methodologies and their Challenges

Blood pressure and heart rate are critical vital signs to health care practitioners. Measurements are taken under various conditions, from routine check ups to emergency evacuations. The blood pressure technique, which is the golden standard by health care practitioners, is the auscultatory method that incorporates a stethoscope and a mercury sphygmomanometer and consists of three basic steps that are schematically illustrated by the following three figures. During the first step, as shown in Figure 16.17, the cuff is inflated to a level one would consider higher than the systolic pressure. The artery is occluded and blood flow is stopped, therefore the artery is still silent. The pressure vs. time illustrated in Figure 16.17 shows the pressure in the cuff during the period that the cuff deflates. As the cuff pressure is slowly deflated (i.e., Step 2, Figure 16.18) a point is

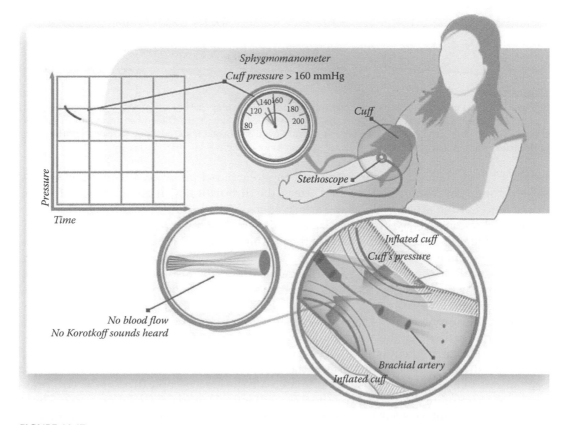

FIGURE 16.17
Schematic representation of auscultatory blood pressure method. Step 1: The cuff is inflated to a level one would consider higher that the subject's systolic pressure. The artery is occluded and blood flow is stopped.

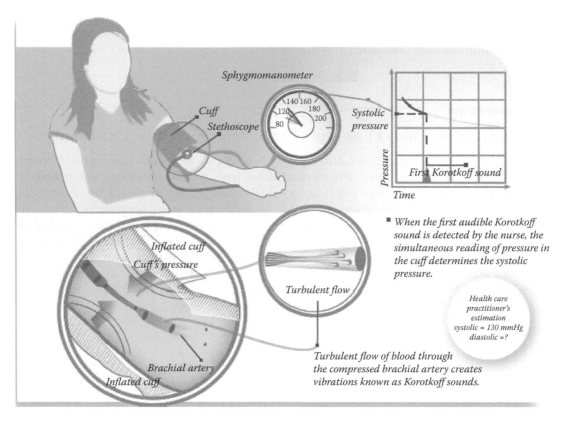

Sphygmomanometer

Cuff
Stethoscope

Systolic
pressure

First Korotkoff sound

Pressure

Time

■ *When the first audible Korotkoff*
sound is detected by the nurse, the
simultaneous reading of pressure in
the cuff determines the systolic
pressure.

Health care
practitioner's
estimation
systolic = 130 mmHg
diastolic =?

Inflated cuff
Cuff's pressure

Turbulent flow

Turbulent flow of blood through
the compressed brachial artery creates
vibrations known as Korotkoff sounds.

Brachial artery
Inflated cuff

FIGURE 16.18

Schematic representation of auscultatory blood pressure method. Step 2: The cuff pressure is slowly deflated, a point is reached where the cuff pressure equals the systolic blood pressure. Blood begins to jet through the compressed brachial artery creating audible turbulent flow (i.e., Korotkoff sounds) due to friction of the blood at the artery wall, as the artery opens while the cuff pressure changes.

reached where the cuff pressure equals the systolic blood pressure. When blood begins to jet through the compressed brachial artery, the turbulent flow is audible due to friction of the blood at the artery wall, as the artery opens while the cuff pressure changes. The audible vibrations created from the artery walls are called the Korotkoff sounds [12,13]. Once the deflated cuff's pressure reaches the person's diastolic blood pressure, the artery is widely open. Then, the flow of blood in the brachial artery becomes laminar and smooth, and the Korotkoff sounds disappear. The moment that the last audible Korotkoff sound is heard by the health care practitioner, a person's diastolic blood pressure is defined (i.e., Step 3, Figure 16.19).

Most of the automated blood pressure systems, however, are based on the oscillometric technique rather that the auscultatory method. The oscillometric technique is simple and no stethoscope is required between the cuff and the subject's skin [12]. While blood is flowing through the patient's artery, the throbbing of the artery causes microscopic oscillations that are superimposed (modulated) on the macroscopic drop of the cuff pressure as illustrated in Figure 16.20 by the the oscillometric pressure deflation curve. These oscillations are interpreted in lieu of Korotkoff sounds. To estimate the systolic and diastolic blood pressure, the peak oscillation signifies the mean pressure. The systolic, diastolic pressure estimates are derived from the largest positive and negative derivatives

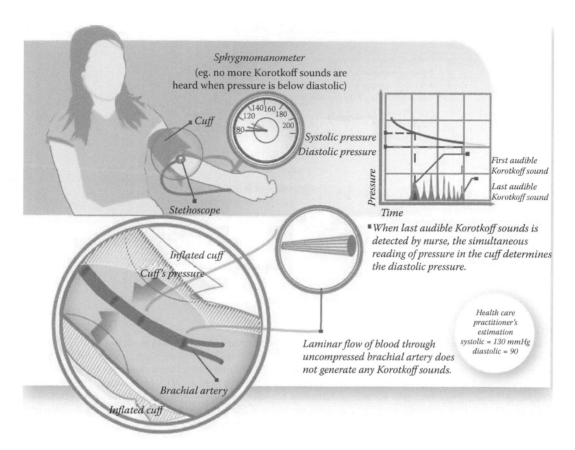

FIGURE 16.19

Schematic representation of auscultatory blood pressure method. Step 2: Once the deflated cuff's pressure reaches the person's diastolic blood pressure, the artery is widely open. Then, the flow of blood in the brachial artery becomes laminar and smooth, and the Korotkoff sounds disappear. The moment that the last audible Korotkoff sound is heard by the health care practitioner, a person's diastolic blood pressure is defined as illustrated by the pressure vs time curve.

of the envelope of the oscillatory response relevant with the Korotkoff sounds, as illustrated in Figure 16.20.

Due to the low signal-to-noise ratio (SNR) inherent by the oscillometric method, when measuring blood pressure, any minor movement by the patient or speaking activity will induce random noise on the deflated pressure signals, that would compromise the measurement and mask the actual event. Thus, the above low SNR properties make the oscillometric method sensitive to movement and as a result the automated blood pressure systems that use this approach tend to be sensitive to patient's movement and environmental vibrations. Thus, the use of these systems is restricted to motionless environments to ensure reliable measurements [12].

Furthermore, in the noisy environments of ambulances, helicopters, airplanes or naval vessels, the environmental noise frequently overwhelms the acoustic signals of interest, making almost impossible to hear the patient's Korotkoff sounds and measure a patient's blood pressure when the traditional auscultatory method is used. Thus, in noisy and vibration intensive environments the pressure fluctuations caused by these disturbances are sizable compared to the pressure fluctuations that need to be detected

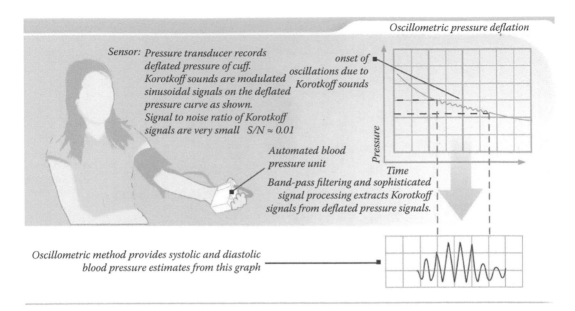

FIGURE 16.20
Schematic representation of the oscillometric technique. The technique is simple and no stethoscope is required between the cuff and the subject's skin. While blood is flowing through the patient's artery, the throbbing of the artery causes microscopic oscillations that are superimposed (modulated) on the macroscopic drop of the cuff pressure as illustrated by the oscillometric pressure deflation curve. These oscillations are interpreted in lieu of Korotkoff sounds. To estimate the systolic and diastolic blood pressure, the peak oscillation signifies the mean pressure. The systolic, diastolic pressure estimates are derived from the largest positive and negative derivatives of the envelope of the oscillatory response relevant with the Korotkoff sounds.

for the proper operation of both the auscultatory and oscillometric techniques, thereby reducing their accuracy.

There are, however, alternative blood pressure estimation techniques that can have limited success in these challenging environments. For example, the Propaq® [14], a widely accepted vital signs monitor for use in helicopters and emergency environments, synchronizes the oscillometric data with that of an ECG (i.e., ECG gating) to remove random noise that is not related to blood pressure flow signals. This approach, however, makes the blood pressure measurements much more complex. For a standard diagnostic ECG, up to 15 leads may be required. The minimum number of leads required is two. Affixing the leads may require cleaning the site, and shaving or clipping hair. The patient is usually required to remain still and hold his or her breath occasionally, and all jewelry must be removed. However, in a medical emergency, it is critical that medical staff can determine such standard information as blood pressure and heart rate with minimal hassle and in the least time. A technique is needed which accounts for noise and vibrations, yet, which is simple to operate and not time consuming.

Potential system solutions to this challenging problem will have to employ a range of advanced signal processing techniques to improve the detection of the Korotkoff sounds in noisy and vibration intense environments. These techniques can include band pass filtering, adaptive interference cancellation and peak discrimination by means of pattern recognition. The use of these advanced signal processing techniques leads to a feasible system for providing vital signs measurements in challenging environments. However the tradeoff for a system that involves this type of signal processing is the greatly increased computational load that is brought on by the addition of the signal processing modules. This means a far more complex system design that is powered by a digital signal processor (DSP) at the core, and includes all of the necessary supporting peripherals and components. The block diagram in Figure 16.21 shows a generic layout for an advanced blood pressure measurement system. The complexity of the system design that is required is immediately evident from Figure 16.21 and the details will be discussed later in the section addressing the computing architecture requirements for advanced blood pressure devices.

Although, the signal processing challenges for robust and reliable electronic blood pressure devices have not been fully addressed yet, many investigations have proposed interesting approaches [13,16] that attempt to reduce the impact of noise effects on blood pressure measurements. One of them [16] introduced the concept of strategic placement of microphones along the brachial artery to reject random noise effects and amplify the Korotkoff signals of interest by simple subtraction of signals from two sets of microphones.

It is the authors' of this chapter strong belief that promising solutions, that can address the signal processing challenges associated with blood pressure systems, will have to include adaptive interference cancellation (AIC) techniques that have been proven to be highly successful in numerous other very complex system applications [9].

Along these lines, it has been chosen to present in this chapter the concept of an adaptive auscultatory method [13] for noninvasive blood pressure monitoring that is based on the implementation of the AIC to remove the effects of noise and vibration from the audible Korotkoff sounds. The system makes use of two sets of acoustic sensors in the pressure cuff, as shown in Figure 16.22. One set of sensors, consisting of a microphone array, is placed over the brachial artery, where an examiner would place the bell of a stethoscope if the blood pressure were being measured conventionally. The other sensor is placed on

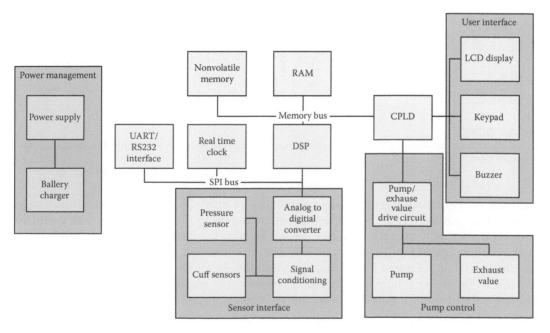

FIGURE 16.21
A generic layout of a computing architecture for an advanced blood pressure measurement system.

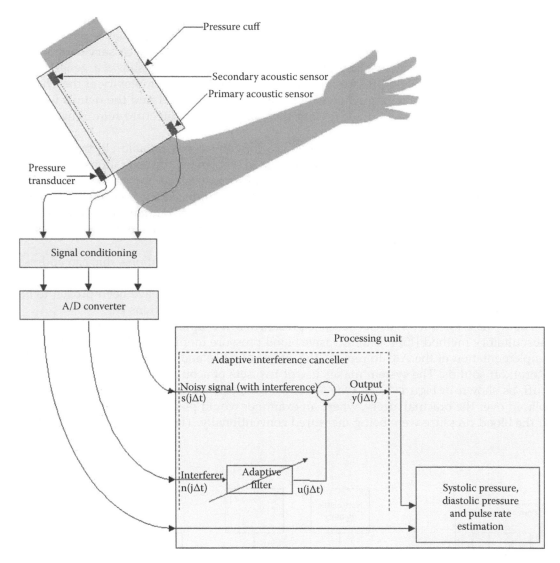

FIGURE 16.22
Schematic representation of a system concept employing adaptive interference cancellation for automated blood pressure monitoring systems.

the back of the patient's arm, where it detects noise only. Then, the adaptive interference canceller is used as a nonlinear filter to remove the noise (interference) from the audible Korotkoff sounds received by the first sensor, as depicted in the "processing unit" shown in the bottom half of Figure 16.22. The AIC is useful in any situation where nonlinear noise interference that is embedded in the signal of interest, can be accurately measured, as is the case in the field measurement of blood pressure. The use of a sound transducer (microphone) in the pressure cuff has been found very effective in measuring blood pressure when the detection of the acoustic signal by the unaided human ear is extremely difficult [13,15].

16.3.1 System Concept of an Adaptive Auscultatory Blood Pressure Technology

The concept of an adaptive auscultatory blood pressure methodology, discussed in this section, is based on relevant developments defined elsewhere [13,17–20]. The objective of the system is to acquire the Korotkoff sounds emanating from the occluded artery of an individual, while reducing all extraneous noises, artifacts and vibration effects by using adaptive interference cancellation and artifact removal procedures. The proposed system detects the Korotkoff sounds with an array of sensors that would enhance detection of very weak Korotkoff sounds and would minimize interferences by applying the following three stages of signal processing. Firstly, adaptive multisensor signal procesing is combined with an array of microphones properly placed inside the pressure cuff. Secondly, adaptive interference cancellation is implemented by using as a reference signals obtained from an external microphone placed on the pressure cuff. Furthermore, an additional third stage of processing is included for the reduction of remaining motion and other physiological artifacts from the signal prior to estimating the blood pressure. In this respect, this approach is unique compared to other blood pressure measuring systems which also employ the auscultatory method to determine blood pressure readings [18–20].

16.3.1.1 Signal Acquisition

The source of the signal of interest is along the brachial artery in the upper arm, hence the signal is travelling along a line. This is depicted in Figure 16.23, which shows also the arrangement of the microphones into a planar array placed at the interior of the pressure cuff. The pressure transducer measures the applied cuff pressure. The external microphone provides a reference signal for noise cancellation. In principle, the microphones could be beamformed using a linear focused beamformer which requires, for each microphone, knowledge of the steering angle directed to the artery, as well the appropriate focal distance to the artery. However, in general the range of angles and focal distances are unknown due to the changing location and the curvature created by placing the sensors on the curved arm every time the pressure cuff is placed on the upper arm. To overcome this problem adaptive beamforming is applied as shown in Figure 16.24 [9,21]. The adaptive beamformer is basically an adaptive multisensor combiner and attempts to achieve

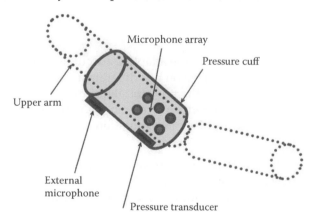

FIGURE 16.23
Position of different sensors on upper arm for the adaptive auscultatory blood pressure technology.

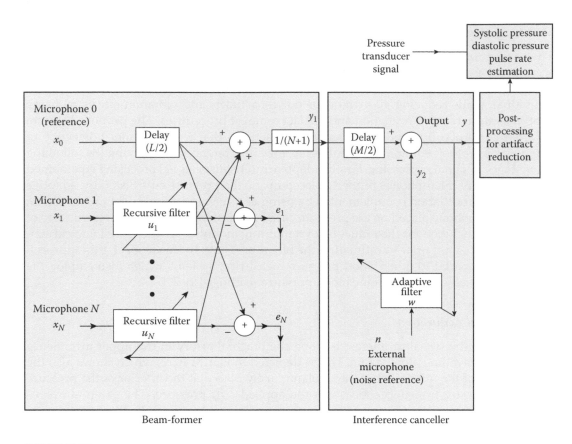

FIGURE 16.24
Signal processing structure for the adaptive auscultatory blood pressure technology.

optimal signal localization and detection, simultaneously. The obtained signals from all sensors are digitized and then processed with the system of Figure 16.24.

16.3.1.2 Signal Processing

The AIC signal processing structure for the adaptive auscultatory blood pressure technology [17] is depicted in Figure 16.24. Based on this diagram, next we provide a brief description of the basic processing stages of the relevant algorithms, which are: adaptive linear beamforming, adaptive noise/interference cancellation, postprocessing for additional artifact reduction and blood pressure/pulse rate estimation.

Adaptive multisensor combining
Assuming that the length of each of the recursive filters is $L+1$, at iteration $k+1$, the adjustment of the filter coefficient vector $\underline{u}_i^k = [u_i^k(-L/2), ..., u_i^k(0), ..., u_i^k(L/2)]$ for each of the N recursive filters $i=1, ..., N$ is carried out using a gradient optimization according to the following equation [19]:

$$\underline{u}_i^{k+1} = \underline{u}_i^k + \mu_1(k) \cdot e_i(k) \cdot \underline{x}_i^k \tag{16.19}$$

where,

$$\underline{x}_i^k = [x_i(k - L/2), ..., x_i(k), ..., x_i(k + L/2)] \tag{16.20a}$$

and

$$e_i(k) = x_0(k - L/2) - \sum_{j=-L/2}^{L/2} x_i(k + j) \cdot u_i^k(j) \tag{16.20b}$$

and the step size $\mu_1(k) = 1/((\underline{x}_i^k)(\underline{x}_i^k)^H)$, where, H denotes conjugate transpose operation. The output of the beamformer/multisensor combiner at iteration k is

$$y_1(k) = \frac{1}{N+1}\left(x_0(k - L/2) - \sum_{i=1}^{N} \underline{x}_i^k \cdot \underline{u}_i^{kT} \right) \tag{16.21}$$

where T denotes transpose operation. In other words, the adaptive beamformer/multisensor combiner tries to minimize the difference between the signals at the reference microphone and each of the other N microphones and add them coherently to achieve a total array gain of $10 \times \log(N)$ at its output.

Adaptive interference canceller
The adaptive interference canceller (AIC) has been shown as an effective method to remove any interference from a signal interest, provided that the interference is measured independently [21]. The use of the AIC in the blood pressure measurement process has been previously described (Figure 16.22) [13,18–20]. With the new sensor configuration described here, there are various options for implementing the AIC. The AIC may be used on each of the individual microphones, or at the output of the multisensor combiner. Performing the AIC processing at the last step is the most efficient implementation, and does not sacrifice any accuracy since the sensor combining process is a linear operation. Thus, given the external microphone interference measurement, the adaptive interference canceller estimates and cancels noisy contributions from the combiner output as follows:

Assuming that the length of the recursive filter is $M+1$, at iteration $k+1$, the adjustment of the canceller coefficient vector $\underline{w}^k = [w^k(-M/2), ..., w^k(0), ..., w^k(M/2)]$ is carried out using a gradient optimization according ... to Equation 16.22,

$$\underline{w}^{k+1} = \underline{w}^k + \mu^2(k).y(k).\underline{n}^k \tag{16.22}$$

where,

$$\underline{n}^k = [n(k - M/2), ..., n(k), ..., n(k + M/2)], \tag{16.23}$$

and the step size is $\mu_2(k) = 1/((\underline{n}^k)(\underline{n}^k)^H)$. The output of the canceller at iteration k is

$$y(k)\cdot = y_1(k - M/2) - y_2(k) = y_1(k - M/2) - \sum_{j=-M/2}^{M/2} n(k + j) \cdot w^k(j). \tag{16.24}$$

Postprocessing for artifact reduction

The goal at this stage is the reduction of artifacts which cannot be removed by the adaptive interference section. Such artifacts arise either during the travelling of the acoustic signal through the body or from physiological motion of the user. Here we propose a novel idea that exploits the repeatability and periodicity properties of the desired signal. The algorithm proceeds as follows:

1. Obtain an estimate of basic period, P, of the signal by calculating the autocorrelation function and finding the second major peak. In other words, given $y(k)$, $k=0,1, …, K-1$ from step (5), calculate

$$R_y(m) = \frac{1}{K} \sum_{k=\max(0,-m)}^{\min(K,K-m)} y(k)\cdot y(k+m), \; m = 0, 1, …, M \tag{16.25}$$

where, $M \approx K/3$ in practice. Since the measured signal has an inherent periodicity corresponding to the heart pulse period, P, the $R_y(m)$ will exhibit peaks at multiples of the pulse period P. Also, since the autocorrelation function is always maximum at $m=0$, it is the second peak of the autocorrelation that correspond to P.

Note: In case of small to moderate arrhythmia, Equation 16.25 will still provide an average value of P. If, however, the arrhythmia is severe then the process will fail. This can trigger an alarm which will prompt the user or physician to extract the blood pressure measurements from the graphical interface, as defined in Section 17.3.3.

2. Remove the mean and divide the entire signal into nonoverlapping segments of length P, that is

Calculate,

$$m_y = \frac{1}{K} \sum_{k=0}^{K-1} y(k) \tag{16.26}$$

$$\tilde{y}(k) = y(k) - m_y, k = 0,1, …, K-1 \tag{16.27}$$

$$z_i(j) = y[(i-1)\cdot P + j], \; j = 0, 1, …, P-1, i = 1, …, [K/P] \tag{16.28}$$

where, $[K/P]$ indicates the integer part of K/P.

3. For each segment, calculate a metric, (e.g., the magnitude of the second coefficient of the P point FFT of the segment envelope), that is

$$Z_i(1) = \left| \sum_{n=0}^{P-1} |z_i(n)| \cdot e^{-j2\pi n/P} \right|, i = 1, …, [K/P] \tag{16.29}$$

The motivation for calculating such a metric stems from the periodic nature of the "desired" signal which is expected to be the same over intervals of one period of P samples. The above metric corresponds to the average power of the first harmonic of the signal.

4. Calculate the mean and variance of the obtained metrics and devise a process to characterize each metric and therefore its corresponding segment either as "normal" or "outlier".

In other words, segments of length P that are either artifacts or significantly contaminated by artifacts will exhibit a significant different value of $Z(1)$ compared to average segment value and thus will be considered as outliers. Based on our experience with real heartbeat pulse data, outliers should be considered those $Z_i(1)$ that are greater or smaller from the mean by at least one standard deviation, that is

$$\text{Obtain the mean: } m_z = \frac{1}{[K/P]} \sum_{i=1}^{[K/P]} Z_i(1) \tag{16.30}$$

$$\text{Obtain the variance: } \sigma_z^2 = \frac{1}{[K/P]} \sum_{i=1}^{[K/P]} (Z_i(1) - m_z)^2 \tag{16.31}$$

Designate as "outlier segment" any segment, $z_i(j)$, with $Z_i(1)$: $|Z_i(1) - m_z| \geq \sigma_z$

5. In the original signal, replace each value of "outlier segments" by the mean of the signal m_y. Keep "normal" segments unchanged.

This process is nothing else but a peak discriminator. The recorded and final "clean" signals from a two sensor implementation of the proposed system are depicted in Figure 16.25.

16.3.2 Pulse Rate and Blood Pressure Estimator

The pulse rate is directly obtained from the "clean" signal obtained at the output of the postprocessing unit described in the previous subsection. The systolic pressure is obtained at the point corresponding to either the greatest magnitude of the positive derivative of the signal's envelope or by the first detected periodic pulse (e.g., Korotkoff sound). The diastolic pressure is obtained either at the point corresponding to the greatest magnitude of the negative derivative of the signal's envelope or by the last detected periodic pulse (e.g., Korotkoff sound). The occurrences of these two points which basically indicate the first and last audible heartbeats, respectively, are referenced to the data acquired by the pressure transducer, which measures the pressure of the deflating pressure cuff as a function of time, as defined by the pressure diagrams of Figures 16.18, 16.19 and 16.26. The corresponding measured pressures, in Figure 16.26 are the systolic and diastolic blood pressure.

16.3.3 Blood Pressure Results Using a Computer User Interface Graph

An automated blood pressure system that will be based on the signal processing structure defined in the previous sections, uses the auscultatory method to determine blood pressure readings. It employs acoustics sensors in the electronic cuff to detect the Korotkoff sounds of a patient during a blood pressure measurement. A unique graphical computer user interface, that can be integrated with the device, has the potential to assist medical practitioners and users to "identify" graphically the Korotkoff sounds in the same way as the medical practitioners "hear" the same sounds using a mercury sphygmomanometer. The Korotkoff sounds, that are sensed by the microphones housed in stethoscope bell-shape sensors, which is the signal after beamforming, noise and artifact reduction, are plotted against the pressure deflation curve on the interface, as shown in Figure 16.26. In this figure, the acquired Korotkoff sounds, are represented by pulses along the time axis.

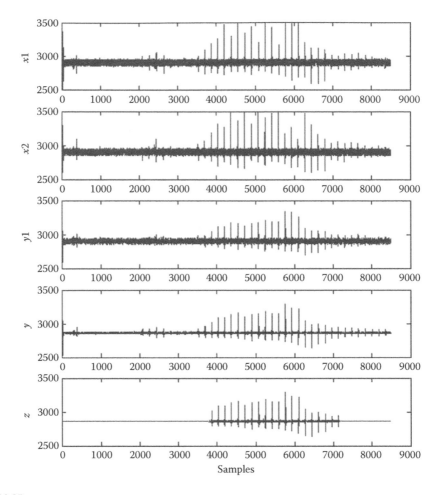

FIGURE 16.25
Examples of signals in different stages of a two sensor implementation of proposed system. From top to bottom, digitized signal at microphone 1, digitized signal at microphone 2, output of beamformer, signal after noise cancellation, signal after artifact reduction.

With the graphic interface depicted in Figure 16.26, a trained user can graphically verify the systolic and diastolic blood pressure and heart rate values of each measurement. The visualization process of the blood pressure graph can aid the health care professionals to accurately determine and verify the blood pressure results provided by an automated blood pressure system and to detect potential abnormalities of the patient's heart beat pattern. Furthermore, the user can check the blood pressure values at any point in the pressure deflation curve, as shown in Figure 16.26. The left hand-side vertical bar shows the position in time of the first audible Korotkoff sound and the corresponding pressure in the deflated cuff, which defines an estimate of a patient's systolic blood pressure. This is synonymous with or in close proximity to the blood pressure when the first heartbeat is heard, at the marked increase in the systolic envelope on the same graph, as detailed graphically by Figure 16.18.

The right hand-side vertical bar in Figure 16.26 identifies the position in time of the last audible Korotkoff sound and the corresponding pressure in the deflated cuff, which

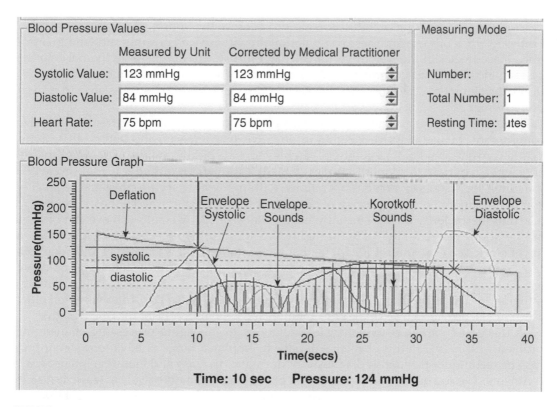

FIGURE 16.26
Graphical user interface with example sample blood pressure record.

defines an estimate of a patient's diastolic blood pressure, as illustrated schematically by Figure 16.19.

16.4 Computing Architecture Requirements

Advanced adaptive algorithms is extremely computationally intensive, requiring the execution of billions of operations per second (GOPS), and relying heavily on the use of fast Fourier transforms (FFTs) for frequency domain operations, such as time delays and spectral analysis. Until recently, conventional DSP architectures were not up to the task. However, most recent innovations in new DSP architectures are fostering the development of advanced medical diagnostic and monitoring equipment that simply was not possible even a few years ago. Support of complex arithmetic, 40-bit or higher extended precision, and DSP architectures that can support 1 GFLOPS at 100 MHz allow the execution of advanced adaptive algorithms for portable, noninvasive diagnostic and monitoring devices that have been discussed in this chapter. The DSP's low clock rate and integration with general purpose processors (e.g., ARM processor) and other peripherals (e.g., A/DC) can help conserve power and space in these kind of portable end-products. The block diagram in Figure 16.21 shows a generic layout for an advanced blood pressure measurement system that is based on the most recent advancements in DSP architecture designs.

Fixed-point computing architectures do not possess the appropriate processing capabilities to efficiently execute these computationally intensive algorithms and tend to introduce inaccuracies in the weight vector calculation that actually increase the noise in the system. Floating-point arithmetic enables much more accurate calculations and provides faster development cycles because the C-code does not have to be translated to a fixed point format.

Thus, 40-bit or higher extended precision architecture can permit extremely accurate computation of the adaptive weights in adaptive interference cancellation algorithms. While conventional floating-point DSP architectures typically require two cycles to execute an FFT, options to support complex domain arithmetic, required for single cycle FFT execution for frequency domain operations, will highly advance progress in the level of sophistication in the system design of portable noninvasive medical diagnostic devices.

While the majority of DSPs usually have a clock of 250 MHz, to achieve the billions of operations per second (GOPS) throughput required for similar type of applications, a lower clock rate (e.g., around 100 MHz) that can sustain also the same level of throughput requirements, is highly preferable. Lower clock rates can keep power consumption down (e.g., around 750 milliWatts), an important consideration for portable monitors. Finally, integration of a DSP with an ARM processor, multichannel ADC/DAC interfaces, and other peripherals, provide a true single chip solution and a small footprint for DSP designs.

As discussed earlier, potential system solutions to the signal processing challenging problems discussed in this chapter will have to employ a range of advanced signal processing techniques to improve diagnostic performance in noisy and vibration intense environments. The use of these advanced signal processing techniques leads to feasible system designs for providing vital signs measurements in challenging environments. However the tradeoff for a system that involves this type of signal processing is the greatly increased computational load that is brought on by the addition of the signal processing modules.

16.5 Conclusion

This chapter describes signal processing structures, in the field of noninvasive vital signs diagnostic applications, that consist of adaptive noise cancellation techniques combined with FIR filters, pattern recognition algorithms, correlation functions and time delay estimation techniques using FFT implementation.

The relevant system concepts, that can be derived from the discussion of this chapter, can address most of the medical requirements for noninvasive monitoring of vital signs in hospital emergency departments, ambulances, hospital intensive and after care, home care and applications for family medical clinics, insurances and old-age nursing homes.

In particular, the concept of the adaptive auscultatory blood pressure system, has been proven successful in noisy and vibration intensive conditions of helicopters and the test results have been reported in Pinto et al. [13].

On the concept of the dispersive ultrasound, it is important to note that similar investigations in the same field are based on time delay estimation techniques by probing the brain with ultrasound pulses at a specific frequency region. In doing so, they disregard information essential to accurately account for the dispersive properties of the human brain that would allow the diagnosis of a wide variety of brain injuries. Thus, a noninvasive ultrasound technology, that takes into consideration the dispersive properties of the human brain (e.g., such as the one discussed in this chapter), has the potential to create a leading technological edge in noninvasive diagnostic applications on traumatic brain injury (TBI), and on security issues relevant with nondestructive identification of contained fluids.

References

1. Obusek, Colonel J.P. Force Maintainer, Military Medical Technology Online, http://www.mmt-kmi.com

2. European Commission, IST Program, http://www.cordis.lu/ist/ka1/health

3. Mayer, S.A., and Dennis, L.L. 1998. Management of increased intracranial pressure. The Neurologist, 4, 2–12.

4. Heifetz, M., and Weiss, M.. 1981. Detection of skull expansion with increased intracranial pressure. *Journal of Neurosurgery*, 55, 811–812.

5. Ragauskas, A., Daubaris, G., and Dziugys, A. 1999. Method and apparatus for determining the pressure inside the brain. US Patent, 5,951,477.

6. Ragauskas, et al. 1995. Method and apparatus for non-invasively deriving and indicating of dynamic characteristics of the human and animal intracranial media. US Patent, 5,388,583.

7. Stergiopoulos, S., and Wrobel, M. 2004. Non-invasive diagnostic ultrasound system monitoring brain abnormalities. Assignee: Defence R&D Canada, US Patent Application: 10/898,208.

8. Freibert, A., Wrobel, M., and Pawlak, M. 2001. Ethics Committee Protocol # 621-2531.01-47/00 Of the State of Bavaria of the Federal Republic of Germany, final report for DRDC-Toronto, Contract W7711-007672/001/TOR.

9. Stergiopoulos, S. 1998. Implementation of adaptive and synthetic aperture beamformers in sonar systems. *Proceedings of the IEEE*, 86(2), 358–396.

10. Stergiopoulos, S., and Ashley, A.T, 1997. An experimental evaluation of split-beam processing as a broadband bearing estimators for line array sonar systems. *The Journal of the Acoustical Society of America*, 102(6), 3556–3563.

11. Szustakowski, M., Jodlowski, L., and Piszczek, M. 1999. Passage time measurement of medium probe signal with the use of phase method with quasi-continuous wave. *Molecular & Quantum Acoustics, Polish Academy of Science*, 20, 279–289.

12. Geddes, L.A., and Moore, A.G. 1968. The efficient detection of Korotkoff sounds. *Medical and Biological Engineering*, 6, 603–609. Also Pergamon Press, UK.

13. Pinto, L., Dhanantwari, A., Wong, W., and Stergiopoulos, S. 2002. Blood pressure monitoring system using adaptive interference cancellation for applications in noisy and vibration intense environments. *Annals of Biomedical Engineering*, 30, 657–670.

14. http://www.welchallyn.com

15. Geddes, L.A., Voelz, M., Combs, C., Reiner, D., Babbs, C.F. 1982. Characterization of the oscillometric method for measuring indirect blood pressure. *Annals of Biomedical Engineering*, 10, 271–280. Also Pergamon Press, West Lafayette, IN.

16. Khan, A.R., Bahr, D.E., and Allen, K.W. 1999. Blood pressure monitoring with improved noise rejection. US Patent, 5,873,836.

17. Stergiopoulos, S., and Hatzinakos, D. 2005. Method and device for measuring systolic and diastolic blood pressure and heart rate. US Patent Application, 60/656,382.

18. Stergiopoulos, S., and Dhanantwari, A. 2003. Method and device for measuring systolic and diastolic blood pressure and heart rate in an environment with extreme levels of noise and vibrations. Assignee: Defence R&D Canada, US Patent: 6,520,918.

19. Stergiopoulos, S., Dhanantwari, A., Pinto, L., Zachariah, R., and Wong, M.Y.W. 2004. Method and device for measuring systolic and diastolic blood pressure and heart rate in an environment with extreme levels of noise and vibrations. Assignee: Defence R&D Canada, US Patent: 6,705,998.

20. Stergiopoulos, S., Dhanantwari, A., Pinto, L., Zachariah, R., and Wong, M.Y.W. 2004. Method and device for measuring systolic and diastolic blood pressure and heart rate in an environment with extreme levels of noise and vibrations. Assignee: Defence R&D Canada, US Patent: 6,805,671.

21. Widrow, B., and Stearns, S. 1985. *Adaptive Signal Processing*. Prentice Hall, Englewood Cliffs, NJ.

17

Examination of Contrast Agents in Medical Imaging Using Physiologically-Constrained Deconvolution

Robert Z. Stodilka
University of Western Ontario

Benoit Lewden
Lawson Health Research Institute

Michael Jerosch-Herold
Brigham & Women's Hospital

Eric Sabondjian
University of Western Ontario

Frank S. Prato
University of Western Ontario

CONTENTS

17.1 Introduction and Background

Contrast agent kinetics is the study of the spatial and temporal changes of contrast agent concentration, as the contrast agent enters into and perfuses a tissue of interest. Contrast agent kinetic analysis has applications in medical diagnostics by helping to characterize the functional state of a tissue, and applications in drug discovery by offering insight into the behavior of the contrast agent itself. Contrast agent kinetic analysis is used often in conjunction with an imaging device that can measure noninvasively the concentration of the contrast agent, at one or more locations, as a function of time.

One fundamental quantity which characterizes the passage of a contrast agent or tracer through tissue is the (mean) transit time, measured between arterial input and venous output. The transit time can be viewed as an average of the transit times of individual transit times. For a region of interest (ROI) where contrast concentration is detected with external detectors it is often more appropriate to consider the residence time in the ROI. The fundamental behavior of the contrast agent in a tissue can be characterized by the distribution of retention times. Conceptually this corresponds to a setting where the tracer is introduced instantaneously in the ROI. The tissue impulse retention function (TIRF), $R(t)$, represents the temporal evolution of contrast agent concentration within a tissue if an ideal bolus—a Dirac delta function—of contrast agent had entered into that tissue without delay [1]. $R(t)$ depends on both the contrast agent properties and the underlying tissue properties, and it can be used to calculate many useful parameters such as blood flow, volume of distribution, extraction fraction, mean transit time, and standard deviation of transit time [2]. From basic physiologic considerations, the expected shape of $R(t)$ is as follows: zero until the contrast agent enters the tissue, at which point $R(t)$ "instantaneously" achieves its maximum value, and then remains constant at this value until the contrast agent begins to leave the tissue. Finally, $R(t)$ returns to zero over a period of time determined by the distribution of retention times of the contrast agent molecules.

In practice, $R(t)$ cannot be observed directly because an ideal bolus of contrast agent cannot be directly injected into the tissue of interest. What is more practical and less invasive is that a nonideal bolus is introduced into the blood at a location upstream to the tissue of interest. As it moves through the blood, the contrast agent concentration is measured by the medical imaging device, as a function of time, in two locations: one after the location of injection but upstream to the tissue of interest, $I(t)$; and one within the tissue of interest $J(t)$. $J(t)$ represents a combination of both $R(t)$ and $I(t)$: the TIRF whose shape represents the sum-total of the response to the bolus. The bolus can be decomposed into a train or comb of ideal impulse inputs, and the response of the system corresponds to a sum of time-shifted and scaled impulse responses, with a scaling factor that corresponds to the amplitude of the actual bolus input at each time point. Mathematically, $J(t)$ is the convolution of $R(t)$ and $I(t)$. The effect of a nonideal $I(t)$ must be corrected in order to calculate $R(t)$ [3]. This correction is known as deconvolution.

Given that $I(t)$ and $J(t)$ are both corrupted by noise, the TIRF can be only estimated as $R'(t)$. There are three major strategies for calculating $R'(t)$: frequency domain strategy, analytical model strategy, and spatial-domain strategy. Deconvolution analysis is difficult because it is inherently a differentiating process that is mathematically unstable and therefore very susceptible to noise in either $I(t)$ or $J(t)$ [4]. The three strategies have evolved from differing opinions regarding stability of numerical methods.

This chapter is divided into three sections. In the first section, we describe the three major strategies for deconvolution, including strengths and weaknesses of each. In the second section, we proceed to develop a novel approach to deconvolution termed physiologically-constrained deconvolution (PCD). PCD is based on the matrix strategy, but overcomes some of the limitations of this strategy by incorporating *a priori* physiologic information. In the third section, we demonstrate the application of our novel approach to measuring cardiac perfusion with both computer simulations and experimental data.

The frequency domain strategy tackles the problem of deconvolution using Fourier transforms. deconvolving $I(t)$ from $J(t)$ in the time domain is a difficult and unstable process. Theoretically, deconvolution could be simplified by using Fourier theory to transform the two functions into their frequency domain counterparts: $FI(v)$ and $FJ(v)$, where v denotes frequency. The frequency domain counterpart of $R(t)$, $FR'(v)$, can now be written as the ratio of $FJ(v)$ and $FI(v)$: $FR'(v) = FJ(v)/FI(v)$. Finally, $FR'(v)$ can be transformed back into the spatial domain to yield $R'(t)$ [5–8]. In practice, the fast Fourier transform (FFT) [9] and the inverse FFT are used to move data between the spatial and frequency domains. FFT algorithms assume that I(t) and J(t) are periodic and without discontinuities [10]. Unfortunately, significant discontinuities appear if the data acquisition is terminated early because of time limitations, while a significant amount of contrast agent remains within the tissue of interest and J(t) has not yet dropped to zero [2]. Such an abrupt end to J(t), coupled with inherently noisy measurements, produces high-frequency oscillations in the calculated R'(t) [8] that have no underlying physiologic basis [4]. To address this concern, some authors have developed techniques to temporally extrapolate the I(t) and J(t) to zero [2,8,10]; however the validity domain of temporal extrapolations of data has not been fully explored.

The analytical model strategy assumes that the TIRF can be described by an analytical model that is derived from assumptions about the behavior of the contrast agent within the biological system under investigation [4,7]. The analytical model of the TIRF has parameters that are initially unknown and must be calculated by fitting the model to the data. In general, reducing the number of parameters improves the stability of the model in the presence of noise. But reducing the number of parameters can limit the model: the final shape of the TIRF is restricted to the family of curves that can be realized by manipulating the parameters of the analytical model. The analytical model strategy is a good choice for situations where an analytical model is known to characterize adequately healthy and pathologic tissue. Models can be formulated in a way such that they provide a direct correspondence between model parameters and the physiological quantities of interest. Analytical models work best in specific well-defined biological systems, and should not be used beyond such systems since it is difficult to know *a priori* the appropriate functional form [8].

The matrix strategy is based on an equivalency between discrete convolution and matrix multiplication: J(t)=I(t)⊗R'(t) can be re-written as **MJ**=**TMI**×**MR'**, where × denotes matrix multiplication; **MJ** and **MR'** are vectors representing the I(t) data and R'(t) values; and **TMI** is a lower-triangular Toeplitz matrix of the I(t) data, also known as a convolution matrix. (See, for example, Equation 5 in Jerosch-Herold et al. [12].) Thus, **MR'** can be expressed as the solution to this linear set of equations [13–15]. The matrix strategy has the advantages of neither requiring assumptions of an underlying analytical model nor requiring curves to go to zero. Unfortunately, this strategy can produce widely-oscillating R'(t) that have no physiological significance. To minimize these effects, I(t) and/or J(t) can be preprocessed or R'(t) can be postprocessed to reduce their oscillatory nature [16,17]. Furthermore, R'(t) derived from matrix strategy methods can lead to R'(t) having values less than zero, which have no physical significance and must be set to zero [17]. Finally, the physiological information which can be derived by the matrix strategy is limited to a few quantities, namely flow and distribution volume.

17.2 PCD: Theory and Methods

The frequency domain, analytical model, and matrix strategies each have limitations. In general, they have a trade-off between potentially nonapplicable physiologic assumptions and numerical stability. In principal, a more ideal strategy would maximize flexibility concerning physiologic considerations while maintaining numerical stability. In laying the groundwork for our novel approach, PCD [18], we first re-visit some basic assumptions about the properties of contrast agents, the measurement system, and the physiologic behavior of the biological system under interrogation.

The first assumption is that of stationarity (or time-invariance), which is required for the validity of the convolution and deconvolution operations [17]. It is assumed that the characteristic behavior of the contrast agent and the tissue do not change during the course of the data acquisition by the medical imaging device.

The second assumption is that MI and MJ are accurately known. In practice, the anatomic region where MI is measured should be in close proximity to the anatomic region where MJ is measured. Referring to Figure 17.1, if Location I is too far upstream from Location J, then the contrast agent will become dispersed prior to entering the tissue at Location J, and MI will not accurately represent the true distribution of contrast agent input before it enters the tissue ROI [1,2]. Ideally, a method should account for dispersion by allowing MR' to begin before its local maximum, to account for any delay of the true input to the ROI, relative to the input measured up-stream, and have freedom in how it increases to that local maximum. The size of Location I should also be small, since measurements over large regions will temporally smooth MI thus decreasing the accuracy of MR'. The second assumption also requires that the medical imaging device be capable of making accurate measurements of the concentration of the contrast agent. In practice, this may require corrections for contaminating signals and effects of limited spatial resolution.

The third assumption concerns contrast agent re-circulation. Referring to Figure 17.1, it is assumed that the contrast agent passes through Location I and subsequently Location J, and that no part of the contrast agent enters Location J without first passing through Location I. If the contrast agent re-circulates after leaving Location J, then any contrast agent molecules that enter Location J for the second time have first passed through Location I a second time. This assumption forms the basis for constraining MR' to have a single local maximum.

The fourth assumption is that the system under observation is linear over the dynamic range of contrast agent concentration [8,17]. That is, an increase in injected contrast agent concentration will result in the same relative increase in the observed contrast agent concentration. From a physiologic perspective, linearity does not exist for some contrast agents

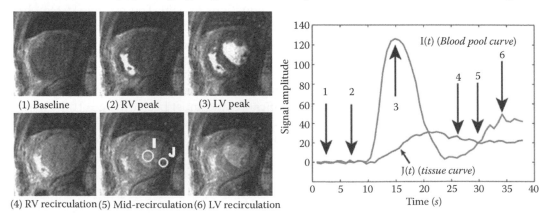

(1) Baseline   (2) RV peak   (3) LV peak
(4) RV recirculation (5) Mid-recirculation (6) LV recirculation

FIGURE 17.1

Cross-section through the short-axis of a canine heart (left images). A contrast agent (Gd-DTPA) was injected, followed by acquisition of dynamic T1-weighted MRI. The images show the progress of the contrast agent first through the right-ventricle (RV), then the left ventricle (LV), and finally into the myocardial tissue. To quantify the passage of this contrast agent, two regions-of-interest (ROIs) were drawn: one over the blood pool (label I) and another over a segment of myocardial tissue (label J). The right image shows a plot of the contrast agent vs time for the two ROIs, where the numbers correspond to the images.

administered at high levels, where receptors or transport mechanisms may become saturated. From an imaging physics perspective, linearity can be compromised by the effects of limited spatial or temporal resolution or nonlinearity in the physical effect which is to be measured [19–21]. However, for most medical imaging device contrast agents, the contrast agents are administered at doses low enough to avoid saturation effects.

With the above stated assumptions in hand, we can proceed with developing a mathematical formalism for an algorithm to calculate an $R'(t)$ in a stable manner. $J(t)$ is formally defined as [1]:

$$J(t) = I(t) \otimes R(t) = \int_0^t I(t - \xi) R(\xi) d\xi, \tag{17.1}$$

where t is a continuous variable of time and $\otimes$ denotes convolution. However, in practice, $I(t)$ and $J(t)$ are measured at discrete points in time at intervals of Δt, for a total of N measurements each. Using zero-order numerical integration, Equation 17.1 is commonly approximated as a discrete time convolution:

$$J(n \cdot \Delta t) = \sum_{m=1}^{N} I(n \cdot \Delta t + 1 - m \cdot \Delta t) R(m \cdot \Delta t) \Delta t, \tag{17.2}$$

where the summation starts at measurement $m=1$ which is taken at time $t=0$. Without loss of generality, herein it is assumed that $\Delta t=1$, recognizing that the time interval between points in $R'(t)$ will be the same as the original interval at which $I(t)$ and $J(t)$ were sampled.

We re-introduce the following variables: **MI** is a vector describing the discrete time measurements of $I(t)$, with elements mi_x, where $x=1, 2, ..., N$; **MJ** is a vector describing the discrete time measurements of $J(t)$, with elements mj_x, where $x=1, 2, ..., N$; **MR** is a vector describing the discrete time function $R(t)$, with elements mr_x, where $x=1, 2, ..., N$; **MR'** is a vector describing the discrete time estimate of the TIRF, with elements mr'_x, where $x=1, 2, ..., N$; and **TMI** as the lower-triangular Toeplitz matrix of **MI** elements. These variables allow us to recast Equation 17.2 as [13]:

$$\begin{bmatrix} mj_1 \\ mj_2 \\ mj_3 \\ ... \\ mj_N \end{bmatrix} = \begin{bmatrix} mi_1 & 0 & 0 & ... & 0 \\ mi_2 & mi_1 & 0 & ... & 0 \\ mi_3 & mi_2 & mi_1 & ... & 0 \\ ... & ... & ... & ... & ... \\ mi_N & mi_{N-1} & mi_{N-2} & ... & mi_1 \end{bmatrix} \times \begin{bmatrix} mr_1 \\ mr_2 \\ mr_3 \\ ... \\ mr_N \end{bmatrix}, \tag{17.3}$$

or more compactly as

$$\mathbf{MJ} = \mathbf{TMI} \times \mathbf{MR}, \tag{17.4}$$

where $\times$ denotes matrix multiplication. The goal is to find an **MR'** such that **TMI** $\times$ **MR'** is the maximum-likelihood estimate of **MJ**. The maximum-likelihood estimate is dependent upon the noise properties of the measured data. For illustrative purposes, we consider two specific cases. First, the maximum-likelihood estimate associated with the Poisson probability density function, which is characteristic of nuclear medicine imaging modalities, such as single photon emission computed tomography (SPECT) and positron emission tomography (PET). If **MI** and **MJ** are subject to Poisson statistics, the maximum-likelihood estimate satisfies the objective function [22]:

$$\max \sum [\mathbf{MJ} \cdot \log (\mathbf{TMI} \times \mathbf{MR'}) - \mathbf{TMI} \times \mathbf{MR'}], \tag{17.5a}$$

where $\cdot$ denotes element-by-element multiplication. Second, the maximum-likelihood estimate associated with the Rician probability density function, which is characteristic of magnetic resonance imaging (MRI) is the objective function [23]:

$$\max \sum \left(\log \left\{ I_0 \left[\frac{(\mathbf{TMI} \times \mathbf{MR'}) \cdot \mathbf{MJ}}{\sigma^2} \right] \right\} - \frac{(\mathbf{TMI} \times \mathbf{MR'})^2}{2\sigma^2} \right), \tag{17.5b}$$

where I_0 is the modified zeroth-order Bessel function of the first kind, and σ is noise. In both Poisson and Rician cases, **TMI** is poorly conditioned in practice; thus there are many **MR'** that satisfy Equation 17.5a and b almost equally well. Having identified this indeterminacy, additional conditions can now be imposed on the family of solutions to Equation 17.5a or b in order to select a preferred **MR'**. These conditions are defined by linear inequality constraints. Three linear inequality constraints on Equation 17.5a and b are introduced to account for the physical and physiological considerations discussed earlier. The first linear inequality constraint ensures the first and last elements of **MR'** are zero or larger (ensuring nonnegativity for all **MR'**):

$$\mathbf{L1} \times \mathbf{MR'} \geq 0, \tag{17.6}$$

where **L1** is a $2 \times N$ matrix with the first element on the first row being 1, the last element on the second row being 1, and the other elements being 0.

Similarly, the second and third linear inequality constrains act together to ensure **MR'** has a single local maximum by enforcing a first derivative greater than zero prior to the maximum and a first derivative less than zero after the local maximum. These can be written as:

$$\mathbf{L2} \times \mathbf{MR'} \geq 0, \tag{17.7}$$

and

$$\mathbf{L3} \times \mathbf{MR'} \geq 0, \tag{17.8}$$

where **L2** and **L3** are numerical differentiation matrices, preferably using Richardson's extrapolation. **L2** is a $P \times N$ matrix where P is the starting location of the local maximum of **MR'**, and $P < N$; and it constrains **MR'** elements mr_x for $x \leq P$. The **L3** restriction operates for $x > P$.

Having defined the variables and functions, we finally express the preferred **MR′** as that which solves:

$$\max \sum [\mathbf{MJ} \cdot \log(\mathbf{TMI} \times \mathbf{MR'}) - \mathbf{TMI} \times \mathbf{MR'}]$$

or

$$\max \sum \left(\log \left\{ I_0 \left[\frac{(\mathbf{TMI} \times \mathbf{MR'}) \times \mathbf{MJ}}{\sigma^2} \right] \right\} - \frac{(\mathbf{TMI} \times \mathbf{MR'})^2}{2\sigma^2} \right),$$

subject to

$$\begin{bmatrix} \mathbf{L1} \\ \mathbf{L2} \\ \mathbf{L3} \end{bmatrix} \times \mathbf{MR'} \geq 0. \tag{17.9}$$

17.3 PCD: Evaluation and Discussion

We sought to evaluate PCD from several perspectives, for the case of data subject to Rician noise.

First, given experimentally measured **MI** and **MJ**, can we produce a reasonable **MR′**? To answer this question, we tested the technique using data previously collected by one of us [24]. Referring to Figure 17.2, middle image: the delay in peak presentation (label A) indicates the contrast agent's transit time between blood pool and tissue curves, the finite

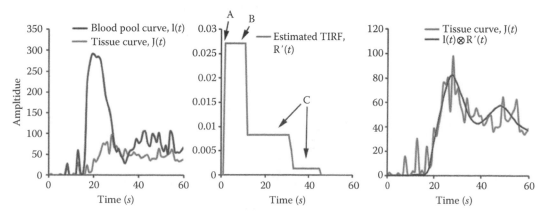

FIGURE 17.2
Left image: contrast agent amplitude vs. time curves for blood pool I(*t*) and tissue J(*t*) [24]. The curves are subject to Rician noise. Middle image: Tissue impulse retention function (TIRF) estimate R′(*t*) calculated using the procedure of Equation 17.9. Quantitatively, the estimate is optimal in the maximum-likelihood sense, assuming a Rician distribution of noise. Right image: the accuracy of R′(*t*) can be inspected visually by convolving I(*t*) with R′(*t*), and superimposing over the original tissue curve J(*t*).

rise time is indicative of contrast agent dispersion between the blood pool and tissue curves, the length of the plateau (label B) indicates the contrast agent's minimum transit time within the tissue ROI. The step-like appearance of R′(t) (label C) indicates quantized transit times of the contrast agent within the tissue, which suggests the presence of discrete capillary populations with differing transit times.

The second method of evaluating PCD was an examination of precision. Since deconvolution is a poorly-conditioned process, we sought to determine the precision in R′(t) as a function of time t. To accomplish this, a synthetic TIRF curve was created, R(t), and convolved with a noise-free ("smoothed") version of the blood pool curve I(t) of Figure 17.2 to produce a noise-free tissue curve. Noise was then added to both the blood pool and tissue curves, such that the final curves had a contrast:noise of 5:1, which is typical of clinically acquired data [24]. Then, from those two noisy curves, R′(t) was calculated according to Equation 17.9. This was repeated for 100 noise realizations. The results of this procedure are shown in Figure 17.3.

The third method of evaluating PCD considered the dynamic range of perfusion estimates. Synthetic R(t)s were constructed corresponding to 1–4 ml/min/g—which is the range of perfusions observed clinically [25,26]. Each of these R(t)s was convolved with a clinically measured blood pool curve (not shown), and clinically-typical Rician noise was added (CNR 5:1). One hundred noise realizations were produced for each perfusion level. Equation 17.9 was employed to estimate the TIRF for each noise realization and perfusion level. The results of this investigation are shown in Figure 17.4.

The fourth method of evaluation was to compare perfusion estimates obtained from PCD with those obtained from a spline-constrained deconvolution technique developed by one of us (MJ-H) [12]. For this evaluation, three volunteers were injected with MRI contrast (Gd-DTPA, 0.04 mmol/kg, 7 ml/s power injected, 10 ml saline flush), and T_1-weighted MRI

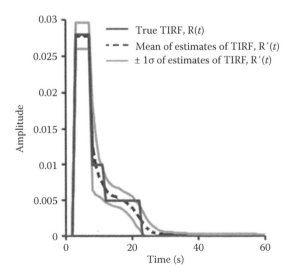

FIGURE 17.3
To evaluate precision as a function of time, a true R(t) was created (solid black), convolved with a blood pool curve, Rician noise added, and then deconvolved. This process was repeated 100 times. The mean R′(t) (100 R′(t) s averaged) is shown as the black dashed curve, along with ± one standard deviation (gray curves). The peak of the mean R′(t), which is defined as the tissue perfusion, coincides with the peak of the true R(t), which suggests that perfusion estimates should be unbiased.

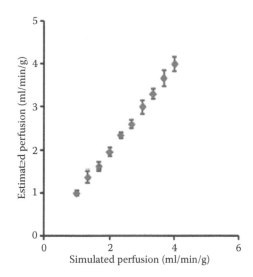

FIGURE 17.4

Results show correlation between simulated perfusion levels and estimates calculated by Equation 17.9 over a range of clinically observed perfusion levels. The diamond symbols indicate means of estimates at each perfusion level, and the bars indicate ± one standard deviation.

(fast gradient echo sequence, three short-axis 8 mm thick slices) acquired for 50 heartbeats to track the contrast through first pass and recirculation. MRI scans were acquired at rest and, for two volunteers, also acquired at adenosine-induced maximal vasodilation. Blood pool and tissue regions-of-interest were defined, from which blood pool $I(t)$ and tissue $J(t)$ curves were generated. For blood pool curves: a region was defined in the center of the left ventricle. For tissue curves: endo- and epicardial contours were traced manually, divided into eight transmural equi-circumferential sectors (not shown). Perfusion was calculated two ways: using the technique of Jerosch-Herold et al. [12] (gold standard), and Equation 17.9. Results are shown in Figure 17.5.

Simulations demonstrate that PCD is numerically stable and estimates perfusion accurately at clinical contrast:noise levels for typical cardiac perfusion values (Figure 17.4). Theoretically, we expect maximum-likelihood estimates of perfusion to be unbiased (example shown in Figure 17.3), and have minimum variance (nearing the Cramer–Rao bound [27]). Perfusion values derived from $R'(t)$s calculated from Equation 17.9 are similar to estimates provided by a widely used clinical perfusion package (Figure 17.5).

Previous deconvolution techniques stabilized deconvolution via smoothing penalty functions or constraining TIRFs to smooth piece-wise functions (such as splines). For example, the method of Jerosch-Herold et al. [12] for determining blood flow quantification uses model-free deconvolution, the matrix strategy and Tikhonov regularization. However, their TIRF is constrained to be a sequence of B-splines (smoothly varying linked polynomial curves); and their implementation of the matrix strategy solves directly for the B-spline coefficients and not the TIRF. Additionally, they do not constrain the shape of their TIRF to have a single local maximum—which is central to our method, nor do they solve for an optimal location of a local maximum.

Our technique removes these constraints, and interestingly, TIRFs calculated by PCD appear step-like, which may indicate discrete capillary populations with differing transit times. Our technique intrinsically models dispersion of the contrast agent between the blood pool and tissue curves, which is not accounted for by the clinical package and may

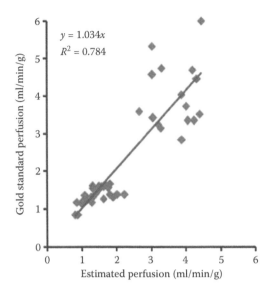

FIGURE 17.5
The method of Equation 17.9 (*x*-axis) was compared with a widely-used technique (*y*-axis, gold-standard) in its ability to estimate perfusion from blood pool curves and tissue curves. The results of the comparison demonstrate general agreement between the two methods, although dispersion is noted visually at higher flow rates.

partially explain discrepancies. Future work could focus on validating PCD against microsphere measurements of perfusion in animal models of heart disease [24,26]. Whether the distribution of transit times indeed allows identification of discrete capillary populations, and whether this is consistent with the observed distribution of blood flows in the myocardium remains to be investigated.

17.4 Conclusions

In this chapter, by introducing the concept of PCD, we have reviewed the different methods in use to extract information of physiological importance from data collected as noninvasive bio-medical imaging methods are used to record the passage of a contrast agent through a tissue or organ. The PCD method is a hybrid of past methods as it is both model independent but also subject to physiological and physical considerations. Only through further research will it be known if this hybrid method is superior.

One of the long-lasting debates between the researcher and the clinician has been for the need for absolute verses relative quantification in functional imaging. This has certainly been the case in the determination of myocardial blood flow by MRI and nuclear medicine. Although both imaging modalities can be used to determine absolute quantification of myocardial blood flow the majority of clinical studies remain qualitative. However, we predict that in the near future this will change as new hybrid imaging platforms start to dominate and molecular imaging matures. The current success of PET/CT and the expected even greater impact of MRI/PET and MRI/SPECT will both make it easier to extract quantitative functional information but also make it more important to provide information complimentary to molecular imaging. For example, as we have referenced in this chapter, contrast-enhanced MRI and CT [28] can be used to measure tissue organ blood flow while, with hybrid platforms, at the same time PET and/or SPECT would extract molecular information. This information collected at the same time would be synergistic. For example, PET could provide information on cardiac receptor activity but extraction of that information would be improved by *a priori* knowledge of the blood flow to the myocardium whereas blood flow to the heart is important information on its own right. It is through such simultaneous functional and molecular imaging information that we will better understand the biomarkers and targets for disease and capture events on appropriate time scales.

We look forward to a new era where: (a) better data (greater CNR) will result from improved bio-medical imaging technology including hybrid technology, (b) better deconvolution methods will further improve accuracy and precision in the determination of quantitative function imaging information, and (c) these will be combined on hybrid imaging platforms to extract new information regarding the molecular and functional basis of disease.

Acknowledgments

This work was supported by grants from the Canadian Institutes of Health Research, the Natural Sciences and Engineering Research Council, the Canadian Foundation for Innovation, the Ontario Research and Development Challenge Fund, Bayer Healthcare and Multi Magnetics Inc.

References

1. Bassingthwaighte, J., and Goresky, C. 1984. Modeling in the analysis of solute and water exchange in the microvasculature. In *Handbook of Physiology Section 2: The Cardiovascular System*, Renkin, E., and Geiger, S. (Eds). Bethesda MD: American Physiological Society, 549–626.
2. Gobbel, G.T., and Fike, J.R. 1994. A deconvolution method for evaluating indicator-dilution curves. *Phys. Med. Biol.*, 39, 1833–1854.
3. Zierler, K. 1965. Equations for measuring blood flow by external monitoring of radioisotopes. *Circ. Res.*, 16, 309–321.
4. Bronikowski, T., Dawson, C., and Linehan, J. 1983. Model-free deconvolution techniques for estimating vascular transport function. *Int. J. Biomed. Comput.*, 14, 411–429.
5. Alderson, P.O., Douglass, K.H., Mendenhall, K.G., Guadini V.A., Watson D.C., Links J.M., and Wagner, H.N. 1979. Deconvolution analysis in radionuclide quantitation of left-to-right cardiac shunts. *J. Nucl. Med.*, 20, 502–506.
6. Gamel, J., Rousseau, W.F., Katholi, C.R., and Mesel, E. 1973. Pitfalls in digital computation of the impulse response of vascular beds from indicator-dilution curves. *Circ. Res.*, 32, 516–523.
7. Kuruc, A., Treves, S., and Parker, J.A. 1983. Accuracy of deconvolution algorithms assessed by simulation studies: concise communication. *J. Nucl. Med.*, 24, 258–263.
8. Juni, J.E., Thrall, J.H., Froelich, J.W., Wiggins, R.C., Campbell, D.A., and Tuscan, M. 1988. The appended curve technique for deconvolution analysis – method and validation. *Eur. J. Nucl. Med.*, 14, 403–407.
9. Cooley, J., and Tukey, J. 1965. An algorithm for the machine calculation of complex Fourier series. *Math. Comput.*, 19, 297–301.

10. Wall, C., Borovetz, H.S., Murphy, J.J., and Hardesty, R.L. 1980. System parameter identification in transport models using the fast Fourier transform (FFT). *Comput. Biomed. Res.*, 14, 570–581.
11. Nakamura, M., Suzuki, Y., Nagasawa, T., Sugihara, M., and Takahashi, T. 1982. Detection and quantitation of left-to-right shunts from radionuclide angiography using the homomorphic deconvolution technique. *IEEE BME*, 23, 192–201.
12. Jerosch-Herold, M., Swingen, C., and Seethamraju, R.T. 2002. Myocardial blood flow quantification with MRI by model-independent deconvolution. *Med. Phys.*, 29, 886–897.
13. Valentinuzzi, M.E. and Montaldo Volachec, E.M. 1975. Discrete deconvolution. *Med. Biol. Eng.*, 13, 123–125.
14. Ham, H.R., Dobbeleir, A., Viart, P., Piepsz, A., and Lenaers, A. 1981. Radionuclide quantitation to left-to-right cardiac shunts using deconvolution analysis: concise communication. *J. Nucl. Med.*, 22, 688–692.
15. Cosgriff, P.S., and Berry, J.M. 1982. A comparative assessment of deconvolution and diuresis renography in equivocal upper urinary tract obstruction. *Nucl. Med. Commun.*, 3, 377–384.
16. Basic, M., Popovic, S., Mackovic-Basic, M., Sedlak-Vadoc, V., and Bajc, M. 1988. Extravascular background subtraction using deconvolution analysis of the renogram. *Phys. Med. Biol.*, 33, 1065–1073.
17. Gonzalez, A., Puchal, R., Bajen, M.T., Mairal, L., Prat, L., and Martin-Comin, J. 1994. 99Tcm-MAG3 renogram deconvolution in normal subjects and in normal functioning kidney grafts. *Nucl. Med. Commun.*, 15, 680–684.
18. Stodilka, R.Z., and Prato, F.S. 2007. Method and apparatus for quantifying the behavior of an administered contrast agent. USPTO 60/972,633.
19. McKenzie, C.A., Pereira, R.S., Prato, F.S., Chen, Z., and Drost, D.J. 1999. Improved contrast agent bolus tracking using T1 FARM. *Magn. Reson. Med.*, 41, 429–435.
20. Moran, G.R., and Prato, F.S. 2004. Modeling (1H) exchange: an estimate of the error introduced in MRI by assuming the fast exchange limit in bolus tracking. *Magn. Reson. Med.*, 51, 816–827.
21. Köstler, H., Ritter, C., Lipp, M., Beer, M., Hahn, D., and Sandstede, J. 2004. Prebolus quantitative MR heart perfusion imaging. *Magn. Reson. Med.*, 52, 296–299.
22. Stodilka, R.Z., Blackwood, K.J., and Prato, F.S. 2006. Tracking transplanted cells using dual-radionuclide SPECT. *Phys. Med. Biol.*, 51, 2619–2632.
23. Sijbers, J., den Dekker, A.J., Scheunders, P., and Van Dyck, D. 1998. Maximum-likelihood estimation of Rician distrbution parameters. *IEEE Trans. Med. Imag.*, 17, 357–361.
24. Bellamy, D.D., Pereira, R.S., McKenzie, C.A., Prato, F.S., Drost, D.J., Sykes, J., and Wisenberg, G. 2001. Gd-DTPA bolus tracking in the myocardium using T1 fast acquisition relaxation mapping (T1 FARM). *Magn. Reson. Med.*, 46, 555–564.
25. Parodi, O., De Maria, R., Oltrona, L., Testa, R., Sambuceti, G., Roghi, A., Merli, M., Belingheri, L., Accinni, R., Spinelli, F., et al. 1993. Myocardial blood flow distribution in patients with ischemic heart disease or dilated cardiomyopathy undergoing heart transplantation. *Circulation*, 88, 509–522.
26. Prato, F.S., McKenzie, C.A., Thornhill, R.E., and Moran, G.R. 2001. Functional imaging of tissues by kinetic modeling of contrast agents in MRI. In *Advanced Signal Processing Handbook*, Stergiopoulos, S. (Ed.). CRC Press, Boca Raton, FL.
27. Cramér, H. 1946. *Mathematical Methods of Statistics*. Princeton University Press, Princeton, NJ.
28. Murphy, B.D., Fox, A.J., Lee, D.H., Sahlas, D.J., Black, S.E., Hogan, M.J., Coutts, S.B., Demchuk, A.M., Goyal, M., Aviv, R.I., Symons, S., Gulka, I.B., Beletsky, V., Pelz, D., Hachinski, V., Chan, R., and Lee, T.Y. 2006. Identification of penumbra and infarct in acute ischemic stroke using computed tomography perfusion-derived blood flow and blood volume measurements. *Stroke*, 37, 1771–1777 (Epub 2006 Jun 8).

18

Arterial Spin Labeling: A Magnetic Resonance
Imaging Technique for Measuring Cerebral Perfusion

Keith St. Lawrence and Daron G. Owen

Lawson Health Research Institute
University of Western Ontario

Frank S. Prato

Lawson Health Research Institute
University of Western Ontario
St Joseph's Health Care

CONTENTS

18.1 Introduction

The brain comprises only 2% of the total body weight, yet consumes 20% of the oxygen intake of the body. Due to its high energy demands and lack of appreciable energy stores, the brain is critically dependent on blood flow for a continuous supply of oxygen and glucose. Disruptions in blood supply can cause a rapid cessation of synaptic activity and, if of sufficient severity, lead to cellular membrane failure [1]. Likewise, increased synaptic activity requires a greater metabolic rate to maintain cellular ionic gradients, and this requires a concomitant increase in cerebral blood flow (CBF) [2]. This coupling of brain function to its blood supply is why measuring CBF forms the basis of many current functional neuroimaging techniques. Such techniques have proven extremely useful in a diverse range of applications from mapping brain regions associated with functional tasks to assessing tissue viability following stroke [3,4].

Techniques for measuring absolute CBF can be broadly divided into three classifications depending on the characteristics of the flow tracer implemented. Conceptually, the simplest approach involves tracers that are trapped in tissue at amounts proportional to blood flow. This approach is considered a gold standard in animal experiments and is the primary method used in human studies involving single photon emission computed tomography (SPECT) [4,5]. The second approach, which was developed primarily by Zierler, determines blood flow using a model-independent description of tracer transport times [6]. Commonly referred to as black-box analysis [7], this approach requires rapid imaging techniques to characterize the passage of a nondiffusible or intravascular tracer as it flows through the tissue microcirculation. For clinical studies it has been adapted to both computed tomography (CT) and magnetic resonance imaging using blood-pool contrast agents [8–11]. The third and perhaps the most established approach is the use of freely diffusible tracers combined with the compartmental analysis developed by Kety [12]. Kety and Schmidt obtained the first quantitative measurements of global CBF in humans over 50 years ago using nitrous oxide as a freely diffusible tracer [13,14]. Modifications of the Kety–Schmidt approach are the principal methods of measuring cerebral blood flow with positron emission tomography (PET) using radio-labeled water, $H_2^{15}O$, as a tracer [15,16]. Although water is not a completely freely diffusible tracer [17], PET combined with $H_2^{15}O$ is commonly considered the gold standard for measuring CBF in humans, as evidenced by the number of validation studies comparing this technique to others [4].

Arterial spin labeling (ASL) methods can be considered the MRI equivalent of PET/ $H_2^{15}O$ since water is used as a tracer and the ASL signal can be interpreted using the same classical tracer kinetic theory to generate quantitative CBF images [18]. However, ASL methods differ significantly from PET/$H_2^{15}O$ in many respects due to the unique labeling procedure, which involves manipulating the longitudinal magnetization (spin) of endogenous water in arterial blood [19–22]. Since this labeling method is completely noninvasive, ASL is well suited to perfusion studies in healthy subjects and to serial studies involving patient populations. This chapter will outline the principles underlying ASL, discuss issues confounding flow quantification and sensitivity, as well as methods developed to address these issues, and finally review some of the applications of ASL.

18.2 Theory: Tissue Compartment

18.2.1 The Kety Model

The ingeniousness of Kety was to recognize that the Fick principle, which had been previously used to measure cardiac output in man, could be adapted to measure CBF. His approach was to introduce into the blood a chemically inert tracer that would readily diffuse across the blood–brain barrier (BBB) and accumulate in tissue [13]. Based on the conservation of mass, the Fick principle states that the amount of tracer in brain at time t equals the difference between the amount entering and leaving by blood flow [12]:

$$C_b(t) = F \cdot \int_0^t \left(C_a(u) - C_v(u)\right) du \tag{18.1}$$

In this equation, $C_b(t)$, $C_a(t)$ and $C_v(t)$ are the concentrations of tracer in brain, arterial blood and venous blood, respectively, and F is cerebral blood flow, which is typically expressed as ml of blood/g of tissue/min or ml of blood/100 g of tissue/min. The first application of this approach required subjects to inhale nitrous oxide while blood samples were acquired from a peripheral artery and the jugular vein to determine $C_a(t)$ and $C_v(t)$, respectively [13]. If the inhalation period is long enough to allow the cerebral tissue and blood tracer concentrations to equilibrate, denoted as $t \rightarrow \infty$, then $C_b(\infty)$ can be defined by:

$$C_b(\infty) = \lambda C_v(\infty) \tag{18.2}$$

where λ is the partition coefficient and is the ratio between the tissue and blood tracer concentrations at equilibrium [12]. The partition coefficient is generally expressed in units of ml blood/g tissue and the values for water in whole brain, grey matter and white matter are 0.9, 0.98 and 0.82 ml/g, respectively [23]. With the tissue concentration defined by Equation 18.2, CBF can be determined from Equation 18.1.

Recognizing that the N_2O method could only measure global CBF since $C_v(t)$ must be determined, Kety expanded this approach to describe the equilibration of a freely diffusion tracer in tissue. In essence, if the diffusion of the tracer is extremely rapid, the tissue concentration can be assumed to be in equilibrium with the venous blood concentration at all times:

$$C_b(t) = \lambda C_v(t) \tag{18.3}$$

Using this relationship, the Fick principle can be written in terms of the instantaneous change in the tissue concentration as:

$$\frac{dC_b(t)}{dt} = FC_a(t) - \frac{F}{\lambda} C_b(t) \tag{18.4}$$

The solution to this differential equation, assuming the tracer concentration in tissue is zero at $t=0$, is:

$$C_b(t) = F \int_0^t C_a(u) e^{-\frac{F}{\lambda}(t-u)} du = F \cdot C_a(t) \otimes e^{-\frac{F}{\lambda}t} \tag{18.5}$$

This expression is commonly referred to as the Kety equation and it is the basis of all imaging techniques that use a freely diffusible tracer to measure CBF.

The key assumption of the Kety model is that blood flow is the rate limiting step governing tracer transport. This requires rapid tracer diffusion to instantaneously dissipate the tracer throughout the tissue compartment. Furthermore, the exchange of tracer across the BBB must not retard tracer movement. Inert gases and other lipophilic substances that have high diffusion coefficients can be considered freely diffusible in brain [7]. As mentioned above, water is not a true "freely diffusible" tracer and its diffusion limitations have been shown to cause a progressive underestimation of CBF at higher values [17,24,25]. The impact of this complication on CBF measurements acquired with ASL will be discussed in more detail in Section 18.4.3.

18.2.2 The Kety Model Adapted to Arterial Spin Labeling

The signal change in the brain caused by the inflow of magnetically labeled water can be described by combining the Fick principle, which accounts for blood flow, with the Block equation, which accounts for the relaxation of the magnetization [18]:

$$\frac{dM_b(t)}{dt} = \frac{M_b^0 - M_b(t)}{T_1} + F \cdot \left(M_a(t) - M_v(t) \right) \tag{18.6}$$

where $M_b(t)$ is the longitudinal magnetization of water in tissue [per unit of tissue], M_b^0 is the equilibrium magnetization, $M_a(t)$ and $M_v(t)$ are the longitudinal magnetizations of water in arterial and venous blood, respectively (per ml of blood), and T_1 is the longitudinal relaxation time in tissue. Assuming the magnetically labeled water in tissue is in equilibrium with the labeled water in venous blood (i.e., the Kety model), $M_v(t)$ can be replaced in Equation 18.6 by $M_b(t)$:

$$\frac{dM_b(t)}{dt} = \frac{M_b^0 - M_b(t)}{T_1} - FM_a(t) - F\frac{M_b(t)}{\lambda} \tag{18.7}$$

The definition of $M_a(t)$ in Equation 18.7 will be defined by the labeling procedure, and these are discussed in the next section. Regardless of the procedure implemented, the signal change caused by the inflow of labeled water is quite small due to T_1 relaxation. To isolate this small signal change, two images are collected: one with labeling (i.e., the label image) and one without (i.e., the control image). The general solution to Equation 18.7 is

$$M_b^i(t) = M_b^i(0)e^{-R_1^* t} + \frac{R_1}{R_1^*} M_b^0 \left(1 - e^{-R_1^* t} \right) + FM_a^i(t) \otimes e^{-R_1^* t} \tag{18.8}$$

where the superscript "i" refers to either label (L) or control (C) condition, $M_b^i(0)$ is the magnetization at $t=0$, $R_1 = 1/T_1, R_1^* = R_1 + F/\lambda$. The last term is a convolution of the arterial input function with the exponential decay function for labeled water in tissue. In an ASL experiment, only the last term will be different between the label and control conditions. Consequently, the signal contribution from blood flow can be isolated by subtracting label from control images:

$$\Delta M_b(t) = M_b^L(t) - M_b^C(t) \tag{18.9}$$

Using Equation 18.8 to define $M_b^L(t)$ and $M_b^C(t)$, the ASL signal is

$$\Delta M(t) = F \cdot \Delta M_a(t) \otimes e^{-R_1^* t} \tag{18.10}$$

where $\Delta M_a(t)$ is the difference in arterial blood magnetization between control and label procedures (see Section 18.3):

$$\Delta M_a(t) = M_a^L(t) - M_a^C(t) \tag{18.11}$$

The similarities between Equations 18.5 and 18.10 should come as no surprise considering the ASL signal equation was derived using the Kety model. The principal difference between these equations is the definition of the rate constant defining the loss of tracer from tissue. With the original Kety model, the rate constant is defined by F/λ, whereas for ASL it is necessary to include T_1 (longitudinal) relaxation [26]. It fact, the loss of the magnetic label is overwhelmingly dominated by longitudinal relaxation, as evident by comparing R_1^* to R_1. At a magnetic field strength of 1.5 T, $R_1 \sim 1$ s^{-1} and, assuming global $F = 50$ ml/100 g/min and $\lambda = 90$ ml/100 g, the corresponding $R_1^* = 1.01$ s^{-1}! This rapid decay allows successive CBF images to be acquired within a few seconds with ASL; in contrast, the quickest acquisition time with PET is approximately 1 min [27]. On the other hand, the dominance of T_1 relaxation is the root cause of the poor sensitivity of ASL.

18.3 Labeling Procedures

18.3.1 Continuous Arterial Spin Labeling

The initial ASL experiments were conducted in rats and labeling was accomplished by continuously saturating (90°) the magnetization of the arterial blood in the carotid arteries [18]. The ASL signal can be effectively doubled by using a continuous inversion (180°) method to label the arterial blood [19]. This and future continuous arterial spin labeling (CASL) studies use a flow-induced adiabatic inversion technique based on the principles of adiabatic fast passage to continuously invert arterial blood magnetization at a plane proximal to the imaging slice [28]. This is accomplished by continuously applying low-power radiofrequency (RF) irradiation in the presence of a magnetic field gradient applied in the direction of flow (Figure 18.1a). In order for spins to be inverted under these circumstances, the following condition must be met:

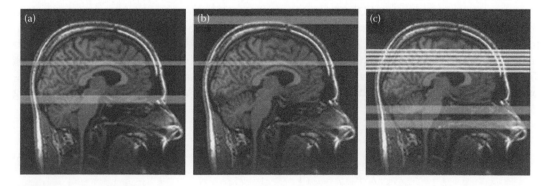

FIGURE 18.1
Sagittal images of a human head showing the location of the imaging slice(s) (white bars). (a) Continuous adiabatic inversion shown proximal to the imaging slice (lower grey bar). (b) Off-resonance irradiation applied distally to imaging slice during control period to balance magnetization transfer effects (upper grey bar). (c) Double inversion of arterial spins using amplitude-modulated CASL for multislice imaging.

$$\frac{1}{T_1}, \frac{1}{T_2} \ll (1/H_1)G\nu \ll \gamma H_1 \tag{18.12}$$

where T_2 is the transverse relaxation time, H_1 is the RF magnetic field strength, G is the gradient field strength, ν is the linear velocity of the inflowing spins, and γ is the gyromagnetic ratio for protons. Assuming that Equation 18.12 is satisfied, the arterial magnetization under label and control conditions is given by:

$$M_a^L = \frac{M_b^0}{\lambda}\left(1 - 2e^{-R_{1a}\tau_a}\right), \quad M_a^C = \frac{M_b^0}{\lambda} \tag{18.13}$$

where R_{1a} is longitudinal relaxation rate of water in blood ($R_{1a}=1/T_{1a}$). In these equations, the blood magnetization is expressed in terms of the equilibrium tissue magnetization since the water concentrations in the two spaces are in equilibrium. The exponential term in brackets accounts for the arterial transit time, τ_a, which is the delay between the inversion of the arterial magnetization and its arrival at the tissue in the imaging slice. From Equation 18.11, $\Delta M_a(t)$ is

$$\Delta M_a(t) = \begin{matrix} 0 & 0 < t < \tau_a \\[6pt] \dfrac{-2\alpha\, M_b^0}{\lambda}e^{-R_{1a}\tau_a} & \tau_a \le t \le \Delta_{\text{label}} + \tau_a \\[6pt] 0 & \Delta_{\text{label}} + \tau_a < t \end{matrix} \tag{18.14}$$

where Δ_{label} is the duration of the labeling period. The labeling efficiency, α, accounts for the fact that not all spins are completely inverted as they pass through the labeling plane. Factors that can reduce the labeling efficiency include duty-cycle limitations of the applied RF field, relaxation effects, and variations in blood velocities [19,29,30]. Computer simulations and experimental results, measured in both flow phantoms and a rat carotid artery, indicate that with the proper labeling conditions the efficiency can be as great as 90% [31,32].

An inadvertent effect of the off-resonance RF pulse used for continuous labeling is the perturbation of tissue macromolecules in the imaging slice, resulting in a decrease in the tissue water magnetization [33]. This magnetization transfer effect is much greater than the signal change caused by spin labeling and therefore careful attention must be taken to ensure that the ΔM signal is isolated [34,35]. One approach is to apply the same off-resonance RF pulse distally to the imaging slice during the control period (Figure 18.1b). This pulse will have the same magnetization transfer effect as the labeling pulse but without affecting arterial spins. The $\Delta M(t)$ signal defined by CASL is obtained by combining Equations 18.10 and 18.14:

$$\frac{\Delta M_b(t)}{M_b^0} = \begin{matrix} 0 & t < \tau_a \\[8pt] \left(\dfrac{-2\alpha F}{\lambda \cdot R_1^{\text{rf}}}\right)e^{-R_{1a}\tau_a}\left\{1 - e^{-R_1^{\text{rf}}[t-\tau_a]}\right\} & \tau_a \le t \le \tau_a + \Delta_{\text{label}} \end{matrix} \tag{18.15}$$

where, R_1^{rf} is the tissue relaxation rate in the presence of off-resonance RF irradiation [36]. Figure 18.2 illustrates a typical time course of $\Delta M_b(t)$ predicted from this equation. If the labeling period is sufficiently long, a steady state is reached between the delivery of label

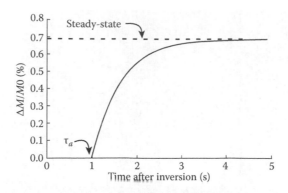

FIGURE 18.2

Time course of $\Delta M_b(t)$ as predicted by Equation 18.15. The signal begins to rise when labeled water reaches the tissue (i.e., $t > \tau_a$). A steady-state signal change is reached after approximately 4 s in this example ($F = 70$ ml/100 g/min, $T_1 = 1$ s, $T_{1a} = 1.2$ s, $\tau_a = 1$ s, $R_1^{rf} = 1.639$ s^{-1}).

by blood flow and the loss of label by relaxation. This Figure illustrates the low sensitivity of ASL as the maximum signal never reaches 1% of M_b^0 due to the rapid loss of the magnetic label.

The disadvantage with applying off-resonance RF irradiation distal to the imaging slice during the control period is that only one imaging slice can be collected at a time since magnetization transfer effects are balanced at only a position equidistant from the labeling and control planes [34]. Multislice imaging can be achieved by using amplitude-modulated off-resonance RF irradiation to invert the arterial spins twice as they pass through the labeling plane, which returns the spins to their equilibrium state (Figure 18.1c) [37]. Since the off-resonance RF irradiation in the control and label conditions are applied at the same centre frequency, magnetization transfer effects are balanced across multiple slices. The trade-off is a 30% reduction in labeling efficiency due to imperfections in the amplitude-modulated control pulse [37,38]. An alternative approach is to avoid magnetization transfer effects completely by using a separate RF coil to label arterial blood in the carotid arteries [39–42]. This approach has the added advantage of providing whole-brain coverage, but requires specialized hardware—primarily the labeling coil and an additional RF channel—that are not readily available on clinical MRI scanners. A recently proposed alternative, which is referred to as pseudo continuous arterial spin labeling (pCASL), employs a series of RF pulses for flow-driven inversion [43]. It has the advantages of not requiring any specialized hardware and magnetization transfer effects are balanced over all slices since the label and control pulses are applied at the same location. Its labeling efficiency was 80% compared to 68% for amplitude-modulated CASL [44]. Figure 18.3 shows an example of perfusion-weighted images obtained with pCASL.

18.3.2 Pulsed Arterial Spin Labeling

The other main labeling approach is pulsed arterial spin labeling (PASL), which involves using a single, short-duration (5–20 ms) RF pulse to invert the magnetization of a large volume of inflowing arterial blood (Figure 18.4) [20–22]. The inversion pulse is followed by a delay, TI, prior to imaging to allow labeled water to flow into the tissue. The difference in arterial blood magnetization, $\Delta M_a(t)$, for PASL is defined as

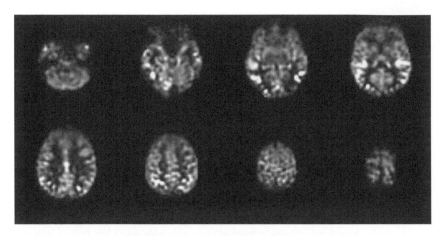

FIGURE 18.3
Mean ΔM images acquired using a pseudo-CASL technique with a labeling duration of 1.5 s and a postlabeling delay of 1 s. Full-brain coverage was obtained by labeling below the petrous portion of carotid arteries. Eight of 16 images are shown. Mean ΔM images were calculated from 32 ΔM volumes. Data were acquired from a healthy volunteer on a clinical 1.5 T scanner using a single-shot spiral imaging technique.

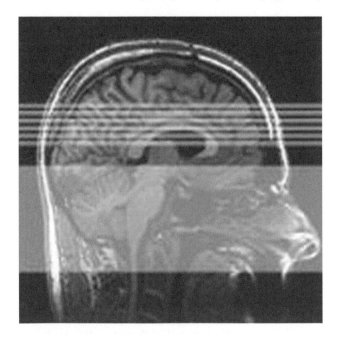

FIGURE 18.4
Sagittal images of a human head showing the location of the imaging slices (white bars) and the spatially selective inversion slab for PASL (lower grey bar).

$$\Delta M_a(t) = \begin{cases} 0 & 0 < t < \tau_a \\ \dfrac{-2\alpha\, M_b^0}{\lambda} e^{-R_{1a}t} & t \geq \tau_a \end{cases} \tag{18.16}$$

The time dependence shown in Equation 18.16 results in reduced labeling efficiency compared to CASL. Since all of the label is created at the same time, label more distal to the imaging slices undergoes greater relaxation prior to reaching them compared to label more proximal. The $\Delta M(t)$ signal for PASL is obtained by combining Equations 18.10 and 18.16:

$$\frac{\Delta M_b(t)}{M_b^0} = \begin{cases} 0 & 0 < t < \tau_a \\ \left(\frac{-2\alpha F}{\lambda \cdot \Delta R}\right) e^{-R_{1a}t}\left\{1 - e^{-\Delta R[t-\tau_a]}\right\} & t \leq \tau_a \end{cases} \tag{18.17}$$

where, $\Delta R = R_{1a} - R_1^*$. Figure 18.5 illustrates a typical time course of $\Delta M_b(t)$ predicted from this equation. Unlike CASL in which the arterial label is continually refreshed, the time course of $\Delta M_b(t)$ for PASL rises to a peak and then subsequently decays due to the relaxation of the arterial magnetization.

Magnetization transfer effects are generally considered negligible with PASL because only a single RF pulse is used for labeling. PASL is also easier to implement than CASL and the RF power deposition is considerably less, which is an important issue when conducting studies at higher magnetic field strengths. The reduction in the signal-to-noise ratio (SNR) due to decreased labeling efficiency can be compensated for by employing a shorter repetition time in PASL, allowing for more signal averaging [45]. There are several variations on the general PASL method, due mainly to differences in the RF pulses used to create the control and label conditions. With all PASL methods, careful attention must be paid to ensure that the spatial profile of the inversion pulse does not interfere with the imaging slices [46,47]. One of the most commonly used methods is referred to as flow sensitive alternating inversion recovery (FAIR) [22]. The review papers by Calamante et al. and Barbier et al. offer thorough descriptions of the many versions of PASL [48,49].

18.4 Issues Regarding Perfusion Quantification

18.4.1 Transit Time Effects

Equations 18.15 and 18.17 reveal that regardless of the labeling technique, the ASL signal is dependent on the time required for the labeled water to reach the tissue of interest (i.e., the arterial transit time τ_a). It is not unreasonable to ignore τ_a in studies involving small

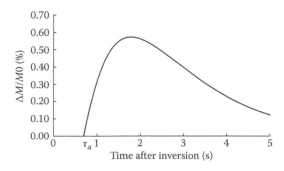

FIGURE 18.5
Time course of $\Delta M_b(t)$ as predicted by Equation 18.17 ($F=70$ ml/100 g/min; $T_1=1$ s; $T_{1a}=1.2$ s; $\tau_a=0.7$ s).

animals since the distance between the labeling location and the imaging slices can be quite short [31]. In contrast, τ_a is not negligible in humans and can significantly affect flow quantification at rest and during neural activation (see Figure 18.6). A number of studies have shown that the fractional decrease in τ_a during activation can be as large as the associated increase in CBF [50,51]. The most direct method for dealing with arterial transit time effects is to measure $\Delta M_b(t)$ at multiple time points and determine both CBF and τ_a by fitting the signal equation to the resulting time series [36,52]. However, this approach can be quite time consuming and typically requires signal averaging over large regions to achieve a reasonable SNR [53]. An intriguing method for reducing acquisition times is to collect PASL signals at multiple inversion times using a Look–Locker approach [54].

An alternative approach is to reduce the sensitivity of the ΔM_b signal to τ_a. With CASL, this is accomplished by introducing a postlabeling delay between the end of the labeling period and image acquisition [55]. Provided the delay is long enough to allow all of the labeled water to reach the tissue, the ΔM_b signal will only depend on the delivery rate (i.e., CBF) and T_1 relaxation. In this case, $\Delta M_b(t)$ is determined for $t > \Delta_{\text{label}} + \tau_a$ [56]:

$$\frac{\Delta M_b(t)}{M_b^0} = \left(\frac{-2\alpha F}{\lambda \cdot R_1^*}\right) e^{-R_1^*(t-\Delta_{\text{label}})} \left\{1 - \Psi_1 - \Psi_2\right\} \left\{e^{\left[R_1^* - R_{1a}\right]\tau_a}\right\} \tag{18.18}$$

where

$$\Psi_1 = \left(1 - \frac{R_1^*}{R_1^{\text{rf}}}\right) e^{-R_1^* \tau_a}, \qquad \Psi_2 = \left(\frac{R_1^*}{R_1^{\text{rf}}}\right) e^{-\left[R_1^* - R_1^{\text{rf}}\right]\tau_a} e^{-R_1^{\text{rf}} \Delta_{\text{label}}} \tag{18.19}$$

The transit time insensitivity of Equation 18.18 can be more easily observed under the assumption that all relaxation rates are equal:

$$\frac{\Delta M_b(t)}{M_b^0} = \left(\frac{-2\alpha F}{\lambda \cdot R_1}\right) e^{-R_1^*(t-\Delta_{\text{label}})} \left\{1 - e^{-R_1^* \Delta_{\text{label}}}\right\} \tag{18.20}$$

Equation 18.20 demonstrates that this is an effective method for reducing the sensitivity to transit times provided the relaxation rates in blood and tissue are similar. An example

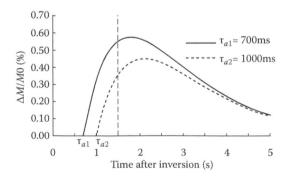

FIGURE 18.6
Time course of $\Delta M_b(t)$ generated by Equation 17 with two transit times but the same CBF value ($F=70$ ml/100 g/min, $T_1=1$ s and $T_{1a}=1.2$ s, $\tau_{a1}=0.7$ s and $\tau_{a2}=1$ s). The vertical dashed line represents a typical acquisition time (TI). Clearly, the calculated CBF values from the two ΔM_b curves at $t=$TI will not be the same due to the difference in τ_a.

of the predicted $\Delta M_b(t)$ signal for CASL including a postlabeling delay is shown in Figure 18.7. At $t=\Delta_{\text{label}}$, the off-resonance RF irradiation is switched off causing the ASL signal to increase as the saturated macromolecules return to their equilibrium state. This, in turn, causes the tissue relaxation rate to revert to R_1. At $t=\Delta_{\text{label}}+\tau_a$, the signal decays according to R_1 since all of the labeled water has reached the tissue. The penalty with this approach is a significant loss of signal if the chosen delay is considerably longer than τ_a.

Reducing transit-time sensitivities with PASL is accomplished by applying a saturation pulse proximal to the imaging slices at a specific time after the initial inversion pulse. This pulse reduces the volume of the inverted arterial blood magnetization to a well-defined "slab" [57]. The τ_a dependency is diminished if the delay after the application of the saturation pulse is long enough to allow the entire slab to reach the tissue. In this case, $\Delta M_a(t)$ is defined as:

$$\Delta M_a(t) = \begin{array}{ll} 0 & 0 < t < \tau_a \\ \dfrac{-2\alpha M_b^0}{\lambda}e^{-R_{1a}t} & \tau_a \leq t \leq \Delta_{\text{slab}} + \tau_a \\ 0 & \Delta_{\text{slab}} + \tau_a < t \end{array} \tag{18.21}$$

where Δ_{slab} is the duration of the label defined by the time between the initial inversion and the application of the saturation pulse. If the postsaturation delay is greater than $\Delta_{\text{slab}} + \tau_a$, the labeled spins all experience the same transit delay and $\Delta M_b(t)$ is given by:

$$\frac{\Delta M_b(t)}{M_b^0} = \left(\frac{-2\alpha F}{\lambda \cdot \Delta R}\right)e^{-R_{1a}t}\left\{e^{-\Delta R[t-\Delta_{\text{slab}}-\tau_a]} - e^{-\Delta R[t-\tau_a]}\right\} \qquad t > \Delta_{\text{slab}} + \tau_a \tag{18.22}$$

Again, the insensitivity to τ_a is more readily observable if $R_{1a} \sim R_1^*$:

$$\frac{\Delta M_b(t)}{M_b^0} = \left(\frac{-2\alpha F}{\lambda}\right)e^{-R_1^*t}\left\{\Delta_{\text{slab}}\right\} \tag{18.23}$$

The PASL signal with and without the application of a saturation pulse to define Δ_{slab} is shown in Figure 18.8. Analogous to the use of a postlabeling delay with CASL, this Figure

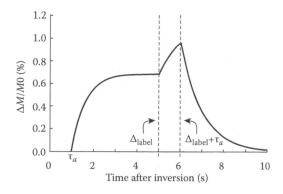

FIGURE 18.7
Time course of $\Delta M_b(t)$ as predicted by Equation 18.7 for CASL with a postlabeling delay ($F=70$ ml/100 g/min, $T_1=1$ s, $R_1^{\text{rf}}=1.639$ s^{-1}, $T_{1a}=1.2$ s, $\tau_a=1$ s and $\Delta_{\text{label}}=4.5$ s).

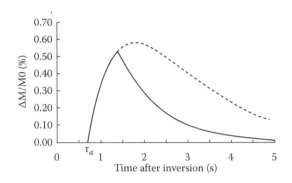

FIGURE 18.8
Time course of $\Delta M_b(t)$ as predicted by Equations 18.17 and 18.22 without (dashed line) and with (solid line) the application of a saturation pulse to define the length of the label ($F=70$ ml/100 g/min, $T_1=1$ s, $T_{1a}=1.2$ s, $\tau_a=0.7$ s and $\Delta_{slab}=0.7$ s).

illustrates that the transit-time insensitive condition (i.e., $t > \Delta_{slab} + \tau_a$) can result in a considerable loss of signal if t is much greater than $\Delta_{slab} + \tau_a$. The postlabeling delay approach works well with healthy subjects since the typical range of τ_a values has been characterized. Consequently, a delay time can be chosen that will accommodate τ_a variations across regions, conditions (ex. activation versus rest) and subjects. However, caution must be exercised when applying this approach to patient groups with extremely long arterial transit delays, such as those encountered in cerebrovascular disease (see Section 18.7.2).

18.4.2 Vascular Artifacts

Most diffusible tracers, such as the PET tracer $H_2{}^{15}O$, have a relatively long relaxation rate or half-life, allowing for a sizable build-up of tracer in the tissue. In such cases, the signal contribution originating from tracer in blood is minimal since the blood volume is only 5% of the total brain volume [58]. With ASL, the concentration of label in the tissue never builds up beyond a few percent of the tissue equilibrium magnetization due to the rapid relaxation of the magnetic label. As a result, the relative signal contribution from arterial blood compared to tissue is far greater than the ratio of their respective volumes. This additional signal can lead to substantial overestimations of blood flow since the model outlined in Section 18.2 only accounts for labeled water in the tissue. Two possible methods have been proposed to correct for the arterial signal contribution. The first is to eliminate this additional signal by using diffusion-weighted gradients to suppress the faster moving spins in the arteries [36,52]. The disadvantages with this approach are the significant loss in ΔM signal, since the arterial contribution can be as large as 50% of the total signal, and the uncertainty in knowing what portion of the vascular bed is suppressed by the diffusion-weighted gradients [59]. The other approach is to use a model with two compartments in series: the first compartment accounts for the arterial signal and the second compartment accounts for the tissue signal as described above [26,55].

Modelling blood flow through the arterial vasculature as plug flow, the arterial contribution to the ASL signal, $\Delta M_{\text{Art}}(t)$, can be written as:

$$\Delta M_{\text{Art}}(t) = F \int_{u=0}^{u=t} \Delta M_a(t) e^{-R_{1a}[t-u]} du \tag{18.24}$$

where the exponential term accounts for the relaxation of the label prior to reaching the tissue compartment. The arterial input function, $\Delta M_a(t)$, has the same form as for the tissue compartment (Equation 18.14 for CASL and Equation 18.21 for PASL), except the transit time to the arterial compartment, τ_a', is shorter than τ_a. Since the $\Delta M_{Art}(t)$ signal equations for PASL and CASL are similar, only the former is presented here. The input function is:

$$\Delta M_a(t) = \begin{cases} 0 & 0 < t < \tau_a' \\ \dfrac{-2\alpha\, M_b^0}{\lambda} e^{-R_{1a}t} & \tau_a' \le t \le \tau_a \\ \dfrac{-2\alpha\, M_b^0}{\lambda} e^{-R_{1a}t} & \tau_a \le t \le \Delta_{slab} + \tau_a' \\ 0 & \Delta_{slab} + \tau_a' < t \end{cases} \tag{18.25}$$

The corresponding arterial compartment signal for PASL is

$$\frac{\Delta M_{Art}(t)}{M_b^0} = \begin{cases} 0 & 0 < t < \tau_{a'} \\ \left(\dfrac{-2\alpha F}{\lambda}\right) e^{-R_{1a}t} \left\{ t - \tau_{a'} \right\} & \tau_{a'} < t < \tau_a \\ \left(\dfrac{-2\alpha F}{\lambda}\right) e^{-R_{1a}t} \left\{ \tau_a - \tau_{a'} \right\} & \tau_a < t < \Delta_{slab} + \tau_{a'} \\ \left(\dfrac{-2\alpha F}{\lambda}\right) e^{-R_{1a}t} \left\{ \Delta_{slab} + \tau_a - t \right\} & \Delta_{slab} + \tau_{a'} < t < \Delta_{slab} + \tau_a \\ 0 & \Delta_{slab} + \tau_a < t \end{cases} \tag{18.26}$$

A schematic of the two-compartment model for PASL is shown in Figure 18.9. In this diagram, arterial blood magnetization proximal to the imaging slice is inverted (labeled) at $t=0$, followed by a delay, TI, to allow the labeled spins to travel from the inversion

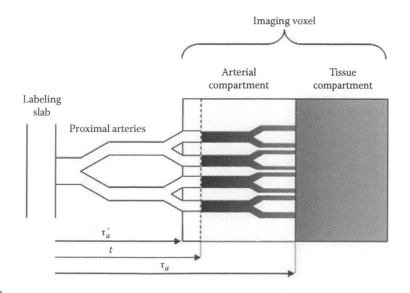

FIGURE 18.9
Two-compartment model for PASL consisting of arterial and tissue compartments in series. Each compartment has its own arterial transit time. τ_a for tissue and τ_a' for arterial.

region, through the proximal arteries, to the arterial compartment in the imaging plane. Increasing the TI value allows sufficient time for the labeled water to pass through the arterial compartment into the tissue compartment. The relative weighting of the two compartments depends on the two transit times and TI (Figure 18.10). Figure 18.11 illustrates the arterial and tissue signals, along with the total ΔM signal, as determined from Equations 18.17, 18.22 and 18.26. Because of the addition of the arterial compartment, the maximum ΔM signal is greater than shown previously in Figure 18.8.

In principle, quantifying blood flow using a two-compartment model requires knowing the transit times to the arterial and tissue compartments. However, if the relaxation rates in blood and tissue are similar, the ASL signal is fairly insensitive to the arterial/tissue signal ratio provided the end of the labeling bolus has reached the arterial compartment [55]. This approach works well in grey matter regions; however, the difference in the T_1 values between blood and white matter can affect the accuracy of CBF measurements. Another potential artifact associated with the two-compartment model is labeled water in arteries that traverse a voxel, but do not supply the local tissue (i.e., nonfeeding arteries). This additional signal will cause an overestimation of CBF, although the magnitude of this artifact diminishes with longer postlabeling delays [57]. As a final note, combining the two-compartment model with diffusion-weighting gradients has been proposed as an alternative method for measuring the arterial transit time to tissue, τ_a [60]. With this technique, τ_a can be determined from the ratio of the ΔM signal with ($\sim\Delta M_b$) and without ($\sim\Delta M_{ART}+\Delta M_b$) diffusion-weighting gradients.

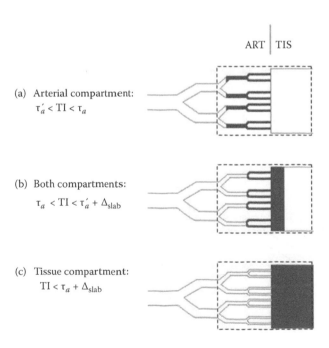

ART | TIS

(a) Arterial compartment:
    $\tau_a' < TI < \tau_a$

(b) Both compartments:
    $\tau_a < TI < \tau_a' + \Delta_{slab}$

(c) Tissue compartment:
    $TI < \tau_a + \Delta_{slab}$

FIGURE 18.10
Schematic diagram showing the time dependency of ΔM signals in the arterial and tissue compartments of the imaging slice (dashed box). (a) At an early acquisition time (TI), labeled water is only in the arterial compartment (denoted ART). (b) If more time is permitted, labeled water will be in both compartments. (c) For long TI values, labeled water is only located in the tissue (TIS) compartment.

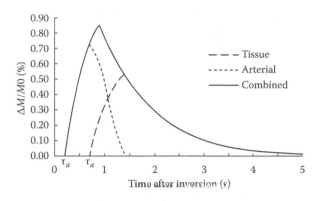

FIGURE 18.11

Time course of $\Delta M(t)$ as defined by combining a tissue compartment, Equations 18.17 and 18.20, with an arterial compartment, Equation 18.26. This example includes the application of a saturation pulse to define the length of the label ($F=70$ ml/100 g/min, $T_1=1$ s, $T_{1a}=1.2$ s, $\tau_a'=0.2$ s, $\tau_a=0.7$ s and $\Delta_{slab}=0.7$ s).

18.4.3 Water Exchange between Capillaries and Tissue

With the Kety model, the capillaries and the extravascular space (EVS) are treated as one homogeneous compartment since the exchange across the BBB is assumed to occur on a time scale much shorter than blood flow. In comparison to lipophilic substances, water transport is slower since it occurs via filtration and diffusion through endothelial membranes [61]. The permeability of the BBB to a substance can be characterized by measuring the unidirectional extraction into the EVS during a single transcapillary passage [12,62]:

$$E = 1 - e^{(-PS/F)} \qquad (18.27)$$

where, PS is the capillary permeability-surface area product for a given substance. The extraction fraction, E, for water is close to 1 for CBF values less than 50 ml/100 g/min, but falls as CBF increases [17]. One of the first adaptations of the Kety model to characterize the uptake of $H_2^{15}O$ in brain included the extraction fraction [16]:

$$C_b(t) = EF \cdot C_a(t) \otimes e^{-\frac{EF}{\lambda}t} \qquad (18.28)$$

It is evident from this equation that CBF can no longer be measured independently since it is coupled to E. Consequently, CBF will be underestimated if E is less than 1, as has been shown to occur at higher flow values [24,25,27].

Analogous to Equation 18.28, the Bloch equation including flow effects, Equation 18.7, can be modified to include E [63–65]. However, the expected progressive underestimation of CBF at higher values has not, in general, been observed in ASL experiments (Section 18.5). This can be explained by the significant signal contribution from labeled water in the capillaries. Even though the capillary blood volume is much smaller than the volume of the EVS, the fraction of labeled water in the capillaries can be substantial due to T_1 relaxation and restricted water extraction. A number of tracer kinetic models have been proposed to account for the capillary contribution [66–70]. All of these models expand on the basic Kety model to include labeled water in capillaries and water exchange across the BBB (Figure 18.12). In the most general form [66], the amount of labeled water in the tissue volume is given by:

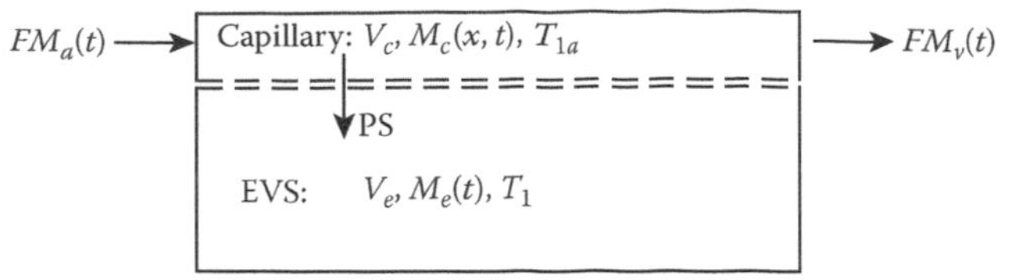

FIGURE 18.12
General capillary-tissue exchange model. The magnetization in the capillary and extra-vascular spaces are defined by $M_c(x,t)$ and $M_e(t)$, respectively (t represents time and x represents position down the length of the capillary). The blood–brain barrier is represented by the double dashed line.

$$\Delta M(t) = F\Delta M_a(t) \otimes (q_c(t) + q_e(t)) \tag{18.29}$$

where $q_c(t)$ and $q_e(t)$ represent the capillary and EVS residual concentrations, respectively, following an idealized impulse input:

$$q_c(t) = \begin{matrix} e^{-\alpha t} & t \le \tau_c \\ 0 & t > \tau_c \end{matrix} \tag{18.30}$$

where $\alpha = \dfrac{PS}{V_c} + R_{1a}$, $\tau_c = \dfrac{V_c}{F}$ and V_c is the capillary blood volume.

$$q_e(t) = \begin{matrix} \beta\left(e^{-R_1 t} - e^{-\alpha t}\right) & t \le \tau_c \\ \beta E_R e^{-R_1 t} & t > \tau_c \end{matrix} \tag{18.31}$$

where $\beta = 1 \Big/ \left(1 + \dfrac{\Delta R_1 V_c}{PS}\right)$ and $E_R = 1 - e^{-\left(\frac{PS}{F} - \Delta R_1 \cdot \tau_c\right)}$.

Equation 18.30 characterizes the loss of labeled water in the capillary space during a single capillary transit due to extraction into tissue, which is governed by PS/V_c, and T_1 relaxation (see Figure 18.13). Equation 18.31 shows that the amount of labeled water in the EVS initially rises due to the influx from blood and then falls due to T_1 relaxation. These equations can be used to explain the effects of the capillary contributions by generating theoretical $\Delta M(t)$ data from Equation 18.29 and analyzing these "data" with the standard one-compartment Kety model, Equation 18.10. To derive a more complete representation of the ASL signal, models have been expanded to include contributions from nonexchanging vessels, both arterial and venous, and to account for transverse relaxation times (T_2 and T_2^*) [66,71].

Figure 18.14 shows the observed CBF values for CASL experiments at 1.5 and 4 T. Included in the Figure is a plot of the product EF, which is the flow value measured by PET/$H_2^{15}O$. These plots suggest that ASL measurements of CBF obtained with the Kety model should be relatively insensitive to restricted water exchange due to the signal contributions from capillaries [66–68]. These predictions are in good agreement with validation studies (Section 18.5). These simulations also demonstrate that the accuracy of the standard ASL equation is reasonable provided the relaxation times in blood and tissue are similar. For example, the reduction in T_2^* of blood at higher magnetic fields will attenuate the capillary and venous contributions, leading to an underestimation of CBF when ASL

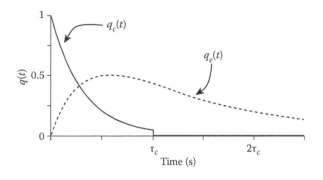

FIGURE 18.13
Concentration profiles in the capillary, $q_c(t)$, and extravascular, $q_e(t)$ spaces for an idealized impulse input. Data were generated from Equations 18.30 and 18.31. $F=60$ ml/100 g/min, $PS=150$ ml/100 g/min, $V_c=1.0$ ml/100 g, $\tau_c=V_c/F$, $T_1=1.26$ s and $T_{1a}=1.49$ s (these T_1 relaxation values represent a magnetic field strength of 3 T).

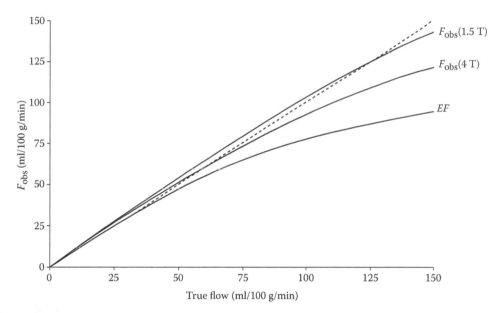

FIGURE 18.14
Observed CBF (F_{obs}) obtained using the one-compartment Kety model to interpret ASL data generated from the blood-tissue exchange model. Dotted line represents true CBF (F) and the solid black line represents the product of the extraction fraction (E) and F. F_{obs} was determined for $PS=150$ ml/100 g/min, $V_c=1.5$ ml/100 g and a nonexchanging venous blood volume of 2 ml/100 g. Data were generated for two magnetic field strengths and one echo time (TE = 15 ms). At 1.5 T, $T_1=1.0$ s, $T_{1a}=1.2$ s, $T_2^*=100$, 185 and 80 ms for tissue, arterial blood and venous blood, respectively. At 4 T, $T_1=1.4$ s, $T_{1a}=1.6$ s, $T_2^*=40$, 60 and 15 ms for tissue, arterial blood and venous blood, respectively.

data are analyzed with the Kety model (Figure 18.14). These predictions were confirmed by comparing ASL data acquired at 1.5 and 4 T [71,72].

A number of studies have analyzed ASL data acquired at multiple acquisition times with two-compartment exchange models [68–70]. It has been argued that accounting for restricted water exchange should improve the accuracy of CBF measurements. However, given the relative insensitivity of the ASL signal to water exchange, as discussed above, and its poor intrinsic SNR, the feasibility of this approach has been questioned [73].

Measuring water permeability remains an intriguing marker for assessing pathological alterations in BBB integrity and improving the quantification of cerebral hemodynamics by dynamic contrast-enhanced MRI [74–76]. One approach for enhancing the sensitivity of ASL to water exchange is to use diffusion-weighted gradients to separate the blood and tissue signal contributions [63,77].

18.5 Validation Studies

As is evident by the previous sections, there are a number of factors that can reduce the accuracy of ASL-CBF measurements, including labeling efficiencies, magnetization transfer effects, transit time uncertainties, vascular artifacts and restricted water exchange across the BBB. Nevertheless, accurate CBF measurements can be obtained with ASL methods provided steps are taken to minimize these confounding factors. This has been demonstrated in a number of validation studies comparing ASL measurements of CBF to measurements obtained with established methods. In rat studies, excellent agreement was observed between ASL-CBF measurements and those obtained with invasive techniques: radioactive microspheres [78] and the hydrogen clearance method [79]. Although in both of these studies, the comparisons were only conducted over large tissue volumes due to technical issues.

In a rat occlusion study, CASL images of CBF were compared to CBF images generated by quantitative autoradiography (QAR) [80]. Although a strong correlation between ASL and QAR measurements was observed, CBF values obtained with ASL consistently overestimated those obtained with QAR. As suggested by the authors, the most likely explanation for these findings is the significant signal contribution from labeled water in the capillary bed. Because of the magnetization transfer effects associated with CASL, there was a sizable difference in T_1 values of blood and tissue. The slower relaxation of the capillary contribution caused an overestimation of CBF when analyzing the data with the one-compartment Kety model. This study clearly demonstrated the importance of conducting validation studies since each ASL technique has its own unique issues regarding flow quantification.

Cerebral blood flow measured in cats by the FAIR technique underestimated CBF measured by radiolabeled microspheres for values greater than ~70 ml/100 g/min [69]. This underestimation was attributed to restricted water exchange at higher flow rates. These results would appear to be at odds with studies in rats that did not observe an underestimation of CBF within the same flow range and with modelling studies that suggest that ASL is relatively insensitive to restricted water exchange (Section 18.4.3). An alternative explanation is that the limited spatial coverage of the implemented surface coil prevented global inversion, which is required for FAIR. Consequently, the arterial input function may have been shorter than expected, particularly at higher flow rates.

The first validation study in humans compared measurements of basal CBF obtained with CASL to CBF measurements obtained with PET using $H_2^{15}O$ as a tracer [81]. Good agreement in terms of grey matter CBF was observed, but ASL consistently underestimated CBF in white matter regions, which was attributed to the substantially longer arterial transit times in white matter compared to grey matter. Recently, a PET/ASL comparison was performed in patients with chronic occlusive cerebrovascular disease [82]. Similar to Ye et al. [81], grey matter CBF values were in good agreement between the two techniques, while the white matter values were underestimated by ASL. The ASL approach also underestimated CBF in occluded regions due to the longer arterial tissue transit times in these regions.

Direct comparisons between ASL techniques and dynamic susceptibility contrast (DSC)-MRI have been performed [83–85]. Since both are MRI techniques, these studies do not require moving subjects between imaging modalities, which avoids errors associated with repositioning and registering datasets. One limitation is the inability to quantify CBF with DSC-MRI due to difficulties in accurately measuring the arterial concentration curve of the contrast agent. Nevertheless, a linear relationship between CBF measurements from ASL and DSC-MRI was reported, along with a significant correlation between the white-to-grey matter CBF ratios from the two methods. Based on the same principles underlying DSC-MRI, quantitative images of CBF can be obtained with perfusion CT (pCT) using an iodinated contrast agent [4,10]. In piglets, good agreement was observed between CBF measurements from pCT and PASL provided the inversion time for PASL was sufficient to allow the labeled water to reach the tissue [86]. Vascular artifacts, which were identified in the high-resolution pCT images, were observed in the ASL-CBF images acquired at a relatively short inversion time (TI=1200 ms) and caused CBF to be overestimated.

18.6 Challenges

Assuming that the appropriate steps are taken to address issues regarding CBF quantification, the primary challenge facing ASL applications is the inherently low signal. Since ASL signals are generally on the order of one percent of the equilibrium magnetization, signal averaging over extended periods of time, typically 5–10 min, is required to achieve an adequate SNR. Fast imaging sequences, such as echo-planar imaging (EPI) or spiral imaging techniques, are typically used to reduce motion artifacts between control and label images, and fairly thick slices are acquired (5–8 mm) to improve the SNR [20,21,52,87]. Recent advances in MRI technology have certainly benefited ASL methods, in particular the increased prevalence of higher field strength magnets and phased array RF coils [38,88]. Increasing the field strength not only improves image SNR, but also enhances the ASL signal due to the increase in T_1 relaxation times [72,89]. Two-to-fourfold improvements in SNR have been reported using receive-only phased array coils instead of conventional volumetric transmit/receive coils [90,91]. Combining high field imaging (7 T) with localized receiver coils has resulted in perfusion-weighted images with considerably higher spatial resolutions than typically obtained at 1.5 T with a conventional volume RF coil ($\sim$1$\times$1$\times$3 cm^3 vs 3.5$\times$3.5$\times$5 mm^3, respectively) [92,93].

Further SNR improvements can be achieved by optimizing the MRI sequences for particular ASL applications. For functional activation studies that do not require quantitative maps of CBF, the temporal resolution of ΔM images can be increased by using an inversion time longer than the repetition time, which is referred to as turbo ASL, or by reducing the length of the postlabeling delay [42,50,94]. Another approach for reducing noise levels is to use multiple inversion pulses to suppress static water magnetization during image acquisition [95]. Background suppression was originally proposed for three-dimensional ASL imaging to reduce phase noise caused by inter-acquisition fluctuations [96]. It has subsequently been used to double the temporal resolution in activation studies by eliminating the need for control images [97]. The number, RF power and type of inversion pulse must be selected carefully when implementing background suppression in order to minimize the attenuation of the ΔM signal caused by pulse imperfections [96,98]. Recently, a six-fold increase in sensitivity was reported by combining a single-shot 3D imaging sequence with background suppression and pCASL [99]. Because physiological noise has a large component that is proportional to the magnitude of the magnetization [100], background suppression also improves the reproducibility of ASL images (see Figure 18.15) [101]. Another method for reducing physiological noise is retrospective image-based correction, although this requires measuring the cardiac and respiratory fluctuations during the study [102].

An additional challenge with ASL is its poor temporal resolution compared to blood oxygenation level dependent (BOLD) imaging. Because of the need to collect both label and control images, ΔM images are typically acquired every 4–8 s, whereas, BOLD images are acquired every 1–3 s. The superior temporal resolution of BOLD imaging is well suited to detecting activation related to individual stimuli (i.e., event-related functional MRI) [103]. There have been few event-related ASL studies, in part due to issues regarding temporal resolution and sensitivity [104–106]. A potential complication with using ASL techniques to measure rapid changes in CBF is that the tracer kinetic models outlined in Sections 18.2 through 18.4 are based on the implicit assumption that CBF remains constant during the labeling period. Consequently, a rapidly changing ΔM signal may not be properly characterized by these models.

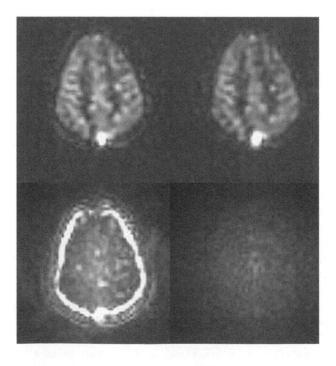

FIGURE 18.15

Top: Mean ΔM images from one slice location acquired by a PASL technique without (left) and with (right) background suppression. Bottom: Corresponding temporal standard deviation (SD) images, which have been scaled by a factor of two. The bright "ring" evident in the SD image without background suppression is due to motion at the edge of the brain. Mean and SD images were calculated from a time series of 80 ΔM volumes acquired with a single-shot spiral imaging technique at 1.5 T.

18.7 Applications

Despite its intrinsic SNR limitations, applications of ASL techniques have continued to expand. Compared to other functional MRI techniques—BOLD contrast and DSC-MRI—ASL techniques have the advantage of measuring absolute CBF. Since the labeling schemes are completely noninvasive, ASL techniques have been used in both short-term studies, such as functional activation experiments, and long-term studies, such as measuring perfusion changes associated with chronic diseases. This section will provide a brief overview of the applications in humans, highlighting the challenges and advantages of using ASL.

18.7.1 Neuroscience

One of the initial interests with ASL techniques was in functional MRI (fMRI) studies to measure regional CBF changes in response to activation tasks or stimuli. Although not quantitative, Kwong et al. published the first such study, and the results compared well with BOLD data acquired during the same visual stimulus [107]. This was followed by a number of fMRI studies using both PASL and CASL techniques to map regional changes in CBF associated with sensorimotor tasks and visual stimulation paradigms [20,22,47,52,56,108–110]. Compared to BOLD imaging, these ASL activation studies had limited spatial coverage and exhibited smaller activation regions. Furthermore, these types of paradigms are also associated with large changes in regional CBF. Detecting smaller perfusion changes associated with other activation paradigms, such as a working memory task, required relatively long scan times [96,111,112]. Due to these sensitivity issues and the lack of ASL sequences on most commercial MRI scanners, there have been far fewer fMRI studies using ASL techniques compared to those based on BOLD imaging.

Despite these challenges, fMRI applications of ASL techniques have demonstrated a number of advantages compared to BOLD imaging. The ASL signal has been shown to have a lower inter-subject and inter-session variability compared to the BOLD signal [113,114]. This has been attributed to the complexity of the BOLD signal as it depends on cerebral blood flow, blood volume and oxygen extraction. Arterial spin labeling has also been shown to better localize the site of neuronal activation than BOLD due to weighting of the BOLD signal by susceptibility effects in draining veins [115]. Although the benefits of this are likely to be small at typical field strengths used in human studies (1.5 and 3 T) considering the relatively poor spatial resolution of ASL images. A more practical advantage is that ASL can be incorporated with imaging methods that are less sensitive to susceptibility artifacts and, therefore, activation maps can be obtained in brain regions associated with large field heterogeneities [112,116].

Recent advances in MRI technology—namely, increased field strengths and phased array RF coils—have considerably improved the sensitivity of ASL and this has lead to its employment in studies of more "complex" activation paradigms. These include studies of learning [117], stress [118], sustained attention [119], speech production [120], memory encoding [99] and acute pain [121]. Increased sensitivity at higher fields has also been used to improve spatial coverage [38,122]. Using a 3 T scanner and a separate labeling coil positioned on the neck, Garraux et al. were able to measure regional CBF changes throughout the whole brain during a motor task [123].

A unique feature of ASL techniques is the insensitivity to task frequency. Low frequency drifts that are present in fMRI time series diminish the sensitivity of BOLD imaging when the duration of the stimulus epochs extends beyond a minute [113]. Due to the pair-wise subtraction used to create ΔM images, low-frequency drifts are removed from ASL time series, and as a result, task intervals can be extended to several minutes, hours or even days [124]. A number of recent ASL-fMRI studies used this feature to study slowly evolving processes, such as learning a motor skill and the psychological stress caused by performing a persistent mental arithmetic task [117,118]. Figure 18.16 illustrates an example in which the changes in regional CBF associated with sustained muscular pain were measured. Tonic pain was induced by the continuous infusion of hypertonic saline (5% NaCl) into the brachioradialis muscle of healthy volunteers. A PASL technique was used to track the progressive decline in pain-induced CBF changes as the effects of hypertonic saline diminished [121].

Although ASL techniques may be superior to BOLD imaging for certain applications, these two techniques are often combined [101,125–128]. One of the main purposes is to measure CBF and the cerebral metabolic rate of oxygen ($CMRO_2$) simultaneously. There has been a great deal of interest in understanding the coupling of blood flow and oxidative metabolism since the seminal work of Fox et al. showing that the CBF increase during somatosensory stimulation exceeded the concomitant increase in $CMRO_2$ [129]. The BOLD signal can be calibrated by concurrent measurements with CBF during a stimulus that does not affect oxidative metabolism, such as a mild hypercapnic challenge. It is then possible to measure the CBF and $CMRO_2$ changes during a functional task by combining ASL and BOLD data [130]. A number of studies have used calibrated BOLD to investigate the blood flow/metabolism coupling in the healthy brain [130–137]. Values for the ratio of $CMRO_2$ to CBF during visual or sensorimotor activation have ranged between 25 and 50%. This technique has also been used to study regional differences [137,138], differences in coupling during activation and deactivation [131,136,139], and the effects of pharmacological agents [135,140]. Although promising, calibrated BOLD remains in its infancy and there are a number of outstanding issues regarding (i) the accuracy of the model proposed by Davis et al., (ii) the validity of the assumed blood flow/blood volume ratio, and (iii) the optimization of the calibration step [137,141]. Nevertheless, calibrated BOLD has the potential to become an extremely powerful technique for studying flow/metabolism coupling, in particular during disease processes that are linked to impaired energy metabolism [142].

18.7.2 Clinical

The most immediate clinical application of any imaging method for measuring CBF is the identification of perfusion deficits related to acute cerebrovascular disease (stroke and transient ischemic attack) [143]. Studies of acute stroke patients have reported reasonable agreement between ASL and DSC-MRI in the assessment of relative perfusion and flow asymmetries [84,144–148]. However, long arterial transit times associated with cerebrovascular disease have hindered the ability to quantify CBF [82]. Postlabeling delays can reduce the sensitivity to transit times, but transit times greater than 2 s have been observed in patients with acute ischemic stroke due to low perfusion and/or extensive collateral supply [145,146]. In such cases, SNR considerations would question the ability to obtain meaningful perfusion images with extended postlabeling delays. A potential solution is the use of velocity-selective labeling (VSL) instead of spatially selective ASL techniques

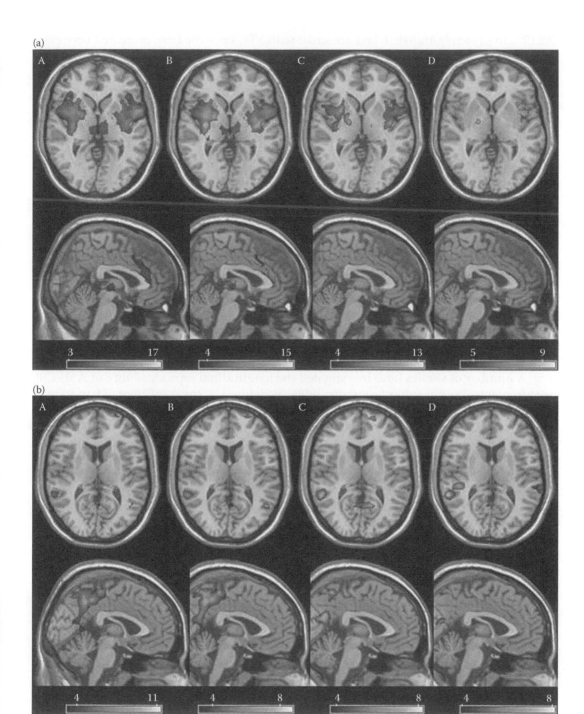

FIGURE 18.16

Group activation maps ($N=12$) showing areas of significant CBF increases (a) and decreases (b) during prolonged muscular pain. (A) 1st 2.5 min; (B) 1st 5 min; (C) 2nd 5 min; (D) 3rd 5 min. Corrected for multiple comparisons (FDR, $p<0.05$). Images displayed in neurological convention.

[149,150]. In principle, transit delays are small with VSL since the label is created very close to the tissue of interest. However, the labeling efficiency is less than with conventional ASL as the blood magnetization is saturated not inverted.

Perhaps a more promising role for ASL techniques in cerebrovascular disease is in the assessment of perfusion reserve in patients with chronic conditions, such carotid artery occlusion [151,152]. A unique feature of ASL is the ability to obtain perfusion maps of vascular territories by selectively labeling individual arteries [40,153–157]. These vessel-selective techniques have been used to map the effects of stenosis on vascular supply and to evaluate clinical outcome of carotid endarterectomy. Due to the ability to obtain serial CBF images, ASL techniques can be used to measure the response of CBF to a vasodilator, such as carbon dioxide, breath holding and acetazolamide [158–160]. Reduced CBF response indicates impaired cerebrovascular reserve, which is a significant risk factor of stroke [161].

Another clinical area that is particularly well-suited to ASL is the assessment of functional deficits associated with neurodegenerative diseases and other forms of dementia. Focal areas of hypoperfusion in patients with Alzheimer's disease, fronto-temporal dementia, and mild cognitive impairment have been reported, and the locations of these areas are consistent with previous nuclear medicine studies [162–164]. The advantages of ASL over nuclear medicine techniques are the relative ease of obtaining serial measurements; the ability to combine functional and structural data, such as morphological measures of brain atrophy; and potential to assess "cognitive" reserve by performing functional tasks [165]. A number of studies have investigated the longitudinal reproducibility of ASL-CBF measurements. Within-subject reproducibility, as defined by the coefficient of variation of whole brain and territorial CBF, has ranged from 10 to 14%, which is good agreement with PET and SPECT studies [166–168].

At the opposite end of the age spectrum, ASL techniques are well suited to paediatric studies, particularly since ethical concerns regarding the use of ionizing radiation in this age group are avoided [169]. In general, the SNR of ASL techniques will be higher in paediatric studies due to the higher blood flow in children, increased water concentration in the brain and longer relaxation times [170]. Clinical applications have included assessment of CBF in infants with congenital heart defects and children with sickle cell disease [171–173]. Measuring CBF could potentially have a significant impact on neonatal health care, since ischemia is a major contributor to preterm brain injury [174]. The challenge to using ASL techniques with this age group will be to overcome SNR limitations due to the extremely low blood flow values in the preterm brain [175,176]. A recent study demonstrated that CBF images in preterm neonates could be obtained on a 3 T scanner using a PASL technique [177]. Although the CBF values were higher than reported in previous studies—possibly due to long transit times—this study demonstrated the feasibility of using ASL to assess CBF in high-risk neonates.

There has been an interest in using ASL techniques to assess blood flow in brain tumors since blood flow changes are believed to be a marker of tumor grade or aggressiveness. Although, DSC-MRI is more commonly used for this propose, there have been several studies using ASL techniques to measure tumor blood flow in humans [178–184]. These studies demonstrated that perfusion images obtained by ASL were able to delineate regional variations in tumor blood flow and distinguish tumor grades. In combination with routine MRI and magnetic resonance spectroscopy, ASL improved the sensitivity and specificity of grading gliomas. However, DSC-MRI remains the more commonly used technique because of superior SNR and the ability to acquire images of tumor hemodynamics and permeability.

18.8 Summary

As evidenced by the variety of ASL applications outlined in Section 18.7, there is a tremendous need—both clinically and in the basic neurosciences—for techniques that can produce images of cerebral blood flow. Numerous methods have been developed on nuclear medicine, CT and MRI platforms, each with its own advantages and disadvantages. Arterial spin labeling can be considered an emerging technique as it has yet to become a standard MRI sequence on clinical scanners. However, the interest in ASL continues to grow because it is completely noninvasive, repeatable and easily adapted to a variety of applications. The significant improvements in sensitivity due to recent developments in MRI hardware will only heighten this interest.

References

1. Astrup J. 1982. Energy-requiring cell functions in the ischemic brain. Their critical supply and possible inhibition in protective therapy. *J. Neurosurg.*, 56:482.
2. Sokoloff L. 1981. Relationships among local functional activity, energy metabolism, and blood flow in the central nervous system. *Fed. Proc.*, 40:2311.
3. Villringer A, Dirnagl U. 1995. Coupling of brain activity and cerebral blood flow: basis of functional neuroimaging. *Cerebrovasc. Brain Metab. Rev.*, 7:240.
4. Wintermark M, Sesay M, Barbier E, Borbely K, Dillon WP, Eastwood JD, Glenn TC, Grandin CB, Pedraza S, Soustiel JF, Nariai T, Zaharchuk G, Caille JM, Dousset V, Yonas H. 2005. Comparative overview of brain perfusion imaging techniques. *Stroke*,36:e83.
5. Heymann MA, Payne BD, Hoffman JI, Rudolph AM. 1977 Blood flow measurements with radionuclide-labeled particles. *Prog. Cardiovasc. Dis.*, 20:55.
6. Zierler KL. 1965. Equations for measuring blood flow by external monitoring of radioisotopes. *Circ. Res.*, 16:309.
7. Lassen NA, Perl W. 1979. *Tracer Kinetic Methods in Medical Physiology*. New York, NY: Raven Press.
8. Ostergaard L, Weisskoff RM, Chesler DA, Gyldensted C, Rosen BR. 1996. High resolution measurement of cerebral blood flow using intravascular tracer bolus passages. Part I: Mathematical approach and statistical analysis. *Magn. Reson. Med.*, 36:715.
9. Axel L. 1980. Cerebral blood flow determination by rapid-sequence computed tomography: theoretical analysis. *Radiology*, 137:679.
10. Cenic A, Nabavi DG, Craen RA, Gelb AW, Lee TY. 1999. Dynamic CT measurement of cerebral blood flow: a validation study. *Am. J. Neuroradiol.*, 20:63.
11. Rosen BR, Belliveau JW, Vevea JM, Brady TJ. 1990. Perfusion imaging with NMR contrast agents. *Magn. Reson. Med.*, 14:249.
12. Kety SS. 1951. The theory and applications of the exchange of inert gas at the lungs and tissues. *Pharmacol. Rev.*, 3:1.
13. Kety SS, Schmidt CF. 1948. The nitrous oxide method for quantitative determinations of cerebral blood flow in man: theory, procedure, and normal values. *J. Clin. Invest.*, 27:476.
14. Kety SS, Schmidt CF. 1945. Determination of cerebral blood flow in man by the use of nitrous oxide in low concentrations. *Am. J. Physiol.*, 143:53.
15. Jones T, Chesler DA, Ter-Pogossian MM. 1976. The continuous inhalation of oxygen-15 for assessing regional oxygen extraction in the brain of man. *Br. J. Radiol.*, 49:339–343.
16. Herscovitch P, Markham J, Raichle ME. 1983. Brain blood flow measured with intravenous H2(15)O. I. Theory and error analysis. *J. Nucl. Med.*, 24:782.

17. Eichling JO, Raichle ME, Grubb RL, Jr., Ter-Pogossian MM. 1974. Evidence of the limitations of water as a freely diffusible tracer in brain of the rhesus monkey. *Circ. Res.* 35:358.

18. Detre JA, Leigh JS, Williams DS, Koretsky AP. 1992. Perfusion imaging. *Magn. Reson. Med.,* 23:37.

19. Williams DS, Detre JA, Leigh JS, Koretsky AP. 1992. Magnetic resonance imaging of perfusion using spin inversion of arterial water. *Proc. Natl. Acad. Sci. USA,* 89:212.

20. Edelman RR, Siewert B, Darby DG, Thangaraj V, Nobre AC, Mesulam MM, Warach S. 1994. Qualitative mapping of cerebral blood flow and functional localization with echo-planar MR imaging and signal targeting with alternating radio frequency. *Radiology,* 192:513.

21. Kwong KK, Chesler DA, Weisskoff RM, Donahue KM, Davis TL, Ostergaard L, Campbell TA, Rosen BR. 1995. MR perfusion studies with T1-weighted echo planar imaging. *Magn. Reson. Med.,* 34:878.

22. Kim SG. 1995. Quantification of relative cerebral blood flow change by flow-sensitive alternating inversion recovery (FAIR) technique: application to functional mapping. *Magn. Reson. Med.,* 34:293.

23. Herscovitch P, Raichle ME. 1985. What is the correct value for the brain--blood partition coefficient for water? *J. Cereb. Blood Flow Metab.,* 5:65.

24. St.Lawrence KS, Lee TY. 1998. An adiabatic approximation to the tissue homogeneity model for water exchange in the brain: II. Experimental validation. *J. Cereb. Blood Flow Metab.,* 18:1378.

25. Herscovitch P, Raichle ME, Kilbourn MR, Welch MJ. 1987. Positron emission tomographic measurement of cerebral blood flow and permeability-surface area product of water using [15O]water and [11C]butanol. *J. Cereb. Blood Flow Metab.,* 7:527.

26. Buxton RB, Frank LR, Wong EC, Siewert B, Warach S, Edelman RR. 1998. A general kinetic model for quantitative perfusion imaging with arterial spin labeling. *Magn. Reson. Med.,* 40:383.

27. Raichle ME, Martin WR, Herscovitch P, Mintun MA, Markham J. 1983. Brain blood flow measured with intravenous H2(15)O. II. Implementation and validation. *J. Nucl. Med.,* 24:790.

28. Dixon WT, Du LN, Faul DD, Gado M, Rossnick S. 1986. Projection angiograms of blood labeled by adiabatic fast passage. *Magn. Reson. Med.,* 3:454.

29. Roberts DA, Bolinger L, Detre JA, Insko EK, Bergey P, Leigh JS, Jr. 1993. Continuous inversion angiography. *Magn. Reson. Med.,* 29:631.

30. Gach HM, Kam AW, Reid ED, Talagala SL. 2002. Quantitative analysis of adiabatic fast passage for steady laminar and turbulent flows. *Magn. Reson. Med.,* 47:709.

31. Zhang W, Williams DS, Detre JA, Koretsky AP. 1992. Measurement of brain perfusion by volume-localized NMR spectroscopy using inversion of arterial water spins: accounting for transit time and cross-relaxation. *Magn. Reson. Med.,* 25:362.

32. Maccotta L, Detre JA, Alsop DC. 1997. The efficiency of adiabatic inversion for perfusion imaging by arterial spin labeling. *NMR Biomed.,* 10:216.

33. Zhang W, Silva AC, Williams DS, Koretsky AP. 1995. NMR measurement of perfusion using arterial spin labeling without saturation of macromolecular spins. *Magn. Reson. Med.,* 33:370.

34. Pekar J, Jezzard P, Roberts DA, Leigh JS, Jr., Frank JA, McLaughlin AC. 1996. Perfusion imaging with compensation for asymmetric magnetization transfer effects. *Magn. Reson. Med.,* 35:70.

35. McLaughlin AC, Ye FQ, Pekar JJ, Santha AK, Frank JA. 1997. Effect of magnetization transfer on the measurement of cerebral blood flow using steady-state arterial spin tagging approaches: a theoretical investigation. *Magn. Reson. Med.,* 37:501.

36. Ye FQ, Mattay VS, Jezzard P, Frank JA, Weinberger DR, McLaughlin AC. 1997. Correction for vascular artifacts in cerebral blood flow values measured by using arterial spin tagging techniques. *Magn. Reson. Med.,* 37:226.

37. Alsop DC, Detre JA. 1998. Multisection cerebral blood flow MR imaging with continuous arterial spin labeling. *Radiology,* 208:410.

38. Wang J, Zhang Y, Wolf RL, Roc AC, Alsop DC, Detre JA. 2005. Amplitude-modulated continuous arterial spin-labeling 3.0-T perfusion MR imaging with a single coil: feasibility study. *Radiology,* 235:218.

39. Silva AC, Zhang W, Williams DS, Koretsky AP. 1995. Multislice MRI of rat brain perfusion during amphetamine stimulation using arterial spin labeling. *Magn. Reson. Med.,* 33:209.

40. Zaharchuk G, Ledden PJ, Kwong KK, Reese TG, Rosen BR, Wald LL. 1999. Multislice perfusion and perfusion territory imaging in humans with separate label and image coils. *Magn. Reson. Med.*, 41:1093.

41. Mildner T, Trampel R, Moller HE, Schafer A, Wiggins CJ, Norris DG. 2003. Functional perfusion imaging using continuous arterial spin labeling with separate labeling and imaging coils at 3 T. *Magn. Reson. Med.*, 49:791.

42. Hernandez-Garcia L, Lee GR, Vazquez AL, Noll DC. 2004. Fast, pseudo-continuous arterial spin labeling for functional imaging using a two-coil system. *Magn. Reson. Med.*, 51:577.

43. Garcia DM, de Bazelaire C, Alsop DC. 2005. Pseudo-continuous flow driven adiabatic inversion for arterial spin labeling. *Int. Soc. Magn. Reson. Med.*, 13:37.

44. Wu WC, Fernandez Seara M, Detre JA, Wehrli FW, Wang J. 2007. A theoretical and experimental investigation of the tagging efficiency of pseudocontinuous arterial spin labeling. *Magn. Reson. Med.*, 58:1020–1027.

45. Wong EC, Buxton RB, Frank LR. 1998. A theoretical and experimental comparison of continuous and pulsed arterial spin labeling techniques for quantitative perfusion imaging. *Magn. Reson. Med.*, 40:348.

46. Yongbi MN, Yang Y, Frank JA, Duyn JH. 1999. Multislice perfusion imaging in human brain using the C-FOCI inversion pulse: comparison with hyperbolic secant. *Magn. Reson. Med.*, 42:1098.

47. Wong EC, Buxton RB, Frank LR. 1997. Implementation of quantitative perfusion imaging techniques for functional brain mapping using pulsed arterial spin labeling. *NMR Biomed.* 10:237.

48. Calamante F, Thomas DL, Pell GS, Wiersma J, Turner R. 1999. Measuring cerebral blood flow using magnetic resonance imaging techniques. *J. Cereb. Blood Flow Metab.*, 19:701.

49. Barbier EL, Lamalle L, Decorps M. 2001. Methodology of brain perfusion imaging. *J. Magn. Reson. Imaging*, 13:496.

50. Gonzalez-At JB, Alsop DC, Detre JA. 2000 Cerebral perfusion and arterial transit time changes during task activation determined with continuous arterial spin labeling. *Magn. Reson. Med.*, 43:739.

51. Yang Y, Engelien W, Xu S, Gu H, Silbersweig DA, Stern E. 2000. Transit time, trailing time, and cerebral blood flow during brain activation: measurement using multislice, pulsed spin-labeling perfusion imaging. *Magn. Reson. Med.*, 44:680.

52. Yang Y, Frank JA, Hou L, Ye FQ, McLaughlin AC, Duyn JH. 1998. Multislice imaging of quantitative cerebral perfusion with pulsed arterial spin labeling. *Magn. Reson. Med.*, 39:825.

53. Figueiredo PM, Clare S, Jezzard P. 2005. Quantitative perfusion measurements using pulsed arterial spin labeling: effects of large region-of-interest analysis. *J. Magn. Reson. Imaging*, 21:676.

54. Gunther M, Bock M, Schad LR. 2001. Arterial spin labeling in combination with a look-locker sampling strategy: inflow turbo-sampling EPI-FAIR (ITS-FAIR). *Magn. Reson. Med.*, 46:974.

55. Alsop DC, Detre JA. 1996. Reduced transit-time sensitivity in noninvasive magnetic resonance imaging of human cerebral blood flow. *J. Cereb. Blood Flow Metab.*, 16:1236.

56. Ye FQ, Smith AM, Yang Y, Duyn J, Mattay VS, Ruttimann UE, Frank JA, Weinberger DR, McLaughlin AC. 1997. Quantitation of regional cerebral blood flow increases during motor activation: a steady-state arterial spin tagging study. *Neuroimage*, 6:104.

57. Wong EC, Buxton RB, Frank LR. 1998. Quantitative imaging of perfusion using a single subtraction (QUIPSS and QUIPSS II). *Magn. Reson. Med.*, 39:702.

58. Grubb RL, Jr., Raichle ME, Eichling JO, Ter-Pogossian MM. 1974. The effects of changes in $PaCO_2$ on cerebral blood volume, blood flow, and vascular mean transit time. *Stroke*, 5:630.

59. Kennan RP, Gao JH, Zhong J, Gore JC. 1994. A general model of microcirculatory blood flow effects in gradient sensitized MRI. *Med. Phys.*, 21:539.

60. Wang J, Alsop DC, Song HK, Maldjian JA, Tang K, Salvucci AE, Detre JA. 2003. Arterial transit time imaging with flow encoding arterial spin tagging (FEAST). *Magn. Reson. Med.*, 50:599.

61. Paulson OB. 2002. Blood-brain barrier, brain metabolism and cerebral blood flow. *Eur. Neuropsychopharmacol.*, 12:495.

62. Crone C. 1963. The permeability of capillaries in various organs as determined by use of the "indicator diffusion" method. *Acta Physiol. Scand.*, 58:292.

63. Silva AC, Williams DS, Koretsky AP. 1997. Evidence for the exchange of arterial spin-labeled water with tissue water in rat brain from diffusion-sensitized measurements of perfusion. *Magn. Reson. Med.*, 38:232.

64. Silva AC, Zhang W, Williams DS, Koretsky AP. 1997. Estimation of water extraction fractions in rat brain using magnetic resonance measurement of perfusion with arterial spin labeling. *Magn. Reson. Med.*, 37:58.

65. Zaharchuk G, Bogdanov AA, Jr., Marota JJ, Shimizu-Sasamata M, Weisskoff RM, Kwong KK, Jenkins BG, Weissleder R, Rosen BR. 1998. Continuous assessment of perfusion by tagging including volume and water extraction (CAPTIVE): a steady-state contrast agent technique for measuring blood flow, relative blood volume fraction, and the water extraction fraction. *Magn. Reson. Med.*, 40:666.

66. St. Lawrence KS, Frank JA, McLaughlin AC. 2000. Effect of restricted water exchange on cerebral blood flow values calculated with arterial spin tagging: a theoretical investigation. *Magn. Reson. Med.*, 44:440.

67. Ewing JR, Cao Y, Fenstermacher J. 2001. Single-coil arterial spin-tagging for estimating cerebral blood flow as viewed from the capillary: relative contributions of intra- and extravascular signal. *Magn. Reson. Med.*, 46:465.

68. Parkes LM, Tofts PS. 2002. Improved accuracy of human cerebral blood perfusion measurements using arterial spin labeling: accounting for capillary water permeability. *Magn. Reson. Med.*, 48:27.

69. Zhou J, Wilson DA, Ulatowski JA, Traystman RJ, van Zijl PC. 2001. Two-compartment exchange model for perfusion quantification using arterial spin tagging. *J. Cereb. Blood Flow Metab.*, 21:440.

70. Li KL, Zhu X, Hylton N, Jahng GH, Weiner MW, Schuff N. 2005. Four-phase single-capillary stepwise model for kinetics in arterial spin labeling MRI. *Magn. Reson. Med.*, 53:511.

71. St Lawrence KS, Wang J. 2005. Effects of the apparent transverse relaxation time on cerebral blood flow measurements obtained by arterial spin labeling. *Magn. Reson. Med.*, 53:425.

72. Wang J, Alsop DC, Li L, Listerud J, Gonzalez-At JB, Schnall MD, Detre JA. 2002. Comparison of quantitative perfusion imaging using arterial spin labeling at 1.5 and 4.0 Tesla. *Magn. Reson. Med.*, 48:242.

73. Carr JP, Buckley DL, Tessier J, Parker GJ. 2007. What levels of precision are achievable for quantification of perfusion and capillary permeability surface area product using ASL? *Magn. Reson. Med.*, 58:281.

74. Barbier EL, St Lawrence KS, Grillon E, Koretsky AP, Decorps M. 2002. A model of blood-brain barrier permeability to water: accounting for blood inflow and longitudinal relaxation effects. *Magn. Reson. Med.*, 47:1100.

75. Schwarzbauer C, Morrissey SP, Deichmann R, Hillenbrand C, Syha J, Adolf H, Noth U, Haase A. 1997. Quantitative magnetic resonance imaging of capillary water permeability and regional blood volume with an intravascular MR contrast agent. *Magn. Reson. Med.*, 37:769.

76. Yankeelov TE, Rooney WD, Huang W, Dyke JP, Li X, Tudorica A, Lee JH, Koutcher JA, Springer CS, Jr. 2005. Evidence for shutter-speed variation in CR bolus-tracking studies of human pathology. *NMR Biomed.*, 18:173.

77. Wang J, Fernandez-Seara MA, Wang S, St Lawrence KS. 2007. When perfusion meets diffusion: in vivo measurement of water permeability in human brain. *J. Cereb. Blood Flow Metab.*, 27:839.

78. Walsh EG, Minematsu K, Leppo J, Moore SC. 1994. Radioactive microsphere validation of a volume localized continuous saturation perfusion measurement. *Magn. Reson. Med.*, 31:147.

79. Pell GS, King MD, Proctor E, Thomas DL, Lythgoe MF, Gadian DG, Ordidge RJ. 2003. Comparative study of the FAIR technique of perfusion quantification with the hydrogen clearance method. *J. Cereb. Blood Flow Metab.*, 23:689.

80. Ewing JR, Wei L, Knight RA, Pawa S, Nagaraja TN, Brusca T, Divine GW, Fenstermacher JD. 2003. Direct comparison of local cerebral blood flow rates measured by MRI arterial spin-tagging and quantitative autoradiography in a rat model of experimental cerebral ischemia. *J. Cereb. Blood Flow Metab.*, 23:198.

81. Ye FQ, Berman KF, Ellmore T, Esposito G, van Horn JD, Yang Y, Duyn J, Smith AM, Frank JA, Weinberger DR, McLaughlin AC. 2000. H(2)(15)O PET validation of steady-state arterial spin tagging cerebral blood flow measurements in humans. *Magn. Reson. Med.*, 44:450.

82. Kimura H, Kado H, Koshimoto Y, Tsuchida T, Yonekura Y, Itoh H. 2005. Multislice continuous arterial spin-labeled perfusion MRI in patients with chronic occlusive cerebrovascular disease: a correlative study with CO2 PET validation. *J. Magn. Reson. Imaging*, 22:189–198.

83. Li TQ, Guang CZ, Ostergaard L, Hindmarsh T, Moseley ME. 2000. Quantification of cerebral blood flow by bolus tracking and artery spin tagging methods. *J. Magn Reson. Imaging*, 18:503.

84. Hunsche S, Sauner D, Schreiber WG, Oelkers P, Stoeter P. 2002. FAIR and dynamic susceptibility contrast-enhanced perfusion imaging in healthy subjects and stroke patients. *J. Magn. Reson. Imaging*, 16:137.

85. Weber MA, Gunther M, Lichy MP, Delorme S, Bongers A, Thilmann C, Essig M, Zuna I, Schad LR, Debus J, Schlemmer HP. 2003. Comparison of arterial spin-labeling techniques and dynamic susceptibility-weighted contrast-enhanced MRI in perfusion imaging of normal brain tissue. *Invest. Radiol.*, 38:712.

86. Koziak AM, Winter J, Lee TY, Thompson RT, St Lawrence KS. 2007. Validation study of a pulsed arterial spin labeling technique by comparison to perfusion computed tomography. *Magn. Reson. Imaging.*, 26(4):543–553.

87. Ye FQ, Pekar JJ, Jezzard P, Duyn J, Frank JA, McLaughlin AC. 1996. Perfusion imaging of the human brain at 1.5 T using a single-shot EPI spin tagging approach. *Magn. Reson. Med.*, 36:217.

88. Roemer PB, Edelstein WA, Hayes CE, Souza SP, Mueller OM. 1990. The NMR phased array. *Magn. Reson. Med.*, 16:192.

89. Yongbi MN, Fera F, Yang Y, Frank JA, Duyn JH. 2002. Pulsed arterial spin labeling: comparison of multisection baseline and functional MR imaging perfusion signal at 1.5 and 3.0 T: initial results in six subjects. *Radiology*, 222:569.

90. Wang Z, Wang J, Connick TJ, Wetmore GS, Detre JA. 2005. Continuous ASL (CASL) perfusion MRI with an array coil and parallel imaging at 3T. *Magn. Reson. Med.*, 54:732.

91. de Zwart JA, Ledden PJ, van Gelderen P, Bodurka J, Chu R, Duyn JH. 2004. Signal-to-noise ratio and parallel imaging performance of a 16-channel receive-only brain coil array at 3.0 Tesla. *Magn. Reson. Med.*, 51:22.

92. Pfeuffer J, Adriany G, Shmuel A, Yacoub E, Van de Moortele PF, Hu X, Ugurbil K. 2002. Perfusion-based high-resolution functional imaging in the human brain at 7 Tesla. *Magn. Reson. Med.*, 47:903.

93. Duyn JH, van Gelderen P, Talagala L, Koretsky A, de Zwart JA. 2005. Technological advances in MRI measurement of brain perfusion. *J. Magn. Reson. Imaging*, 22:751.

94. Wong EC, Luh WM, Liu TT: Turbo ASL. 2000. arterial spin labeling with higher SNR and temporal resolution. *Magn. Reson. Med.*, 44:511.

95. Mani S, Pauly J, Conolly S, Meyer C, Nishimura D. 1997. Background suppression with multiple inversion recovery nulling: applications to projective angiography. *Magn. Reson. Med.*, 37:898.

96. Ye FQ, Frank JA, Weinberger DR, McLaughlin AC. 2000. Noise reduction in 3D perfusion imaging by attenuating the static signal in arterial spin tagging (ASSIST). *Magn. Reson. Med.*, 44:92.

97. Duyn JH, Tan CX, van Gelderen P, Yongbi MN. 2001. High-sensitivity single-shot perfusion-weighted fMRI. *Magn. Reson. Med.*, 46:88.

98. Garcia DM, Duhamel G, Alsop DC. 2005. Efficiency of inversion pulses for background suppressed arterial spin labeling. *Magn. Reson. Med.*, 54:366.

99. Fernandez-Seara MA, Wang J, Wang Z, Korczykowski M, Guenther M, Feinberg DA, Detre JA. 2007. Imaging mesial temporal lobe activation during scene encoding: comparison of fMRI using BOLD and arterial spin labeling. *Hum. Brain Map.*, 28:1391.

100. Kruger G, Glover GH. 2001. Physiological noise in oxygenation-sensitive magnetic resonance imaging. *Magn. Reson. Med.*, 46:631.

101. St Lawrence KS, Frank JA, Bandettini PA, Ye FQ. 2005. Noise reduction in multislice arterial spin tagging imaging. *Magn. Reson. Med.*, 53:735.

102. Restom K, Behzadi Y, Liu TT. 2006. Physiological noise reduction for arterial spin labeling functional MRI. *Neuroimage* 31:1104.

103. Rosen BR, Buckner RL, Dale AM. 1998. Event-related functional MRI: past, present, and future. *Proc. Natl. Acad. Sci. USA*, 95:773.

104. Liu TT, Wong EC, Frank LR, Buxton RB. 2002. Analysis and design of perfusion-based event-related fMRI experiments. *Neuroimage*, 16:269.

105. Yang Y, Engelien W, Pan H, Xu S, Silbersweig DA, Stern E. 2000. A CBF-based event-related brain activation paradigm: characterization of impulse-response function and comparison to BOLD. *Neuroimage*, 12:287.

106. Huppert TJ, Hoge RD, Diamond SG, Franceschini MA, Boas DA. 2006. A temporal comparison of BOLD, ASL, and NIRS hemodynamic responses to motor stimuli in adult humans. *Neuroimage*, 29:368.

107. Kwong KK, Belliveau JW, Chesler DA, Goldberg IE, Weisskoff RM, Poncelet BP, Kennedy DN, Hoppel BE, Cohen MS, Turner R, et al. 1992. Dynamic magnetic resonance imaging of human brain activity during primary sensory stimulation. *Proc. Natl. Acad. Sci. USA*, 89:5675.

108. Talagala SL, Noll DC. 1998. Functional MRI using steady-state arterial water labeling. *Magn. Reson. Med.*, 39:179.

109. Sanes JN, Donoghue JP, Thangaraj V, Edelman RR, Warach S. 1995. Shared neural substrates controlling hand movements in human motor cortex. *Science*, 268:1775.

110. Hoge RD, Atkinson J, Gill B, Crelier GR, Marrett S, Pike GB: Stimulus-dependent BOLD and perfusion dynamics in human V1. *Neuroimage*, 9:573.

111. Ye FQ, Smith AM, Mattay VS, Ruttimann UE, Frank JA, Weinberger DR, McLaughlin AC. 1999. Quantitation of regional cerebral blood flow increases in prefrontal cortex during a working memory task: a steady-state arterial spin-tagging study. *Neuroimage*, 8:44.

112. Kemeny S, Ye FQ, Birn R, Braun AR. 2005. Comparison of continuous overt speech fMRI using BOLD and arterial spin labeling. *Hum. Brain Map.*, 24:173.

113. Aguirre GK, Detre JA, Zarahn E, Alsop DC. 2002. Experimental design and the relative sensitivity of BOLD and perfusion fMRI. *Neuroimage*, 15:488.

114. Tjandra T, Brooks JC, Figueiredo P, Wise R, Matthews PM, Tracey I. 2005. Quantitative assessment of the reproducibility of functional activation measured with BOLD and MR perfusion imaging: implications for clinical trial design. *Neuroimage*, 27:393.

115. Kim SG, Duong TQ. 2002. Mapping cortical columnar structures using fMRI. *Physiol. Behav.*, 77:641.

116. Wang J, Li L, Roc AC, Alsop DC, Tang K, Butler NS, Schnall MD, Detre JA. 2004. Reduced susceptibility effects in perfusion fMRI with single-shot spin-echo EPI acquisitions at 1.5 Tesla. *Magn. Reson. Imaging*, 22:1–7.

117. Olson IR, Rao H, Moore KS, Wang J, Detre JA, Aguirre GK. 2006. Using perfusion fMRI to measure continuous changes in neural activity with learning. *Brain Cogn.*, 60(3):262–271.

118. Wang J, Rao H, Wetmore GS, Furlan PM, Korczykowski M, Dinges DF, Detre JA. 2005. Perfusion functional MRI reveals cerebral blood flow pattern under psychological stress. *Proc. Natl. Acad. Sci. USA*, 102:17804.

119. Kim J, Whyte J, Wang J, Rao H, Tang KZ, Detre JA. 2006. Continuous ASL perfusion fMRI investigation of higher cognition: Quantification of tonic CBF changes during sustained attention and working memory tasks. *Neuroimage.*, 31(1):376–385.

120. Troiani V, Fernandez-Seara MA, Wang Z, Detre JA, Ash S, Grossman M. 2007. Narrative speech production: An fMRI study using continuous arterial spin labeling. *Neuroimage.*, 40(2):932–939.

121. Owen DG, Bureau Y, Thomas AW, Prato FS, St Lawrence KS. 2007 Quantification of pain-induced changes in cerebral blood flow by perfusion MRI. *Pain.*, 136(1–2):85–96.

122. Talagala SL, Ye FQ, Ledden PJ, Chesnick S. 2004. Whole-brain 3D perfusion MRI at 3.0 T using CASL with a separate labeling coil. *Magn. Reson. Med.*, 52:131.

123. Garraux G, Hallett M, Talagala SL. 2005. CASL fMRI of subcortico-cortical perfusion changes during memory-guided finger sequences. *Neuroimage*, 25:122–132.

124. Wang J, Aguirre GK, Kimberg DY, Detre JA. 2003. Empirical analyses of null-hypothesis perfusion FMRI data at 1.5 and 4 T. *Neuroimage*, 19:1449.
125. Yongbi MN, Fera F, Mattay VS, Frank JA, Duyn JH. 2001. Simultaneous BOLD/perfusion measurement using dual-echo FAIR and UNFAIR: sequence comparison at 1.5T and 3.0T. *Magn. Reson. Imaging*, 19:1159.
126. Schwarzbauer C. 2000. Simultaneous detection of changes in perfusion and BOLD contrast. *NMR Biomed.*, 13:37.
127. van Gelderen P, C WHW, de Zwart JA, Cohen L, Hallett M, Duyn JH. 2005. Resolution and reproducibility of BOLD and perfusion functional MRI at 3.0 Tesla. *Magn. Reson. Med.*, 54:569.
128. Kim SG, Ugurbil K 1997. Comparison of blood oxygenation and cerebral blood flow effects in fMRI: estimation of relative oxygen consumption change. *Magn. Reson. Med.*, 38:59.
129. Fox PT, Raichle ME. 1986. Focal physiological uncoupling of cerebral blood flow and oxidative metabolism during somatosensory stimulation in human subjects. *Proc. Natl. Acad. Sci. USA*, 83:1140.
130. Davis TL, Kwong KK, Weisskoff RM, Rosen BR. 1998. Calibrated functional MRI: mapping the dynamics of oxidative metabolism. *Proc. Natl. Acad. Sci. USA*, 95:1834.
131. Stefanovic B, Warnking JM, Pike GB. 2004. Hemodynamic and metabolic responses to neuronal inhibition. *Neuroimage*, 22:771.
132. Hoge RD, Atkinson J, Gill B, Crelier GR, Marrett S, Pike GB. 1999. Linear coupling between cerebral blood flow and oxygen consumption in activated human cortex. *Proc. Natl. Acad. Sci. USA*, 96:9403.
133. Kim SG, Rostrup E, Larsson HB, Ogawa S, Paulson OB. 1999. Determination of relative CMRO2 from CBF and BOLD changes: significant increase of oxygen consumption rate during visual stimulation. *Magn. Reson. Med.*, 41:1152.
134. Kastrup A, Kruger G, Neumann-Haefelin T, Glover GH, Moseley ME. 2002. Changes of cerebral blood flow, oxygenation, and oxidative metabolism during graded motor activation. *Neuroimage*, 15:74.
135. St.Lawrence KS, Ye FQ, Lewis BK, Frank JA, McLaughlin AC. 2003. Measuring the effects of indomethacin on changes in cerebral oxidative metabolism and cerebral blood flow during sensorimotor activation. *Magn. Reson. Med.*, 50:99.
136. Uludag K, Dubowitz DJ, Yoder EJ, Restom K, Liu TT, Buxton RB. 2004. Coupling of cerebral blood flow and oxygen consumption during physiological activation and deactivation measured with fMRI. *Neuroimage*, 23:148.
137. Chiarelli PA, Bulte DP, Gallichan D, Piechnik SK, Wise R, Jezzard P. 2007. Flow-metabolism coupling in human visual, motor, and supplementary motor areas assessed by magnetic resonance imaging. *Magn. Reson. Med.*, 57:538.
138. Stefanovic B, Warnking JM, Rylander KM, Pike GB. 2006. The effect of global cerebral vasodilation on focal activation hemodynamics. *Neuroimage*, 30:726.
139. Stefanovic B, Warnking JM, Kobayashi E, Bagshaw AP, Hawco C, Dubeau F, Gotman J, Pike GB. 2005. Hemodynamic and metabolic responses to activation, deactivation and epileptic discharges. *Neuroimage*, 28:205.
140. Perthen JE, Lansing AE, Liau J, Liu TT, Buxton RB. 2007. Caffeine-induced uncoupling of cerebral blood flow and oxygen metabolism: A calibrated BOLD fMRI study. *Neuroimage.*, 40(1):237–247.
141. Li TQ, Haefelin TN, Chan B, Kastrup A, Jonsson T, Glover GH, Moseley ME. 2000. Assessment of hemodynamic response during focal neural activity in human using bolus tracking, arterial spin labeling and BOLD techniques. *Neuroimage*, 12:442.
142. Blass JP. 2003. Cerebrometabolic abnormalities in Alzheimer's disease. *Neurol. Res.*, 25:556.
143. Davis SM, Donnan GA. 2004. Advances in penumbra imaging with MR. *Cerebrovasc. Dis.*, 17 Suppl 3:23.
144. Detre JA, Alsop DC, Vives LR, Maccotta L, Teener JW, Raps EC. 1998. Noninvasive MRI evaluation of cerebral blood flow in cerebrovascular disease. *Neurology*, 50:633.

145. Siewert B, Schlaug G, Edelman RR, Warach S. 1997. Comparison of EPISTAR and T2*-weighted gadolinium-enhanced perfusion imaging in patients with acute cerebral ischemia. *Neurology,* 48:673.
146. Chalela JA, Alsop DC, Gonzalez-Atavales JB, Maldjian JA, Kasner SE, Detre JA. 2000. Magnetic resonance perfusion imaging in acute ischemic stroke using continuous arterial spin labeling. *Stroke,* 31:680.
147. Yoneda K, Harada M, Morita N, Nishitani H, Uno M, Matsuda T. 2003. Comparison of FAIR technique with different inversion times and post contrast dynamic perfusion MRI in chronic occlusive cerebrovascular disease. *Magn. Reson. Imaging,* 21:701.
148. Wolf RL, Alsop DC, McGarvey ML, Maldjian JA, Wang J, Detre JA. 2003. Susceptibility contrast and arterial spin labeled perfusion MRI in cerebrovascular disease. *J. Neuroimaging,* 13:17.
149. Wong EC, Cronin M, Wu WC, Inglis B, Frank LR, Liu TT. 2006. Velocity-selective arterial spin labeling. *Magn. Reson. Med.,* 55:1334.
150. Duhamel G, de Bazelaire C, Alsop DC. 2003. Evaluation of systematic quantification errors in velocity-selective arterial spin labeling of the brain. *Magn. Reson. Med.,* 50:145.
151. Hendrikse J, van Osch MJ, Rutgers DR, Bakker CJ, Kappelle LJ, Golay X, van der Grond J. 2004. Internal carotid artery occlusion assessed at pulsed arterial spin-labeling perfusion MR imaging at multiple delay times. *Radiology,* 233:899.
152. Ances BM, McGarvey ML, Abrahams JM, Maldjian JA, Alsop DC, Zager EL, Detre JA. 2004. Continuous arterial spin labeled perfusion magnetic resonance imaging in patients before and after carotid endarterectomy. *J. Neuroimaging,* 14:133.
153. Gunther M. 2006. Efficient visualization of vascular territories in the human brain by cycled arterial spin labeling MRI. *Magn. Reson. Med.,* 56:671.
154. Jones CE, Wolf RL, Detre JA, Das B, Saha PK, Wang J, Zhang Y, Song HK, Wright AC, Mohler EM, III, Fairman RM, Zager EL, Velazquez OC, Golden MA, Carpenter JP, Wehrli FW. 2006. Structural MRI of carotid artery atherosclerotic lesion burden and characterization of hemispheric cerebral blood flow before and after carotid endarterectomy. *NMR Biomed.,* 19:198.
155. Werner R, Norris DG, Alfke K, Mehdorn HM, Jansen O: Continuous artery-selective spin labeling (CASSL). *Magn. Reson. Med.,* 53:1006.
156. Davies NP, Jezzard P. 2003. Selective arterial spin labeling (SASL): perfusion territory mapping of selected feeding arteries tagged using two-dimensional radiofrequency pulses. *Magn. Reson. Med.,* 49:1133.
157. Golay X, Hendrikse J, Van Der Grond J. 2005. Application of regional perfusion imaging to extra-intracranial bypass surgery and severe stenoses. *J. Neuroradiol.,* 32:321.
158. Kastrup A, Li TQ, Glover GH, Moseley ME. 1999. Cerebral blood flow-related signal changes during breath-holding. *Am. J. Neuroradiol.,* 20:1233.
159. St.Lawrence KS, Ye FQ, Lewis BK, Weinberger DR, Frank JA, McLaughlin AC. 2002. Effects of indomethacin on cerebral blood flow at rest and during hypercapnia: an arterial spin tagging study in humans. *J. Magn. Reson. Imaging,* 15:628.
160. Detre JA, Samuels OB, Alsop DC, Gonzalez-At JB, Kasner SE, Raps EC. 1999. Noninvasive magnetic resonance imaging evaluation of cerebral blood flow with acetazolamide challenge in patients with cerebrovascular stenosis. *J. Magn. Reson. Imaging,* 10:870.
161. Yonas H, Smith HA, Durham SR, Pentheny SL, Johnson DW. 1993. Increased stroke risk predicted by compromised cerebral blood flow reactivity. *J. Neurosurg.,* 79:483.
162. Johnson NA, Jahng GH, Weiner MW, Miller BL, Chui HC, Jagust WJ, Gorno-Tempini ML, Schuff N. 2005. Pattern of cerebral hypoperfusion in Alzheimer disease and mild cognitive impairment measured with arterial spin-labeling MR imaging: initial experience. *Radiology,* 234:851.
163. Alsop DC, Detre JA, Grossman M. 2000. Assessment of cerebral blood flow in Alzheimer's disease by spin-labeled magnetic resonance imaging. *Ann. Neurol.,* 47:93.
164. Sandson TA, O'Connor M, Sperling RA, Edelman RR, Warach S. 1996. Noninvasive perfusion MRI in Alzheimer's disease: a preliminary report. *Neurology,* 47:1339.
165. Stern Y. 2006. Cognitive reserve and Alzheimer disease. *Alzheimer Dis. Assoc. Disord.,* 20:S69.

166. Parkes LM, Rashid W, Chard DT, Tofts PS. 2004. Normal cerebral perfusion measurements using arterial spin labeling: reproducibility, stability, and age and gender effects. *Magn. Reson. Med.*, 51:736.

167. Yen YF, Field AS, Martin EM, Ari N, Burdette JH, Moody DM, Takahashi AM. 2002. Test-retest reproducibility of quantitative CBF measurements using FAIR perfusion MRI and acetazolamide challenge. *Magn Reson. Med.*, 47:921.

168. Jahng GH, Song E, Zhu XP, Matson GB, Weiner MW, Schuff N. 2005. Human brain: reliability and reproducibility of pulsed arterial spin-labeling perfusion MR imaging. *Radiology*, 234:909.

169. Wang J, Licht DJ. 2006. Pediatric perfusion MR imaging using arterial spin labeling. *Neuroimaging Clin. N. Am.*, 16:149.

170. Wang J, Licht DJ, Jahng GH, Liu CS, Rubin JT, Haselgrove J, Zimmerman RA, Detre JA. 2003. Pediatric perfusion imaging using pulsed arterial spin labeling. *J. Magn. Reson. Imaging*, 18:404.

171. Oguz KK, Golay X, Pizzini FB, Freer CA, Winrow N, Ichord R, Casella JF, van Zijl PC, Melhem ER. 2003. Sickle cell disease: continuous arterial spin-labeling perfusion MR imaging in children. *Radiology.*, 227(2):567–574.

172. Licht DJ, Wang J, Silvestre DW, Nicolson SC, Montenegro LM, Wernovsky G, Tabbutt S, Durning SM, Shera DM, Gaynor JW, Spray TL, Clancy RR, Zimmerman RA, Detre JA. 2004. Preoperative cerebral blood flow is diminished in neonates with severe congenital heart defects. *J. Thorac Cardiovasc. Surg.*, 128:841.

173. Strouse JJ, Cox CS, Melhem ER, Lu H, Kraut MA, Razumovsky A, Yohay K, van Zijl PC, Casella JF. 2006. Inverse correlation between cerebral blood flow measured by continuous arterial spin-labeling (CASL) MRI and neurocognitive function in children with sickle cell anemia (SCA). *Blood*, 108:379.

174. Volpe JJ. 2001. *Neurology of the Newborn*. Philadelphia, PA: W.B. Saunders Company.

175. Altman DI, Powers WJ, Perlman JM, Herscovitch P, Volpe SL, Volpe JJ. 1988. Cerebral blood flow requirement for brain viability in newborn infants is lower than in adults. *Ann. Neurol.*, 24:218.

176. Greisen G, Borch K. 2001. White matter injury in the preterm neonate: the role of perfusion. *Dev. Neurosci.*, 23:209.

177. Miranda MJ, Olofsson K, Sidaros K. 2006. Noninvasive measurements of regional cerebral perfusion in preterm and term neonates by magnetic resonance arterial spin labeling. *Pediatr. Res.*, 60:359.

178. Gaa J, Warach S, Wen P, Thangaraj V, Wielopolski P, Edelman RR. 1996. Noninvasive perfusion imaging of human brain tumors with EPISTAR. *Eur. Radiol.*, 6:518.

179. Warmuth C, Gunther M, Zimmer C. 2003. Quantification of blood flow in brain tumors: comparison of arterial spin labeling and dynamic susceptibility-weighted contrast-enhanced MR imaging. *Radiology*, 228:523.

180. Kimura H, Takeuchi H, Koshimoto Y, Arishima H, Uematsu H, Kawamura Y, Kubota T, Itoh H. 2006. Perfusion imaging of meningioma by using continuous arterial spin-labeling: comparison with dynamic susceptibility-weighted contrast-enhanced MR images and histopathologic features. *Am. J. Neuroradiol.*, 27:85.

181. Wolf RL, Wang J, Wang S, Melhem ER, O'Rourke DM, Judy KD, Detre JA. 2005. Grading of CNS neoplasms using continuous arterial spin labeled perfusion MR imaging at 3 Tesla. *J. Magn. Reson. Imaging*, 22:475.

182. Weber MA, Thilmann C, Lichy MP, Gunther M, Delorme S, Zuna I, Bongers A, Schad LR, Debus J, Kauczor HU, Essig M, Schlemmer HP. 2004. Assessment of irradiated brain metastases by means of arterial spin-labeling and dynamic susceptibility-weighted contrast-enhanced perfusion MRI: initial results. *Invest. Radiol.*, 39:277.

183. Weber MA, Zoubaa S, Schlieter M, Juttler E, Huttner HB, Geletneky K, Ittrich C, Lichy MP, Kroll A, Debus J, Giesel FL, Hartmann M, Essig M. 2006. Diagnostic performance of spectroscopic and perfusion MRI for distinction of brain tumors. *Neurology*, 66:1899.

184. Kim HS, Kim SY. 2007. A prospective study on the added value of pulsed arterial spin-labeling and apparent diffusion coefficients in the grading of gliomas. *Am. J. Neuroradiol.*, 28:1693.

19

Computer Aided Diagnosis in Mammography with Emphasis on Automatic Detection of Microcalcifications

Yong Man Ro and Sung Ho Jin
Information Communication University

Konstantinos N. Plataniotis
University of Toronto

CONTENTS

19.1 Background

19.1.1 Breast Cancer and Mammography

Breast cancer is one of the major causes of increasing mortality in women, especially in developed countries including the USA and Europe. The World Health Organization's International Agency for Research on Cancer in Lyon, France, estimates that more than 150,000 women in the world die of breast cancer each year [1]. In order to decrease the death rate of breast cancer, many diagnostic and treatment methods have been studied and developed [1–11]. Among them, mammography, which is known as a low-dose X-ray imaging system, has emerged as an effective diagnostic technique in the early phases of breast cancer [2,3]. In mammogram, radiologists define a tiny abnormal deposit of calcium salts, known as microcalcification, as one of the early warning sign for breast cancer [4]. Unfortunately, the variability in type, size and distribution of microcalcification makes the early diagnosis of microcalcification a difficult task [5,6]

Mammography is typically characterized in terms of the acquisition mode. For example, digital database for screening mammography (DDSM) is obtained by scanning the mammography film. Obviously in such a system the obtained digital image is corrupted by noise augmented from the converter. On the other hand, full-field digital mammography (FFDM) utilizes solid-state detectors, similar to those found in digital camera, that convert X-rays into electrical signals producing images with better SNR compared to DDSM images [7]. Figure 19.1 depicts two test images so that a visual assessment of the relative merits of DDSM and FFDM can be performed. It is obvious that the blur intensity observed in the background regions of the DDSM image can easily make abnormalities resulting in misdiagnosis [8]

19.1.2 Microcalcification

Breast cancer is typically characterized and defined by small and bright spots corresponding to small calcium deposits, generally know as calcifications [9]. Since calcifications are difficult to be visually detected without instruments digital mammograms, with a high spatial resolution, are usually utilized for their detection [10]. Depending on its size, the calcification could be categorized into two types: macrocalcifications and microcalcifications. Macrocalcifications are large calcium deposits in the breast tissue. Most of them are benign and usually not considered to be cancers. On a mammogram, the macrocalcifications appear as spots within the breast tissue. Most often they form round shapes, and do not require a biopsy. In contrast, microcalcifications are tiny specks of calcium that may be found in an area of rapidly dividing cells. When many are seen in a cluster, they may indicate a small cancer [11]. About half of the cancers detected appear as these clusters [12].

Due to the microcalcifications' variability in shape, brightness and size, it is difficult to distinguish them from neighboring breast tissues [6,13]. According to Breast Imaging and

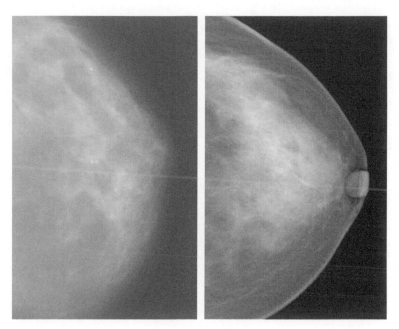

FIGURE 19.1
Examples of DDSM and FFDM image.

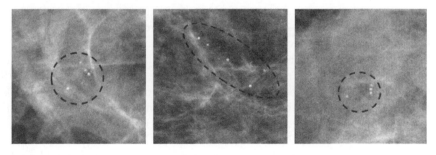

FIGURE 19.2
Examples of microcalcifications.

Reporting Data System (BI-RDS) in the American College of Radiology (ACR), the types and arrangements of microcalcifications are categorized into typically benign, intermediate concern, and a higher probability of malignancy [6]. If microcalcifications are easily found, most cases are benign, but it could occasionally be a sign of breast cancer. It is known that microcalcifications consist of brighter spots compared with neighboring normal regions. The range of the size of spots is typically observed between 0.1 mm and 1 mm. If more than four spots are found in an area of (1 cm×1 cm), the area can be judged as an abnormal region, i.e., a doubtful cluster as a microcalcification (Figure 19.2) [1]. Small size and variable shape of microcalcifications cause a difficulty achieving an automatic detection.

19.1.3 Computer Aided Diagnosis (CAD) System for Mammogram

The importance of automatic diagnosis of microcalcifications in early detection and diagnosis procedures led to the development of computer aided diagnosis (CAD) in

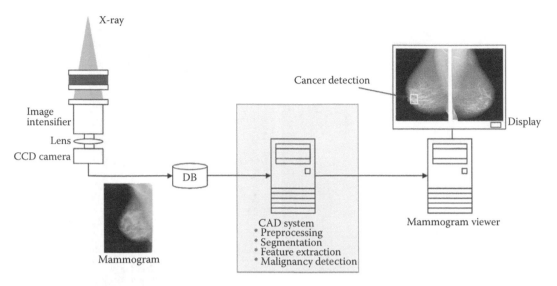

FIGURE 19.3
CAD system for mammogram.

mammography. CAD is a procedure in medical science that provides medical doctors with additional information allowing them to have a second opinion as well as double reading to correctly identify regions with a high suspicion of malignancy [14]. Much research has been carried out for developing accurate CAD to help indicate suspicious locations with great accuracy and reliability [15–17]. CAD uses image processing and pattern recognition techniques to automatically detect the suspected regions.

Figure 19.3 shows a mammography with a CAD system. As seen, intensity is recorded on the image from X-ray and mammogram database is acquired (FFDM with CCD camera shown in Figure 19.3 or DDSM from digitizing mammogram films). Digital mammogram goes through CAD module to find the cancer region. Through the CAD module, abnormalities are detected to draw the attention of medical doctors to those legions. This process helps medical doctors interpret the cancer possibility more precisely. As seen in the right side of the figure, the mammogram is displayed to show the highlighted suspicious regions so that medical doctor can perform a biopsy as a precaution.

Inside CAD modules, preprocessing techniques are applied to improve the performance of microcalcification detection as well as to provide doctors with enhanced images when difficult with the naked eye.

Most of the CAD systems utilize procedures that can be used to identify lesions and malignant tumors [17]. These detection processes are composed of segmentation, feature extraction and classification steps. Segmentation is the process of separating suspected malignancy from the original image. In feature extraction process, the feature values, which take characteristics of malignancy into consideration, are extracted from the segmented region. Lastly, based on the feature values extracted, the classification process decides whether suspicious regions are malignant or not.

It is known that medical doctors' chance of successful treatment is improved when they are assisted by CAD. Unfortunately, CAD currently is not perfect so that it could give a number of normal areas on mammograms as malignant, which causes unnecessary biopsies.

19.1.4 Related Works for Microcalcification Detection

The high attenuation properties of the mammogram and the small sizes of microcalcifications are the major causes of poor visibility and small lesions. In order to improve the contrast and to eliminate the noise of a mammogram, there were many preprocessing techniques. Laine et al. [18] developed image contrast enhancement using multiresolution representation of the dyadic wavelet. Strickland and Hahn [19] used undecimated wavelet transform. Gordon and Rangayyan [20] performed adaptive neighborhood image processing to enhance the contrast of features relevant to mammography noise and digitization effects. Dhawan et al. [21–23] developed an adaptive neighborhood-based image processing technique that utilized low-level analysis and knowledge about a desired feature in the design of a contrast enhancement function to improve the contrast of specific features. Tahoces et al. [24] developed an enhancement technique by automatic spatial filtering.

For the segmentation of the mammogram, techniques presented by Chan et al. [25] and Wu et al. [26] compared each pixel to a local threshold computed within a square neighborhood around the pixel. The method proposed by Davies and Dance [27,28] divided an image into small square subimages and set individual threshold for each subimage. The method that was based on localized region growing in Kallergi et al. [29] required the selection of a neighborhood, and it regulated the growth of region by comparing the value of the pixel to be appended to the value of the seed pixel.

There also were many CAD algorithms which were for microcalcification detection based on various classifiers. Karssemeijer [30] developed a statistical Bayesian image analysis model for detection of microcalcifications. Chan et al. [31] adopted linear discriminant analysis for classification of microcalcifications and Dhawan [32] developed a radial basis function neural network for classifying hard-to-diagnose cases. Kang et al. [33] and Kouskas et al. [34] also utilized artificial neural networks as a core classifier and Cheng et al. [35] applied fuzzy logic for microcalcification detection. In the previous works, even though they needed a multidimensional feature vector to represent the characteristics of microcalcification, the features showed to cause high false positive rate originated from image noise components and ambiguous boundary decisions. As a reliable solution, a support vector machine (SVM) classifier has emerged for detecting microcalcifications, which maximizes the margin between the data located closely to the boundaries [36,37].

19.2 Overview of Microcalcification Detection System

In this section, we provide an overview of a detection system for microcalcifications. The system is typically divided into two parts: preprocessing and detection. Figure 19.4 depicts a generic representation of the detection system.

Preprocessing includes background segmentation and image enhancement, including image normalization, enhancement, and noise reduction, which handle the visibility of a mammogram. In a mammogram, we can observe that background regions occupy large areas unrelated to microcalcification. The regions should be segmented and eliminated to minimize the search range of the microcalcification. The background segmentation could use various morphological image processing techniques such as dilation and erosion, opening and closing.

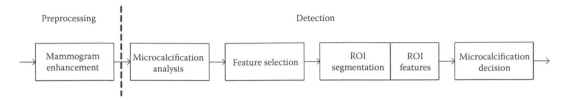

FIGURE 19.4
Layout of a generic detection system for microcalcifications.

The visibility of an input mammogram could be improved by image processing techniques, e.g., contrast enhancement, normalization resolution or pixel depth, and histogram equalization, etc. The contrast enhancement affects the detection performance because the contrast of a pixel belonging to regions of microcalcifications could be a dominant factor to find the microcalcifications. In addition, noise component needs to be eliminated or reduced in the preprocessing phase. Image acquisition is the major source of noise in a mammogram. The performance of imaging sensors is affected by a diversity of factors, such as environmental conditions at the image acquisition, and by the quality of the sensing elements themselves [38].

The microcalcification detection stage includes suitable feature extraction, regions of interest (ROI) segmentation and classification of microcalcifications. The characteristics of microcalcifications, e.g., size, shape, intensity are utilized in the detection system. The features should sufficiently reflect the characteristics of microcalcification with mathematical descriptions which discriminate between benign and malignant. However, it is difficult to predict whether a specific feature or feature combination gives good performance. The extracted features are used for ROI segmentation and microcalcification detection. The last procedure in the detection part is microcalcification decision which determines whether ROI is benign and malignant by using classifiers with extracted features on ROI.

In this chapter, the preprocessing stage is described in Section 19.3. Information on technologies and implementation details regarding the background segmentation, contrast enhancement and denoising steps are provided [33,38]. Through combining homomorphic filtering with subband decomposition of wavelet transform, we establish nonlinear contrast enhancement and effective denoising scheme. Section 19.4 addresses the microcalcification detection procedures. SVM-based detection systems for microcalcifications and associated detail algorithms are described in detail. In this approach, the system forms a hierarchical structure combined by multiple classifiers. The aim of the hierarchical structure is to diminish search range of a full image to minimize the processing time. The system firstly seeks ROIs by detecting candidate pixels, in other words, one classifier is required to filter ROIs from whole pixels of the image. The next classifier is utilized to obtain the regions of microcalcifications from the selected ROIs. In Section 19.4, experimental results are described to show the demonstration of microcalcification enhancement and microcalcification detection mentioned in this chapter.

19.3 Preprocessing

A mammogram image can be divided into three distinctive regions: the breast region, background (nonbreast) region, and the regions of artifacts. The breast region is created

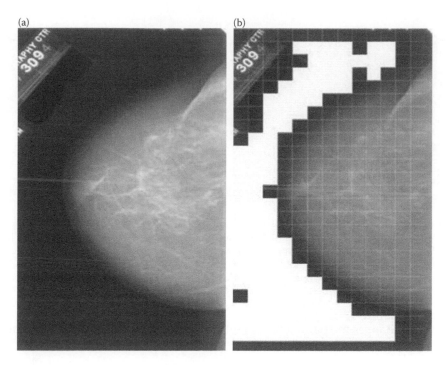

FIGURE 19.5

Example of background segmentation (a) original image and (b) background segmentation. (From Ho-Kyung Kang, Nguyen N. Thahh, Sung-Min Kim, and Yong Man Ro., *Robust Contrast Enhancement for Microcalcification in Mammography*. LNCS, Springer Berlin, Vol. 3045, pp. 604, 2004. With permission.)

when the X-ray is absorbed in the breast; the background is a region where the X-ray has no obstacle; and artifacts are objects such as labels. The first work of the preprocessing is to reduce the search range from an entire mammogram. CAD systems on microcalcifications need segmentation of the background from a mammogram because eliminating the background leads to a significant reduction in terms of computational cost. If the main processing, i.e., the detection of microcalcifications, deals with only the breast region, the entire processing time will be minimized. Therefore, preprocessing is required to limit regions for the main processing and to normalize or enhance the characteristics of an input image. Figure 19.5 shows an example of a mammogram with the background removed from an original image. In the figure, the breast is shown in the right side of image (a) and (b), whereas the top left shows a tag with the patient's information, and the dark part on the left side is for the background.

Following the successive completion of the first step, noise removal and signal enhancement are attempted. The original image has relatively low performance in the microcalcification detection because of low contrast and high noise. Typically, various kinds of noise components from heterogeneous environments exist in mammograms such as acquisition devices, X-ray intensities, and concentrations of sensitizers of mammogram films. Some images often show the background and breast area diffused by the noise components. Here, we can see that the background can play a role as an indicator and give the characteristics of the noise components. Figure 19.6 gives two examples of a mammogram, in which (a) is a mammogram with a high noise condition, whereas (b) shows a low noise case.

Even though previous contrast enhancement methods usually used fixed parameters, e.g., gains of filtering and thresholds of denoising, for all input mammograms, the approach

(a)   (b)

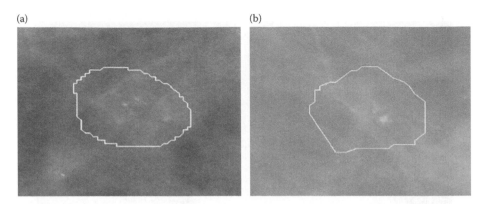

FIGURE 19.6
Examples of two different noise condition in microcalcification areas: (a) shows the breast with high noise
($var_b > 40$); (b) shows the backgrounds and breast area with low noise ($var_b < 20$) (12 bit gray level). Note that the
marks in (a) and (b) show the microcalcification. (See Equation 19.9 for the definition for var_b.) (From Ho-Kyung
Kang, Nguyen N. Thahh, Sung-Min Kim, and Yong Man Ro., *Robust Contrast Enhancement for Microcalcification
in Mammography*. LNCS, Springer Berlin, Vol. 3045, pp. 604, 2004. With permission.)

of this chapter changes the parameters adapting to noise characteristics of an input mammogram by using homomorphic filtering.

19.3.1 Background Segmentation

The background is segmented by using differences of mean and variance of the gray levels
between neighboring pixels. The background segmentation employs a block with $N \times N$
size to calculate gray level histogram, where the noise variance of the block can be utilized
in the contrast enhancement and noise reduction parts followed by the background segmentation. In Figure 19.5, the boundary of the background is unclear and the sharpness
of the boundary depends on the type of digitization method used. The procedure of the
segmentation is carried out by the following steps:

Step 1: Make histogram ($H(x)$) of mammogram by using gray value.

Step 2: Find a gray value range (μ_l, μ_h) as shown in Figure 19.7, where μ_l and μ_h can
be obtained as

$$H(\mu_l) = H(\mu_h) = \frac{1}{3}\max \text{ (lower part of } H(x)) \tag{19.1}$$

The background area of a mammogram has unique properties representing gray
value, mean, and variance. The peak of the lower part of the histogram is located
in the background. If a block has a mean value in the range of (μ_l, μ_h) it can be a
candidate block of the background.

Step 3: Select candidate blocks in which mean gray values are between μ_l and μ_h.

Step 4: Calculate the variance of the selected blocks and make a histogram of the
variance ($H_s(x)$).

Step 5: Find the variance range (σ_l, σ_h) as shown in Figure 19.8, where σ_l and σ_h can
be obtained as

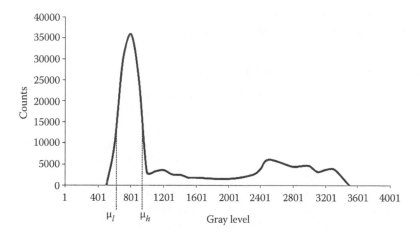

FIGURE 19.7
Example of histogram $H(x)$ and mean range (μ_l, μ_h). (From Kokyung Kang, Yong Man Ro, and Sung Min Kim., *A Microcalculation Detecion Using Adaptive Contrast Enhancement on Wavelet Transform and Neutral Network.* IEICE Transactions on Information and Systems. The Institute of Electronics, Information and Communication Engineers, Vol. 1389-D, No. 3, pp. 12382, 2006. With permission.)

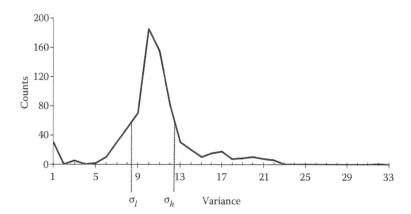

FIGURE 19.8
Example of histogram on variances $H_s(x)$ in the selected block and variance range (σ_l, σ_h). (From Kokyung Kang, Yong Man Ro, and Sung Min Kim., *A Microcalculation Detecion Using Adaptive Contrast Enhancement on Wavelet Transform and Neutral Network.* IEICE Transactions on Information and Systems. The Institute of Electronics, Information and Communication Engineers, Vol. 1389-D, No. 3, pp. 12382, 2006. With permission.)

$$H_s(\sigma_l) = H_s(\sigma_h) = \frac{1}{3}\max\,(H_s(x)) \tag{19.2}$$

The variance is calculated within a block ($N \times N$) whose mean gray value is in the range of (μ_l, μ_h). In Equations 19.1 and 19.2, 1/3 value is obtained heuristically from the mammogram database used in the experiment.

Step 6: If a block has variance which belongs to the variance range of (σ_l, σ_h), the block is determined to belong to the background, finally.

Figure 19.5b is the segmentation result of a block based on the background detection. The white blocks show the background area that is finally selected.

19.3.2 Contrast Enhancement and Denoising

There are generally two types of techniques to enhance the image contrast: linear and nonlinear enhancement [39]. The linear enhancement technique tends to emphasize edges strongly. The strong edges cause to reduce the dynamic range of small edges of the cancer so that it can deteriorate displaying of the mammogram on a display screen. So, it is important to keep subtle features of the mammogram to describe and represent the cancer. On the other hand, a nonlinear method is not always predictable but it sometimes offers better performance. In this section, a nonlinear enhancement technique is explored, which combines with two different nonlinear methods: one is a nonlinear mapping function on wavelet subbands and the other is to use adaptive channel gains using the homomorphic filter.

19.3.2.1 Dyadic Wavelet Transform

The one dimensional discrete dyadic wavelet transform [18,40,41] used in this section is illustrated in Figure 19.9. The left box shows the part of signal decomposition whereas the right box represents the part of signal reconstruction.

In Figure 19.9, $H(\omega)$ and $L(\omega)$ denote a high pass and a lowpass filter respectively. $H^*(\omega)$ is a reconstruction filter of high pass components whereas $L^*(\omega)$ is a reconstruction filter of low pass components. The variable 2ω indicates a subsampling of input signal [40]. As shown in the figure, a signal can be decomposed into three high pass channels ($H(\omega)$, $H(2\omega)$, $H(4\omega)$) and one low pass channel $L(4\omega)$. Each channel corresponds to a wavelet coefficient [41,42].

Let $s(x, y)$ be an input signal. In order to enhance image contrast, then, a generic unsharp masking can be written as:

$$\tilde{s} = s(x, y) - k\nabla^2 s(x, y) \tag{19.3}$$

where Laplacian operator, ∇^2 can be written as $\nabla^2 = \partial^2/\partial x^2 + \partial^2/\partial y^2$ and k is a constant.

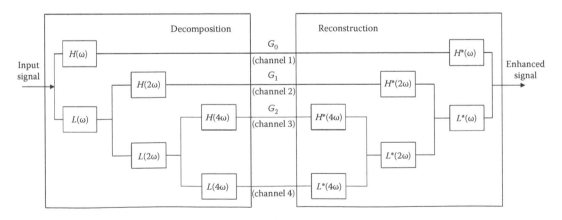

FIGURE 19.9
One dimensional dyadic discrete wavelet transform (three levels shown. Note G_m is channel gain.). (From Ho-Kyung Kang, Nguyen N. Thahh, Sung-Min Kim, and Yong Man Ro., *Robust Contrast Enhancement for Microcalcification in Mammography*. LNCS, Springer Berlin, Vol. 3045, pp. 603, 2004. With permission.)

If a discrete image of $s(x, y)$ is $s(i, j)$, the discrete form of the Laplacian operator can be written as:

$$\nabla^2 s(i, j) = [s(i+1, j) - 2s(i, j) + s(i-1, j)] + [s(i, j+1) - 2s(i, j) + s(i, j-1)]$$

$$= -5\left\{ s(i, j) - \frac{1}{5}[s(i+1, j) + s(i-1, j) + s(i, j) + s(i, j+1) + s(i, j-1)] \right\} \qquad (19.4)$$

where i, j are discrete variables corresponding to x, y of Equation 19.3. This formula shows that the discrete Laplacian operator can be established by subtracting the average value of neighborhood pixels from the value of a central pixel. Then, the unsharp masking in discrete form can be rewritten as

$$\tilde{s}(i, j) = s(i, j) + k[s(i, j) - s(i, j) * h(i, j)] \qquad (19.5)$$

where $h(i, j)$ is the discrete averaging filter and $*$ is convolution operator.

On the other hand, in a dyadic wavelet framework as shown in Figure 19.9, N channels have equal gain such that $G_m = G_0 > 1$, where $0 \leq m \leq N-1$ and m denotes the channel index. Then system frequency response $V(\omega)$ can be described as [18]:

$$V(\omega) = \sum_{m=0}^{N-1} G_m C_m(\omega) + C_N(\omega)$$

$$= 1 + (G_0 - 1)[1 - C_N(\omega)] \qquad (19.6)$$

where $C_m(\omega)$ is a frequency response for each channel. As shown in Equation 19.6, the gains of all channels are same, i.e., $G_m = G_0$. In spatial domain, Equation 19.6 can be written as $v(i) = 1 + (G_0 - 1)[1 - c_N(i)]$. If a signal $s(i)$ comes into the wavelet framework of Figure 19.9, the output of the system is then simplified as:

$$\tilde{s}(i) = s(i) + (G_0 - 1)[s(i) - s(i) * c_N(i)] \qquad (19.7)$$

Then, the form of Equation 19.7 is similar to Equation 19.5 which is the signal after the unsharp masking. Equation 19.7 forms the 1-D representation of Equation 19.5, and $c_N(i)$ of Equation 19.7 is a Gaussian lowpass filter which is replaced instead of an averaging filter of Equation 19.5. Note that gains of channels have the same value in the unsharp masking of wavelet domain.

On this subband decomposition, a nonlinear function varying gains of the given channels will be described to improve the visibility of breast cancer.

19.3.2.2 Homomorphic Filtering

Homomorphic filtering is one of the image processing techniques which are developed for improving the appearance of an image by simultaneous gray-level range compression and contrast enhancement [39]. In the mammographic images, the homomorphic filtering leads to the contrast stretching into the lower gray level thereby enhancing the contrast. The homomorphic filter gives varying gains to wavelet channels so that the contrast of a malignant tumor is improved. The enhancement approach involves a nonlinear mapping to a different domain in which linear filter techniques are applied, followed by mapping back to the original domain. The concept of this approach can be summarized in Figure 19.10.

Through the above procedure, the homomorphic filter function $H(u, v)$ can control and affect the low- and high-frequency components of the Fourier transform in various ways. Figure 19.11 illustrates the homomorphic filter function where K_L and K_H denote the gain of the low- and high-frequency components, respectively.

If the parameters K_L and K_H are chosen so that $K_L < 1$ and $K_H > 1$, the filter function shown in Figure 19.11 tends to decrease the low frequency components of an image and to amplify the high frequency components. Thus, they achieve contrast enhancement as well as dynamic range compression at the same time.

19.3.2.3 Adaptive Contrast Enhancement and Denoising

In the 1-D dyadic discrete wavelet transform of Figure 19.9, linear enhancement can be represented as a linear mapping of a wavelet coefficient u of a given subband channel, i.e., $E_m(u) = G_m \cdot u$, where $E_m(u)$ denotes a linear function. A nonlinear mapping function employed in this chapter adopts the piecewise linear functions, so that it can adapt to the corresponding wavelet subband channels. More details are explained as follows.

Taking into account microcalcification characteristics mentioned in Section 19.1, a nonlinear enhancement technique used this chapter is designed with the following guidelines [18,38]:

1. Low contrast area should be enhanced more than the high contrast area. In other words, the small values of wavelet coefficient, u, should be mapped into large values.

2. A sharp edge should not be blurred.

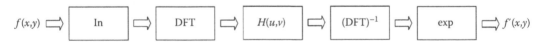

FIGURE 19.10
Homomorphic filtering approach for image enhancement. (Reprinted from Rafael C. Gonzalez and Richard E. Woods., *Digital Processing Book*. Prentice Hall, pp. 193, 2002. Pearson Education, Inc., Upper Saddle River, NJ. With permission.)

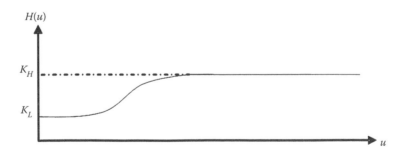

FIGURE 19.11
Cross section of homomorphic filter function. (From Jeong Hyun Yoon and Yong Man Ro., *Enhancement of the Contrast in Mammographic Images Using the Homomorphic Filter Method*. IEICE Transactions on Information and Systems. The Institute of Electronic, Information and Communication Engineers, Vol. 1085, No. 1. pp. 300, 2002. With permission.)

3. Mapping function should be monotonically increasing.
4. Mapping function should be antisymmetry such that $E_m(-u) = -E_m(u)$, in order to preserve the phase polarity.

   On top of above, one requirement is added in this work.
5. Mapping function should give different responses depending on the corresponding channels.

Thus, a piecewise linear function which meets the above requirements can be defined as

$$E_m(u) = \begin{cases} u - (K_m - 1)'T'_m & \text{if } u < -T'_m \\ K_m \cdot u & \text{else if } |u| \le T_m \\ u + (K_m - 1)T_m & \text{otherwise} \end{cases} \quad (19.8)$$

where $K_m > 1$ and m is a channel number. For each channel m, an enhancement mapping function is denoted as E_m. Two parameters of the function, threshold and gain are denoted as T_m and K_m, respectively. To strengthen the weak features, threshold T_m is set by $t \times \max\{|u|\}$, where $0 < t < 1$. By setting a small t across all levels, weak features at distinct scales are favored and effectively enhanced.

To improve the visibility of the mammogram, a homomorphic filter is used to find the gain K_m [18] in Figure 19.12. The homomorphic filter function decreases the energy of low frequencies while increasing those of high frequencies in the image. Therefore, increasing the homomorphic filter shows contrast stretching for the lower gray level by compressing the dynamic range of the gray level. Based on the characteristics of the homomorphic filter function, we determine the gain of the mapping function, i.e., the weighting wavelet coefficients of channels corresponding to the homomorphic filter function. Figure 19.12 represents the gain K_m that is determined according to discrete homomorphic filtering. Here, taking logarithmic function for an input signal is the first work of homomorphic filtering process. It also inverts the exponential operation caused by the radioactive absorption, which is generated in the process of obtaining a mammographic image. Note that the dotted line in Figure 19.12 plots a continuous homomorphic filter function.

In mammogram contrast enhancement, noise reduction is an important issue. The wavelet shrinkage that was presented in Laine et al. [18] is one method of denoising. Therefore, applying the same parameters in noise reduction and the gain K_m for every mammogram

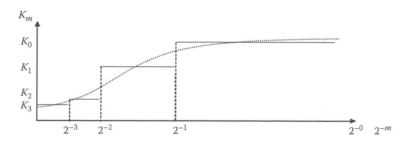

FIGURE 19.12
Homomorphic filter function for applying to wavelet coefficients. (From Ho-Kyung Kang, Nguyen N. Thahh, Sung-Min Kim, and Yong Man Ro., *Robust Contrast Enhancement for Microcalcification in Mammography*. LNCS, Springer Berlin, Vol. 3045, pp. 608, 2004. With permission.)

are not efficient. In order to take into account the noise properties of each mammogram, a robust method for mammogram enhancement is essential.

To obtain the noise characteristics of the mammogram, the background obtained in the above section is used with the values combining the gray-level, the mean, and the variance of pixels. The segmented background areas (blocks) are supposed to contain the noise of the image. Thus, we can take noise characteristics from this area and they are measured by background noise variance, var_b, which can be written as

$$\mathrm{var}_b = \frac{1}{N_b} \sum_{(x,y) \in \text{background}} \{I(x, y) - \mathrm{mean}(x, y)\}^2, \qquad (19.9)$$

where N_b is the number of pixels in the background area. $I(x,y)$ and $\mathrm{mean}(x,y)$ are calculated using background pixels only. If background noise has a high variance, we need to reduce the gain of the homomorphic filter in the high frequency domain. The gain of the homomorphic filter can be written adaptively as

$$K'_m = K_m \times \frac{A}{\mathrm{var}_b + A}, \quad \text{if } m = 0 \quad \text{or} \quad m = 1 \qquad (19.10)$$

where A is the constant value to normalize noise variance, which is the mean value of the background noise variance in the training mammogram database, m is the level of wavelet, and K_m means the gain of each wavelet level. In Equation 19.10, $m=0$ means the highest frequency level in the wavelet, and $m=1$ means the second highest wavelet level. The gains are reduced in high noise mammograms and then a higher gain of contrast enhancement is acceptable in low noise mammograms.

Figure 19.13 represents the modified homomorphic filtering approach with denoising as well as contrast enhancement. In the figure, three-level wavelet decomposition and reconstruction with 1-D signal is shown. Here, we first take the logarithmic function for an input signal. This also inverts the exponential operation caused by radioactive absorption, which is generated in the process of obtaining a mammography image. K'_m is the linear enhancement gain of each wavelet channel. This is well-suited for enhancement of microcalcification because it emphasizes weak edges with strong edges at the same time.

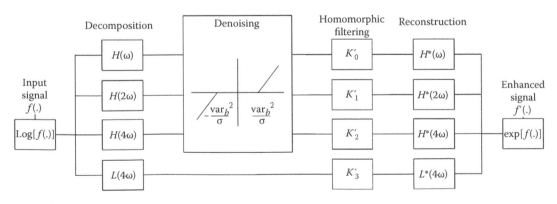

FIGURE 19.13
Entire scheme for mammogram contrast enhancement with denoising and modified homomorphic filtering. (From Ho-Kyung Kang, Nguyen N. Thahh, Sung-Min Kim, and Yong Man Ro., *Robust Contrast Enhancement for Microcalcification in Mammography*. LNCS, Springer Berlin, Vol. 3045, pp. 606, 2004. With permission.)

Further, an adaptive denoising is included in the enhancement process in the wavelet domain shown in the denoising block of Figure 19.13. To achieve edge-preserved denoising, a nonlinear wavelet shrinkage method is applied. In denoising, wavelet coefficient values are reduced to zero according to a level-dependent threshold. The noise adaptive shrink operator $S(u)$ for denoising can be written as

$$S(u) = \begin{cases} \text{sgn}(u) \times \left(|u| - \text{var}_b^2 / \sigma \right) & \text{if } |u| > \text{var}_b^2 / \sigma \\ 0 & \text{otherwise,} \end{cases} \tag{19.11}$$

where u is the wavelet coefficient, and σ is a variance of the reconstructed image using wavelet coefficients in a subband. The sgn(u) means a sign function which determines the positive or negative sign of u. The threshold in this wavelet shrinkage is referred to as a nearly optimal threshold [43].

By taking the modified homomorphic filter gains of the high frequency area in the wavelet domain and the optimal denoising operators, a microcalcification can be enhanced while noise is reduced in the breast area.

19.4 Microcalcification Detection

The microcalcification detection system in this chapter consists of two processes: the first process is a pixel-based segmentation which extracts pixels of interest (POI) from a digital mammogram (Figure 19.14). Note that the POI is presumed to contain microcalcifications. In addition, the extracted POIs are clustered into ROI. In this process, suspected regions are classified based on pixel characteristics of a microcalcification. The next process, the region-based microcalcification detection, categorizes the clustered ROIs into normal and abnormal regions and determines microcalcifications.

Regarding the classifier, the two processes adopt appropriate SVM depending on their aims. A linear SVM gives fast processing with acceptable classification performance, so that it can minimize the search range on an entire mammogram. On the other hand, because a nonlinear SVM shows higher classification accuracy than the linear one, it is suitable for decisions regarding microcalcifications where ROI classification should be performed in greater detail. Therefore, the pixel-based segmentation adopts a linear kernel-based SVM and the region-based microcalcification detection uses a nonlinear kernel-based SVM.

The detection system utilizes multiple classifiers which combines SVMs and binary tree. We define two classes for a mammogram: normal pixels and POIs; and we also categorize POIs into two subclasses: normal regions and microcalcifications. Then, the two cascaded processes of the detection system including the respective classifiers (SVM in this document) are illustrated in Figure 19.15.

Therefore, the first classifier, which is a linear kernel-based SVM of the pixel-based segmentation, accepts pixel characteristics-based features and filter POIs from a digital mammogram. In region-based microcalcification detection, the second classifier, which is a nonlinear kernel-based SVM, distinguishes microcalcifications from suspected ROIs by using features based on ROI characteristics of microcalcifications.

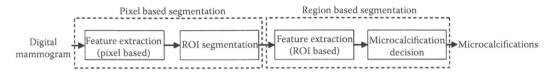

FIGURE 19.14
Layout of the detection system.

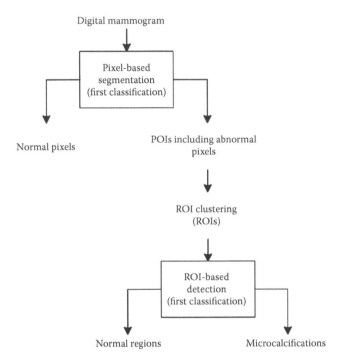

FIGURE 19.15
A hierarchical structure combined by multiple classifications.

19.4.1 Pixel-Based Segmentation to Get ROIs

The goal of this process is to obtain all of the ROIs from an entire digital mammogram. Thus, pixel-based segmentation is divided into three successive substeps as shown in Figure 19.16. The first substep, feature extraction, uses the fact that a microcalcification has abrupt change of intensity compared to normal regions (e.g., tissues). Then, the process seeks POIs by using a linear kernel-based SVM and represents them as the form of a group, i.e., an ROI.

19.4.1.1 Feature Extraction Based on Pixel Characteristics

In order to find a set of pixels which are supposed to be the area of a doubtful tumor, a windowing technique is used to catch intensity changes existing between adjacent pixels of a mammogram image. Suppose that a window W with $n \times n$ has a center point (i, j). Then, two statistical values, intensity difference (ID) and contrast-to-noise ratio (CNR) for gray levels of an input mammogram are calculated within the window. The values of ID

FIGURE 19.16
Layout of pixel-based segmentation process to obtain ROIs.

and CNR represent the intensity differences between pixels and then filter dominant pixels, relatively. They are defined as,

$$f_{ID} = I(i, j) - \mu \tag{19.12}$$

$$f_{CNR} = \frac{I(i, j) - \mu}{\sigma} \tag{19.13}$$

where $I(i, j)$ is the intensity of the pixel coordinated at (i, j) in an input mammogram. μ and σ are mean and standard deviation values for the pixels located in the window W, respectively.

The first feature extraction performs spatial filtering using the two different windows: a global window W_g and a local window W_l. Each window size adapts to the resolution of a digital mammogram. The motivation behind this is to solve constraints on different resolutions of mammograms due to the diagnosis device or an environment with a diversity of noise conditions, X-ray intensities, and concentrations of sensitizers of mammogram films. Let L_g and L_l be the lengths of sides of W_g and W_l, respectively. This leads to

$$L_g = \begin{cases} W_g + 1 & \text{if } W_g(\text{mod}\,2) \equiv 0 \\ W_g & \text{otherwise} \end{cases} \tag{19.14}$$

and

$$L_l = \begin{cases} W_l + 1 & \text{if } W_l(\text{mod}\,2) \equiv 0 \\ W_l & \text{otherwise} \end{cases} \tag{19.15}$$

where

$$W_g = [\text{resolution}^{-1} \times \gamma_g], \quad W_l = [\text{resolution}^{-1} \times \gamma_l]$$

and 'resolution' is the resolution of an input mammogram and the unit is expressed by μm/pixel. In this work, 1.97 and 0.47 mm are utilized as the parameter γ_g, and γ_l for W_g and W_l, respectively.

Suppose that a mammogram with 94.09 μm/pixel resolution is fed into the detection system. Then, W_g and W_l are calculated as 21 and 5 pixels separately and the spatial filtering is performed with 21×21 and 5×5 windows. From Equations 19.12 through 19.15, we can see that f_{ID_g} and f_{CNR_g} become the features obtained from the global window, and f_{ID_l} and f_{CNR_l} are from the local window. In this process, then, the pixel characteristics of the mammogram are represented by the 4-D feature vector: f_{ID_g}, f_{CNR_g}, f_{ID_l}, f_{CNR_l}.

19.4.1.2 Extraction of Candidate Pixels: A First Classification to Obtain POIs

As an input vector, the extracted four features are fed into a linear kernel-based SVM classifier in order to find POIs. In this case, a linear kernel, $K(\mathbf{x}_i, \mathbf{x}) = <\mathbf{x}_i, \mathbf{x}>$, measures the similarity between feature vectors. Substituting the kernel for a hyperplane decision function, the SVM permits the computation of dot products in high-dimensional feature space [44–46]. Thus, this leads to

$$f(\mathbf{x}) = \text{sgn}\left(\sum_{i=1}^{N_s} \alpha_i y_i \mathbf{x}_i^T \mathbf{x} + b\right) \tag{19.16}$$

where sgn is a sign function, that extracts the sign of a real number. α_i is a *Lagrange* multiplier, y_i is a class label corresponding to a support vector $\mathbf{x}_i$, and b is an optimal bias. N_s denotes the number of support vectors. $\mathbf{x}$ is a feature vector for a given POI, i.e., an input feature vector consisting of the previous four features.

The POIs can be determined as the linear SVM classifier outputs a numeric value representing how confident each point (i.e., a pixel) in the input mammogram is to be a part of microcalcification. In the standard formulation [47], the threshold λ for $\sum \alpha_i y_i x_i^T x + b$ (i.e., $\sum \alpha_i y_i x_i^T x + b > \lambda$) is set to zero so that the decision rule simply corresponds to determining which side of the separating hyperplane (defined by $\sum \alpha_i y_i x_i^T$ and b) the vector $\mathbf{x}$ is located at. After training decides $\sum \alpha_i y_i x_i^T$ and b, however, the value of λ can be changed adaptively to bias a given mammogram database [48]. In other words, b, the threshold for defining the hyper plane, is adjusted at the time of training. Individual training of the SVM requires the normalization of contrast differences existing among heterogeneous mammogram databases. This work establishes λ by using an average of CNR values through empirical cross-validation. λ is, thereupon, replaced by

$$\lambda = \frac{2\omega}{f_{\text{CNR}\_g} + f_{\text{CNR}\_l}} \tag{19.17}$$

where, ω is a weighting value and set as 1.5 in this document.

19.4.1.3 ROI Generation

A candidate pixel p extracted in the section above at coordinates (i, j) has eight horizontal, vertical, and diagonal neighbors. If another candidate, pixel q, is among the neighbors, then the two pixels p and q form eight connectivity and establish a region (R). After establishing regions in the entire mammogram, the process needs to filter only the microcalcification ROIs from the connected regions, e.g., a set of $R_1, R_2, R_3, \ldots, R_i$. The criterion for the filtering is determined by the characteristics of the size of a microcalcification noted above. The criterion that the ROI could be a microcalcification ROI can be written as

$$N_{\min} < N(R_i) < N_{\max} \tag{19.18}$$

where $N_{\max}$ and $N_{\min}$ are the maximum and minimum number of pixels of the real microcalcification ROI, respectively. $N(R_i)$ denotes the total number of pixels contained in the ith region R_i. $N_{\max}$ and $N_{\min}$ are observed and measured from experiments with the microcalcification database.

Since the criterion of Equation 19.18 does not reflect the shape of the region, the results could include both ROIs which could be both normal areas and ROIs which could be microcalcifications. The next process, henceforth, will distinguish the microcalcifications from the ROIs depending on the characteristics of their shapes.

19.4.2 Microcalcification Detection Based on ROIs

In the previous module, i.e., ROI generation, we obtained clusters of suspicious pixels with abrupt changes of gray level, or intensity. Additionally, this second module is the process of microcalcification decision using a second SVM classifier as shown in Figure 19.17. Firstly, it extracts and uses a feature vector related to the size of the extracted ROI, and then it seeks ROIs that include only a part of microcalcifications from the clusters through the SVM.

19.4.2.1 Feature Extraction Based on ROI Characteristics

The previous pixel-based segment utilized a pixel intensity-based approach to catch spots with the characteristics of microcalcifications, whereas this process mainly employs a region shape-based approach. In order to classify ROIs into microcalcifications, we adopt a nonlinear kernel SVM and require a feature vector representing the characteristics of the ROI [49]. Table 19.1 illustrates a summary of the features.

The feature f_{num} is the total number of pixels that belong to the corresponding ROI and represents the size of the ROI. Density, f_{den}, is based on the observation that the regions belonging to a microcalcification have high correlation with each other [50]. In other words, the microcalcification forms a colony of ROIs within a specific domain. Let the center coordinates of a given ROI and a neighboring ROI be (Cx, Cy) and (Cx_n, Cy_n), respectively. Then, f_{den}, is equal to the total number of neighboring ROIs satisfying the following condition,

$$\sqrt{(Cx - Cx_n)^2 + (Cy - Cy_n)^2} < T_d \tag{19.19}$$

where T_d denotes a threshold value which is a radius from the center of the corresponding ROI.

FIGURE 19.17
Layout of region-based detection process.

TABLE 19.1

Summary of the Features Based on ROI Characteristics for the Nonlinear Kernel SVM

Features	Description
Number of pixels in ROI, f_{num}	Total number of pixels contained in a ROI
Density, f_{den}	Degree of distribution of the neighbor ROIs within given area
Variance, f_{var}	Variance value for the intensities of the pixels in an ROI
Gradient, f_{grad}	Mean value of 3×3 Sobel gradient for each pixel in an ROI
Direction of gradient, f_{dir}	Angle value of a gradient vector
Directional edges, $f_0, f_{45}, f_{90}, f_{135}$	Summation values of directional edge components in a block
Nondirectional edge, f_{ND}	Summation value of nondirectional edge components in a block
f_{diff} Difference	Minimum value among directional edges between a block enclosing an ROI and eight adjacent blocks

From the observation that pixels of a microcalcification carry similar intensities in a mammogram, f_{var}, the variance value for the intensities of the pixels existing in an ROI, is adopted to classify ROIs. Several kinds of high frequency components in a mammogram describe a diversity of shapes on microcalcifications, as mentioned in Section 19.1. *Gradient* and *direction of gradient* in Table 19.1 originate from the first-derivatives of a digital mammogram. The two quantities give the sums of the magnitudes and the direction angles of gradient vectors with respect to the pixels of ROIs, respectively; and are defined as

$$f_{grad} = \sum_{\forall (i,j) \in ROI} \sqrt{S_V^2 + S_H^2} \tag{19.20}$$

and

$$f_{dir} = \sum_{\forall (i,j) \in ROI} \tan^{-1} \left(\frac{S_V}{S_H} \right) \tag{19.21}$$

where S_V and S_H are the vertical and horizontal directional gradient components after *Sobel* mask performs at coordinates (i, j) of a given ROI, respectively. The vertical and horizontal directional Sobel masks are illustrated in Figure 19.18 [39].

Directional edges, *nondirectional edge*, and *difference* in Table 19.1 are also based on the shape and edge information of a microcalcification. Tissue or mammary gland tends to emphasize directional edges components whereas a microcalcification generally shows high nondirectional edge value [33]. To obtain such features, as shown in Figure 19.19, an ROI is segmented as the rectangular block at the points of the exterior boundary.

(a)			(b)		
−1	0	1	−1	−2	−1
−2	0	2	0	0	0
−1	0	1	1	2	1

FIGURE 19.18
Sobel masks in an (a) vertical and (b) horizontal direction.

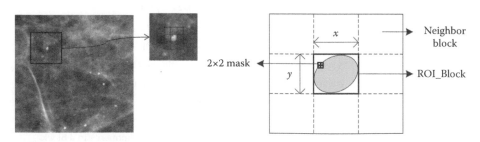

FIGURE 19.19
The block enclosing ROI, eight adjacent blocks, and 2 × 2 mask.

For *directional edges* and *nondirectional edge* features ($f_0, f_{45}, f_{90}, f_{135}, f_{ND}$) in Table 19.1, spatial filtering is performed by a filter mask sized 2×2 as shown in Figure 19.19. The mask moves from point to point within a rectangular block in a process called *ROI_Block*. The spatial filtering uses the filter masks in an MPEG-7 edge histogram: four directional masks (0°, 45°, 90°, 135°) and one nondirectional mask with a 2×2 size [51]. Given a particular mask for a direction, it only generates a feature for that direction, i.e., the 0° mask only generates a feature f_0 for 0°. Let z_1, z_2, z_3, z_4 be components of a given mask which are arranged in zigzag order from top-left to bottom-right. Then, the responses of the filtering are calculated by

$$f_0 = \frac{1}{M} \sum_{\forall (i,j) \in \text{ROI\_Block}} \{z_1 \cdot I(i,j) - z_2 \cdot I(i,j+1) + z_3 \cdot I(i+1,j) - z_4 \cdot I(i+1,j+1)\}, \quad (19.22)$$

$$f_{45} = \frac{\sqrt{2}}{M} \sum_{\forall (i,j) \in \text{ROI\_Block}} \{z_1 \cdot I(i,j) - z_4 \cdot I(i+1,j+1)\}, \quad (19.23)$$

$$f_{90} = \frac{1}{M} \sum_{\forall (i,j) \in \text{ROI\_Block}} \{z_1 \cdot I(i,j) + z_2 \cdot I(i,j+1) - z_3 \cdot I(i+1,j) - z_4 \cdot I(i+1,j+1)\}, \quad (19.24)$$

$$f_{135} = \frac{\sqrt{2}}{M} \sum_{\forall (i,j) \in \text{ROI\_Block}} \{z_3 \cdot I(i+1,j) - z_2 \cdot I(i,j+1)\} \quad (19.25)$$

and

$$f_{ND} = \frac{2}{M} \sum_{\forall (i,j) \in \text{ROI\_Block}} \{z_1 \cdot I(i,j) - z_2 \cdot I(i,j+1) - z_3 \cdot I(i+1,j) + z_4 \cdot I(i+1,j+1)\}, \quad (19.26)$$

where, M is the number of available pixels for spatial filtering in ROI_Block.

For the eight-adjacent blocks defined by the given block of Figure 19.19, difference (f_{diff}) seeks the difference of high frequency components between neighboring blocks. Note that microcalcifications are brighter spots compared with neighboring normal regions. The distribution of the high frequency components in a microcalcification is relatively higher than that in a normal area. This tendency can be also revealed in the four directional features mentioned in Equations 19.22 through 19.25. Then, the difference of the high frequency components is measured by

$$f_{\text{diff}} = \min \{e_0, e_{45}, e_{90}, e_{135}\} \quad (19.27)$$

where,

$$e_0 = \sum_{\forall (i,j) \in \text{ROI\_Block}} [2 \cdot I(i,j) - I(i,j-y) - I(i,j+y)], \quad (19.28)$$

$$e_{45} = \sum_{\forall (i,j) \in \text{ROI\_Block}} [2 \cdot I(i,j) - I(i+x,j-y) - I(i-x,j+y)] \quad (19.29)$$

$$e_{90} = \sum_{\forall (i,j) \in \text{ROI\_Block}} \left[2 \cdot I(i,j) - I(i-x,j) - I(i+x,j) \right], \tag{19.30}$$

$$e_{135} = \sum_{\forall (i,j) \in \text{ROI\_Block}} \left[2 \cdot I(i,j) - I(i-x,j-y) - I(i+x,j+y) \right], \tag{19.31}$$

and ROI_Block is the rectangular block which encloses at the points coordinated at the outer boundaries of an ROI as illustrated in Figure 19.19. x and y are offsets of x- and y-axis directions, i.e., x and y denote the width and height of the block that covers an ROI in Figure 19.19.

This feature describes the minimum value of four-directional high frequency components existing between a given ROI and its neighboring regions. It compares the minimum high frequency components among ROIs. Thus, the extracted feature vector for the nonlinear kernel SVM as described in Table 19.1 are denoted as $f_{num}, f_{den}, f_{var}, f_{grad}, f_{dir}, f_0, f_{45}, f_{90}, f_{135}, f_{ND}$, and f_{diff}.

19.4.2.2 Classification of ROIs: A Second Classification to Detect Microcalcifications

This substep employs an SVM classifier using a Gaussian radial basis function (RBF) kernel to measure the similarity of features. A Gaussian RBF monotonically decreases with distance from the center and gives a significant response only in a neighborhood near the center, i.e., its response is finite [52]. The RBF kernel, therefore, utilizes the basic assumption that it gives maximum similarity when support vectors and input feature vectors are identical. The Gaussian RBF is given by

$$K(\mathbf{x}_i, \mathbf{x}) = \exp\left(-\frac{\|\mathbf{x} - \mathbf{x}_i\|^2}{2\sigma} \right) \tag{19.32}$$

where σ is chosen so that the smallest distance between any pair of training points is much larger than the width σ. Equation 19.32 yields a nonlinear decision function with the kernel [44–46].

$$f(\mathbf{x}) = \text{sgn}\left(\sum_{i=1}^{N_s} \alpha_i y_i \exp\left(-\frac{\|\mathbf{x} - \mathbf{x}_i\|^2}{2\sigma} \right) + b \right) \tag{19.33}$$

where $\mathbf{x}$ denotes a feature vector corresponding to a given ROI.

The feature vector of Equation 19.33 is composed of eleven features noted in the previous section. The classification result of this Gaussian RBF SVM classifier indicates how close an ROI is to the subsets of a microcalcification. Based on these results, the next substep performs ROI clustering and establishes a microcalcification.

19.4.2.3 Detection of Microcalcifications

In most case, a microcalcification is formed by the connection of multiple adjacent regions as mentioned earlier [5,53]. Therefore, ROIs, that might be a part of a microcalcification, should be clustered into one group. If the distance between two adjacent ROIs is smaller than a predefined distance threshold value, then they are merged into a new cluster. The merging criterion is defined as,

$$\sqrt{(Cx_i - Cx_j)^2 + (Cy_i - Cy_n)^2} < T_{d\_\text{cluster}},\qquad(19.34)$$

where (Cx_i, Cy_i) and (Cx_j, Cy_j) are the center coordinates of ROI$_i$ and ROI$_j$, respectively, and T_{d_cluster} is a predefined distance threshold.

Through this merging substep, finally, the detection of microcalcifications is achieved. Here, the substep does not perform connection of the pixels of ROIs practically, and it informs us the number of microcalcifications (i.e., groups of ROIs) that exist in the mammogram.

19.5 Experimental Results

To verify the effectiveness of the microcalcification contrast enhancement and detection system mentioned in the previous sections, some experiments are performed; the first experiment is for the microcalcification enhancement and the other is for the microcalcification detection.

19.5.1 Experiments for Contrast Enhancement

To verify the usefulness of the microcalcification enhancement method mentioned in the previous section, an experiment was performed with DDSM database. In order to focus on the effect of the contrast enhancement and denoising, we selected and performed the experiments with the DDSM mammogram database which contains high noise components and has low resolution. The DDSM database was obtained from Samsung medical center. The resolution of a mammogram was 50 μm/pixel and gray level depths were 12 bits and 16 bits with various kinds of noise characteristics.

In this experiment, three contrast enhancement methods were performed and compared: linear enhancement in wavelet domain (unsharp marking) [18]; homomorphic filter in wavelet domain (homomorphic filtering) [38]; and the adaptive image enhancement using the modified homomorphic filter in wavelet domain. Since the adaptive method was to enhance the contrast of microcalcifications and reduce noise, parameters of enhancement methods were chosen that are well-suited to this purpose.

In order to compare contrast enhancement, we introduce one of the measurement methods. A quantitative measurement of contrast improvement can be estimated by using the contrast improvement index (CII). Laine et al. [18], defined CII as

$$\text{CII} = \frac{C_\text{enhanced}}{C_\text{original}},\qquad(19.35)$$

where $C_{enhanced}$ and $C_{original}$ denote the contrast values of microcalcifications in the enhanced and original images, respectively. The contrast C of a microcalcification in the image is defined as

$$C = \frac{f - b}{f + b}, \tag{19.36}$$

where f is a mean value of the microcalcification, and b is a mean value of background. The standard deviation (Std.) of pixels in the background region is also measured in order to represent a noise level.

Figure 19.20 shows enhancement and analysis results for a high noise image which contains microcalcifications in the center area. Figure 19.20e through h draw the profiles of the enhanced mammograms. Note that the vertical axis of the profiles represents a gray level of the corresponding pixel and the horizontal axis shows pixel positions in the lines shown in Figure 19.20a through d. The contrast improvements of both homomorphic filtering and the adaptive contrast enhancement methods are equivalent to the center peak of profile in Figure 19.20, and the CII value of Table 19.2. However, Figure 19.20g and h show that the adaptive enhancement is much better than homomorphic filtering in denoising. This is also indicated by the standard deviation of noise Std. in Table 19.2. The Std. of noise of homomorphic filtering is 32.5, whereas the Std. of noise of the adaptive enhancement is 12.3. The experiment shows that the adaptive enhancement is more effective in denoising when compared with the previous enhancement methods in a high noise condition.

Figure 19.21 provides another example which shows the original image with low noise. Figure 19.21 and Table 19.3 show that noise is reduced in all enhanced images. However, the adaptive enhancement increases the contrast of microcalcification better than the others. In Table 19.3, the adaptive enhancement obtains 3.8858 of CII while homomorphic filtering obtains 3.2860. This example indicates that in case of a low noise condition, the adaptive enhancement is better than homomorphic filtering in contrast enhancement with similar denoising. This is due to higher gains in the high frequency channels.

As seen in Figure 19.21e, high noise mammograms have much fluctuation in the breast region shown in the original image. Unsharp masking and homomorphic filtering enhance the contrast of the microcalcification with relatively high noise. On the other hand, the adaptive image enhancement modifies wavelet gains and increases the denoising threshold in high noise cases. Therefore, noise is reduced in the breast area with high CII value. In low noise cases, the adaptive image enhancement is superior to other methods in denoising and improving the contrast of microcalcifications.

The experimental results reveal that the subband decomposition using homomorphic filtering gives effective denoising as well as contrast enhancement.

19.5.2 Experiments for Detection Performance

To verify the performance of the detection system, the experiment was performed with two different mammogram databases offered from Samsung medical center: the first contain mammograms acquired by a mammography platform from General Electronic (GE) Co., and the other is comprised of mammograms obtained by a device from Lorad Co. In this work, we refer to them as DB 1 (mammograms from GE systems) and DB 2 (mammograms from Lorad systems), respectively. The mammograms from DB 1 show a 94.09 μm/

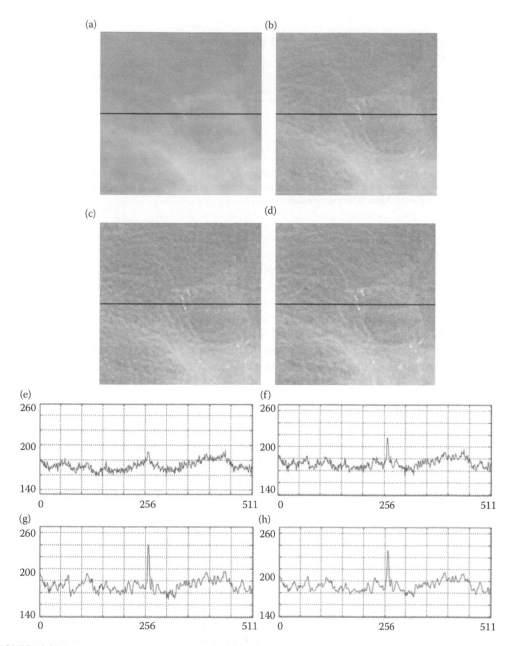

FIGURE 19.20
Contrast enhancement for high noise mammography image and profiles of one line: (a) original image, (b) unsharp masking, (c) homomorphic filtering, (d) adaptive enhancement method, (e) profile of the line in (a), (f) profile of the line in (b), (g) profile of the line in (c), and (h) profile of the line in (d). (From Ho-Kyung Kang, Nguyen N. Thahh, Sung-Min Kim, and Yong Man Ro., *Robust Contrast Enhancement for Microcalcification in Mammography*. LNCS, Springer Berlin, Vol. 3045, pp. 607, 2004. With permission.)

pixel resolution and 12 bits gray level depth. They had collected mammograms from 753 Korean women from 2001 to 2004. The mammograms from DB 2 have 70 μm/pixel resolution and 16 bits gray level depth. They are mammograms from 800 Korean women from 2006 to 2007. Figure 19.22 illustrates the examples of mammograms from both databases.

TABLE 19.2

Contrast Improvement Index and Standard Deviation of Noise for a Mammogram with High Noise (Figure 19.20)

	Original Image	Unsharp Masking	Homomorphic Filtering	Adaptive Enhancement
C	0.0490	0.0871	0.1310	0.1103
CII_n	N/A	1.7761	2.6714	2.2500
Std.	41.4	40.04	32.5	12.3

Source: From Ho-Kyung Kang, Nguyen N. Thahh, Sung-Min Kim, and Yong Man Ro., *Robust Contrast Enhancement for Microcalcification in Mammography.* LNCS, Springer Berlin, Vol. 3045, pp. 608, 2004. With permission.

Both databases are digital mammograms obtained from digital mammography called full-field digital mammography (FFDM).

From DB 1, in the first experiment we randomly selected 60 images as the training data set and another 100 images as a test data set. Half of the training set includes microcalcifications which are confirmed by medical doctors and the others are normal mammograms. For the test database to evaluate the detection system, 50 images contain microcalcifications and the remainder consists of normal images. 92 mammograms in DB 2 are used for training and evaluating in the second experiment, respectively. In the training database, the half of the mammograms shows microcalcification regions, and the rest are normal. The number of positive data (i.e., a mammogram with microcalcifications) in the test database is equal to that of negative data. The databases we utilized consist of mammograms with mediolateral-oblique (MLO) and cranio-caudal (CC) views.

The detection result of each process for the detection system is shown in Figure 19.23. Figure 19.23a is the original input mammogram and the clusters of a microcalcification are located at the center of the image. Marked regions in Figure 19.23b represent all doubtful regions, ROIs as the output of the first SVM classifier and ROI generation. The used thresholds should be adaptive to differences in mammogram resolutions. Similar to the establishment of window sizes in Equations 19.14 and 19.15, they are calibrated depending on the ratio of mammogram resolutions used in both databases. From the fact that DB 1 and DB 2 used 94.09 μm/pixel and 70 μm/pixel, respectively, the ratio can be calculated. For the mammograms from DB 1, ROI generation employs ROI sizes ranging from larger than 5 (N_{min}) pixels to smaller than 400 (N_{max}). On the other hand, the ROIs of the mammograms from DB 2 are decided between 7 and 536 pixels. Finally, in Figure 19.23c the regions corresponding to microcalcification remain from the second SVM. In order to merge the regions into microcalcifications, the detection system uses 70 pixels (for mammograms in DB 1) and 94 pixels (for mammograms in DB 2), and $T_{d\_cluster}$ as the maximum distance between ROIs. A mark in the form of a triangle is located at the center coordinates of detected potential regions and reveals that the regions are organized into one group, i.e., one microcalcification.

In order to measure the performance of detecting microcalcifications, a free-response operating characteristics (FROC) curve was used [54], which draws the true-positive fraction (TPF) versus an allowed number of false-positive (FP) per image as illustrated in Figure 19.24. The TPF refers to the true-positive detection ratio, which shows how many microcalcifications are recognized correctly; and it is equal to TP/(TP + FN). The false-positive (FP) is the number of samples containing no microcalcifications but having been incorrectly

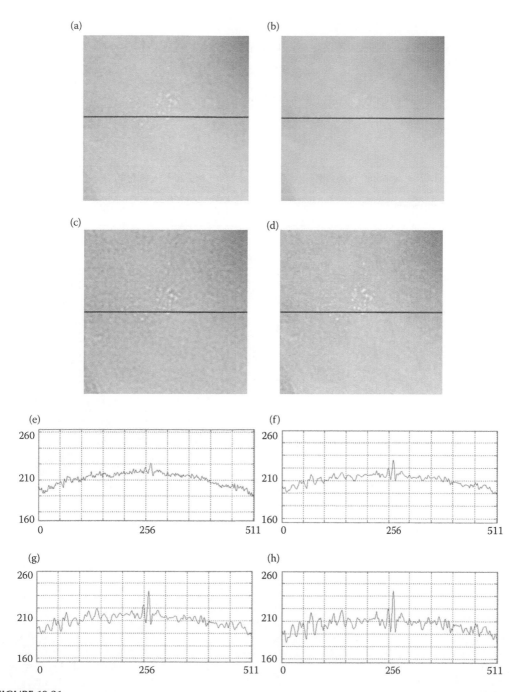

FIGURE 19.21

Contrast enhancement for low noise mammography image and profiles of one line: (a) original image, (b) unsharp masking, (c) homomorphic filtering, (d) adaptive enhancement method, (e) profile of the line in (a), (f) profile of the line in (b), (g) profile of the line in (c), and (h) profile of the line in (d). (From Ho-Kyung Kang, Nguyen N. Thahh, Sung-Min Kim, and Yong Man Ro., *Robust Contrast Enhancement for Microcalcification in Mammography.* LNCS, Springer Berlin, Vol. 3045, pp. 608, 2004. With permission.)

TABLE 19.3

Contrast Improvement Index and Standard Deviation of Noise for a High Noise Mammogram (Figure 19.21)

	Original Image	Unsharp Masking	Homomorphic Filtering	Adaptive Enhancement
C	0.0183	0.0382	0.0600	0.0710
CII_n	N/A	2.0906	3.2860	3.8858
Std.	23.1	11.3	11.5	10.3

Source: From Ho-Kyung Kang, Nguyen N. Thahh, Sung-Min Kim, and Yong Man Ro., *Robust Contrast Enhancement for Microcalcification in Mammography.* LNCS, Springer Berlin, Vol. 3045, pp. 608, 2004. With permission.

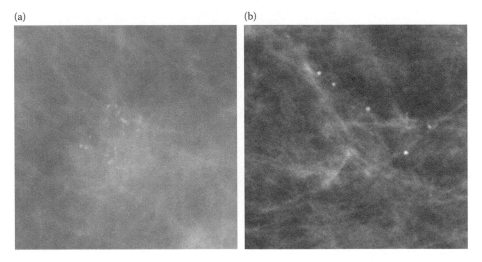

FIGURE 19.22
Examples of mammograms from (a) DB 1 and (b) DB 2.

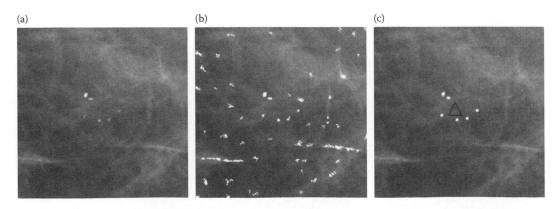

FIGURE 19.23
Mammograms at detection processes: (a) original, (b) extracted ROIs, and (c) regions decided as a microcalcification.

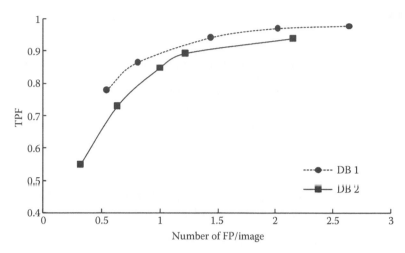

FIGURE 19.24
FROC curve on microcalcification decision results with DB 1 and DB 2.

classified as having microcalcifications. The true-positive (TP) is the number of spots that belong to true microcalcifications and have also been detected correctly as having microcalcifications, and the false-negative (FN) is the number of samples that contain microcalcifications but have been determined incorrectly as having no microcalcifications. Figure 19.24 shows the microcalcification decision results with the two different databases: DB 1 and DB 2.

In Figure 19.24, the detection system gives reliable detection performance in DB 1 as well as in DB 2. When the number of FP per image is equal to one, the system shows detection accuracy of more than 82%. And besides, if the system allows two false-positive detections in an image, it can show over 92% detection performance regardless of which database it is applied to. These results signify that the adaptive calibration of parameters on the first SVM is effective and useful in individual training for a given mammogram database. They also indicate that the system can maintain reliable detection performance regardless of database replacement.

19.6 Conclusion

The accurate detection of microcalcifications has been increasingly demanded in the field of CAD in the past decade. Due to various types, shapes, and distribution of microcalcifications, it is difficult to detect them. In this chapter, we have presented an automatic microcalcification detection system as a computer aided diagnosis in mammography. Detailed algorithms including the image enhancement and the microcalcification classification were presented. As a result, we obtained more enhanced mammograms in cooperation with the homomorphic filtering and the adaptive denoising method. In the detection of microcalcification, we discussed an automatic detection system using SVM classifiers. The experimental results showed that the multiple classifiers detection system gave robust and reliable performance in detecting microcalcifications, regardless of the size and type of data sets.

References

1. S. Detounis. 2004. Computer-aided detection and second reading utility and Implementation in a high-volume breast clinic. *Appl. Radiol.*, 33, 8–15.
2. S. K. Moore. 2001. Better breast cancer detection. *IEEE Spectrum*, 50–54.
3. K. Thangavel, M. Karnan, R. Sivakumar, and A. Kaja Mohideen. 2005. Automatic detection of microcalcification in mammograms – a review. *The International Congress for Global Science and Technology*, 5, 31–61.
4. Imagins: The Breast Cancer Resource [available online] 2006. Mar. http://www.imaginis.com/breasthealth/.
5. B. Rothenberg, K. Ziegler, N. Aronson. 2006. Technology evaluation center assessment synopsis: full-field digital mammography. *J. Am. Coll. Radiol.*, 3(8), 586–588.
6. ACR National Mammography Database (NMD) [available online] 2004. Jan. http://www.acr.org/.
7. S. Vedantham, A. Karellas, S. Suryanarayanan, D. Albagli, S. Han, E. J. Tkaczyk, C. E. Landberg, B. Opsahl-Ong, P. R. Granfors, I. Levis, C. J. D'Orsi, and R. E. Hendrick. 2000. Full breast digital mammography with an amorphous silicon-based flat panel detector: physical characteristics of a clinical prototype. *Med. Phys.*, 27, 558–567.
8. R. M. Nishikawa and M. J. Yaffe. 1985. Signal-to-noise properties of mammographic film-screen systems. *Med. Phys.*, 12, 32–39.
9. B. S. Monsees. 1995. Evaluation of breast microcalcifications. *Radiol. Clinics N. Am.*, 33(6), 1109–1121.
10. E. A. Sickles. 1986. Mammographic features of 300 consecutive nonpalpable breast cancers. *Am. J. Roentgenol.*, 146, 661–663.
11. D. B. Kopans. 1989. *Breast Imaging.* Lippincott Williams & Wikins, Philadelphia, PA, 81–95.
12. M. Lanyi. 1985. Morphological analysis of microcalcifications. In: *Early Breast Cancer: Histopathology, Diagnosis and Treatment.* Springer-Verlag, Berlin.
13. S. G. Komen. Breast Cancer Foundation [available online] 2006. Sep. http://www.komen.org.
14. S. M. Astley. 2004. Computer-based detection and prompting of mammographicabnormalities. *Br. J. Radiol.*, 77, 194–200.
15. K. Engan, T. Gulsrud, K. Fretheim, B. Iversen, and L. Eriksen. 2007. A computer aided detection (CAD) system for microcalcifications in mammograms—MammoScan μCaD. *Int. J. Biomed. Sci.*, 2(3) 168–179.
16. M. L. Giger. 2002. Computer-aided diagnosis in radiology. *Acad. Radiol.*, 9, 1–3.
17. J. S. Suri and R. M. Rangayyan. 2006. *Recent Advances in Breast Imaging, Mammography, and Computer-Aided Diagnosis of Breast Cancer.* SPIE Press, NJ, 130–131.
18. A. Laine, J. Fan, and W. Yang. 1995. Wavelets for contrast enhancement of digital mammography. *IEEE Eng. Med. Biol. Mag.*, 145, 536–550.
19. R. N. Strickland and H. I. Hahn. 1996. Wavelet transform for detecting microcalcifications in mammograms. *IEEE Trans. Med. Image*, 15, 218–229.
20. R. Gordon, R. M. Rangayyan. 1984. Feature enhancement of film mammograms using fixed and adaptive neighborhoods. *Appl. Optics*, 23, 560–564.
21. A. P. Dhawan, G. Buelloni, and R. Gordon. 1986. Enhancement of mammographic features by optimal adaptive neighborhood image processing. *IEEE Trans. Med. Imaging*, MI-5, 8–15.
22. A. P. Dhawan and R. Gordon. 1987. Reply to comments on enhancement of mammographic features by optimal adaptive neighborhood image processing. *IEEE Trans. Med. Imaging*, MI-6, 82–83.
23. A. P. Dhawan and E. Le Royer. 1988. Mammographic feature enhancement by computerized image processing. *Comput. Methods Programs Biomed.*, 27, 23.

24. P. G. Tahoces, J. Correa, M. Souto, C. Gonzalez, L. Gomez, and J. Vidal. 1991. Enhancement of chest and breast radiographs by automatic spatial filtering, *IEEE Trans. Med. Imaging*, MI-10(3), 330–335.

25. H. P. Chan, K. Doi, C. J. Vyborny, K. L. Lam, and R. A. Schmidt. 1988. Computer-aided detection of microcalcifications in mammograms. *Investigative Radiol.*, 9, 664–671.

26. Y. Wu, K. Doi, M. L. Giger, and Robert M. Nishikawa. 1992. Computerized detection of clustered microcalcification in digital mammograms: applications of artificial neural networks. *Med. Phys.*, 19, 555–560.

27. D. H. Davies and D. R. Dance. 1990. Automated computer detection of clustered microcalcifications in digital mammograms. *Phys. Med. Biol.*, 35, 1111–1118.

28. D. H. Davies and D. R. Dance. 1992. The automated computer detection of subtle calcifications in radio graphically dense breasts. *Phys. Med. Biol.*, 37, 1385–1390.

29. M. Kallergi, K. Woods, L. P Clarke, W. Qian, and R. Clark. 1992. Image segmentation in digital mammography: comparison of local thresholding and region growing algorithms. *Comput. Med, Imaging Graphics*, 16, 323–331.

30. N. Karssemeijer. 1991. A stochastic model for automated detection calcifications in digital mammograms. In *Proc. 12th Int. Conf. Information Medical Imaging*, Wye, UK, 227–238.

31. H. P. Chan, K. Doi, C. J. Vyborny, K. L. Lam, and R. A. Schmidt. 1988. Computer-aided detection of microcalcifications in mammograms. *Investigative Radiol.*, 9, 664–671.

32. A. P. Dhawan and E. Le Royer. 1988. Mammographic feature enhancement by computerized image processing. *Comput. Methods Programs Biomed.*, 27, 23–25.

33. H. Kang, Y. M. Ro, and S. M. Kim. 2005. A microcalcification detection using adaptive contrast enhancement on wavelet transform and neural network. *IEICE Trans. on Inform. Syst.*, E89-D(3), 1280–1289.

34. E. Kouskos, C. Markopoulos, K. Revenas, K. Koufopoulos, V. Kyriakou, and J. Gogas. 2003. Computer-aided preoperative diagnosis of microcalcifications on mammograms. *Acta Radiologica*, 44, 43–46.

35. H. Cheng, Y. M. Liu, and R. I. Freimanis. 1998. A novel approach to microcalcifications detection using fuzzy logic techniques. *IEEE Trans. Medical Imaging*, 17, 442–450.

36. J. S. Taylor and N. Cristianini. 2000. Support vector machines and other Kernel-based learning methods. Cambridge University Press.

37. K. R. Muller, S. Mika, G. Ratsch, K. Tsuda, and B. Scholkopf. 2001. An introduction to kernel-based learning algorithms. *IEEE Trans. Neural Networks*, 12, 181–201.

38. J. H. Yoon and Y. M. Ro. 2002. Enhancement of the contrast in mammographic images using the hommorphic filter method. *IEICE Trans. on Information and Systems*, E85-D(1), 298–303.

39. R. C. Conzale and R. E. Woods. 2002. *Digital Image Processing*. Prentice Hall, NJ.

40. M. Vetterli and J. Kovacevic. 1995. *Wavelets and Subband Coding*. Prentice Hall PTR.

41. S. Mallat and S. Zhong. 1992. Characterization of signals from multiscale edges. *IEEE Trans. Pattern Anal. Machine Intell.*, PAMI-14, 710–732.

42. S. Mallat. 1989. A theory for multiresolution signal decomposition: the wavelet representation, *IEEE Trans. Pattern Anal. Machine Intell.*, PAMI-11, 674–693.

43. S. G. Chang, B. Yu, and M. Vetterli. 2000. Adaptive wavelet thresholding for image denoising and compression. *IEEE Trans. Image Proc.*, 9(9), 1532–1546.

44. B. Scholkopf and A. J. Smola. 2002. *Learning with Kernels: Support Vector Machines, Regularization, Optimization, and Beyond*. The MIT Press, London, UK.

45. C. J. C. Burges. 1998. A tutorial on support vector machines for pattern recognition. In *Data Mining and Knowledge Discovery*. Kluwer Academic Publishers, Boston, MA, 121–167.

46. C.-W. Hsu, C.-C. Chang, and C.-J. Lin. 2003. *A practical guide to support vector classification*. Technical Report, Department of Computer Science and Information, Engineering, National Taiwan University, Taiwan.

47. K. Schutte and J. Glass. 2005. Robust detection of sonorant landmarks. *Proc. Interspeech*, 1005–1008

48. V. Vapnik. 1982. *Estimation of Dependencies Based on Empirical Data*. Springer-Verlag, NY.

49. Y. Sun, C. F. Babbs, and E. J. Delp. 2005. A comparison of feature selection methods for the detection of breast cancers in mammograms: adaptive sequential floating search vs. genetic algorithm. In *Proc. 27th Annu. Conf. IEEE Engineering in Medicine and Biology Society*, 6532–6535.

50. RadiologyInfo, [available online] 2007. Oct. http://www.radiologyinfo.org/

51. C. J. Cao and A. Cai. 2005. A method for classification of scenery documentary using MPEG-7 edge histogram descriptor. In *Proc. Inter. Workshop on VLSI Design and Video Technology*, 105–108.

52. T. Joachims, SVMlight, 2007. Oct. http://svmlight.joachims.org.

53. C. M.-Thoms, S. M. Dunn, C. F. Nodine, and H. L. Kundel. 2003. The perception of breast cancers – A spatial frequency analysis of what differentiates missed from reported cancers. *IEEE Trans. Medical Imaging*, 22(10) 1297–1306.

54. J. S. Suri and R. M. Rangayyan. 2006. *Recent Advances in Breast Imaging, Mammography, and Computer-Aided diagnosis of Breast Cancer*. SPIE Press, NJ, 230–262.

Index